An Introduction to the
World's Oceans

An Introduction to the
World's Oceans

Tenth Edition

Keith A. Sverdrup
University of Wisconsin–Milwaukee

E. Virginia Armbrust
University of Washington

Boston   Burr Ridge, IL   Dubuque, IA   New York   San Francisco   St. Louis
Bangkok   Bogotá   Caracas   Kuala Lumpur   Lisbon   London   Madrid   Mexico City
Milan   Montreal   New Delhi   Santiago   Seoul   Singapore   Sydney   Taipei   Toronto

AN INTRODUCTION TO THE WORLD'S OCEANS, TENTH EDITION

Published by McGraw-Hill, a business unit of The McGraw-Hill Companies, Inc., 1221 Avenue of the Americas, New York, NY 10020. Copyright © 2009 by The McGraw-Hill Companies, Inc. All rights reserved. Previous editions © 2008, 2005, 2003, 2000, 1997, 1994, and 1991. No part of this publication may be reproduced or distributed in any form or by any means, or stored in a database or retrieval system, without the prior written consent of The McGraw-Hill Companies, Inc., including, but not limited to, in any network or other electronic storage or transmission, or broadcast for distance learning.

Some ancillaries, including electronic and print components, may not be available to customers outside the United States.

Printed in China

2 3 4 5 6 7 8 9 0 CTP/CTP 0 9

ISBN 978–0–07–337670–7
MHID 0–07–337670–1

Publisher: *Thomas D. Timp*
Executive Editor: *Margaret J. Kemp*
Director of Development: *Kristine Tibbetts*
Senior Marketing Manager: *Lisa Nicks*
Senior Project Manager: *Kay J. Brimeyer*
Lead Production Supervisor: *Sandy Ludovissy*
Lead Media Project Manager: *Stacy A. Patch*
Associate Design Coordinator: *Brenda A. Rolwes*
Cover Designer: *Studio Montage, St. Louis, Missouri*
(USE) Cover Image: *Royalty-Free/CORBIS*
Lead Photo Research Coordinator: *Carrie K. Burger*
Photo Research: *Pam Carley/Sound Reach*
Compositor: *S4Carlisle Publishilng Services*
Typeface: *10/12 Times Roman*
Printer: *CTPS*

The credits section for this book begins on page 483 and is considered an extension of the copyright page.

Library of Congress Cataloging-in-Publication Data

Sverdrup, Keith A.
    An introduction to the world's oceans. — 10th ed. / Keith Sverdrup, E. Virginia Armbrust.
        p. cm.
    Includes index.
    ISBN 978–0–07–337670–7 — ISBN 0–07–337670–1 (hard copy : alk. paper)
    1. Oceanography—Textbooks. I. Armbrust, E. Virginia (Elisabeth Virginia) II. Title.

GC11.2.D89 2009
551.46—dc22   2008032035

www.mhhe.com

Dedicated to

Barbara J. and Stephanie J. Sverdrup Stone
and
Bob and Charles Armbrust

About the Authors

Keith A. Sverdrup is a Professor of Geophysics at the University of Wisconsin–Milwaukee [UWM] where he has taught oceanography for twenty-five years and conducts research in tectonics and seismology. He is a recipient of UWM's Undergraduate Teaching Award. Keith received his B.S. in Geophysics from the University of Minnesota and his Ph.D. in Earth Science with a dissertation on seismotectonics in the Pacific Ocean basin from the Scripps Institution of Oceanography.

Keith has been active in educational programs of the American Geophysical Union [AGU], the American Institute of Physics [AIP], and the Geological Society of America [GSA]. He was a member of AGU's Education and Human Resources Committee for twelve years [chairing it for four years], and also chaired AGU's Excellence in Geophysics Education Award Committee, the Editorial Advisory Committee for Earth and Space, and the Sullivan Award Committee for excellence in science journalism. Keith served as a member of AIP's Physics Education Committee for six years. He is a member of the American Geophysical Union, American Society of Limnology and Oceanography, Oceanography Society, Geological Society of America, National Association of Geology Teachers, National Science Teachers Association, and Sigma Xi. Keith was the Geosciences Program Officer at the National Science Foundation's Division of Undergraduate Education from 2005–2007.

Dr. Virginia (Ginger) Armbrust is a Professor in the School of Oceanography at the University of Washington, where she teaches and conducts research on marine phytoplankton. She received her A.B. in Human Biology from Stanford University and her Ph.D. in Biological Oceanography from Massachusetts Institute of Technology and Woods Hole Oceanographic Institution.

Ginger directs research projects that examine the biodiversity, physiology, and ecology of marine phytoplankton, to understand how these organisms shape and respond to changes in their habitats. She was the lead scientist for an international project to determine the entire DNA sequence of a marine diatom to better understand how these organisms function in their environment. She is the co-director of the Pacific Northwest Center for Human Health and Ocean Studies, created in response to the need to understand the links between ocean processes and human health. She is the director of the Center for Environmental Genomics in the College of Ocean and Fisheries Science and is a Gordon and Betty Moore Foundation Investigator in Marine Microbiology.

Ginger teaches at both the undergraduate and graduate levels, and she has twice received her college's Distinguished Graduate Teaching Award. She has supervised the research of thirty undergraduates in her laboratory and has served as either the chair or a member of the Research Advisory Committee for forty graduate students. As a mentor for "Partners in Science," she has supervised the summer research of high school teachers in her lab.

Brief Contents

Contents

Chapter 4

The Sea Floor and Its Sediments   90

Chapter 5

The Physical Properties of Water   124

Chapter 6

The Chemistry of Seawater   148

Chapter 7

The Structure and Motion of the Atmosphere   167

Chapter 8

Circulation and Ocean Structure   200

Chapter 9

The Surface Currents   218

Chapter 10

The Waves   239

Going to Sea   266

Chapter 11

The Tides   270

Chapter 12

Coasts, Beaches, and Estuaries   288

Chapter 13

Environmental Issues and Concerns   319

Chapter 14

The Living Ocean   343

Chapter 15

Production and Life   359

A Note to Students

Human beings have been curious about the oceans since they first walked along their shores. As people have learned more about the oceans, they have come to understand more fully and appreciate the tremendous influence these bodies of salt water have on our lives. The oceans cover over 70% of Earth's surface, creating a habitat for thousands of known species and countless others still to be discovered. The sea contains vast quantities of diverse natural resources in the water and on the sea floor; some are actively exploited today, and many more may be recovered in the future with improved technology and greater demand. Global climate and weather are strongly influenced by the oceans as they interact with the atmosphere through the transfer of moisture and heat energy. The ocean basins also serve as the location of great geologic processes and features such as earthquakes, volcanoes, massive mountain ranges, and deep trenches, all of which are related to the creation and destruction of sea floor in the process of plate tectonics.

Much of what happens in the oceans and on the sea floor is hidden from direct observation. Although the *Hubble Space Telescope* can form images from light that has traveled over 10 billion trillion kilometers, we cannot see more than a few tens of meters below the ocean's surface even under the most favorable conditions because of the efficient scattering and absorption of light by seawater. Consequently, most of what we know about the oceans comes from indirect, or remote, methods of observation. With constantly improving technology and innovative applications of that technology, we continue to learn more about the geological, physical, chemical, and biological characteristics of the oceans.

Although careful scientific study of the oceans is often difficult and challenging, it is both necessary and rewarding. Our lives are so intimately tied to the oceans that we benefit from each new fact that we discover. Continued research and a better understanding of the oceans become increasingly important, as the population of this planet grows ever larger. Early in the new millennium, there is both good news and bad news concerning global population growth. The rate of population increase has slowed with falling birth rates, and there is some indication that the human population will level off by the end of this century. But even if the human population does stabilize, it will not do so before there is an increase of several billion people over today's population. We clearly will continue to face difficult environmental decisions affecting the oceans as well as the land in the foreseeable future. Our best chance of dealing wisely and effectively with these challenges is to promote more widespread understanding of the oceans.

Although it is critical that we continue to train marine scientists to study the oceans, it is no less important for people in all walks of life to develop a basic understanding of how the oceans influence our lives and how our actions influence the oceans. In studying oceanography, you are preparing yourself to be an informed global citizen. It is likely that at some point in the future you will have the opportunity to voice your concern about the health of the oceans, either directly or through the governmental process. Your interest in and study of oceanography will help you participate in future discussions and decision-making processes in an informed manner.

The book's website at www.mhhe.com/sverdrup10e provides you with links to Internet addresses relevant to this text. To expand your knowledge of oceanography, Internet exercises for many of these sites are found on the website. Also included is a comprehensive student study guide that includes detailed outlines of the chapters and questions to test your understanding.

A Note to Instructors

A major objective of this text is to stimulate student interest and curiosity by blending contemporary information and research with basic principles in order to present an integrated introduction to the many and varied sciences used in the study of the oceans. To do so, we have extensively reviewed and rewritten material from the ninth edition to produce this new tenth edition. In the face of constant and rapid change, we have added new material for both content and interest. We have also invited six scientists to write guest essays in their fields of specialization. There is also a seventh essay written by a chief scientist and a ship's captain on planning and executing an oceanographic expedition.

We realize that the students who use this book come from diverse backgrounds and that for many of them this is an elective course. The content continues to be reasonably rigorous, but we have chosen to use simple algebra rather than advanced mathematics. For instance, we use centrifugal force to explain tidal principles because most students do not have much background in vectors.

An ecological approach and descriptive material are used to integrate the biological chapters with the other subject fields. We strive to emphasize oceanography as a cohesive and united whole rather than a collection of subjects gathered under a marine umbrella.

In order to understand the constant barrage of information concerning our planet and marine issues, students must have a basic command of the language of marine science in addition to mastering processes and principles. For this reason we maintain an emphasis on critical vocabulary. All terms are defined in the text; terms that are particularly important are printed in boldface. A list of important terms appears at the end of each chapter, with a glossary included at the end of the book. The website for this text also hosts interactive flashcards of key terms for student study.

End-of-chapter Summaries provide quick reviews of key concepts. Study Problems are included in many chapters, and Study Questions are at the end of each chapter. The Study Questions are not intended merely for review, but also to challenge students to think further about the lessons of the chapter.

This book may be used in a one-quarter or one-semester course. Because the experience and emphasis of faculty using this book will differ, it is expected that each instructor will emphasize and elaborate on some topic at the expense of other topics. We continue to make each chapter stand as independently as possible and encourage instructors to use the chapters in the order that best suits their purposes. Cross-references from one chapter to another indicate discussion of topics elsewhere in the text. Faculty wishing to use a more quantitative approach in some areas are encouraged to make use of appendix C, Equations and Quantitative Relationships. The answers to the Study Questions and Study Problems from the text appear in the Instructor's Manual, within the password-protected instructor's area of the website.

Changes to the Tenth Edition

In this edition of the book we introduce learning outcomes for each chapter. These can be used to guide both the instructor and the students as they study the material. In **Chapter 1** we have included historical information about Thomas Huxley and his identification of *Bathybius haeckellii* as a suspected primordial form of deep-sea life and we have updated the information about the Argo program. A new, updated "Field Notes: New Approaches to Exploring the Oceans" has been prepared for **Chapter 3.** Several figures in **Chapter 4** have been modified for greater clarity. In **Chapter 5** the discussion of light attenuation in water has been modified and a new figure added to illustrate the difference in attenuation coefficients in open ocean water and coastal water. In addition, information on massive icebergs in the Antarctic has been updated. Details concerning El Niño conditions have been updated in **Chapter 7** along with an expanded discussion of Hurricane Katrina. We have also included a new "Field Notes: The Oceans and Climate Change." In **Chapter 8** we have expanded the discussion of OTEC plants. In **Chapter 9** we have included a section on current volume transport. We have updated the information about the DART program of buoys in **Chapter 10** and we have expanded our discussion of internal waves. In **Chapter 12** we have updated information on National Marine Sanctuaries and rising sea level. In **Chapter 13** we have extensively updated information on marine pollution, the Gulf of Mexico dead zone, and oil spills. **Chapter 15** has been modified to include a new figure and a new table illustrating the concepts of trophic transfers. A new figure has been added to **Chapter 16** to better illustrate the complexity of the microbial food web. The section Practical

Considerations in **Chapter 17** has been updated with information on the world fish catch over past decades. **Chapters 16, 17, and 18** have been modified to better illustrate the concepts of "top-down" and "bottom-up" controls on food webs.

Instructor Supplements

McGraw-Hill offers a variety of supplements to assist instructors with both preparation and classroom presentation.

The 10th Edition Website (www.mhhe.com/sverdrup10e)

The 10th edition website offers a wealth of teaching and learning tools for instructors and students. Instructors will appreciate:

- A password-protected Instructor's Manual with answers to the study questions and study problems in the text
- PowerPoint lecture outlines
- Scripps videos
- Animations
- Access to the online **Presentation Center,** including most of the illustrations, photographs, and tables from the text in convenient jpeg format
- A student center with multiple-choice quizzes, a student study guide, key term flashcards, Internet exercises, and web links to chapter-related material
- A test bank utilizing McGraw-Hill's EZ Test software. EZ Test is a flexible and easy-to-use electronic testing program that allows instructors to create tests in a wide variety of question types.

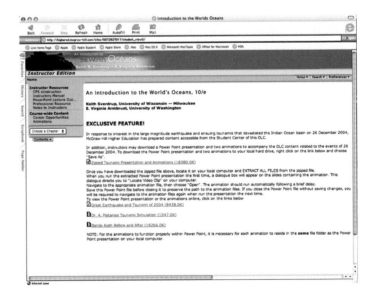

Presentation Center

Build instructional materials wherever, whenever, and however you want!

The **Presentation Center** is an online digital library containing assets such as photos, artwork, animations, PowerPoints, and other media types that can be used to create customized lectures, visually enhanced tests and quizzes, compelling course websites, or attractive printed support materials.

Access to your book, access to all books!
The **Presentation Center library** includes thousands of assets from many McGraw-Hill titles. This ever-growing resource gives instructors the power to utilize assets specific to an adopted textbook as well as content from all other books in the library.

Nothing could be easier!
Accessed from the instructor side of the website, the **Presentation Center's** dynamic search engine allows you to explore by discipline, course, textbook chapter, asset type, or keyword. Simply browse, select, and download the files you need to build engaging course materials. All assets are copyright McGraw-Hill Higher Education but can be used by instructors for classroom purposes.

Student Supplements

The Internet makes oceanographic information and data available to researchers and it also provides images and information in many forms to instructors and students. Public agencies and museums, universities and research laboratories, satellites and oceanographic projects, interest groups, and individuals all over the planet provide information that can be publicly accessed.

The website for *An Introduction to the World's Oceans* is a great place to review chapter material and enhance your study routine. Visit www.mhhe.com/sverdrup10e for access to the following online study tools:

- Multiple-choice quizzes
- Student study guide
- Key term flashcards
- Internet exercises
- Web links to chapter-related material

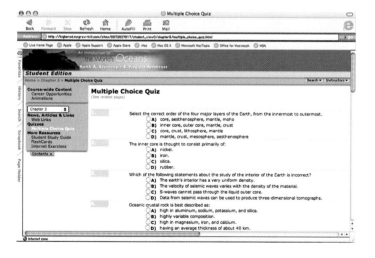

Acknowledgments

As a book is the product of many experiences, it is also the product of people other than the authors. We extend many thanks to our friends and colleagues who have graciously answered our questions and provided us with information and access to their photo files. We owe very special thanks to faculty, staff, and students of the School of Oceanography and to students from the School of Aquatic and Fisheries Science, both within the College of Ocean and Fishery Sciences, University of Washington. They have answered numerous questions, shared data, and provided insights into how to convey concepts presented in this edition. We are also grateful to Scripps Institution of Oceanography, which has allowed us the privilege of providing their videotape series as an instructor ancillary to this tenth edition of the text.

We would particularly like to thank the following people who authored the Field Notes boxes in this edition:

Field Notes for the Tenth Edition

Virginia Armbrust, *University of Washington*

Eddie Bernard, *National Oceanic and Atmospheric Administration/Pacific Marine Environmental Laboratory*

Christopher Brown, *National Oceanic and Atmospheric Administration/National Environmental Satellite, Data and Information Service*

Francisco Chavez, *Monterey Bay Aquarium Research Institute*

David Clague, *Monterey Bay Aquarium Research Institute*

John Delaney, *University of Washington*

Marcia McNutt, *Monterey Bay Aquarium Research Institute*

LuAnne Thompson, *The Oceans and Climate Change*

Ian Young, *Monterey Bay Aquarium Research Institute*

Thanks are also extended to the reviewers who provided their thoughtful comments and suggestions for the tenth edition:

Reviewers for the Tenth Edition

Douglas Biggs, *Texas A & M University*

Istvan Csato, *Collin College*

Huan Feng, *Montclair State University*

William Hoyt, *University of Northern Colorado*

Kevin Johnson, *Florida Institute of Technology*

Greta Mackenzie, *University of Miami*

We thank all members of the team at McGraw-Hill, without whose help, enthusiasm, and coordinated efforts this tenth edition could not have been completed.

Guided Tour

A variety of tools within this textbook have been designed to assist with chapter review and critical analysis of chapter topics.

Chapter Outline

Each chapter begins with an outline of the subsections and boxed readings within each chapter.

Learning Outcomes

Learning outcomes appear at the onset of each chapter to help instructors better facilitate course management and set goals for each chapter topic.

Learning Outcomes

After studying the information in this chapter students should be able to:

1. review the evolution of methods to measure ocean depth with time up to the present,
2. construct a simple cross section of an ocean basin, including both a passive and active continental margin,
3. discuss the formation of atolls,
4. sketch the location of ocean ridges and trenches,
5. explain three different ways to classify sediments,
6. list the organisms that contribute the majority of calcareous and siliceous sedimentary particles,
7. identify where biogenous and lithogenous sediments are dominant on the sea floor,
8. define isotopes and describe how they can be used with marine sediments as historical records,
9. list multiple seabed resources and appraise the extent to which they are currently being recovered, and
10. write a short history of the evolution of the Law of the Sea.

Field Notes Boxes

The essays represented within these boxes are written by oceanographers in the field. These readings highlight relevant oceanographic topics and provide insights into engaging oceanographic careers.

Field Notes

The Oceans and Climate Change
by LuAnne Thompson

LuAnne Thompson is an Associate Professor of Oceanography, Adjunct Associate Professor of Atmospheric Sciences, and Interim Director of the Program on Climate Change at the University of Washington. Her research interest is the role of the ocean in climate variability and change, which she explores using numerical models of ocean circulation, biogeochemical cycles, and atmospheric circulation.

The ocean serves as the memory of the climate system, storing heat, fresh water, and chemicals over time spans from decades to millennia. This memory results from the chemical and physical properties of seawater. First, water is heavier than air. A 10 meter column of water is heavier than a column of air that extends to the top of the Earth's atmosphere. Second, water has four-fold higher specific heat than air and five-fold higher specific heat than soil. Because it takes vastly more energy to heat up the ocean, ocean temperature is much more resistant to change than air or land temperature. Although the temperature of surface waters of the ocean varies according to latitude, the temperature of deep water in all oceans is almost always close to freezing. Finally, gases such as carbon dioxide dissolve in water making the ocean a major storage depot. In fact, fifty-fold more carbon dioxide is stored in the ocean than in the atmosphere. Without absorption of carbon dioxide by the ocean, atmospheric concentrations of this greenhouse gas would be even higher. Thus the oceans provide a damper that keeps the Earth's climate relatively constant and benign.

Over the past fifty years, large research programs have greatly informed how this ocean damper works. In 1957–58, under the auspices of the International Geophysical Year, detailed profiles of temperature and salinity were done in the Atlantic Ocean. These surveys allowed oceanographers to see clearly that the sources of the deep (below 1500 m depth) and intermediate water (between 500 and 1500 m depth) of the world's oceans come from geographically isolated regions, with deep water forming in the northern North Atlantic and near Antarctica, and intermediate water originating at high latitudes and in marginal seas.

In the 1970s, data collected during the Geosecs (Geochemical Ocean Sections Study) Program greatly added to our knowledge of the role of the ocean in global cycles of nutrients and gases such as carbon dioxide and oxygen. By this time, a fairly complete picture of the full three-dimensional structure of the ocean water properties was developed and there was general acceptance that the modern ocean system was relatively stable. In the 1980s a simple cartoon of the global ocean circulation called the great ocean conveyor belt (see Figure 9.14) was popularized. This cartoon suggests that cold water created in the high northern latitudes of the North Atlantic sinks to the abyss and moves slowly toward the North Pacific and Indian Oceans where it rises to the surface, warms, and then makes its way back to the Nordic Seas via surface currents and ultimately the Gulf Stream.

In the mid-1980s through the 1990s, extensive and detailed ocean surveys were conducted as part of the World Ocean Circulation Experiment (WOCE). These studies generated the most comprehensive observational analyses of the ocean general circulation and indicated that the conveyer belt circulation model is an oversimplification. Analyses of WOCE data clearly emphasize the significance of deep water formed near Antarctica, and suggest that much of the upwelling of deep water may occur in the Southern Ocean, and that there are multiple routes by which water is exchanged among the various ocean basins.

The role the ocean plays in climate variability and change is becoming clearer. Geochemical tracers show that deep-water cycles in the ocean take several centuries, while water cycles through waters above the thermocline much more quickly, on the order of decades. Analyses of satellite measurements suggest that the oceans carry the bulk of the excess heat away from the equator in the tropics. However, the amount of heat transferred by the oceans' amount falls off at higher latitudes, and poleward of 40° of the atmosphere carries the lion's share. These results have called into question how important the conveyer belt circulation is to maintaining the climate of Europe, and whether we would expect large and abrupt climate changes in the future if there were a large influx of fresh water into the high latitude North Atlantic Ocean from the rapid melting of Greenland ice, for example.

At the same time that more comprehensive observations of the oceans were being made, the first general circulation models of the ocean and the climate system were constructed. In the early 1970s numerical modeling of ocean circulation was born with the construction of the first general circulation model of the ocean that qualitatively reproduced the major ocean currents. Shortly thereafter, this model was coupled to a model atmosphere and exchange of heat and water across the air-sea interface was modeled. These models are the predecessors of the models used today. Written in Fortran, a scientific programming language, they divided the ocean up into boxes, with the sides of each box about 200–300 km, and the depth about 100–500 m. The original models were written on paper computer cards. Each model run took weeks or more of computer time even though necessarily restricted to poor resolution of features and physical processes in the ocean and the atmosphere.

The development of comprehensive climate models has allowed testing of hypotheses of how the climate system would respond to changes in both the greenhouse effect and to other climatic perturbations. For example, these models can be used to ask if Europe would rapidly cool if a significant fraction of

192

Chapter Summary

Each chapter's summary provides a quick review of key concepts.

Key Terms

Key terms are boldfaced and defined within the text, and end-of-chapter key terms listings indicate the most important terms and their locations within each chapter.

Study Questions and Problems

Study Questions and Study Problems serve not only as a concept review, but challenge students to think further about the lessons within each chapter.

Visit Our Website:
www.mhhe.com/sverdrup10e

The website offers a wealth of teaching and learning tools for instructors and students.

Instructors will appreciate:

- A password-protected Instructor's Manual with answers to the study questions and study problems in the text
- PowerPoint lecture outlines
- Scripps videos
- Animations
- Access to the new online **Presentation Center,** including all of the illustrations, photographs, and tables from the text in convenient jpeg format
- A test bank utilizing McGraw-Hill's EZ Test Software

Students will find:

- A student center with multiple-choice quizzes
- A student study guide
- Key term flashcards
- Internet exercises
- Web links to chapter-related material

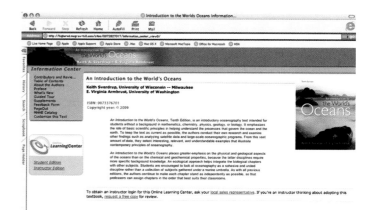

Summary

Phytoplankton are the dominant photosynthetic autotrophs in the sea. These microscopic organisms use sunlight and inorganic compounds to generate the organic matter that serves as food for life in the sea. The process of generating organic carbon from carbon dioxide is commonly referred to as carbon fixation. All photosynthetic organisms use the pigment chlorophyll a to absorb sunlight. Oxygen is formed as a by-product of photosynthesis. Organic matter is broken down through respiration to yield chemical energy, water, and carbon dioxide.

Gross primary productivity is the total amount of organic matter produced by photosynthesis per volume of seawater per unit of time. Net primary production is the gain in organic matter from photosynthesis by phytoplankton minus the reduction in organic matter due to respiration by phytoplankton. Primary productivity can be determined by measuring the rate of uptake of carbon dioxide or the rate of production of oxygen.

Phytoplankton remove required inorganic nutrients from seawater in a ratio that reflects their biological demands. When organisms die and decompose, the nutrients are released back into seawater in a similar ratio. Nutrients cycle between the land and the sea and through organic and inorganic compounds. Inorganic nutrient concentrations, sunlight, and temperature influence the rate of primary production. Phytoplankton blooms occur when phytoplankton reproduce more rapidly than they are consumed by zooplankton and other heterotrophs. Standing stock is the total phytoplankton biomass present at a given site at a given instant in time and is related to chlorophyll a concentrations.

Key Terms

All key terms from this chapter can be viewed by term or definition when studied as flashcards on this book's website at www.mhhe.com/sverdrup10e.

density, 51
seismic wave, 51
P-wave, 51
S-wave, 51
refract, 52
inner core, 54
outer core, 54
mantle, 54
seismic tomography, 54
crust, 54
Mohorovičić discontinuity, 54
Moho, 54
granite, 54

basalt, 54
lithosphere, 55
asthenosphere, 55
mesosphere, 55
isostasy, 55
Gondwanaland, 56
continental drift, 56
Pangaea, 56
Laurasia, 56
ridge, 57
rise, 57
rift valley, 57
trench, 57
convection cell, 59
seafloor spreading, 59
spreading center, 59
subduction zone, 59

epicenter, 59
focus, 59
hypocenter, 59
sediment, 60
core, 60
dipole, 61
Curie temperature, 62
paleomagnetism, 62
polar reversal, 62
polar wandering curve, 65
inertia, 65
plate tectonics, 65
divergent plate boundary, 66
convergent plate boundary, 66
transform boundary, 66
transform fault, 66
rift zone, 67

graben, 68
pillow basalt, 68
gabbro, 69
escarpment, 69
fracture zone, 71
andesite, 73
island arc, 73
Wadati-Benioff zone, 73
passive/trailing margin, 73
active/leading margin, 74
spreading rate, 74
hot spot, 74
guyot, 79
transverse/aseismic ridge, 79
craton, 82
terrane, 82
hydrothermal vent, 83

Study Questions

1. What is meant by the term *polar wandering?* Have the magnetic poles actually wandered?
2. Describe the three types of plate boundaries. What processes take place at each type of boundary? In what direction do the plates move at each boundary?
3. What mechanisms have been proposed to account for plate motion?
4. What is the difference between the leading edge and the trailing edge of a continent? Between a divergent plate boundary and a convergent plate boundary?
5. If the ability of the oceanic crust to transmit heat were uniform, the rate of heat flow through the ocean floor would depend only on the temperature change across the oceanic crust. Under such a condition, how would the heat flow measurements in figure 3.12 indicate the presence of ascending convection cells in the asthenosphere?
6. If the polar wandering curves for North America and Europe are made to coincide, how will these continents move relative to each other?
7. Using the techniques and reasoning employed to discover the properties of the interior of Earth, explain how you would determine what is inside a sealed box (for example, measuring the box, weighing it, spinning it, balancing it on different axes, sampling its exterior). What clue would

each of these measurements give you to the contents of the box?
8. What had to be learned about Earth before Alfred Wegener's ideas could be accepted?
9. Why does a newly formed mid-ocean volcanic island gradually subside?
10. Explain the formation and symmetry of the magnetic stripes found on either side of the mid-ocean ridge system. What is their significance when the magnetic information is correlated with the age of the crust?
11. Under what conditions will a convergent boundary form a mountain range? An island arc system? Why do volcanoes associated with subduction zones usually erupt more explosively than mid-ocean volcanoes associated with hot spots and spreading centers?
12. On an outline map of the world draw in (a) earthquake belts, (b) mid-ocean ridges, and (c) trenches. Relate your map to figure 3.8. What do you conclude?
13. How have recent advances in seismic tomography modified our ideas of Earth's internal layers, as shown in figure 3.3?
14. Why do P-waves pass through Earth's outer core?
15. What is a terrane? What role do terranes play in our understanding of today's continents?

Study Problems

1. If a plate moves away from a spreading center at the rate of 5 cm/yr, what is the displacement of a landmass carried by that plate after 180×10^6 years?
2. Magnetic stripes with the same magnetic orientation are measured on either side of a ridge crest. The stripe on the west side

of the ridge is displaced 11 km from the crest; the stripe on the east side is displaced 9 km from the crest. The age of the rock in both stripes is 4×10^5 years. Calculate the average spreading rate at this ridge.

Presentation Center

Build instructional materials wherever, whenever, and however you want!
The **Presentation Center** is an online digital library containing assets such as photos, artwork, animations, PowerPoints, and other media types that can be used to create customized lectures, visually enhanced tests and quizzes, compelling course websites, or attractive printed support materials.

Access to your book, access to all books!
The **Presentation Center library** includes thousands of assets from many McGraw-Hill titles. This ever-growing resource gives instructors the power to utilize assets specific to an adopted textbook as well as content from all other books in the library.

Nothing could be easier!
Accessed from the instructor side of the website, the **Presentation Center's** dynamic search engine allows you to explore by discipline, course, textbook chapter, asset type, or keyword. Simply browse, select, and download the files you need to build engaging course materials. All assets are copyright McGraw-Hill Higher Education but can be used by instructors for classroom purposes.

Electronic Textbook

CourseSmart is a new way for faculty to find and review eTextbooks. It's also a great option for students who are interested in accessing their course materials digitally and saving money. CourseSmart offers thousands of the most commonly adopted textbooks across hundreds of courses from a wide variety of higher education publishers. It is the only place for faculty to review and compare the full text of a textbook online, providing immediate access without the environmental impact of requesting a print exam copy. At CourseSmart, students can save up to 50% off the cost of a print book, reduce their impact on the environment, and gain access to powerful Web tools for learning, including full text search, notes and highlighting, and email tools for sharing notes between classmates. **www.CourseSmart.com**

A sextant and marine charts. The sextant is an early navigational aid first constructed by John Bird in 1759.

Chapter 1

The History of Oceanography

Chapter Outline

Learning Outcomes

After studying the information in this chapter students should be able to:

1. *describe* the difference between a scientific hypothesis and a theory,
2. *discuss* the interaction of early civilizations with the oceans,
3. *sketch* the major seafaring routes of the great voyages of discovery in the fifteenth and sixteenth centuries, James Cook's voyages of discovery, and the scientific voyages of Charles Darwin and the *Challenger* expedition.
4. *list* the major discoveries of the *Challenger* expedition,
5. *compare* and *contrast* the methods of making scientific measurements in the nineteenth and twentieth centuries, and
6. *describe* the difference in both the quantity of oceanographic data and the density of that data available to oceanographers now compared to the nineteenth century.

Oceanography is a broad field in which many sciences are focused on the common goal of understanding the oceans. Geology, geography, geophysics, physics, chemistry, geochemistry, mathematics, meteorology, botany, and zoology have all played roles in expanding our knowledge of the oceans. Oceanography today is usually broken down into a number of subdisciplines because the field is extremely interdisciplinary.

Geological oceanography includes the study of Earth at the sea's edge and below its surface, and the history of the processes that form the ocean basins. Physical oceanography investigates the causes and characteristics of water movements such as waves, currents, and tides and how they affect the marine environment. It also includes studies of the transmission of energy such as sound, light, and heat in seawater. Marine meteorology (the study of heat transfer, water cycles, and air-sea interactions) is often included in the discipline of physical oceanography. Chemical oceanography studies the composition and history of the water, its processes, and its interactions. Biological oceanography concerns marine organisms and the relationship between these organisms and the environment in the oceans. Ocean engineering is the discipline that designs and plans equipment and installations for use at sea.

Scientists make discoveries about the natural world, both as it is now and as it has been throughout its history, by gathering data through observation and experimentation. Data are the "facts" used by scientists. Scientific data are reproducible and accompanied by an estimate of error. If an observation is not reproducible and does not include an error estimate, it is not a *scientific* datum. Repeated observations and measurements by independent scientists are expected to yield data whose error regions overlap with the error regions around the original data.

A scientific **hypothesis** is an initial explanation of data that is based on well-established physical or chemical laws. If the data are quantitative (i.e., if they are represented as numbers), the hypothesis is often expressed as a mathematical equation. In order to be considered a *scientific* hypothesis, it must be subject to testing and falsification (demonstration that something is not true). A scientific hypothesis that has been repeatedly tested and found to be in agreement with observed facts is provisionally considered to be a valid hypothesis, recognizing that it may be replaced by a more complete hypothesis in the future.

If a hypothesis is consistently supported by repeated, different experiments, then it may be advanced to the level of a **theory**. The great value of a theory is its ability to predict the existence of phenomena or relationships that had not previously been recognized. Scientists use the word "theory" in a much more restrictive sense than the general public, who use the word in the same way the word "speculation" is used. A scientific theory is not an idle speculation, however. It is a tested, reliable, and precise statement of the relationships among reproducible observations.

A collection of hourly measurements of sea surface elevation at a specific point would comprise a set of scientific data or facts. An initial explanation of these data might be the hypothesis that sea surface elevation varies in response to tidal forces. This hypothesis could be expressed as a mathematical equation. If repeated measurements elsewhere in the oceans yielded reproducible data that continued to be accurately explained by the hypothesis, it would rise to the level of tidal theory (discussed in chapter 11).

Even when a hypothesis is elevated to the status of a theory, the scientific investigation will not necessarily stop. Scientists do not discard accepted theories easily; new discoveries are first assumed to fit into the existing theoretical framework. It is only when, after repeated experimental tests, the new data cannot be explained that scientists seriously question a theory and attempt to modify it.

The study of the oceans was promoted by intellectual and social forces as well as by our needs for marine resources, trade and commerce, and national security. Oceanography started slowly and informally; it began to develop as a modern science in the mid-1800s and has grown dramatically, even explosively, in the last few decades. Our progress toward the goal of understanding the oceans has been uneven and progress has frequently changed direction. The interests and needs of nations as well as the scholarly curiosity of scientists have controlled the ways we study the oceans, the methods we use to study them, and the priority we give to certain areas of study. To gain perspective on the current state of knowledge about the oceans, we need to know something about the events and incentives that guided people's previous investigations of the oceans.

1.1 The Early Times

People have been gathering information about the oceans for millennia, accumulating bits and pieces of knowledge and passing it on by word of mouth. Curious individuals must have acquired their first ideas of the oceans from wandering the seashore, wading in the shallows, and gathering food from the ocean's edges. During the Paleolithic period, humans developed the barbed spear, or harpoon, and the gorge. The gorge was a double-pointed stick inserted into a bait and attached to a string. At the beginning of the Neolithic period, the bone fishhook was developed and later the net (fig. 1.1). By 5000 B.C., copper fishhooks were in use.

As early humans moved slowly away from their inland centers of development, they were prepared to take advantage of the sea's food sources when they first explored and later settled along the ocean shore. The remains of shells and other refuse,

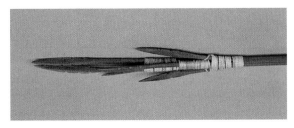

(a)

(b)

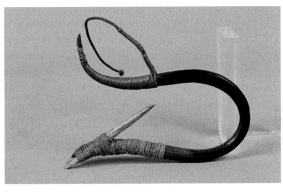

(c)

Figure 1.1   Traditional fishing and hunting implements from coastal Native American cultures of the Pacific Northwest. (a) A duck spear made of cedar. (b) A bone harpoon point and lanyard made of sinew, hemp, and twine. (c) A fishhook made of bone and steam-bent cedar root.

in piles known as kitchen middens, have been found at the sites of ancient shore settlements. These remains show that our early ancestors gathered shellfish, and fish bones found in some middens suggest that they also used rafts or some type of boat for offshore fishing. Some scientists think that many more artifacts have been lost or scattered as a result of rising sea level. The artifacts that have been found probably give us only an idea of the minimum extent of ancient shore settlements. Drawings on ancient temple walls show fishnets; on the tomb of the Egyptian Pharaoh Ti, Fifth Dynasty (5000 years ago), is a drawing of the poisonous pufferfish with a hieroglyphic description and warning. As long ago as 1200 B.C. or earlier, dried fish were traded in the Persian Gulf; in the Mediterranean, the ancient Greeks caught, preserved, and traded fish, while the Phoenicians founded fishing settlements, such as "the fisher's town" Sidon, that grew into important trading ports.

Early information about the oceans was mainly collected by explorers and traders. These voyages left little in the way of recorded information. Using descriptions passed down from one voyager to another, early sailors piloted their way from one landmark to another, sailing close to shore and often bringing their boats up onto the beach each night.

Some historians believe that seagoing ships of all kinds are derived from early Egyptian vessels. The first recorded voyage by sea was led by Pharaoh Snefru about 3200 B.C. In 2750 B.C., Hannu led the earliest documented exploring expedition from Egypt to the southern edge of the Arabian Peninsula and the Red Sea.

The Phoenicians, who lived in present-day Lebanon from about 1200 to 146 B.C., were well-known as excellent sailors and navigators. While their land was fertile it was also densely populated, so they were compelled to engage in trade to acquire many of the goods they needed. They accomplished this by establishing land routes to the east and marine routes to the west. The Phoenicians were the only nation in the region at that time that had a navy. They traded throughout the Mediterranean Sea with the inhabitants of North Africa, Italy, Greece, France, and Spain. They also ventured out of the Mediterranean Sea to travel north along the coast of Europe to the British Isles and south to circumnavigate Africa in about 590 B.C. In 1999, the wreckage of two Phoenician cargo vessels circa 750 B.C. was explored using remotely operated vehicles (ROVs) that could dive to the wreckage and send back live video images of the ships. The ships were discovered about 48 km (30 mi) off the coast of Israel at depths of 300–900 m (roughly 1000–3000 ft).

Extensive migration throughout the Southwestern Pacific may have begun by 2500 B.C. These early voyages were relatively easy because of the comparatively short distance between islands in the far Southwestern Pacific region. By 1500 B.C., the Polynesians had begun more extensive voyages to the east, where the distance between islands grew from tens of kilometers in the western Pacific to thousands of kilometers in the case of voyages to the Hawaiian Islands. They reached and colonized the Hawaiian Islands sometime between A.D. 450 and 600. By the eighth century A.D., they had colonized every habitable island in a triangular region roughly twice the size of the

Figure 1.2 On Satawal Island master navigator Mau Piailug teaches navigation to his son and grandson with the help of a star compass. The compass consists of an outer ring of stones, each representing a star or a constellation when it rises or sets on the horizon, and an inner ring of pieces of palm leaf representing the swells which travel from set directions and which together with the stars help the navigator find his way over the sea. In the center of the ring, the palm leaves serve as a model outrigger canoe.

Figure 1.3 A navigational chart (*rebillib*) of the Marshall Islands. Sticks represent a series of regular wave patterns (swells). Curved sticks show waves bent by the shorelines of individual islands. Islands are represented by shells.

United States bound by Hawaii on the north, New Zealand in the southwest, and Easter Island to the east.

A basic component of navigation throughout the Pacific was the careful observation and recording of where prominent stars rise and set on the horizon. Observed near the equator, the stars appear to rotate from east to west on a north-south axis. Some rise and set farther to the north and some farther to the south, and they do so at different times. Navigators created a "star structure" by dividing the horizon into thirty-two segments where their known stars rose and set. These directions form a compass and provide a reference for recording information about the direction of winds, currents, waves, and the relative positions of islands, shoals, and reefs (fig. 1.2). The Polynesians also navigated by making close observations of waves and cloud formations. Observations of birds and distinctive smells of land such as flowers and wood smoke alerted them to possible landfalls. Once islands were discovered, their locations relative to one another and to the regular patterns of sea swell and waves bent around islands could be recorded with stick charts constructed of bamboo and shells (fig. 1.3).

As early as 1500 B.C., Middle Eastern peoples of many different ethnic groups and regions were exploring the Indian Ocean. In the seventh century A.D., they were unified under Islam and controlled the trade routes to India and China and consequently the commerce in silk, spices, and other valuable goods. (This monopoly wasn't broken until Vasco da Gama defeated the Arab fleet in 1502 in the Arabian Sea.)

The Greeks called the Mediterranean "Thalassa" and believed that it was encompassed by land, which in turn was surrounded by the endlessly circling river Oceanus. In 325 B.C., Alexander the Great reached the deserts of the Mekran Coast, now a part of Pakistan. He sent his fleet down the coast in an apparent effort to probe the mystery of Oceanus. He and his troops had expected to find a dark, fearsome sea of whirlpools and water

spouts inhabited by monsters and demons; they did find tides that were unknown to them in the Mediterranean Sea. His commander, Nearchus, took the first Greek ships into the ocean, explored the coast, and brought them safely to the port of Hormuz eighty days later. Pytheas (350–300 B.C.), a navigator, geographer, astronomer, and contemporary of Alexander, made one of the earliest recorded voyages from the Mediterranean to England. From there, he sailed north to Scotland, Norway, and Germany. He navigated by the Sun, stars, and wind, although he may have had some form of sailing directions. He recognized a relationship between the tides and the Moon, and made early attempts at determining latitude and longitude. These early sailors did not investigate the oceans; for them, the oceans were only a dangerous road, a pathway from here to there, a situation that continued for hundreds of years. However, the information they accumulated slowly built into a body of lore to which sailors and voyagers added each year.

While the Greeks traded and warred throughout the Mediterranean, they observed the sea and asked questions. Aristotle (384–322 B.C.) believed that the oceans occupied the deepest parts of Earth's surface; he knew that the Sun evaporated water from the sea surface, which condensed and returned as rain. He also began to catalog marine organisms. The brilliant Eratosthenes (c. 264–194 B.C.) of Alexandria, Egypt, mapped his known world and calculated the circumference of Earth to be 40,250 km, or 25,000 mi (today's measurement is 40,067 km, or 24,881 mi). Posidonius (c. 135–50 B.C.) reportedly measured an ocean depth of about 1800 m (6000 ft) near the island of Sardinia, according to the Greek geographer Strabo (c. 63 B.C.–A.D. 21). Pliny the Elder (c. A.D. 23–79) related the phases of the Moon to the tides and reported on the currents moving through the Strait of Gibraltar. Claudius Ptolemy (c. A.D. ~85–161) produced the first world atlas and established world boundaries: to the north, the British Isles, Northern Europe, and the unknown lands of Asia; to the south, an unknown land, "Terra Australis Incognita," including Ethiopia, Libya, and the Indian Sea; to the east, China; and to the west, the great Western Ocean reaching around Earth to China. His atlas listed more than 8000 places by latitude and longitude, but his work contained a major flaw. He had accepted a value of 29,000 km (18,000 mi) for Earth's circumference. This value was much too small and led Columbus, more than 1000 years later, to believe that he had reached the eastern shore of Asia when he landed in the Americas.

1.2 The Middle Ages

After Ptolemy, intellectual activity and scientific thought declined in Europe for about 1000 years. However, shipbuilding improved during this period; vessels became more seaworthy and easier to sail, so sailors could make longer voyages. The Vikings (Norse for *piracy*) were highly accomplished seamen who engaged in extensive exploration, trade, and colonization for nearly three centuries from about 793 to 1066 (fig. 1.4). During this time, they journeyed inland on rivers through Europe and western Asia, traveling as far as the Black and Caspian Seas. The Vikings are probably best known for their voyages across the North Atlantic Ocean. They sailed to Iceland in 871 where as many as 12,000 immigrants eventually settled. Erik Thorvaldsson (known as Erik the Red) sailed west from Iceland in 982 and discovered Greenland. He lived there for three years before returning to Iceland to recruit more settlers. Icelander Bjarni Herjolfsson, on his way to Greenland to join the colonists in 985–86, was blown off course, sailed south of Greenland, and

is believed to have come within sight of Newfoundland before turning back and reaching Greenland. Leif Eriksson, son of Erik the Red, sailed west from Greenland in 1002 and reached North America roughly 500 years before Columbus.

To the south, in the region of the Mediterranean after the fall of the Roman Empire, Arab scholars preserved Greek and Roman knowledge and continued to build on it. The Arabic writer El-Mas'údé (d. 956) gives the first description of the reversal of the currents due to the seasonal monsoon winds. Using this knowledge of winds and currents, Arab sailors established regular trade routes across the Indian Ocean. In the 1100s, large Chinese junks with crews of 200 to 300 sailed the same routes (between China and the Persian Gulf) as the Arab dhows.

During the Middle Ages, while scholarship about the sea remained primitive, the knowledge of navigation increased. Harbor-finding charts, or *portolanos,* appeared. These charts carried a distance scale and noted hazards to navigation, but they did not have latitude or longitude. With the introduction of the magnetic compass to Europe from Asia in the thirteenth century, compass directions were added. A Dutch navigational chart from Johannes van Keulen's *Great New and Improved Sea-Atlas or Water-World* of 1682–84 is shown in figure 1.5. The compass directions follow the pattern used in early fourteenth-century *portolanos.*

Although tides were not understood, the Venerable Bede (673–735) illustrated his account of the tides with data from the British coast. His calculations were followed in the tidal observations collected by the British Abbot Wallingford of Saint Alban's Monastery in about 1200. His tide table, titled "Flod at London Brigge," documented the times of high water. Sailors made use of Bede's calculations until the seventeenth century.

As scholarship was reestablished in Europe, Arabic translations of early Greek studies were translated into Latin and thus became available to European scholars. The study of tides continued to absorb the medieval scientists, who were also interested in the saltiness of the sea. By the 1300s, Europeans had established successful trade routes, including some partial ocean crossings. An appreciation of the importance of navigational techniques grew as trade routes were extended.

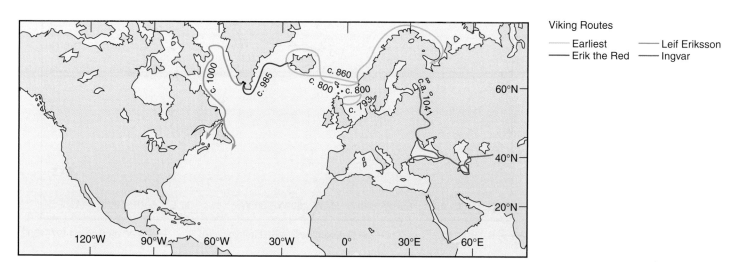

Figure 1.4 Major routes of the Vikings to the British Isles, to Asia, and across the Atlantic to Iceland, Greenland, and North America.

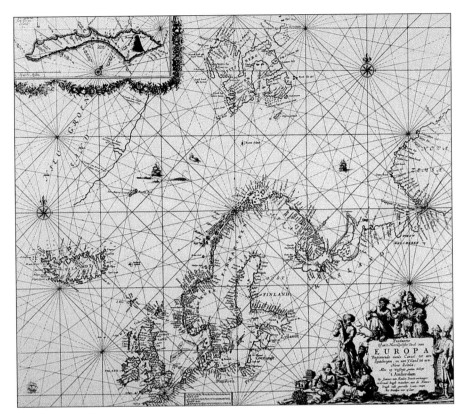

Figure 1.5 A navigational chart of northern Europe from Johannes van Keulen's *Sea-Atlas* of 1682–84.

1.3 Voyages of Discovery

From 1405 to 1433, the great Chinese admiral Zheng He conducted seven epic voyages in the western Pacific Ocean and across the Indian Ocean as far as Africa. Zheng He's fleet consisted of over 300 ships. The fleet is believed to have included as many as sixty-two "treasure ships" thought to have been as much as 122 m (400 ft) long and 52 m (170 ft) wide; this was ten times the size of the ships used for the European voyages of discovery during this period of time. The purpose of these voyages remains a matter of a debate among scholars. Suggested reasons include the establishment of trade routes, diplomacy with other governments, and military defense. The voyages ended in 1433, when their explorations led the Chinese to believe that other societies had little to offer, and the government of China withdrew within its borders, beginning 400 years of isolation.

In Europe, the desire for riches from new lands persuaded wealthy individuals, often representing their countries, to underwrite the costs of long voyages to all the oceans of the world. The individual most responsible for the great age of European discovery was Prince Henry the Navigator (1394–1460) of Portugal. In 1419, his father, King John, made him governor of Portugal's southernmost coasts. Prince Henry was keenly interested in sailing and commerce, and studied navigation and mapmaking. He established a naval observatory for the teaching of navigation, astronomy, and cartography about 1450. From 1419 until his death in 1460, Prince Henry sent expedition after expedition south along the west coast of Africa to secure trade routes and establish colonies. These expeditions moved slowly due to the mariners' belief that waters at the equator were at the boiling point and that sea monsters would engulf ships. It wasn't until twenty-seven years after Prince Henry's death that Bartholomeu Dias (1450?–1500) braved these "dangers" and rounded the Cape of Good Hope in 1487 in the first of the great voyages of discovery (fig. 1.6). Dias had sailed in search of new and faster routes to the spices and silks of the East.

Portugal's slow progress along the west coast of Africa in search for a route to the east finally came to fruition with Vasco da Gama (1469–1524) (fig. 1.6). In 1498, he followed Bartholomeu Dias's route to the Cape of Good Hope and then continued beyond along the eastern coast of the African continent. He successfully mapped a route to India but was

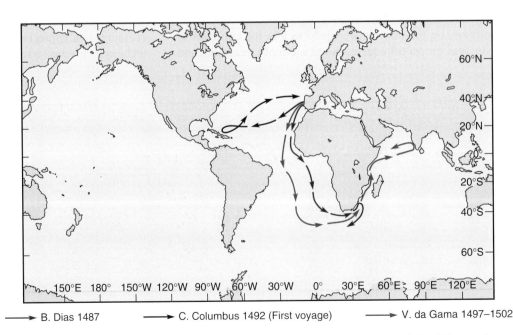

— B. Dias 1487   → C. Columbus 1492 (First voyage)   → V. da Gama 1497–1502

Figure 1.6 The routes of Bartholomeu Dias and Vasco da Gama around the Cape of Good Hope and Christopher Columbus's first voyage.

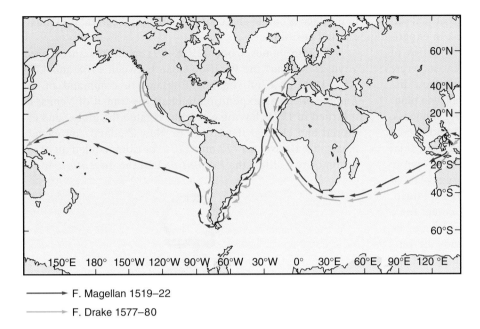

——▶ F. Magellan 1519–22

——▶ F. Drake 1577–80

Figure 1.7 The sixteenth-century circumnavigation voyages by Magellan and Drake.

challenged along the way by Arab ships. In 1502, da Gama returned with a flotilla of fourteen heavily armed ships and defeated the Arab fleet. By 1511, the Portuguese controlled the spice routes and had access to the Spice Islands. In 1513, Portuguese trade extended to China and Japan.

Christopher Columbus (1451–1506) made four voyages across the Atlantic Ocean in an effort to find a new route to the East Indies by traveling west rather than east. By relying on inaccurate estimates of Earth's size, he badly underestimated the distances involved and believed he had found islands off the coast of Asia when, in fact, he had reached the New World (fig. 1.6).

Italian navigator Amerigo Vespucci (1454–1512) made several voyages to the New World (1499–1504) for Spain and Portugal, exploring nearly 10,000 km (6000 mi) of South American coastline. He accepted South America as a new continent not part of Asia, and in 1507, German cartographer Martin Waldseemüller applied the name "America" to the continent in Vespucci's honor. Vasco Núñez de Balboa (1475–1519) crossed the Isthmus of Panama and found the Pacific Ocean in 1513, and in the same year, Juan Ponce de León (1460?–1521) discovered Florida and the Florida Current. All claimed the new lands they found for their home countries. Although these men had sailed for fame and riches, not knowledge, they more accurately documented the extent and properties of the oceans, and the news of their travels stimulated others to follow.

Ferdinand Magellan (1480–1521) left Spain in September 1519 with 270 men and five vessels in search of a westward passage to the Spice Islands. The expedition lost two ships before finally discovering and passing through the Strait of Magellan and rounding the tip of South America in November 1520. Magellan crossed the Pacific Ocean and arrived in the Philippines in March 1521, where he was killed in a battle with the natives on April 27, 1521. Two of his ships sailed on and reached the

Spice Islands in November 1521, where they loaded valuable spices for a return home. In an attempt to guarantee that at least one ship made it back to Spain, the two ships parted ways. The *Victoria* continued sailing west and successfully crossed the Indian Ocean, rounded Africa's Cape of Good Hope, and arrived back in Spain on September 6, 1522, with eighteen of the original crew. This was the first circumnavigation of Earth (fig. 1.7). Magellan's skill as a navigator makes his voyage probably the most outstanding single contribution to the early charting of the oceans. In addition, during the voyage, he established the length of a degree of latitude and measured the circumference of Earth. It is said that Magellan tried to test the mid-ocean depth of the Pacific with a hand line, but this idea seems to come from a nineteenth-century German oceanographer; writings from Magellan's time do not support this story.

By the latter half of the sixteenth century, adventure, curiosity, and hopes of finding a trading shortcut to China spurred efforts to find a sea passage around North America. Sir Martin Frobisher (1535?–94) made three voyages in 1576, 1577, and 1578, and Henry Hudson (d. 1611) made four voyages (1607, 1608, 1609, and 1610), dying with his son when set adrift in Hudson Bay by his mutinous crew. The Northwest Passage continued to beckon, and in 1615 and 1616, William Baffin (1584–1622) made two unsuccessful attempts.

While European countries were setting up colonies and claiming new lands, Francis Drake (1540–96) set out in 1577 with 165 crewmen and five ships to show the English flag around the world (fig. 1.7). He was forced to abandon two of his ships off the coast of South America. He was separated from the other two ships while passing through the Strait of Magellan. During the voyage Drake plundered Spanish shipping in the Caribbean and in Central America and loaded his ship with treasure. In June 1579, Drake landed off the coast of present-day California and sailed north along the coast to the present United States–Canadian border. He then turned southwest and crossed the Pacific Ocean in two months' time. In 1580, he completed his circumnavigation and returned home in the *Golden Hind* with a cargo of Spanish gold, to be knighted and treated as a national hero. Queen Elizabeth I encouraged her sea captains' exploits as explorers and raiders because, when needed, their ships and knowledge of the sea brought military victories as well as economic gains.

1.4 The Beginnings of Earth Science

New ideas and new knowledge had stimulated the practical exploration of the oceans during the fifteenth and sixteenth centuries, but most of the thinking about the sea was still rooted in the ideas of Aristotle and Pliny. In the seventeenth century, although the practical needs of commerce, national security, and economic and political expansionism guided events at sea, scientists on land were beginning to show an interest in experimental science and the study of the specific properties of substances. Curiosity about Earth flourished; scientists wrote pamphlets and formed societies in which to discuss their discoveries. The works of Johannes Kepler (1571–1630) on planetary motion and those of Galileo Galilei (1564–1642) on mass, weight, and acceleration would in time be used to understand the oceans. Although Kepler and Galileo had theories of tidal motion, it was not until Sir Isaac Newton (1642–1727) wrote his *Principia* in 1687 and gave the world the unifying law of gravity that the processes governing the tides were explained. Edmund Halley (1656–1742), the astronomer of comet fame and a friend of Newton, also had an interest in the oceans; he made a voyage in 1698 to measure longitude and study the variation of the compass, and he suggested that the age of the oceans could be calculated by determining the rate at which rivers carry salt to the sea. The physicist John Joly followed Halley's suggestion and in 1899 reported an age for the oceans of 90–100 million years based on the mass of salt in the oceans and the annual rate of addition of salt to the oceans. This estimate was far too young because he did not account for the recycling of salt, the incorporation of salt into seafloor sediments, or marine salt deposits.

1.5 The Importance of Charts and Navigational Information

As colonies were established far from their home countries and as trade, travel, and exploration expanded, interest was renewed in developing more accurate charts and navigational techniques. The first hydrographic office dedicated to mapping the oceans was established in France in 1720 and was followed in 1795 by the British Admiralty's appointment of a hydrographer.

As early as 1530, the relationship between time and longitude had been proposed by Flemish astronomer Gemma Frisius, and in 1598, King Philip III of Spain had offered a reward of 100,000 crowns to any clockmaker building a clock that would keep accurate time onboard ship (see the discussion of longitude and time in chapter 2, section 2.4). In 1714, Queen Anne of England authorized a public reward for a practical method of keeping time at sea, and the British Parliament offered 20,000 pounds sterling for a seagoing clock that could keep time with an error not greater than two minutes on a voyage to the West Indies from England. A Yorkshire clockmaker, John Harrison, built his first chronometer (high-accuracy clock) in 1735, but not until 1761 did his fourth model meet the test, losing only fifty-one seconds on the eighty-one-day voyage. Harrison was awarded only a portion of the prize after his success in 1761, and it was not until 1775, at the age of eighty-three, that he received the remainder from the reluctant British government. In 1772, Captain James Cook took a copy of the fourth version of Harrison's chronometer (fig. 1.8)

to produce accurate charts of new areas and correct previously charted positions.

Captain James Cook (1728–79) made his three great voyages to chart the Pacific Ocean between 1768 and 1779 (fig. 1.9). In 1768, he left England in command of the *Endeavour* on an expedition to chart the transit of Venus; he returned in 1771, having circumnavigated the globe, and explored and charted the coasts of New Zealand and eastern Australia. Between 1772 and 1775, he commanded an expedition of two ships, the *Resolution* and the *Adventure,* to the

Figure 1.8 John Harrison's fourth chronometer. A copy of this chronometer was used by Captain James Cook on his 1772 voyage to the southern oceans.

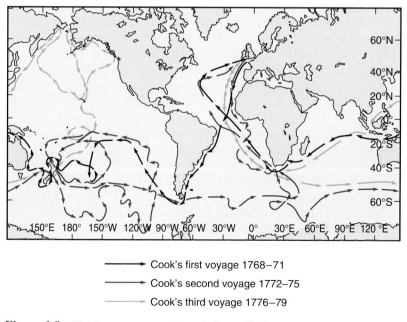

→ Cook's first voyage 1768–71
→ Cook's second voyage 1772–75
→ Cook's third voyage 1776–79

Figure 1.9 The three voyages of Captain James Cook.

South Pacific. On this journey, he charted many islands, explored the Antarctic Ocean, and, by controlling his sailors' diet, prevented vitamin C deficiency and scurvy, the disease that had decimated crews that spent long periods of time at sea. Cook sailed on his third and last voyage in 1776 in the *Resolution* and *Discovery*. He spent a year in the South Pacific and then sailed north, discovering the Hawaiian Islands in 1778. He continued on to the northwest coast of North America and into the Bering Strait, searching for a passage to the Atlantic. He returned to Hawaii for the winter and was killed by natives at Kealakekua Bay on the island of Hawaii in 1779. Cook takes his place not only as one of history's greatest navigators and seamen but also as a fine scientist. He made soundings to depths of 400 m (1200 ft) and accurate observations of winds, currents, and water temperatures. Cook's careful and accurate observations produced much valuable information and made him one of the founders of oceanography.

In the United States, Benjamin Franklin (1706–90) became concerned about the time required for news and cargo to travel between England and America. With Captain Timothy Folger, his cousin and a whaling captain from Nantucket, he constructed the 1769 Franklin-Folger chart of the Gulf Stream current (fig. 1.10). When published, the chart encouraged captains to sail within the Gulf Stream en route to Europe and return via the trade winds belt and follow the Gulf Stream north again to Philadelphia, New York City, and other ports. Since the Gulf Stream carries warm water from low latitudes to high latitudes, it is possible to map its location with satellites that measure sea surface temperature. Compare the Franklin-Folger chart in figure 1.10 to a map of the Gulf Stream shown in figure 1.11 based on the average sea surface temperature during 1996. In 1802, Nathaniel Bowditch (1773–1838), another American, published the *New American Practical Navigator*. In this book, Bowditch made the techniques of celestial navigation available for the first time to every competent sailor and set the stage for U.S. supremacy of the seas during the years of the Yankee clippers. When Bowditch died, his copyright was bought by the U.S. Navy, and his book continued in print. Its information was

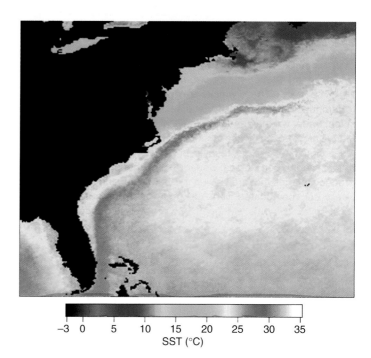

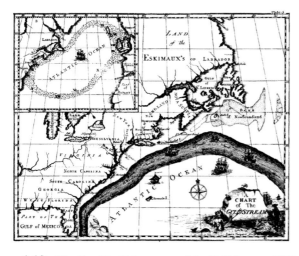

Figure 1.10 The Franklin-Folger map of the Gulf Stream, 1769. Compare this map with figure 1.11.

Figure 1.11 Annual average sea surface temperature for 1996. The *red-orange streak* of 25° to 30°C water shows the Gulf Stream. Compare with figure 9.8.

updated and expanded with each edition, serving generations of mariners and navigators.

In 1807, the U.S. Congress, at the direction of President Thomas Jefferson, formed the Survey of the Coast under the Treasury Department, later named the Coast and Geodetic Survey and now known as the National Ocean Survey. The U.S. Naval Hydrographic Office, now the U.S. Naval Oceanographic Office, was set up in 1830. Both were dedicated to exploring the oceans and producing better coast and ocean charts. In 1842, Lieutenant Matthew F. Maury (1806–73), who had worked with the Coast and Geodetic Survey, was assigned to the Hydrographic Office and founded the Naval Depot of Charts. He began a systematic collection of wind and current data from ships' logs. He produced his first wind and current charts of the North Atlantic in 1847. At the 1853 Brussels Maritime Conference, Maury issued a plea for international cooperation in data collection, and from the ships' logs he received, he produced the first published atlases of sea conditions and sailing directions. His work was enormously useful, allowing ships to sail more safely and take days off their sailing times between major ports around the world. The British estimated that Maury's sailing directions took thirty days off the passage from the British Isles to California, twenty days off the voyage to Australia, and ten days off the sailing time to Rio de Janeiro. In 1855, he published *The Physical Geography of the Sea*. This work includes chapters on the Gulf Stream, the atmosphere, currents, depths, winds, climates, and storms, as well as the first bathymetric chart of the North Atlantic with contours at 6000, 12,000, 18,000, and 24,000 ft. Many marine scientists consider Maury's book the first textbook of what we now call oceanography and consider Maury the first true oceanographer. Again, national and commercial interests were the driving forces behind the study of the oceans.

Marine Archaeology

More than 2000 years of exploration, war, and trading have left thousands of wrecks scattered across the ocean floors. These wrecks are great storehouses of information for archaeologists and historians searching for information about ships, trade, warfare, and the details of human life in ancient times. Marine archaeologists use techniques developed for oceanographic research to find, explore, recover, and preserve wrecks and other artifacts lying under the sea. Sound beams that sweep the sea floor produce images that are viewed onboard ship, and an object of interest can be located very precisely with computer-controlled positioning systems. Magnetometers are also used in these searches to detect iron from sunken vessels. Divers can be sent down to verify a wreck in shallow water, and a research submersible (submarine) carrying observers can be used in deeper water. Unstaffed, towed camera and instrument sleds or remotely operated vehicles (ROVs) equipped with underwater video cameras explore in deep water or areas that are difficult or unsafe for divers and submersibles. ROVs and submersibles also collect samples to help identify wrecks.

The oldest known shipwreck, from the fourteenth century B.C., a Bronze Age merchant vessel, was discovered in 1983 more than 33 m (100 ft) down in the Mediterranean Sea off the Turkish coast. Divers have recovered thousands of artifacts from its cargo, including copper and tin ingots, pottery, ivory, and amber. Using these items, archaeologists have been able to learn about the life and culture of the period, trace the ship's trade route, and understand more about Bronze Age people's shipbuilding skills.

The waters around northern Europe have claimed thousands of wrecks since humans began to trade and voyage along these coasts. Two vessels, found in shallow water, have provided quantities of information for historians and archaeologists. In the summer of 1545, the English warship *Mary Rose* sank as it sailed out to engage the French fleet. The ship was studied in place and raised in 1982. More than 17,000 objects were salvaged from it, giving archaeologists and naval historians insights into the personal and working lives of the officers and crew of a naval vessel at that time. The portion of the hull that was buried in the mud was preserved and is on display at Portsmouth, England (box fig. 1). The Swedish man-of-war *Vasa* sank in 1628 in Stockholm Harbor at the beginning of its maiden voyage. The ship was located in 1956 and raised in 1961. The *Vasa* was a great ship, over 200 ft long and, like others of the period, fantastically decorated with carvings and statues. Divers searched the seabed every summer from 1963–67 and recovered sculptures and carved details that had adorned the ship. Because the water of the Baltic Sea is much less salty than the open ocean, there were no shipworms to destroy the wooden hull, and the *Vasa* is now on exhibit in Stockholm, Sweden (box fig. 2).

Another fighting ship of this period, the Spanish galleon *San Diego,* sank in 1600 as it engaged two Dutch vessels off the Philippine coast. An intensive, two-year archaeological excavation began in 1992. Relics recovered include hundreds of pieces of intact Chinese porcelain, a bronze astrolabe for determining latitude, and a bronze and glass compass. Most of the hull had been destroyed by shipworms and currents; surviving pieces were measured and then covered with sand for protection.

Wrecks that lie in deep water are initially much better preserved than those in shallow water because they lie below the depths of the strong currents and waves that break up most shallow-water

Box Figure 1 The hull of the *Mary Rose* in its display hall at Portsmouth Dockyard, England. To preserve and stabilize it, the remaining wood is constantly sprayed with a preservative solution.

Box Figure 2 The *Vasa,* a seventeenth-century man-of-war, is on permanent display in Stockholm, Sweden. Most of the *Vasa* is original, including two of the three masts and parts of the rigging.

wrecks. Shallow-water wrecks are also often the prey of treasure hunters who destroy the history of the site while they search for adventure and items of market value. Over long periods, the cold temperatures and low oxygen content of deep water favor preservation of wooden vessels by slowing decomposition and excluding the marine organisms that bore into wood in shallow areas. Also, the muds and sands falling from above cover objects on the deep-sea floor much more slowly than they do in shallow water.

In 1685, the *Belle,* the ship of the explorer Robert Cavelier, Sieur de la Salle, sank in a storm in Matagordo Bay, along the Texas coast. In 1995, this wreck was located by scanning the bay for magnetic anomalies caused by iron in the wreckage. Reaching the vessel, under 4 m (12 ft) of water and tons of mud required a cofferdam made of two concentric octagonal walls of interlocking steel plates. It was constructed 12 m (40 ft) below the bay floor and 6 m (18 ft) above sea level. The water and mud were then pumped out of the cofferdam to allow access to the *Belle* (box fig. 3). The wreck yielded a rich harvest of artifacts, including cannons, brass pots, candlesticks, coils of rope and brass wire, personal belongings, and trade goods such as rings, glass beads, and combs.

Robert Ballard of the Institute for Exploration in Mystic, Connecticut, has led several expeditions using sophisticated electronic and robotic equipment to find both ancient and modern vessels. He discovered the wreck of the *Titanic* in 1985, 4790 m (15,700 ft) down in the North Atlantic. In 1989, he searched 518 km^2 (200 mi^2) of ocean to find the German battleship *Bismarck,* sunk in 1941 after one of the most famous sea battles of World War II. The ship lies 4750 m (15,600 ft) below the surface (box fig. 4). Both ships were located using a towed underwater camera sled, controlled from the vessel at the surface. Once the wrecks were found, observers descended via submersible for direct observation and further inspection of the vessels. Photographic surveys using ROVs that could be maneuvered to take pictures inside and outside the wrecks documented both vessels.

In 1997, a team of oceanographers, engineers, and archaeologists led by Ballard discovered a cluster of five Roman ships, along with thousands of artifacts, on the bottom of the Mediterranean Sea. The ships lie in 762 m (2500 ft) of water beneath an ancient trade route, where they were lost sometime between about 100 B.C. and A.D. 400. Prior to this discovery, no major ancient shipwrecks had been discovered and explored by archaeologists in water deeper than 61 m (200 ft). The ships were remarkably well

Box Figure 4 The World War II German battleship *Bismarck* was sunk in 1941 and lies 4790 m (15,700 ft) below the surface in the North Atlantic. In 1989, the towed camera sled Argo photographed part of the ship's superstructure.

preserved, having been protected from looting and encrustation with coral by the deep water in which they sank. The vessels were spread over an area of about 52 km^2 (20 mi^2) and probably sank when they were caught in sudden, violent storms. The ships were initially located using the U.S. Navy's nuclear submarine *NR-1.* The *NR-1* is capable of searching over large areas for extended periods of time, and its long-range sonar can detect objects at a much greater distance than conventional sonar systems typically used by oceanographers. Once the site had been identified, the small ROV *Jason* was used for detailed mapping, observation, and the recovery of over 100 artifacts selected to help the archaeologists date the ancient wrecks.

These expeditions have moved marine archaeology from the shallow waters into the deep sea. Today's oceanographic instrumentation makes all shipwrecks available to archaeologists.

To Learn More About Marine Archaeology

Ballard, R. D. 1985. How We Found *Titanic. National Geographic* 168 (6): 698–722.

Ballard, R. D. 1986. A Long Last Look at *Titanic. National Geographic* 170 (6): 698–727.

Ballard, R. D. 1989. Find the *Bismarck. National Geographic* 176 (5): 622–37.

Ballard, R. D. 1998. High-Tech Search for Roman Shipwrecks. *National Geographic* 193 (4): 32–41.

Bass, G. F. 1987. Oldest Known Shipwreck Reveals Splendors of the Bronze Age. *National Geographic* 172 (6): 693–734.

Gerard, S. 1992. The Carribean Treasure Hunt. *Sea Frontiers* 38 (3): 48–53.

Goddio, F. 1994. The Tale of the *San Diego. National Geographic* 186 (1): 35–56.

LaRoe, L. 1997. La Salle's Last Voyage. *National Geographic* 191 (5): 72–83.

Oceanus 28 (4) 1985-86. Winter issue devoted to the discovery of the *Titanic.*

Roberts, D. 1997. Sieur de la Salle's Fateful Landfall. *Smithsonian* 28 (1): 40–52.

Box Figure 3 Surrounded by a cofferdam to keep out mud and water, archaeologists excavate amidships of the wreck of the *Belle.* Shop vacuums and water hoses are used to expose the ship's structure and cargo, including casks and boxes of muskets.

1.6 Ocean Science Begins

As charts became more accurate and as information about the oceans increased, the oceans captured the interest of naturalists and biologists. Baron Alexander von Humboldt (1769–1859) made observations on a five-year (1799–1804) cruise to South America; he was particularly fascinated with the vast numbers of animals inhabiting the current flowing northward along the western coast of South America, the current that now bears his name. Charles Darwin (1809–82) joined the survey ship *Beagle* and served as the ship's naturalist from 1831–36 (fig. 1.12). He described, collected, and classified organisms from the land and sea. His theory of atoll formation is still the accepted explanation.

At approximately the same time, another English naturalist, Edward Forbes (1815–54), began a systematic survey of marine life around the British Isles and in the Mediterranean and Aegean seas. He collected organisms in deep water and, on the basis of his observations, proposed a system of ocean depth zones, each characterized by specific animal populations. However, he also mistakenly theorized that there was an azoic, or lifeless, environment below 550 m (1800 ft). His announcement is curious, because twenty years earlier, the Arctic explorer Sir John Ross (1777–1856), looking again for the Northwest Passage, had taken bottom samples at over 1800 m (6000 ft) depth in Baffin Bay with a "deep-sea clamm," or bottom grab, and had found worms and other animals living in the mud. Ross's nephew, Sir James Clark Ross (1800–62), took even deeper samples from Antarctic waters and noted their similarity to the Arctic species recovered by his uncle. Still, Forbes's systematic attempt to make orderly predictions about the oceans, his enthusiasm, and his influence make him another candidate as a founder of oceanography.

Christian Ehrenberg (1795–1876), a German naturalist, found the skeletons of minute organisms in seafloor sediments and recognized that the same organisms were alive at the sea surface; he concluded that the sea was filled with microscopic life and that the skeletal remains of these tiny organisms were still being added to the sea floor. Investigation of the minute drifting plants and animals of the ocean was not seriously undertaken until German scientist Johannes Müller (1801–58) began his work in 1846. He used an improved fine-mesh tow net similar to that used by Charles Darwin to collect these organisms, which he examined microscopically. This work was continued by Victor Hensen (1835–1924), who improved the Müller net, introduced the quantitative study of these minute drifting sea organisms, and gave them the name *plankton* in 1887.

Although science blossomed in the seventeenth and eighteenth centuries, there was little scientific interest in the sea except as we have seen for the practical reasons of navigation, tide prediction, and safety. In the early nineteenth century, ocean scientists were still few and usually only temporarily attracted to the sea. Some historians believe that the subject and study of the oceans were so vast, requiring so many people and such large amounts of money, that government interest and support were required before oceanography could grow as a science. This did not happen until the nineteenth century in Great Britain.

In the last part of the nineteenth century, laying transatlantic telegraph cables made a better knowledge of the deep sea a necessity. Engineers needed to know about seafloor conditions, including bottom topography, currents, and organisms that might

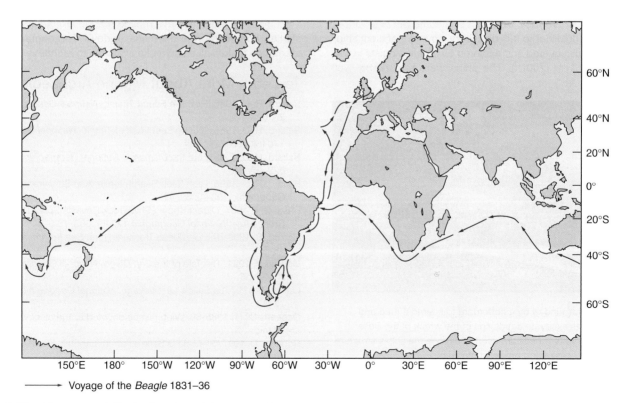

———→ Voyage of the *Beagle* 1831–36

Figure 1.12   The voyage of Charles Darwin and the survey ship HMS *Beagle,* 1831–36.

dislodge or destroy the cables. The British began a series of deep-sea studies stimulated by the retrieval of a damaged cable from more than 1500 m (5000 ft) deep, well below Forbes's azoic zone. When the cable was brought to the surface, it was found to be covered with organisms, many of which had never been seen before. In 1868, the *Lightning* dredged between Scotland and the Faroe Islands at depths of 915 m (3000 ft) and found many animal forms. The British Admiralty continued these studies with the *Porcupine* during the summers of 1869 and 1870, dredging up animals from depths of more than 4300 m (14,000 ft). Charles Wyville Thomson (1830–82), like Forbes, a professor of natural history at Edinburgh University, was one of the scientific leaders of these two expeditions. On the basis of these results, he wrote *The Depths of the Sea,* published in 1873, which became very popular and is regarded by some as the first book on oceanography.

English biologist Thomas Henry Huxley (1825–95), a close friend and supporter of Charles Darwin, was particularly interested in studying the organisms that inhabit the deep sea. Huxley was a strong supporter of Darwin's theory of evolution and believed that the organisms of the deep sea could supply evidence of its validity. In 1868, while examining samples of mud recovered from the deep-sea floor of the Atlantic eleven years before and preserved in alcohol, Huxley noticed that the surface of the samples was covered by a thick, mucus-like material with small embedded particles. Under the microscope it appeared as if these particles moved, leading him to conclude that the mucus was a form of living protoplasm. Huxley named this "organism" *Bathybius haeckelii* after the noted German naturalist Ernst Haeckel (fig. 1.13*b*). Haeckel, also a strong supporter of the theory of evolution, viewed this protoplasm as the primordial ooze from which all other life evolved. He believed it blanketed the deep-sea floor and provided an inexhaustible supply of food for higher order organisms of the deep ocean. One of the primary scientific objectives of the *Challenger* expedition, described in the next section, was to study the distribution of *Bathybius haeckelii.*

1.7 The *Challenger* Expedition

With public interest running high, the Circumnavigation Committee of the British Royal Society was able to persuade the British Admiralty to organize the most comprehensive oceanographic expedition yet undertaken. The Society obtained the use of the naval corvette *Challenger,* a sailing vessel with auxiliary steam power. All but two of the corvette's guns were removed, and the ship was refitted with laboratories, winches, and equipment, including 232 km (144 mi) of sounding rope. The leadership was offered to Charles Wyville Thomson, whose assistant was a young geologist, John Murray (1841–1914). The *Challenger* sailed from Portsmouth, England, on December 21, 1872, for a voyage that was to last nearly three-and-a-half years, during which time the vessel logged 110,840 km (68,890 mi) (fig. 1.13). The first leg of the voyage took the vessel to Bermuda, then to the South Atlantic island of Tristan da Cunha, around the Cape of Good Hope, and east across the southernmost part of the Indian Ocean; this vessel was the first steamship to cross the Antarctic Circle. It continued on to Australia, New Zealand, the Philippines, Japan, and

China. Turning south to the Marianas Islands, the vessel took its deepest sounding at 8180 m (26,850 ft). Sailing across the Pacific to Hawaii, Tahiti, and through the Strait of Magellan, it returned to England on May 24, 1876. Queen Victoria conferred a knighthood on Thomson, and the *Challenger* expedition was over.

The *Challenger* expedition's purpose was scientific research; during the voyage, the crew took soundings at 361 ocean stations, collected deep-sea water samples, investigated deepwater motion, and made temperature measurements at all depths. Thousands of biological and sea-bottom samples were collected. The cruise brought back evidence of an ocean teeming with life at all depths and opened the way for the era of descriptive oceanography that followed.

The numerous dredges of deep-sea sediments obtained during the expedition failed to find any evidence of *Bathybius haeckelii* in fresh samples. One of the naturalists noticed, however, when alcohol was added to a sample, something similar to *Bathybius haeckelii* was produced. Rather than being the primitive life form from which all other organisms evolved, *Bathybius haeckelii* was shown to be a chemical precipitate produced by the reaction of the sediment with alcohol.

Although the *Challenger* expedition ended in 1876, the work of organizing and compiling information continued for twenty years, until the last of the fifty-volume *Challenger Reports* was issued. John Murray (later Sir John Murray) edited the reports after Thomson's death and wrote many of them himself. He is considered the first geological oceanographer. William Dittmar (1833–92) prepared the information on seawater chemistry for the *Challenger Reports.* He identified the major elements present in the water and confirmed the findings of earlier chemists that in a seawater sample, the relative proportions of the major dissolved elements are constant. Oceanography as a modern science is usually dated from the *Challenger* expedition. The *Challenger Reports* laid the foundation for the science of oceanography.

The *Challenger* expedition stimulated other nations to mount ocean expeditions. Although their avowed purpose was the scientific exploration of the sea, in large measure prestige was at stake. Norway explored the North Atlantic with the *Voringen* in the summers of 1876–78; Germany studied the Baltic and North Seas in the SS *Pomerania* in 1871 and 1872 and in the *Crache* in 1881, 1882, and 1884. The French government financed cruises by the *Travailleur* and the *Talisman* in the 1880s. The Austrian ship *Pola* worked in the Mediterranean and Red seas in the 1890s. The U.S. vessel *Enterprise* circumnavigated Earth between 1883 and 1886, as did Italian and Russian ships between 1886 and 1889.

1.8 Oceanography as Science

During the late nineteenth and early twentieth centuries, intellectual interest in the oceans increased. Oceanography was changing from a descriptive science to a quantitative one. Oceanographic cruises now had the goal of testing hypotheses by gathering data. Theoretical models of ocean circulation and water movement were developed. The Scandinavian oceanographers were particularly active in the study of water movement. One of them, Fridtjof Nansen (1861–1930) (fig. 1.14*a*), a well-known

Figure 1.13 The *Challenger* expedition: December 21, 1872, to May 24, 1876. Engraving from the *Challenger Reports,* volume 1, 1885.
(a) "H.M.S. *Challenger*—Shortening Sail to Sound," decreasing speed to take a deep-sea depth measurement. (b) *Bathybius haeckelii,* the supposed primordial protoplasm believed to cover the ocean floor. (c) "Dredging and Sounding Arrangement on board *Challenger.*" Rigging is hung from the ship's yards to allow the use of the over-the-side sampling equipment. A biological dredge can be seen hanging outboard of the rail. The large cylinders in the rigging are shock absorbers. (d) Sieving bottom samples for organisms. (e) "Deep Sea Deposits." This plate shows the shells of microscopic organisms making up different kinds of muds and clays from the floor of the deep sea. (f) "H.M.S. *Challenger* at St. Paul's Rocks," in the equatorial mid-Atlantic. (g) "Zoological Laboratory on the Main Deck." (h) "Chemical Laboratory." (i) A biological dredge used for sampling bottom organisms. Note the frame and skids that keep the mouth of the net open and allow it to slide over the sea floor.

14

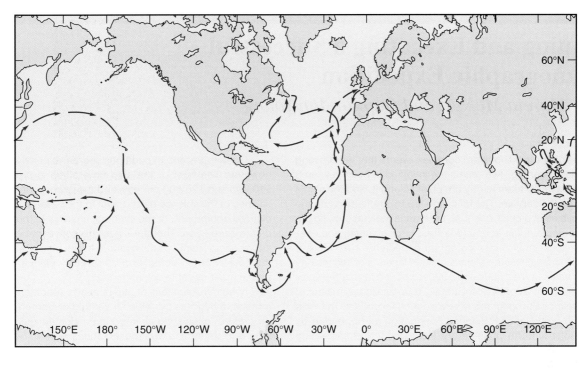

(j) ——————▶ Cruise of the *Challenger*, 1872–76

Figure 1.13 (continued) (j) The cruise of the *Challenger,* the first major oceanography research effort.

(a)

(b)

Figure 1.14 (a) Fridtjof Nansen, Norwegian scientist, explorer, and statesman (1861–1930), using a sextant to determine his ship's position. (b) The *Fram* frozen in ice. As the ice pressure increased, it lifted its specially designed and strengthened hull so that the ship would not be crushed.

athlete, explorer, and zoologist, was interested in the currents of the polar seas. This extraordinary man decided to test his ideas about the direction of ice drift in the Arctic by freezing a vessel into the polar ice pack and drifting with it to reach the North Pole. To do so, he had to design a special vessel that would be able to

survive the great pressure from the ice; the 39 m (128 ft) wooden *Fram* ("to push forward"), shown in figure 1.14*b,* was built with a smoothly rounded hull and planking over 60 cm (2 ft) thick.

Nansen departed with thirteen men from Oslo in June 1893. The ship was frozen into the ice nearly 1100 km (700 mi) from

Field Notes

Planning and Executing a Successful Oceanographic Expedition

by Dr. Marcia McNutt and Captain Ian Young

Oceanographic expeditions are complex events that require long and detailed planning. Two teams are involved in every research cruise: the science team led by the chief scientist, and the ship's crew, led by the captain. The following are two reflections on the responsibilities of a chief scientist, Dr. Marcia McNutt (box fig. 1), and a ship captain, Ian Young (box fig. 2), for planning and executing a successful oceanographic expedition. Dr. McNutt served on the faculty at the Massachusetts Institute of Technology for fifteen years in the 1980s and 1990s, where she was the Griswold Professor of Geophysics. Captain Young sailed as master of the R/V *Maurice Ewing* for many years during the 1990s. This ship is owned by the National Science Foundation and operated by Lamont-Doherty Earth Observatory. Both Dr. McNutt and Young are employed at the Monterey Bay Aquarium Research Institute (box fig. 3), where Young sails on the R/V *Western Flyer* and Dr. McNutt is the president and CEO.

The Role of the Chief Scientist

Box Figure 1 Dr. Marcia McNutt.

All scientists are motivated by the thrill of discovery. For oceanographers such as myself, discovery does not lie in a test tube in the lab but in the ocean itself. Most of the deep sea remains either completely unexplored or observed in only a cursory sense with modern technology, and thus there are ample opportunities to go where no one has gone before and to see what is there through new sets of technological "eyes." My own expeditions invariably return with wondrous new discoveries that were not predicted by any prevailing theories. For that reason, I live for going to sea.

Oceanographic expeditions are an expensive undertaking, however. The cost of the ship time alone is typically between $10,000 and $20,000 per day, and those costs do not include the salaries of the science party, travel to the vessels, use of very specialized equipment, or the eventual analysis of the data collected on the expedition. Therefore, it is imperative that each expedition address important scientific problems, use the best technology for the data to be collected, and be carefully planned and executed by the most capable scientific team.

I begin the process of planning an expedition by formulating a testable hypothesis, which if verified or refuted, would lead to progress in answering an important scientific question. The next step is to determine what observations need to be made or what experiment needs to be performed to test the hypothesis. From this consideration naturally flows the equipment that must be assembled and the expertise needed from my team. I select other team members primarily based on their scientific ability and expertise, but personal considerations (such as do you want to be confined for 30–60 days on 250 ft of floating steel with this person) weigh in as well.

The team then writes a proposal to a funding agency, commonly the National Science Foundation, to support the expedition and waits. And waits. The proposal is reviewed by other scientists who make recommendations on which projects should be funded with the limited money available. Typically, about six months later a decision is reached as to which expeditions will be supported. If my team is one of the lucky ones, we are then placed on the schedule for one of the ships in the national fleet of vessels operated by academic institutions. To which ship we are assigned depends on the type of equipment needed on the ship and where the research area is. Much of my own work is specialized enough that only one or two ships can conduct the mission. Therefore, we may have to wait another year or more for that ship to eventually sail to the right part of the world to conduct our experiment.

Once the ship is determined, I write up a summary of the proposed work for the benefit of the captain and technical personnel on the designated ship. Unlike the science proposal, which is meant to excite other scientists about the importance of the proposed expedition, this summary is a detailed account of where we will go, what equipment we will use, and how we will use it. Several months before the expedition, I meet with the operators of the ship and its captain to discuss the mission and answer questions they may have about the summary. This is our chance to make sure that no unpleasant surprises surface after we set sail (e.g., "We thought that you were bringing liners for the piston core." "No, I thought the ship already stocked liners for the core!"). This is also the time when we make sure that we have the necessary permits to work in the territorial waters of other nations.

If this is my first meeting with the captain, I want very much to impress him with my knowledge of marine operations and thoroughness of preparation. Our teamwork must be built on a foundation of mutual respect: I must trust him not to place limitations on our scientific operations at sea unless he truly feels that the safety of the vessel and its occupants are at risk. On the other hand, he must feel confident that I would not ask for anything extra from the crew or the ship unless the success of the science mission depended on it.

Once at sea, we make the most of every minute. Operations go around the clock, with subsets of the crew and science party assigned to pairs of four-hour watches. As chief scientist, it is my responsibility to make sure that the captain and mates on the bridge know well in advance what the science plan is for the next watch. From a science perspective, the very best expeditions are those in which the cruise plan is constantly in flux as we follow up on new discoveries. But such expeditions are not always the favorite on the bridge, nor do they allow the chief scientist to get eight hours of uninterrupted sleep!

In the final analysis, the success of any expedition can often be attributed to the willingness of the captain and his crew to put in the extra effort necessary to achieve what are invariably ambitious objectives in the face of bad weather, temperamental equipment, and Murphy's laws. That is why the last line in so many oceanographic research publications is some variant of the following: "Special thanks to the captain and crew of the R/V _____ for making this expedition such a success."

The Role of the Ship's Captain

I've sailed professionally for more than twenty years on various types of vessels, with different levels of responsibility. Whether signed on a freighter as third mate bound for Africa or wandering the world's oceans on a tramp research vessel, I've always loved being "safely back to sea."

As captain of a modern research vessel, I am primarily concerned with the readiness of the ship. Usually, no two expeditions are alike, and each requires unique preparations. Information about each scientific voyage is provided by the chief scientist either through the ship's marine office or directly to the captain one to three months in advance.

The captain starts planning for the cruise months before it begins. The five Ps apply at all times: **P**roper **P**lanning **P**revents **P**oor **P**erformance. This includes ordering charts for the research site, preparing equipment, and performing needed maintenance on the ship's machinery and systems (electrical and mechanical) that will be needed during the next cruise for around the clock operations.

The captain relies on the crew—science technician, chief mate, chief engineer, and steward—to make all preparations for each cruise. The ship's science technician is responsible for making sure that the ship's data collection systems are ready and compatible with science computers. The chief mate will perform deck maintenance to have equipment ready and deck space clear for storing, staging, launching, and recovering equipment. All safety equipment has to be in good working order as required by the U.S. Coast Guard. The chief engineer plans ahead to order fuel, lubricating oil, hydraulic oil, and spare parts for all the systems under his care. These systems are all important for the success of the

Box Figure 2 Captain Ian Young.

science mission. A failure in any one of these systems can delay, slow, or stop an expedition. The ship's steward must choose and stock adequate provisions for forty people for a thirty- to fifty-day voyage, a job that requires experience and attention to detail to ensure quality meals.

Ideally, clearances from the State Department will arrive prior to the vessel sailing. These clearances give the vessel legal permission to work in the territorial waters of foreign sovereign nations and are required to avoid conflict with foreign governments. Depending on the proposed location for the research, multiple countries may need to be contacted for the same expedition. The ship's marine office typically applies for the necessary clearances six to twelve months in advance of the cruise. A delay in the receipt of a clearance can have serious consequences because it can force a cruise to be rescheduled for a later date. Rescheduling can result in the expedition missing the optimum weather window for the research site, which can degrade the quality and quantity of data collected.

The captain has to work closely with the chief scientist to make sure the cruise objectives are achieved. This starts with the first science meeting held between them. Here, the captain gets the first full view of the cruise for which he has been preparing the vessel. The cruise can have many different aspects: station work, towed gear deployments, bottom instrument launches, buoy recoveries, and grid surveys. This meeting provides an opportunity for the captain to contribute operations suggestions to help the scientists make better use of time and equipment because he knows the capabilities of the ship and personnel.

(continued)

(concluded)

The captain is always watching the weather. Sometimes work can be scheduled around certain weather patterns. Quite often there are "weather days" when work has to stop for the safety of the people on deck and the vessel. This may require the vessel heaving-to for a night or retrieving gear and leaving the area due to a hurricane or typhoon. This is a difficult decision for the captain to make, because the chief scientist never wants to stop collecting data and lose time from the cruise. This is one of the many situations in which the captain's decision has to be right; an error in judgment could jeopardize the safety of the vessel and the people on board or cost him his job.

At the end of the voyage, there is a time-honored ritual involving the captain and chief scientist: negotiating the break-off time for the "end of science." The distance to port from the work area may be measured by a few hours or a few weeks. The captain and chief scientist will estimate the time required to return to port; frequently, both will come up with different answers to this question. The captain will allow for weather and currents, while the chief scientist generally will not. At this point in the cruise, the chief scientist likes to state: "If I can't have time to collect the last bit of data, then the whole expedition will be for nothing." Of course, this isn't true, but it contributes to the drama required to end a successful cruise. Usually, the time requested to finish the scientific

Box Figure 3 Aerial view of the Monterey Bay Aquarium Research Institute.

mission is within reason, and the captain is able to safely return to port on schedule. Even though there are days of transit to the next port, in the captain's mind, this trip is over, and he is already preparing for the next expedition.

the North Pole and remained in the ice for thirty-five months. During this period, measurements were made through holes in the ice that showed that the Arctic Ocean was a deep-ocean basin and not the shallow sea that had been expected. Water and air temperatures were recorded, water chemistry was analyzed, and the great plankton blooms of the area were observed. Nansen became impatient with the slow rate of drift and, with F. H. Johansen, left the *Fram* locked in the ice some 500 km (300 mi) from the pole. They set off with dogsleds toward the pole, but after four and a half weeks, they were still more than 300 km (200 mi) from the pole, with provisions running low and the condition of their dogs deteriorating. The two men turned away from the pole and spent the winter of 1895–96 on the ice, living on seals and walrus. They were found by a British polar expedition in June 1896 and returned to Norway in August of that year. The crew of the *Fram* continued to drift with the ship until they freed the vessel from the ice in 1896 and returned home. Nansen's expedition had laid the basis for future Arctic work.

After the expedition's findings were published, Nansen continued to be active in oceanography, and his name is familiar today from the Nansen bottle, which he designed to collect water samples from deep water. In 1905, he turned to a career as a statesman, working for the peaceful separation of Norway from Sweden. After World War I, he worked with the League of Nations to resettle refugees, for which he received the 1922 Nobel Peace

Prize. The well-designed *Fram* paid another visit to polar waters, carrying the Norwegian explorer Roald Amundsen (1872–1928) to the Antarctic continent on his successful 1911 expedition to the South Pole. It was also Amundsen who finally made a Northwest Passage entirely by water in the *Gjoa,* leaving Norway in 1903 and arriving in Nome, Alaska, three years later (fig. 1.15).

Fluctuations in the abundance of commercial fish in the North Atlantic and adjacent seas, and the effect of these changes on national fishing programs, stimulated oceanographic research and international cooperation. As early as 1870, researchers began to realize their need for knowledge of ocean chemistry and physics to understand ocean biology. The study of the ocean and its fisheries required crossing national boundaries, and in 1902, Germany, Russia, Great Britain, Holland, and the Scandinavian countries formed the International Council for the Exploration of the Sea to coordinate and sponsor research in the ocean and in fisheries.

Advances in theoretical oceanography sometimes could not be verified with practical knowledge until new instruments and equipment were developed. Lord Kelvin (1824–1907) invented a machine in 1872 that made it possible to combine tidal theory with astronomical predictions to predict the tides. Deep-sea circulation could not be systematically explored until approximately 1910, when Nansen's water-sampling bottles were combined with thermometers designed for deep-sea temperature

Figure 1.15 (a) Roald Amundsen, Norwegian explorer of the Arctic and Antarctic (1872–1928). (b) The *Gjoa* preparing for its journey through the Northwest Passage (1903).

measurements and an accurate method for measuring salinity was devised by the chemist Martin Knudsen (1871–1949). The reliable and accurate measurement of ocean depths had to wait until the development of the echo sounder, which was given its first scientific use on the 1925–27 German cruise of the *Meteor.* Although the *Meteor* expedition was supposedly sent out for purely oceanographic reasons, it was also an attempt by the German government to find an affordable way to separate dissolved gold from seawater. Although the expedition failed to find a cheap way to produce gold, it did accumulate a great deal of information about the South Atlantic.

1.9 Oceanography in the Twentieth Century

In the United States, government agencies related to the oceans proliferated during the nineteenth century. These agencies were concerned with gathering information to further commerce, fisheries, and the navy. After the Civil War, the replacement of sail by steam lessened government interest in studying winds and currents and in surveying the ocean floor. Private institutions and wealthy individuals took over the support of oceanography in the United States. Alexander Agassiz (1835–1910), mining engineer, marine scientist, and Harvard University professor,

financed a series of expeditions that greatly expanded knowledge of deep-sea biology. Agassiz served as the scientific director on the first ship built especially for scientific ocean exploration, the U.S. Fish Commission's *Albatross,* commissioned in 1882. He designed and financed much of the deep-sea sampling equipment that enabled the *Albatross* to recover more specimens of deep-sea fishes in one haul than the *Challenger* had collected during its entire three-and-a-half years at sea.

One of Agassiz's students, William E. Ritter, became a professor of zoology at the University of California–Berkeley. From 1892–1903, Ritter conducted summer field studies with his students at various locations along the California coast. In 1903, a group of business and professional people in San Diego established the Marine Biological Association and invited Ritter to locate his field station in San Diego permanently. With financial support from members of the Scripps family, who had made a fortune in newspaper publishing, Ritter was able to do this. This was the beginning of the University of California's Scripps Institution of Oceanography (fig. 1.16*a*). The property and holdings of the Marine Biological Association were formally transferred to the University of California in 1912.

Also, in the first twenty years of the century, the Carnegie Institute funded a series of exploratory cruises, including investigations of Earth's magnetic field, and maintained a

(a)

(b)

Figure 1.16 (a) The Scripps Institution of Oceanography in La Jolla, California. Established in 1903 by William Ritter, a zoologist at the University of California–Berkeley, with financial support from E. W. Scripps and his daughter Ellen Browning Scripps. The first permanent building was erected in 1910. (b) The Woods Hole Oceanographic Institution in Woods Hole, Massachusetts. In a rare moment, all three WHOI research vessels are in port. *Knorr* is in the foreground, with *Oceanus* bow forward and *Atlantis* stern forward, on the opposite side of the pier.

biological laboratory. In 1927, a National Academy of Sciences committee recommended that ocean science research be expanded by creating a permanent marine science laboratory on the East Coast. This led to the establishment of the Woods Hole Oceanographic Institution in 1930 (fig. 1.16*b*). It was funded largely by a grant from the Rockefeller Foundation. The Rockefeller Foundation allocated funds to stimulate marine research and to construct additional laboratories, and oceanography began to move onto university campuses. Teaching oceanography required that the subject material be consolidated, and in 1942, *The Oceans,* by Harald U. Sverdrup, Martin W. Johnson, and Richard H. Fleming, was published. It captured nearly all the world's knowledge of oceanographic processes and was used to train a generation of ocean scientists.

Oceanography mushroomed during World War II, when practical problems of military significance had to be solved quickly. The United States and its allies needed to move personnel and materials by sea to remote locations, to predict ocean and shore conditions for amphibious landings, to know how explosives behaved in seawater, to chart beaches and harbors from aerial reconnaissance, and to use underwater sound to find submarines. Academic studies ceased as oceanographers pooled their knowledge in the national war effort.

After the war, oceanographers returned to their classrooms and laboratories with an array of new, sophisticated instruments, including radar, improved sonar, automated wave detectors, and temperature-depth recorders. They also were aided with large-scale government funding for research and education. The Earth sciences in general and oceanography in particular blossomed during the 1950s. The numbers of scientists, students, educational programs, research institutes, and professional journals all increased.

Major funding for applied and basic ocean research was supplied by both the Office of Naval Research and the National Science Foundation. The Atomic Energy Commission financed oceanographic work at the west-central Pacific atoll sites of atomic tests. During the 1950s, the Coast and Geodetic Survey expanded its operations and began its seismic sea wave (tsunami) warning system. International cooperation brought about the 1957–58 International Geophysical Year (IGY) program, in which sixty-seven nations cooperated to explore the sea floor and made discoveries that changed the way geologists thought about continents and ocean basins. As a direct result of the IGY program, special research vessels and submersibles were built to be used by federal agencies and university research programs.

The decade of the 1960s brought giant strides in programs and equipment. In 1963–64, another multinational endeavor, the Indian Ocean Expedition, took place. In 1965, a major reorganization of governmental agencies occurred. The Environmental Science Services Administration (ESSA) was formed by consolidating the Coast and Geodetic Survey and the Weather Bureau, among others. Under ESSA, federal environmental research institutes and laboratories were established, and the use of satellites to obtain data became a major focus of ocean research. In 1968, the Deep Sea Drilling Program, a cooperative venture between research institutions and universities, began to sample Earth's crust beneath the sea (fig. 1.17*a*) (see chapter 3) using the specially built drill ship *Glomar Challenger.* It was finally retired in 1983 after fifteen years of extraordinary service. The *Glomar Challenger* was named after the ship used during the *Challenger* expedition. Electronics developed for the space program were applied to ocean research. Computers went aboard research vessels, and for the first time, data could be sorted, analyzed, and interpreted at sea. This made it possible for

(a)

(b)

Figure 1.17 (a) The *Glomar Challenger,* the Deep Sea Drilling Program drill ship used from 1968–83. (b) The *JOIDES Resolution,* the Ocean Drilling Program drill ship in use since 1985.

scientists to adjust experiments while they were in progress. Government funding allowed large-scale ocean experiments. Fleets of oceanographic vessels from many institutions and nations carried scientists studying all aspects of the oceans.

In 1970, the U.S. government reorganized its earth science agencies once more. The National Oceanic and Atmospheric Administration (NOAA) was formed under the Department of Commerce. NOAA combined several formerly independent agencies, including the National Ocean Survey, National Weather Service, National Marine Fisheries Service, Environmental Data Service, National Environmental Satellite Service, and Environmental Research Laboratories. NOAA also administers the National Sea Grant College Program. This program consists of a network of twenty-nine individual programs located in each of the coastal and Great Lakes states. The Sea Grant College Program encourages cooperation in marine science and education among government, academia, and industry.

The International Decade of Ocean Exploration (IDOE) in the 1970s was a multinational effort to survey seabed mineral resources, improve environmental forecasting, investigate coastal ecosystems, and modernize and standardize the collection, analysis, and use of marine data.

At the end of the 1970s, oceanography faced a reduction in funding for ships and basic research, but the discovery of deep-sea hot-water vents and their associated animal life and mineral deposits renewed excitement over deep-sea biology, chemistry, geology, and ocean exploration in general. Instrumentation continued to become more sophisticated and expensive as deep-sea mooring, deep-diving submersibles, and the remote sensing of the ocean by satellite became possible. Increased cooperation among institutions led to the integration of research at sea between subdisciplines and resulted in large-scale, multifaceted research programs. Although collection of oceanographic data from vessels at sea expanded, data collected by satellites since the 1970s have increasingly presented researchers with the ability to observe the sea surface on a global scale. Currents, eddies, algae production, sea-level changes, waves, thermal properties, and air-sea interactions are all monitored via satellite, allowing scientists to develop computerized prediction models and to test them against natural phenomena.

During these years of expanding programs, Earth scientists began to recognize the signs of global degradation and the need for progressive policies and management of living and nonliving resources. Students were attracted to the programs, and ocean management courses were added to curricula. As more and more nations were turning to the sea for food and as technology was increasing our ability to exploit sea resources, problems of resource ownership, dwindling fish stocks, and the need for fishery management had to be faced.

In 1983, the Deep Sea Drilling Project became the Ocean Drilling Program (ODP). The ODP was managed by an international partnership of fourteen U.S. science organizations and twenty-one international organizations called the Joint Oceanographic Institutions for Deep Earth Sampling (JOIDES). The ODP replaced the retired drilling ship *Glomar Challenger* with a larger vessel, the *JOIDES Resolution* (fig. 1.17*b*), named after the HMS *Resolution,* used by Captain James Cook to explore the Pacific Ocean basin over 200 years ago. The *JOIDES Resolution* started drilling operations in 1985 and was in continuous service until ODP ended in 2003. The total crew included fifty scientists and technicians and sixty-five ship crew members.

NASA's *NIMBUS-7* satellite, launched in 1978, carried a sensor package called the Coastal Zone Color Scanner (CZCS) that detected multiband radiant energy from chlorophyll in sea and land algae. The sensor operated from 1978–86. CZCS images can be used to determine levels of biological productivity in the oceans. They can also indicate how and where physical processes in the oceans influence the distribution and health of marine biological communities, particularly the small marine plants called phytoplankton, which are discussed in detail in chapter 16 (fig. 1.18). The CZCS sensor was the predecessor to the ocean color measuring sensors SeaWiFS (Sea-viewing Wide Field-of-view Sensor) and MODIS (Moderate Resolution Imaging Spectroradiometer) which are in orbit today. SeaWiFS

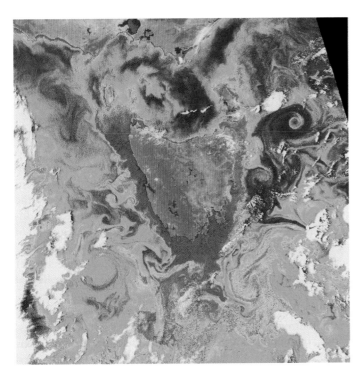

Figure 1.18 False color image centered on the island of Tasmania. Tasmania is located south of the eastern coast of Australia. *Yellow* and *reds* indicate high concentrations of phytoplankton, *greens* and *blues* low concentrations, and *dark blue* and *purple* very low concentrations. The complex current interactions, indicated by swirling color patterns, around the island have significant influence on the distribution of phytoplankton.

has now operated continuously for more than ten years (1997–2007). MODIS began recording data in 2002. SeaWiFS is carried by the *SEASTAR* satellite and MODIS is onboard the *Aqua* satellite. These are discussed further in section 1.10.

In 1985, the U.S. Navy *Geodynamic Experimental Ocean Satellite (GEOSAT)* was launched. It was designed to collect data for military purposes, but its orbit was changed to replace the failed *SEASAT*. From 1986–90, it monitored topography, surface winds and waves, local gravity changes, and sea surface "boundaries" caused by abrupt changes in salinity and temperature. *GEOSAT* was a predecessor to the *TOPEX* and *Jason* missions discussed in section 1.10.

1.10 The Recent Past, the Present, and the Future of Oceanography

Today's scientists see Earth not as a single entity but as a complex of systems and subsystems acting as a whole. Projects that emerged in the 1990s and continue in the 2000s require that scientists cross from one discipline to another and share information for common goals. Satellites are used for global observation. Earth and ocean scientists are able to retrieve and manipulate data quickly by computer at sea or on land, and they share it rapidly over the Internet. In addition, successful integrated approaches to Earth studies require that governments, agencies, universities, and national and international programs agree to set common priorities and share program results.

Several large-scale oceanographic programs have been developed to better understand the role of the oceans in processes of the atmosphere-ocean-land system. These programs provide data for models that scientists use to predict the evolution of Earth's environment as well as the consequences of human-influenced changes. The World Ocean Circulation Experiment (WOCE) studies the world oceans using computers and chemical tracers to model the present state of the oceans and predict ocean evolution in relation to long-term changes in the atmosphere. This effort combines sampling by ship, satellites, and floating buoys with sensors. The U.S. Joint Global Ocean Flux Study (JGOFS) is the largest and most complex ocean biogeochemical research program ever organized. Begun in 1988, the JGOFS program resulted in over 3000 ship days (more than eight years) of research and 343,000 nautical miles of ship travel (almost sixteen times around the globe) before ending in 2003. The goal of JGOFS research programs was to measure and understand on a global scale the processes controlling the cycling of carbon and other biologically active elements between the ocean, atmosphere, and land. This knowledge is needed to better predict the ocean's response to change, especially global climate change. The Global Ocean Atmosphere-Land System (GOALS) studies the energy transfer between the atmosphere and the tropical oceans to better understand El Niño and its effects and to provide improved large-scale climate prediction.

In August 1992, the satellite *TOPEX/Poseidon* was launched in a joint U.S.–French mission to explore ocean circulation and its interaction with the atmosphere. *TOPEX/Poseidon* measures sea level along the same path every ten days. This information is used to relate changes in ocean currents with atmospheric and climate patterns. The measurements allow scientists to chart the height of the seas across ocean basins with an accuracy of 3 cm (1.1 in). *TOPEX/Poseidon*'s three-year prime mission ended in the fall of 1995 and is now in its extended observational phase. A major follow-on mission to continue these studies began in December 2001, with the launch of *Jason-1*. *Jason-1*'s mission is the same as *TOPEX/Poseidon*'s, but the satellite is designed to acquire continuous data over longer periods of time to monitor global climate interactions between the atmosphere and the oceans (see the discussion of rising sea level in chapter 12).

In 1991, the Intergovernmental Oceanographic Commission (IOC) recommended the development of a Global Ocean Observing System (GOOS) to include satellites, buoy networks, and research vessels. The goal of this program is to enhance our understanding of ocean phenomena so that events such as El Niño (see section 7.8 in chapter 7) and its impact on climate can be predicted more accurately and with greater lead time. The successful prediction of the 1997–98 El Niño six months in advance made planning possible prior to its arrival.

An integral part of the GOOS is an international project called Argo, named after the mythical vessel used by the ancient Greek seagoing hero Jason. Argo consists of an array of 3000 independent instruments, or floats, throughout the oceans (fig. 1.19*a,b*). Each float is programmed to descend to a depth of up to 2000 m (6560 ft, or about 1.25 mi), where it remains for ten to fourteen days (fig. 1.19*c*). It then ascends to the surface, measuring temperature and salinity as it rises. When it reaches the surface, it relays the

(a)

data to shore via satellite and then descends once again and waits for its next cycle. The floats drift with the currents as they collect and transmit data. In this fashion, the entire array will provide detailed temperature and salinity data of the upper 2000 m of the oceans every ten to fourteen days. Argo floats have a life expectancy of four years. Roughly one quarter of the floats will have to be replaced each year.

In 1997 NASA launched the *SeaStar* satellite for a planned five-year mission to record ocean color with the Sea-WiFS color scanner. Now, ten years after launch, SeaWiFS is still operational providing information about basic marine biology, including oceanic primary productivity, plant biomass, and plant diversity.

The United Nations designated 1998 as the Year of the Ocean. Goals included (1) a comprehensive review of national ocean policies and programs to ensure coordinated advancements leading to beneficial results and (2) raising public awareness of the significance of the oceans in human life and the impact that human life has on the oceans. Also in 1998, the National Research Council's Ocean Studies Board released a report highlighting three areas that are likely to be the focus of future research: (1) improving the health and productivity of the coastal oceans, (2) sustaining ocean ecosystems for the future, and (3) predicting ocean-related climate variations.

Figure 1.19 (a) The Japanese coast guard cutter *Takuyo* prepares to retrieve an Argo float (photo courtesy of the International Argo Steering Team). (b) Schematic of an Argo float. The float's buoyancy is controlled by a hydraulic pump that moves hydraulic fluid between an internal reservoir and an external flexible bladder that expands and contracts. (c) A single measurement cycle involves the slow descent of the float to a depth of up to 2000 m (6560 ft) where it remains for about ten days before hydraulic fluid is pumped into the external bladder and the float rises again to the surface to transmit data before descending again.

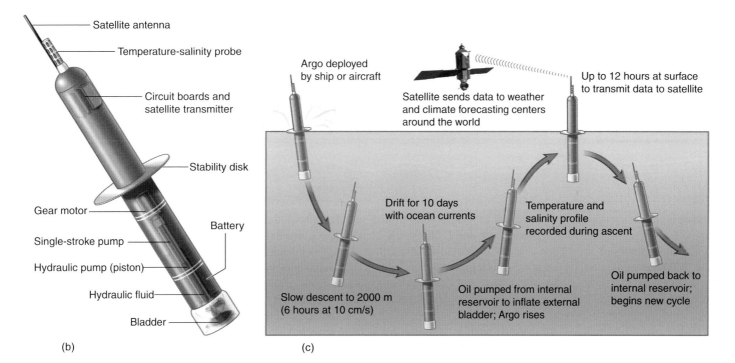

(b)   (c)

Figure 1.20 The drill ship *Chikyu* from its launching ceremony in January 2002.

maximum water depth the *Chikyu* is able to drill in is 7000 m (23,000 ft). The *Chikyu*'s derrick height is 110 m (361 ft) above water level. State-of-the-art drilling technology allows the new vessel to drill safely in areas with gas or hydrocarbon deposits along continental margins as well as into regions with thick sediment deposits or fault zones.

Future research priorities of ocean drilling are currently envisioned to investigate global climate change, to assist in the emplacement of geophysical and geochemical observatories on the sea floor, and to explore the deep structure of continental margins and oceanic crust.

One of the largest and most ambitious research and education programs under development is Project NEPTUNE. An international, multi-institutional project, NEPTUNE is part of a worldwide effort to develop regional, coastal, and global ocean observatories. Project NEPTUNE is scheduled to last at least thirty years. It is described in more detail in chapter 3.

Until the late nineteenth century, the great oceanographic voyages were largely voyages of exploration. Explorers such as James Cook and scientists such as those on the *Challenger* set sail into unknown waters to discover what they could find. With the beginning of the twentieth century, modern oceanographic science matured to become a process of hypothesis testing. Modern-day oceanographers typically first form an idea based on existing knowledge and then carefully design an experiment to test it. The sense of exploration has largely been in the background in recent decades with a few clear exceptions, such as the discovery of hydrothermal vent systems on mid-ocean ridges. The Office of Ocean Exploration (OOE) at NOAA was created in 2001 to encourage and fund new exploratory missions in the oceans. The OOE will also fund the development of new technology to support underwater exploration, such as manned or robotic submersibles and underwater-imaging systems.

Although large-scale, federally funded studies are presently in the forefront of ocean studies, it is important to remember that studies driven by the specific research interests of individual scientists are essential to point out new directions for oceanography and other Earth sciences. In the following chapters, you will follow the development of the ideas that have enabled us to build an understanding of the dynamic and complex systems that are Earth's oceans.

In 2002, NASA launched the Earth Observing System (EOS) satellite *Aqua* with the color sensor MODUS to collect information about Earth's water cycle, including evaporation from the oceans, the extent of sea ice, radiative energy flux from the oceans, water temperature, and the distribution of phytoplankton and dissolved organic matter in the oceans.

Following the end of the ODP in 2003, a new ocean drilling program began. The Integrated Ocean Drilling Program (IODP) is an international marine drilling program involving sixteen countries and hundreds of scientists. The U.S. involvement is directed by a consortium called the Joint Oceanographic Institutions (JOI). IODP is the third phase of scientific ocean drilling succeeding the Deep Sea Drilling Project (DSDP) and Ocean Drilling Program (ODP). IODP utilizes two major drilling vessels, the *JOIDES Resolution*, operated by the United States, and a new vessel, the *Chikyu* (fig. 1.20), built and operated by Japan.

The *Chikyu* hull was launched in January 2002, and initial sea trials began in December 2004. Scientific drilling cruises began in 2007. The deepest hole drilled by the *JOIDES Resolution* to date is 2994 m (9823 ft) in 5333 m (17,498 ft) of water. The *Chikyu* is designed to drill up to 7500 m (24,600 ft) beneath the sea floor initially, with plans to increase that to 8000 m (26,250 ft). The

Summary

Oceanography is a multidisciplinary field in which geology, geophysics, chemistry, physics, meteorology, and biology are all used to understand the oceans. Early information about the oceans was collected by explorers and traders such as the Phoenicians, the Polynesians, the Arabs, and the Greeks. Eratosthenes calculated the first accurate circumference of Earth, and Ptolemy produced the first world atlas.

During the Middle Ages, the Vikings crossed the North Atlantic, and shipbuilding and chartmaking improved. In the

fifteenth and sixteenth centuries, Dias, Columbus, da Gama, Vespucci, and Balboa, as well as several Chinese explorers, made voyages of discovery. Magellan's expedition became the first to circumnavigate Earth. In the sixteenth and seventeenth centuries, some explorers searched for the Northwest Passage, while others set up trading routes to serve developing colonies.

By the eighteenth century, national and commercial interests required better charts and more accurate navigation techniques. Cook's voyages of discovery to the Pacific produced much valuable information, and Franklin compiled a chart of the Atlantic's Gulf Stream. A hundred years later, the U.S. Navy's Maury collected wind and current data to produce current charts and sailing directions and then wrote the first book on oceanography.

Ocean science began with the nineteenth-century expeditions and research of Darwin, Forbes, Müller, and others. The three-and-a-half-year *Challenger* expedition laid the foundation for modern oceanography with its voyage, which gathered large quantities of data on all aspects of oceanography. Exploration of the oceans in Arctic and Antarctic regions was pursued by Nansen and Amundsen into the beginning of the twentieth century.

In the twentieth century, private institutions played an important role in developing U.S. oceanographic research, but the largest single push came from the needs of the military during World War II. After the war, large-scale government funding and international cooperation allowed oceanographic projects that made revolutionary discoveries about the ocean basins. Development of electronic equipment, deep-sea drilling programs, research submersibles, and use of satellites continued to produce new and more detailed information of all kinds. At present, oceanographers are focusing their research on global studies and the management of resources as well as continuing to explore the interrelationships of the chemistry, physics, geology, and biology of the sea.

Key Terms

All key terms from this chapter can be viewed by term or definition when studied as flashcards on this book's website at **www.mhhe.com/sverdrup10e**.

hypothesis, 2
theory, 2

Study Questions

1. Eratosthenes estimated the circumference of Earth at approximately 40,250 km (25,000 mi). Compare this estimate with the circumference used by Ptolemy. What difference would it have made to later voyages of discovery if Eratosthenes's measurement had been used rather than that of Ptolemy?
2. Who first assigned the name "America" to the New World? For whom was it named?
3. Why was there such great interest in finding and establishing a Northwest Passage?
4. Who first understood the tides and published an explanation of them?
5. What were Captain James Cook's contributions to our understanding of the oceans?
6. Why did Benjamin Franklin consider it so important to chart the Gulf Stream?
7. Who was Matthew F. Maury, and why is he considered by many to be the founder of oceanography?
8. Why do you think Edward Forbes concluded that there was no life in the oceans below 550 m (1800 ft)?
9. What did the engineers who laid the first transatlantic cable need to know about the oceans?
10. The *Challenger* and its expedition are often called unique. Why is this term used? What were the benefits of this expedition to the science of the oceans?
11. What was Fridtjof Nansen trying to prove by freezing the *Fram* into the polar ice?
12. The amount of ocean data has been expanding at an ever-increasing rate since the early years of ocean exploration. Why?
13. How has each of the following affected twentieth-century oceanography? (a) Economics. (b) Commerce and transportation. (c) Military needs.
14. In what ways have computers altered oceanography?
15. What are the reasons for the increased interest in resources of the sea? What types of management do you think may be required for these resources in the future?

Earth from space.

Chapter 2

The Water Planet

Chapter Outline

Learning Outcomes

After studying the information in this chapter students should be able to:

1. *explain* the "Big Bang" theory of the origin of the universe, and *describe* its structure,
2. *describe* the origin of the solar system,
3. *list* two possible sources of the water in the oceans,
4. *review* how we have come to estimate Earth's age as 4.5 to 4.6 billion years,
5. *rank* the eons and eras of geologic time in chronological order,

6. *list* and *date* the three major mass extinctions,
7. *define* and *sketch* lines of latitude and longitude,
8. *calculate* the difference in time between two locations of known longitude,
9. *diagram* the hydrologic cycle, and
10. *calculate* the mean depth of the oceans using data in table 2.4.

Questions concerning the origin and the age of the planet we call Earth are related to the origin of the universe and the beginnings of our solar system. Many theories have been suggested for these origins. In this chapter, we consider the most widely accepted hypotheses for the birth of our universe and our planet as well as the methods now used to measure time and calculate the age of Earth.

Water did not exist on Earth in the beginning, but its formation on a planet that was neither too far from nor too close to the Sun, neither too cold nor too hot, changed Earth and allowed the development of life. In this chapter, we begin to investigate our water planet (fig. 2.1)—cycling of water, its distribution, and its largest reservoirs, known as the oceans. We will begin to understand how scientists and mariners find their way about the oceans, and we will learn some of the basic principles required for the study of the oceans, which we call oceanography.

2.1 Beginnings

Origin of the Universe

For centuries, our concept of the nature of the universe was governed by visual observations from Earth's surface. Our current understanding of its history and structure has been greatly enhanced by observations made with instruments that are sensitive to energy across the **electromagnetic spectrum,** from radio waves to gamma rays, such as the *Hubble Space Telescope (HST)* (fig. 2.2).

The *HST* was deployed in April 1990, in low-Earth orbit at an altitude of 595 km (370 mi), and it circles Earth every ninety-seven minutes. Because of its location above the atmosphere, the 2.4 m (94.5 in) reflecting telescope (the size of a reflecting telescope refers to the diameter of its mirror) has an optical resolution, or image clarity, that is about ten times better than the best ground-based telescopes can achieve. It can detect objects one-billionth as bright as the human eye is capable of detecting.

In recent years, observational data have provided increasing evidence that the universe originated in an event known as the **Big Bang.** The Big Bang model envisions all energy and

Figure 2.1 The water planet. Earth as seen from space.

matter in the universe as having initially been concentrated in an extremely hot, dense singularity much smaller than an atom. Roughly 13.7 billion years ago, this singularity experienced a cataclysmic explosion that caused the universe to rapidly expand and cool as it grew larger. One second after the Big Bang, the temperature of the universe was about 10 billion K (roughly 1000 times the temperature of the Sun's interior). The Kelvin (K) temperature scale is an absolute temperature scale, so 0 K is absolute zero, the coldest possible temperature. At this temperature, all atoms and molecules would stop moving.

Figure 2.2   The *Hubble Space Telescope (HST)* is a joint venture between the European Space Agency and the National Aeronautics and Space Administration. It was proposed in the 1940s, designed and constructed in the 1970s and 1980s, and began its operational life with its launch in 1990.

Room temperature is about 300 K (see appendix B for conversions from K to °C and °F). At this time, the universe consisted mostly of elementary particles, light, and other forms of radiation. The elementary particles, such as protons and electrons, were too energetic to combine into atoms. One hundred seconds after the Big Bang, the temperature had cooled to about 1 billion K (roughly the current temperature in the centers of the hottest stars).

The universe was now cool enough for protons and neutrons, the nuclei of hydrogen atoms, as well as nuclei of deuterium, helium, and lithium, to begin to form. While the temperature was still very high, the universe was dominated by the radiation. Later, as the universe cooled, matter took over. Eventually, when the temperature had dropped to a few thousand

degrees K, electrons and nuclei would have started to combine to form atoms, and strong interactions between matter and radiation ceased. It was then possible for small concentrations of matter to begin to grow gravitationally. Denser, cooler regions pulled in additional matter gravitationally, increasing their density even further. An extraordinary composite image of the early universe is shown in figure 2.3. This image was compiled from data obtained by the *Wilkinson Microwave Anisotropy Probe (WMAP)* satellite over a period of more than one year. It is essentially a temperature map of the universe 380,000 years after the Big Bang, over 13 billion years ago. The spatial variation in temperature reflects the clumping of mass in the universe at that time. The *WMAP* is so sensitive that it can resolve differences in temperature of only a millionth of a degree. Roughly 200 million years after the Big Bang, gravity began to pull matter into the structures we see in the universe today, and the first stars were formed.

The universe has a distinct structure. On a small scale, there are individual stars. Stars are responsible for the formation of elements heavier than lithium. Stars fuse hydrogen and helium in their interiors to form heavier elements such as carbon, nitrogen, and oxygen. The higher temperatures of more massive stars continue the nuclear fusion process to create elements as heavy as iron. These elements, so important in oceanographic processes, are created in stars and were not made by the Big Bang at the formation of the universe. Some of these stars are at the center of solar systems like our own, with planets that orbit them. Detecting planets orbiting other stars is extremely difficult because they are small and dark, and the brilliance of their parent stars commonly conceals their presence. Consequently, the first confirmation of planets outside our solar system came from indirect observations. Both a large planet and the star it orbits will rotate around their common center of mass, causing a wobble in the star's motion. The existence of planets outside our solar system was first demonstrated several years ago by the detection of just such a wobble in certain stars. Astronomers have directly observed a number of faint points of light in the constellation Orion that they believe are planetlike objects a few times more massive than Jupiter. These objects can only be seen because they are

Figure 2.3   The structure of the early universe about 380,000 years after the Big Bang, as seen by the *Wilkinson Microwave Anisotropy Probe* satellite. Colors indicate "warmer" (red) and "cooler" (blue) spots. These patterns are extremely small temperature differences within an extraordinarily evenly dispersed microwave light bathing the universe, which now averages only 2.73 K. The difference in temperature between the warmest and coolest regions in this image is only 400 μK (400×10^{-6} K).

very young and still warm after the process of formation, and they do not orbit a nearby star that would mask the light they radiate.

Galaxies are composed of clumps of stars. Our galaxy, the Milky Way, is composed of about 200 billion stars. It is shaped like a flattened disk, with a thickness of about 1000 **light-years** and a diameter of about 100,000 light-years. A light-year is equal to the distance light travels in one year, which is 9.46×10^{12} km (5.87×10^{12} mi). The observable universe contains from 10 billion to 100 billion galaxies. Galaxies are preferentially found in groups called **clusters.** A single cluster may contain thousands of galaxies. Clusters typically have dimensions of 1 million to 30 million light-years. Individual clusters tend to group in long, stringlike or wall-like structures called superclusters. Superclusters may contain tens of thousands of galaxies. The largest supercluster known is about 500 million light-years across. At very large scales, the universe looks something like a sponge, with galaxies arranged in interconnected lines and sheets interspersed with huge regions in which very few galaxies are seen.

Throughout the universe, some stars are burning out or exploding; others are still forming, incorporating original matter from the Big Bang and recycling matter from older generations of stars. A widely accepted model of the universe is one in which the vast majority of the energy is in a form that cannot be detected. Roughly 30% of all energy is thought to be dark matter that does not interact with light. Sixty-five percent is thought to consist of an even more mysterious dark energy that is causing the expansion of the universe to accelerate. The familiar matter, the kind found in planets and stars, makes up only 5% of the energy in the universe.

Origin of Our Solar System

Present theories attribute the beginning of our solar system to the collapse of a single, rotating interstellar cloud of gas and dust that included material that was produced within older stars and dispersed into space when the old stars exploded. This rotating cloud, or **nebula,** appeared about 5 billion years ago. The shock wave from a nearby exploding star, or supernova, is thought to have imparted spin to the cloud, pushing it together and causing it to compress from its own gravitational pull. As the nebula collapsed, its speed of rotation increased, and, heated by compression, its temperature rose. The gas and dust, spinning faster and faster, flattened perpendicular to the axis of spin, forming a disk. At the center of the disk a star, our Sun, was formed. Self-sustaining nuclear reactions kept the Sun hot, but the outer regions began to cool. In this cooler outer portion of the rotating disk, molecules of gas and dust began to collide, accrete (or stick together), and chemically interact. The collisions and interactions produced particles that grew from further accretion and became large enough to have sufficient gravity to attract still other particles. The planets of our solar system had begun to form. After a few million years, the Sun was orbited by nine planets (in order from the Sun): Mercury, Venus, Earth, Mars, Jupiter, Saturn, Uranus, Neptune, and Pluto.

If Mercury, Venus, Earth, and Mars are compared to Jupiter, Saturn, Uranus, and Neptune, the four planets closer to the Sun are seen to be much smaller in diameter and mass. (See table 2.1. Note use of metric units; see appendix B for further information.) These four inner planets are rich in metals and rocky materials. The four outer planets are cold giants, dominated by ices of water, ammonia, and methane. Their atmospheres are made up of helium and hydrogen; the planets located nearer the Sun lost these lighter gases because the higher temperature and intensity of solar radiation tend to push these gases out and away from the center of the solar system. In addition, these inner planets are not massive enough for their gravitational fields to prevent these lighter gases from escaping. If the mass of each planet in table 2.1 is divided by its volume, the results will show that the outer planets are composed of lighter, or less dense, materials than the inner planets.

Pluto, at least 500 times less massive than the Earth, has an elliptic orbit that takes it inside the orbit of Neptune. It has a

Table 2.1	**Features of the Planets in the Solar System**						
Planet	**Mean Distance from Sun (10^6 km)**	**Diameter (km)**	**Mass Relative to Earth Mass**	**Rotation Period[1] (hours, days)**	**Orbit Period (years)**	**Mean Temperature of Surface (°C)**	**Principal Atmospheric Gases[2]**
Mercury	57.9	4,878	0.055	58.6 d	0.24	−170 night 430 day	Essentially a vacuum
Venus	108.2	12,104	0.815	243 d*	0.62	−23 clouds 480 surface	CO_2, N_2
Earth	149.6	12,756	1.000	23.94 h	1.00	16	N_2, O_2
Mars	227.9	6,787	0.107	24.62 h*	1.88	−50 (average)	CO_2, N_2, Ar
Jupiter	778.3	142,800	317.8	9.93 h	11.86	−150	H_2, He, CH_4, NH_3
Saturn	1429	120,000	95.2	10.5 h	29.48	−180	H_2, He, CH_4, NH_3
Uranus	2875	50,800	14.5	17.24 h*	84.01	−210	H_2, He, CH_4
Neptune	4504	48,600	17.2	16 h	164.8	−220	H_2, He, CH_4
Pluto	5900	2,245	0.002	6.4 d	247.71	−230	CH_4, N_2

1. * by rotation period indicates rotation in a direction opposite that of Earth.

2. CO_2 = carbon dioxide; N_2 = nitrogen; O_2 = oxygen; Ar = argon; H_2 = hydrogen; He = helium; CH_4 = methane; NH_3 = ammonia.

Origin of the Oceans

The oldest sedimentary rocks found on Earth, rocks that formed by processes requiring liquid water at the surface, are about 3.9 billion years old. This indicates that there have been oceans on Earth for approximately 4 billion years. Where did the water in the oceans come from? There are two possible sources for this water, the interior of Earth and outer space.

Traditionally, scientists have suggested that the water in the oceans and atmosphere originated in the interior of Earth in a region called the mantle (discussed in more detail in chapter 3, section 3.1) and was brought to the surface by volcanism, a process that continues to this day (box fig. 1). The rock that makes up the mantle is thought to be similar in composition to meteorites, which contain from 0.1%–0.5% water by weight. The total mass of rock in the mantle is roughly 4.5×10^{27} g; thus, the original mass of water in the mantle would have been approximately 4.5×10^{24} to 2.25×10^{25} g. This is from three to sixteen times the amount of water currently in the oceans, so it is clear that the mantle is an adequate source for the water in the oceans; but is enough water brought to the surface through volcanism to actually fill the oceans? Magmas erupted by volcanoes contain dissolved gases that are held in the molten rock by pressure. Most magmas consist of 1%–5% dissolved gas by weight, most of which is water vapor. The gas that escapes from Hawaiian magmas is about 70% water vapor, 15% carbon dioxide, 5% nitrogen, and 5% sulfur dioxide, with the remainder consisting mostly of chlorine, hydrogen, and argon. It is estimated that thousands of tons of gas are ejected in volcanic eruptions each day. Undoubtedly, the rate of volcanic eruptions on Earth has varied with time, probably being much greater earlier in Earth's history when the planet was still very hot. However, if we conservatively assume that the present rate of ejection of water vapor by volcanism has been roughly constant over the last 4 billion years, then the volume of water expelled by volcanoes would have produced roughly 100 times the volume of water in the oceans.

The traditional view of the interior of Earth serving as the source of ocean water has recently been challenged by a bold new suggestion that large volumes of water are continually being added from outer space. Evidence for this idea comes from data collected by a polar-orbiting satellite called the *Dynamics Explorer 1 (DE-1)* (box fig. 2). The *DE-1* carried an ultraviolet photometer capable of taking pictures of Earth's **dayglow.** Dayglow is ultraviolet light, invisible to the naked eye, emitted by atomic oxygen in the upper atmosphere when it absorbs and reradiates electromagnetic energy from the Sun. In many of the dayglow images of Earth obtained by the satellite, there are distinct dark spots, roughly 48 km (30 mi) in diameter, that appear to move across the face of Earth, suggesting

Box Figure 1 Volcanic eruptions release gases including water vapor, carbon dioxide, and sulfur dioxide to the atmosphere and oceans. Even if it is not erupting, an active volcano can release thousands of tons of sulfur per day.

that they were caused by moving objects (box fig. 3). The direction of motion of the dark spots matches the direction of motion of meteoritic material as it approaches Earth. Atmospheric physicist Louis Frank has suggested that these dark spots are created when small icy comets vaporize in the outer atmosphere, creating clouds of water vapor that absorb the ultraviolet radiation of Earth's dayglow over a small area, thus creating a dark spot in the bright ultraviolet background. The size of the spots implies that the average mass of the comets is about 10 kg (22 lb). He estimates that an average of twenty of these comets enter the atmosphere each minute, or a staggering 10 million each year. If all of the water in these comets condensed to form a layer on the surface of Earth, it would be reflective surface that may be primarily composed of frozen methane.

Extraterrestrial Oceans

Data obtained by NASA's *Voyager* and *Galileo* spacecraft indicate that two of Jupiter's moons, Europa and Callisto, may have oceans beneath their ice-covered surfaces (fig. 2.4*a*). Liquid oceans are believed to be possible despite extremely cold surface temperatures, $-162°C$ ($-260°F$) on Europa, because of heat generated by friction due to the continual tidal deformation by Jupiter's strong gravitational force.

Some of the most compelling evidence for the presence of these oceans has come from magnetic measurements made by the *Galileo* spacecraft. Neither moon has a strong internal magnetic field of its own, but *Galileo* detected induced magnetic fields around both moons, indicating that they both consist partly of strongly conducting material.

It is unlikely that the ice covering the moons can account for the induced magnetic fields because ice is a poor electrical

Box Figure 2 A prelaunch photo of the *Dynamics Explorer 1* and *2 (DE-1/DE-2)* spacecraft stack before being covered by their fairings and mated with the Delta Launch vehicle. *DE-1* is on the bottom. *DE-1* was launched into a high-altitude elliptical orbit, while *DE-2* was launched into a lower orbit.

roughly 0.0025 mm (0.0001 in) deep. While this doesn't seem like a significant amount of water, over 4 billion years this rate of accumulation would fill the oceans two to three times.

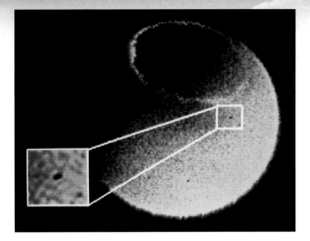

Box Figure 3 Image of Earth's dayglow at ultraviolet wavelengths taken from an altitude of 18,500 km (11,500 mi). The dayglow is due to the excitation of atomic oxygen by solar radiation. Inset shows a magnified view of a dark spot, or "hole," in the dayglow thought to be caused by the vaporization of small cometlike balls of ice.

Some debate continues about the role of comet impacts in the formation of the oceans. Additional study may give us further insight to the relative importance of volcanism and the impact of extraterrestrial objects in creating the oceans. It is very likely that both processes have contributed to their formation.

To Learn More About the Formation of the Oceans

Delsemme, A. H. 2001. An Argument for the Cometary Origin of the Biosphere. *American Scientist* 89 (5): 432–42.

Frank, L. A., J. B. Sigwarth, and J. D. Craven. 1986. On the Influx of Small Comets into Earth's Upper Atmosphere. *Observations and Interpretations Geophysical Research Letters* 13 (4): 303–10.

Kasting, J. F. 1998. The Origins of Water on Earth. *Scientific American Presents: The Oceans* 9 (3): 16–21.

Vogel, S. 1996. Living Planet. *Earth* 5 (2): 26–35.

Weisburd, S. 1985. Atmospheric Footprints of Icy Meteors. *Science News* 128: 391.

conductor. Fresh water is also a relatively poor conductor, but water with a high concentration of dissolved ions, such as seawater, is a very good conductor. The most plausible explanation for the observed magnetic effect is that Europa and Callisto have liquid oceans containing electrolytic salts beneath their surfaces. It is believed that magnesium sulfate might be a major component of Europa's water rather than sodium chloride, as is the case in Earth's oceans.

One proposed model for Europa includes a surface ice layer 15 km (10 mi) thick, covering a 100 km (62 mi) deep ocean. If this is the case, the Europan ocean would contain twice as much water as Earth's oceans, and it would be roughly ten times deeper than the greatest depths below sea level on Earth. In contrast, one model proposed for Callisto is a surface ice layer 100 km (62 mi) thick, covering a 10 km (6.2 mi) deep ocean. If such oceans exist, they may provide a possible environment for life.

In January 2004, two robotic rovers, *Spirit* and *Opportunity*, landed on opposite sides of the planet Mars. Their mission has been to probe the rocks of Mars for signs of past or current deposits of water, and they have succeeded in collecting both

(a)

(b)

Figure 2.4   (a) View of a small region of the thin, disrupted ice crust of Jupiter's moon Europa. North is to the top of the picture. The image covers an area approximately 70 km (43 mi) east-west by 30 km (19 mi) north-south. The *white* and *blue colors* outline areas that have been blanketed by a fine dust of ice particles ejected at the time of formation of a large impact crater roughly 1000 km (620 mi) to the south. A few small craters of less than 500 m (1600 ft) in diameter can be seen. These were probably formed by large, intact blocks of ice thrown up in the impact explosion that formed the large crater to the south. (b) Image of small holes or cavities taken by the microscopic imager on the Mars Exploration Rover *Opportunity* in a region called "El Capitan" on a rock outcrop at Meridiani Planum, Mars. Several cavities have disk-like shapes with wide midpoints and tapered ends. This feature is consistent with salt minerals that crystallize within a rock matrix, either pushing the matrix grains aside or replacing them. These crystals are then either dissolved in water or eroded by wind activity to produce the cavities.

chemical and physical data that strongly support the hypothesis that water was once present.

Chemical analyses have discovered various kinds of salts in some Martian rocks. On Earth, rocks that contain equally large amounts of these salts either formed in water or, after formation, were altered by long exposures to water. Photographs of some rocks show the presence of small holes or indentations similar to those found in some Earth rocks when crystals of salt minerals that were originally formed in the rocks were later dissolved by the flow of fresh water through the rocks (fig. 2.4*b*). The rover *Opportunity* detected the presence of an iron-bearing mineral called gray hematite in some Martian rocks; on Earth, hematite containing crystalline grains of the size found in the Martian rocks typically forms in the presence of water.

Of particular interest was *Opportunity*'s discovery of sedimentary rocks with ripple marks, structures that are formed by waves or currents of water. Taken together, the data accumulated by the rovers strongly support the hypothesis that there was once a large body of saltwater on the Martian surface and that *Opportunity*'s landing site was once the shoreline of a salty sea.

In August 2007, NASA launched the *Phoenix Probe* on a mission to sample ice beneath the surface in the north polar region of Mars and test it for the presence of water. The craft carried an oven-like instrument called the "Thermal and Evolved Gas Analyzer (TEGA)," designed to bake soil and ice samples and analyze the chemistry of released vapors. The *Phoenix Probe* landed on Mars in May 2008 and, in July 2008, confirmed the presence of water on Mars. This is the first time water has been directly identified on another planet.

Early Planet Earth

During the first billion years of Earth's existence, Earth is thought to have been a mixture of silicon compounds, iron and magnesium oxides, and small amounts of other elements. According to this model, Earth formed originally from cold matter, but events occurred that raised Earth's temperature and initiated processes that obliterated its earlier history and resulted in its present form. Early Earth was bombarded by particles of all sizes, and a portion of their energy was converted into heat on impact. Each new layer of accumulated material buried the material below it, trapping the heat and raising the temperature of Earth's interior. At the same time, the growing weight of the accumulating layers compressed the interior, and the energy of compression was converted to heat, raising Earth's internal temperature to approximately 1000°C. Atoms of radioactive elements, such as uranium and thorium, disintegrated by emitting subatomic particles that were absorbed by the surrounding matter, further raising the temperature.

Some time during the first few hundred million years after Earth formed, its interior reached the melting point of iron and nickel. When the iron and nickel melted, they migrated toward the center of the planet. Frictional heat was generated, and lighter substances were displaced. In this way, the temperature of Earth was raised to an average of 2000°C. The less dense material from the partially molten interior moved upward and spread over the surface, cooling and solidifying. The melting and solidifying probably happened repeatedly, separating the lighter, less dense compounds from the heavier, denser substances in the interior of the planet. In this way, Earth became completely reorganized and differentiated into a layered system, which is explored in greater detail in chapter 3.

Earth's oceans and atmosphere are probably both, at least in part, by-products of this heating and differentiation. As Earth warmed and partially melted, water, hydrogen, and oxygen locked in the minerals were released and brought to the surface along with other gases in volcanic eruptions. As Earth's surface cooled, water vapor condensed to form the oceans. Another possible source of water for the oceans is from space objects, such as cometlike balls of ice or ice meteorites, that have collided with Earth throughout its history. The origin of the oceans is discussed further in the box titled "Origin of the Oceans."

At first, Earth must have had too little gravity to have accumulated an atmosphere. It is generally believed that, during the process of differentiation, gases released from Earth's hot, chemically active interior formed the first atmosphere, which was primarily made up of water vapor, hydrogen gas, hydrogen chloride, carbon monoxide, carbon dioxide, and nitrogen. Any free oxygen present would have combined with the metals of the crust to form compounds such as iron oxide. Oxygen gas could not accumulate in the atmosphere until its production exceeded its loss by chemical reactions with Earth's crust. Oxygen production did not exceed loss until life evolved to a level of complexity in which photosynthetic organisms could convert carbon dioxide and water with the energy of sunlight into organic matter and free oxygen. This process and its significance to life are discussed in chapters 6 and 15.

2.2 Age and Time

Age of Earth

Over the centuries, people have asked the question, How old is Earth? In the seventeenth century, Archbishop Ussher of Ireland attempted to answer the question by counting the generations listed in the Bible; he determined that the first day of creation was Sunday, October 23, 4004 B.C. In 1897, the English physicist Lord Kelvin calculated the time necessary for a molten Earth to cool to present temperatures and dated Earth as 20 million to 40 million years old. In 1899, John Joly (see chapter 1, section 1.4) calculated the age of Earth to be 90–100 million years based on the rate of addition of salt to the oceans from rivers. In 1896, Antoine Henri Becquerel discovered radioactivity. With this discovery and an understanding of radioactive decay, scientists were able to accurately date rocks and minerals. In 1905, Ernest Rutherford and Bertrum Boltwood used radioactive decay to date rock and mineral samples 500 million years old. Two years later, in 1907, Boltwood

calculated an age of 1.64 billion years for a mineral sample rich in uranium.

The method pioneered by Rutherford and Boltwood, known as **radiometric dating,** uses radioactive **isotopes.** Each atom of a radioactive isotope has an unstable nucleus. This unstable nucleus changes, or decays, and emits one or more subatomic particles plus energy. For example, the radioactive isotope carbon-14 decays or changes to nitrogen-14; uranium-235 decays to lead-207; and potassium-40 decays to argon-40. The time at which any single nucleus will decay is unpredictable, but if large numbers of atoms of the same radioactive isotope are present, it is possible to predict that a certain fraction of the isotope will decay over a certain period of time. In this process of decay, the atom changes from one element (the parent element) to another (the daughter product). The time over which one-half of the atoms of a radioactive isotope decay is known as the isotope's **half-life** (fig. 2.5). The half-life of each radioactive isotope is characteristic and constant. For example, the half-life of carbon-14 is 5730 years; that of uranium-235 is 704 million years; and that of potassium-40 is 1.3 billion years. Therefore, if a substance is found that was originally made up only of atoms of uranium-235, after 704 million years, the substance is one-half uranium-235 and one-half lead-207. In another 704 million years, three-quarters of the substance will be lead-207 and only one-quarter uranium-235. Because each radioactive isotopic system behaves uniquely in nature, data must be carefully tested, compared, and evaluated. The best data are those in which different radioactive isotopic systems give the same date.

The most widely accepted age of Earth is between 4.5 billion and 4.6 billion years. This age is based on lead isotope studies of meteorites and samples of lead minerals from rocks. Heat generated by the formation of Earth probably created a molten surface. The original solid surface that formed as Earth cooled is believed

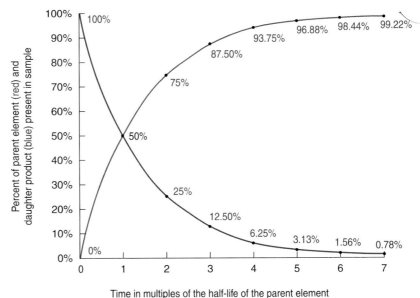

Figure 2.5 The half-life for a radioactive isotope is the time in which half of the parent element atoms decay to daughter atoms. The radioactive decay curve is exponential. It displays the percentage of original parent element atoms and daughter product atoms as a function of time (measured in multiples of the half-life of the parent element).

to have been destroyed by a variety of geologic processes. The oldest minerals found are between 4.1 billion and 4.2 billion years old, providing a minimum age for Earth. The accepted age of Earth is based on dating objects in the solar system that are believed to have formed at the same time as the planets but are not geologically active (they have not changed significantly over time), such as meteorites. Several different dating methods on multiple meteorites have all resulted in estimated ages of about 4.5 billion years.

Geologic Time

To refer to events in the history and formation of Earth, scientists use geologic time (table 2.2). The principal divisions are the four eons: the Hadean (4.6 to 4.0 billion years ago), the Archean (4.0 to 2.5 billion years ago), the Proterozoic (2.5 billion to 570 million years ago), and the Phanerozoic (since 570 million years ago). The first three eons are known as the Precambrian; it represents nearly 88% of all geologic time. Fossils are known from other eons but are common only from the Phanerozoic. The Phanerozoic eon is divided into three eras: the Paleozoic era of ancient life; the Mesozoic era of middle life (popularly called the Age of Reptiles); and the Cenozoic era of recent life (the Age of Mammals). Each of these eras is subdivided into periods and epochs; today, for instance, we live in the Holocene epoch of the Quaternary period of the Cenozoic era. The appearance or disappearance of fossil types was used to set the boundaries of the time units before radiometric dating enabled scientists to set these timescale boundaries more accurately. The accurate calibration of radiometric dates and relative time determined from fossils is an ongoing process, and the exact dates defining the time units are constantly being adjusted with the acquisition of new data. Dating time units in the Precambrian eons is particularly difficult because of the lack of very old marine sediments and fossils.

Very long periods of time are incomprehensible to most of us. We often have difficulty coping with time spans of more than ten years. What were you doing exactly ten years ago today? We have nothing with which to compare the 4.6-billion-year age of Earth or the 500 million years since the first **vertebrates** (animals with a spinal column) appeared on this planet. To place geologic time in a framework we can understand, let us divide Earth's age by 100 million. If we do so, then we can think of Earth as being just forty-six years old. What has happened over that forty-six years?

No record remains of events during the first three years. The earliest history preserved can be found in some rocks of Canada, Africa, and Greenland that formed forty-three years ago. Sometime between thirty-five and thirty-eight years ago, the first primitive living cells of bacteria-type organisms appeared. Oxygen production by living cells began about twenty-three years ago, half the age of the planet. Most of this oxygen combined with iron in the early oceans and did not accumulate in the atmosphere. It took about eight years, or until roughly fifteen years ago, for enough oxygen to accumulate in the atmosphere to support significant numbers of complex oxygen-requiring cells. Oxygen reached its present concentration in the atmosphere approximately eleven years later. The first invertebrates (animals without backbones) developed seven years ago, and two years later, the first vertebrates appeared. Primitive fish first swam in the oceans, and corals appeared just five years

ago. Three years, eight and one-half months ago, the first sharks could be found in the oceans, and roughly five months later, reptiles could be found on land. A massive extinction struck the planet just two and one-half years ago, killing 96% of all life. Following this catastrophic event, the dinosaurs appeared just two years, three months, and ten days ago. Three and one-half weeks later, the first mammals developed. A second major extinction occurred roughly two years ago. This event killed over half of the species on Earth, leaving mostly dinosaurs on land. Just eighteen and one-half months ago, the first birds flew in the air, and they would have seen the first flowering plants a little less than five months later. A third major extinction struck Earth 237 days ago, killing off the dinosaurs as well as many other species. Two hundred eleven days ago the mammals, birds, and insects became the dominant land animals. Our first human ancestors (the first identifiable member of the genus *Homo*) appeared just a little less than six days ago. About half an hour ago, modern humans began the long process we know as civilization, and only one minute ago, the Industrial Revolution began changing the Earth and our relationship with it.

Natural Time Periods

People first defined time by the natural motions of the Earth, Sun, and Moon. Later, people grouped natural time periods for their convenience, and still later, artificial time periods were created for people's special uses. Time is used to determine the starting point of an event, the event's duration, and the rate at which the event proceeds. An accurate measurement of time is required to determine location or position; this use of time is discussed in section 2.4.

The year is the time required by Earth to complete one orbit about the Sun. This time is 365¼ days, adapted for convenience to 365 days with an extra day added every four years, except for years ending in hundreds and not divisible by 400. As Earth orbits around the Sun, those who live in temperate zones and polar zones are very conscious of the seasons and of the differences in the lengths of the periods of daylight and darkness. The reason for these seasonal changes is seen in figure 2.6. Earth moves along its orbit with its axis tilted 23½° from the vertical. During the year, Earth's North Pole is sometimes tilting toward the Sun and sometimes tilting away from it. The Northern Hemisphere receives its maximum hours of sunlight when the North Pole is tilted toward the Sun; this is the Northern Hemisphere's summer. During the same period, the South Pole is tilted away from the Sun, so the Southern Hemisphere receives the least sunlight; this period is winter in the Southern Hemisphere (note in fig. 2.6 that summer in the Northern Hemisphere and winter in the Southern Hemisphere occur when Earth is farthest from the Sun). When Earth is closest to the Sun, the North Pole is tilted away from the Sun, creating the Northern Hemisphere's winter, and the South Pole is inclined toward the Sun, creating the Southern Hemisphere's summer. Note also that during summer in the Northern Hemisphere, the periods of daylight are longer around the North Pole and shorter around the South Pole; the opposite is true during the Northern Hemisphere's winter.

As we live on the spinning Earth orbiting the Sun, we are not conscious of any movement. What we sense is that the Sun rises

Table 2.2 The Geologic Timescale

Eon	Era	Period	Epoch	Time (millions of years ago)	Life Forms/Events
Phanerozoic	Cenozoic "Age of Mammals"	Quaternary	Holocene	0.0	Modern humans
			Pleistocene	0.01	Earliest humans
		Tertiary	Pliocene	1.6	
			Miocene	5.3	Earliest hominids
			Oligocene	23.7	Flowering plants
			Eocene	36.6	Earliest grasses
					Mammals, birds, and insects dominant
			Paleocene	57.8	
				65	

Cretaceous-Tertiary boundary: The extinction of 50% of all species, including the dinosaurs, at the end of the Mesozoic era (65 million years ago).

Eon	Era	Period	Epoch	Time (millions of years ago)	Life Forms/Events
	Mesozoic "Age of Reptiles"	Cretaceous			Earliest flowering plants (115)
					Dinosaurs in ascendence
		Jurassic		144	First birds (155)
					Dinosaurs abundant
				208	

Triassic-Jurassic boundary: The extinction of over 50% of all species, including the last of the mammal-like reptiles, leaving mainly dinosaurs on land (208 million years ago).

Eon	Era	Period	Epoch	Time (millions of years ago)	Life Forms/Events
		Triassic			First turtles (210)
					First mammals (221)
					First dinosaurs (228)
					First crocodiles (240)
				245	

Permian-Triassic boundary: The greatest mass extinction of all time; 96% of all species perish at the end of the Paleozoic era (245 million years ago).

Eon	Era	Period	Epoch		Time (millions of years ago)	Life Forms/Events
	Paleozoic	Permian		"Age of Amphibians"		Extinction of trilobites and many other marine animals
		Carboniferous	Pennsylvanian		286	Large coal swamps
						First reptiles (330)
			Mississippian		320	Amphibians abundant
		Devonian		"Age of Fishes"	360	First seed plants (365)
						First sharks (370)
						First insect fossils (385)
						Fishes dominant
		Silurian			408	First vascular land plants (430)
		Ordovician		"Age of Marine Invertebrates"	438	First land plants similar to lichen (470)
						First fishes (505)
						Earliest corals
						Marine algae
		Cambrian			505	Abundant shelled invertebrates
						Trilobites dominant
					570	
Proterozoic						First invertebrates (700)
						Earliest shelled organisms (~ 750)
						Oxygen begins to accumulate in the atmosphere (1500).
						First fossil evidence of single-celled life with a cell nucleus: eukaryotes (1500)
Archean					2500	First evidence of by-products of eukaryotes (2700)
						Earliest primitive life, bacteria and algae: prokaryotes (3500–3800)
						These will dominate the world for the next 3 billion years.
Hadean					4000	Oldest surface rocks (4030)
						Oldest single mineral (~ 4200)
						Oldest Moon rocks (4440)
						Oldest meteorites (4560)
					~ 4600	

Collectively, these are popularly known as the Precambrian.

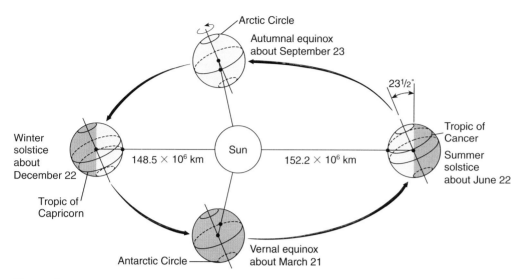

Figure 2.6   Earth's yearly seasons. The fixed orientation of Earth's axis during its orbit of the Sun causes different portions of Earth to remain in shadow at different seasons. The mean distance between Earth and the Sun is 149×10^6 km, but note the change in this distance during the year.

in the east and sets in the west daily and slowly moves up in the sky from south to north and back during one year. The periods of daylight in the Northern Hemisphere increase as the Sun moves north to stand above 23½°N, the **Tropic of Cancer.** It reaches this position at the **summer solstice,** on or about June 22, the day with the longest period of daylight and the beginning of summer in the Northern Hemisphere. On this day, the Sun does not sink below the horizon above the **Arctic Circle,** 66½°N latitude, nor does it rise above the **Antarctic Circle,** 66½°S. Following the summer solstice, the Sun appears to move southward until on or about September 23, the **autumnal equinox,** when it stands directly above the equator. On this day, the periods of daylight and darkness are equal all over the world. The Sun continues its southward movement until about December 21, when it stands over 23½°S, the **Tropic of Capricorn;** this position marks the **winter solstice** and the beginning of winter in the Northern Hemisphere. On this day, the daylight period is the shortest in the Northern Hemisphere; above the Arctic Circle, the Sun does not rise, and south of the Antarctic Circle, the Sun does not set. The Sun then begins to move northward, and on about March 21, the **vernal equinox,** it stands again above the equator; spring begins in the Northern Hemisphere, and the periods of daylight and darkness are once more equal around the world. Follow figure 2.6 around again, checking the position of the South Pole, and note how the seasons of the Southern Hemisphere are reversed from those of the Northern Hemisphere.

The greatest annual variation in the intensity of direct solar illumination occurs in the temperate zones. In the polar regions, the seasons are dominated by the long periods of light and dark, but the direct heating of Earth's surface is small because the Sun is always low on the horizon. Between the Tropics of Cancer and Capricorn, there is little seasonal change in solar radiation because the Sun is always nearly directly overhead at midday hours.

Weeks and months, as they presently exist, modify natural time periods. The Moon requires 27⅓ days to orbit Earth, but a period of 29½ days defines the **lunar month.** In the lunar month

the Moon passes through four phases: new moon, first quarter, full moon, and last quarter. The four phases approximately match the four weeks of the month. Days are grouped into twelve months of unequal length in order to form one calendar year. The present arrangement is known as the Gregorian calendar after Pope Gregory XIII, who in the sixteenth century made the changes necessary to correct the old Julian calendar, adopted in 46 B.C. and named after Julius Caesar. The Gregorian calendar was adopted in the United States in 1752, by which time the Julian calendar was eleven days in error. In that year, by parliamentary decree, in both Great Britain and the United States, September 2 was followed by September 14. People rioted in protest, demanding their eleven days back. Also in 1752, the beginning of the calendar year was changed from the original date of the vernal equinox, March 21, to January 1. The year 1751 had no months of January and February.

The day is derived from Earth's rotation. The average time for Earth to make one rotation relative to the Sun is twenty-four hours; this is the average, or mean, **solar day**—our clock day. Another measure of a day is the time required for Earth to make a complete rotation with respect to a far-distant point in space. This is known as the **sidereal day** and is about four minutes shorter than the mean solar day; it gives the true rotational period of Earth. The sidereal day is useful in astronomy and navigation.

Living organisms respond to these natural cycles. In temperate zones, as the periods of daylight lengthen and the temperatures increase, flowers bloom and trees produce new leaves; then, as periods of darkness increase and temperatures fall, flowers die back, and trees lose their leaves and enter dormancy. In tropical areas, however, forests remain lush year-round. Some animals migrate and alternate periods of activity and hibernation or estivation with the seasons. Other animals set their internal clocks to the day-night pattern, hunting in the dark and sleeping in the light; still others do the reverse. Plants and animals of the sea also react to these rhythms, as do the physical processes that stir the atmosphere and circulate the water in the oceans. Understanding these cycles helps us understand processes that occur at the ocean surface and are discussed in later chapters: climate zones, winds, currents, vertical water motion, plant life, and animal migration.

2.3 Shape of Earth

As Earth cooled and turned in space, gravity and the forces of rotation produced its nearly spherical shape. The Earth's mean radius can be defined as the radius of a sphere having the same volume as Earth. This is called the volumetric radius (V_r) and

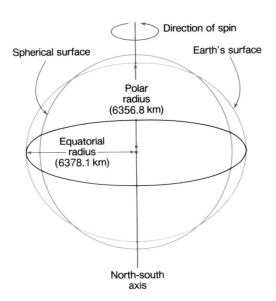

Figure 2.7   Rotation makes Earth bulge outward at the equator. Notice that the equatorial radius is larger than the polar radius.

it is easily calculated from the equatorial radius (a) and the polar radius (b) using:

$$V_r = \sqrt[3]{a^2b} = 6371 \text{ km (3959 mi)}$$

It has a shorter polar radius (6356.8 km; 3950 mi) and a longer equatorial radius (6378.1 km; 3963 mi). This difference of 21.3 km, or about 13 mi, occurs because Earth is not a rigid sphere. As Earth spins, it tends to bulge at the equator (fig. 2.7). Because the landmasses are presently concentrated in the middle region of the Northern Hemisphere and centered on the South Pole in the Southern Hemisphere, Earth's surface is depressed slightly in these areas and is elevated at the North Pole and in the middle region of the Southern Hemisphere. This land distribution causes Earth to have a very slight pear shape, about 15 m (50 ft) between depressions and elevations. However, Earth is a nearly perfect sphere.

Earth is also quite smooth. The top of Earth's highest mountain, Mount Everest in the Himalayas, is about 8840 m (29,000 ft) above sea level; the deepest ocean depth, the bottom of the Challenger Deep in the Mariana Trench of the Pacific Ocean, is about 11,000 m (36,000 ft). If these measurements are divided by the mean radius of Earth (6371 km, or 20,896,000 ft), the resulting elevation-to-radius ratios are 0.00139 for the mountain and 0.00173 for the trench. On a scale model of Earth with a radius of 50 cm (20 in), Mount Everest would be about 0.07 cm (0.027 in) high and the Challenger Deep would be about 0.086 cm (0.034 in) in depth. The Earth model's surface would feel rather like the skin of a grapefruit or the surface of a basketball. The topographic relief of Earth's surface, its high mountains and deep oceans, is minor compared to the size of the planet.

2.4 Location Systems

Latitude and Longitude

To find our way on the surface of our planet, we need a reference (or location) system. Most of us use such a system daily:

city name, street name or number, and building number. Armed with a city map, we confidently navigate to areas never visited before. Most of Earth's surface, however, is not provided with streets and building numbers, and we must use another system. To determine the location of a position on Earth, we use a grid of reference lines that are superimposed on Earth's surface and cross at right angles. These grid lines are called lines of **latitude** and **longitude.** Lines of latitude, also known as **parallels,** are referenced to the **equator.** The equator is created by passing a plane through Earth halfway between the poles and at right angles to Earth's axis. This process is much like cutting an orange in two pieces halfway between the depressions marking the stem and the navel. The equator is marked as 0° latitude, and other latitude lines are drawn parallel to the equator, northward to 90°N, or the North Pole, and southward to 90°S, or the South Pole (fig. 2.8a). Notice that the parallels of latitude describe increasingly smaller circles as the poles are approached. Notice

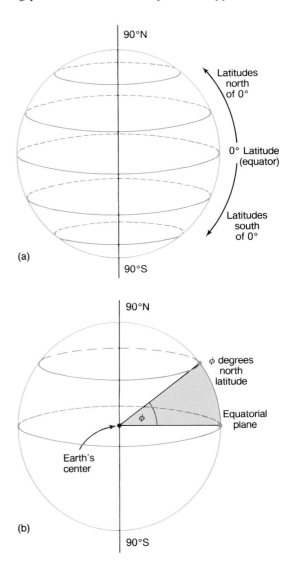

Figure 2.8   (a) Latitude circles are drawn parallel to the equator. (b) The value of a latitude circle is expressed in degrees determined by the angle between the equatorial plane and a line from Earth's center to a surface point on the latitude circle. This is the angle φ (phi). The degree value of φ must be noted as north or south of the equator.

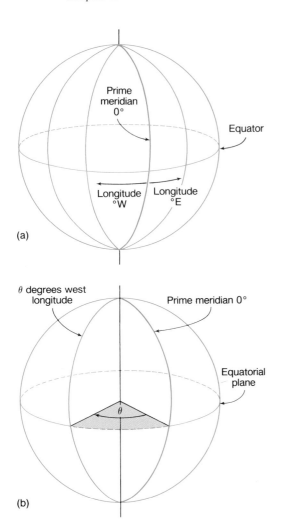

(a)

(b)

Figure 2.9   (a) Longitude lines are drawn with reference to the prime meridian. (b) The value of a longitude line is expressed in degrees determined by the arc between the prime meridian and the longitude line. This is the angle θ (theta). The value of θ is given in degrees east or west of the prime meridian.

Figure 2.10   The Royal Naval Observatory at Greenwich, England. The fiber optic line running into the observatory's door marks the prime meridian, the division between east and west longitudes. The prime meridian is defined by the position of the telescope in the Observatory's Meridian Building. This was built by Sir George Biddell Airy, the seventh Astronomer Royal, in 1850. The crosshairs in the eyepiece of the telescope define longitude 0° for the world. In this night image of the observatory, its roof has been opened to allow use of the telescope.

also that latitudes are designated as either north or south of the equator. The latitude value is determined by the angle (φ, or phi) between the latitude line and the equatorial plane at Earth's center (fig. 2.8*b*). The Tropics of Cancer and Capricorn (see fig. 2.6) correspond to latitudes 23½°N and S, respectively. Latitudes 66½°N and S, respectively, correspond to the Arctic and Antarctic Circles.

Lines of longitude, or **meridians,** are formed at right angles to the latitude lines (fig. 2.9). Longitude is referenced to an arbitrarily chosen point: the 0° longitude line on Earth's surface extends from the North Pole to the South Pole and passes directly through the Royal Naval Observatory in Greenwich, England, just outside London (fig. 2.10). On the other side of Earth, 180° longitude is directly opposite 0°. The 0° longitude line is known as the **prime meridian.** The 180° longitude line approximates the **international date line.** Longitude lines are identified by their angular displacement (θ, or theta) to the east and west of 0° longitude, as shown in figure 2.9*b*. Thus, longitude may be reported as either 0° to 180° east (also known

as positive longitude) and 0° to 180° west (also known as negative longitude) or 0° to 360° east (always positive). In this manner −90°, 90°W, and 270° longitude all mark the same meridian. Note in figure 2.9*a* that meridians are the same size, much like the lines marking the segments of an orange. The meridians mark the intersection of Earth's surface with a plane passing through Earth's rotational axis at right angles to the parallels of latitude. Any circle at Earth's surface with its center at Earth's center is a **great circle.** All longitude lines form great circles; only the equator is a great circle of latitude. A great circle connecting any two points on Earth's surface defines the shortest distance between them. See appendix C for an example calculation of the great circle distance between two points.

To identify any location on Earth's surface, we use the crossing of the latitude and longitude lines; for example, 158°W, 21°N is the approximate location of the Hawaiian Islands, and 20°E, 33°S identifies the Cape of Good Hope at the southern tip of Africa. Because the distance expressed in whole degrees is large (1 degree of latitude equals 60 nautical miles), each degree (1°) of arc is divided into 60 minutes (60'), each minute into 60 seconds (60"), and each second into tenths of a second. The location of Honolulu Bay, Hawaii, is 21° 18' 34"N, 157° 52' 21"W. One **nautical mile** is equal to one minute of arc length of latitude or longitude at the equator, or 1852 m (1.15 land mi; see appendix B). Positions on Earth can be specified with great accuracy when this system is used.

Charts and Maps

Charts and maps show Earth's surface on a flat, or two-dimensional, sheet. Maps usually show Earth's land features or land and sea relationships, whereas charts depict the sea and sky. Any chart or map produces a distorted image of the curved surface of Earth. The task of the mapmaker, or cartographer, is to produce the most accurate, most convenient, and least distorted picture for the task for which the map or chart is to be used.

Maps are made by projecting Earth's features and the latitude-longitude system onto a surface; the resulting picture is a map or chart **projection.** Imagine a transparent globe with the continents and the latitude and longitude lines painted on its surface. Place a light in the center of the globe and the light rays will shine out through it. The light will project the shapes and locations of the continents and the latitude and longitude lines onto a piece of paper held up to the outside of the globe. Different projections are obtained by varying the position of the light and the type of surface on which the projection is made. Many chart and map projections have been constructed, but most are modifications of three basic types: cylindric, conic, and tangent plane, as shown in figure 2.11. In these projections, the surface that is to be the map is rolled around the globe as a cylinder (fig. 2.11*a*), made into a cone (fig. 2.11*b*), or laid flat (tangent) against the sphere (fig. 2.11*c*). Although the cylindric and conic surfaces may be placed around, over, or tangent to Earth at any location, they are usually placed so that the cylinder touches the equator and the cone is centered on the polar axis.

Compare the three parts of figure 2.11 and notice that distortion on a map or chart increases at greater distances from the point or line in contact with the globe. Consider Greenland. In the tangent plane projection (fig. 2.11*c*), its size and shape are very close to its true form. In the conic projection (fig. 2.11*b*), the island is shown larger than its true size. And in the cylindric projection (fig. 2.11*a*), both its size and its shape are greatly distorted. The traditional and familiar world map used in many books and school classrooms is the **Mercator projection,** a form of the cylindric type shown in

figure 2.11*a*. Although distortion is great at high latitudes and poles cannot be shown, the Mercator projection, unlike other projections, has the advantage that a straight line as drawn on the map is a line of true direction or constant compass heading; therefore, the Mercator projection is useful in navigation. Each type of chart or map has its own characteristics. The user must select the projection with the least distortion and best properties for his or her purpose.

Maps that show lines connecting points of similar elevation on land (known as **contours** of elevation) show Earth's **topography;** they are topographic maps. Charts of the ocean showing contour lines connecting points of the same depth below the sea surface depict the area's **bathymetry;** they are bathymetric charts (fig. 2.12). Color, shading, and perspective drawings may be added to simulate topography and produce visual representations or bird's-eye views, called **physiographic maps.** See figure 2.13 for an example of a physiographic map and compare it to the bathymetric chart in figure 2.12. Today, computers use

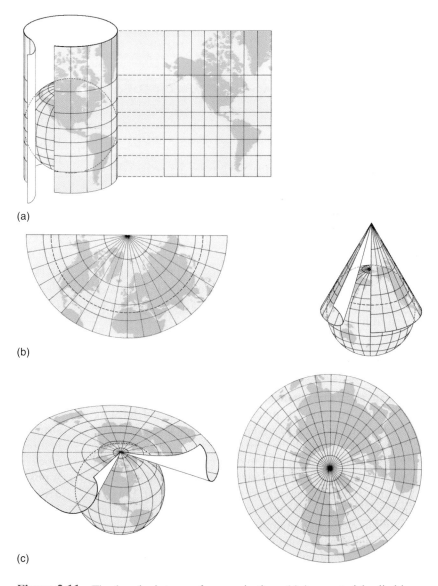

(a)

(b)

(c)

Figure 2.11 The three basic types of map projections. (a) An equatorial cylindric projection. (b) A simple polar conic projection. (c) A polar tangent plane projection.

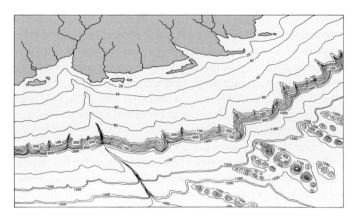

Figure 2.12   A bathymetric, or contour, chart of the sea floor along a section of generalized coast. Changes in the pattern of and spacing between contour lines indicate changes in depth.

Figure 2.13   A physiographic map of the area shown in figure 2.12. Shading and perspective have been added.

electronic ocean-depth measurements to form detailed bathymetric charts that are rapidly converted into simulated three-dimensional, color images of the sea floor that can be viewed from any angle (fig. 2.14).

Measuring Latitude

Early maps show that the first cartographers and navigators had considerable difficulty in precisely describing and locating known features. When measurements were not accurate, artistic license appeared to fill the gaps. The major problem was that early navigators were not able to determine their position accurately. As navigational techniques improved, so did the maps. It was known by early navigators that the North Star, **Polaris,** appeared to hang in the sky above the North Pole and did not appreciably move from that spot. In the Northern Hemisphere, therefore, measuring the angle of elevation of Polaris above the horizon gave a good estimate of one's latitude. Once a ship was out of sight of land, it sailed north or south until reaching the desired latitude, then sailed east or west along that line of latitude until reaching land. Adjustments north and south to reach the

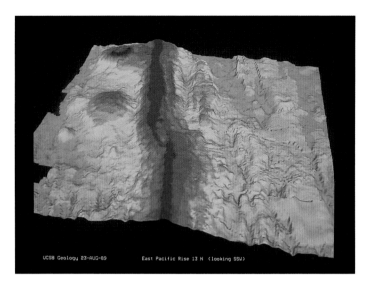

Figure 2.14   Three-dimensional computer-processed image of a section of the East Pacific Rise. Data for this image were obtained by using the Sea Beam echo sounder.

desired landfall were made close to shore and according to visible landmarks.

Longitude and Time

Determining longitude was a much more difficult task. Because the longitude lines rotate with Earth, 360° in twenty-four hours, it becomes necessary to know the time of day and the position of the Sun or the stars relative to one's longitude line. Although the theory for using time to determine longitude had been proposed by Flemish astronomer Gemma Frisius in 1530, early clocks did not work satisfactorily on ships, and precise longitude measurements were not possible until the construction of an accurate ship's chronometer in the eighteenth century. See chapter 1 for the history of this achievement.

If a clock is set to exactly noon at some initial location when the Sun is at its **zenith,** or highest elevation above a reference longitude, and if that clock is then carried to a new location and the time on the clock is noted at the zenith time (noon) of the Sun at the new location, the difference in time between the time on the clock and noon at the new location can be used to determine the difference in longitude between the initial and new locations. When this technique is used, a position that is 15° of longitude west of the reference longitude is directly *under the Sun* one hour later, or at 1 P.M. by the clock, because Earth has turned eastward 15° during that hour. A position 15° east of the reference longitude is directly under the Sun at 11 A.M., because an hour is required to turn the 15° to bring the reference longitude directly under the Sun at its zenith (fig. 2.15).

The reference longitude in use today is the prime meridian, or 0° longitude. The clock time is set to noon when the Sun is at its zenith above the prime meridian. This is **Greenwich Mean Time (GMT),** now **Universal Time** or **Zulu time** (zero meridian time). Because Sun time changes by one hour for each

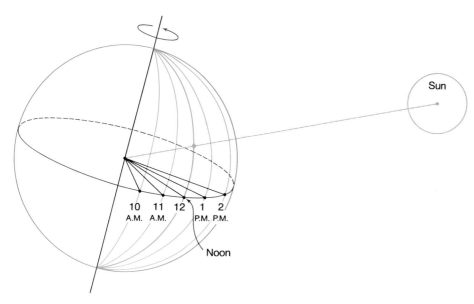

Figure 2.15 The time on a meridian relative to the Sun changes by one hour for each 15° change in longitude.

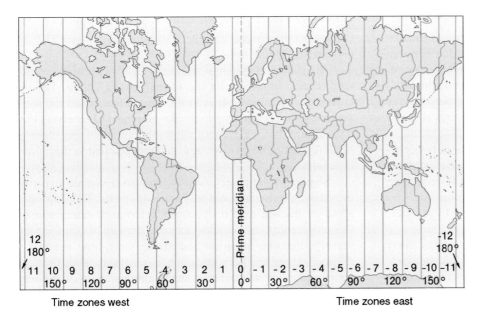

Figure 2.16 Distribution of the world's time zones. Degrees of longitude are marked along the bottom. Time zones are positive (west zones) or negative (east zones). Time at Greenwich is determined by adding the zone number to the local time.

15° of longitude, Earth has been divided into time zones that are 15° of longitude wide. The time zones do not exactly follow lines of longitude; they follow political boundaries when necessary for the convenience of the people living in those zones (fig. 2.16).

2.5 Modern Navigational Techniques

Modern navigators still use chronometers and wait for clear skies to "shoot" the Sun or stars with a sextant to determine positions at sea, but such measurements are used primarily to check their

modern electronic navigational equipment. Chronometers are calibrated, or reset, by broadcast time signals. When vessels are near land, **radar** (radio detecting and ranging) bounces radio pulses off a target such as the shoreline or another vessel. The image on the radar screen is formed by radiation energy that is sent out by a transmitter, reflected from an object, returned to the antenna, and then displayed on a screen. **Loran** (long-range navigation) can be used farther out at sea. This electronic timing device measures the difference in arrival time of radio signals from pairs of land stations. The position of the receiving ship is plotted on a chart that shows the time delay lines for these stations. Recent improvements in loran include receivers with computers that can be programmed with the latitude and longitude of the desired destination. The loran receiver monitors the signals from the stations and directly reads out the course to be sailed and the distance to the destination. The computer can also monitor the signals and continuously calculate the latitude and longitude of the vessel, enabling the ship's personnel and ocean scientists to know their position at all times.

The **satellite navigation system** is an accurate and sophisticated navigational aid. Satellites orbiting Earth emit signals of a precise frequency that are picked up by a receiver on the ship. The ship's receiver monitors the frequency shift of the signal as the satellite passes and determines the exact instant in time at which the frequency is correct. At this instant, the ship's path and the satellite's orbit are at right angles. Given this information, a computer programmed with the satellite's orbital properties can determine the ship's position to within 30 m (100 ft) or less.

A more versatile and accurate method of finding one's position uses the U.S. Navstar **Global Positioning System (GPS).** GPS is a worldwide radio-navigation system consisting of twenty-four navigational satellites—twenty-one operational and three active spares—and five ground-based monitoring stations (fig. 2.17). The satellites orbit at an altitude of about 20,165 km (12,500 mi) and repeat the same track and relative configuration over any point about every twenty-four hours. At any given time, from five to eight satellites are visible from any point on Earth. The system uses this constellation of satellites as reference points for calculating positions on the surface with an accuracy of a few meters for commercial and private users and to better than a centimeter with advanced forms of GPS.

Each GPS satellite transmits a unique digital code that is sent as a radio signal to ground receivers. GPS receivers

(a)

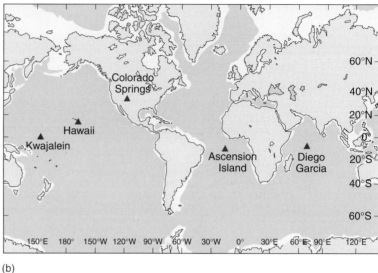

(b)

Figure 2.17 The Global Positioning System (GPS) consists of (a) twenty-four satellites and (b) five monitoring stations. The main control station is located at Falcon Air Force Base in Colorado Springs, Colorado.

generate codes that are identical to those sent by the satellites at exactly the same time the satellites do. Because of the distance the satellite signals travel, there is a time lag between when the receiver generates a specific signal and when it receives the same signal generated at the same time by the satellite. This time delay is a function of the distance between the receiver and the satellite. The satellite signals travel at the speed of light, or approximately 300,000 km/s (186,000 mi/s), so measuring the arrival time of the signal accurately is critically important. The signal from a satellite directly overhead would reach a ground receiver in roughly 0.06 s. Precise measurements of distance between a receiver and four satellites will pinpoint the exact location of the receiver. In practice, there are a variety of sources of error in the measurements, so measurements are made between the receiver and as many satellites as can be detected to maximize location accuracy. GPS can be used to measure the velocity of a moving receiver as well as its location.

GPS is used in a wide variety of applications, including positioning and navigation, mapping land and sea surface features, monitoring tides and currents, measuring the motion of tectonic plates, and monitoring strain along faults. It is even used to study the viscosity of Earth's mantle (see chapter 3); it observes three-dimensional crustal velocities as regions previously depressed under heavy ice sheets during the last glacial period continue slowly to rise as the ice melts and releases its tremendous weight. Oceanographers are developing new techniques for using GPS in underwater applications. A major problem in the use of GPS underwater is that GPS signals are radio waves, which are quickly absorbed by water. Underwater GPS systems that are being tested involve surface buoys containing GPS receivers that communicate with underwater targets using acoustic (sound) energy. A further challenge is that it is difficult to rapidly transmit the large amounts of information that are contained in GPS radio signals in an acoustic signal.

Solutions to these problems will likely be developed rapidly because of the enormous benefit of being able to use GPS in underwater applications.

The U.S. GPS system is similar to a global position system, GLONASS, employed by the former Soviet Union. The increasing cooperation between the United States and the Russia Federation is pointing to the development of receivers that will accept both systems, allowing increased positioning accuracy, better Earth coverage, and independent verifications of results.

Shipboard computers can now store an electronic atlas that includes surface charts and seafloor bathymetry. The ship's position is tracked and its position determined continuously. Oceanographic vessels use devices that draw charts showing the vessel's changing position and at the same time conduct a seafloor survey and keep track of water measurements (such as temperature and salt content) made automatically as the ship moves along. Today's oceanographers are able to return more and more accurately to the same place at sea to repeat measurements and follow changes in ocean processes. Scientists now have the ability to evaluate data as they are being taken, allowing changes to be made in a vessel's sampling pattern and thereby ensuring that the data satisfy a project's research requirements.

2.6 Earth: The Water Planet

Water on Earth's Surface

As Earth and the other planets of the solar system cooled, the Sun's energy gradually replaced the heat of planet formation in maintaining their surface temperatures. Earth developed a nearly circular orbital path at a mean distance of 149 million km (149×10^6 km), or 93 million mi (93×10^6 mi) from the Sun. (If you are unfamiliar with scientific notation to express very large numbers, see appendix A.) Moving along this orbit Earth is 152.2×10^6 km (94.5×10^6 mi) from the Sun in June and 148.5×10^6 km (92.2×10^6 mi) away in December, as shown in figure 2.6. At these distances from the Sun, Earth's orbit keeps the annual heating and cooling cycle within moderate limits. Earth's mean surface temperature is about 16°C (61°F), which allows water to exist as a gas, as a liquid, and as a solid.

The rotation of Earth on its axis is also important in moderating temperature extremes. Earth completes one rotation, turning from west to east, in twenty-four hours. If Earth rotated more slowly, the side of Earth toward the Sun would be exposed to the Sun's energy for a longer period than at present and would become very hot, while the side in darkness would lose heat and become very cold. Temperature changes from day to night would be large. By contrast, a shorter period of rotation would decrease the present variation from day to night.

Earth's solar orbit, its rotation, and its blanket of atmospheric gases produce surface temperatures that allow the existence of liquid water. The atmosphere acts as a protective shield for Earth against ultraviolet radiation from the Sun. Compare Earth's distance to the Sun, the period of rotation, and the time required for Earth to complete one orbit of the Sun with those of other planets, as shown in table 2.1. Note the surface temperature of Earth compared to that of other planets.

The amount of water on Earth's surface can be expressed in several ways. For example, the oceans cover 362 million km² (362×106 km²), or about 140 million mi² (140×106 mi²). Because these numbers are so large, they do not convey a clear idea of size; therefore, an easier concept is to remember that 71% of Earth's surface is covered by the oceans, and only 29% of the surface area is land above sea level.

The volume of water in the oceans is enormous: 1.35 billion km³ (1.35×10^9 km³, or 0.324×10^9 mi³). Another way to express the oceanic volume is to think of a smooth sphere with exactly the same surface area as Earth (510×10^6 km², or 197×10^6 mi²), uniformly covered with the water from Earth's oceans. The ocean water would be 2646 m (8682 ft) deep, a depth of about 1.6 mi. If the water from all other sources in the world were added, the depth would rise 75 m to 2721 m (8928 ft). When water volumes are considered as depths over a smooth sphere, they are referred to as **sphere depths.** The ocean sphere depth is 2646 m, and the total water sphere depth of all Earth's water is 2721 m.

Hydrologic Cycle

Earth's water occurs as a liquid in the oceans, rivers, lakes, and below the ground surface; it occurs as a solid in glaciers, snow packs, and sea ice; it occurs as water droplets and vapor in the atmosphere. The places in which water resides are called **reservoirs,** and each type of reservoir, when averaged over the entire Earth, contains a fixed amount of water at any one instant. But water is constantly moving into and out of reservoirs. This movement of water through the reservoirs, diagrammed in figure 2.18, is called the **hydrologic cycle.**

Water is taken out of the oceans and moved into the atmosphere by evaporation. Most of this water returns directly to the sea by precipitation, but air currents carry some water vapor over the continents. Precipitation in the form of rain and snow transfers this water from the atmosphere to the land surface, where it percolates into the soil; is taken up by plants; runs off into rivers, streams, and lakes; or remains for longer periods as snow and ice in some areas. Some of this water returns to the atmosphere by evaporation, **transpiration** (the

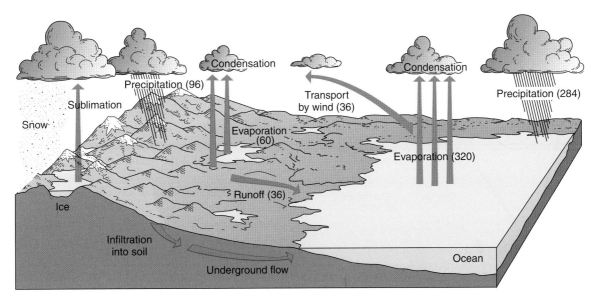

Figure 2.18 The hydrologic cycle and annual transfer rates for Earth as a whole. Precipitation transfer rate includes both snow and rain. Evaporation transfer rate from the continents includes evaporation of surface water, transpiration (the release of water to the atmosphere by plants), and sublimation (the direct change in state from ice to water vapor). Runoff from the continents includes both surface flow and underground flow. Annual transfer rates in thousands of cubic kilometers (10^3 km³).

Table 2.3	Earth's Water Supply		
	Approximate Water Volume		**Approximate Percent of Total Water**
Reservoir	**(km³)**	**(mi³)**	
Oceans and sea ice	1,349,929,000	323,866,000	97.26
Ice caps and glaciers	29,289,000	7,000,000	2.11
Groundwater	8,368,000	2,000,000	0.60
Freshwater lakes	125,500	30,000	0.009
Saline lakes and inland seas	105,000	25,000	0.008
Soil moisture	67,000	16,000	0.005
Atmosphere	13,000	3,100	0.0009
Rivers	1,250	300	0.0001
Total water volume	1,387,897,750	332,940,400	100

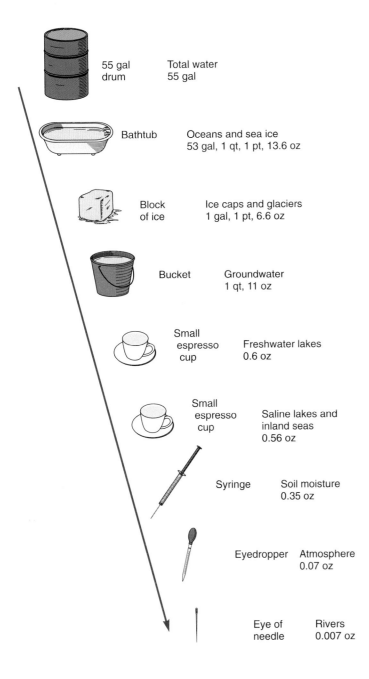

55 gal drum	Total water 55 gal
Bathtub	Oceans and sea ice 53 gal, 1 qt, 1 pt, 13.6 oz
Block of ice	Ice caps and glaciers 1 gal, 1 pt, 6.6 oz
Bucket	Groundwater 1 qt, 11 oz
Small espresso cup	Freshwater lakes 0.6 oz
Small espresso cup	Saline lakes and inland seas 0.56 oz
Syringe	Soil moisture 0.35 oz
Eyedropper	Atmosphere 0.07 oz
Eye of needle	Rivers 0.007 oz

release of water by plants), and **sublimation** (the conversion of ice directly to water vapor). Melting snow and ice, rivers, groundwater, and land runoff move the water back to the oceans to complete the cycle and maintain the oceans' volume. For a comparison of the water stored in Earth's reservoirs, see table 2.3 and figure 2.19.

The properties of climate zones are principally determined by their surface temperatures and their evaporation-precipitation patterns: the moist, hot equatorial regions; the dry, hot subtropical deserts; the cool, moist temperate areas; and the cold, dry polar zones. Differences in these properties, coupled with the movement of air between the climate zones, move water through the hydrologic cycle from one reservoir to another at different rates. The transfer of water between the atmosphere and the oceans alters the salt content of the oceans' surface water, and, with the seasonal and latitudinal changes in surface temperature, determines many of the characteristics of the world's oceans. These characteristics are explored in chapters 8 to 10.

Reservoirs and Residence Time

Because the total amount of water on Earth is essentially constant, the hydrologic cycle must maintain a balance between the addition and removal of water from Earth's water reservoirs. The rate of removal of water from a reservoir must equal the rate of addition to it, for if the balance is disturbed, one reservoir will gain at the expense of another. The average length of time that a water molecule spends in any one reservoir is called the **residence time** in that reservoir. Water's residence time can be calculated by dividing the volume of water in the reservoir by the rate at which the water is replaced. A large reservoir generally has a long residence time because of its large volume, while

Figure 2.19 Comparison of the amount of the world's water supply held in each of the major water reservoirs. For purpose of illustration, Earth's total water supply has been scaled down to the volume of a 55 gal drum.

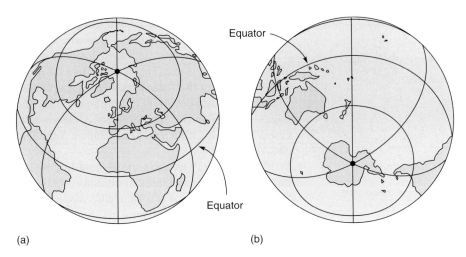

(a)   (b)

Figure 2.20   Continents and oceans are not distributed uniformly over Earth. The Northern Hemisphere (a) contains most of the land; the Southern Hemisphere (b) is mainly water.

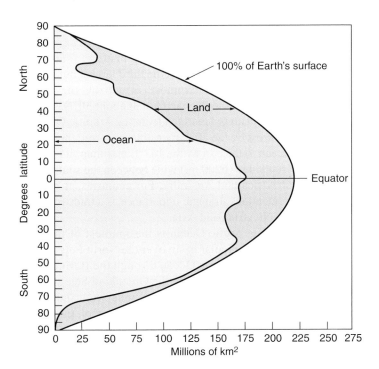

Figure 2.21   Distribution of land and ocean by latitude. In the Northern Hemisphere, middle-latitude areas of land and ocean are nearly equal. Land is almost absent at the same latitude in the Southern Hemisphere. The areas are calculated on the basis of 5° latitude intervals. The oceans cover 71% of Earth's surface and the remaining 29% is covered by land.

small reservoirs generally have a short residence time and the water in them can be replaced comparatively quickly. The size also determines how a reservoir reacts to changes in the rate at which the water is gained or lost. Large reservoirs show little effect from small changes, whereas small reservoirs may alter substantially when exposed to the same gain or loss. For example, if the ocean volume decreased by 6½% and that volume of water were added to the land ice, the result would be a 300%

increase in the present volume of land ice and sea level would drop by 172 m (564 ft). This example reflects the changes that have occurred on Earth during the major ice ages, when more water was stored on land as ice and sea level was lower.

About 380,000 km^3 (91,167 mi^3) of water move through the atmosphere each year. Because the atmosphere holds the equivalent of 13,000 km^3 (3119 mi^3) of liquid water at any one time, a little arithmetic shows that the water in the atmosphere can be replaced twenty-nine times each year. Atmospheric water has a very short residence time. The residence time for water in the other, larger reservoirs is much longer. For example, it would take 4219 years to evaporate and pass all of the water in the oceans through the atmosphere, to the land as precipitation, and back to the oceans via rivers. Further study of water's movement shows us that annually 320,000 km^3 (76,772 mi^3) is evaporated from the oceans, and 60,000 km^3 (14,395 mi^3) is evaporated from land. When the water returns as precipitation, 284,000 km^3 (68,135 mi^3) is returned directly to the sea surface and 96,000 km^3 (23,032 mi^3) to the land. However, the excess gained by the land (36,000 km^3 or 8637 mi^3) flows back to the oceans in rivers, streams, and groundwater (see fig. 2.18).

Distribution of Land and Water

To understand the present distribution of land and water on Earth, consider Earth viewed from the north (fig. 2.20a) and from the south (fig. 2.20b). About 70% of Earth's landmasses are in the Northern Hemisphere, and most of this land lies in the middle latitudes. The Southern Hemisphere is the water hemisphere, with its land located mostly in the tropical latitudes and in the polar region. The details of this land-water-latitude distribution are presented in figure 2.21.

Oceans

The distribution and shape of the continents divide the world ocean into four (fig. 2.22), and some would argue five, individual oceans. Oceanographers all recognize the three major oceans—Pacific, Atlantic, and Indian Oceans—that extend northward from the continuous body of water surrounding Antarctica. A fourth smaller ocean—Arctic Ocean—is located at the North Pole. The Arctic Ocean has sometimes been considered an extension of the North Atlantic, but because of its size and relative isolation, most people consider it to be an independent ocean basin. Each of these oceans has its own characteristic surface area, volume, and mean depth (table 2.4). In 2000, the International Hydrographic Organization defined the seawater between 60°S and Antarctica to be an independent fifth ocean called the Southern Ocean, thus significantly reducing the size of the three major oceans by eliminating their far southern regions.

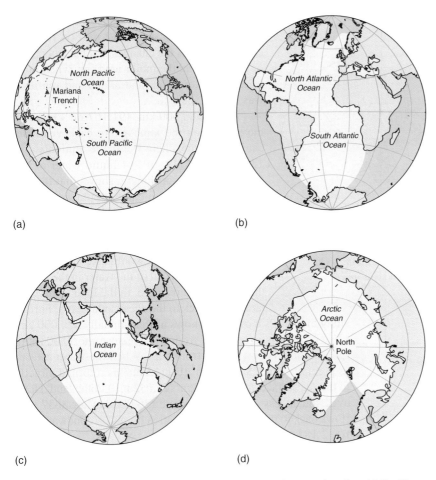

North Pacific
Ocean

Mariana
Trench

South Pacific
Ocean

(a)

North Atlantic
Ocean

South Atlantic
Ocean

(b)

Indian
Ocean

(c)

Arctic
Ocean

North
Pole

(d)

Figure 2.22 The world's four major oceans in order of decreasing size: (a) Pacific Ocean (the deepest spot in the oceans is located in the Mariana Trench); (b) Atlantic Ocean; (c) Indian Ocean; and (d) Arctic Ocean.

The Pacific Ocean has greater surface area, volume, and mean depth than any of the other oceans. The Pacific was named by Ferdinand Magellan in 1520 for the calm weather he and his crew enjoyed while crossing it (*paci* = peace). The Pacific covers a little over a third of Earth's surface and just over half of the world ocean's surface. At its maximum width near 5°N, the Pacific stretches 19,800 km (12,300 mi) from Indonesia to Colombia. There are roughly 25,000 islands in the Pacific, the majority of which are south of the equator. This is more than the number of islands in the rest of the oceans combined. There are a number of marginal seas along the edges of the Pacific, including the Celebes Sea, Coral Sea, East China Sea, Sea of Japan, Sulu Sea, and Yellow Sea.

The Atlantic Ocean is the second largest. Its name is derived from Greek mythology and means "Sea of Atlas" (Atlas was a Titan who supported the heavens by means of a pillar on his shoulders). The land area that drains into the Atlantic is four times larger than that of either the Pacific or Indian Ocean. There are relatively few islands given the size of the basin. The irregular coastline of the Atlantic includes a number of bays, gulfs, and seas. Some of the larger ones include the Caribbean Sea, Gulf of Mexico, Mediterranean Sea, North Sea, and Baltic Sea.

The Indian Ocean is primarily a Southern Hemisphere ocean; it is the third largest but is quite deep. The northernmost extent of the Indian Ocean is in the Persian Gulf at about 30°N. The Indian Ocean is separated from the Atlantic Ocean to the west by the 20°E meridian and from the Pacific Ocean to the east by the 147°E meridian. It is nearly 10,000 km (6200 mi) wide between the southern tip of Africa and Australia. For centuries, it has had tremendous strategic importance as a trade route between Africa and Asia.

The Arctic Ocean is the smallest of the four, occupying a roughly circular basin over the North Pole region. It is connected to the Pacific Ocean through the Bering Strait and to the Atlantic Ocean through the Greenland Sea. Its floor is divided into two deep basins by an underwater mountain range. The major flow of water into and out of the Arctic Ocean is through the North Atlantic Ocean.

Table 2.4	**Ocean Depths, Areas, and Volumes**				
Ocean	**Average Depth**	**Area**	**Volume**	**Percent of Ocean Surface Area**	**Percent of Earth's Surface Area**
Pacific	3940 m	181.3×10^6 km^2	714.4×10^6 km^3	50.1	35.6
	12,927 ft	70.0×10^6 mi^2	171.4×10^6 mi^3		
Atlantic	3575 m	94.3×10^6 km^2	337.2×10^6 km^3	26.0	18.5
	11,730 ft	36.4×10^6 mi^2	80.9×10^6 mi^3		
Indian	3840 m	74.1×10^6 km^2	284.6×10^6 km^3	20.5	14.5
	12,599 ft	28.6×10^6 mi^2	68.3×10^6 mi^3		
Arctic	1117 m	12.3×10^6 km^2	13.7×10^6 km^3	3.4	2.4
	3665 ft	4.7×10^6 mi^2	3.3×10^6 mi^3		
All Oceans	3729 m	362.0×10^6 km^2	1349.9×10^6 km^3	100	71.0
	12,235 ft	139.8×10^6 mi^2	323.9×10^6 mi^3		

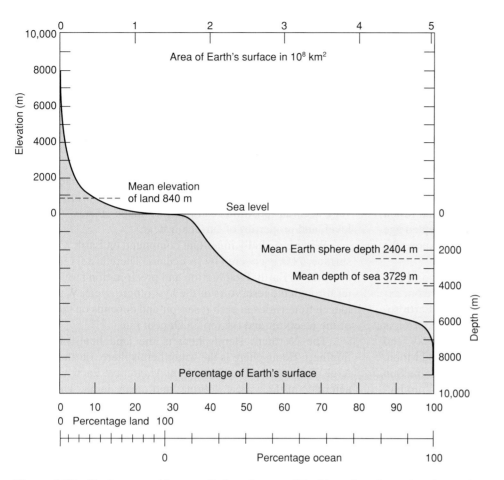

Figure 2.23 The hypsographic curve displays the area of Earth's surface above elevations and water depths.

that the areas well below the sea surface are much greater than the areas well above it; there are basins beneath the sea that are about four times greater in area than the area of land in mountains. Mount Everest, the highest land peak, reaches 8.84 km (5.49 mi) above sea level, whereas the ocean's deepest trench descends 11.02 km (6.84 mi) below sea level.

Because the hypsographic curve is constructed as a plot of area versus height, an area of the diagram shows volume, which is the product of area times height. The mean elevation of the land is 840 m (2750 ft), and the entire land volume above sea level fits within a box 840 m high, covering 29% of Earth's surface. The mean depth of the ocean can be calculated from the average depth of the individual oceans and the percent of total ocean area they cover (see table 2.4) as being (3940 m × 50.1%) + (3575 m × 26%) + (3840 m × 20.5%) + (1117 m × 3.4%) or 3729 m (12,235 ft).

Because humans are land-dwellers, their reference is sea level. If we are considering the hypsographic curve for Earth, a mean Earth elevation becomes more suitable for describing the location of Earth's solid surface; this value is called the **mean Earth sphere depth.** This level is (840 m × 29%) + (−3729 m × 71%) or 2404 m (7887 ft) below present sea level, as calculated from the mean elevations, mean depths, and areas given in figure 2.23. This value represents the level at which the volume of crust above and water volume below the Earth sphere depth are equal. To arrive at the mean Earth sphere depth, think of moving all the elevated land and some of the sea bottom down into the ocean depressions until Earth is a perfectly smooth ball. When this is accomplished, the mean Earth sphere depth reference level would stand at the 2403 m (7884 ft) depth. Such an operation would also result in the displacement of seawater upward, until a new sea level 243 m (797 ft) above the present sea level was reached. This depth of water is the **mean ocean sphere depth,** 2646 m (8682 ft). It is the depth that the oceans would have if the volume of ocean water were spread over an area equal to the total Earth area rather than over 71% of Earth's area. The oceans swallow the land, with only a comparatively small rise in water level, emphasizing again that Earth is the water planet (or the ocean planet) and not the land planet.

Hypsographic Curve

Another method used by oceanographers to depict land-water relationships is shown in figure 2.23. This graph of depth or elevation versus Earth's area is called a **hypsographic curve.** Find the line indicating sea level and note that the elevation of land above sea level is given in meters along the left margin; the depth below sea level is given in meters along the right margin. The scale across the top of the figure indicates total Earth area in 100 million km² (108 km²). The scales along the bottom of the figure indicate percentages of Earth's surface area; note that the curve crosses sea level at the 29% mark, showing that 29% of Earth's surface is above sea level and 71% is below sea level. The lower of the two scales gives land and ocean areas as separate percentages. Referring to the land area percentage scale, note that only 20% of all land areas are at elevations above 2 km. The ocean area scale shows that approximately 85% of the ocean floor is deeper than 2 km. The hypsographic curve helps us to see not only that Earth is 71% covered with water but also

Summary

The beginning expansion of the universe was followed by the first stars, the reactions that produced the elements, and billions of galaxies. Our solar system is part of the Milky Way galaxy; it began as a rotating cloud of gas. A series of events produced nine planets orbiting the Sun, each planet having unique characteristics. Over approximately 1.5 billion years, Earth heated, cooled, changed, and accumulated a gaseous atmosphere and liquid water.

Reliable age dates for Earth rocks, meteorites, and Moon samples are obtained by radiometric dating. The accepted age of Earth is 4.6 billion years. Geologic time is used to express the timescale of Earth's history.

The distance between Earth and the Sun, Earth's orbit, its period of rotation, and its atmosphere protect Earth from extreme temperature change and water loss. Because Earth rotates, its shape is not perfectly symmetrical. Its exterior is relatively smooth. Natural time periods (the year, day, and month) are based on the motions of the Sun, Earth, and Moon. Because of the tilt of Earth's axis as it orbits the Sun, the Sun appears to move annually between 23½°N and 23½°S, producing the seasons.

Latitude and longitude are used to form a grid system for the location of positions on Earth's surface. Different types of map and chart projections have been developed to show Earth's features on a flat surface. These projections distort Earth's features to some extent. Bathymetric and physiographic charts and maps use elevation and depth contours to depict Earth's topography.

To determine longitudinal position, one must be able to measure time accurately. This need required the development of accurate seagoing clocks for celestial navigation.

Modern navigational techniques make use of radar, radio signals, computers, and satellites. A satellite network provides very accurate position readings and maps storms, tides, sea level, and properties of surface waters.

Water is a vitally important compound on Earth. Of Earth's surface, 71% is covered by its oceans. There is a fixed amount of water on Earth. Evaporation and precipitation move the water through the reservoirs of the hydrologic cycle. Water's residence time varies in each reservoir and depends on the volume of the reservoir and the replenishment rate.

The Northern Hemisphere is the land hemisphere; the Southern Hemisphere is the water hemisphere. Earth has three large oceans extending north from Antarctica. Each has a characteristic surface area, volume, and mean depth. The hypsographic curve is used to show land-water relationships of depth, elevation, area, and volume. It is also used to determine mean land elevation, mean ocean depth, Earth sphere depth, and ocean sphere depth.

Key Terms

All key terms from this chapter can be viewed by term or definition when studied as flashcards on this book's website at www.mhhe.com/sverdrup10e.

electromagnetic spectrum, 27
Big Bang, 27
galaxy, 29
light-year, 29
cluster, 29
nebula, 29
dayglow, 30
radiometric dating, 33
isotope, 33
half-life, 33
vertebrate, 34

Tropic of Cancer, 36
summer solstice, 36
Arctic Circle, 36
Antarctic Circle, 36
autumnal equinox, 36
Tropic of Capricorn, 36
winter solstice, 36
vernal equinox, 36
lunar month, 36
solar day, 36
sidereal day, 36
latitude, 37
longitude, 37
parallel, 37
equator, 37
meridian, 38

prime meridian, 38
international date line, 38
great circle, 38
nautical mile, 39
projection, 39
Mercator projection, 39
contour, 39
topography, 39
bathymetry, 39
physiographic map, 39
Polaris, 40
zenith, 40
Greenwich Mean Time
   (GMT), 40
Universal Time, 40
Zulu Time, 40

radar, 41
loran, 41
satellite navigation system, 41
Global Positioning System
   (GPS), 41
sphere depth, 43
reservoir, 43
hydrologic cycle, 43
transpiration, 43
sublimation, 44
residence time, 44
hypsographic curve, 47
mean Earth sphere depth, 47
mean ocean sphere depth, 47

Study Questions

1. How and why have estimates of the age of Earth changed over the past few hundred years? Do you think the present estimate of Earth's age will change in the future?
2. Describe the distribution of water and land on Earth.
3. Why does Earth's average surface temperature differ from the surface temperature of other planets in the solar system?
4. Why is the twilight period at sunset shorter at low latitudes than it is at high latitudes?
5. The route of a ship sailing a constant compass course on a Mercator projection is indicated by a straight line that cuts all longitude lines at the same angle. This is a rhumb line. Discuss how this line appears (a) on a polar conic projection, (b) on a globe, and (c) on a tangent plane projection centered on the polar axis.
6. Discuss how the hypsographic curve is used to determine the mean depth and sphere depth of the oceans.
7. Why are the Arctic and Antarctic Circles located at 66½°N and 66½°S, respectively?
8. What are some advantages of using satellites for oceanographic research? Are there any disadvantages?
9. How will the seasons change over a calendar year at each of these latitudes: (a) 10°N; (b) 70°N; (c) 30°S? Make a simple diagram for each latitude to show why the seasonal pattern occurs.
10. Explain why Earth sustains a wide variety of life forms but the other planets do not.
11. Trace several possible routes for a water molecule moving between a mountain lake and an ocean. In which reservoirs would the molecule spend the greatest amount of time and in which the least?
12. Use an atlas to find the appropriate latitudes and longitudes for each of the following:
    a. St. John's, Newfoundland, and London, England
    b. Cape Town, South Africa, and Melbourne, Australia
    c. Anchorage, Alaska, and Moscow, Russia
    d. Strait of Gibraltar, Strait of Magellan, and Straits of Florida
    e. Galápagos Islands, Tristan da Cunha, and Reykjavík, Iceland
13. Although latitude and longitude were used on very early charts, navigators continued to use charts with many compass direction lines (*portolano* type) well into the seventeenth century. Why?
14. If the lunar month were used as the length of a month, what would happen to the Gregorian calendar year relative to the Sun?

Study Problems

1. Determine the distance between two locations: 110°W, 38½°N and 110°W, 45°N. Express this distance in nautical miles and kilometers.
2. The contour interval on a bathymetric chart is equal to 100 m. Graph the slope of the sea floor across four evenly spaced contour lines if the distance between the first line and the fourth line is 2.5 km.
3. A plane leaves Tokyo, Japan, on June 6, at 0800 hours local Tokyo time and flies for nine hours, landing in San Francisco, California. Give the local time and date of arrival in San Francisco.
4. Show that the annual net evaporative loss of water from the world's oceans equals the annual net gain of water by precipitation on the land. Why does the ocean volume not decrease?
5. Use the volume of the oceans and Earth's area to determine the sphere depth of the oceans.

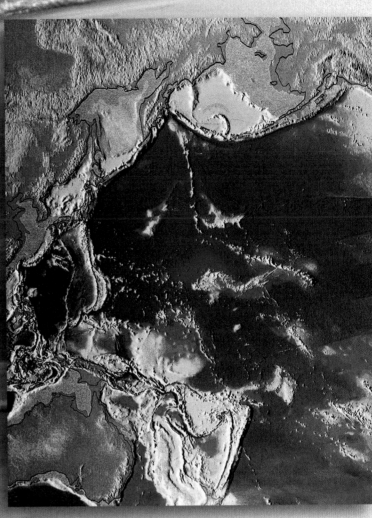

The convergent plate boundaries and island chains of the western Pacific Ocean.

Chapter 3

Plate Tectonics

Chapter Outline

Learning Outcomes

After studying the information in this chapter students should be able to:

1. *sketch* Earth's internal structure and label the thickness of each region,
2. *explain* how seismology has provided critical data for modeling Earth's interior,
3. *differentiate* between the lithosphere, asthenosphere, and mesosphere,

4. *diagram* the three types of plate boundaries,
5. *review* the evidence in a written summary Alfred Wegener used to support his hypothesis of continental drift and the evidence supporting plate tectonics,
6. *distinguish* between continental drift and plate tectonics,
7. *describe* the formation of hydrothermal vents, and
8. *illustrate* the formation of magnetic stripes on the sea floor.

Many features of our planet have presented Earth scientists with contradictions and puzzles. For example, the remains of warm water coral reefs are found off the coast of the British Isles, marine fossils occur high in the Alps and the Himalayas, and coal deposits that were formed in warm, tropical climates are found in northern Europe, Siberia, and northeastern North America. In addition, similar patterns were observed in widely scattered places but for no known reason. Great mountain ranges divide the oceans, the volcanoes known as the ring of fire border both the east and west coasts of the Pacific Ocean, and deep-ocean trenches are found adjacent to long island arcs. No single coherent theory explained all these features, until the technology and scientific discoveries in the 1950s and early 1960s combined to trigger a complete reexamination of Earth's history.

In this chapter, we investigate Earth's interior, and we explore the history of the theory of plate tectonics, as well as the evidence that allowed this theory to become accepted. We review the research that continues to provide us with new insights into Earth's past, to comprehend and appreciate its present, and even to look forward into its future.

3.1 Earth's Interior

Investigating Earth's Structure

Although we cannot directly observe the interior of the planet, scientists have been able to learn a great deal about the structure, composition, physical state, and behavior of Earth's interior using indirect methods. Earth's spherical shape, its mean (or average) radius, and its mass can be used to determine the average density of Earth, 5.51 g/cm³. **Density** is a measure of mass per unit volume and is usually given in grams per cubic centimeter, written g/cm³. This calculated density is considerably greater than the average density of surface rocks, which is about 2.7 g/cm³. Earth's high average density requires that the material below the surface have a much greater density. Because Earth wobbles only slightly as it rotates and the acceleration due to gravity over its surface is relatively uniform, Earth's mass must be distributed fairly uniformly about its center as a series of concentric layers. Gravity, density, and Earth's dimensions enable us to calculate the pressures within Earth and the

temperatures that can be reached under these pressures. Because Earth has a magnetic field, we conclude that its central part must include materials that produce magnetic fields.

Another clue to Earth's structure is furnished by meteorites that occasionally hit Earth and are considered to be the remains of planets. More than half of the meteorites that have been found are "stony" silicate or rocky lumps; another large group is made mainly of iron and nickel; and a few are "stony-iron" with metal inclusions. Radiometric dating of meteorites gives a maximum age of 4.6 billion years, the same as the age of the solar system and Earth. These fragments allow us to directly analyze the density, chemistry, and mineralogy of the nickel-iron cores and stony shells of bodies that we believe to have a composition similar to that of Earth.

The most detailed information we have about the interior has come from roughly a century of recording and studying the passage of **seismic waves** through Earth. Geologists and geophysicists monitor recording stations all over the surface of Earth that measure the type, strength, and arrival time of seismic waves generated by earthquakes, volcanic eruptions, and humanmade explosions.

Two basic kinds of seismic waves occur: surface waves travel relatively slowly along the surface of Earth, and body waves travel at higher speeds through Earth's interior (fig. 3.1). Most of the information we have about Earth's interior comes from the study of body waves. There are two kinds of body waves: **P-waves,** or primary waves (so called because they travel faster than any other seismic waves and are the first to arrive at a recording station), and **S-waves,** or secondary waves (so called because they travel more slowly than P-waves and are the second waves to arrive at a station).

P-waves and S-waves produce different types of motion in the material they travel through. P-waves, also known as compressional waves, alternately compress and stretch the material they pass through, causing an oscillation in the same direction as they move. P-waves can travel through all three states of matter: solid, liquid, and gas (sound propagates through the air and the oceans as a compressional wave). S-waves, also known as shear waves, oscillate at right angles to their direction of motion (similar to a plucked string). S-waves propagate only through solids. The motion generated in materials by these two types of waves is shown in figure 3.2.

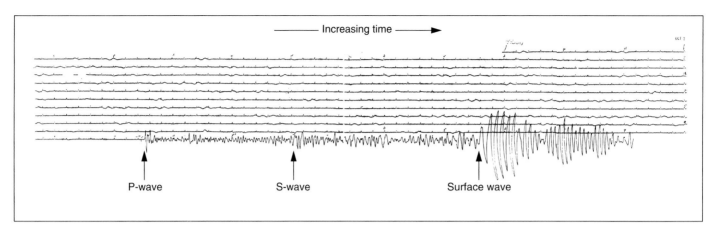

Figure 3.1   A seismogram of an earthquake that occurred in Taiwan recorded in Berkeley, California, 10,145 km (6300 mi) away. The faster body waves (P- and S-waves) arrive before the slower surface waves. Time increases from top to bottom and from left to right on the seismogram.

The speed and direction of seismic waves depend on the characteristics of the material, including its chemistry, its density, and changes in the physical state of the material (solid, partially molten, or molten) caused by variations in pressure and temperature with depth. Detailed modeling of the paths taken by seismic waves and their expected travel times through Earth as a function of distance has produced an Earth model consisting of four major layers: the inner core, the outer core, the mantle, and the crust (fig. 3.3). These are discussed in detail in "Internal Layers." As seismic body waves move through one layer and into another, their speeds change and the waves bend, or **refract,** as shown in figure 3.4*a* and *b*. The paths taken by P- and S-waves through Earth provide information about the dimensions, structure, and physical properties of each of the internal layers. The outer core is a unique region where P-waves are strongly refracted (fig. 3.4*a*) and S-waves cannot propagate (fig. 3.4*b*), indicating that the outer core behaves like a liquid. The presence of the outer core creates both P- and S-wave shadow zones (areas on the surface of Earth away from an earthquake where P- and S-waves generated by the event will not be recorded). The size of these shadow zones is determined by the depth to the top of the outer core. If the interior of Earth had uniform properties, seismic body waves would follow straight lines, their speed would be constant, and there would be no shadow zones.

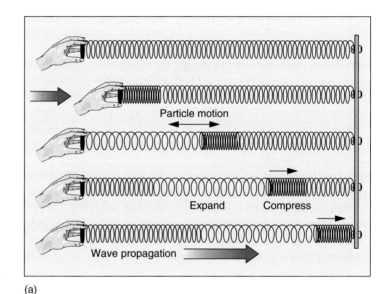

(a)

(b)

Figure 3.2   Particle motion in seismic waves. (a) P-wave motion can be illustrated with a sudden push on the end of a stretched spring. Vibration is parallel to the direction of propagation. (b) S-wave motion can be illustrated by shaking a rope to transmit a deflection along its length. Vibration is perpendicular to the direction of propagation.

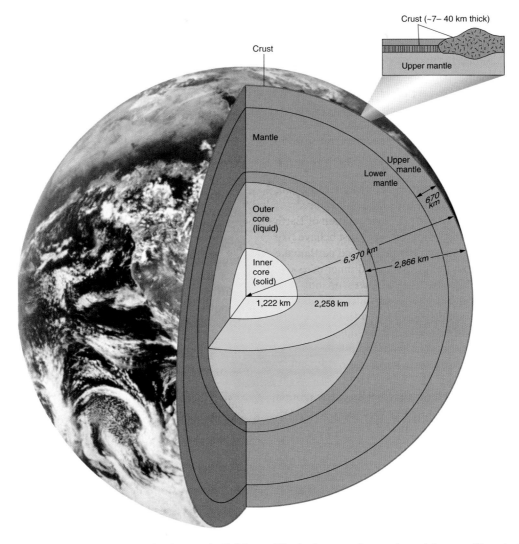

Figure 3.3 Earth's interior. Seismic waves show the three main divisions of Earth: the crust, the mantle, and the core. Photo by NASA.

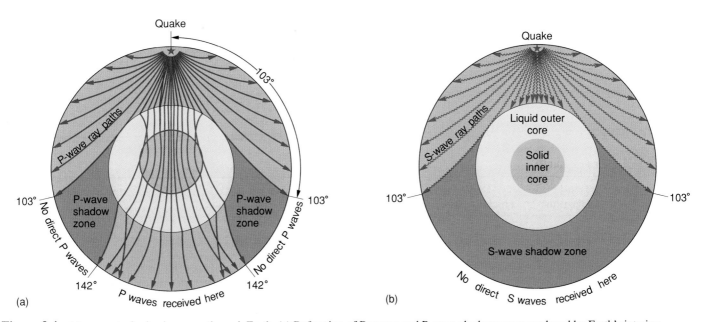

Figure 3.4 Movement of seismic waves through Earth. (a) Refraction of P-waves and P-wave shadow zones produced by Earth's interior structure. (b) Refraction of S-waves and S-wave shadow zone produced by the inability of S-waves to travel into the liquid outer core.

Internal Layers

At the planet's center is the **inner core;** its radius is 1222 km (759 mi). It is solid and nearly five times as dense as common surface rocks such as granite partly because of the tremendous pressure at that depth. The inner core is composed primarily of iron with lesser amounts of other elements that most likely include nickel, sulfur, and oxygen. The temperature ranges from about 4000°–5500°C, far above the melting temperature of iron at Earth's surface, but the inner core remains solid because of the high pressure. The inner core is surrounded by a shell, 2258 km (1402 mi) thick, of similar composition and lower temperature (3200°C) and pressure. This shell is called the **outer core.**

As early as 1926, studies of the tidal deformation of Earth made it clear that at least a portion of the core must behave like a fluid. In 1936, seismologist Inge Lehmann used earthquake data to establish the fact that there was a solid inner core surrounded by a "fluid" outer core that S-waves do not pass through. Although it behaves like a fluid, the outer core may not be completely molten. It would behave like a fluid even if as much as 30% were composed of suspended crystals, most probably of iron oxides and sulfides, that had formed from the surrounding liquid. Recent studies have determined that the inner core rotates about 1° per year faster than the mantle. This increase in eastward rotation is thought to be caused by motion in the fluid outer core. The fluid motion in the outer core moves at a speed that is probably on the order of kilometers per year, generating Earth's magnetic field.

Detailed study of P-waves traveling in the vicinity of the core-mantle boundary has revealed that the upper surface of the core is not smooth but has peaks and valleys. These features extend as much as 11 km (7 mi) above and below the mean surface of the outer core. It is thought that the regions above a peak are areas where the mantle has excess heat and mantle material rises, drawing the core upward. Cooler, denser, and more viscous mantle material sinks to cause depressions in the core's surface. It is likely that these peaks and valleys last only as long as it takes a rising plume to lose its excess heat and sink back toward the core, perhaps a hundred million years.

The **mantle** makes up about 70% of Earth's volume. The mantle is 2866 km (1780 mi) thick and is less dense and cooler (1100°–3200°C) than the core. It is composed of magnesium-iron silicates (rocky material rather than metallic like the core). Earthquake data indicate the presence of a distinct change in the speed of P- and S-waves at a depth of about 670 km (416 mi). It is thought that this is caused by a collapse of the internal structure of certain minerals into denser minerals due to rising pressure. Although the mantle is solid, some parts are weaker than others. Material in the mantle flows very slowly in response to variations in temperature, which create changes in density. Warmer, more buoyant material rises toward the surface, while cooler, denser material sinks. The velocity of this motion is generally on the order of centimeters per year, which is much slower than the flow in the liquid outer core.

Data from an increasingly densely spaced and sophisticated array of seismic-recording stations and the computer capacity to analyze the travel times of the thousands of seismic waves generated by the world's earthquakes are used to produce three-dimensional maps of the interior of Earth. This process, known as **seismic tomography,** is giving a detailed description of Earth's interior layers and demonstrating that these layers, especially the mantle, are less homogeneous than once thought. Three-dimensional tomographic images of the mantle show large regions of the mantle with higher-than-average and lower-than-average seismic wave velocities, indicating regions of colder and warmer rock, respectively.

Earth's outermost layer is the cool, rigid, thin surface layer called the **crust.** The boundary between the crust and the mantle is a chemical boundary; the rocks on either side have different chemical compositions. We know this by studying rocks that originated in the mantle and are now exposed at the surface. This boundary is called the **Mohorovičić discontinuity** in honor of its discoverer, Andrija Mohorovičić. It is more commonly known simply as the **Moho.** There are two kinds of crust, continental and oceanic. Continental crust is relatively low in density and averages about 40 km (25 mi) in thickness. Its composition and structure are highly variable, but it consists primarily of **granite**-type rock, which has a high content of sodium, potassium, aluminum, and silica. Oceanic crust is relatively more dense, with an average thickness of about 7 km (4.3 mi). It is more homogeneous both chemically and structurally than continental crust and is composed primarily of **basalt**-type rock, which is low in silica and high in iron, magnesium, and calcium. See table 3.1 for a comparison of these layers and their properties.

Table 3.1	**Layers of Earth**					
Layer	**Depth Range (km)**	**Layer Thickness (km)**	**State**	**Composition**	**Density (g/cm³)**	**Temperature (°C)**
Crust						
Continental	0–65	40 (average)	Solid	Silicates rich in sodium, potassium, and aluminum	2.67	~0–1000
Oceanic	0–10	7 (average)	Solid	Silicates rich in calcium, magnesium, and iron	3.0	~0–1100
Mantle	Base of crust–2891	2866	Solid and mobile	Magnesium-iron silicates	3.2–5.6	1100–3200
Outer core	2891–5149	2258	Liquid	Iron, nickel	9.9–12.2	3200
Inner core	5149–6371	1222	Solid	Iron, nickel	12.8–13.1	4000–5500

3.2 Lithosphere and Asthenosphere

The Layers

More detailed study of the uppermost part of Earth, the upper mantle and the crust, has shown that, independent of the sharp chemical boundary marked by the Moho, it is possible to identify a different layered structure characterized by changes in the mechanical properties of the rock from rigid to ductile behavior. Rocks that behave rigidly do not permanently deform when a force less than the breaking strength of the rock is applied. Rocks that have ductile behavior will deform, or flow, in response to an applied force. This has led to the identification of a strong, rigid surface shell called the **lithosphere,** which consists of crust and upper mantle material fused together. It is the lithosphere that comprises the plates of plate tectonics, discussed in detail in section 3.4. In oceanic regions, the lithosphere thickens with increasing age of the sea floor. It reaches a maximum thickness of about 100 km (62 mi) at an age of 80 million years. In continental regions, the thickness of the lithosphere is slightly greater than in the ocean basins, varying from about 100 km (62 mi) beneath the geologically young margins of continents to about 150 km (93 mi) beneath old continental crust. The base of the lithosphere corresponds roughly to the region in the man-

tle where temperatures reach 650°C ± 100°C. At the higher temperatures found at these depths, mantle rock begins to lose some of its strength.

The lithosphere is underlain by a weak, deformable region in the mantle called the **asthenosphere** where the temperature and pressure conditions lead to partial melting of the rock and loss of strength. The asthenosphere is often equated with a region of low seismic velocity called the low velocity zone (LVZ). Seismic waves travel more slowly through the asthenosphere, indicating that it may be as much as 1% melt. The asthenosphere behaves in a ductile manner, deforming and flowing slowly when stressed. It behaves roughly the way hot asphalt does.

The increase in pressure with increasing depth results in greater strength in the lower mantle, sometimes called the **mesosphere,** where the material is all solid but will convect slowly, moving upward in some regions and downward in others, because of temperature gradients and density differences. The depth to the base of the asthenosphere remains a matter of scientific debate. If it does correspond roughly to the LVZ, it would extend from the base of the lithosphere to about 350 km (217 mi). Some scientists believe it may be as shallow as 200 km (122 mi), while others think it may extend to depths to 700 km (435 mi). The lithosphere and underlying asthenosphere are shown in figure 3.5 and are compared in table 3.2.

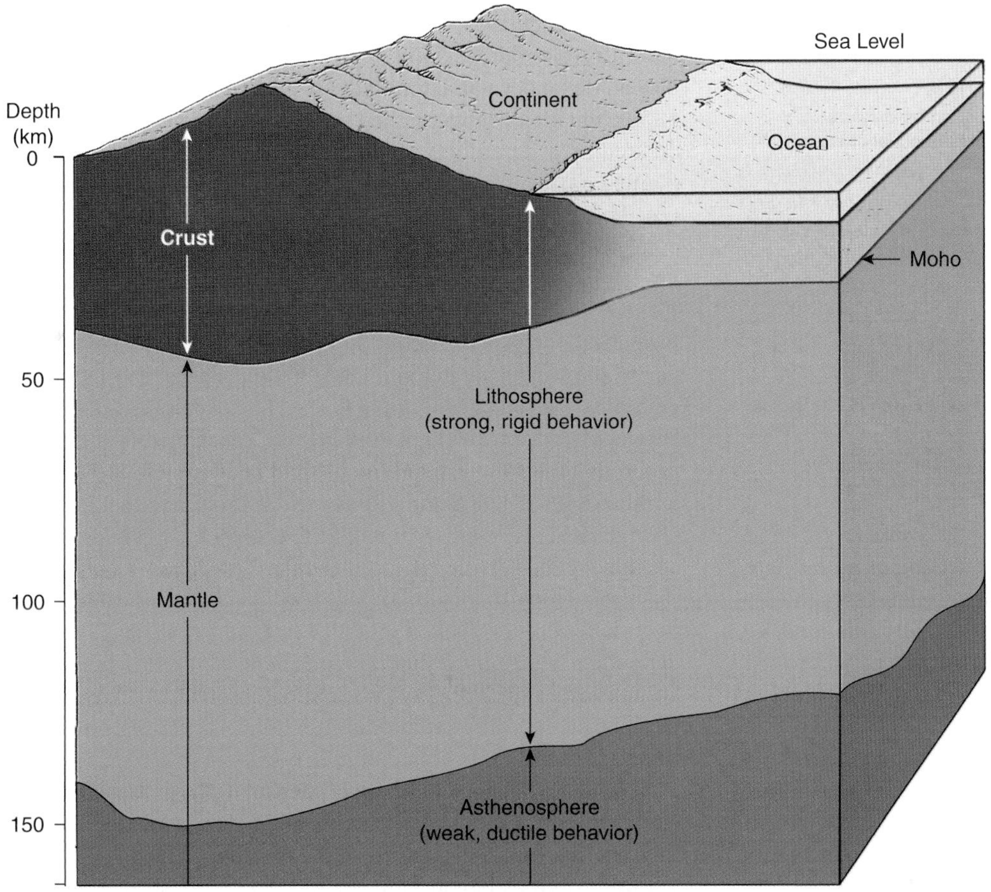

Isostasy

The distribution of elevated continents and depressed ocean basins requires that a balance be kept between the internal pressures under the land blocks and those under the ocean basins. This is the principle of **isostasy.** The balance is possible because the greater thickness of low-density granitic crust in the continental regions is compensated for by the elevated higher-density mantle material under the thinner crust of the oceans (fig. 3.6). The situation is often compared to the floating of an iceberg. The top of the iceberg is above the sea surface, supported by the buoyancy of the displaced water below the surface. The deeper the ice extends below the surface, the higher the iceberg reaches above the water. The less dense continental land blocks float on the denser mantle in the same way, with most of the continental volume below sea level.

In other words, the asthenosphere offers buoyant support to a section of lithosphere sagging under the weight of a mountain range.

Figure 3.5 The lithosphere is formed from the fusion of crust and upper mantle. It varies in thickness; it is thinner in ocean basins and thicker in continental regions. The lithosphere rides on the weak, partially molten asthenosphere. Notice that the Moho is relatively close to Earth's surface under the ocean's basaltic crust but is depressed under the granitic continents.

Layer	Depth Range (km)	Layer Thickness (km)	Characteristics
Lithosphere	0–100 (oceanic regions) 0 to 100–150 (continental regions)	0–150	Solid Rigid response
Asthenosphere	Base of lithosphere—350	200–350	Solid (~1% melt) Ductile response
Mesosphere	350–2891 (core-mantle boundary)	2541	Solid mobile

Table 3.2 **Upper Layers of Earth, Based on Their Response to Applied Stress**

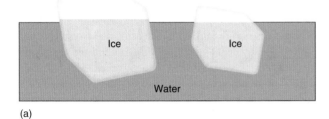

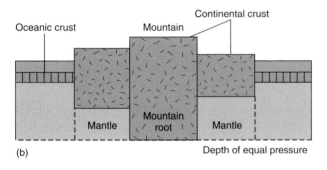

Figure 3.6 Isostasy. (a) Blocks of ice float in water with most of their volume submerged. The proportion of the ice block above water (roughly 10% of the volume) compared to the proportion of the block submerged (roughly 90% of the volume) is constant, regardless of the size of the block. Consequently, larger blocks will extend farther above and below the surface than smaller blocks. (b) Relatively light crustal blocks "float" on the denser underlying mantle in a similar way. The thicker the continental crust (or block), the further it extends above sea level and the deeper it extends into the mantle.

Because the lithosphere is cooler, it is more rigid and stronger than the asthenosphere. If a thick section of lithosphere has a large area and is mechanically strong, it depresses the asthenosphere slightly and is able to support a mountain range such as the Himalayas or the Alps. In other places where the crust is fractured and mechanically weak, a mountainous region must penetrate deeper into the mantle to provide buoyancy. The Andes are a mountain range with roots deep in the mantle.

If material is removed from or added to the continents, isostatic adjustment occurs. For example, parts of North America and Scandinavia continue to rise as the continents readjust to the lost weight of the ice sheets that receded at the close of the last ice age 10,000 years ago. Newly formed volcanoes protruding above the sea surface as islands often subside or sink back under the sea as their weight depresses the oceanic crust. The upper mantle gradually changes its shape in response to weight changes in the overlying, more rigid crust. In the middle 1800s, the concept of isostasy was firmly rooted in geologic studies. It was thought that landmasses separated and oceans changed boundaries in response to changes in isostasy as sections of Earth's crust moved up or down.

3.3 Movement of the Continents

History of a Theory: Continental Drift

As world maps became complete and more accurate, observant individuals were intrigued by the shapes of the continents on either side of the Atlantic Ocean. The possible "fit" of the bulge of South America into the bight of Africa was noted by English scholar and philosopher Francis Bacon (1561–1626), French naturalist George Buffon (1707–88), German scientist and explorer Alexander von Humboldt (1769–1859), and others. In the 1850s, the idea was expressed that the Atlantic Ocean had been created by a separation of two landmasses during some unexplained cataclysmic event early in the history of Earth.

Thirty years later, it was suggested that a portion of Earth's continental crust had been torn away to form the Moon, creating the Pacific Ocean and triggering the opening of the Atlantic. As scientific studies of Earth's crust continued, patterns of rock formation, fossil distribution, and mountain range placement began to show even greater similarities between the lands now separated by the Atlantic Ocean. In a series of volumes published between 1885 and 1909, Austrian geologist Edward Suess proposed that the southern continents had been joined into a single continent he called **Gondwanaland.** He assumed that isostatic changes had allowed portions of the continents to sink and create the oceans between the continents. This idea was known as the subsidence theory of separation. At the beginning of the twentieth century, Alfred L. Wegener and Frank B. Taylor independently proposed that the continents were slowly drifting about Earth's surface. Taylor soon lost interest, but Wegener, a German meteorologist, astronomer, and Arctic explorer, continued to pursue this concept until his death in 1930.

Wegener's theory, called **continental drift,** proposed the existence of a single supercontinent called **Pangaea** (fig. 3.7). He thought that forces arising from the rotation of Earth combined with tidal forces began Pangaea's breakup. First, the northern portion composed of North America and Eurasia, which he called **Laurasia,** separated from the southern portion formed of Africa, South America, India, Australia, and Antarctica, for which he retained the earlier name Gondwanaland. The continents as we

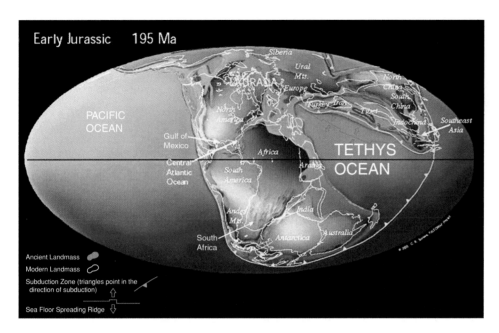

Figure 3.7 Pangaea in the early Jurassic, 195 Ma (million years before the present). Pangaea is composed of the two subcontinents, Laurasia and Gondwanaland, which will split apart with the opening of the central Atlantic Ocean. There is one great ocean, Panthalassa, which was the ancestral Pacific Ocean, and a second small ocean, the Tethys, which would eventually close with the opening of the Indian Ocean.

know them today then gradually separated and moved to their present positions. Wegener based his ideas in part on the geographic fit of the continents and the way in which some of the older mountain ranges and rock formations appeared to be related when the landmasses were assembled in Pangaea. He also noted that fossils more than 150 million years old collected on different continents were remarkably similar, implying the ability of land organisms to move freely from one landmass to another. Fossils more recent than 150 million years old showed quite different forms in different places, suggesting that the continents and their evolving populations had separated from one another.

Wegener's theory provoked considerable debate in the 1920s. The most serious weakness in the theory was Wegener's inability to identify a mechanism that could cause the continents to break apart and drift through the ocean basins. When challenged on this point, he suggested that there were two possible driving mechanisms, centrifugal force from the rotation of Earth and the same tidal forces that raise and lower sea level. Both of these were quickly demonstrated to be too small to be capable of moving continental rock masses through the rigid basaltic crust of the ocean basins. With the lack of a viable driving mechanism, much of the scientific community remained skeptical of the continental drift theory, and little attention was given to it after Wegener's death.

Evidence for a New Theory: Seafloor Spreading

Armed with new sophisticated instruments and technology developed during World War II, Earth scientists returned to their studies of the sea floor in the 1950s. Although the use of sound to examine the sea floor began in the 1920s, by the 1950s the equipment was greatly improved and much more readily available. (See chapter 4 for a discussion of echo sounders and depth recorders.) In 1947, geologists Bruce Heezen and Marie Tharp of Columbia University's Lamont-Doherty Geological Laboratory began working to systematically map Earth's seafloor features. Before they began their work, maps of the sea floor were limited mainly to shallow coastal areas necessary for the safe navigation of ships. Heezen researched and recorded ocean floor data during 33 cruises between 1947 and 1965, while Tharp remained on shore constructing the maps because women were not allowed onboard research cruises at the time. The physiographic maps that Tharp constructed from Heezen's surveys represented the first comprehensive study of ocean floor bathymetry. After 1965, Tharp was able to accompany Heezen on cruises. The culmination of their work was the publication of the map of the *World Ocean Floor* by Heezen and Tharp in 1974, a year after Heezen's death (fig. 3.8). This map revealed the presence of a series of mountain ranges through the ocean basins (fig. 3.8). These ranges are 65,000 km (40,000 mi) long; they rise 2–3 km (1.2–1.9 mi) above the adjacent sea floor and typically are between 1000 and 3000 km (600 and 1800 mi) wide. These are the deep seas' mid-ocean **ridge** and **rise** systems. If the slopes of these mountain ranges are steep and the width of the ranges is narrow, they are referred to as ridges (such as the Mid-Atlantic Ridge and the Mid-Indian Ridge); if the slopes are gentler and their width is broad, they are called rises (such as the East Pacific Rise). The ocean ridges typically have a central **rift valley,** a depression along the axis of the ridge, 50–3000 m (165–9850 ft) deep and 20–50 km (12.5–31 mi) wide. Ocean rises typically do not have a central rift valley; instead, they have their highest elevation (their shallowest depth) along the center. Along the axes of both ridges and rises is a narrow zone roughly 2 km (1.2 mi) wide that is volcanically active. Ocean ridges and rises are discussed in more detail in chapter 4, and the locations of specific ridges and rises are illustrated in figure 4.14.

Another dominant feature of the ocean floor was found to be narrow, steep-sided **trenches,** 6000–11,000 m (20,000–36,000 ft) deep, that are most characteristic of the Pacific Ocean. Some trenches are located seaward of chains of volcanic islands; Japan, Indonesia, the Philippines, and the Aleutian Islands are all associated with trenches. Other deep-sea trenches follow the edges of South and Central America (fig. 3.8). Older theories had not predicted the existence of such extensive mid-ocean features and could not explain them. However, a new theory was advanced that made the old idea of continental drift seem a possible and plausible explanation.

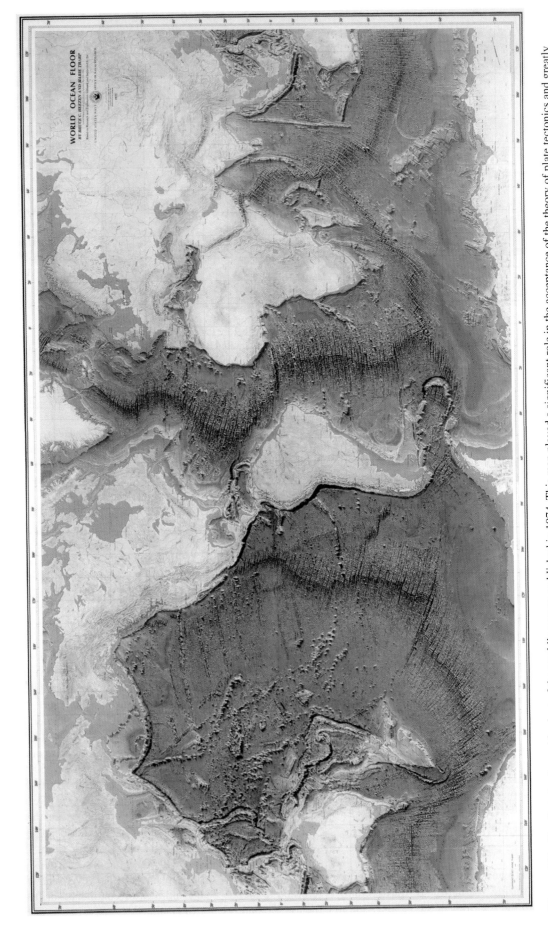

Figure 3.8 The first physiographic chart of the world's oceans, published in 1974. This map played a significant role in the acceptance of the theory of plate tectonics and greatly expanded public awareness and understanding of the major geologic features of the sea floor.

In the early 1960s, H. H. Hess of Princeton University promoted the concept that deep within Earth's mantle are currents of low-density molten material heated by Earth's natural radioactivity. When the upward-moving mantle rock reaches the base of the lithosphere, it moves horizontally beneath the lithosphere, cooling and increasing in density. Hess suggested that the lithosphere rides atop the convecting mantle rock. At some point, the mantle rock is cool and dense enough to sink once again. These patterns of moving mantle material are called **convection cells** (fig. 3.9). There are two proposed models of mantle convection. Some scientists believe that two regions of convection exist, one cell confined to the upper mantle above a depth of 700 km (435 mi) and the other in the lower mantle. Other scientists believe that convection occurs throughout the entire mantle to the core-mantle boundary; this model is called whole-mantle convection.

In Hess's model, the upward-moving mantle rock would carry heat with it toward the surface. This heat would cause the overlying oceanic crust and shallow mantle rock to expand, thereby creating the mid-ocean ridge. Active volcanism along the axis of the ridge results in the extrusion of basaltic magma onto the sea floor, where it cools, hardens, and creates new sea floor and oceanic crust. If new sea floor is being produced in this manner, there must be some mechanism for removing old sea floor because no measurable change occurs in the total surface area of Earth. Hess proposed that the great, deep ocean trenches are areas where relatively old, cold, and dense oceanic lithosphere descends back into Earth's interior. Figure 3.9 shows this process of producing new lithosphere at the ridges and losing old lithosphere at the trenches in a system driven by the motion of convection cells.

Although some of the ascending molten material breaks through the crust and solidifies, most of the rising material is turned aside under the rigid lithosphere and moves away toward the descending sides of the convection cells, dragging pieces of the lithosphere with it. This lateral movement of the oceanic lithosphere produces **seafloor spreading** (fig. 3.9). Areas in which new sea floor and oceanic lithosphere are formed above rising magma are **spreading centers;** areas of descending older oceanic lithosphere are **subduction zones.** The seafloor spreading mechanism provides the forces causing movement of the lithosphere. The continents are not moving through the basalt of the sea floor; instead, they are being carried as passengers on the lithosphere, similar to boxes on a conveyor belt.

Hess's general model of seafloor spreading in which the sea floor is created along ocean ridges, moves away from the ridges, and eventually is subducted into the mantle at ocean trenches

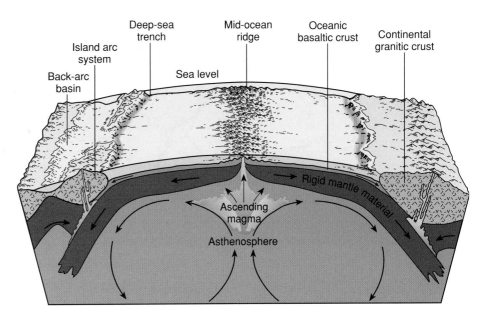

Figure 3.9 The model of mantle convection and seafloor spreading proposed by Harry Hess. Seafloor spreading leads to the creation of ocean crust at mid-ocean ridges and its destruction at deep-sea trenches. This process is shown being driven by convection cells in the atmosphere. The actual driving mechanisms of the plates is thought to be more complex than this simple model (see the discussion in section 3.5 of this chapter).

was correct. His proposal that there is convection in the mantle was also correct. However, his suggestion that mantle convection is the driving force for seafloor spreading is now believed to be wrong. The actual forces driving the motion of the sea floor are discussed in section 3.5 following the discussion of plate tectonics in section 3.4.

Evidence for Crustal Motion

Additional evidence was needed to support the idea of seafloor spreading. As oceanographers, geologists, and geophysicists explored Earth's crust on land and under the oceans, evidence began to accumulate.

Epicenters of earthquakes were known to be distributed around Earth in narrow and distinct zones. Epicenters are the points on Earth's surface directly above the actual earthquake location, which is called the earthquake **focus** or **hypocenter.** These zones were found to correspond to the areas along the ridges, or spreading centers, and the trenches, or subduction zones. Earthquakes that are shallower than 100 km (60 mi) are prevalent in the ocean basins along ridges and rises and at trenches where the lithosphere bends and fractures as it is subducted into the mantle. These relatively shallow events are also found in continental regions that are undergoing significant deformation. Earthquakes deeper than 100 km are generally associated with the subduction of oceanic lithosphere. The deeper the earthquakes are, the farther they are displaced inland away from the trench beneath continents or island arcs (fig. 3.10).

Researchers sank probes into the sea floor to measure the heat flowing from the interior of Earth through the oceanic

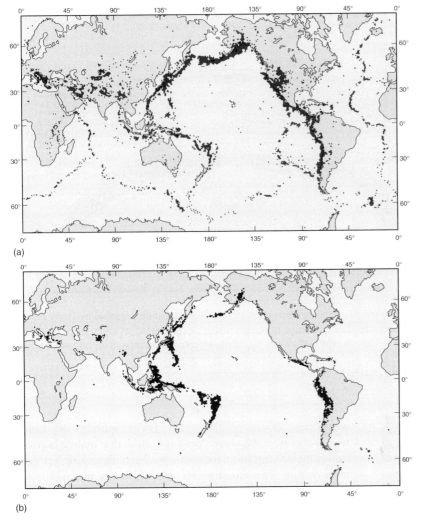

(a)

(b)

Figure 3.10 Earthquake epicenters, 1961–67.
(a) Epicenters of earthquakes with depths less than 100 km
(60 mi) outline regions of crustal movement. (b) Epicenters
of earthquakes with depths greater than 100 km are related
to subduction.

older. The sea floor formed by convection-cell
processes at the spreading centers is young and short-
lived, for it is lost at subduction zones, where it plunges
back down into the mantle.

Vertical, cylindric samples, or **cores,** were obtained
by drilling through the sediments that cover the ocean
bottom and into the rock of the ocean floor. (A detailed
discussion of sediments is found in chapter 4.) Drilling
cores through, on average, 500–600 m (1600–2000 ft) of
sediments and into the ocean floor rock required the development of a new technology and a new kind of ship.
In the late summer of 1968, the specially constructed
drilling ship *Glomar Challenger* (see fig. 1.17*a*) was
used for a series of studies. This ship, 122 m (400 ft)
long, with a beam of 20 m (65 ft) and a draft of 8 m
(27 ft), displaced 10,500 tons when loaded. Its
specialized bow and stern thrusters and its propulsion
system responded automatically to computer-
controlled navigation, using acoustic beacons on the
sea floor. This setup enabled the ship to remain for
long periods of time in a nearly fixed position over a
drill site in water too deep to anchor. An acoustic
guidance system made it possible to replace drill bits
and place the drill pipe back in the same bore holes in
water about 6000 m (20,000 ft) deep. This method is

crust (fig. 3.11). The measured heat flow shows a pattern with
a high degree of variability, even over closely spaced intervals.
In part, this variation is attributed to the seeping of seawater
down through porous or fractured portions of the crust at one
location along a ridge system and rising of heated water at another. This circulation does not occur over regions of the sea
floor that are sealed by a thick layer of loose particles, or
sediment. In these areas, the measured heat flow shows a regular pattern. It is highest in the vicinity of the mid-ocean ridges
over the ascending portion of a mantle convection cell, where
the crust is more recent, and it decreases as the distance from
the ridge center and the crustal age increases (fig. 3.12).

Radiometric age dating of rocks from the land and sea floor
shows that the oldest rocks from the oceanic crust are only about
200 million years old, whereas rocks from the continents are much

Figure 3.11 Recovering a heat flow probe from the deep-sea floor. The probe consists
of a cylinder holding the measuring and recording instruments above a thin probe that
penetrates the sediment. Thermisters (devices that measure temperature electronically) at the
top and bottom of the probe measure the temperature difference over a fixed distance equal to
the length of the probe. This temperature change with depth is used to compute the heat flow
through the sediment.

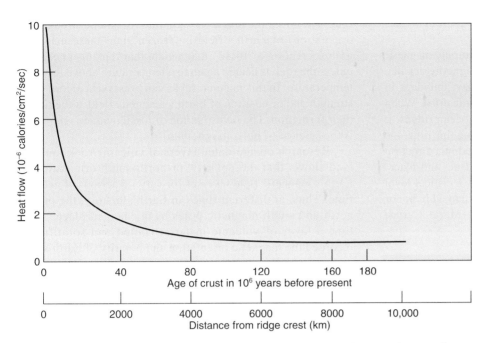

Figure 3.12 Heat flow through the Pacific Ocean floor. Values are shown against age of crust and distance from the ridge crest.

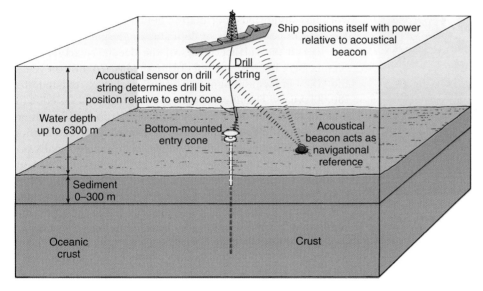

Figure 3.13 Deep-ocean drilling technique. Acoustical guidance systems are used to maneuver the drilling ship over the bore hole and to guide the drill string back into the bore hole.

illustrated in figure 3.13. A more general discussion of cores and coring methods is found in chapter 4.

In 1983, the *Glomar Challenger* was retired, after logging 600,000 km (375,000 mi), drilling 1092 holes at 624 drill sites, and recovering a total of 96 km (60 mi) of deep-sea cores for study. A new deep-sea drilling program with a new vessel, the *JOIDES Resolution* (fig. 3.14), began in 1985. The drill pipe comes in sections 9.5 m (31 ft) long that can be screwed together to make a single pipe up to 8200 m (27,000 ft) long.

The cores taken by the *Glomar Challenger* provided much of the data needed to establish the existence of seafloor

spreading. No ocean crust older than 180 million years was found, and sediment age and thickness were shown to increase with distance from the ocean ridge system (fig. 3.15). Note that the sediments closest to the ridge system are thin over the new crust, which has not had as long to accumulate its sediment load. The crust farther away from the ridge system is older and is more heavily loaded with sediments.

Although many pieces of evidence fit the theory that Earth's crust produced at the ridge system is new and young, the most elegant proof for seafloor spreading came from a study of the magnetic evidence locked into the ocean's floors.

The familiar geographic North (90°N) and South (90°S) Poles mark the axis about which Earth rotates. Earth also behaves as if a giant bar magnet is embedded in its interior and tilted roughly 11.5° away from the axis of rotation (fig. 3.16). A magnetic field like Earth's, with two opposite poles, is called a **dipole.** Earth's magnetic field consists of invisible lines of magnetic force that are parallel to Earth's surface at the magnetic equator and converge and dip toward Earth's surface at the magnetic poles. A small, freely suspended magnet (such as a compass needle) will align itself with these lines of magnetic force, with the north-seeking end pointing to the north magnetic pole and the south-seeking end pointing to the south magnetic pole. In addition, a suspended magnet dips toward Earth's surface by differing amounts depending on its distance from the magnetic equator. At the magnetic equator a suspended magnet would be horizontal, or parallel to Earth's surface. In the north magnetic hemisphere, the north-seeking end of the magnet would point downward at an increasing angle until it pointed vertically downward at the north magnetic pole. In the south magnetic hemisphere, the north-seeking end of the magnet would point upward at an increasing angle until it pointed vertically upward at the south magnetic pole. The task of locating the magnetic poles is difficult for many reasons: the area over which the dip is nearly 90° is often relatively large, the pole can move because of daily variations and magnetic storms, and it is difficult for survey crews to work in polar regions. Repeated surveys by the Canadian Geological Survey have shown that the north magnetic pole has been moving approximately northwest at 40 km (25 mi) per year in recent years. The estimated location for the north magnetic pole in 2005 was 82.7°N, 114.4°W, just north of the Sverdrup Islands in the Canadian Arctic. The location of the south magnetic pole in 2001 was determined to be 64.6°S, 138.3°E, near the coast of Wilkes Land in Antarctica.

Most igneous rocks contain particles of a naturally magnetic iron mineral called magnetite. Particles of magnetite act like small magnets. These particles are particularly abundant in basalt, the type of rock that makes up the oceanic crust. When basaltic magma erupts on the sea floor along ocean ridges, it cools and solidifies to form basaltic rock. During this time, the magnetite grains in the rock become magnetized in a direction parallel to the existing magnetic field at that time and place. When the temperature of the rock drops below a critical level called the **Curie temperature,** or Curie point (named in honor of the twice Nobel Prize–winning physicist Marie Curie),

Figure 3.14 The deep-sea drilling vessel *JOIDES Resolution.*

roughly 580°C, the magnetic signature (magnetic strength and orientation) of Earth's field is "frozen" into these magnetite grains, creating a "fossil" magnetism that remains unchanged unless the rock is heated again to a temperature above the Curie temperature. In this manner, rocks can preserve a record of the strength and orientation of Earth's magnetic field at the time of their formation. The investigation of fossil magnetism in rocks is the science of **paleomagnetism.**

Research on age-dated layers of volcanic rock found on land shows that the polarity, or north-south orientation, of Earth's magnetic field reverses for varying periods of geologic time. Thus, at different times in Earth's history, the present north and south magnetic poles have changed places. Each time a layer of volcanic material cooled and solidified, it recorded the magnetic orientation and polarity of Earth's field at that time. The dating and testing of samples taken through a series of volcanic layers have enabled scientists to build a calendar of these events (fig. 3.17). During these **polar reversals,** Earth's magnetic field gradually decreases in strength by a factor of about ten over a period of a few thousand years. Some evidence exists that during a reversal, Earth's field may not remain a simple dipole. The collapse of the magnetic field during reversals enables more intense cosmic radiation to penetrate the planet's surface, and recent work has suggested a correlation between such periods and a decrease in populations of delicate, single-celled organisms that live in the surface layers of the ocean. Nearly 170 reversals have been identified during the last 76 million years, and the present magnetic orientation has existed for 780,000 years. The interval between reversals has varied, with some notable periods of constant orientation for long intervals. During the Cretaceous period (from roughly 120 million to 80 million years ago), the field was stable and normally polarized. The field was stable and reversely polarized for about 50 million years in the Permian period, roughly 300 million years before present. The cause of reversals is unknown but is thought to be associated with changes in the motion of the material in Earth's liquid outer core. How long our current period of normal polarity will last we do not know.

During the 1950s, oceanographers began to measure the strength of the magnetism of the oceanic crust during research cruises by towing marine magnetometers behind the ship, away from the magnetic noise of the vessel. From 1958–61, the results of several marine magnetic surveys were published as maps of high and low magnetic intensity that resembled the ridges and valleys of a topographic map. These alternating areas of high and low intensity came to be known as "magnetic stripes." The stripes varied in width from a few kilometers to many tens of kilometers and were as much as several thousand kilometers long. When the stripes were compared to marine bathymetry, it was noted that the maps revealed a mirror image

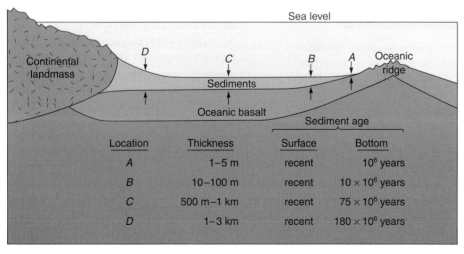

Figure 3.15 Variation in age and thickness of seafloor sediments with distance from the mid-ocean ridge.

Location	Thickness	Sediment age	
		Surface	Bottom
A	1–5 m	recent	10^6 years
B	10–100 m	recent	10×10^6 years
C	500 m–1 km	recent	75×10^6 years
D	1–3 km	recent	180×10^6 years

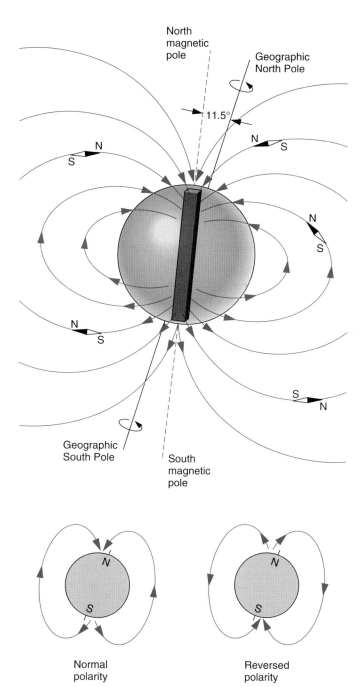

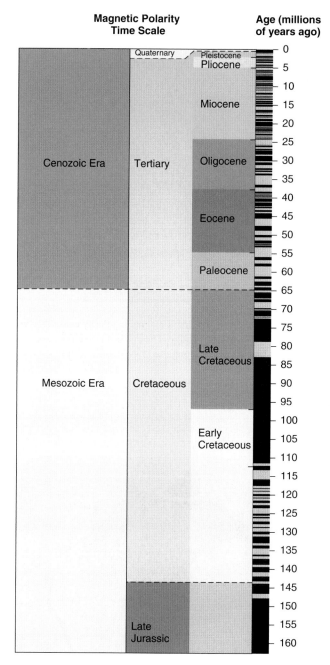

Figure 3.16   Lines of magnetic force surround Earth and converge at the magnetic poles. A freely suspended magnet aligns itself with these lines of magnetic force, with the north-seeking end of the magnet pointing to the north magnetic pole and the south-seeking end pointing to the south magnetic pole. The magnet hangs parallel to Earth's surface at the magnetic equator and dips at an increasing angle as it approaches the magnetic poles. N and S in the two small figures indicate the *geographic* poles.

Figure 3.17   Worldwide magnetic polarity time scale for the Cenozoic and Mesozoic eras. *Black* indicates positive anomalies (and therefore normal polarity). *Tan* indicates negative anomalies (reverse polarity).

pattern of roughly parallel magnetic stripes on each side of the mid-ocean ridge (fig. 3.18). The significance of this pattern was not understood until 1963, when F. J. Vine and D. H. Matthews of Cambridge University proposed that these stripes represented a recording of the polar reversals of the vertical component of Earth's magnetic field, frozen into the rocks of the sea floor. As the molten basalt rose along the crack of the ridge system and solidified, it locked in the direction of the prevailing magnetic field. As seafloor spreading moved this material to both flanks of the ridge, it was replaced at the ridge crest by more molten materials. Each time Earth's magnetic field reversed, the direction of the magnetic field was recorded in the new crust. Vine and Matthews proposed that if such were the case, there should be a symmetric pattern of magnetic stripes centered at the ridges and becoming older away from the ridges. Their ideas were confirmed; the polarity and age of

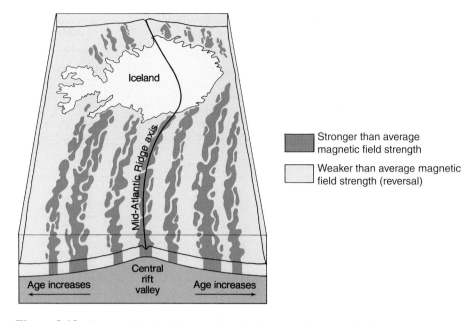

Figure 3.18   Reversals in Earth's magnetic polarity cause the symmetrically striped pattern centered on the Mid-Atlantic Ridge. The age of the sea floor increases with the distance from the ridge. The spreading rate along the Mid-Atlantic Ridge is about 2–4 cm per year.

these stripes corresponded to the same magnetic field changes found in the dated layers on land. Their discovery dramatically demonstrated seafloor spreading.

The magnetic stripe data in the world's oceans has been used to produce a map of the age of the sea floor (fig. 3.19). Seafloor age increases away from the oceanic spreading centers, providing another verification of seafloor spreading. The present ocean basins are young features, created during the past 200 million years, or the last 5% of Earth's history.

Polar Wandering Curves

Paleomagnetic studies of fossil magnetism in continental rocks provide evidence of relative motion between tectonic plates and Earth's magnetic poles through time. By carefully measuring the fossil magnetism in a rock, the location of the north magnetic pole at the time of the rock's formation can be calculated. This is done

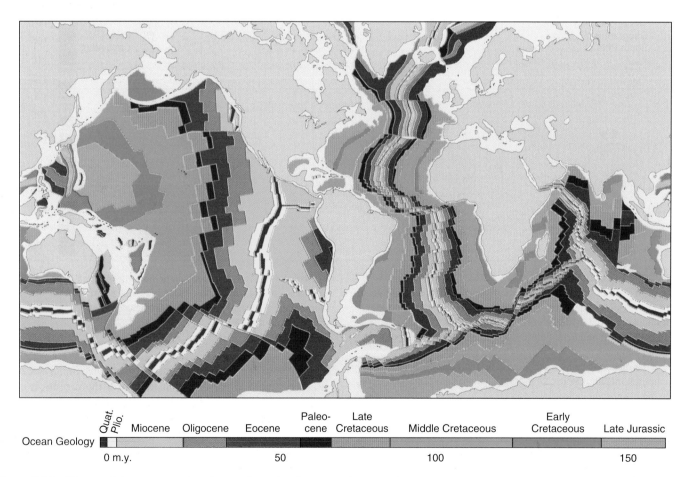

Figure 3.19   The age of the ocean crust based on seafloor spreading magnetic patterns is shown in this color-shaded image. The age of the ocean floor increases with increasing distance from the mid-ocean ridge as a result of seafloor spreading. Oceanic fracture zones offset the ridge and the age patterns.

by determining the direction and distance to the pole's position. The horizontal component of the rock's magnetism points in the direction of the pole. The angle of dip of the magnetic signature reveals the magnetic latitude at which the rock formed, and the magnetic latitude is directly related to the distance between the north magnetic pole and the rock's formation site.

By measuring the fossil magnetism in continental rocks of different ages found on the same tectonic plate, it is possible to create a plot of the apparent location of the north magnetic pole through time. Such a plot is known as a **polar wandering curve** (fig. 3.20). A polar wandering curve is evidence that the relative positions of the tectonic plate, or the continent embedded in it, and the north magnetic pole have changed with time. This may be the result of the movement of either the pole or the plate. While the magnetic poles do move a little with time, we know that their average location remains near the geographic poles. Consequently, a polar wandering curve records the motion of the plate with respect to a nearly stationary north magnetic pole.

Because all plates do not move along the same path, each plate produces its own polar wandering curve, which converges with all other polar wandering curves at the current location of the north magnetic pole. If we look at the polar wandering curves constructed for the North American and Eurasian plates, we see that they are very similar from 500 to 200 million years ago. Over the past 200 million years, the curves have parted as the American and Eurasian plates have diverged from each other.

The divergence of these two polar wandering paths is a record of the opening of the North Atlantic Ocean and the drifting apart of the North American and European Plates. When these two polar wandering curves are superimposed, the movements of these plates reconstruct the formation of the two-lobed portion of Pangaea proposed by Wegener, with Laurasia to the north, Gondwanaland to the south, and the Sea of Tethys in between. On the basis of this method, the present landmasses join nicely at the edges of the continental blocks, at a boundary on the continental slope that is about 2000 m (6560 ft) below present sea level. When the plates and their landmasses are moved together, major ancient mountain ranges and fault systems spanning several of our modern continents join; for example, the fault through the Caledonian Mountains of Scotland joins the Cabot Fault extending from Newfoundland to Boston. Also, the lands occupied by countries of western Europe and Great Britain that are presently located in temperate and high latitudes were once found in the equatorial zone; their previous locations explain their fossil coral reefs and desert-type sandstones.

Current research leads scientists to believe that the assumption that the rotational polar axis has not varied its location over time may not be correct. The rotating Earth possesses tremendous **inertia,** which keeps it spinning on its axis. However, as the lithospheric plates migrate about on Earth's surface, Earth may tend to roll about its center slightly, trying to adjust so that the majority of the crustal mass is centered about the equator. If Earth does roll about its center, the location of the pole points marking Earth's rotational axis relative to the lithospheric plates may also shift.

3.4 Plate Tectonics

The theory of **plate tectonics** incorporates the ideas of continental drift and seafloor spreading in a unified model. Drifting continents and the motion of the sea floor both result from the fragmentation and movement of the outer rigid shell of Earth called the lithosphere (you may want to review section 3.2 and figure 3.5). The lithosphere is fragmented into seven major plates along with a number of smaller ones (fig. 3.21 and table 3.3). The plates are outlined by the major earthquake belts of the world (see fig. 3.10); this relation was first pointed out in 1965 by J. T. Wilson, a geophysicist at the University of Toronto.

Plates and Their Boundaries

Each lithospheric plate consists of the upper roughly 80–100 km (50–62 mi) of rigid mantle rock capped by either oceanic or continental crust. Lithosphere capped by oceanic crust is often simply called oceanic lithosphere, and, in a similar fashion, lithosphere capped by continental crust is referred to as continental lithosphere. Some plates, such as the Pacific Plate, consist entirely of oceanic lithosphere, but most plates, such as the South American Plate, consist of variable amounts of both oceanic and continental lithosphere with a transition from one to the other along the margins of continents. The plates move with respect to

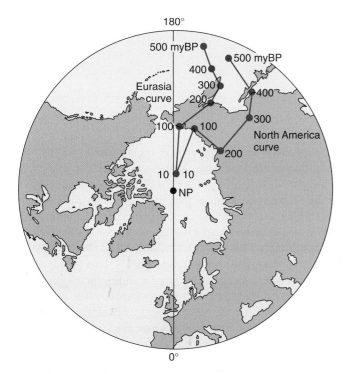

Figure 3.20 The positions of the north magnetic pole millions of years before the present (myBP), judged by the magnetic orientation and the age of the rocks of North America and Eurasia. The divergence of the two paths indicates that the landmasses of North America and Eurasia have been displaced from each other as the Atlantic Ocean opened.

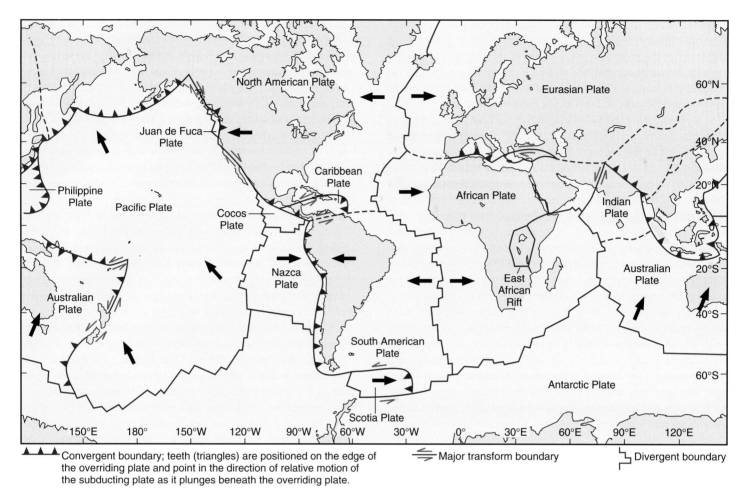

▲▲▲ Convergent boundary; teeth (triangles) are positioned on the edge of the overriding plate and point in the direction of relative motion of the subducting plate as it plunges beneath the overriding plate.

⇇ Major transform boundary

⌐⌐ Divergent boundary

Figure 3.21 Earth's lithosphere is broken into a number of individual plates. The edges of the plates are defined by three different types of plate boundaries. *Arrows* indicate direction of plate motion.

Table 3.3	Approximate Plate Area (10^6 km^2)
Major Plates	
Pacific	105
African	80
Eurasian	70
North American	60
Antarctic	60
South American	45
Australian	45
Smaller Plates	
Nazca	15
Indian	10
Arabian	8
Philippine	6
Caribbean	5
Cocos	5
Scotia	5
Juan de Fuca	2

one another on the ductile asthenosphere below. As the plates move, they interact with one another along their boundaries, producing the majority of Earth's earthquake and volcanic activity.

The three basic kinds of plate boundaries are defined by the type of relative motion between the plates. Each of these boundaries is associated with specific kinds of geologic features and processes (table 3.4 and fig. 3.22). Plates move away from each other along **divergent plate boundaries.** Ocean basins and the sea floor are created along marine divergent boundaries marked by the mid-ocean ridges and rises, while continental divergent boundaries are often marked by deep rift zones. Plates move toward one another along **convergent plate boundaries.** Marine convergent boundaries are associated with deep-ocean trenches, the destruction of sea floor, and the closing of ocean basins, while continental convergent boundaries are associated with the creation of massive mountain ranges. The third type of plate boundary occurs where two plates are neither converging nor diverging but are simply sliding past one another. These are called **transform boundaries** and are marked by large faults called **transform faults.**

Table 3.4	**Types and Characteristics of Plate Boundaries**					
Plate Boundary	**Types of Lithosphere**	**Geologic Process**	**Geologic Feature**	**Earthquakes**	**Volcanism**	**Examples**
Divergent (move apart)	Ocean-Ocean	New sea floor created, ocean basin opens	Mid-ocean ridge	Yes, shallow	Yes	Mid-Atlantic Ridge, East Pacific Rise
	Continent-Continent	Continent breaks apart, new ocean basin forms	Continental rift, shallow sea	Yes, shallow	Yes	East African Rift, Red Sea, Gulf of Aden, Gulf of California
Convergent (move together)	Ocean-Ocean	Old sea floor destroyed by subduction	Ocean trench	Yes, shallow to deep	Yes	Aleutian, Mariana, and Tonga Trenches (Pacific Ocean)
	Ocean-Continent	Old sea floor destroyed by subduction	Ocean trench	Yes, shallow to deep	Yes	Peru-Chile and Middle-America Trenches (Eastern Pacific Ocean)
	Continent-Continent	Mountain building	Mountain range	Yes, shallow to intermediate	No	Himalaya Mountains, Alps
Transform (slide past each other)	Ocean	Sea floor conserved (neither created nor destroyed)	Transform fault (offsets segments of ridge crest)	Yes, shallow	No	Mendocino and Clipperton (Eastern Pacific Ocean)
	Continent	Sea floor conserved (neither created nor destroyed)	Transform fault (offsets segments of ridge crest)	Yes, shallow	No	San Andreas Fault, Alpine Fault (New Zealand), North and East Anatolian Faults (Turkey)

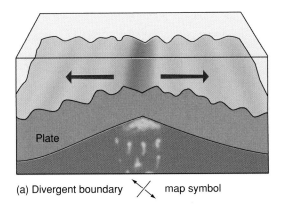

(a) Divergent boundary   map symbol

(b) Convergent boundary   map symbol

(c) Transform fault boundary   map symbol

Figure 3.22   The three basic types of plate boundaries include (a) divergent boundaries where plates move apart, (b) convergent boundaries where plates collide, and (c) transform boundaries where plates slide past one another.

Divergent Boundaries

Ocean basins are created along divergent boundaries that break apart continental lithosphere (fig. 3.23). A series of successive divergent boundaries were responsible for the breakup of Pangaea, producing the individual continents and ocean basins we see today. Upwelling of hot mantle rock is thought to be responsible for the creation of divergent boundaries in continental lithosphere. Mantle upwelling heats the base of the lithosphere, thinning it and causing it to dome upward (fig. 3.23a). This weakens the lithosphere and produces extensional forces and stretching that lead to faulting and volcanic activity. Faulting along the boundary creates rifting and subsidence in the continental crust, forming a **rift zone**

Figure 3.23 (a) Continental rifting begins as rising magma heats the overlying continental crust, making it dome upward and thin as tensional forces cause it to stretch. (b) Stretching and pulling apart of the crust produce a rift valley with active volcanism. (c) Continued spreading results in the formation of new sea floor along the boundary and the creation of a young, shallow sea. (d) Eventually, a mature ocean basin and spreading ridge system are created.

A Continent undergoes extension. The crust is thinned and a rift valley forms (East African Rift Valleys).

B Continent tears in two. Continent edges are faulted and uplifted. Basalt eruptions form oceanic crust (Red Sea).

C Continental sediments blanket the subsiding margins to form continental shelves and rises. The ocean widens and a mid-oceanic ridge develops (Atlantic Ocean).

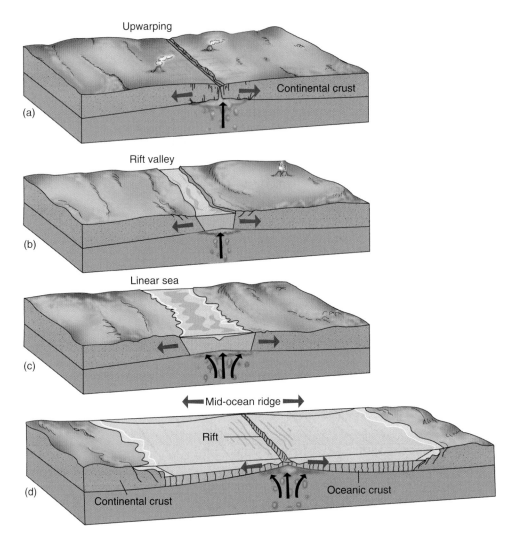

along the length of the boundary (fig. 3.23*b*). A present-day example of this early rifting is the continuing development of the East African Rift valley, stretching from Mozambique to Ethiopia (fig. 3.24). Volcanoes in the rift include Kilimanjaro and Mount Kenya. Such a sunken rift zone is called a **graben.** The next "stage" in continental rifting can be seen farther to the north, where two shallow seas, the Red Sea and the Gulf of Aden, have formed (see fig. 3.23*c*). If this rift system remains active in the future, East Africa will separate from the rest of the continent and a new ocean basin will form with seafloor spreading and a central spreading ridge (see fig. 3.23*d*).

Most divergent boundaries are located along the axis of a mid-ocean ridge system (fig. 3.25). Mantle upwelling beneath the ridge heats the overlying oceanic lithosphere, causing it to expand, and creates a submarine mountain range. As the plates on either side of the boundary move apart, cracks form along the crest of the ridge, allowing molten rock to seep up from the mantle and flow onto the ocean floor. In the early stages of a volcanic eruption, basaltic magma can flow rapidly onto the sea floor in relatively flat flows called sheet flows that may extend

several kilometers away from their source. Basaltic magma has a low viscosity, allowing it to flow freely, because the magma is very hot and has a relatively low content of silicon dioxide. As the eruption rate decreases, the magma is often extruded more slowly onto the sea floor, creating rounded flows called pillow lavas, or **pillow basalts** (fig. 3.26). The magma solidifies to form new ocean crust along the edges of each diverging plate. The thickness of the oceanic crust remains relatively constant from the time of its formation at the ridge. However, as the two plates diverge, the mantle rock immediately beneath the crust cools, fuses to the base of the crust, and behaves rigidly, causing the thickness of the oceanic lithosphere to increase with age and distance from the ridge. Plates move apart at different rates along the length of the ridge system (see section 3.5).

Detailed seismic studies, coring, and the direct observation of large transform faults from submersibles have provided ocean scientists with considerable information about the process of rifting, the formation of oceanic crust, and the structure of ridges and rises. A generalized section across a ridge is shown in

Figure 3.24   An active continent-continent divergent boundary, still in its early stage of development, has created the East African Rift system. This same boundary is more fully developed to the north, where it has created two shallow seas, the Red Sea and the Gulf of Aden.

figure 3.27. This figure shows the four basic layers that make up the oceanic lithosphere.

The topmost layer (layer 1 on fig. 3.27) is composed of sediment; it may not be present at the ridge axis if the sea floor is too young to have acquired a covering of this loose material. Layer 2 is made up of two forms of basalt. The surface portion of this layer is the result of the rapid cooling of magma to form fine-grained, or glassy, lava. The deeper part is formed from magma that solidifies in a series of vertically oriented basalt dikes. Layer 3 is also basaltic, but it has cooled more slowly, forming a granular, dark-colored igneous rock called **gabbro.** The Moho is found at the bottom of this layer. Layers 2 and 3 together are usually 6–7 km (about 4 mi) thick, but they are thinner at the rift zones. The lithosphere contains these layers and a thick fourth layer of rock known as peridotite that forms the rigid upper mantle and the base of the lithospheric plate. The partly molten asthenosphere portion of the mantle lies below this fourth layer.

The mechanism whereby peridotite produces basalt and gabbro was determined by subjecting laboratory samples of peridotite to the high temperature and pressure found 100 km (60 mi) below the surface of the oceanic crust. When hot peridotite is decompressed, it partially melts and produces a liquid similar in composition to the basalt found in the upper layers of oceanic crust.

Transform Boundaries

The ocean ridge system is divided into segments that are separated by transform faults (fig. 3.28). As two plates move away from the ridge crest, they simply slide past one another along the transform fault. In this manner, the ridge crest segments and transform faults form a continuous plate boundary that alternates between being a divergent boundary and a transform boundary. Differences in the age and temperature of the plates across a transform boundary can create significant and rapid changes in the elevation of the sea floor called **escarpments.** These changes in elevation are

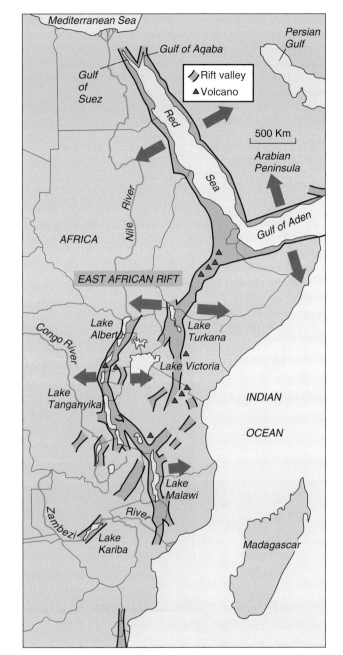

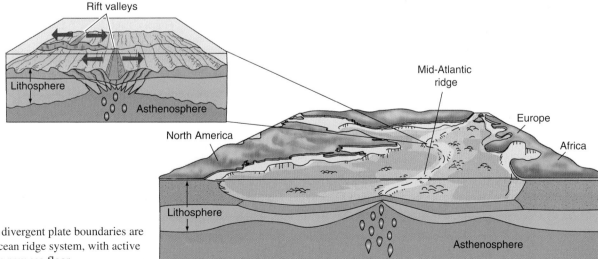

Figure 3.25   Most divergent plate boundaries are marked by the mid-ocean ridge system, with active volcanism that creates new sea floor.

Figure 3.26 Pillow lavas are mounds of elongated lava "pillows" formed by repeated oozing and quenching of extruding basalt magma on the sea floor. First, a flexible, glassy crust forms around the newly extruded lava, forming an expanded pillow. Next, pressure builds within the pillow until the crust breaks and new basalt extrudes like toothpaste, forming another pillow.

Figure 3.27 A mid-ocean ridge cross section showing the four layers of the oceanic lithosphere.

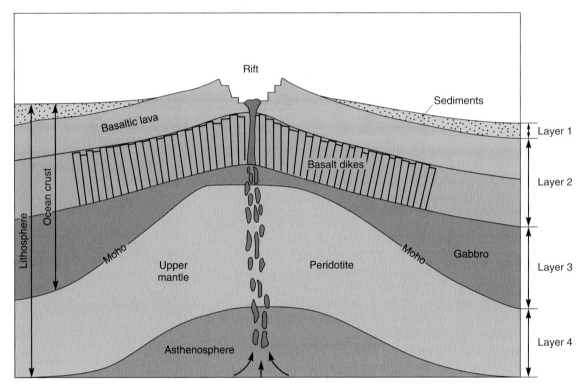

Figure 3.28 Transform boundaries are marked by transform faults. Transform faults usually offset segments of ocean ridges (divergent boundaries). Plates slide past one another along transform boundaries. The direction of motion is opposite the sense of displacement of the ridge segments.

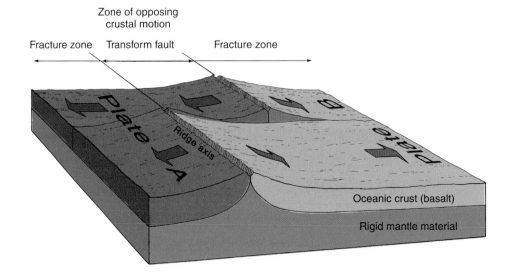

propagated into each plate by seafloor spreading, where they are preserved as "fossil" transform faults. These fossil faults, along with the active transform fault between the two segments of ridge crest, form a continuous linear feature called a **fracture zone.** The longest fracture zones can be up to 10,000 km (6200 mi) long, extending deep into each plate on either side of the ridge. It is important to remember that no movement or earthquake activity occurs along the fracture zone where it extends into either plate. However, there is relative motion and seismicity along the transform fault between adjacent segments of ridge crest. The direction of motion of the plates on either side of a transform fault is determined by the direction of seafloor spreading in each plate and is opposite the sense of displacement of the ridge segments.

While most transform faults join two divergent boundaries (segments of ridge crest), they may also join different combinations of other types of plate boundaries. Transform faults can join two convergent boundaries (ocean trenches), as they do between the Caribbean and American Plates, or a convergent boundary and a divergent boundary, as they do on either side of the Scotia Plate (see fig. 3.21). The diverse settings in which transform faults play a role in marking the boundaries of plates can be seen along the west coast of the United States (fig. 3.29). Much of the boundary between the North American and Pacific Plates is marked by a long transform fault; its southern portion is known as the San Andreas Fault, cutting through continental crust from the Gulf of California to Cape Mendocino, California. Offshore of Cape Mendocino, a transform fault known as the Mendocino Fault links a spreading center and a trench, forming a boundary between the Pacific and Juan de Fuca Plates. Farther to the north, additional transform faults offset segments of the Juan de Fuca Ridge.

The San Andreas Fault allows crustal sections on either side to move horizontally relative to each other. The land on the Pacific side, including the coastal area from San Francisco to the tip of Baja California, is actively moving northward with respect to the land on the east side of the fault system. The motion is not uniform; when the accumulated stress exceeds the strength of the rock, a sudden movement and displacement occur along the fault, causing an earthquake. Movement along this fault system caused the famous 1906 and 1989 (Loma Prieta) earthquakes in the San Francisco Bay area.

The reason for the many transform faults that are found associated with the ridge system is related to changes in the speed and direction of the plates as they move apart on a spherical surface. Variations in the strength or location of convection cells and collisions between sections of lithosphere may also result in transform faults.

Convergent Boundaries

Plates move toward each other along convergent plate boundaries. This is caused by the subduction, or sinking, of one plate (often referred to as the downgoing plate) into the mantle beneath the other plate (called the overriding plate). Subduction of the downgoing plate typically creates, and continues to occur along, an ocean trench. Subduction also usually results in both volcanoes and earthquakes (refer to fig. 3.10 for a map of the distribution of earthquakes). The specific kinds of geologic features and processes associated with subduction depend in large part on whether the downgoing lithosphere is oceanic or continental. This is because of the difference in density between continental and oceanic lithosphere. Three possible combinations occur: ocean-continent convergence, ocean-ocean convergence, and continent-continent convergence (fig. 3.30).

Roughly 20%–40% of the thickness of continental lithosphere consists of continental crust. Because the density of continental crust is relatively low compared to the density of the mantle (see table 3.1), continental lithosphere is buoyant and will not subduct into the mantle. The density of oceanic lithosphere is greater than continental lithosphere and increases with age as seafloor spreading carries the oceanic lithosphere away from the ridge crest and the plate cools and thickens. The thickness of oceanic lithosphere increases at a

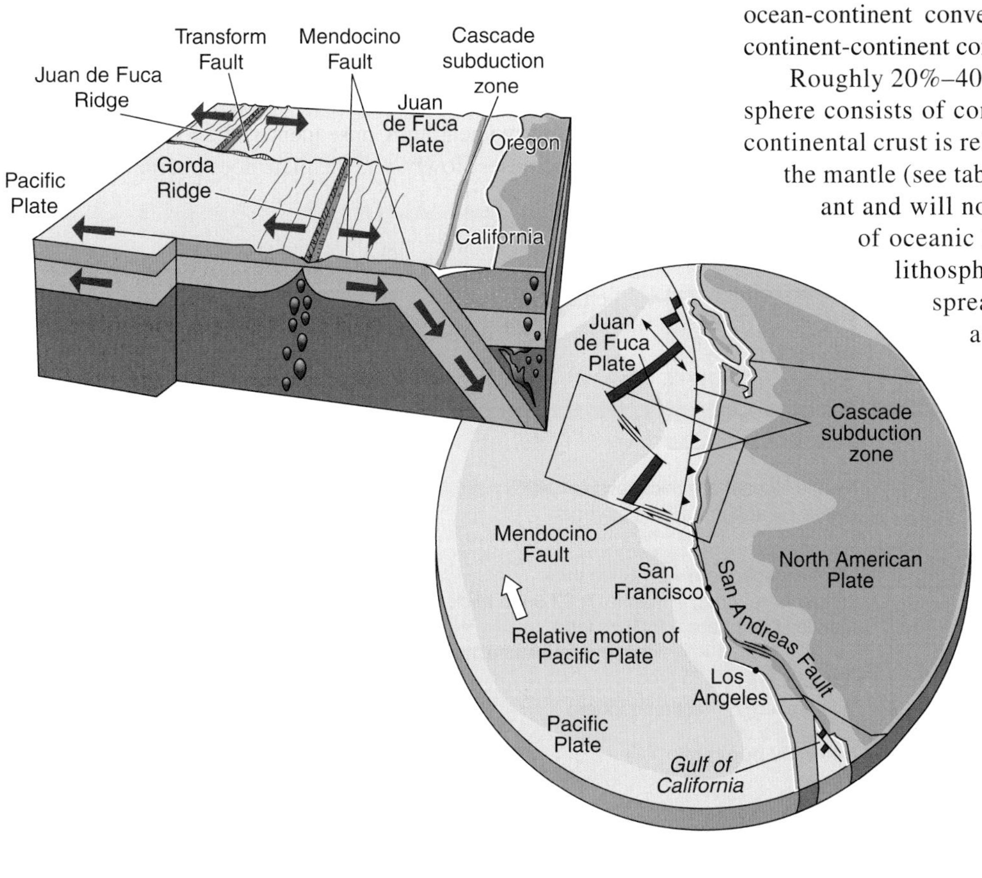

Figure 3.29 Transform boundaries can occur in continental or oceanic crust. In addition, they may join either divergent or convergent boundaries together. The San Andreas Fault cuts through continental crust, joining a divergent boundary in the Gulf of California with a subduction zone off the northern coast of California. Other transform faults offset segments of ridge crest where there is spreading between the Pacific and Juan de Fuca Plates.

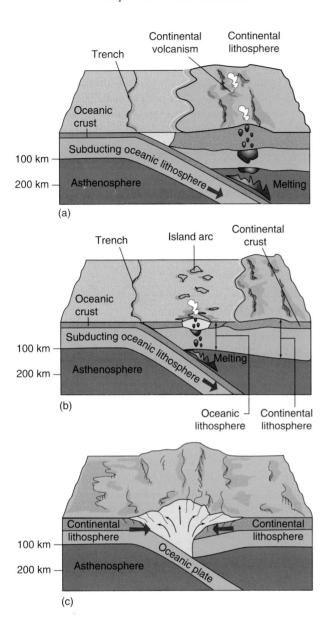

(a)

(b)

(c)

Figure 3.30   The three different types of convergent plate boundaries are (a) ocean-continent convergence, (b) ocean-ocean convergence, and (c) continent-continent convergence.

rate proportional to the square root of age, as shown by the following equation and illustrated in figure 3.31.

$$\text{Thickness (km)} = 10 \times \sqrt{\text{(age in millions of years)}}$$

The increase in thickness of the oceanic lithosphere is due to an increase in the thickness of the mantle portion of the lithosphere. The thickness of the crustal portion of the oceanic lithosphere remains constant. Roughly 95% of old oceanic lithosphere consists of relatively cold, dense mantle rock. Old oceanic lithosphere is about 1% more dense than the underlying asthenosphere, making it possible for oceanic lithosphere to be subducted. Subducted lithosphere always consists of oceanic lithosphere.

If oceanic lithosphere is being actively subducted along an ocean trench, volcanism occurs in the overriding plate along a line roughly parallel to the trench. The source of the magma for the volcanism depends on the age, and hence the temperature, of the oceanic lithosphere subducted. If subduction is occurring close to the spreading center where the sea floor was created, the subducting oceanic lithosphere will be relatively young (less than about 50 million years), warm, and buoyant. This lithosphere tends to resist subduction because of its buoyancy, so it descends into the mantle at a shallow angle. During descent, the temperature of the plate increases even more by conductive heating from the mantle. The plate will partially melt and produce magma. One area where magma may be generated this way is along the Pacific Northwest of the United States (fig. 3.29). More commonly, the subducting oceanic lithosphere is relatively old (greater than about 100 million years), cold, and dense. Old oceanic lithosphere is subducted easily, descending into the mantle at a steep angle. The rate of subduction of old oceanic lithosphere is relatively fast compared to the rate of conductive heating. Consequently, the temperature of the cold plate increases slowly, and the plate does not melt. However,

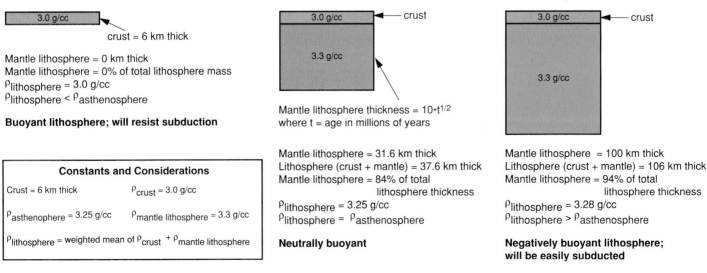

(a) Age = 0

3.0 g/cc

crust = 6 km thick

Mantle lithosphere = 0 km thick
Mantle lithosphere = 0% of total lithosphere mass
$\rho_{lithosphere}$ = 3.0 g/cc
$\rho_{lithosphere}$ < $\rho_{asthenosphere}$

Buoyant lithosphere; will resist subduction

Constants and Considerations	
Crust = 6 km thick	ρ_{crust} = 3.0 g/cc
$\rho_{asthenophere}$ = 3.25 g/cc	$\rho_{mantle\ lithosphere}$ = 3.3 g/cc
$\rho_{lithosphere}$ = weighted mean of ρ_{crust} + $\rho_{mantle\ lithosphere}$	

(b) Age = 10 Ma

3.0 g/cc ← crust

3.3 g/cc

Mantle lithosphere thickness = $10 \cdot t^{1/2}$
where t = age in millions of years

Mantle lithosphere = 31.6 km thick
Lithosphere (crust + mantle) = 37.6 km thick
Mantle lithosphere = 84% of total
                                    lithosphere thickness
$\rho_{lithosphere}$ = 3.25 g/cc
$\rho_{lithosphere}$ = $\rho_{asthenosphere}$

Neutrally buoyant

(c) Age = 100 Ma

3.0 g/cc ← crust

3.3 g/cc

Mantle lithosphere = 100 km thick
Lithosphere (crust + mantle) = 106 km thick
Mantle lithosphere = 94% of total
                                    lithosphere thickness
$\rho_{lithosphere}$ = 3.28 g/cc
$\rho_{lithosphere}$ > $\rho_{asthenosphere}$

**Negatively buoyant lithosphere;
will be easily subducted**

Figure 3.31   Diagrammatic representation of how the thickness and density (ρ) of oceanic lithosphere increase with age.

the temperature does rise high enough to drive water and other volatiles out of the plate and into the surrounding mantle. The addition of water to the mantle reduces the melting temperature of the mantle rock it is moving through by about 200°C (360°F), causing that mantle to partially melt, producing a basaltic magma. The formation of this magma typically begins at a depth of about 100–150 km (roughly 60–100 mi). The composition of erupted lava will depend on whether the overriding plate consists of continental or oceanic lithosphere.

Along ocean-continent convergent boundaries, the magma produced by subduction must rise through the continental rocks. This magma may partially melt the overlying continental crust, resulting in a magma with a mixed composition between basalt and granite called **andesite.** Eventually, this rising magma produces active volcanoes along the edge of the continental plate (fig. 3.30*a*). Such volcanic action is distinctly different from the volcanism associated with spreading centers, where basaltic magma is extruded on the sea floor in nonviolent eruptions. These andesitic volcanoes often erupt explosively, due to high concentrations of volatiles (or gases) and higher concentrations of silicon dioxide in the melt than are found in basaltic magmas. When andesitic volcanoes erupt they eject large amounts of rock, ash, and gas that can rise high into the atmosphere. Examples of active continental volcanic chains include the Andes along the west coast of South America and the Cascade Range of the Pacific Northwest, including Mount St. Helens (fig. 3.32).

Ocean-ocean convergent boundaries are also marked by ocean trenches (fig. 3.30*b*). The plate that is subducted generally is the one whose convergent edge is the farthest distance from the spreading center where it was created, and hence the oldest and most dense. The magma produced by ocean-ocean convergence will rise beneath the overriding oceanic lithosphere and produce a line of active volcanoes on the ocean floor of the overriding plate. These volcanoes may eventually grow large

enough to reach the sea surface and form a volcanic **island arc.** Island arcs generally erupt basalt that contains more volatiles than the basalt erupted along ocean ridges. As a result, island arc volcanism can be explosive. Island arcs are most commonly found in the Pacific Ocean. Good examples include the Mariana Islands, Tonga, the Aleutian Islands, and the Philippines. There are only two island arcs in the Atlantic Ocean: the Sandwich Islands in the South Atlantic and the Lesser Antilles on the eastern side of the Caribbean Sea. In some cases, island arcs can form on pieces of continental crust that have broken away from the mainland. Examples of this include New Zealand and Japan.

Subducting oceanic lithosphere is also associated with intense and unusually deep earthquake activity. Dipping zones of earthquakes, called **Wadati-Benioff zones** in honor of the geophysicists who first observed them, mark the approximate location of the subducted plate. The deepest earthquakes in the world occur in Wadati-Benioff zones extending to depths of as much as 650 km (400 mi) in the western Pacific Ocean. Earthquakes outside Wadati-Benioff zones are typically no more than 30 km deep.

Continent-continent convergent boundaries are the end result of the closing of an ocean basin by ocean-continent convergence (fig. 3.30*c*). As the ocean basin closes, the continental lithosphere of the downgoing plate moves closer to the trench. When the sinking oceanic lithosphere of the downgoing plate is completely subducted and the ocean basin has closed, the continental lithosphere of the downgoing plate will collide with the continental lithosphere of the overriding plate, forming a suture zone that may include marine sediments scraped off the oceanic lithosphere. This suture zone marks the site of the former ocean basin. The plates will continue to move together as the dense oceanic lithosphere of the downgoing plate continues to pull the plate toward the overriding plate. Along the edges of the two converging continents, the continental crust buckles, fractures, and thickens. What was once a well-defined convergent boundary marked by a narrow ocean trench becomes a broad zone of intense deformation with no clear, discrete boundary. This produces a great mountain range that can have old marine sediments and fossils raised to high elevation, as in the Alps and Himalayas.

Continent-continent convergence created the Himalayas as India collided with Asia. India continues to collide with Asia at the rate of about 5 cm (2 in) per year, and the Himalayas are still rising in elevation. Other examples of mountain ranges created in this manner are the Appalachians, Alps, and Urals.

Continental Margins

When a continent rifts and moves away from a spreading center, the resultant continental margin is known as a **passive,** or **trailing, margin.** Such margins are also referred to as Atlantic-style margins since they are found on both sides of the Atlantic Ocean as well as around Antarctica, the Arctic Ocean, and the Indian Ocean. Continental and oceanic lithosphere are joined along passive margins so there is no plate boundary at the margin. As passive margins move away from the ridge, the oceanic lithosphere cools, increases its density, thickens, and subsides. This causes the edge of the continent to slowly subside as well. Passive margins are modified by waves and currents; they may also be modified by

Figure 3.32 Mount St. Helens erupted violently on May 18, 1980. The mountain lost nearly 4.1 km³ (1 mi³) from its once symmetrical summit, reducing its elevation from 2950 to 2550 m (a change of 1312 ft). The force of the lateral blast blew down forests over a 594 km² (229 mi²) area. Huge mud flows of glacial meltwater and ash moved down the mountain.

coral reefs. These margins accumulate sediment eroded from the continents to a depth of about 3 km (1.8 mi). Currents move the sediments downslope to the sea floor, where thick deposits of sediments build on top of the oceanic crust. Old passive margins are not greatly modified by tectonic processes because of their distance from the ridge. These margins are broad, the water is shallow, and the sedimentary deposits are thick, as along the eastern coast of the United States. While passive margins begin at a divergent plate boundary, they move to a midplate position as a result of seafloor spreading and the opening of the ocean basin.

When a plate boundary is located along a continental margin, the margin is called an **active,** or **leading, margin.** Active continental margins are often marked by ocean trenches where oceanic lithosphere is subducted beneath the edge of the continent. These margins are typically narrow and steep with volcanic mountain ranges, as along the west coast of South America as well as Oregon and Washington. Because sediments moving from the land into the coastal ocean move directly downslope into deeper water, into trenches, or into adjacent ocean basins, thick sediment deposits do not routinely accumulate to form broad shallow shelves at active margins. Active margins are found primarily in the Pacific Ocean.

3.5　Motion of the Plates

Mechanisms of Motion

The mechanism that drives the plates is still not fully understood. The plates could be forced apart at the ridges by the formation of new crust, moved sideways, and then thrust downward at the trenches. Another possibility is that the thick, dense, oceanic crust and its sediment load sink into the mantle and the resulting tension drags the remainder of the plate with it. This latter mechanism could result in tension cracks through which the mantle material escapes upward to form the ridges and new crust. In this theory, the weight of the dense, cold, descending slab of oceanic lithosphere pulls the remainder of the plate.

The fact that the speed of a plate seems to increase as the amount of subducting lithosphere along the edge of the plate increases further supports theories that slab pull is an important driving force. Other factors that may be involved include changes in the weight of the plates resulting from land erosion or the accumulation of sediments, the rate at which magma wells up into the ridges, and thickening of the older lithosphere as it cools.

Rates of Motion

The sea floor, acting like a conveyor belt, moves away from the ridge system where it is created by volcanic activity. The rate at which each plate moves away from the axis of the ridge is known as the half spreading rate, while the rate at which the two plates move away from each other is called the full spreading rate, or simply the **spreading rate.** Spreading rates vary between about 1 and 20 cm (0.4 and 8 in) per year but are generally between about 2 and 10 cm (0.8 and 4 in) per year. The average spreading rate is about 5 cm (2 in) per year, roughly the rate at which fingernails grow. In the course of a human

lifetime of seventy-five years, a plate moving at an average spreading rate would travel 3.75 m (12.3 ft), or roughly the length of an automobile. Although these rates are slow by everyday standards, they produce large changes over geologic time. For example, if a plate moved at the rate of just 1.6 cm (0.6 in) per year, it would take 100,000 years for the plate to travel 1.6 km (1 mi); therefore, in the 200 million years since the breakup of Pangaea, it could move more than 3200 km (2000 mi), which is more than half the distance between Africa and South America. The spreading rate affects the physical structure of divergent plate boundaries. Slow spreading rates are found along ridges that have steep profiles and deep central valleys, such as the Mid-Atlantic Ridge. Fast spreading rates produce ridges with gentler slopes and shallower, or nonexistent, central valleys, such as the East Pacific Rise. Spreading rates are estimated at about 2.5–3 cm (1–1.2 in) per year for the Mid-Atlantic Ridge and about 8–13 cm (3–5 in) per year for the East Pacific Rise. Keep in mind that the process of spreading does not occur smoothly and continuously but goes on in increments, with varying time periods between occurrences. Compare the width of the age band centered on the East Pacific Rise with the width of the band centered on the Mid-Atlantic Ridge in figure 3.19; the wider the age band, the faster the spreading rate. Spreading rate is also believed to be related to the frequency of volcanic eruptions along the ridge. While very little is actually known about eruption frequency, it has been estimated that eruptions will occur approximately every 50–100 years at any given location along a fast-spreading ridge, while the rate may be only once every 5000–10,000 years on slow-spreading ridges.

Although seafloor spreading can be observed directly with great expense and difficulty at sea, there is one place where many of the processes can be seen on land—in Iceland, the only large island lying across a mid-ocean ridge and rift zone. Spreading in Iceland occurs at rates similar to those found at the crest of the Mid-Atlantic Ridge. Northeastern Iceland had been quiet for 100 years, until a volcanic rift opened in 1975; in six years, this rift widened by 5 m (17 ft) along an 80 km (50 mi) long stretch of the ridge's crest. Over 100 years, this spreading rate is 5 cm (2 in) per year, which is within the typical range. The spreading motion between the tectonic plates can now be monitored by satellite. The Global Positioning System of satellites (see chapter 2, section 2.5) can be used to determine the relative motion of tectonic plates directly by measuring the distance between two widely separated points to an accuracy of 1 cm (0.4 in) (fig. 3.33).

Investigations of ancient granitic rocks, formed during the Archean eon and exposed in West Greenland, eastern Labrador, Wyoming, western Australia, and southern Africa, indicate that, 3.5 billion years ago, crustal plates existed and moved granitic continental blocks at an average rate of about 1.7 cm (0.67 in) per year, again within today's average rates of plate motion.

Hot Spots

Scattered around Earth are approximately forty areas of isolated volcanic activity known as **hot spots** (fig. 3.34). They are found under continents and oceans, in the center of

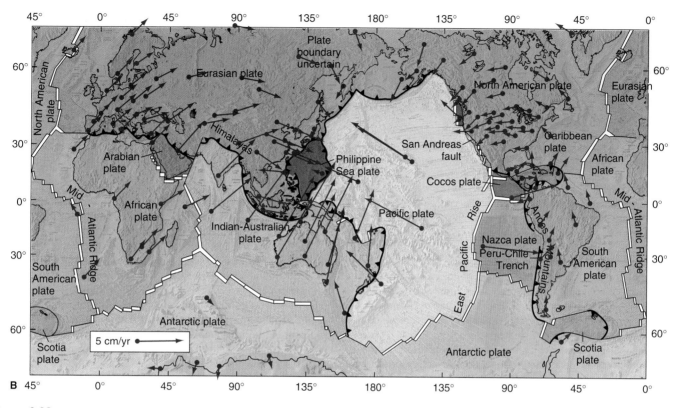

Figure 3.33 Annual plate motion vectors measured by GPS at stations around the world. From NASA.

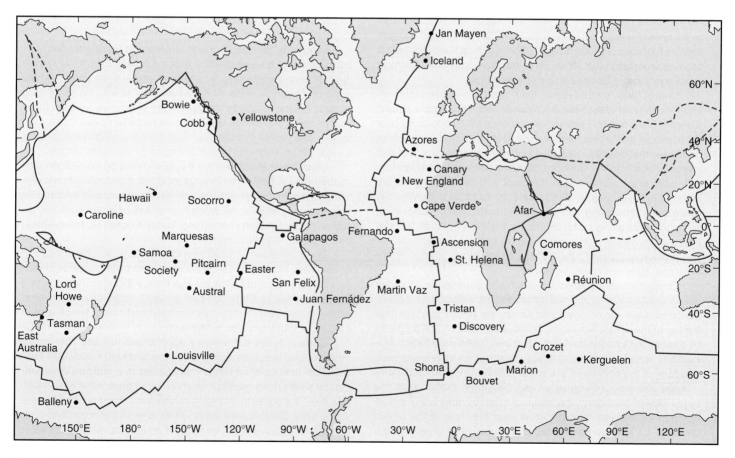

Figure 3.34 Location of major hot spots.

Field Notes

New Approaches to Exploring the Oceans

by Dr. John Delaney

Dr. John Delaney is a professor of oceanography at the University of Washington, specializing in marine geology. His research focuses on the deep-sea volcanic activity of the Juan de Fuca Ridge in the northeast Pacific Ocean.

In the toxic superheated brew of water and dissolved gases that circulates through the rocks beneath the sea floor, microbes thrive. Here in this "ocean beneath the sea floor" what would be poison to many of Earth's residents is nutrition for an ancient form of microbial life called *Archaea*. When the rocks crack or shift, due perhaps to an earthquake or a submarine volcanic eruption, *Archaea* are released onto the sea floor, making an as yet unevaluated, but potentially significant contribution to Earth's carbon cycle.

This deep, sub-seafloor biosphere is largely unexplored and was unknown only 20 years ago. For the past decade and a half, however, when marine scientists have visited erupting volcanoes on the sea floor of the Pacific Ocean, they have found billowing clouds of microbial material issuing vigorously from seafloor cracks. After viewing and sampling these eruptive venting phenomena, scientists concluded there is a deep, hot microbial biosphere that resides in the pores and cracks of the rocks making up the volcanic portion of the oceanic crust. Some of the microbes are adapted to especially high temperatures and do not begin to reproduce until the temperatures are higher than 60°C. Many of these life forms do not require sunlight as a source of energy, but instead use volcanic chemicals to energize their life functions, a process called chemosynthesis. Indeed there is growing speculation that life on Earth may have originated billions of years ago under submarine volcanic conditions.

One of today's most compelling questions in planetary science focuses on identifying and quantifying the linkages among a host of plate-tectonic processes and microbial productivity near the sea floor and within Earth's crust. Deformational and thermal processes that operate most intensely near plate margins, and less intensely within plate interiors, result in forced fluid migration within the crust and between the crust and its overlying ocean. These fluid fluxes may be steady state, episodic, or both. Whether operating at spreading centers, in mid-plate, within subduction complexes, or along transform faults the moving fluids transfer heat and chemically active organic and inorganic compounds that can provide nutritional support for a widespread, but poorly understood, deep microbial biosphere.

Oceanographers have considerable experience studying ocean productivity supported by photosynthesis and its dependence on mass and energy fluxes across the air-sea interface near the top of the ocean. Scientists, however, are currently ill prepared to assess the importance of fluid-driven, plate-modulated productivity linked to mass and energy fluxes across the water-rock interface at the bottom of the ocean. Unlike the predictably distributed character of irradiance and gas-liquid exchange at the upper-ocean boundary, the input of chemosynthetically active compounds derived from the crust near the base of the ocean tends to be localized along faults, fissures, and other venting structures at a variety of scales and in patterns that cannot be predicted, but must be identified. Without such studies, it is not possible to assess the relative proportions of planetary carbon fixation at the top of the ocean versus that near the bottom of the ocean.

This tectonically forced fluid-expulsion/microbial-bloom theme represents a new class of geobiological processes that operate steadily and episodically in a "seasonal" rhythm attuned to plate dynamics. Knowing where, how, and when these features channel nutrient-laden fluids from the crust into the overlying ocean requires new approaches within the ocean sciences and represents the initial step toward quantification of a process that may operate on other bodies in our solar system and beyond. Focusing on these processes at the scale of a single plate could give rise to a new class of research designed to conduct quantitative plate-scale studies of heat, chemical, and biomass fluxes within a well-defined plate-tectonic framework. Such an effort would be the basis for launching a deep-sea, long-term ecological study of plate tectonically modulated biogeocomplexity. The local, regional, and global importance and pervasiveness of these processes on Earth and on other active, water-bearing planets will be a major focus of research for decades.

In order to fully explore and understand these linkages, scientists must enter and interact with the full volume of the marine environment—water-column, sea floor, and sub–sea floor—at the scale of a tectonic plate. And they must be able to do so for long periods of time, enabling the generation of time-series data to reveal patterns and trends that occur on decadal scales rather than for just one month in a summer field season using the traditional expeditionary methods of a ship and an underwater vehicle.

Such new capabilities are being fostered by convergent advancements in fields that include robotics, communications, distributed power, computing, sensor development, and information management. Indeed, an entire new phase of oceanographic research and education is emerging. Today's ocean scientists stand on the edge of a revolution.

The integration of these novel technological approaches can enable vast, efficient three-dimensional arrays of thousands of sensors, on, above, and below the sea floor. Advances in computing technology will facilitate comparing results from these distributed sensor arrays with increasingly sophisticated computer-generated models of many oceanic processes.

Certain ocean observatory approaches will use fiber-optic/power cables to supply significant power and high-bandwidth, two-way, real-time communication capabilities to a network of experimental sites on the sea floor. Such observatories will enable rapid, adaptive responses to changing conditions in remote or hostile environments. Shore-based users will have real-time command-and-control of ocean-based sensors, instruments, and underwater vehicles. Within the next decade, users on land will have electronic

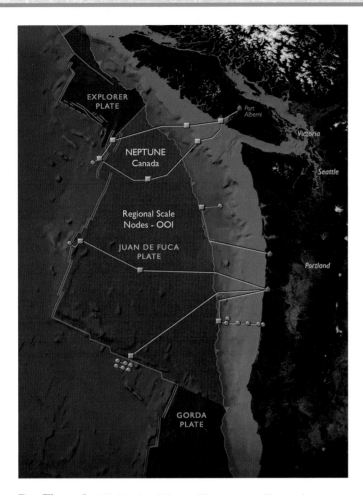

Box Figure 1 The Regional Ocean Observatory will extend continuous high-bandwith (tens of gigabits/second) and power (tens of kilowatts) to a network of instruments widely distributed across, above, and below the sea floor in the northeast Pacific Ocean. As the world's first ocean observatory to span a tectonic plate, this facility will provide a constant stream of data in real time from throughout the water, on the sea floor, and below the sea floor associated with the Juan de Fuca plate.

"ringside seats" during submarine volcanic eruptions using HDTV and autonomous underwater vehicles.

The development and construction of cabled ocean observatories are under way in many countries, including the U.S., Canada, Japan, and the European Union. Within the U.S., a major National Science Foundation program—the Ocean Observatories Initiative—includes cabled observatories at coastal and regional scales. The regional component will be built off the coast of Oregon and Washington and will be modeled after the NEPTUNE system developed by the University of Washington and partners. Associated with the Juan de Fuca tectonic plate, where a global suite of ocean processes is represented, the observatory will use a heavily instrumented network of 750 miles of fiber-optic/power cable to enable long-term, real-time measurements of physical, chemical, geological, and biological variables in the ocean and sea floor. The West Coast coastal component of the NSF Initiative includes a cabled line of instruments running out from Newport, Oregon. A complementary Canadian effort—NEPTUNE Canada—is building a 480-mile loop of instrumented cable on the northern portion of the plate (box figure 1).

Experimental sites will be established at nodes along the cables and will be instrumented to interact with physical, chemical, and biological processes that operate within the ocean and in the underlying crust across multiple scales of space and time.

These new capabilities will provide novel and enduring ways to study the oceans (box figure 2). The data will be distributed worldwide and in near real time via the Internet. Students, educators, decision makers, and the general public will be able to join scientists on their journeys of exploration and discovery. These new approaches will revolutionize not only how we humans look at oceans and the earth, but eventually the way we learn to manage our entire planet. In September 2005, the VISIONS05 expedition to the underwater volcanoes of the Northeast Pacific previewed a cabled observatory's ability to provide stunning live imagery from the deep sea when live high-definition video from the sea floor was broadcast for the first time. Images were transmitted from a high-definition underwater video camera deployed on a robot on the sea floor. Data traveled from the robot to ship by fiber-optic cable and from ship to the University of Washington via high-bandwidth satellite connection. From the

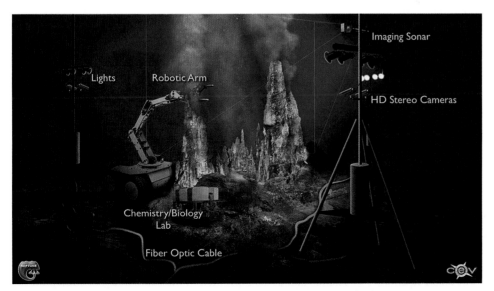

Box Figure 2 Conceptual representation of a future seafloor laboratory on the Regional Ocean Observatory of the Ocean Observatories Initiative network in the NE Pacific ocean. The cable will provide continuous power and high-bandwidth communications to instruments on the sea floor, including lights and high-definitions (HD) cameras, underwater robots, and *in situ* laboratories for chemical and biological analyses. HD video and other data will be transmitted via fiber-optic cable in real and near real time to land-based laboratories, classrooms, and science centers. Because these images and data will be available via the Internet, learners of all ages will be able to participate in this journey of exploration and discovery.

(continued)

university, the transmission went out over Internet2 to viewers in the U.S., Canada, Australia, and Tokyo.

Funding support for VISIONS05 came from many sources, including the W. M. Keck Foundation. In 2001, the Keck Foundation granted $5 million to the University of Washington for a five-year experiment designed to explore the basic premise that when rock deforms, the nutrient-rich fluids set in motion are capable of supporting microbial blooms in adjacent portions of the crust or within the overlying ocean. It is not possible to fully test such a hypothesis without establishing a permanent presence on the sea floor that can continuously observe, document, and interact with co-varying processes of deformation, fluid flow, chemistry, and microbial activity.

The Keck study focused on the establishment of well-instrumented proto-observatories at three major plate boundaries in the Northeast Pacific to investigate linkages among episodic deformation, fluid venting, and microbial productivity. Observatory sites were selected along the likely path of the regional observatory cable, so that experience gained during the study would allow early experiments with the full cabled observatory system to be optimally designed and effectively implemented.

By the end of the five-year study, forty new instruments had been developed and deployed at the undersea observatories.

These successes relied on the efforts of multiple interdisciplinary teams to design the instruments, map the detailed topography of the plate, and coordinate the complex system design and instrument deployment

Keck funding was also used to develop an *in situ* microbial incubator that examines the extreme conditions under which microbial communities thrive, survive, and expire within the walls of actively venting sulfide chimneys. More can be learned about microorganisms in these extreme environments at: http://www.visions05.washington.edu/science/investigations/micro_survival.html. Analysis of biofilms that grew during this experiment indicated that life may thrive at temperatures above 150°C.

In addition, a long-term seafloor seismic network was deployed on plate boundaries to characterize seismicity over a one- to three-year interval. This was one of the first deployments of a seafloor seismic network with sensor installation and data quality that are comparable to a good land network.

Websites

www.ooi.washington.edu
www.neptune.washington.edu
www.joiscience.org/ocean_observing
www.visions05.washington.edu

---

plates, and at the mid-ocean ridges. These hot spots periodically channel hot material to the surface from deep within the mantle, possibly from the core-mantle boundary. Above these sites, a plume of mantle material may force its way through the lithosphere and form a volcanic peak, or seamount. If the hot spot does not break through, it may produce a broad swell on the ocean floor or the continent that subsides as the plate moves over and away from the magma source. Hot spots may also resupply magma to the asthenosphere, which cools and becomes attached to the base of the lithosphere, thickening the plates. Some people believe that the breakup of Pangaea began when a chain of hot spots developed under the supercontinent.

Hot-spot plumes of magma are not uniform; they differ in chemistry, suggesting that they come from different mantle depths, and it has been suggested that their discharge rates may also vary. Hot spots may fade away and new ones may form. The life span of a typical hot spot appears to be about 200 million years. Although their positions may change slightly, hot spots tend to remain relatively stationary in comparison to moving plates, and they can be useful in tracing plate motions.

As the oceanic lithosphere moves over a hot spot, successive eruptions can produce a linear series of peaks, or seamounts, on the sea floor. The youngest peak is above the hot plume, and the seamounts increase in age as their distance from the hot spot increases (fig. 3.35). For example, in the islands and seamounts of the Hawaiian Islands system, the island of Hawaii, with its active volcanoes, is presently located over the hot spot. The newest volcanic seamount in the series is Loihi, found in 1981 45 km (28 mi) east of Hawaii's southernmost tip and rising 2450 m (8000 ft) above the sea floor but still under water. At its current rate of growth, it should become an island in 50,000–100,000 years.

The land features of the island of Hawaii show little erosion, for it is comparatively young. To the west are the islands of Maui, Oahu, and Kauai, which have been displaced away from the magma source. Kauai's canyons and cliffs are the result of erosion over the longer period of time that it has been exposed to the winds and rains. Although these four islands are the most familiar, other islands and atolls attributed to the same hot spot stretch farther

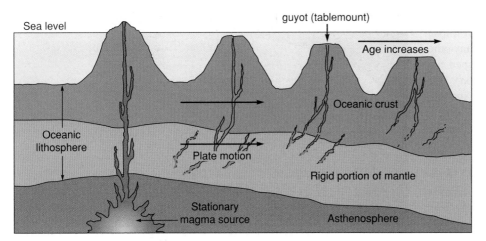

Figure 3.35 A chain of islands is produced when oceanic lithosphere moves over a stationary hot spot.

west across the Pacific. They are peaks of eroded and subsided seamounts formed above the same hot spot. Subsidence occurs when the heated and expanded plate moves slightly downhill and away from the mantle bulge over the hot spot. As the plate moves away from the hot spot, it cools and contracts, and combined with the weight of the seamount it is carrying, the result is a depression in the mantle that gradually carries the seamount below the ocean surface. When seamounts in tropical areas sank slowly, coral grew upward and coral atolls were the result. The formation of atolls is discussed further in chapter 4.

West of Midway Island, the chain of peaks changes direction and stretches to the north, indicating that the Pacific Plate moved in a different direction some 40 million years ago. This line of peaks is the Emperor Seamount Chain, volcanic peaks that once were above the sea surface as islands but have since eroded and subsided over time, resulting in many flat-topped seamounts known as **guyots** (or tablemounts) 1000 m (0.6 mi) below the surface (fig. 3.35). The northern end of the Emperor Seamount Chain is estimated to be 75 million years old, and Midway Island itself may be 28 million years old. Refer to figure 3.8 to follow this seamount chain. Notice in figure 3.35 that the peaks formed by the hot spot get older in the direction in which the plate is moving. The plate is presently moving westward; Midway Island is northwest of the main Hawaiian Islands, and it is also older.

It is possible to check the rate at which the plate is moving by using the distance between seamounts in conjunction with radiometric dating. For example, the distance between the islands of Midway and Hawaii is 2430 km (1509 mi). Midway was an active volcano 28 million years ago, when it was located above the hot spot currently occupied by Hawaii. In other words, Midway has moved 2430 km (1509 mi) in 28 million years, or 8.7 cm (3.4 in) per year.

There is another hot spot at 37° 27'S, in the center of the South Atlantic, marked by the active volcanic island of Tristan da Cunha (see fig. 3.34). Volcanic activity at this more slowly diverging plate boundary produces seamounts that can be carried either to the east or to the west, depending on the side of the

spreading center on which the seamount was formed. The hot spot produces a continuous series of seamounts very close together, forming a **transverse,** or **aseismic, ridge:** the Walvis Ridge to the east, between the Mid-Atlantic Ridge and Africa, and the Rio Grande Rise to the west, between the Mid-Atlantic Ridge and South America. Refer to figures 3.8 and 4.14.

If a hot spot is located at a spreading center, the flow of material to the surface is intensified. The crust may thicken and form a platform. Iceland is an extreme example of this process in which the crust has become so thick that it stands above sea level.

Seamount chains, plateaus, and swellings of the sea floor, all products of hot spots, are being used to trace the motion of Earth's plates over known hot-spot locations. Recently, hot-spot tracks have been used to reconstruct the opening of the Atlantic and Indian Oceans. Although the exact mechanism is not known, hot spots also appear to have formed huge basaltic plateaus: the Kerguelen Plateau in the southern Indian Ocean about 110 million years ago, and the Ontong Java Plateau in the western Pacific about 122 million years ago.

3.6 History of the Continents

The Breakup of Pangaea

Figure 3.36 traces the recent plate movements that led to the configuration of the continents and oceans as we know them today. In the early Triassic, when the first mammals and dinosaurs appeared, the continents were all joined in a single landmass called Pangaea (fig. 3.36a). Most of the rest of the globe at this time was covered by the massive Panthalassic Ocean, also known as Panthalassa. A second, much smaller ocean called the Tethys occupied an indentation in Pangaea between what would eventually become present-day Australia and Asia. About 200 million years ago, Pangaea began to break apart into Laurasia and Gondwana. By about 150 million years ago, when the dinosaurs were flourishing, Laurasia and Gondwana were separated by a narrow sea that would grow to be the central Atlantic Ocean (fig. 3.36b). At the same time, India and Antarctica were beginning to move away from South America and Africa. Water flooded the spreading rift between South America and Africa about 135 million years ago, and the South Atlantic Ocean began to form. As seafloor spreading opened the Atlantic Ocean, Panthalassa (what we now call the Pacific Ocean) was growing progressively smaller. At the same time, India was moving north and the southern Indian Ocean was forming as the Tethys Ocean was closing.

By the late Cretaceous, 97 million years ago, the North Atlantic Ocean was opening and the Caribbean Sea was beginning to form. India was continuing to move toward Asia, and the Tethys Ocean was being consumed from the north as the Indian

Ocean was expanding from the south (fig. 3.36*c*). By the end of the Cretaceous, shortly before the meteor impact that led to the extinction of the dinosaurs and many other species, the Atlantic Ocean was well developed and Madagascar had separated from India (fig. 3.36*d*). During the middle Eocene, 50 million years ago, Australia had separated from Antarctica, India was on the verge of colliding with Asia, and the Mediterranean Sea had formed (fig. 3.36*e*). About 40 million years ago, India collided with Asia and initiated a continent-continent convergent boundary, destroying the last of the Tethys Ocean and beginning the formation of the Himalayas. Twenty million years ago, Arabia moved away from Africa to form the Gulf of Aden and the new, still-opening Red Sea.

Before Pangaea

Because Earth is about 4.6 billion years old, there is no reason to believe that Pangaea and its breakup into to-day's continental configuration was the only time during which the sea floor spread or the continents changed position and recombined. Scientists are now searching for new and more extensive evidence to indicate the positions of the landmasses before the breakup of Pangaea. Because there is no record of pre-Pangaean relationships in the present 200-million-year-old oceanic crusts, scientists must depend on evidence from the continents, where the oldest rocks are found.

Radiometric dating and magnetic and mineral analyses indicate that ancient granitic continental core regions are common to all landmasses; they have been present in continental crust for more than 3 billion years. Ancient mountain ranges located in the interiors of to-day's continents are evidence of collisions between old plates; seismic evidence has been used to mark old plate edges below the Ural and Appalachian Mountains. The magnetic and fossil evidence frozen into continental rocks during the 350 million years prior to Pangaea have been used to propose a series of pre-Pangaea plate movements during the Paleozoic era, which began about 570 million years ago.

Six major continents are recognized from the Paleozoic era: Gondwana (Africa, South America, India, Australia, Antarctica), Baltica (Scandinavia), Laurentia (North America), Siberia, China, and Kazakhstania. Approximately 555 million to 540 million years ago, these landmasses were strung along Earth's equator; there was no land above 60°N and below 60°S, and the polar

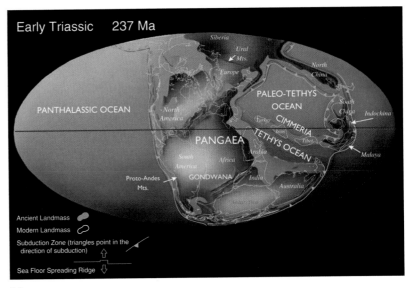

(a)

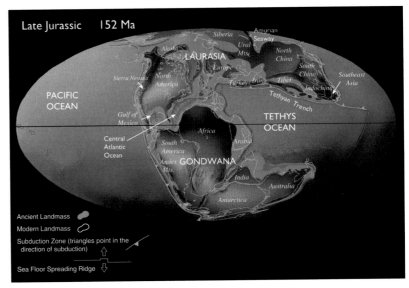

(b)

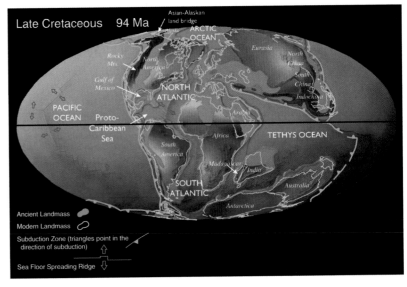

Figure 3.36 The configuration of landmass and oceans from Pangaea to the present. Geologic period (or epoch) and time before the present in millions of years (Ma) are given in the upper left-hand corner of each illustration.

(c)

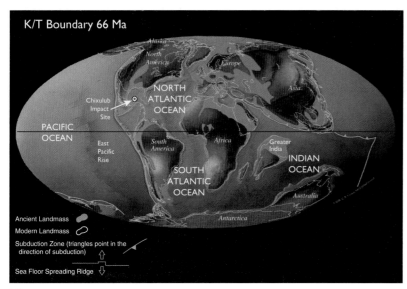

(d)

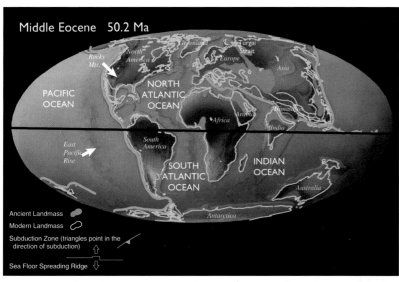

(e)

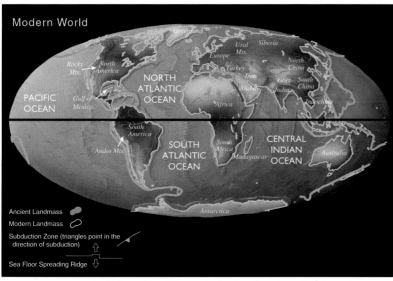

(f)

Figure 3.36 (continued)

regions were wide expanses of ocean. Gondwana and Baltica had shifted to the east and south by 514 million years ago (fig. 3.37). Gondwana's southward motion carried it across the South Pole 435 million to 430 million years ago and then north again on the opposite side of Earth to a position that would eventually make it a part of Pangaea about 350 million years ago. This movement completely reversed the north-south orientation of modern Africa and South America. Their present-day southern tips pointed to the north at the beginning of the journey, but as the landmass drifted across the South Pole to the other side of the world, they pointed to the south, as we know them today. About 310 million years ago, the ocean between Gondwana and the other continental fragments began to close. Siberia moved from low to high latitudes and merged with Kazakhstania, while China moved westward. The assembly of Pangaea 260 million to 250 million years ago resulted from a series of continental collisions that formed great mountain belts that are still detectable. This scenario is not the only possible solution to the continent puzzle. As time passes, new evidence may point to a different fit and movements of the puzzle pieces through time.

Because the tectonic processes are continuous, there probably has never been a stable geography of Earth. The modern configuration is no more stable than the one in the Paleozoic era. They are both steps in an ever-changing pattern. We have followed this changing pattern of Earth's landmasses from wide dispersal 555 million years ago, to a single joined continental mass 260 million years ago, and then through another period of dispersal to the present time.

It has been suggested that this is an orderly, cyclic, 500- million-year pattern powered by the heat from radioactive decay in Earth's interior. Although the production of heat is continuous, this model proposes that the heat is released in periodic bursts that are related to the movement of the continents. When sufficient heat accumulates under the continental mass, the rifting process begins fragmenting the continents, and as the continents move apart, they cool and subside, then sea level rises and covers their border lands.

Using this model, we see the Atlantic Ocean opening and closing as the continents move apart and then reassemble over the 500-million-year cycle, while the Pacific Ocean boundaries, which have remained more stable, approximate the wide, ancient hemispheric ocean of that period. We can even look forward in time to see the Mediterranean Sea being closed and lost as Africa continues to move northward, and the Atlantic and Indian Oceans continuing to expand while the Pacific Ocean narrows as North and South America collide with Asia. Australia will continue to move northward, eventually colliding with Eurasia, while Los Angeles and coastal southern California will pass San Francisco on their way toward the Aleutian Trench.

Terranes

Studies of ancient crustal rocks from the interior of North America indicate that the core of the continent was assembled about 1.8 billion years ago from several large pieces of granitic crust more than 3 billion years old. It appears that four or five large pieces of crust called **cratons** collided and joined over a 100-million-year period. Mixtures of sediments from the craton margins and the deep-sea floor, the remains of ancient island arc systems, and other small pieces of continental crust were trapped between the colliding blocks of continent. Radiometric dating has identified boundaries between North American cratons at 1.78 billion and 1.65 billion years old.

Geologic studies of Alaska show that the characteristics of one area are not necessarily related to the history, age, structure, or mineral composition of an adjacent area. Alaska is made up of crustal fragments, some showing the characteristic homogeneity and higher density of oceanic crust and others showing the highly varied mineral content and lower density of continental crust. Smaller crustal fragments associated with craton margins, bounded by faults, and with a history distinct from adjoining crustal fragments are known as **terranes.** See figure 3.38 for a map of North American terranes.

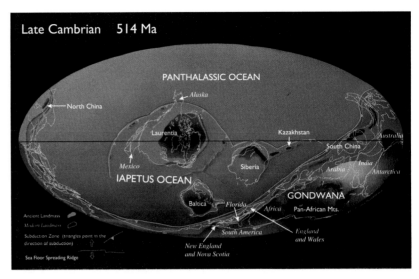

Figure 3.37 The configuration of landmass and oceans in the Late Cambrian, 514 million years ago, before the convergence of landmass to form Pangaea. The largest continent at this time was Gondwana which comprised present day Africa, South America, India, Australia, and Antarctica. Additional continents included Laurentia, Baltica, Siberia, Kazakhstania, and China.

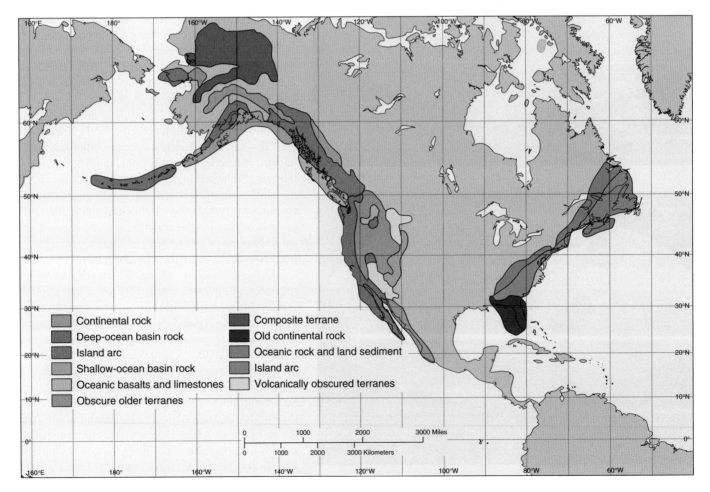

Figure 3.38 Examples of North American terranes, crustal fragments with histories distinct from adjoining fragments.

Terranes are often elongate, as if produced by an island arc system that has collided with a craton, or by faulting, which has cut off a sliver of continent similar to the land west of California's San Andreas Fault. The terranes of Alaska appear to have arrived from the south, rotated clockwise, and were faulted and stretched as the Pacific and North American Plates moved relative to each other. A terrane known as Wrangellia originated at or to the south of the equator, moved northward, and crashed into western North America about 70 million years ago. Faulting has spread its fragments throughout eastern Oregon, Vancouver Island, the Queen Charlotte Islands, and the Wrangell Mountains of southeast Alaska.

India is considered a single giant terrane by some; terranes north of the Himalayas predate the arrival of India. The area of North America west of Montana, south into New Mexico, and north into Canada and Alaska appears to be an assemblage of terranes that have been modified by faulting and vulcanism. In Oregon, some volcanic peaks appear to be basalt seamounts that have moved landward from relatively short distances offshore. Rock formations found in the San Francisco area are typical of rock formed in the South Pacific. Fossils collected between Virginia and Georgia indicate that a long north-south section of the East Coast was formed somewhere adjacent to an island arc system; the fossils point to a past European connection (fig. 3.38).

3.7 Exploring Divergent Boundaries

The exploration of the sea floor that began in the 1950s has continued for fifty years, providing oceanographers with the data to understand the basic features of ocean crust and plate tectonics. The present scientific effort is more specialized and seeks to understand ongoing seafloor changes as well as the processes of the past. Using more and more sophisticated technology, the search is on for the details, particularly along divergent plate boundaries, and numerous programs and projects are underway to these ends.

Project FAMOUS

Oceanographers made their first descent into the rift valley of the mid-ocean ridge in 1973. The previous year, more than twenty U.S., French, Canadian, and British oceanographic vessels had mapped a small section of the Mid-Atlantic Ridge. In the summer of 1973, in an area centered at 36° 50'N, southwest of the Azores, French scientists aboard the submersible *Archimede* made the first visual study of a rift valley, or spreading center. The following summer, the French submersible *Cyana* and the U.S. submersible *Alvin* (fig. 3.39) joined the dive program. This effort was known as the French-American Mid-Ocean Undersea Study, or Project FAMOUS.

The submersibles descended nearly 3000 m (9800 ft) to the floor of the huge rift valley. In a series of fifty dives, they took more than 50,000 photographs and made 100 hours of film recording. They also recovered 150 rock samples and seventy water samples. The cracks and fissures running parallel to the direction of the rift indicated continuous spreading with episodic, localized volcanic eruptions occurring in the central rift area.

Seafloor Spreading and Hydrothermal Vents

In 1972, measurements of water temperature and water chemistry were made in the Galápagos Rift. This spreading area lies between the East Pacific Rise and the South American mainland, 964 km (599 mi) west of Ecuador. Measurement data showed active circulation of seawater through newly formed oceanic crust. This circulation appeared to be the cause of plumes of hot water rising from **hydrothermal vents** along the rift. Water issuing from these vents had temperatures of 17°C (63°F), compared with 2°C (36°F) for the surrounding seawater. The hot water mixed with the cold bottom water to produce shimmering, upward-flowing streams rich in silica, barium, lithium, manganese,

Figure 3.39 Oblique view of the research vessel *Atlantis* with the submersible *Alvin* hanging on an A-frame.

84   *Chapter* 3   Plate Tectonics

hydrogen sulfide, and sulfur. Rocks in the vicinity of the vents were coated with chemical deposits rich in metals precipitated out of the vent waters.

In 1979, the Mexican-French-American Riviera Submersible Experiment (RISE) research program investigated ocean-floor hot springs near the tip of Baja California. Two thousand meters (6000 ft) down, *Alvin* found mounds and chimney-shaped vents 20 m (65 ft) high that were ejecting hot (350°C, 660°F), black streams of particles containing sulfides of lead, cobalt, zinc, silver, and a variety of other minerals.

Hydrothermal vent activity is now known to occur at seafloor spreading centers worldwide. Cold, dense seawater circulates near magma chambers close to the sea floor; there, it is heated and then released through cracks in the newly created upper lithosphere. Some of these vents discharge low temperature, clear waters up to 30°C (86°F); others, known as white smokers, give rise to milky

discharges with temperatures ranging from 200°–330°C (392°–626°F); and still others, black smokers, release jets of sulfide-blackened water at temperatures between 300° and 400°C (572° and 752°F). This hydrothermal vent activity is thought to go on for years, perhaps for decades. Rapid, unpredictable temperature fluctuations have been recorded on a timescale of days to seconds, indicating a very unstable environment surrounding the vents.

During a dive on the Juan de Fuca Ridge, a camera carried on the *Alvin* recorded the features of massive sulfide structures with large, protruding flanges that trap beneath them layers of hot vent water (350°C, 660°F) (fig. 3.40).

Researchers investigating a section of the East Pacific Rise in the winter of 1993–94 discovered an area of extreme volcanic activity. Scientists mapping the sea floor northwest of Easter Island in the South Pacific found about 1200 seamounts in a region of about 21,000 km^2 (8100 mi^2). No other region of Earth is known

(a)

(b)

(c)

Figure 3.40   (a) A large sulfide structure in the main vent field of the Endeavour Segment of the Juan de Fuca Ridge. The structure is about 10–15 m (33–49 ft) wide and 15–20 m (49–66 ft) high. The *Alvin* carried the life-sized plywood cutout down to the sea floor and placed it beside the structure. The cutout points to a large flange on the side of the structure. The light color of the flange is caused by organisms growing on the flange's upper surface. (b) A composite illustration of an Endeavour sulfide structure demonstrates a variety of features. Flanges protrude from the sides. Vented hot water (350°C) is often trapped under these flanges. As this water leaks around the flange edge and rises, it forms a deposit that increases the flange's outward growth. Some flanges have vent water channels within them and form smokers on upper surfaces. Organisms may also be found on flange surfaces. (c) A closeup of a flange protruding from the side of a large sulfide structure.

to have so many volcanoes. The lack of sediment in the area indicates that eruptions are frequent and recent. The area lies at the boundary between the Pacific and Nazca Plates and is a zone of rapid spreading. This same region was explored in 1993 with the French submersible *Nautile.* Many lavas appeared fresh—less than fifty years old—and many hydrothermal vents were noted, as was the presence of warm water rising off the lava beds.

Until recently, it was thought that nearly all hydrothermal vent systems shared some basic characteristics. Most of the known vent systems were located on the axis of ocean ridges on very young sea floor right along the plate boundary. The fluids emanating from them passed through basaltic rock, and the resultant minerals that precipitated from the fluids were iron- and sulfide-rich, especially at the dramatic black smokers. There is a growing body of evidence, however, that hydrothermal systems with a different chemical signature located off the axis of the ridge system in older sea floor may be more common than previously thought. The discovery and investigation of a spectacular new type of hydrothermal vent field located just off the axis of the Mid-Atlantic Ridge at 30°N, which has been named "Lost City," is a case in point. Lost City is located nearly 15 km (9 mi) away from the axis of the ridge on sea floor that is roughly 15 million years old. The Mid-Atlantic Ridge is a slow-spreading ridge. Along slow-spreading ridges, hydrothermal fluids not only react with shallow basaltic rock but may also circulate through deeper rocks that have been chemically altered. This is apparently the case at Lost City, where the venting fluids are very cool (40°–75°C, 104°–167°F) and abnormally alkaline (pH 9.0–9.8 compared with the average pH of seawater of 7.8; see the discussion of pH in chapter 6). The low-temperature venting at Lost City has created numerous chimneys, one of which is 60 m (197 ft) tall. These chimneys are composed of carbonate and magnesium-rich minerals, making them chemically very different from the other well-studied vent systems on young sea floor.

At present, two major U.S. programs are associated with identifying and understanding processes along the mid-ocean ridges: NOAA's VENTS program and the National Science Foundation's Ridge Inter-Disciplinary Global Experiments (RIDGE). The VENTS program is exploring the source and strength of hydrothermal discharges as well as the time intervals between discharges. The program is interested in the pathways followed by the materials issuing from vents, and its researchers have set up monitors to discover the impact of vent discharges on the chemistry of the water.

The RIDGE program's mission is to integrate observations, experimental results, and theoretical studies in a decade-long program to understand the geological, geochemical, and biological processes responsible for creating new oceanic crust along the mid-ocean ridge systems.

Detailed mapping of the East Pacific Rise has revealed a significant asymmetry to the sea floor in the study area. The sea floor is subsiding more slowly and the underlying mantle is warmer on the Pacific Plate side to the west than on the Nazca Plate side to the east. In addition, the Pacific Plate is moving much faster and has more seamounts and recent lava flows away from the axis of the ridge.

Fifty-one seismometers were deployed in two linear arrays about 800 km (500 mi) long across the East Pacific Rise. The seismometers were used to record seismic waves generated by regional and distant earthquakes. The velocity of seismic waves decreases with decreasing density and increasing temperature, characteristics that would be found where magma is being generated in the mantle. The first results from the seismic experiment indicate that there is a broad region of melting in the mantle beneath the ridge. The magma generation zone is thought to be several hundred kilometers wide and extend to a depth of 100–150 km (60–90 mi). The magma generation zone is also asymmetric. It extends farther to the west of the ridge axis. This difference would explain why the mantle temperatures are higher west of the ridge and why the sea floor is shallower there. The seismic data also indicate that the mantle is relatively homogeneous below depths of about 300 km (190 mi), indicating that the geographic location of the ridge is not controlled by deep mantle processes. An unexpected result was the observation that the lowest seismic velocities are not directly beneath the axis of the ridge but to the west of the ridge beneath the Pacific Plate. This finding suggests that the center of magma generation and upwelling may be west of the ridge axis.

Hydrothermal Vent Communities

The submersible *Alvin* carries two scientists and a pilot. It is equipped with cameras inside and out, as well as baskets for samples taken with its mechanical claws (fig. 3.41). It dives at

Figure 3.41 *Alvin's* mechanical arm collects a sample.

*I*n 1998, an ambitious expedition was undertaken by scientists at the University of Washington and the American Museum of Natural History in cooperation with the Canadian Coast Guard to recover black smokers from an active hydrothermal vent system on the Endeavour Segment of the Juan de Fuca Ridge. The Endeavour, located about 300 km (180 mi) off Washington and Vancouver Island, is one of the most active hydrothermal vent regions ever discovered. There are at least four major hydrothermal vent fields along the Endeavour Segment with hundreds of black smokers, some of which emit fluids at temperatures of 400°C (750°F). One giant smoker, Godzilla, stood roughly 45 m (150 ft) tall before it collapsed in 1996. The southernmost field in this area, the Mothra Hydrothermal Field, was chosen as the recovery site because it contains abundant steep-sided smokers that host diverse macrofaunal communities. Venting sulfide structures in this area are up to 24 m (79 ft) high and extend linearly for over 400 m (1300 ft). The tops of black smokers targeted for recovery were initially identified during a preliminary survey conducted in 1997 using the ROV *Jason* and the staffed submersible *Alvin.* The smokers identified as possible targets for retrieval were relatively small; the largest was about 3 m (10 ft) tall and weighed roughly 6800 kg (15,000 lb).

The expedition used the Canadian Remotely Operated Platform for Ocean Science (ROPOS) to recover four sulfide chimneys from a depth of about 2250 m (7400 ft). ROPOS works out of a cage that acts as a garage for the vehicle while it is being lowered to the bottom or brought back to the surface (box fig. 1). The fiberoptic tether that connects the cage to the ship is also used to transmit images from cameras on ROPOS and other data, including piloting commands, to the vehicle. After the cage is lowered to the target area, ROPOS "swims" out on a separate tether connected to the cage. Because ROPOS is on its own tether, it is unaffected by any motion of the cage caused by ship motion at the surface.

Once the expedition had arrived at the site, ROPOS was lowered from the University of Washington's research vessel *Thomas G. Thompson* to the bottom, where it moved to the first target, an inactive black smoker called Phang. Its initial task was to take pictures of this structure to document its characteristics for before-and-after studies. It then cinched cables around the smoker using a recovery cage that had been specially designed to fit over it. The next step was for ROPOS to approach the smoker and cut into its base using a chain saw with carbide and diamond-embedded blocks in the chain (box fig. 2). After a series of cuts had been made, ROPOS was backed away from the chimney. The recovery cage and cables were attached to a line brought over by ROPOS from a previously deployed recovery basket on the bottom holding 2400 m (8000 ft) of line. ROPOS was then brought on deck, and the recovery line was floated to the surface, where it was brought onboard the Canadian Coast Guard's vessel *John P. Tully* and attached to a winch. The *Tully* then took up slack on the line, broke the smoker off its base, and brought it to the surface (box fig. 3), where the structure was cut in half in preparation for geologic and microbiological studies.

This process was repeated three times during the course of the expedition to recover a total of four sulfide chimneys (box fig. 4). The sulfide edifices were chosen in part for their diversity. Phang,

Box Figure 1   The Canadian ROPOS (Remotely Operated Platform for Ocean Science) in its deployment cage.

Box Figure 2   With the recovery cage in place, ROPOS approaches the base of the chimney in preparation for cutting with a carbide and diamond-embedded chain saw.

Box Figure 3 A sulfide chimney breaking the surface during recovery, with the recovery cage surrounding it.

(a)

(b)

Box Figure 4 (a) Several segments of recovered chimneys secured on deck. (b) Section of a chimney that was cut in half for study of its interior.

the dead chimney, did not have any water flowing through it, but another chimney, Finn, was very active, with 304°C (579°F) hydrothermal fluid flowing out. Neither chimney had many large organisms attached. The remaining two chimneys recovered, Roane and Gwenen, were venting fluids at temperatures of 20°–200°C (68°–392°F) and had diverse communities of organisms living on them, including tube worms, limpets, and snails (box fig. 5). These two structures remained hot when brought up from a water depth of over 2000 m (nearly 1.5 mi); Roane had temperatures of 90°C (194°F), and Gwenen had temperatures of 60°C (140°F). Their temperatures were measured from within the structures when they were on deck. The largest chimney recovered is 1.5 m (5 ft) tall and weighs 1800 kg (4000 lb); it is now on display at the American Museum of Natural History in New York City.

The smokers will be studied in detail by geologists, biologists, and chemists to learn more about the extreme chemical and thermal gradients that characterize these environments, the conditions under which sulfide structures grow and evolve, and how nutrients may be delivered to the organisms that live on and within the structures. Such studies are likely to find new species of microorganisms that thrive in these high-temperature, sunlight-free environments. Preliminary studies on these samples indicate that microorganisms living within the structures at temperatures of 90°C (194°F) derive their energy from carbon-bearing species in the hydrothermal fluids and from mineral-fluid reactions within the rocks. These and similar findings are exciting in that they show that microorganisms are capable of living in the absence of sunlight in volcanically active water-saturated regions of our planet. Other hydrothermally active planets may harbor similar life forms.

Box Figure 5 Tube worm colony attached to one of the recovered chimneys.

a speed of 2 knots (2.3 mph) to a depth of 3000 m (10,000 ft) and spends four to five hours on the bottom. In 1977 and 1979, expeditions from Woods Hole Oceanographic Institution (WHOI) returned with the submersible *Alvin* to explore the vent areas along the Galápagos Rift. An underwater camera assembly called ANGUS (Acoustically Navigated Underwater Survey System) was slowly towed above the ocean floor to pinpoint the vent area, and the *Alvin* made twenty-four dives to the rift, which is at a depth of 2500 m (8000 ft).

The most surprising discovery of these dives was the perplexing presence of large communities of animals so far away from surface food sources. Clams, mussels, limpets, tube worms, and crabs were viewed, photographed, and collected. New species were identified; giant tube worms and clams with red blood and flesh similar to beef were collected for laboratory analysis. The presence of so many large animals at such depths immediately brought up the question of their source of food. Such dense communities could not be supported by the fall of organic matter from the sea surface. Instead of depending on plants using the Sun's energy to produce organic material, these ocean-bottom populations rely on bacteria that rely on the hydrogen sulfide and particulate sulfur in the hot vent water to provide their food. The vent animals feed on the bacteria, which were calculated to reach concentrations as high as 0.1–1 g/L. No sunlight is necessary in these chemosynthetic (rather than photosynthetic) communities, and no food is needed from the surface; the populations are sustained by the vents themselves. In areas where the vents have become inactive, the animal communities have died. Since the initial discoveries along the Galápagos Rift, communities of vent organisms have been discovered in association with the Gorda Plate's Juan de Fuca Rift off the coasts of Oregon and Washington, in sites along the Mid-Atlantic Ridge, along the East Pacific Rise off Central America, and at several places in the western Pacific. Further discussion of these communities, the organisms, and their food supply is found in chapter 18.

Summary

Earth is made up of a series of concentric layers: the crust, the mantle, the liquid outer core, and the solid inner core. The evidence for this internal structure comes indirectly from studies of Earth's dimensions, density, rotation, gravity, and magnetic field and of meteorites. It also comes from the ways in which seismic waves change speed and direction as they move through Earth. Seismic tomography is being used to describe Earth's interior layers.

Continental crust is formed from granitic rocks, which are less dense than oceanic crust (formed of basalt). The top of the mantle is fused to the crust to form the rigid lithosphere. The lithosphere floats on the deformable upper mantle, or asthenosphere. The pressures beneath the elevated continents and depressed ocean basins are kept in balance by vertical adjustments of the crust and mantle, a process known as isostasy.

Alfred Wegener's theory of drifting continents was based on the geographic fit of the continents and the similarity of fossils collected on different continents. His ideas were ignored until the discovery of the mid-ocean ridge system and until the proposal of convection cells in the asthenosphere led to the concept of seafloor spreading. New lithosphere is formed at the ridges, or spreading centers. Old lithospheric material descends into trenches at subduction zones. Seafloor spreading is the mechanism of continental drift. Evidence for lithospheric motion includes the match of earthquake zones to spreading centers and subduction zones, the greater heat flow along ridges, age measurement of seafloor rocks, age and thickness measurements of sediment from deep-sea cores, and the magnetic stripes in the sea floor on either side of the ridge system. Rates of seafloor spreading are generally 1–20 cm (0.4–8 in) per year, averaging about 5 cm (2 in) per year. Plate tectonics is the unifying concept of lithospheric motion. Plates are made up of continental and oceanic lithosphere bounded by ridges, trenches, and faults. Plates move apart at divergent plate boundaries and together at convergent plate boundaries.

Rift zones separate ocean basins, and, in the past, they have separated landmasses and produced new ocean basins. Oceanic lithosphere is made up of layers of sediment, fine-grained basalt, vertical basalt dikes, igneous rock, and, under the Moho, mantle rock. Subduction produces island arcs, mountain ranges, earthquakes, and volcanic activity. The passive, or trailing, continental margin is closest to the ridge system; the active, or leading, margin borders a subduction, or collision, zone. Hot spots and polar wandering curves are used to trace plate motions. The mechanism that drives the plates is not fully known, but the most widely accepted idea is that cold, dense subducting oceanic lithosphere pulls plates away from divergent boundaries.

Plate movements traced over the last 225 million years show the breakup of Pangaea. Paleozoic plate movements in the 225 million years prior to Pangaea have recently been estimated using climatological evidence. Much of North America appears to be made of continental fragments, or terranes, around a core continent, or craton.

Submersibles have been used to explore the animal communities and hydrothermal vents found in many areas of the world's oceans. Numerous ongoing programs and projects are exploring and monitoring the formation of new oceanic crust; sampling devices and cameras have been installed at several locations. Deep-sea drilling continues in all the oceans, working to understand the formation of seafloor features and looking for evidence of global environmental change.

Key Terms

All key terms from this chapter can be viewed by term or definition when studied as flashcards on this book's website at www.mhhe.com/sverdrup10e.

density, 51
seismic wave, 51
P-wave, 51
S-wave, 51
refract, 52
inner core, 54
outer core, 54
mantle, 54
seismic tomography, 54
crust, 54
Mohorovičić discontinuity, 54
Moho, 54
granite, 54

basalt, 54
lithosphere, 55
asthenosphere, 55
mesosphere, 55
isostasy, 55
Gondwanaland, 56
continental drift, 56
Pangaea, 56
Laurasia, 56
ridge, 57
rise, 57
rift valley, 57
trench, 57
convection cell, 59
seafloor spreading, 59
spreading center, 59
subduction zone, 59

epicenter, 59
focus, 59
hypocenter, 59
sediment, 60
core, 60
dipole, 61
Curie temperature, 62
paleomagnetism, 62
polar reversal, 62
polar wandering curve, 65
inertia, 65
plate tectonics, 65
divergent plate boundary, 66
convergent plate boundary, 66
transform boundary, 66
transform fault, 66
rift zone, 67

graben, 68
pillow basalt, 68
gabbro, 69
escarpment, 69
fracture zone, 71
andesite, 73
island arc, 73
Wadati-Benioff zone, 73
passive/trailing margin, 73
active/leading margin, 74
spreading rate, 74
hot spot, 74
guyot, 79
transverse/aseismic ridge, 79
craton, 82
terrane, 82
hydrothermal vent, 83

Study Questions

1. What is meant by the term *polar wandering?* Have the magnetic poles actually wandered?
2. Describe the three types of plate boundaries. What processes take place at each type of boundary? In what direction do the plates move at each boundary?
3. What mechanisms have been proposed to account for plate motion?
4. What is the difference between the leading edge and the trailing edge of a continent? Between a divergent plate boundary and a convergent plate boundary?
5. If the ability of the oceanic crust to transmit heat were uniform, the rate of heat flow through the ocean floor would depend only on the temperature change across the oceanic crust. Under such a condition, how would the heat flow measurements in figure 3.12 indicate the presence of ascending convection cells in the asthenosphere?
6. If the polar wandering curves for North America and Europe are made to coincide, how will these continents move relative to each other?
7. Using the techniques and reasoning employed to discover the properties of the interior of Earth, explain how you would determine what is inside a sealed box (for example, measuring the box, weighing it, spinning it, balancing it on different axes, sampling its exterior). What clue would each of these measurements give you to the contents of the box?
8. What had to be learned about Earth before Alfred Wegener's ideas could be accepted?
9. Why does a newly formed mid-ocean volcanic island gradually subside?
10. Explain the formation and symmetry of the magnetic stripes found on either side of the mid-ocean ridge system. What is their significance when the magnetic information is correlated with the age of the crust?
11. Under what conditions will a convergent boundary form a mountain range? An island arc system? Why do volcanoes associated with subduction zones usually erupt more explosively than mid-ocean volcanoes associated with hot spots and spreading centers?
12. On an outline map of the world draw in (a) earthquake belts, (b) mid-ocean ridges, and (c) trenches. Relate your map to figure 3.8. What do you conclude?
13. How have recent advances in seismic tomography modified our ideas of Earth's internal layers, as shown in figure 3.3?
14. Why do P-waves pass through Earth's outer core?
15. What is a terrane? What role do terranes play in our understanding of today's continents?

Study Problems

1. If a plate moves away from a spreading center at the rate of 5 cm/yr, what is the displacement of a landmass carried by that plate after 180×10^6 years?
2. Magnetic stripes with the same magnetic orientation are measured on either side of a ridge crest. The stripe on the west side of the ridge is displaced 11 km from the crest; the stripe on the east side is displaced 9 km from the crest. The age of the rock in both stripes is 4×10^5 years. Calculate the average spreading rate at this ridge.

Expanse of clear tropical water and shallow marine sediment in the Maldives, Indian Ocean.

Chapter 4

The Sea Floor and Its Sediments

Chapter Outline

Learning Outcomes

After studying the information in this chapter students should be able to:

1. *review* the evolution of methods to measure ocean depth from the time of the Greeks to the present,
2. *construct* a simple cross section of an ocean basin, including both a passive and active continental margin,
3. *discuss* the formation of atolls,
4. *sketch* the location of ocean ridges and trenches,
5. *explain* three different ways to classify sediments,

6. *list* the organisms that produce the majority of calcareous and siliceous sedimentary particles,
7. *identify* where biogenous and lithogenous sediments are dominant on the sea floor,
8. *define* isotopes and describe how they can be used with marine sediments as historical records,
9. *list* multiple seabed resources and *appraise* the extent to which they are currently being recovered, and
10. *write* a short history of the evolution of the Law of the Sea.

Early mariners and scholars believed that the oceans were large basins or depressions in Earth's crust, but they did not conceive that these basins held features that were as magnificent as the mountain chains, deep valleys, and great canyons of the land. As maps became more detailed and as ocean travel and commerce increased, measurement of water depths and recording of seafloor features in shallower regions became necessary to maintain safe travel and ocean commerce. The secrets of the deeper oceanic areas had to wait for hundreds of years until the technology of the late twentieth century made it relatively easy to map and sample the sea floor. It was only then that large numbers of survey vessels accumulated sufficient data to provide the details of this hidden terrain.

What we know about the sea floor and its covering of sediments comes almost entirely from the observations by surface ships; more recently, submersibles, robotic devices, and satellites have added to our knowledge. Some areas of the sea floor have been measured in great detail; charts of other areas have been made from scanty data. The demand for more measurements to describe and explain the features of the sea floor continues.

In this chapter, we survey the world's ocean floors and discuss their topography and geology. We examine the sources, types, and sampling of sediments and discuss seabed mineral resources.

4.1 Measuring the Depths

In about 85 B.C., a Greek geographer named Posidonius set sail, curious about the depth of the ocean. He directed his crew to sail to the middle of the Mediterranean Sea, where they eased a large rock attached to a long rope over the side. They lowered it nearly 2 km (1.2 mi) before it hit bottom and answered Posidonius's question. Crude as this method was, it continued with minor modifications as the means of obtaining **soundings,** or depth measurements, for the next 2000 years.

An early modification made by nineteenth-century surveyors was the use of hemp line or rope with a greased lead

weight at its end. This line was marked in equal distances (usually **fathoms;** a fathom is the length between a person's fully outstretched hands, standardized at 6 ft). The change in line tension when the weight touched bottom indicated depth, and the particles from the bottom adhering to the grease confirmed the contact and brought a bottom sample to the surface. This method was quite satisfactory in shallow water, and the experienced captain used the properties of the bottom sample to aid in navigation, particularly at night or in heavy fog.

Later, piano wire with a cannonball attached was used in deep water. These nineteenth-century surveyors used a mechanical sounding machine that allowed the rope or wire to free-fall. Using a clock, they carefully timed the rate at which premeasured lengths of the sounding line paid out over the side of the ship. This rate was a constant as long as the weight was free-falling toward the bottom but when the rate of payout decreased abruptly, they knew the weight had hit the bottom. In this way, they were able to obtain reasonably precise soundings in deep water even when the ship was moving at a slow speed (see fig. 1.13a). The time required (eight to ten hours) to let the weight free-fall to the bottom and then winch it back up in deep water was so great that by 1895, only about 7000 measurements had been made in water deeper than 2000 m (6600 ft) and only 550 measurements had been made of depths greater than 9000 m (29,500 ft).

It was not until the 1920s, when acoustic sounding equipment was invented, that deep-sea depth measurements became routine. The **echo sounder,** or **depth recorder,** which measures the time required for a sound pulse to leave the surface vessel, reflect off the bottom, and return, allows continuous measurements to be made easily and quickly when a ship is underway. The behavior of sound in seawater and its uses as an oceanographic tool are discussed in chapter 5. A trace from a depth recorder is shown in figure 4.1. (See figs. 5.17 and 5.18 for an illustration of how sound can be used to measure water depth and a picture of a precision depth recorder.)

In 1925, the German vessel *Meteor* made the first large-scale use of an echo sounder on a deep-sea oceanographic research cruise and detected the Mid-Atlantic Ridge for the first

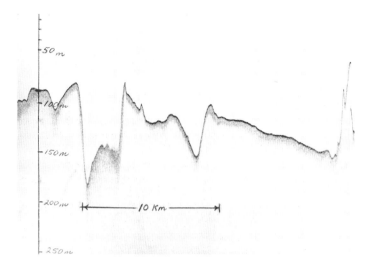

Figure 4.1   Depth recorder trace. A sound pulse reflected from the ocean floor traces a depth profile as the ship sails a steady course. The horizontal scale depends on ship speed.

time. After this expedition, depth measurements gradually accumulated at an ever-increasing rate. As the acoustic equipment improved and was used more frequently, knowledge of the ocean floor's bathymetry expanded and improved, culminating in the 1950s with the first detailed mapping of the mid-ocean ridge and trench systems.

Today, a wide variety of methods are used to obtain even more detailed seafloor bathymetry at scales that range from centimeters (inches) to thousands of kilometers (thousands of miles). The specific technique used depends on the amount of time that can be spent, the scale of the feature that is being examined, and the amount of detail that is required. When necessary, direct observation of small-scale structures is possible with the use of staffed submersibles and remotely operated vehicles (ROVs) carrying video cameras. These images can be transmitted to surface ships and relayed by satellite anywhere in the world in real time (fig. 4.2). Investigations by staffed submersibles or ROVs provide great detail, but they typically cover very small areas and are both time-consuming and expensive for the amount of sea floor surveyed. On large scales of tens or hundreds of square kilometers, sophisticated multibeam sonar systems can rapidly map extensive regions at relatively low cost with great accuracy (see fig. 2.14).

Detailed bathymetric data can be collected in shallow coastal water using an airborne laser. The laser airborne depth sounder (LADS) system is flown in a small fixed-wing aircraft 350–550 m (1200–1800 ft) over the surface; the exact position of the aircraft is determined by the Global Positioning System (GPS). The laser is used to measure the distance between the aircraft and the sea floor. The LADS system can take up to 900 soundings per second (3.24 million soundings per hour). Individual measurements typically are taken on 5 × 5 m (16.5 × 16.5 ft) spacing in a swath 240 m (790 ft) wide along the survey line. For greater detail, the spacing can be reduced to 2 × 2 m (6.5 × 6.5 ft). Because light is rapidly attenuated in water, LADS has an operational depth range of

Figure 4.2   This internally recording television camera is placed on the sea floor, where it automatically photographs events until it is retrieved by the researchers. The *red box* contains the camera's power supply.

0.5–70 m (1.5–230 ft). The actual maximum depth in a specific area depends on water clarity. In pristine coral reef environments, soundings as great as 70 m (230 ft) can be obtained, whereas in clear to moderately turbid coastal waters, the effective depth of penetration decreases to 20–50 m (66–164 ft). In very turbid water, the system is restricted to 0–15 m (0–49 ft).

Very-large-scale seafloor surveys use satellite measurements of changes in sea surface elevation caused by changes in Earth's gravity field due to seafloor bathymetry. These changes in sea surface elevation can be detected by radar altimeters that measure the distance between the satellite and the sea surface. The sea surface is not flat even when it is perfectly calm. Changes in gravity caused by seafloor topography create gently sloping hills and valleys in the sea surface. The excess mass of features such as seamounts and ridges creates a gravitational attraction that draws water toward them, resulting in a higher elevation of the sea surface. Conversely, the deficit of mass along deep-ocean trenches, and subsequent weaker gravitational attraction, results in a depression of the sea surface as water is drawn away

Bathymetrics

Visualizing the sea floor began with single soundings made with a lead line and continued with simple echo sounders and contour maps drawn by hand. In one hour, an individual with a lead line could take twenty measurements in water 10 m (33 ft) deep and in four hours, only one measurement in water 4000 m (13,200 ft) deep; the echo sounder allowed 36,000 measurements to be made each hour in 10 m of water and 680 in the same time in 4000 m of water. Today's multibeam sound systems can take 293,000 measurements per hour in 10 m of water and 20,000 measurements in 4000 m. Advances in multibeam sound system technology and improved computer graphics, combined with satellite navigation for precise positioning, are opening dramatic new windows to the sea floor.

A single sound beam device releases a cone of sound; as the depth of the water increases, the area of the sea floor from which the echo is reflected also increases. Depth is averaged over the "footprint" of the sound beam; therefore, seafloor features smaller than the footprint are difficult to detect and detail is reduced. Two new technologies employing multiple sound devices are being used to produce detailed, high-resolution seafloor maps: (1) side-scan acoustical imaging and (2) swath bathymetry.

Side-scan measurements can be made from either a surface vessel or a submerged system towed behind a vessel. If the ship is pitching and rolling, the path of the sound beam from a surface vessel will be displaced from its intended direction, resulting in inaccurate data. A towed system is below the depth of surface waves and winds; it is also closer to the sea floor, allowing the use of a conical sound beam that produces a smaller sound footprint. The smaller footprint increases the detail that is imaged but decreases the scanned area for each cone of sound. The area surveyed is increased by sending out multiple sound beams obliquely on either side of the sound device; no image is obtained from directly under the side scanner.

Side-scan acoustical images are the product of the reflectivity of the seafloor materials and the angle at which the sound beams strike the sea floor. Changes in the reflection of the sound come from the irregularities and the changing properties of the bottom being scanned. The sides of a seamount, a fault, and other objects with strong topographic relief act as good reflectors.

Side-scan sonar systems that are housed in torpedo-like casings and towed behind ships are known as towfish; GLORIA (geological long-range inclined asdic) is one of the most sophisticated. It is towed at 10 knots (nautical miles per hour), has a depth capability of 5000 m (16,400 ft), and scans the sea floor with two sound beams 30 km (18 mi) wide. The sound beams are composed of sound pulses that last four seconds, with forty-second intervals between pulses to allow the echo to return to the towfish for recording (box fig. 1).

Side-scan acoustical imaging also works very well to detect sunken ships, planes, or other structures, because the reflecting surfaces of these structures are at an angle to the sea floor, and their acoustical properties are very different from those of the sea floor. The object's shape is accompanied by an acoustical shadow (seen behind the plane in box fig. 2) that provides strong image definition and indicates elevation above the sea floor.

Box Figure 2 The image of a plane on the sea floor obtained by side-scan sonar. Note the shadow generated when sound was not returned from the seabed.

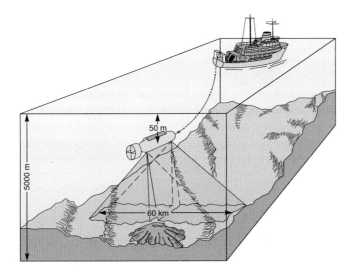

Box Figure 1 A surface vessel tows a side-scan sonar system, or towfish, to acoustically map a swath of sea floor. The base of the *darker blue* triangular prism shows the seafloor area covered by the sound beams. Vertical scale is distorted for clarity.

The GLORIA surveys of the 1980s produced the first accurate maps of the deep-sea floor within U.S. resource limits, but it was too slow and the sound beam was too narrow to obtain detailed information over the shallow continental shelves. In the 1990s, the U.S. Geological Survey (USGS) began a high-resolution survey of U.S. continental shelves using a system based on multiple sound sources known as swath bathymetry. Swath bathymetry determines depth and produces images by analyzing the sound interference patterns between outgoing sound beams and returning echoes. This method uses hull-mounted sonar arrays of 60 to more than 150 electronically separated sound sources mounted along the length of the hull and multiple sound receivers mounted across the hull. Both sound sources and receivers are directed vertically downward. The swath length is the length of the area returning echoes. The width of the swath is in the direction of the vessel's motion and is determined by the sound cone angle. A computer analyzes the sound interference patterns between outgoing sound pulses and their returning echoes. Once a computer image has been assembled, the image can be viewed from any angle and enhanced with color to indicate depth (see fig. 1.13). The ship is positioned by GPS, and measurements can be collected at vessel speeds in excess of 15 knots (17 mi/h).

Between 1994 and 1997, the USGS mapped five areas using swath bathymetry: Massachusetts Bay, parts of the continental shelf off New York, the Hudson River, the coastal margin of Santa Monica, and the central portion of San Francisco Bay. In 1998, the project continued, mapping portions of the slopes of the Hawaiian Islands and the shelf areas of San Diego and Newport, California.

The images obtained with these multibeam systems are providing highly detailed and accurate images of the sea floor, similar to those seen in aerial photographs. These maps are providing basic information, fundamental for geologic research and biological management of the coastal margin.

To Learn More About Bathymetrics

Gardner, J., P. Butman, and L. Mayer. 1998. Mapping U.S. Continental Shelves. *Sea Technology* 39 (6): 10–17.

Pratson, L. F., and W. F. Haxby. 1997. Panoramas of the Seafloor. *Scientific American* 276 (6): 82–87.

toward surrounding areas with greater gravitational attraction. Sea level over large seamounts is elevated by as much as 5 m (16 ft) and over ocean ridges by about 10 m (33 ft); it is depressed over trenches by about 25–30 m (80–100 ft). These changes in elevation occur over tens to hundreds of kilometers, so the slopes are very gentle. The sea surface is always perpendicular to the local direction of gravity, so precise measurements of the slope of the surface can be used to determine the direction and magnitude of the gravitational field at any point. Because these changes in gravity are related to seafloor topography, it is possible to use them to reconstruct a bathymetry that produces the observed variations in sea surface topography (fig. 4.3). Tides, currents, and changes in atmospheric pressure can cause undulations of more than a meter (3 feet) in the ocean surface. These effects are filtered out to produce the bathymetric details. Bathymetric features with "footprints," or horizontal dimensions as small as about 10 km (6.2 mi), can be resolved with satellite altimetry data. Satellite maps are particularly valuable in the Southern Ocean, where the weather and sea conditions are frequently bad and it is difficult to conduct general bathymetric surveys to locate areas of scientific interest.

4.2 Bathymetry of the Sea Floor

The area below the sea surface is as rugged as any land above it. The Grand Canyon, the Rocky Mountains, the desert mesas in the Southwest, and the Great Plains all have their undersea counterparts. In fact, the undersea mountain ranges are longer, the valley floors are wider and flatter, and the canyons are often deeper than those found on land. Features of land topography, such as mountains and canyons, are continually and aggressively eroded by wind, water, ice, changes in temperature, and the chemical alteration of minerals in rocks. The erosion of ocean bathymetry is generally slow. Physical weathering is accomplished primarily by waves and currents, and chemical erosion occurs by the dissolution of minerals. More rapid erosion is generally restricted to the continental margin, as discussed in section 4.3.

The most important agents of physical change on the deep-sea floor are the gradual burial of features by a constant rain of sediments falling from above and volcanism associated with the mid-ocean ridge system, hot spots, island arcs and active seamounts, and some abyssal hills. Movements of Earth's crust may displace features and fracture the sea floor, and the weight of some underwater volcanoes may cause them to subside, but the appearance of the bathymetric features of the ocean basins and sea floor has remained much the same through the last 100 million years. Computer-drawn profiles of crustal elevations across the United States and the Atlantic Ocean are shown in figure 4.4. At 40°N latitude, the height above zero elevation (*dotted line*) and width of mountains in the western United States are about the same as the height and width of the Mid-Atlantic Ridge system on the sea floor. Compare the topography of the Rocky Mountains and the undersea peaks in this figure.

Continental Margin

The edges of the landmasses below the ocean surface and the steep slopes that descend to the sea floor are known as the **continental margin.** There are two basic types of continental margins: passive, or Atlantic, margins and active, or Pacific, margins. Passive margins have little seismic or volcanic activity and

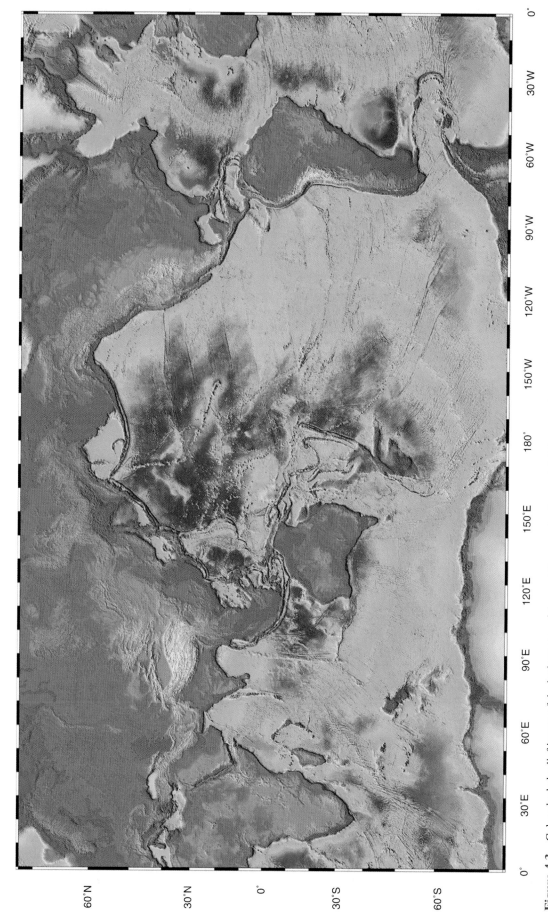

Figure 4.3 Color-shaded relief image of the bathymetry of the world's ocean basins modeled from marine gravity anomalies mapped by satellite altimetry and checked against ship depth soundings.

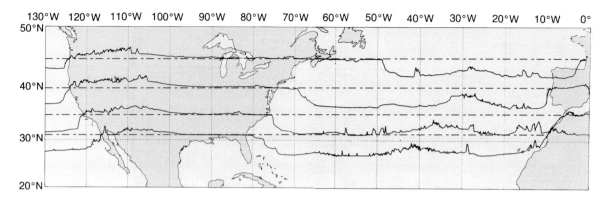

Figure 4.4   Computer-drawn topographic profiles from the western coast of Europe and Africa to the Pacific Ocean. The elevations and depths above and below 0 m are shown along a line of latitude by using the latitude line as zero elevation. For example, the ocean depth at 40°N and 60°W is 5040 m (16,531 ft). The vertical scale has been extended about 100 times the horizontal scale. If both the horizontal and vertical scales were kept the same, a vertical elevation change of 5000 m would measure only 0.05 mm (0.002 in).

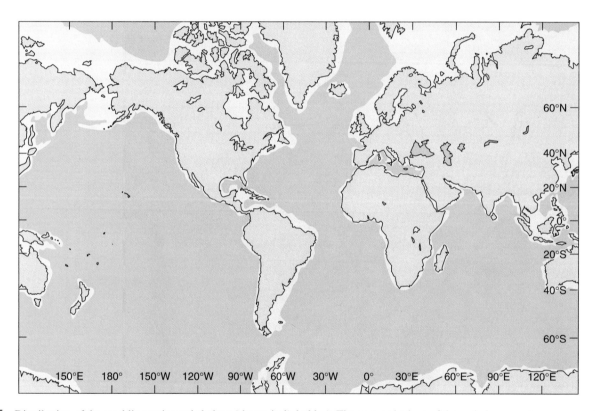

Figure 4.5   Distribution of the world's continental shelves (shown in *light blue*). The seaward edges of these shelves are at an average depth of approximately 130 m (430 ft).

involve a transition from continental crust to oceanic crust in the same lithospheric plate. They form after continents are rifted apart, creating a new ocean basin between them. Passive margins tend to be relatively wide. Active margins are tectonically active and associated with earthquakes and volcanism. Most are associated with plate convergence and subduction of oceanic lithosphere beneath a continent. Active margins are plate boundaries and are frequently relatively narrow. The continental margin is made up of the continental shelf, shelf break, slope, and rise. The **continental shelf** lies at the edge of the continent; continental shelves are the nearly flat borders

of varying widths that slope very gently toward the ocean basins. Shelf widths average about 65 km (40 mi) but are typically much narrower along active margins than passive margins. The width of the continental shelf can be as much as 1500 km (930 mi). Water depth at the outer edge of the continental shelf varies from 20–500 m (65–1640 ft), with an average of about 130 m (430 ft).

The distribution of the world's continental shelves is shown in figure 4.5. The width of the shelf is often related to the slope of the adjacent land; it is wide along low-lying land and narrow along mountainous coasts. Note the narrow shelf along the western coast of South America and the wide

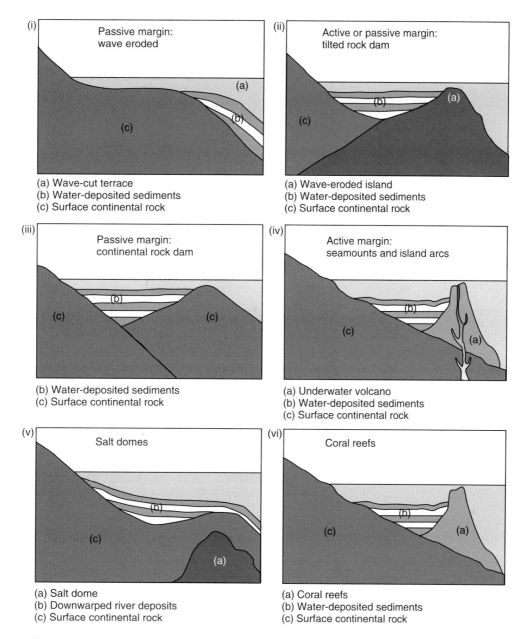

Figure 4.6 Examples of how continental shelves are formed by trapping land-derived sediments at the edge of the continental landmasses.

expanses of continental shelf along the eastern and northern coasts of North America, Siberia, and Scandinavia. The continental shelves are geologically part of the continental crust; they are the submerged seaward edges of the continents. Several processes contribute to the formation of continental shelves. Storm waves may erode continental shelves (fig. 4.6*i*), and, in some areas, natural dams trap sediments between the offshore dam and the coast (fig. 4.6*ii, iii*). Seamounts and island arcs (fig. 4.6*iv*) and coral reefs (fig. 4.6*vi*) also trap sediments. In the Gulf of Mexico, sediments from the land are trapped behind low ridges and salt domes on the sea floor (fig. 4.6*v*). Along the northeastern coast of North America, sediments are trapped behind upturned rock near the outer edges of the continental shelf (fig. 4.6*ii*).

During past ages, the shelves have been covered and uncovered by fluctuations in sea level. During the glacial ages of the Pleistocene epoch, a number of short-term changes occurred in sea level, some of which were greater than 120 m (400 ft). When sea level was low, erosion deepened valleys, waves eroded previously submerged land, and rivers left sediments far out on the shelf. When the glacial ice melted, these areas were flooded, and sediments built up in areas closer to the new shore. At present, although submerged, these areas still show the scars of old riverbeds and glaciers acquired when the land was above water. Today, some continental shelves are covered with thick deposits of silt, sand, and mud sediments derived from the land: examples are offshore from the mouths of the Mississippi and Amazon Rivers, where large amounts of such sediments are deposited annually. Other shelves are bare of sediments, such as where the fast-moving Florida Current sweeps the tip of Florida, carrying the sediments northward to the deeper water of the Atlantic Ocean.

The boundary of the continental shelf on the ocean side is determined by an abrupt change in slope and a rapid increase in depth. This change in slope is referred to as the **continental shelf break;** the steep slope extending to the ocean basin floor is known as the **continental slope.** These features are shown in figure 4.7. The angle and extent of the slope vary from place to place. The slope may be short and steep (for example, the depth may increase rapidly from 200 m [650 ft] to 3000 m [10,000 ft], as in fig. 4.7), or, along an active margin,

it may drop as far as 8000 m (26,000 ft) into a great deep-seafloor depression or trench (for example, off the western coast of South America, where the narrow continental shelf is bordered by the Peru-Chile Trench). The continental slope may show rocky outcroppings and be relatively bare of sediments because of its steepness, tectonic activity, or a low supply of sediments from land.

The most outstanding features of the continental slopes are **submarine canyons.** These canyons sometimes extend up, into, and across the continental shelf. A submarine canyon is steep-sided and has a V-shaped cross section, with tributaries similar to those of river-cut canyons on land. Figure 4.8*a* shows the Monterey and Carmel Canyons off the coast of California.

Figure 4.7   A typical profile of a passive continental margin. Notice both the vertical and horizontal extent of each subdivision. The average slope is indicated for the continental shelf, slope, and rise. The vertical scale is 100 times greater than the horizontal scale (V.E. = 100 times).

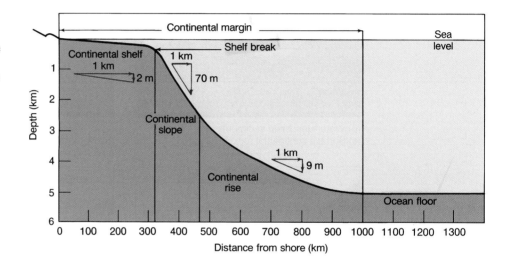

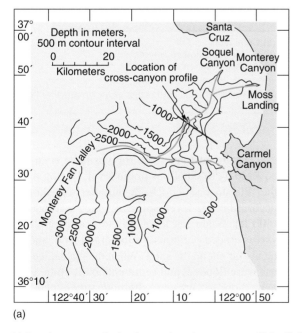

(a)

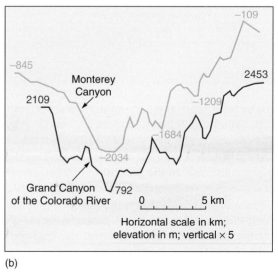

(b)

Figure 4.8   (a) Depth contours depict three submarine canyons off the California coast as they cut across the continental slope and continental shelf. The axes of the canyons, which merge seaward, are indicated by the *blue line.* (b) Cross-canyon profile, along the *red line* in (a), of the Monterey Canyon. Compare this profile to that of the Grand Canyon drawn to the same scale.

Figure 4.8*a* is a bathymetric chart; figure 4.8*b* compares the profile of the Monterey Canyon with the profile of the Grand Canyon of the Colorado River. A submarine canyon is also shown in figures 2.12 and 2.13.

Many of these submarine canyons are associated with existing river systems on land and were apparently cut into the shelf during periods of low sea level, when the glaciers advanced and the rivers flowed across the continental shelves. Ripple marks on the floor of the submerged canyons and sediments fanning out at the ends of the canyons suggest that they were formed by moving flows of sediment and water called **turbidity currents.** These

Figure 4.9 This beach cliff shows an ancient turbidite deposit that has been uplifted and then exposed by wave erosion. Turbidites are graded deposits, with the largest particles in the deposit at the bottom of the turbidite and the smallest at the top.

sediment-laden currents can travel at speeds up to 90 km (56 mi) per hour and carry in suspension up to 300 kg of sediment per cubic meter (18.7 lb/ft^3). Caused by earthquakes or the overloading of sediments on steep slopes, turbidity currents are fast-moving avalanches of mud, sand, and water that flow down the slope, eroding and picking up sediment as they gain speed. In this way, the currents erode the slope and excavate the submarine canyon. As the flow reaches the bottom, it slows and spreads, and the sediments settle. Because of their speed and turbulence, such currents can transport large quantities of materials of mixed sizes. The settling process produces graded beds of coarse material overlain (upward) by smaller particles. These graded deposits are called **turbidites.** Figure 4.9 shows a turbidite preserved in compacted seafloor sediments that have been uplifted and exposed by wave erosion. These large and occasional currents have never been directly observed, although similar but smaller and more continuous flows, such as sand falls, have been observed and photographed (fig. 4.10).

Research on turbidity currents began with laboratory experiments in the 1930s. These experiments were based on earlier observations of the silty Rhone River water moving along the bottom of Lake Geneva in Switzerland. Later analysis of a 1929 earthquake that broke transatlantic telephone and telegraph cables on the continental slope and rise off the Grand Banks of Newfoundland showed a pattern of rapid and successive cable breaks high on the continental slope, followed by a sequence of downslope breaks. These breaks were calculated to have been caused by a turbidity current that ran for 800 km (500 mi) at speeds of 40–55 km (25–35 mi) per hour. Later samples taken from the area showed a series of graded sediments at the end of the current's path. Searches of cable company records showed similar patterns of cable breaks in other parts of the world.

A turbidity flow can be demonstrated by placing water and loose sediments of mixed particle size in a 6 ft section of 2–3 in-diameter clear plastic tube. Cap both ends securely and stand the tube on end for a day or so. Then carefully tilt the tube until it is horizontal and slowly elevate the end with the sediments. Tap the tube gently as you raise it, and the sediments will move down slope as a turbidity flow. When the sediment settles at the other end, a turbidite pattern is formed.

At the base of the steep continental slope may be a gentle slope formed by the accumulation of sediment. This portion of the sea floor is the **continental rise,** made up of sediment deposited by turbidity currents, underwater landslides, and any other processes that carry sands, muds, and silt down the continental slope. Continental rises may be compared to the landforms known as alluvial fans found where sediments from steep canyons

Figure 4.10 Sand fall in San Lucas submarine canyon. Sand moving down the continental slope is directed seaward to the ocean basin floor.

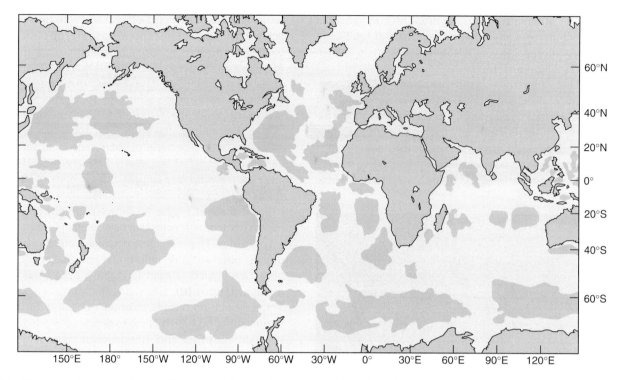

Figure 4.11 The major deep-ocean basins of the world (*dark blue*) are separated by ridges, rises, and continents.

Ocean Basin Floor

spread across a valley floor. The continental rise is a conspicuous feature at passive margins in the Atlantic and Indian Oceans and around the Antarctic continent. Few continental rises occur in the Pacific Ocean, where active margins border the great seafloor trenches located at the base of the continental slope. Refer to figure 4.7 to see the relationship of the continental rise to the continental slope.

Ocean Basin Floor

The true oceanic features of the sea floor occur seaward of the continental margin. The deep-sea floor, between 4000 and 6000 m (13,000 and 20,000 ft), covers more of Earth's surface (30%) than do the continents (29%). In many places, the ocean basin floor is a vast plain extending seaward from the base of the continental slope. It is flatter than any plain on land and is known as the **abyssal plain.** The abyssal plain is formed by sediments that fall from the surface and are deposited by turbidity currents to cover the irregular topography of the oceanic crust. An area of the abyssal plain that is isolated from other areas by continental margins, ridges, and rises is known as a basin, and some basins may be subdivided into subbasins by ridge and rise subsections. The distribution of these basins and subbasins is shown in figure 4.11. Low ridges allow some exchange of deeper water between adjacent basins, but if the ridge is high, both the deep water and the deep-dwelling marine organisms within the basin are effectively cut off from other basins. For example, the deep water of the Angola Basin, in the bight of the western coast of Africa, is cut off from the Brazil Basin to the west by the Mid-Atlantic Ridge and from the Cape Basin to the south by the Walvis Ridge. A physio-

graphic chart of the sea floor (see fig. 3.8) provides another view of these basins.

Abyssal hills and **seamounts** are scattered across the sea floor in all the oceans. Abyssal hills are less than 1000 m (3300 ft) high, and seamounts are steep-sided volcanoes rising abruptly and sometimes piercing the surface to become islands. These features are shown in figure 4.12. Abyssal hills are probably Earth's most common topographic feature. They are found over 50% of the Atlantic sea floor and about 80% of the Pacific floor; they are also abundant in the Indian Ocean. Most abyssal hills are probably volcanic, but some may have been formed by other movements of the sea floor. Submerged, flat-topped seamounts, known as **guyots,** are found most often in the Pacific Ocean; a guyot is also shown in figure 4.12. The tops of Pacific guyots are 1000–1700 m (3300–5600 ft) below the surface; many are at the 1300 m (4300 ft) depth. Many guyots show the remains of shallow-water coral reefs and evidence of wave erosion at their summits. These features indicate that at one time they were warm-water surface features and that their flat tops are the result of wave erosion. They have since subsided owing to their weight, the accumulated rock load bearing down on the oceanic crust, and the natural subsidence of the sea floor with increasing age as the crust cools and grows in density as it moves farther away from the ridge where it formed. They have also been submerged by rising sea level during periods when glacial ice melted on land.

In the warm waters of the Atlantic, Pacific, and Indian Oceans, coral reefs and coral islands are formed in association with seamounts. Reef-building corals are warm-water animals that require a place of attachment and grow in intimate association with a single-cell, plantlike organism; reef-building corals are

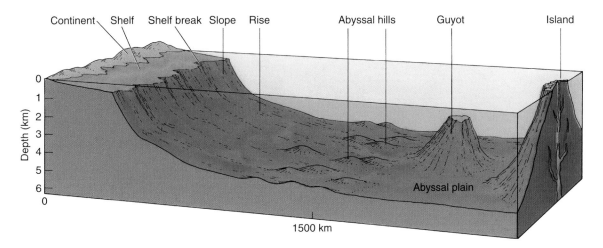

Figure 4.12 An idealized portion of ocean basin floor with abyssal hills (less than 1000 m of elevation), a guyot (a flat-topped seamount), and an island on the abyssal plain. The island was previously a seamount before it reached the surface. Seamounts and guyots are known to be volcanic in origin (vertical × 100).

confined to sunlit, shallow tropical waters. When a seamount pierces the sea surface to form an island, it provides a base on which the coral can grow. The coral grows to form a **fringing reef** around the island. If the seamount sinks or subsides slowly enough, the coral continues to grow upward at a rate that is not exceeded by the rising water, and a **barrier reef** with a lagoon between the reef and the island is formed. If the process continues, eventually the volcanic portion of the seamount disappears below the surface and the coral reef is left as a ring, or **atoll.** This process is illustrated in figure 4.13.

On the basis of the observations made during the voyage of the *Beagle* from 1831–36, Charles Darwin suggested that these were the steps necessary to form an atoll. Darwin's ideas have been proved to be substantially correct by more recent expeditions when drilling through the debris on a lagoon floor found the basalt peak of a seamount that once protruded above the sea surface. The organisms that inhabit coral reefs are discussed in chapter 18.

Ridges, Rises, and Trenches

The most notable features of the ocean floor are the mid-ocean ridge and rise systems stretching for 65,000 km (40,000 mi) around the world and running through every ocean. Their origin and role in plate tectonics were discussed in chapter 3 (see fig. 3.8). Review their distribution using figure 4.14. Recall also the roles played by the rift valleys and transform faults.

The relationship of the deep-sea trenches to plate tectonics was also discussed in chapter 3 (see fig. 3.8). Use figure 4.15 to trace the Japan-Kuril Trench, the Aleutian Trench, the Philippine Trench, and the deepest ocean trench, the Mariana Trench. All these trenches are associated with **island arc systems.** The Challenger Deep, a portion of the Mariana Trench, has a depth of 11,020 m (36,150 ft), making it the deepest known spot in all the oceans. The longest of the trenches is the Peru-Chile Trench, stretching 5900 km (3700 mi) along the western side of South America. To the north, the Middle America Trench borders Central America. The Peru-Chile and Middle America Trenches

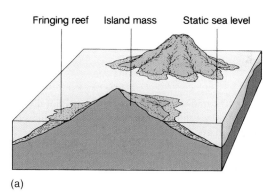

(a)

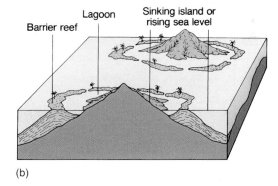

(b)

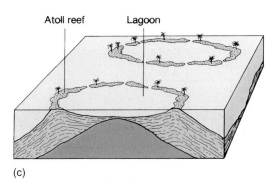

(c)

Figure 4.13 Types of coral reefs and the steps in the formation of a coral atoll shown in profile. (a) Fringing reef. (b) Barrier reef. (c) Atoll reef.

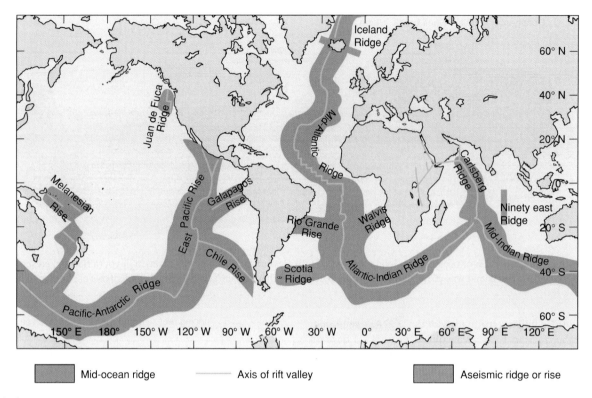

Figure 4.14 The mid-ocean ridge and rise system of divergent plate boundaries. Locations of major aseismic (no earthquakes) ridges and rises are added. Aseismic ridges and rises are elevated linear features thought to be created by hot-spot activity.

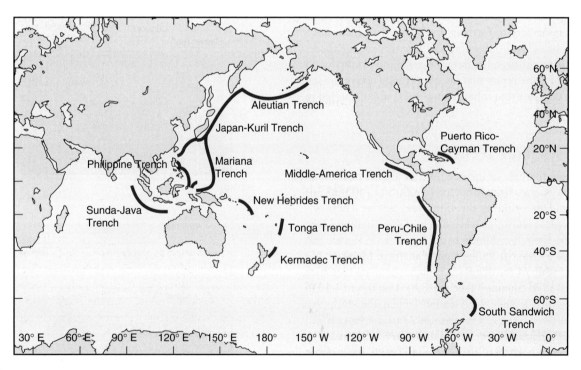

Figure 4.15 Major ocean trenches of the world. The deepest ocean depth is 11,020 m (36,150 ft), east of the Philippines in the Mariana Trench. It is known as the Challenger Deep.

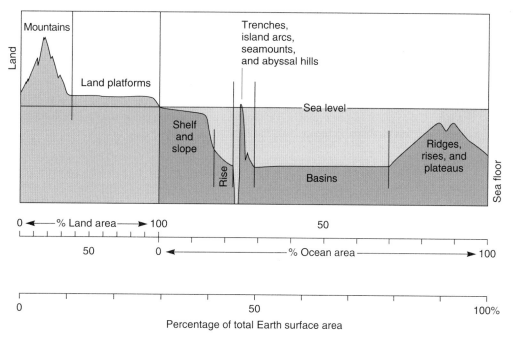

Figure 4.16 Earth's main topographic features shown as percentages of Earth's total surface and as percentages of the land and of the oceans.

are associated with volcanic chains on land. In the Indian Ocean, the great Sunda-Java Trench runs for 4500 km (2800 mi) along Indonesia. In the Atlantic, there are only two comparatively short trenches: the Puerto Rico-Cayman Trench and the South Sandwich Trench, both associated with chains of volcanic islands. To view the bathymetry of the ocean floor as it is known to exist today, see figure 3.8.

In figure 4.16, the topography of the land and the bathymetry of the sea floor are summarized as percentages of Earth's area. Compare the tectonically active areas of trenches and ridges, as well as the area of low-lying land platforms with the area of the ocean basins.

4.3 Sediments

The margins of the continents and the ocean basin floors receive a continuous supply of particles from many sources. Whether these particles have their origin in living organisms, the land, the atmosphere, or the sea itself, they are called sediment when they accumulate on the sea floor. The thickest deposits of sediment are generally found near the continental margins, where sediment is deposited relatively rapidly; in contrast, the deep-sea floor receives a constant but slow accumulation of sediment that produces a thinner layer that varies in thickness with the age of the oceanic crust.

Oceanographers study the rate at which sediments accumulate, the distribution of sediments over the sea bottom, their sources and abundance, their chemistry, and the history they record in layer after layer as they slowly but continuously accumulate on the ocean floors. To describe and catalog the sediments, geologic oceanographers classify sediments by particle size, location, origin, and chemistry.

Table 4.1	Sediment Size Classifications	
Descriptive Name		**Diameter (mm)**
Gravel	Boulder	> 256
	Cobble	64–256
	Pebble	4–64
	Granule	2–4
Sand	Very coarse	1–2
	Coarse	0.5–1
	Medium	0.25–0.5
	Fine	0.125–0.25
	Very fine	0.0625–0.125
Mud	Silt	0.0039–0.0625
	Clay	< 0.0039

Particle Size

Sediment particles are classified by size, as indicated in table 4.1. Familiar terms such as *gravel, sand,* and *mud* are used to identify broad size ranges of large, intermediate, and small particles, respectively. Within each of these ranges, particles are further ranked to produce a more detailed scale from boulders to the very smallest clay-sized particles, which can only be seen with a microscope.

When a sediment sample is collected, it can be dried and shaken through a series of woven-mesh sieves of decreasing opening size. Material that passes through one sieve but not the next is classified by one of the sizes listed in table 4.1.

A sample is said to be "well sorted" if it is nearly uniform in particle size and "poorly sorted" if it is made up of many

Field Notes

Giant Hawaiian Landslides
by Dr. David Clague

Dr. David Clague is a senior scientist at the Monterey Bay Aquarium Research Institute in Moss Landing, California. A marine geologist, Dr. Clague's primary research interests are the formation and degradation of ocean volcanoes.

A short paper was published in 1964 in which some lumpy topography on the deep sea floor northeast of Oahu was hypothesized to be a cluster of landslide blocks. The author of that paper, Dr. James Moore, then the scientist-in-charge of the Hawaiian Volcano Observatory, realized that his idea would be controversial because the largest of the blocks measured nearly 30 km (18.6 mi) long, 5 km (3 mi) wide, and more than 2 km (1.25 mi) tall. The obvious implication was that huge parts of the flanks of Hawaiian volcanoes could fail as giant landslides.

The paper triggered a lively debate for about seven years in which other scientists in general agreed with the observational data but argued that the huge block was a submarine volcano. A period of disinterest in the topic ensued, due mainly to a lack of critical new data or observations; it lasted until 1986. At that time, the U.S. Geological Survey began conducting large-scale sonar mapping of the sea floor around the Hawaiian Islands as part of an effort to map the newly declared U.S. Exclusive Economic Zone that extended 200 miles from all U.S. land. One of the participants on the first cruise around Hawaii was the same James Moore who had made the initial highly controversial suggestion more than twenty years earlier.

The mapping system employed was the GLORIA sonar system because it could create a complete image of the sea floor by combining 30-km wide swaths of coverage (see the box on bathymetrics in this chapter). The ship was driven back and forth on parallel tracks, and slowly the data were acquired and merged with that collected on prior tracks. The image of the region northeast of Oahu showed numerous angular blocks ranging in size from the giant block discovered so many years earlier to blocks as small as a football field—the smallest features that could be resolved by GLORIA. The sea floor was covered with such large and small blocks in a huge region extending roughly 200 km (124 mi) from the shore of Oahu.

The new data clearly demonstrated that the seafloor topography was indeed dominated by the blocky deposit from a giant landslide and that Moore had been correct in his hypothesis in 1964 (box fig. 1). However, in addition to the slide northeast of Oahu, the data also revealed that such blocky deposits were common. We now know they can be found along the entire Hawaiian volcanic chain as far west as Midway Island. In all, seventeen separate, large (more than 20 km, or 12 mi, long) deposits were identified around the main islands (box fig. 2), and another sixty-one were found along the western parts of the chain. The size of the landslides coupled with their apparent frequency, several per

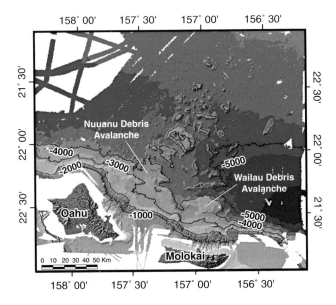

Box Figure 1 Bathymetric map of the region northeast of Oahu showing the blocky deposits of the Nuuanu and Wailau landslides scattered across the sea floor. This map is based on modern swath bathymetry, mainly collected by the Japan Marine Science and Technology Center in 1998 and 1999.

million years, suggested that such giant slides posed an important hazard and warranted much more detailed study to understand their dynamics.

Two main types of slide deposits were recognized: rotational slumps and debris avalanches. The rotational slumps are up to 10 km (6.2 mi) thick, are up to 110 km (68.3 mi) wide, and do not extend far from shore, whereas the debris avalanches are less than 2 km (1.25 mi) thick, are narrow, and extend up to 230 km (143 mi) from shore. The observation that blocks at the outer limits of some avalanche deposits appeared to have run uphill suggested that this type of slide occurred rapidly and that the debris moved down the slopes of the volcanoes at high velocity in order to have enough momentum to eventually run uphill. These characteristics suggest that debris avalanches occur catastrophically and may be a serious hazard in Hawaii and on other volcanic islands.

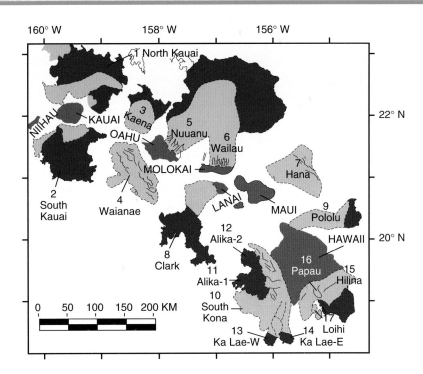

Box Figure 2 Map showing the distribution of landslide debris around the Hawaiian Islands based on GLORIA side-scan imagery. The blocky portions of the debris avalanche are shown in *brown,* and the rotational slump deposits are shown in *yellow.*

Many compelling questions remained to be addressed, with the central one being what causes the slides. These questions are not easy to answer, particularly when one considers the enormous areas that must be mapped and sampled to characterize even a single landslide. The questions scientists have sought to answer are very basic: What slid? When did it slide? How did it slide? Why did it slide? These questions lead to a most important question: Can we identify the conditions that will lead to the next giant slide?

Several different approaches to answering these questions about the landslides were begun in the 1990s. They included collecting high-resolution multibeam bathymetry to better define the distribution and sizes of blocks, amassing deep seismic data to image the subsurface structure of the slides, and employing manned and unmanned submersible dives to determine what types of rocks made up the landslide blocks. An Ocean Drilling Program site south of Oahu and long piston cores collected around the islands sampled layers of volcanic sands interpreted as turbidites deposited from the giant landslides at their outer edges. Beginning in 1998, submersible and remote vehicle surveys took place with increasing frequency. In particular, the Japan Marine Science and Technology Center ran a series of four cruises over a five-year period with a major goal of exploring the blocks from several of the largest landslides around the Hawaiian Islands.

Additional remote vehicle dives were conducted by the Monterey Bay Aquarium Research Institute in 2001 during an expedition to Hawaii. Most of the rocks observed and collected from the blocks are various types of volcanic rocks. These rocks commonly formed where lava flows entered the ocean and the shattered debris accumulated on the flanks of the volcanoes. In the lower parts of several of the slides, sediment from beneath the volcanoes is thrust up as the flanks of the spreading volcano slide over the ocean crust.

We have been successful in determining what slid and, to a lesser degree, when it slid. We are beginning to get a three-dimensional picture of the structure of the slides. However, the more difficult questions of how and why these slides occurred, and what the future holds, remain to be answered.

References

Clague, D. A., and J. G. Moore. 2002. The Proximal Part of the Giant Submarine Wailau Landslide, Molokai, Hawaii. *Journal of Volcanology and Geothermal Research* (113): 259–87.

Moore, J. G., D. A. Clague, R. T. Holcomb, P. W. Lipman, W. R. Normark, and M. E. Torresan. 1989. Prodigious Submarine Landslides on the Hawaiian Ridge. *Journal of Geophysical Research* (94): 17465–84.

different particle sizes (fig. 4.17). Size influences the horizontal distance a particle is transported before settling out of the water and the rate at which it sinks. In general, it takes more energy to transport large particles than it does small particles. In the coastal environment, when poorly sorted sediment is transported by wave or current action, the larger particles will settle out and be deposited first, while the smaller particles may be carried farther away from the coast and deposited elsewhere. In the open ocean, the variation in sinking rate between large and small particles has a tremendous influence on how long it takes for a particle to sink to the deep-sea floor and, hence, how far the settling particle may be transported by deep horizontal currents (table 4.2). A very fine sand-sized particle may settle to the deep-sea floor in a matter of days, where it could come to rest a short horizontal distance away from the point at the surface where it began its journey. In contrast, it may take clay-sized particles over 125 years (nearly 50,000 days) to make the same journey (see Stokes Law for small-particle settling velocity in appendix C). The speed of deep horizontal currents in the oceans is generally quite slow, but even at a speed of 5 cm (2 in) per second, a clay-sized particle could theoretically be transported around the world five times before it reached the deep-sea floor. Smaller soluble particles also have time to dissolve as they slowly sink in the deep ocean. Settling rate is also influenced by particle shape. Stokes Law assumes that the sediment particles are spherical, and the resulting calculated settling rates tend to be maximum rates. Angular grains generate small turbulent eddies that slow their rate of descent. Relatively flat particles such as clays also settle more slowly than spheres of the same density.

Scientists have puzzled over the close correlation between the particle types found in surface waters and those found almost directly below on the sea floor. This observation seems to contradict the large horizontal displacement of very slowly sinking particles due to currents in the water. Some mechanisms must be working to aggregate the tiny particles into larger particles. Scientists have observed that small particles often attract each other owing to their electrical charges. This attraction forms larger particles, which sink more rapidly. This process is important in the formation of the abundant sediment deposits in river deltas. The remains of microscopic organisms known as **phytoplankton** and **zooplankton** can also be found on the sea floor. Phytoplankton are **photosynthetic** and can generate organic matter; zooplankton are **heterotrophic,** which means that, just like humans, they consume organic matter for energy. Phytoplankton and zooplankton are small enough that they can be seen best with the aid of a microscope. Plankton are discussed in more detail in chapter 16. When zooplankton or phytoplankton are consumed by larger organisms, their inorganic cell walls or **tests** are expelled by these organisms in large fecal pellets that sink rapidly. It is estimated that as many as 100,000 tests can be packaged into a single large fecal pellet. This packaging of small particles into larger particles decreases the time for the remains of plankton to sink to the sea floor from years to just ten to fifteen days, minimizing their horizontal displacement by water movements. Once the fecal pellets are deposited on the

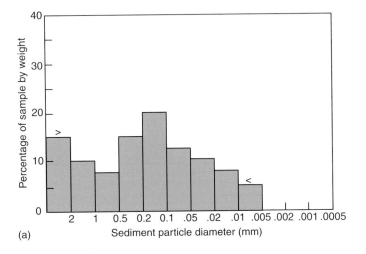

(a)

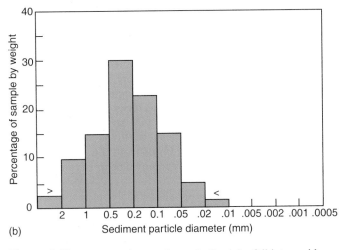

(b)

Figure 4.17 (a) A poorly sorted sample. Particles fall into a wide variety of size ranges in approximately equal amounts. (b) A well-sorted sample. One size range predominates in a limited distribution of sizes.

Table 4.2	Sediment Sinking Rate and Distance Traveled		
Sediment Size	**Approximate Sinking Rate (m/s)**	**Time for a Vertical Fall of 4 km (days)**	**Horizontal Distance Traveled in a 5 cm/s Current (km)**
Very fine sand	9.8×10^{-3}	4.7	20.4
Silt	9.8×10^{-5}	470	2040
Clay	9.8×10^{-7}	47,000	204,000

Note: The sinking rate of a particle depends on its density, shape, and diameter. These rates are based on the assumption that the particles are spherical and have a density similar to that of quartz. Estimates of the speed of deep currents vary. A conservative estimate of 5 cm/s is chosen for purposes of illustration. See appendix C for the formula for small-particle settling velocity.

bottom, breakdown of the remaining organic portion of the pellets liberates the small inorganic particles.

Location

Marine sediments are classified as either **neritic** (*neritos* = of the coast) or **pelagic** (*pelagios* = of the sea) based on where they are found (fig. 4.18). Neritic sediments are found near continental margins and islands and have a wide range of particle sizes. Most neritic sediments are eroded from rocks on land and transported to the coast by rivers. Once they enter the ocean, they are spread across the continental shelf and down the slope by waves, currents, and turbidity currents. The largest particles are left near coastal beaches, while smaller particles are transported farther from shore.

Pelagic sediments are fine-grained and collect slowly on the deep-sea floor. The thickness of pelagic sediments is related to the length of time they have been accumulating or the age of the sea floor they cover. Consequently, their thickness tends to increase with increasing distance from mid-ocean ridges (see fig. 3.15).

Rates of Deposit

The rates at which marine sediments accumulate have a wide range, due to the natural variability of the processes that produce and transport sediments. Accumulation rates of neritic sediments are highly variable. In river estuaries, the rate may be more than 800,000 cm (315,000 in) per 1000 years, or 8 m (over 26 ft) per year. Each year, the rivers of Asia, such as the Ganges, the Yangtse, the Yellow, and the Brahmaputra, contribute more than one-quarter of the world's land-derived marine sediments. In quiet bays, the rate may be 500 cm (197 in)

per 1000 years, and on the continental shelves and slopes, values of 10–40 cm (3.9–15.7 in) per 1000 years are typical, with the flat continental shelves receiving the larger amounts. Many sediments covering the continental shelves away from river mouths are laid down by processes that no longer exist at that location. Such sediments are called **relict sediments** and represent conditions that existed many thousands of years ago, when sea level was lower because of the accumulation of water in ice caps and glaciers.

Accumulation rates for pelagic sediments are much slower than those of typical neritic sediments. An average accumulation rate for deep-ocean pelagic sediment is 0.5–1.0 cm (0.2–0.4 in) per 1000 years. Although deep-sea sedimentation rates are excessively slow, there has been plenty of time during geologic history to accumulate the average pelagic sediment thickness of approximately 500–600 m (1600–2000 ft) on the continental rises and areas of older sea floor. At a rate of 0.5 cm (0.2 in) per 1000 years, it takes only 100 million years to accumulate 500 m (1600 ft) of sediment, and the oldest sea floor is known to be roughly twice that age, or about 200 million years old.

Source and Chemistry

Marine sediments are also classified by the source of the particles that make up the sediment and may be further subdivided by their chemistry. Sedimentary particles may come from one of four different sources: preexisting rocks, marine organisms, seawater, or space.

Sediments derived from preexisting rocks are classified as **lithogenous** (*lithos* = stone, *generare* = to produce) **sediments.** These are also commonly called **terrigenous** (*terri* = land, *generare* = to produce) **sediments.** While terrigenous sediment technically includes any type of material coming off the land, such as rock fragments, wood chips, and sewage sludge, the majority of terrigenous material consists of lithogenous particles. Active volcanic islands in the ocean basins are also an important source of lithogenous sediment. Rocks on land are weathered and broken down into smaller particles by wind, water, and seasonal changes in temperature that result in freezing and thawing. The resulting particles are transported to the oceans by water, wind, ice, and gravity. Windblown dust from the continents, ash from active volcanoes, and rocks picked up by glaciers and embedded in icebergs are additional sources of lithogenous materials.

Lithogenous material can be found everywhere in the oceans. It is the dominant neritic sediment because the supply of lithogenous particles from land simply overwhelms all other types of material. Pelagic lithogenous sediments on the deep-sea floor, called **abyssal clay,** are composed of at least 70% by weight clay-sized particles. Abyssal clay accumulates

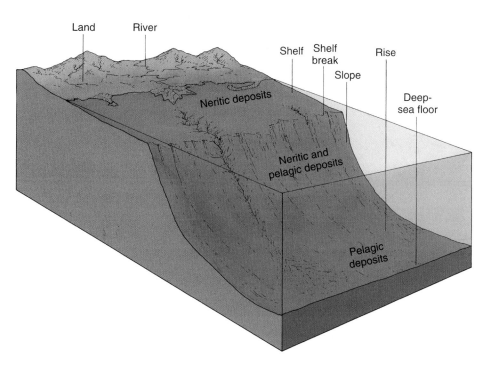

Figure 4.18 Classification of sediments by location of deposit. The distribution pattern is partially controlled by proximity to source and rate of supply.

very slowly at rates that are generally less than 0.1 cm (0.04 in) per 1000 years. Because the accumulation rate is so slow, even a thin deposit represents a very long period of time. It is important to understand that where abyssal clay is the dominant pelagic sediment, it is only because of the lack of other types of material, not because of an increase in the supply of clay-size particles. This is generally the case in regions where there is little marine life in the surface waters above. This fine rock powder, blown out to sea by wind and swept out of the atmosphere by rain, may remain suspended in the water for many years. These clays are often rich in iron, which oxidizes in the water and turns a reddish brown color; hence, they are frequently called **red clay** (fig. 4.19*a*). The distribution of red clay is illustrated in figure 4.20.

(a)

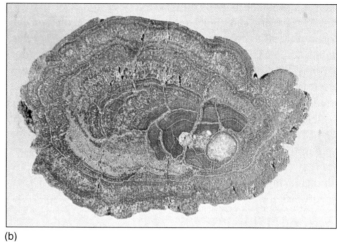

(b)

Figure 4.19   (a) Manganese nodules resting on red clay photographed on deck in natural light. Nodules are 1–10 cm in diameter. (b) A cross section of a manganese nodule showing concentric layers of formation.

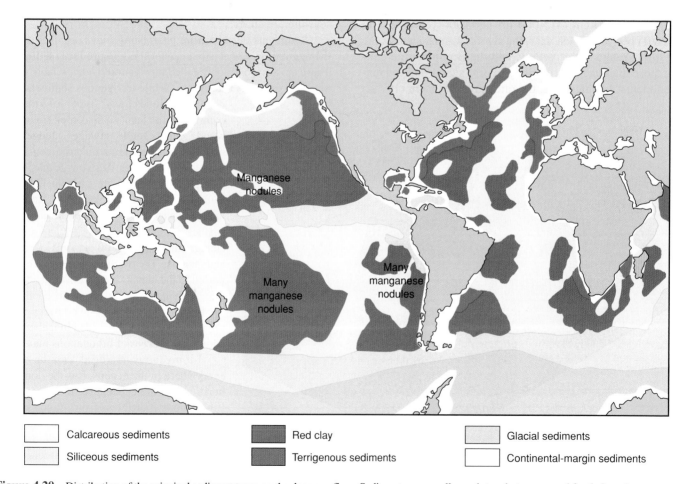

Calcareous sediments	Red clay	Glacial sediments
Siliceous sediments	Terrigenous sediments	Continental-margin sediments

Figure 4.20   Distribution of the principal sediment types on the deep-sea floor. Sediments are usually a mixture but are named for their major component.

The composition of lithogenous sediments, generally various clays and quartz, is controlled by the chemistry of the rocks they came from and their response to chemical and mechanical weathering. Most lithogenous sediments have quartz because it is one of the most abundant and stable minerals in continental rocks. Quartz is very resistant to both chemical and mechanical weathering, so it can easily be transported long distances from its source. The distribution pattern of quartz grains in the sediment can provide important information concerning changes in wind patterns and intensity through time.

Clays are abundant because they are produced by chemical weathering. Four clay minerals make up the deep-sea clays: chlorite, illite, kaolinite, and montmorillonite. The distribution of these four clays reflects different climatic and geologic conditions in the areas where and when they originated as well as along the paths they traveled before settling on the sea floor. These conditions often have a strong dependence on latitude. The warm, moist climate of low latitudes supports strong chemical weathering on land. Mechanical weathering tends to be dominant in the cold, dry climate typical of high latitudes. Chlorite is highly susceptible to chemical weathering and can be altered to form kaolinite. Consequently, chlorite is abundant in deep-sea clays at high latitudes, where chemical weathering is less effective. Kaolinite is produced in the strong chemical weathering of minerals to form soil. It is ten times as abundant in the tropics as in polar regions, where soil-forming processes are very slow. Illite is the most widespread clay mineral. It has a clear hemispheric rather than climatic distribution.

In the Southern Hemisphere, it comprises up to 20%–50% of the clay minerals; in the Northern Hemisphere, it usually accounts for more than 50% of the clay minerals. Illite forms under a variety of conditions that are not dependent on latitude, so its abundance in marine sediment depends on the degree of dilution by other clay minerals. Montmorillonite is produced by the weathering of volcanic material on land and on the sea floor. It is common in regions of low sedimentation near sources of volcanic ash. It is more abundant in the Pacific and Indian Oceans than in the Atlantic Ocean, where there is little volcanic activity along the surrounding coastlines.

Sediments derived from organisms are classified as **biogenous** (*bio* = life, *generare* = to produce) **sediments.** These may include shell and coral fragments as well as the hard skeletal parts of single-celled phytoplankton and zooplankton that live in the surface waters. Pelagic biogenous sediments are composed almost entirely of the shells, or tests, of plankton (fig. 4.21). The chemical composition of these tests is either calcareous (calcium carbonate: $CaCO_3$, as in most seashell material) or siliceous (silicon dioxide: SiO_2, clear and hard). If pelagic sediments are more than 30% biogenous material by weight, the sediment is called an **ooze;** specifically, either a **calcareous ooze** or **siliceous ooze,** depending on the chemical composition of the majority of the tests. The distribution of calcareous and siliceous oozes on the sea floor is related to the supply of organisms in the overlying water, the rate at which the tests dissolve as they descend, the depth at which they are deposited, and dilution with other sediments types (see fig. 4.20).

(a)

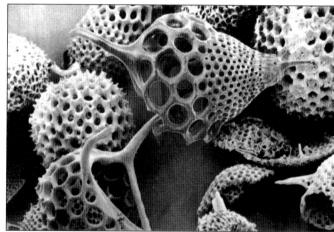

(b)

 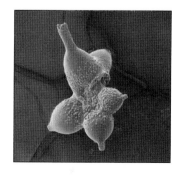

(c)

Figure 4.21 Scanning electron micrographs of biogenous sediments: (a) Diatoms and coccolithophorids. The small disks are detached coccoliths. (b) Radiolarians. (c) Foraminifera. *Figure 4.21c images courtesy of Steve Nathan and R. Mark Leckie.*

Calcareous tests are created by a group of phytoplankton called **coccolithophorids** (covered with calcareous plates called **coccoliths**), snails called **pteropods,** and amoeba-like animals called **foraminifera** (fig. 4.21*a* and *c*). Most coccoliths are smaller than 20 μm. Pteropod tests range from a few millimeters to 1 cm in size, while foraminifera tests range from about 30 μm to 1 mm. These deposits are often named for their principal constituent: coccolithophorid ooze, pteropod ooze, or foraminiferan ooze. Calcareous oozes are the dominant pelagic sediments (see fig. 4.20). The dissolution, or destruction, rate of calcium carbonate varies with depth and temperature and is different in different ocean basins. Calcium carbonate generally dissolves more rapidly in cold, deep water, which characteristically has a higher concentration of CO_2 and is slightly more acidic (this is discussed in detail in the sections on the pH of seawater and dissolved gas in chapter 6). The depth at which calcareous skeletal material first begins to dissolve is called the **lysocline.** Below the lysocline, there is a progressive decrease in the amount of calcareous material preserved in the sediment. The depth at which the amount of calcareous material preserved falls below 20% of the total sediment is called the **carbonate compensation depth (CCD).** The CCD is also commonly defined as the depth at which the rate of accumulation of calcium carbonate is equal to the rate at which it is dissolved. Calcareous ooze tends to accumulate on the sea floor at depths above the CCD and is generally absent at depths below the CCD. The CCD has an average depth of about 4500 m (14,800 ft), or roughly midway between the depth of the crests of ocean ridges and the deepest regions of the abyssal plains. In the Pacific, the CCD is generally at depths of about 4200–4500 m (13,800–14,800 ft). An exception to this is the deepening of the CCD to about 5000 m (16,400 ft) in the equatorial Pacific, where high rates of biological productivity result in a large supply of calcareous material. In the North Atlantic and parts of the South Atlantic, it is at or just below depths of 5000 m (16,400 ft). Calcareous oozes are found at temperate and tropical latitudes in shallower areas of the sea floor such as the Caribbean Sea, on elevated ridge systems, and in coastal regions.

Siliceous tests are created by another group of phytoplankton called diatoms and a type of zooplankton called radiolaria (fig. 4.21*a* and *b*). Their skeletal remains are the dominant components of diatomaceous and radiolarian ooze, respectively. The pattern of dissolution of siliceous tests is opposite to that of calcareous tests. The oceans are undersaturated in silica everywhere, so siliceous material will dissolve at all depths, but it dissolves most rapidly in shallow, warm water. Siliceous oozes are only preserved below areas of very high biological productivity in the surface waters (see fig. 4.20). Even in these areas, an estimated 90% or more of the siliceous tests are dissolved, either in the water or on the sea floor.

Diatomaceous ooze is found at cold and temperate latitudes around Antarctica and in a band across the North Pacific. Because diatoms are photosynthetic, they require sunlight and inorganic **nutrients,** such as nitrate, phosphate, and silicate (similar to what is found in fertilizers) for growth. The sunlight is available at the ocean's surface; the nutrients are produced by the decomposition of organisms in the ocean, and these nutrients are liberated in the deeper water as decomposition takes place. Only at certain locations are these nutrients returned to the surface by the large-scale upward flow of deeper water. Where this upward flow occurs, sunlight combines with the nutrients to produce the conditions needed for high levels of phytoplankton production. Large numbers of diatoms are found in the areas that combine suitable light, nutrients, and the correct temperature.

Radiolarian ooze is found beneath the warm waters of equatorial latitudes. Radiolaria thrive in warm water, producing siliceous outer shells that are often covered with long spines. Figure 4.21*b* shows radiolaria tests at high magnification.

Sediments derived from the water are classified as **hydrogenous** (*hydro* = water, *generare* = to produce) **sediments.** Hydrogenous sediments are produced in the water by chemical reactions. Most are formed by the slow precipitation of minerals onto the sea floor, but some are created by the precipitation of minerals in the water column in plumes of recirculated water at hydrothermal vents along the ocean ridge system. Hydrogenous sediments include some **carbonates** (limestone-type deposits), **phosphorites** (phosphorus in the form of phosphate in crusts and nodules), **salts,** and **manganese nodules.** In addition, hydrothermally generated sulfides rich in iron and other metals form along the axis of spreading centers on young sea floor, and carbonates and magnesium-rich minerals form off the axis of spreading centers on older sea floor, as discussed in chapter 3.

Hydrogenous carbonates are known to form by direct precipitation in some shallow, warm-water environments as a result of an increase in water temperature or a slight decrease in the acidity of the water. In shallow, warm water with high biological productivity, photosynthetic organisms can remove enough dissolved carbon dioxide in the water to decrease the acidity of the water and trigger the precipitation of calcium carbonate (see the discussion of carbon dioxide as a buffer in section 6.3 of chapter 6). The calcium carbonate often precipitates in small pellets called **ooliths** (*oon* = egg) about 0.5–1.0 mm (0.02–0.04 in) in diameter. In the present oceans, there are relatively few places where this is known to be occurring. The largest modern deposits of hydrogenous carbonates are currently forming on the Bahama Banks. Additional deposits are forming on Australia's Great Barrier Reef and in the Persian Gulf.

Phosphorites contain phosphorus in the form of phosphate and are most abundant on the continental shelf and upper part of the continental slope. They are occasionally found as nodules as much as 25 cm (10 in) in diameter or in beds of sand-size grains, but more often they form thick crusts. Most phosphate deposits on continental margins do not appear to be actively accumulating. Phosphorite deposits are currently forming in regions of high biological productivity off the coasts of southwestern Africa and Peru.

Salt deposits occur when a high rate of evaporation removes most of the water and leaves a very salty brine in shallow areas. Chemical reactions occur in the brine, and salts are precipitated or separated from solution and then deposited on the bottom. In such processes, carbonate salts are formed first, followed by sulfate salts, and then chlorides, including sodium chloride. Studies of precipitated material on the floor of the Mediterranean Sea have provided clues to its past isolation from the Atlantic Ocean.

Manganese nodules are composed primarily of manganese and iron oxides but also contain significant amounts of copper, cobalt, and nickel. They were first recovered from the ocean floor in 1873 during the *Challenger* expedition. They are found in a variety of marine environments, including the abyssal sea floor, on seamounts, along active ridges, and on continental margins. Their chemistry is related to the ocean basin they are found in as well as the specific marine environment where they have grown (tables 4.3 and 4.4). Nodules from the Pacific Ocean tend to have the highest concentrations of metals, with the exception of iron. Nodules in the Atlantic Ocean generally have the highest iron concentration. The average weight percent of manganese and iron in nodules is about 18% and 17%, respectively, while the average weight percent of nickel, cobalt, and copper varies from about 0.5% down to 0.2%. Nodules that form on continental margins are very distinct chemically. They have very high manganese concentrations combined with very low iron concentrations. The chemistry of nodules can also be influenced by their position on the sea floor with respect to other sediments. Manganese nodules may lie on top of the other sediment (see fig. 4.19*a*) or be buried at shallow depth in the sediment. Nodules lying on top of the sediment react chemically with the seawater and can become enriched in iron and cobalt. Those that are buried react with both the seawater and the sediment and can become enriched in manganese and copper. The concentric layers in a nodule typically have slightly different chemistries (see fig. 4.19*b*). This chemical layering is the result of changes in the chemistry of the seawater as the nodule grew.

On the deep-sea floor, manganese nodules form black or brown rounded masses typically 1–10 cm (0.5–4 in) in diameter, roughly the size of a golf ball or a little larger. Continental margin manganese and iron oxide deposits can take a variety of forms, from nodules similar to those found on the deep-sea floor to extensive slabs, or crusts. Most manganese nodules grow very slowly: 1–10 mm (0.004–0.04 in) per million years for deep-sea nodules, roughly 1000 times slower than accumulation rates of other pelagic sediments. Nodules grow layer upon layer, often around a hard skeletal piece such as a shark's tooth, rock fragment, or fish bone that acts as a seed, much as a pearl grows around a grain of sand. They generally form in areas of very little sediment supply from other sources or where rapid bottom currents prevent them from being deeply buried. Manganese nodules on continental margins are unique in their rapid growth, having growth rates on the order of 0.01–1 mm per year—from 1000 to 1 million times faster than their deep-sea counterparts. Manganese nodules have been mapped in all oceans except the Arctic. They are most abundant in the central Pacific north and south of the biogenous oozes along the equator (see fig. 4.20). In the Atlantic and Indian Oceans, there are higher rates of lithogenous and biogenous sedimentation and consequently fewer deposits of manganese nodules.

Sediments derived from space are classified as **cosmogenous** (*cosmos* = universe, *generare* = to produce) **sediments.** Particles from space constantly bombard Earth. Most of these particles burn up as they pass through the atmosphere, but roughly 10% of the material reaches the surface of Earth. Cosmogenous particles are generally small, and those that survive the passage through the atmosphere and fall in the ocean stay in suspension in the water long enough usually to dissolve before they reach the sea floor. These iron-rich sediments are found in small amounts in all oceans, mixed in with the other sediments. The pattern of related cosmic materials can indicate the direction of the particle shower that supplied them. The particles become very hot as they pass through Earth's atmosphere and partially melt; this melting gives the particles a characteristic rounded or teardrop shape. Cosmic bodies can disintegrate and melt surface materials as they strike Earth. Their impact can cause a splash of melted particles that spray outward and produce splash-form **tektites** (fig. 4.22). Microtektites are found on the ocean floor and on land.

Table 4.3	Average Chemistry of Manganese Nodules from the Three Ocean Basins			
Element	**Atlantic**	**Pacific**	**Indian**	**Average for All Three Oceans**
Mn	16.18	19.75	18.03	17.99
Fe	21.2	14.29	16.25	17.25
Ni	0.297	0.722	0.510	0.509
Co	0.309	0.381	0.279	0.323
Cu	0.109	0.366	0.223	0.233

Note: The average abundances of manganese (Mn), iron (Fe), nickel (Ni), cobalt (Co), and copper (Cu) in manganese nodules from the Atlantic, Pacific, and Indian Ocean basins. Numbers are weight % of each metal.

Table 4.4	Average Chemistry of Manganese Nodules from Different Environments			
Element	**Seamounts**	**Active Ridges**	**Continental Margins**	**Abyssal Depths**
Mn	14.62	15.51	38.69	17.99
Fe	15.81	19.15	1.34	17.25
Ni	0.351	0.306	0.121	0.509
Co	1.15	0.400	0.011	0.323
Cu	0.058	0.081	0.082	0.233
Mn/Fe	0.92	0.81	28.9	1.04

Note: The average abundances of manganese (Mn), iron (Fe), nickel (Ni), cobalt (Co), and copper (Cu), and the manganese-to-iron ratio in manganese nodules, from different environments. Numbers are weight % of each metal.

6.09 cm (2.4 in)

6.85 cm (2.7 in)

8.33 cm (3.3 in)

6.35 cm (2.5 in)

6.35 cm (2.5 in)

Figure 4.22 These splash-form tektites were collected in Southeast Asia.

Table 4.5	Sediment Summary		
Type	**Source**	**Areas of Significant Deposit**	**Examples**
Lithogenous (terrigenous)	Eroded rock, volcanoes, airborne dust	Dominantly neritic, pelagic in areas of low productivity	Coarse beach and shelf deposits, turbidites, red clay
Biogenous	Living organisms	Regions of high surface productivity, areas of upwelling, dominantly pelagic, some beaches, shallow warm water	Calcareous ooze (above the CCD), siliceous ooze (below the CCD), coral
Hydrogenous	Chemical precipitation from seawater	Mid-ocean ridges, areas starved of other sediment types, neritic and pelagic	Metal sulfides, manganese nodules, phosphates, some carbonates
Cosmogenous	Space	Everywhere but in very low concentration	Meteorites, space dust

A brief summary of the major sediment types is given in table 4.5.

Patterns of Deposit on the Sea Floor

The patterns formed by the sediments on the sea floor reflect both distance from their source and processes that control the rates at which they are produced, transported, and deposited. Seventy-five percent of marine sediments are terrigenous. The majority of terrigenous sediments are initially deposited on the continental margins but are moved seaward by the waves, currents, and turbidity flows that move across the continental shelves and down the continental slopes. The terrigenous sediments of coastal regions are primarily lithogenous, supplied by rivers and wave erosion along the coasts. Worldwide river sediment transport is about $12-15 \times 10^9$ metric tons per year. The majority of this sediment enters the tropical and subtropical oceans.

Coarse sediments are concentrated close to their sources in high-energy environments—for example, beaches with swift currents and breaking waves. The waves and currents move quite large rock particles in the shore zone, but these larger particles settle out quickly. Finer particles are held in suspension and are carried farther away from their source. This pattern results in a gradation by particle size: coarse particles close to shore and to their source, with finer and finer particles predominating as the distance from the source increases.

Finer sediments are deposited in low-energy environments, offshore away from the currents and waves or in quiet bays and estuaries. In higher latitudes, deposits of rock and gravel carried along by glaciers are found in coastal environments, whereas in low latitudes, fine sediments predominate and are considered to be products of large rivers, heavy rainfall, and loose surface soils.

At the present time, most of the land-derived sediments are accumulating off the world's river mouths and in estuaries. Estuaries and river deltas serve as sediment traps, preventing terrigenous sediments from reaching the deep-sea floor in such places as the Chesapeake and Delaware Bay systems along the North Atlantic coast, in the Georgia Strait of British Columbia, and in California's San Francisco Bay along the North Pacific coast. If sediments are supplied to a delta faster than they can be retained, the sediments will move across the shelf into the deeper water environments. This is currently the case with the sediments of the Mississippi River. Much of the thick sediment layer on the outer continental shelf was laid down during the ice ages, when the sea level was lower—for example, at Georges Bank southeast of Cape Cod. Little is currently being added to these outer regions of the continental shelf.

The accumulation of sediments on the passive shelves of continents results in unstable, steep-sided deposits that may slump, sending a flow of terrigenous sediment moving rapidly down the continental slope in a turbidity current (see section 4.2). Turbidity currents move coarse terrigenous materials farther out to sea; in doing so, they distort the general deep-sea sediment pattern and reduce the abundance of pelagic deposits. Near shore, the spring flooding of rivers alternates with periods of low river discharge in summer and fall. The floods bring large quantities of sediment to the coastal waters, and the contributions of this flooding are recorded in the layering of the sediments. Sudden masses of sediment from the collapse of a cliff or the eruption of a volcano are seen in the sediment pattern as specific additions of large quantities of sand or ash.

Oceanic sediments form visually distinct layers characterized by color, particle size, type of particle, and supply rate (fig. 4.23). Seasonal variations and the patterns of long and short growing seasons for marine life can also be determined from the properties and thicknesses of the layers of biogenous material. Over long periods of geologic time, climatic changes such as the ice ages have altered the biological populations that produce sediment and have left a record in the sediment layers.

In shallow coastal areas, cycles of climate change cause variation in rates of sediment production. Along passive continental margins, biogenous sediments may also be diluted by large amounts of lithogenous sediment washing from the land. In coastal areas where marine life is very abundant and river deposits are sparse, biogenous sediments are formed from both shell fragments and broken corals. In the more homogeneous environment of the deep sea, biogenous sediments make up the majority of the pelagic deposits. There is less dilution with terrigenous materials, and few environmental changes disturb the deep bottom deposits, allowing them to remain relatively unchanged for long periods of time. Calcareous oozes are found where the production of organisms is high, dilution by other sediments is small, and depths are less than 4000 m (13,000 ft). (See the areas including the mid-ocean ridges and the warmer shallower areas of the South Pacific in fig. 4.20.) Siliceous oozes cover the deep-sea floor beneath the colder surface waters of 50°–60°N and S latitude and in equatorial regions where cold deeper water is brought to the surface by vertical circulation. Deep basin areas of the Pacific have extensive deposits of red clay (again, see fig. 4.20).

Large rock particles of land origin are also moved out to sea by a process known as **rafting.** Glaciers carry sand, gravel, and rocks embedded in the ice. When the glacier reaches the sea, parts break off and fall into the water as icebergs. The icebergs are carried away from land by the currents and winds, taking the terrigenous materials far from their original sources. As the ice melts, rocks and gravel that were frozen in the ice sink to the sea floor. In addition, sea ice formed in shallow water along the shore can incorporate material from the sea floor and transport it out to sea. Figure 4.20 indicates areas of terrigenous deposits that are affected by ice rafting. It is estimated that ice-rafted material can be found over about 20% of the sea floor. Sometimes large, brown seaweeds known as kelp, which grow attached to rocks in coastal areas, are dislodged by storm waves. The kelp may have enough buoyancy to float away, carrying the attached rock. When the seaweed dies or sinks, the rock is deposited on the ocean floor at some distance from its origin. The deposition of larger rocks by this rafting process is infrequent and irregular.

The wind is an effective agent for moving lithogenous materials out to sea in some parts of the world. Winds blowing offshore from the Sahara Desert or other arid regions transfer sand particles directly from land to sea, sometimes 1000 km (600 mi) or more offshore (fig. 4.24). A similar process can occur between sand dunes and coastal waters. In the open ocean, airborne dust probably supplies much of the deep-sea red clay material. Figure 4.25 indicates the frequency with which winds carry dust, or haze, out to sea. The world's volcanoes are another source of airborne particles. Volcanic ash is present in seafloor sediments and can be found in layers of significant thickness associated with past volcanic events. The annual supply of airborne particles to the sediments is estimated at 100×10^6 metric tons.

Figure 4.23 A deep-sea sediment core obtained by the drilling ship *Glomar Challenger*. Note the layering of the sediments.

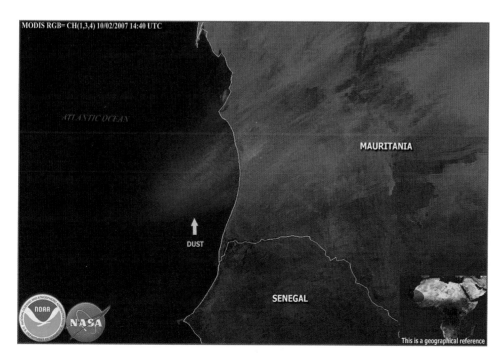

Figure 4.24 Satellite image taken October 2, 2007, showing wind-blown dust from the western Sahara Desert moving over the Atlantic Ocean.

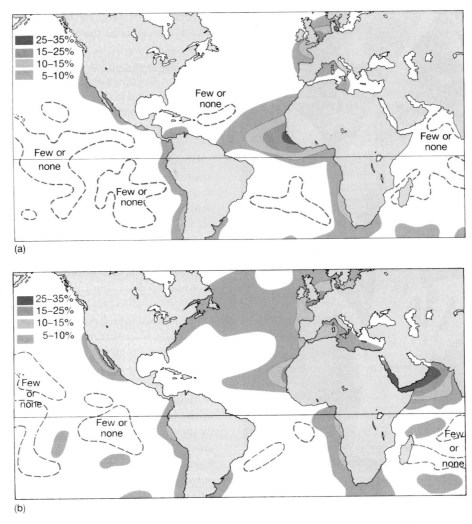

(a)

(b)

Figure 4.25 Frequency of haze as a result of airborne dust during the Northern Hemisphere's (a) winter and (b) summer. Values are given in percentages of total observations.

Formation of Rock

Loose sediments on the sea floor are transformed into **sedimentary rock** in a process known as **lithification.** Lithification can occur through burial, compaction, recrystallization, and cementation. As one layer of sediment covers another, the weight of the sediments puts pressure on the lower sediment layers, and the sediment particles are squeezed more and more tightly together. The particles begin to stick to each other, and the pore water between the sediment particles, with its dissolved solids, moves through the sediments. As it does so, minerals precipitate on the surfaces of the particles and, in time, act to cement the sediment particles together into a mass of sedimentary rock. The sediments in these processes are also exposed to increasing temperature with increasing depth of burial. Chemical changes also occur in sedimentary particles through interaction with seawater and pore water in a process called **diagenesis.** One example of diagenesis is the gradual lithification of calcareous ooze to form chalk or limestone. In this process, calcite particles in the sediments are cemented by calcite precipitated from the pore waters. The transformation of calcareous ooze to chalk occurs at a sediment depth of a few hundred

meters, and the further transformation to limestone occurs with additional cementation under about 1 km burial. Siliceous oozes can be lithified to form a very hard rock called chert.

Sedimentary rock may preserve the layering of the sediments in visually distinct features and strata. Ripple marks from the motion of waves and currents may be seen, and fossils may also be present. Sedimentary rocks are found beneath the sediments of the deep-sea floor, along the passive margins of continents, and on land where they have been thrust upward along active margins or formed in ancient inland seas. Sedimentary rocks include sandstone, shale, and limestone.

If sediments are subjected to greater changes in temperature, pressure, and chemistry, **metamorphic rock** results. Slate is a metamorphic rock derived from shale, and marble is recrystallized metamorphosed limestone.

Sampling Methods

To analyze sediments, the geological oceanographer must have an actual bottom sample to examine. A variety of devices have been developed to take a sample from the sea floor and return it to the laboratory for analysis. **Dredges** are net or wire baskets that are dragged across a bottom to collect loose bulk material, surface rocks, and shells in a somewhat haphazard manner (fig. 4.26). **Grab samplers** are hinged devices that are spring- or weight-loaded to snap shut when the sampler strikes the bottom. See figure 4.27 for examples of this device. Grab samplers sample surface sediments from a fixed area of the sea floor at a single known location.

A **corer** is essentially a hollow pipe with a sharp cutting end. The free-falling pipe is forced down into the sediments by its weight or, for longer cores, by a piston device that enables water pressure to help drive the core barrel into the sediment; coring devices are shown in figure 4.28*a–e*. The product is a cylinder of mud, usually 1–20 m (3.3–65.6 ft) long, that contains undisturbed sediment layers (see fig. 4.23). Box corers (fig. 4.28*e*) are used when a large and nearly undisturbed sample of surface sediment is needed. These corers drive a rectangular metal box into the sediment; they have doors that close over the bottom before the sample is retrieved. Long cores that penetrate the thick sediment overlying older sea floor and reach the older sediment layers nearer the oceanic basalt may be obtained by drilling through both loose sediments and rock. The highly sophisticated drilling techniques used by the research vessel *JOIDES Resolution* are discussed in chapter 3.

Geologic oceanographers and geophysicists also study sediment distribution and seafloor structure with high-intensity sound, a technique known as **acoustic profiling.** Bursts of sound are directed toward the sea floor, where the sound waves either reflect from or penetrate into the sediments. Sound waves that penetrate the sediments are refracted and change speed as they pass through the different layers of sediments. A surface vessel tows an array of underwater microphones, or hydrophones, to sense the returning sound waves, and a recorder plots the returning sound energy to produce a profile of the sediment structure. This technique details the structure of the continental margin and finds buried faults, filled submarine canyons, and clues to oil and gas deposits. Acoustic profiling is very similar to the techniques used to study the interior of Earth (see chapter 3, section 3.1).

Today's ocean scientists are searching for records of Earth's history in the sediment and rock layers of the ocean floor. These

(a)

(b)

Figure 4.26 (a) Rocks can be recovered from the sea floor with a dredge having a chain basket. Sediments and other fine material escape through the chains. (b) Basalt dredged from a depth of about 8 km (5 mi) near the Tonga Trench in the western Pacific Ocean. The dredge is in the *foreground.*

Figure 4.27 Grab samplers: Van Veen (*left*) and orange peel (*right*), both in open positions. Grabs take surface sediment samples.

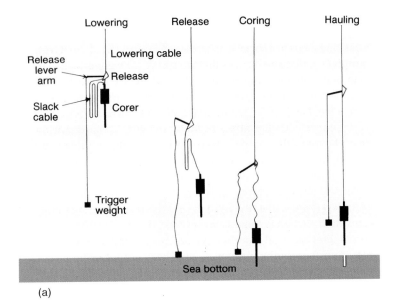

(a)

(c)

(d)

(b)

(e)

Figure 4.28 (a) The Phleger corer is a free-fall gravity corer. The weights help to drive the core barrel into the soft sediments. Inside the corer is a plastic liner. The sediment core is removed from the corer by removing the plastic tube, which is capped to form a storage container for the core. (b) A sketch of a piston corer in operation. The corer is allowed to fall freely to the sea bottom. The action of the piston moving up the core barrel owing to the tension on the cable allows water pressure to force the core barrel into the sediments. (c) Loading a piston corer with weights to prepare it for use. (d) A gravity corer ready to be lowered. (e) A box corer is used to obtain large, undisturbed seafloor surface samples.

116

layers hold evidence for understanding the formation of the ocean basins and continents, changing climate, periods of unusual volcanism, the presence and absence of various life forms, and much more. The information is there, but it requires a combination of sophisticated technical know-how at sea and increasingly detailed scientific research in the laboratory to discern and understand it.

Sediments as Historical Records

Marine sediments and the skeletal materials in them provide important information about processes that have shaped the planet and its ocean basins over the past 200 million years. The study of the oceans through an analysis of sediments is called **paleoceanography.** Two examples of the use of marine sediments to unravel history are (1) the study of the distribution of skeletal remains of marine organisms to date the initiation of the Antarctic Circumpolar Current (ACC) and (2) the study of the relative abundance of different oxygen isotopes in foraminifera tests preserved in the sediment to determine variations in climate and seawater temperature.

Prevailing westerly winds at high southern latitudes cause the ACC to flow continuously from west to east around Antarctica. The interaction of prevailing winds and ocean currents is discussed in detail in chapter 7, and the general pattern of surface currents is discussed in chapter 9. The ACC is a very deep current, extending to depths of 3000–4000 m (9800–13,000 ft), and it is able to flow unimpeded around the globe because there are no shallow seafloor features to block its path. This situation has not always existed, however. The southern continents began to break apart at different times. About 135 million years ago, Africa and India first began to separate from Antarctica, South America, and Australia (see fig. 3.36). As recently as 80 million years ago, South America, Antarctica, and Australia were still effectively one landmass. Sometime around 55 million years ago, some sea floor existed between Australia and Antarctica, and by 35 million years ago, they had separated sufficiently to create a narrow expanse of water called the Austral Gulf. South America had not yet separated from Antarctica. Marine sediments deposited at this time indicate that a small, single-celled, shallow-water marine foraminiferan called *Guembelitria* lived in the restricted waters of the Austral Gulf. The absence of its remains in other Southern Hemisphere sediments of the same age indicates that the organism had not been spread to other areas by ocean currents. Skeletal remains of *Guembelitria* appear quite suddenly in sediments deposited all around Antarctica about 30 million years ago. Even though the Drake Passage between South America and Antarctica did not fully open before 20 million years ago, there must have been a shallow channel a few hundred meters deep as early as 30 million years ago that allowed the ACC to first flow around the continent, carrying *Guembelitria* with it.

Calcite tests found in successive layers of sediment can provide information about changes in climate and seawater temperature over time through a careful analysis of the relative abundance of different oxygen isotopes in the calcite. **Isotopes** are atoms of the same element that have different numbers of neutrons in the nucleus; thus, they have different atomic masses but behave identically chemically. Some marine organisms

remove oxygen from water molecules in the ocean to construct calcareous hard parts (see the discussion of foraminiferans in chapter 16, for example). Water contains the two main isotopes of oxygen: the common ^{16}O and the rarer ^{18}O. These isotopes are stable and do not decay radioactively, so once they have been incorporated into an organism's skeletal material, their relative proportion ($^{18}O:^{16}O$) remains constant even after the organism dies. The $^{18}O:^{16}O$ ratio in a skeletal fragment depends in part on the relative abundance of the isotopes in the seawater at the time the organism formed it. Thus, calcareous biogenous remains record changes in the isotopic chemistry of seawater that are related to changes in global temperature.

Water molecules containing ^{16}O are lighter than molecules containing ^{18}O, so they are more easily removed from the oceans by evaporation. During glacial periods, the water evaporated from the sea surface is trapped in ice sheets; the sea level is lowered, and ^{16}O is removed from the ocean system. This process increases the $^{18}O:^{16}O$ ratio in the seawater and in skeletal parts that organisms are forming at that time. When these organisms die, their skeletal parts sink to the sea floor and are incorporated into the sediment. During warmer, interglacial periods, the melting of ice sheets causes a rise in sea level and returns ^{16}O-enriched fresh water to the oceans. The result is a drop in the $^{18}O:^{16}O$ ratio in the seawater and in the skeletal parts that are being formed. The isotopic composition of skeletal parts is also influenced by seawater temperature. As temperature decreases, organisms preferentially take up more ^{18}O than ^{16}O in their skeletons. The actual $^{18}O:^{16}O$ ratio preserved in a skeletal fragment is primarily due to changes in seawater composition related to the growth and decay of global ice sheets and consequent fall and rise of sea level.

4.4 Seabed Resources

Long ago, people began to exploit the materials of the seabed. The ancient Greeks extended their lead and zinc mines under the sea, medieval Scottish miners followed seams of coal under the Firth of Forth, and, more recently, coal has been mined from undersea strata off Japan, Turkey, and Canada. As technology has developed and as people have become concerned about the depletion of onshore mineral reserves, interest in seabed minerals and mining has grown. At present, the United States is showing little interest in new seabed resources, but international interest remains strong; research continues in exploration, technology development, and environmental studies, especially in Japan, India, China, and South Korea. Keep in mind that each potential deep-sea source is in competition with an onshore supply. Whether the seabed source will be developed depends largely on international markets, needs for strategic materials, and whether offshore production costs can compete with onshore costs.

Sand and Gravel

The largest superficial seafloor mining operation is for sand and gravel, widely used in construction. The technology and cost required to mine sand and gravel in shallow water differ very little from land operations. This is a high-bulk, low-cost material tied to the economics of transport and the distance to market. Annual world production is approximately 1.2 billion metric tons; the reported potential reserve is more than 800 billion metric tons. The United Kingdom and Japan each take 20% of their total annual sand and gravel requirements from the sea floor.

Sand and gravel mining is the only significant seabed mining done by the United States at this time. It is estimated that the United States has a reserve of 450 billion tons of sand off its northeastern coast; there are large deposits of gravel along Georges Bank off New England and in the area off New York City. Along the coasts of Louisiana, Texas, and Florida, shell deposits are mined for use in the lime and cement industries, as a source of calcium oxide used to remove magnesium from seawater as part of the process of making magnesium metal, and, when crushed, as a gravel substitute for roads and highways.

Sands are mined as a source of calcium carbonate throughout the Bahamas, which have an estimated reserve of 100 billion metric tons. Coral sands are mined in Fiji, in Hawaii, and along the U.S. Gulf Coast. Other coastal sands contain iron, tin, uranium, platinum, gold, and diamonds. The "tin belt" stretches for 3000 km (1800 mi) from northern Thailand and western Malaysia to Indonesia. Here, sediments rich in tin have been dredged for hundreds of years and supply more than 1% of the world's market. Iron-rich sediments are dredged in Japan, where the reserve of iron in shallow coastal waters is estimated at 36 million tons. The United States, Australia, and South Africa recover platinum from some sands, and gold is found in river delta sediments along Alaska, Oregon, Chile, South Africa, and Australia. Diamonds, like gold, are found in sediments washed down the rivers in some areas of Africa and Australia; mining of diamonds at depths of 300 m (1000 ft) began off southwestern Africa in 1994. Muds bearing copper, zinc, lead, and silver also occur on the continental slopes, but they lie too deep for exploitation, considering the present demand and their market value.

Phosphorite

Phosphorite, which can be mined to produce phosphate fertilizers, is found in shallow waters as phosphorite muds and sands containing 12%–18% phosphate and as nodules on the continental shelf and slope. The nodules contain about 30% phosphate, and large deposits are known to exist off Florida, California, Mexico, Peru, Australia, Japan, and northwestern and southern Africa. Recently, a substantial source of phosphorite was located in Onslow Bay, North Carolina. Eight beds have been found, and five are thought to be economically valuable; they have been estimated to contain 3 billion metric tons of phosphate concentrates.

The world's ocean reserve of phosphorite is estimated at about 50 billion tons. Readily available land reserves are not in short supply, but most of the world's land reserves are controlled by relatively few nations. Therefore, political considerations may make these marine deposits attractive as mining ventures for some countries. No commercial phosphorite mining occurs at present in the oceans.

Sulfur

Sulfur is necessary for the production of sulfuric acid, which is used in many industrial processes. Presently, the most economical way to acquire sulfur is to recover sulfur from pollution-control equipment. In the past, sulfur has been mined in the Gulf of Mexico by injecting high-pressure steam into wells to melt the sulfur and then pumping it ashore to processing plants. Millions of tons of sulfur reserves exist in the Gulf of Mexico and the Mediterranean Sea.

Coal

Coal is produced by the burial and alteration of large amounts of land plant material in swampy environments with low oxygen concentration. Plant material undergoes a series of changes as more volatiles and impurities are removed at higher temperatures and pressures. Initially, the partially altered plant material forms peat. Peat can then progress through a series of stages to form coal of increasing hardness, from lignite, to bituminous, and finally to anthracite coal. Changes in sea level and land geography over geologic time have caused some coal deposits to be submerged. These are mined when the coal is present in sufficient quantity and quality to make the operation worthwhile. In Japan, the undersea coal deposits are reached by shafts that stretch under the sea from the land or descend from artificial islands.

Oil and Gas

Oil and gas represent more than 95% of the value of all resources extracted from the sea floor or below. Oil and gas deposits are almost always associated with marine sedimentary rocks and are believed to be produced by the slow conversion of marine plant and animal organic matter to hydrocarbons. Conditions must be just right for marine organic material to eventually be converted to oil and gas. It must first accumulate in relatively shallow, quiet water with low oxygen content. Anaerobic bacteria can then utilize the organic matter to produce methane and other light hydrocarbons. As these simple hydrocarbons are buried beneath deeper layers of sediment, they are subjected to higher pressure and temperature. Over a period of millions of years, they can be converted to oil or gas. Oil forms if the depth of burial is on the order of about 2 km (1.2 mi). If the organic material is buried even deeper or cooked for a longer period of time at higher temperature, gas is produced. Oil deposits are generally found at depths less than 3 km (1.9 mi), and below 7 km (4.3 mi) only gas is found.

Because oil and gas are very light, they migrate upward over time, moving slowly out of the source rock and into porous rocks above. This upward migration continues until the fluids reach an impermeable layer of rock. The oil and gas then stop their ascent and fill the pore spaces of the reservoir rock below this impermeable layer.

Petroleum-rich marine sediments are more likely to accumulate during periods of geologic time when sea level is unusually high and the oceans flood extensive low-lying continental regions to create large shallow basins. Much oil and gas are found in marine rocks that formed from sediments deposited during a relatively short period of time during the Jurassic and Cretaceous, between about 85 million and 180 million years ago, when sea level was high.

Roughly 24% of U.S. oil and 25% of U.S. gas production came from offshore areas in 2007. Worldwide, approximately 32% of oil and 24% of gas production in 1998 was from offshore wells. Major offshore oil fields are found in the Gulf of Mexico, the Persian Gulf, and the North Sea, and off the northern coast of Australia, the southern coast of California, and the coasts of the Arctic Ocean. At the present time, many U.S. companies are finding it more profitable to drill for oil and gas in foreign waters and are moving their rigs to the waters of the North Sea, West Africa, and Brazil.

Bringing the offshore oil fields into production has required the development of massive drilling platforms and specialized equipment to withstand heavy seas and fierce storms and to allow drilling and well development at great depth (fig. 4.29). Although the cost of drilling and equipping an offshore well is three to four times greater than that of a similar venture on land,

Figure 4.29 The oil-drilling platform Hibernia, located 320 km (200 mi) offshore Newfoundland, produced its first oil in November 1997.

the large size of the deposits allows offshore ventures to compete successfully. The gas and oil potential in even deeper offshore waters is still unknown, but the deeper the water in which the drilling must be done, the higher the cost. Exxon is studying a project that would require drilling in 1400 m (4600 ft), more than twice the depth of its current deepest well.

The new methods and equipment developed and used for deep-sea oceanographic drilling and research have provided the prototypes for new generations of deep-sea commercial drilling systems. Even though legal restraints, environmental concerns, and worldwide political uncertainties will continue to contribute to the slow development of offshore deposits, petroleum exploration and development will undoubtedly continue to be the main focus of ocean mining in the near future.

Gas Hydrates

In recent years, interest has been growing in gas hydrates trapped in marine sediments. Gas hydrates are a combination of natural gas, primarily methane (CH_4), and water, which forms a solid, icelike structure under pressure at low temperatures. Drill cores of marine sediment have recovered samples of gas hydrates that melt and bubble as the natural gas escapes. These melting samples burn if lit. Gas hydrates are a subject of intense interest for three reasons: they are a potential source of energy, they may contribute to slumping along continental margins, and they may play a role in climate change.

When 1 cubic foot of gas hydrate melts, it releases about 160 cubic feet of gas. A gas hydrate accumulation thus can contain a huge amount of natural gas. Estimates of the amount of natural gas contained in the world's gas hydrate accumulations are speculative and range over three orders of magnitude, from about 2800 to 8,000,000 trillion (2.8×10^{15} to 8×10^{18}) cubic meters of gas. By comparison, in 2000, the U.S. Geological Survey estimated conventional natural gas accumulations for the world at approximately 440 trillion (4.4×10^{14}) cubic meters. Despite the enormous range in the estimated amount of natural gas contained in gas hydrates, even the lowest estimates suggest that gas hydrates are a much greater resource of natural gas than conventional accumulations and may be a substantial source of energy in the future (table 4.6). It is important to note, however, that none of these assessments have predicted how much gas could actually be produced from the world's gas hydrate accumulations given present technology and their location.

A second reason gas hydrates are significant is their effect on seafloor stability. Along the southeastern coast of the United States, a number of submarine landslides, or slumps, have been identified that may be related to the presence of gas hydrates. The hydrates may inhibit normal sediment consolidation and cementation processes, creating a weak zone in the sediments. Alternately, the lowering of sea level during the last glacial period may have reduced the pressure on the sea floor enough to allow some of the gas to escape from the hydrates and accumulate in the sediment, decreasing its strength.

A final reason for studying gas hydrates is their potential link to climate changes. The amount of methane stored in

Table 4.6	Potential Significance of Gas Hydrates		
Estimated Volume of Gas Hydrates, EVGH (10^{12} m^3)	**Ratio of EVGH to World Supply of Natural Gas**	**Ratio of EVGH to Natural Gas Consumption in the:**	
		United States in 2000	**World in 1999**
low of 2800	6.4:1	4375:1	1175:1
high of 8,000,000	18,200:1	12,500,000:1	3,355,700:1

hydrates is believed to be about 3000 times the amount currently present in the atmosphere. Since methane is a greenhouse gas, its release from hydrates could affect global climate.

Manganese Nodules

Manganese nodules are found scattered across the world's deep-ocean floors, with particular concentrations in the red clay regions of the northeastern Pacific (see figs. 4.19 and 4.20). The nodule chemistry varies from place to place, but the nodules in some areas contain 30% manganese, 1% copper, 1.25% nickel, and 0.25% cobalt: these are much higher concentrations than are usually found in land ores. Cobalt is of particular interest, since it is classified as being of "strategic" importance to the United States and hence essential to the national security. Cobalt is an important component in the manufacture of strong alloys used in tools and aircraft engines. The nodules grow very slowly, but they are present in huge quantities. An estimated 16 million additional tons of nodules accumulate each year.

Since the 1960s, large multinational consortia have spent hundreds of millions of dollars to locate the highest nodule concentrations and to develop technologies for their collection. However, their expectations of rapid development have not been realized. In the 1980s, some of these consortia had withdrawn completely and others were dormant. The primary reason for the lack of development of this industry is the presently depressed international market in metals. Another reason involves the history of ownership of the pelagic nodules (see the section in this chapter titled "Laws and Treaties").

Cobalt-enriched manganese crusts, or hard coatings on other rocks, were discovered in relatively shallow water on the slopes of seamounts and islands within U.S. territorial waters in the 1980s. The concentration of cobalt in these deposits is roughly twice that found in typical pelagic manganese nodules and about one and one-half times that found in known continental deposits. These crusts are not being actively mined because of the relatively low cost and continued availability of continental sources.

Progress in mining cobalt from the sea floor continued to be slow in the 1990s and mostly outside the United States. An organization of twelve South Pacific Island nations (Cook Islands, Federated States of Micronesia, Fiji, Guam, Keribati, Marshall Islands, Papua New Guinea, Solomon Islands, Tonga, Tuvalu, Vanuatu, and Indonesia), with Australia and New Zealand as associate members, has supported over twenty-five deep-ocean survey cruises. The Cook Island government is accepting proposals for the mining of cobalt crusts in its coastal waters. These crusts have four to five times the cobalt content of manganese nodules. India has applied to the United Nations to develop manganese nodule deposits in the Indian Ocean and is working to advance mining systems and processing plant designs. Japan continues its research interests.

Sulfide Mineral Deposits

Expeditions to the rift valleys of the East Pacific Rise near the Gulf of California, the Galápagos Ridge off Ecuador, and the Juan de Fuca and Gorda Ridges off the northwestern United States have found sulfides of zinc, iron, copper, and possibly silver, molybdenum, lead, chromium, gold, and platinum. Molten material from beneath Earth's crust rises along the rift valleys, fracturing and heating the rock. Seawater percolates into and through the fractured rock, forming metal-rich hot solutions. When these solutions rise from the cracks and cool, the metallic sulfides precipitate to the sea floor. Deposits may be tens of meters thick and hundreds of meters long. Too little is presently known about these deposits to determine whether they might be of economic importance at some future date. No practical technology exists to sample or retrieve them at this time, and, like the manganese nodules, these deposits are found outside national economic zones, so there are ownership problems (see "Laws and Treaties").

In the 1960s, metallic sulfide muds were discovered in the Red Sea. Deposits of mud 100 m (330 ft) thick were found in small basins at depths of 1900–2200 m (6200–7200 ft). High amounts of iron, zinc, and copper and smaller amounts of silver and gold were found. The salty brines over these muds contained hundreds of times more of these metals than normal seawater.

Laws and Treaties

Because of the potential value of deep-sea minerals, specifically manganese nodules, and because the nodules are found in international waters, outside the usual 200-mile economic zones of coastal nations, the developing nations of the world feel they have as much claim to this wealth as those countries that are presently technically able to retrieve the nodules. The developing nations want access to the mining technology and a share in the profits. For nearly ten years, UN Law of the Sea Conferences worked to produce a treaty to regulate deep-ocean exploitation, including mining. The Law of the Sea Treaty was completed in April 1982. The treaty recognizes deep-sea mineral resources as the heritage of all humankind, to be regulated by a UN seabed authority that would license private companies to mine in tandem with a UN company. The quantities removed would be limited, and the profits would be shared. A UN cartel would both regulate and compete with those mining the sea.

The United States chose not to sign the treaty. In 1984, the United States, Belgium, France, West Germany, Italy, Japan, and the Netherlands signed a separate Provisional Understanding Regarding Deep Seabed Matters, and, under this provisional understanding, four international consortia have been awarded exploration licenses by the United States, West Germany, and the United Kingdom. By 1991, forty-five countries had ratified the Law of the Sea Treaty. Tempers cooled as the years passed, and the UN, realizing that it needs U.S. support to ensure international cooperation for the exploration and exploitation of ocean resources, included the United States in the working groups that met to resolve mining conflicts.

The UN Convention on the Law of the Sea (UNCLOS) came into force in 1994 without U.S. endorsement, but the ongoing negotiations to resolve the issues related to the section governing seabed mining have produced some results. That same year, an agreement that changes the provisions covering deep-seabed mining beyond the 200-mile Exclusive Economic Zone of individual nations was completed by the UN. Both the new agreement and UNCLOS have been submitted to the U.S. Senate for advice and consent. On October 31, 2007, the Senate Foreign Relations Committee voted in favor of sending the treaty to the full U.S. Senate for a ratification vote. In any case, it is unlikely that any deep-sea mining will occur until well into the twenty-first century. The high costs of sea mining, low metal prices, and still-undeveloped land sources combine to make rapid commercialization unlikely under any regulatory system.

Because there may be rich deposits of minerals and oil in Antarctica and its surrounding seas and because both national and private corporations are interested in surveying these areas for possible future mining and drilling, a multinational meeting in June 1988 produced an agreement to regulate mining exploration in and around that continent. This treaty must be ratified by sixteen of the twenty nations that signed the 1959 Antarctica Treaty, which banned all military activity and permitted scientific research. According to the 1988 treaty, no activities will be permitted if they will cause "significant changes" in atmospheric, terrestrial, or marine environments. Because no one really knows for sure how much mineral and oil wealth may be under Antarctica's ice and snow, the extent of future operations is also unknown.

In the United States, there is no offshore mining except for the sand and gravel operations in the state-owned waters of a few coastal states. Mining for manganese nodules, cobalt crusts, metallic sulfides, and phosphorites would be carried out within U.S. waters under provisions of U.S. domestic laws. Under these laws, the Department of the Interior is authorized to issue leases for mineral exploration and development on the continental shelf. No leasing or regulatory program has yet been developed, but federal and state task forces are working to develop programs for cobalt-rich crusts in the Hawaiian Islands, metallic sulfides on the Gorda Ridge off the coast of Oregon and northern California, and phosphorites off North Carolina.

Summary

Ocean-depth measurements were made first with a hand line, then with wire, and, since the 1920s, with echo sounders. Today, they are made with precision depth recorders. Seafloor features can also be sensed by satellites that measure the distance between the satellite and the sea surface.

The bathymetric features of the ocean floor are as rugged as the topographic features of the land but erode more slowly. The continental margin includes the continental shelf, slope, and rise. The continental shelf break is located at the change in steepness between the continental shelf and the continental slope. Submarine canyons are major features of the continental slope and, in some cases, the continental shelf. Some canyons are associated with rivers; others are believed to have been cut by turbidity currents. Turbidity currents deposit graded sediments known as turbidites.

The ocean basin floor is a flat abyssal plain, but it is interrupted by scattered abyssal hills, volcanic seamounts, and flat-topped guyots. In warm shallow water, corals have grown up around the seamounts to form fringing reefs. A barrier reef is formed when a seamount subsides while the coral grows. An atoll results when the seamount's peak is fully submerged. The mid-ocean ridges and rises extend through all the oceans; trenches are associated with island arcs and are found mainly in the Pacific Ocean.

Sediment classifications are based on their size, location, origin, and chemistry. Sediment particles are broadly categorized in order of decreasing size as gravel, sand, and mud. Within each of these categories, particles can be further subdivided by size. The sinking rate and distance traveled in the water column are related to sediment size, shape, and currents. Small particles sink more slowly than large particles. The very smallest particle sizes, silts and clays, sink so slowly that they may be transported large distances while falling to the sea floor. The sinking rate of particles is increased by clumping and incorporation into fecal pellets.

Sediments that accumulate on continental margins and the slopes of islands are called neritic sediments. Sediments of the deep-sea floor are pelagic sediments. In general, pelagic sediments accumulate very slowly and neritic sediments accumulate more rapidly.

Sediments formed from particles of preexisting rocks are called lithogenous sediments. These sediments are also sometimes called *terrigenous sediments*. Since lithogenous sediments are typically derived from the land, they are also known as terrigenous sediments. Pelagic lithogenous sediment is dominated by red clay. Red clay dominates marine sediments only in regions that are starved of other sources of sediment. Biogenous sediments come from living organisms. Sediments composed of at least 30% biogenous material are called oozes; this material accumulates in regions of high biological productivity. Siliceous sediments are subjected to dissolution everywhere in the oceans, while calcareous sediments dissolve rapidly in deep, cold water below the CCD. Sediments that precipitate directly from the water are called hydrogenous sediments. These include manganese nodules on the deep-sea floor and metal sulfides along mid-ocean ridges. Sediments containing particles that originate in space are called cosmogenous sediments.

Patterns of sediment deposit result from the distance from the source area, the abundance of living organisms contributing remains, the seasonal variations in river flow, waves and currents including turbidity currents, the variability in land sources, the prevailing winds, and sometimes rafting.

Coarse sediments are concentrated close to shore; finer sediments are found in quiet offshore or nearshore environments. Terrigenous sediments are found mainly along coastal margins; most deep-sea sediments come from biogenous sources. The distributions of particle sizes reveal the processes that formed the deposit, and the sediment layers provide clues to ancient climate patterns.

In general, sedimentation rates are slowest in the deep sea and greatest near the continents. In some areas, relict sediments were deposited under conditions that no longer exist. Mechanisms that increase the sinking rates of particles include clumping and incorporation of sediment particles into larger fecal pellets of small marine organisms. Loose sediments are transformed into sedimentary rock in which the layering of the sediments may be preserved.

Sediments are sampled with dredges, grabs, and corers; deep-sea drilling takes samples through the sediments and from the seafloor rock below.

Calcareous biogenous sediments preserve records of changes in the oxygen isotopic composition of seawater that are related directly to water temperature and hence can be used to study changes in global climate. Consequently, variations in $^{18}O:^{16}O$ isotopic ratios in calcareous skeletal remains record fluctuations in global coverage by ice sheets and in sea level.

Seabed resources include sand and gravel used in construction and landfills. Sands and muds that are rich in mineral ores are mined. Phosphorite nodules are the raw material of fertilizer. Oil and gas are the most valuable of all seabed resources. Manganese nodules are rich in copper, nickel, and cobalt; they are present on the ocean floor in huge numbers. Retrieval of seafloor mineral resources is slowed by disputes over international law, high mining costs, and low market prices. Sulfide mineral deposits have been discovered along rift valleys; their economic importance is unknown.

Large deposits of gas hydrates are being studied to determine their potential as economically important sources of methane gas. These deposits are icelike accumulations of natural gas and water that form at low temperature and high pressure on the sea floor. Scientists are also studying their possible role in submarine landslides and global climate change.

Key Terms

All key terms from this chapter can be viewed by term or definition when studied as flashcards on this book's website at www.mhhe.com/sverdrup10e.

soundings, 91
fathom, 91
echo sounder, 91
depth recorder, 91
continental margin, 94
continental shelf, 96
continental shelf break, 98
continental slope, 98
submarine canyon, 98
turbidity current, 99
turbidite, 99
continental rise, 99

abyssal plain, 100
abyssal hill, 100
seamount, 100
guyot, 100
fringing reef, 101
barrier reef, 101
atoll, 101
island arc system, 101
phytoplankton, 106
zooplankton, 106
photosynthetic, 106
heterotrophic, 106
test, 106
neritic, 107
pelagic, 107
relict sediment, 107
lithogenous sediment, 107

terrigenous sediment, 107
abyssal clay, 107
red clay, 108
biogenous sediment, 109
ooze, 109
calcareous ooze, 109
siliceous ooze, 109
coccolithophorids, 110
coccolith, 110
pteropod, 110
foraminifera, 110
lysocline, 110
carbonate compensation depth (CCD), 110
nutrient, 110
hydrogenous sediment, 110
carbonate, 110

phosphorite, 110
salt, 110
manganese nodule, 110
oolith, 110
cosmogenous sediment, 111
tektite, 111
rafting, 113
sedimentary rock, 114
lithification, 114
diagenesis, 114
metamorphic rock, 114
dredge, 115
grab sampler, 115
corer, 115
acoustic profiling, 115
paleoceanography, 117
isotope, 117

Study Questions

1. Are calcareous oozes more common in the southern Pacific Ocean or the northern Pacific Ocean? Why?
2. What is a turbidity current? Where would you expect a turbidity current to occur? How does the structure of a sediment

deposit left by a turbidity current differ from that of a shallow-water, nearshore sand deposit?

3. List the four basic sediment types classified by source. Where is each sediment type most likely to be found?

4. Discuss the future of commercial development and exploitation of deep-sea mineral resources.

5. Imagine that you are in a submersible on the ocean bottom. You leave New York and travel across the North Atlantic to Spain. Draw a simple ocean-bottom profile showing each major bathymetric feature you see as you move across the ocean. Name each feature. Do the same for the South Pacific between the coast of Chile and the west coast of Australia. Compare the two profiles. Did your depth scale differ from your horizontal scale? How much?

6. What processes form submarine canyons?

7. What is the continental margin? What pattern of sediment deposit would you expect to find associated with it? What processes produce these patterns of deposit?

8. What combination of factors is required to form a coral atoll?

9. What is a relict sediment? Where would you be likely to find such a deposit, and why would you find it in that place?

10. Describe several ways in which a continental shelf may be formed.

11. Describe methods used to recover sediment samples from the sea floor. Discuss the advantages and disadvantages of each method.

12. What is the average depth of the oceans in meters? In miles? In fathoms?

13. How is particle size used in understanding the pattern of seafloor deposits?

14. What are the implications for the marine environment as exploitation of seabed resources continues? Consider mining and drilling on continental shelves, in Antarctic waters, and in the open ocean.

15. The Grand Banks is an extensive, relatively shallow area southeast of Newfoundland, Canada, and off the U.S. northeastern coast. Why are there large boulders scattered across this area so far from shore?

16. Has the United States ratified the UN Convention on the Law of the Sea yet? If so, when did the ratification occur?

Study Problems

1. If underwater cables are spaced 14 km apart on the sea floor and if monitoring equipment shows that they break in sequence from shallow to deeper water at fifteen-minute intervals, what can you determine about the event causing the breaks?

2. If the average concentration of suspended sediment in the water is 1 g/m^3 and the volume of water in a harbor is 158 km^3, what is the average residence time of sediment in the water of this harbor? The daily sediment supply rate averages 1×10^7 kg. If the harbor has an average depth of 15.8 m, what is the surface area of the harbor? What would be the length of the side of a square having the same area as the surface area of the harbor?

3. In how many days will each of the particles listed in the following table reach the sea floor if the particles fall through 4000 m of seawater? All the particles are derived from land rock of the same density (2.8 g/cm^3). Settling rate is calculated from Stokes Law:

$$V \text{ cm/s} = 2.62 \times 10^4 r^2$$

where r is the radius expressed in centimeters (2.62×10^4 contains gravity, viscosity of water, the difference between the density of the particle and the density of water, and a constant for particle shape). How does the settling rate change if the diameter remains constant but the density of a particle changes?

Particle Type	Particle Diameter (mm)	Settling Rate (V) (cm/s)
Very fine sand	0.1	6.6×10^{-1}
Silt	0.06	2.4×10^{-1}
Clay	0.004	1.05×10^{-3}

4. Assuming a constant sedimentation rate of 0.4 cm per 1000 years, how thick will the sediments be in a portion of an ocean basin where the underlying crust is 130 million years old?

Fog along the coast at Big Sur, California.

Chapter 5

The Physical Properties of Water

Chapter Outline

Learning Outcomes

After studying the information in this chapter students should be able to:

1. *review* the physical properties of water listed in table 5.1,
2. *distinguish* between temperature and heat,
3. *construct* a plot of temperature vs. heat gain in 1 gram of water as it goes from a temperature of –10°C to 110°C,
4. *construct* a plot of density vs. temperature for pure water from a temperature of –2°C to 10°C,

5. *contrast* the change in density of pure water and average salinity seawater as they cool from a temperature of 10°C to –2°C,
6. *illustrate* the attenuation of light in open ocean water and coastal water, and
7. *theorize* how submarines could evade detection by surface ships using acoustic equipment.

---

*W*ater is one of the commonest substances on Earth, yet it is uncommon in many of its properties. Water is a unique liquid. It makes life possible, and its properties largely determine the characteristics of the oceans, the atmosphere, and the land. To understand the oceans, one must examine water as a substance and learn something of its physical and chemical characteristics. In this chapter, we learn about the structure of the water molecule and explore the properties of water. We also review three interesting and potentially hazardous forms of water: sea ice, icebergs, and fog.

5.1 The Water Molecule

The properties of water have excited scientists for over 2000 years. The early Greek philosophers (500 B.C.) counted four basic elements from which they believed all else was made: fire, earth, air, and water. In 1783, more than 2200 years later, English scientist Henry Cavendish determined that water was not a simple element but a substance made up of hydrogen and oxygen. Shortly afterward, another Englishman, Sir Humphrey Davey, discovered that the correct formula for water was two parts hydrogen to one part oxygen, or H_2O.

The chemical properties that make water such a special, useful, and essential substance result from its molecular structure. These properties are the subject of this chapter and are summarized in table 5.1.

The water molecule is deceptively simple, made up of three atoms: two hydrogen atoms and one oxygen atom. An atom is the smallest unit of matter that retains the properties of an element. We can envision an atom as consisting of three distinct types of particles located in two different regions. At the center is the nucleus. The nucleus contains positively charged particles called protons tightly packed together with electrically neutral particles called neutrons. A characteristic number of protons are present in the nucleus of every atom of a given element. For instance, every atom with one proton in the nucleus is an atom of hydrogen, and every atom with eight protons in the nucleus is an atom of oxygen. Negatively charged particles called electrons orbit the nucleus in a series of energy levels. The different energy levels can hold different numbers of electrons. Electrically neutral atoms have the same number of electrons as protons. In the case of hydrogen, there is one electron in the first energy level—an energy level that can hold a maximum of two electrons. In the case of oxygen, two electrons fill the first energy level, and six additional electrons are in the second energy level—a level that can hold a total of eight electrons when it is full. Thus, a hydrogen atom's outermost energy level is one electron short of being full and an oxygen atom's outermost energy level is two electrons short of being full. When two hydrogen atoms and one oxygen atom combine to form a water molecule, each hydrogen atom shares its single electron with the oxygen atom, and the oxygen atom shares one of its electrons with each hydrogen atom. Shared pairs of electrons form **covalent bonds.** The formation of covalent bonds in the water molecule has the effect of filling the outer energy levels of all three atoms in the molecule (fig. 5.1*a*).

The angle between the hydrogen atoms in the water molecule is about 105° (fig. 5.1*b*). A molecule of water is electrically neutral, but the negatively charged electrons within the molecule are distributed unequally. The shared electrons spend more time around the oxygen nucleus than they do around the hydrogen nuclei, giving the oxygen end of the molecule a slightly negative charge. The hydrogen end of the molecule carries a slightly positive charge. As a consequence, the opposite ends of the water molecule have opposite charges, and the molecule is an electrically unbalanced, or **polar, molecule.**

When one end of a water molecule comes close to the oppositely charged end of another water molecule, a bond forms between their positively and negatively charged ends (fig. 5.1*c*). These bonds are known as **hydrogen bonds,** and each water molecule can establish hydrogen bonds with neighboring water molecules. Any single hydrogen bond is weak (less than one-tenth the strength of the covalent bonds between the hydrogen and oxygen atoms), but as one hydrogen bond is broken, another is formed. Consequently, water is characterized by an extensive but ever-changing three-dimensional network of hydrogen-bonded molecules; the result is an atypical liquid with the extraordinary properties that are the subject of this chapter.

Table 5.1	Properties of Water	
Definition	**Comparison**	**Effects**
Physical States Gas, Liquid, Solid Addition or loss of heat breaks or forms bonds between molecules to change from one state to another.	The only substance that occurs naturally in three states on Earth's surface.	Important for the hydrologic cycle and the transfer of heat between the oceans and atmosphere.
Specific Heat One calorie per gram of water per °C.	Highest of all common solids and liquids.	Prevents large variations of surface temperature in the oceans and atmosphere.
Surface Tension Elastic property of water surface.	Highest of all common liquids.	Important in cell physiology, water surface processes, and drop formation.
Latent Heat of Fusion Heat required to change a unit mass from a solid to a liquid without changing temperature.	Highest of all common liquids and higher than most solids.	Results in the release of heat during freezing and the absorption of heat during melting. Moderates temperature of polar seas.
Latent Heat of Vaporization Heat required to change a unit mass from a liquid to a gas without changing temperature.	Highest of all common substances.	Results in the release of heat during condensation and the absorption of heat during vaporization important in controlling sea surface temperature and the transfer of heat in the atmosphere.
Compressibility Average pressure on total ocean volume 200 atmospheres; ocean depth decreased by 37 m (121 ft).	Seawater is only slightly compressible. 4–4.6×10^{-5} cm^3/g for an increase of 1 atmosphere of pressure.	Density changes only slightly with pressure. Sinking water can warm slightly due to its compressibility.
Density Mass per unit volume: grams per cubic centimeter, g/cm^3.	Density of seawater is controlled by temperature, salinity, and pressure.	Controls the ocean's vertical circulation and layering. Affects ocean temperature distribution.
Viscosity Liquid property that resists flow. Internal friction of a fluid.	Decreases with increasing temperature. Salt and pressure have little effect. Water has a low viscosity.	Some motions of water are considered friction free. Low friction dampens motion; retards sinking rate of single-celled organisms.
Dissolving Ability Dissolves solids, gases, and liquids.	Dissolves more substances than any other solvent.	Determines the physical and chemical properties of seawater and the biological processes of life forms.
Heat Transmission Heat energy transmitted by conduction, convection, and radiation.	Molecular conduction slow; convection effective. Transparency to light allows radiant energy to penetrate seawater.	Affects density; related to vertical circulation and layering.
Light Transparency Transmits light energy.	Relatively transparent for visible wavelength light.	Allows plant life to grow in the upper layer of the sea.
Sound Transmission Transmits sound waves.	Transmits sound very well compared to other fluids and gases.	Used to determine water depth and to locate objects.
Refraction The bending of light and sound waves by density changes that affect the speed of light and sound.	Refraction increases with increasing salt content and decreases with increasing temperature.	Makes objects appear displaced when viewed by light and sound.

5.2 Temperature and Heat

The atoms and molecules in any gas, liquid, or solid are always in motion. Thus, each individual atom or molecule has a certain amount of kinetic energy that is equal to one-half its mass times its velocity squared (kinetic energy $= 1/2\ mv^2$). Even in solids, where atoms and molecules are tightly packed and fixed in place, unable to move from one location to another as in a gas or liquid, those atoms and molecules will vibrate around their average position; they will have kinetic energy. The **temperature** of a substance is a measure of the average kinetic energy of the atoms and molecules in the substance. For a homogeneous material consisting of identical atoms or molecules all having the same mass, temperature is simply related to the average velocity of the atoms or molecules. The colder a substance is, the slower the motion of its atoms and molecules. Conversely, the warmer a substance is, the faster the motion of its atoms and molecules. Temperature is measured in **degrees** using one of

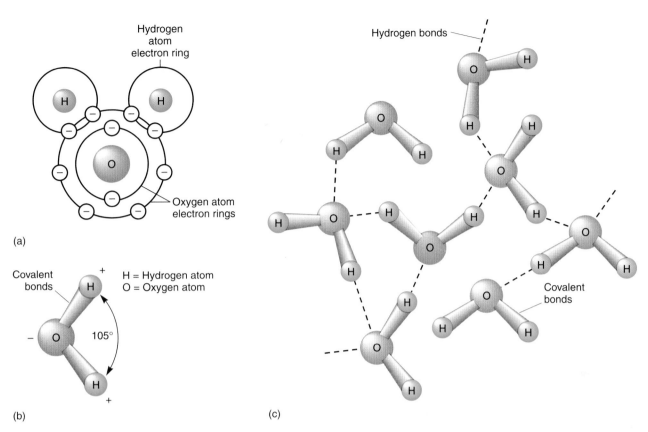

Figure 5.1   The water molecule. (a) The hydrogen atoms share electrons with the outer ring of the oxygen atoms. (b) The angle at which the hydrogen atoms form covalent bonds with the oxygen atom results in a polar molecule. (c) The positive and negative charges allow each water molecule to form hydrogen bonds with other water molecules.

three different scales. The two most commonly used scales are the Fahrenheit (°F) and Celsius (°C) scales (see "Temperature," appendix B). The third scale is the Kelvin (K) scale. It is used to measure extremes of hot and cold and is constructed in such a manner that 0 K corresponds to absolute zero, the temperature at which all atomic and molecular motion ceases. Absolute zero, 0 K, is equal to −273.2°C or −459.7°F; it is a temperature that can never be reached. **Heat** is a measure of the total kinetic energy of the atoms and molecules in a substance. The amount of heat in a substance is determined by the sum of the product of one-half the mass of every atom or molecule in the substance and its velocity squared. Heat is measured in **calories.** One calorie is the amount of heat needed to raise the temperature of 1 g of water by 1°C from 14.5°–15.5°C (see "Energy," appendix B). One thousand calories is equivalent to 1 Calorie, kilocalorie (kcal), or food calorie.

The difference between heat and temperature can be readily understood with a simple example. Imagine comparing boiling water in a pot on your stove to the water in a swimming pool on a warm day. The boiling water in the pot clearly has a higher temperature than the water in the pool; the average kinetic energy, or velocity, of the water molecules in the pot is much faster than in the pool. However, there is far more heat in the pool water. Even though the average kinetic energy, or velocity, of the water molecules in the pool is relatively small, there are so many of them that the total kinetic energy, or heat, is greater.

5.3 Changes of State

Water exists on Earth in three physical states: solid, liquid, and gas (fig. 5.2). When it is a solid, we refer to it as ice, and when it is a gas, we call it water vapor. Pure water ice melts at 0°C and pure water boils at 100°C at standard atmospheric pressure. Pure water is defined as fresh water without suspended particles or dissolved substances, including gases. Standard atmospheric pressure is equal to 76 cm of mercury measured by barometer and is discussed in chapter 7. If you are unfamiliar with the Celsius temperature scale, see appendix B.

Because of the hydrogen bonding between the water molecules, it takes energy (heat) to separate them from each other: that is, for liquid water to evaporate or for ice to melt. In the natural environment, this heat is supplied by the Sun.

When pure water makes any of the changes among liquid, solid, and gas, it is said to change its state. Changes of state are due to the addition or loss of heat. When enough heat is added to solid water or ice, the hydrogen bonds break and the ice melts, forming water. When heat is added to liquid water, the temperature of the water rises and some of the water molecules escape from the liquid or evaporate to form water vapor. When heat is removed from water vapor and the temperature falls below the **dew point,** or temperature of water vapor saturation, the water vapor condenses to liquid. When liquid water loses heat and its temperature is lowered to its freezing point, ice is formed as the water molecules form into a crystalline lattice.

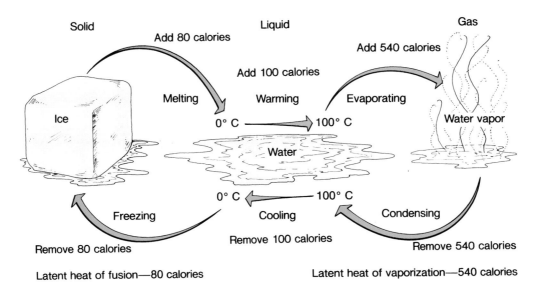

Figure 5.2   Heat energy must be added to convert a gram of ice to liquid water and to convert liquid water to water vapor. The same quantity of heat must be removed to reverse the process.

To change pure water from its solid state (ice) to liquid water at 0°C requires the addition of 80 calories for each gram of ice. There is no change in temperature; there is a change in the physical state of the water as hydrogen bonds break. The reverse of this process is required to change liquid water to ice. For each gram of liquid water that becomes ice, 80 calories of heat must be removed at 0°C. The heat necessary to change the state of water between solid and liquid is known as the **latent heat of fusion.** This addition or loss of heat takes time in nature. A lake does not freeze immediately, even though the surface water temperature is 0°C, nor does the ice thaw on the first warm day. Time is required to remove or add the heat needed for the change of state.

One gram of liquid water requires 1 calorie of heat to raise its temperature 1°C. Therefore, 100 calories are needed to raise the temperature of 1 g of water from 0° to 100°C. In comparison, the 80 calories of heat required per gram to convert ice to and from liquid water at 0°C with no change in temperature is relatively large. Ice is a very stable form of water; a large addition of heat is required to melt ice, and a large removal of heat is required to freeze water.

The change of state between liquid water and water vapor requires 540 calories of heat to convert 1 g of water to water vapor at 100°C. When 1 g of water vapor condenses to the liquid state, 540 calories of heat are liberated. Again, there is no change in temperature; there is only a change in the water's physical state. The heat needed for a change between the liquid and vapor states is known as the **latent heat of vaporization.**

The energy required to change water from solid to liquid to vapor is shown in figure 5.2. Both the latent heat of fusion and the latent heat of vaporization are presented in this diagram, as well as the calories needed for each change of state and the temperatures at which these changes occur. Water converts from the liquid to the vapor state at temperatures other than 100°C; for example, rain puddles evaporate and clothes dry on the clothesline. This change requires slightly more heat to convert the liquid water to a gas at these lower temperatures. Water can remain in the liquid phase at temperatures above 100°C when the pressure is greater than 1 atmosphere. One atmosphere (atm) is equal to the pressure of a column of mercury 76 cm (30 in) high; in English units, 1 atm equals 14.7 pounds per square inch (lb/in²). The conversion from liquid to vapor at temperatures above 100°C requires slightly less heat energy than at lower temperatures (fig. 5.3).

As water is evaporated from the world's lakes, streams, and oceans and returned as precipitation, heat is being removed from Earth's surface and liberated into the atmosphere,

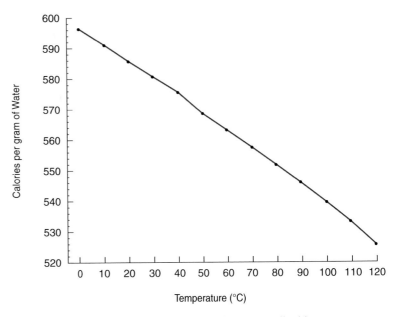

Figure 5.3   The amount of energy required to convert liquid water to water vapor as a function of temperature (the latent heat of vaporization).

where condensation occurs to form clouds. This heat is a major source of the energy used to power Earth's weather systems (see chapter 7).

Under certain conditions it is possible (1) to cool liquid water below 0°C without freezing, (2) to change ice directly to a gas, a process known as **sublimation,** or (3) to boil water at temperatures below 100°C. Cooling below 0°C occurs in some clouds and in the laboratory by controlled cooling of pure water to produce supercooled water. Sublimation is seen in nature when snow or ice evaporates directly under very cold and dry conditions. Anyone living at high altitudes knows that potatoes must be boiled longer and that brewed coffee is cooler there than at sea level because of the decrease in atmospheric pressure and the lower boiling temperature of the water.

The behavior of water, as described, is explained by considering the processes occurring between the water molecules. At the molecular level, the addition of heat energy increases the speed of the molecules, while the loss of heat energy decreases their rate of motion. Heat energy is required to break hydrogen bonds between water molecules, and heat is released when hydrogen bonds are formed. The addition of heat to water causes a relatively small change in the temperature because much of the heat energy is used to disrupt the hydrogen bonds. These bonds must be broken before the molecules are able to move more rapidly. When heat is removed from water, the molecules slow, many additional hydrogen bonds are formed, and considerable energy is released as heat; this process prevents any rapid drop in temperature.

Water molecules stay close together because of their polarity. If the molecules are moving fast enough, they overcome the attractions between them and leave the liquid, entering the air as a gas. Relatively large amounts of heat are needed to evaporate water because hydrogen bonds must first be broken. The greater the addition of heat, the more hydrogen bonds are disrupted and the greater the average energy of motion of the molecules. Under these conditions, more water molecules leave the liquid more quickly. In the same way, large amounts of heat must be extracted from water to form the hydrogen bonds required to freeze water. On Earth, natural temperatures required for boiling are rare and those for freezing are frequent but geographically limited; the average Earth temperature of about 16°C ensures that liquid water is abundant.

Dissolved salts in water change its boiling and freezing points. The boiling temperature is raised and the freezing temperature is lowered. The amount of change is controlled by the amount of salt dissolved. The rise in the boiling temperature is of little consequence to oceanography because seawater does not normally reach such high temperatures in nature, but the lowering of the freezing point is important in the formation of sea ice. Seawater freezes at about −2°C.

5.4 Specific Heat

Of all the naturally occurring earth materials, water changes its temperature the least for the addition or removal of a given amount of heat. The ability of a substance to give up or take in a given amount of heat and undergo large or small changes in temperature is a measure of the substance's **specific heat.** The specific heat of water is

Table 5.2	**Specific Heat of Earth Materials**
Material	**Specific Heat (Calories/g/°C)**
Water	1.00
Air	0.25
Sandstone	0.47
Shale	0.39
Sandy soil	0.24
Basalt	0.20
Limestone	0.17

very high compared to that of soil, rock, and air (table 5.2). For example, summer temperatures in the Libyan desert reach 50°C and temperatures in the Antarctic drop to −50°C, for a worldwide temperature range of 100°C on land. Ocean temperatures vary from a nearly constant high of approximately 28°C in the equatorial areas to a low of −2°C in Antarctic waters, for a worldwide water temperature range of 30°C. The water in a lake changes its temperature very little between noon and midnight, but the adjacent land and air temperature changes are large during this same time. The high specific heat of water and the ability of water to redistribute heat over depth allow the world's lakes and oceans to change temperature slowly, helping to keep Earth's surface temperature stable.

Specific heat of a material is the quantity of heat required to produce a unit change of temperature in a unit mass of that material. The specific heat of water is 1.0 calorie per gram per degree Celsius (cal/g/°C). This is much higher than specific heats of most other liquids because of water's extensive hydrogen bonding. Energy that is used to break hydrogen bonds between water molecules is used in other liquids to directly increase the molecular motion and temperature. The high specific heat of water allows water to gain or lose large quantities of heat with little change in temperature. When salt is added to pure water, the changes in heat capacity, latent heat of fusion, and latent heat of vaporization are small.

The **heat capacity** of a material is the quantity of heat required to produce a unit change of temperature in the material. A material's heat capacity depends on the material's specific heat and the mass of the material. For instance, the heat capacity of 1 kilogram of water is 1000 calories/°C, and the heat capacity of 1 kilogram of sandy soil is 240 calories/°C (see table 5.2). The specific heat of a material is numerically the same as the heat capacity of 1 gram of the material.

5.5 Cohesion, Surface Tension, and Viscosity

In the liquid state, the bonds between water molecules form, break, and re-form with great frequency. Each bond lasts only a few trillionths of a second. However, at any instant, a substantial percentage of all water molecules are bonded to their neighbors. Therefore, water has more structure than other liquids. Collectively, the hydrogen bonds hold water together; this property is known as **cohesion.**

Cohesion is related to **surface tension,** which is a measure of how difficult it is to stretch or penetrate the surface of a liquid. At the surface between air and water, water molecules arrange themselves in an ordered system, hydrogen-bonded to each other laterally and to the water molecules beneath. This arrangement forms a weak elastic membrane that can be demonstrated by filling a water glass carefully; the water can be made to brim above the top of the glass but not overflow. A steel needle can be floated on water; insects such as the water strider walk about on the surface of lakes and streams. These things are possible because water has a high surface tension. This property is important in the early formation of waves. A gentle breeze stretches and wrinkles the smooth water surface, enabling the wind to get a better grip on the water and to add more energy to the sea surface (see chapter 10, section 10.1).

The addition of salt to pure water increases the surface tension. Decreasing the temperature also increases the surface tension, and increasing the temperature decreases it.

Liquid water pours and stirs easily. It has little resistance to motion or internal friction. This property is called **viscosity.** Water has a low viscosity when compared to motor oil, paint, or syrup. Viscosity is affected by temperature. Consider pancake syrup. When it is stored in the refrigerator, it becomes more viscous and slow to pour. It has a high viscosity. When the syrup is returned to room temperature or heated, it becomes more runny; it has a low viscosity. The same is true of water, but the change in viscosity is much less, and it is not noticeable with normal temperature variations. Surface water at the equator is warmer and therefore less viscous than surface water in the Arctic. Minute microscopic organisms find it easier to float in the more viscous polar waters; some of their tropical-water cousins have adapted to the less-viscous water by developing spines and frilly appendages to help keep them afloat. The addition of salt to water increases the viscosity of the water, but the change is small.

5.6 Density

Density is defined as mass per unit volume of a substance. Water density is usually measured in grams per cubic centimeter (g/cm^3). The density of pure water is often determined at 3.98°C, the temperature of maximum density, or approximately 4°C, and is 1 g/cm^3. Therefore, a cube of pure water with 1 cm long edges has a mass of 1 g or a density of 1 g/cm^3. The density of seawater is greater than the density of pure water at the same temperature because seawater contains dissolved salts. At 4°C, the density of seawater of average salinity is 1.0278 g/cm^3. The densities of other substances may be considered in the same way (table 5.3). Less-dense substances will float on denser liquids (for example, oil on water, dry pine wood on water, and alcohol on oil). Ocean water is denser than fresh water; therefore, fresh water floats on salt water.

The Effect of Pressure

Pure water is nearly incompressible, and so is seawater. Pressure in the oceans increases with increasing depth. For every 10 m (33 ft) in depth, the pressure increases by about 1 atmosphere,

or 14.7 lb/in^2. See figure 5.4 for the effect of water pressure at 2000 m (6560 ft). Pressure in the deepest ocean trench, 11,000 m (36,000 ft) deep, is about 1100 atm. These great pressures have only a small effect on the volume of the oceans because of the slight compressibility of the water. A cubic centimeter of seawater at the surface will lose only 1.7% of its volume if it is lowered to 4000 m (13,000 ft), where the pressure is 400 atm. The average pressure acting on the total world's ocean volume results in the reduction of ocean depth by about 37 m (121 ft). In other words, if the ocean water were truly incompressible, sea level would stand about 37 m higher than it does at present. The pressure effect is small enough to be ignored in most instances except when very accurate determination of seawater density is required. Further information on units of pressure can be found in appendix B.

The Effect of Temperature

Water density is very sensitive to temperature changes. When water is heated, energy is added and the water molecules speed up and move apart; therefore, the mass per cubic centimeter

Table 5.3	Densities of Common Materials
Material	**Density (g/cm^3)**
Ice (pure) 0°C	0.917
Water (pure) 0°C	0.99987
Water (pure) 3.98°C	1.0000
Water (pure) 20°C	0.99823
White pine wood	0.35–0.50
Olive oil 15°C	0.918
Ethyl alcohol 0°C	0.791
Seawater 4°C (salt 35 g/kg)	1.0278
Steel	7.60–7.80
Lead	11.347
Mercury	13.6

Figure 5.4   These oceanography students hold research cruise souvenirs—polystyrene coffee cups that were attached to a sampler and lowered 2000 m (6560 ft) into the sea. The water pressure compressed the cups to the size of thimbles.

becomes less because there are fewer molecules per cubic centimeter. For this reason, the density of warm water is less than that of cold water, and warm water floats on cold water. When water is cooled, it loses heat energy, and the water molecules slow down and come closer together; there are then more water molecules, or a greater mass, per cubic centimeter. Because cold water is denser than warm water, it sinks below the warm water. To see these changes, try the following experiment. Chill a small quantity of water in the refrigerator; mix it with a little dye such as ink or food coloring to help see the effect. Allow the water from the tap to run hot and fill a glass half full of the hot water. Now slowly and carefully pour the colored water down the side of the glass on top of the hot water and watch what happens. You should see the cold, colored water sink below the hot water to create two distinct layers: the higher-density cold water will be on the bottom, and the lower-density hot water will be on top.

In pure fresh water, molecules move more slowly and come closer and closer together as water is cooled to 4°C. At this temperature, the water molecules are so close together and moving so slowly that each molecule can form hydrogen bonds with four other molecules. Unlike other fluids that continuously increase in density as they cool, water reaches its greatest density, 1 g/cm³, at 3.98°C, or about 4°C. As the temperature of the water falls below 4°C, the molecules move slightly apart, and at 0°C, they form an open latticework that is the stable structure for ice. Water as a solid takes up more space than water as a liquid; there are fewer water molecules per cubic centimeter, and therefore, ice is less dense than water and floats on water (figs. 5.5 and 5.6). Frozen water's increase in volume is dramatically and often disastrously demonstrated every winter when water pipes freeze and then burst. However, if water continued to contract until it froze, the ice would sink, and lakes would accumulate ice from the bottom up.

When ice absorbs enough heat, the hydrogen bonds between molecules are broken, the lattice starts to collapse, and the molecules move closer together. Above 4°C, the molecules move apart as they increase their motion (fig. 5.6). If enough heat is added, the energy level of the molecules increases until they overcome the attractive forces in the liquid and evaporate into the air as water vapor. Because water vapor is less dense than the mixture of gases that form the atmosphere, a mixture of water vapor and dry air is less dense than dry air alone at the same temperature and pressure.

The Effect of Salt

When salts are dissolved in water, the density of the water increases because the salts have a greater density than water. In other words, they have more mass per cubic centimeter. The average density of seawater at 4°C is 1.0278 g/cm³, compared with 1 g/cm³ for fresh water. Therefore, fresh water floats on salt water. To see this happen, take two small quantities of fresh water; add dye to one and some salt to the other. Carefully and slowly add the salty water to the dyed fresh water. You should see the salty water sink below the dyed fresh water to create two layers: the lower-density fresh water on top and the higher-density salty water on the bottom.

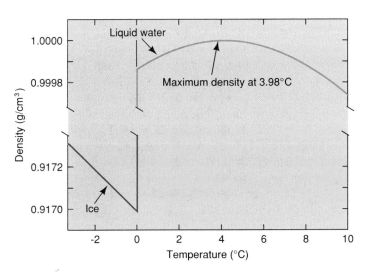

Figure 5.5   The density of pure water reaches its maximum at 3.98°C. Pure water is free of dissolved gases.

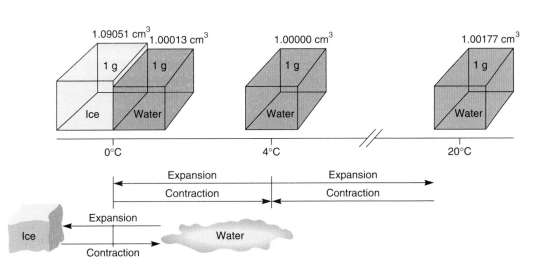

Figure 5.6   Expansion and contraction, or change in volume, of 1 g of water with changes in temperature. At 0°C, water expands during change of state to ice. Between 0° and 4°C, water contracts as it warms and expands as it cools. Above 4°C, water expands as it warms and contracts as it cools.

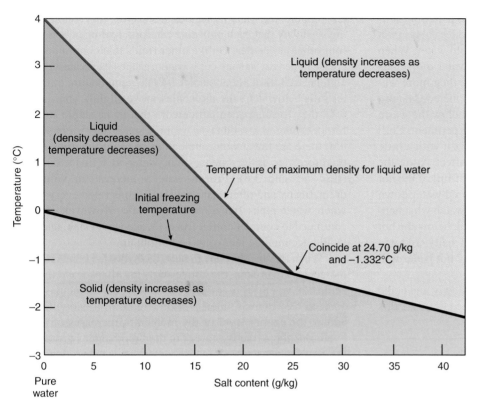

Figure 5.7   The initial freezing point and the temperature of maximum density of water decrease as the salt content increases.

If seawater contains less than 24.7 grams of salt per kilogram, then the seawater, although denser, will behave much like fresh water. Seawater of low salt concentration will reach maximum density before freezing at a temperature less than 0°C. Cooling surface water with a salt content of less than 24.7 g/kg causes the water to increase in density and sink. This process continues until the temperature of maximum density is reached. Further cooling causes this low-salinity surface water to become less dense and remain at the surface.

At a salt content of 24.7 g/kg, the freezing point and temperature of maximum density of seawater coincide at −1.332°C. If the salt content is greater than 24.7 g/kg, freezing occurs before a maximum density is reached. This relationship is shown in figure 5.7. Open-ocean water generally has an average salt content of 36 g/kg; therefore, the density of seawater increases, and the seawater sinks continuously as it is cooled to its freezing point. The effect of temperature and salt content on the density of water is shown in figure 5.8. Notice in this figure that (1) if temperature remains constant, density increases with increasing salt content, and (2) if the salt content remains constant and is greater than 24.7 g/kg (review fig. 5.7), density increases as temperature decreases.

5.7 Dissolving Ability

More substances—solids, liquids, and gases—dissolve in water than in any other common liquid. Water has been called the universal solvent because of its exceptionally good dissolving ability for a wide range of materials.

The polar nature of the molecules makes water an excellent solvent. For example, common table salt, chemically sodium chloride (NaCl), dissolves in water. Each salt molecule is made up of one atom of sodium with a positive charge and one atom of chlorine with a negative charge. These charged atoms are known as **ions.** When sodium chloride is in its natural crystalline lattice, the sodium ions and the chloride ions are held together by the attraction of their opposite electrical charges. To dissolve in water, the sodium and chloride ions must overcome their attraction for each other and interact instead with the molecules of water. Because of their polarity, the water molecules can form spheres around both ions. Figure 5.9a shows water molecules clustered around a sodium ion (Na^+) with their negative (oxygen) ends pointing toward it. Around the chloride ion (Cl^-), the orientation of the water molecules is reversed (fig. 5.9b), with the positive (hydrogen) ends of the solvent molecules pointing toward the ion. The salt ions are surrounded and separated by water molecules. The many dissolved salts in seawater are discussed in chapter 6.

For millions of years, volcanism and rain washing over the land have been supplying the oceans with dissolved salts. Once these reach the ocean basins, most of the dissolved salts stay in the ocean. Water is recycled to the land by oceanic evaporation, but salts remain in the sea and the bottom sediments. Some land salt deposits resulted from geologic uplift processes that bring seafloor deposits above the ocean surface. Other salt deposits on land are the remnants of ancient shallow seas that became isolated over geologic time; the water evaporated, leaving the salt deposits behind. Salts in the seawater and marine sediments are recycled as a result of subduction and subsequent volcanism, as discussed in chapter 3.

5.8 Transmission of Energy

Fresh water and salt water both transmit energy. The energy may be in the form of heat, light, or sound. Because the principles of transmission are the same in fresh water and salt water, our discussion will center on energy transmission in the oceans.

Heat

There are three ways in which heat energy may be transmitted within a material: by **conduction,** by **convection,** and by **radiation.** Conduction is a molecular process. When heat is applied at one location, the molecules move faster because of the addition of energy; gradually, this more rapid molecular motion passes on to the adjacent molecules and the motion spreads. For

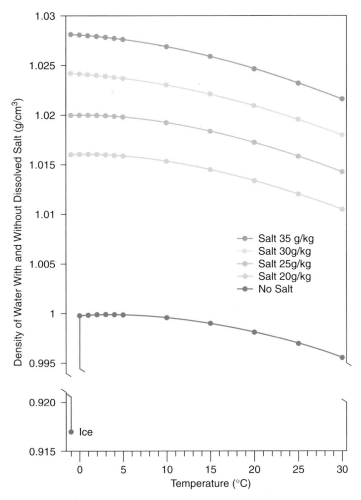

Figure 5.8 The density of seawater increases with increasing salt content at constant temperature and increases with decreasing temperature to the freezing point at constant salt content. The density of pure water with zero salinity increases with decreasing temperature until it reaches a maximum density at about 4°C and then decreases to the freezing point.

example, if a metal spoon is placed in a hot liquid, the handle soon becomes hot. Heat has been conducted from the part of the spoon that is in contact with the heat source to the handle. Metals are excellent conductors; water is a poor conductor and transmits heat slowly in this way.

Convection is a density-driven process in which a heated fluid moves and carries its heat to a new location. In older home heating systems, hot air is supplied through vents at floor level; the hot air rises because it is less dense than the cooler air above it, and the rising air carries the heat with it. When this air is cooled, it sinks toward the floor because it has become denser. In these energy-conscious days, ceiling fans are used to force the less dense hot air down, to keep the room warm at all levels rather than accumulating excess heat at the ceiling. Water behaves in the same way; it rises when heated and its density decreases; it sinks when cooled and its density increases. Remember that convection cells were discussed in chapter 3.

Radiation is the direct transmission of heat from its energy source. Switch on a heat lamp and feel, from a foot away, the heat released. This is radiant energy. The Sun provides Earth with radiant energy that penetrates and warms the surface waters of the planet. Unlike conduction and convection, which require a medium for the transfer of heat, radiated heat can be transferred through the vacuum of space and through transparent materials.

Think about trying to warm the water in each of two containers. If solar radiation is used to heat the surface water of the first container, some of the radiation will be reflected from the water's surface, adding no heat to the water, and some of the radiation will be absorbed by the surface water, raising the water's temperature. Because warm water is less dense, it remains at the surface. A small amount of heat may be transmitted slowly downward by molecular conduction, but unless the container is stirred by mechanical energy from another source, the heat remains at the water's surface. If the second container of water is heated from below, the water at the bottom of the container gains heat, decreases in density, and rises, taking heat with it and distributing the heat through the water volume.

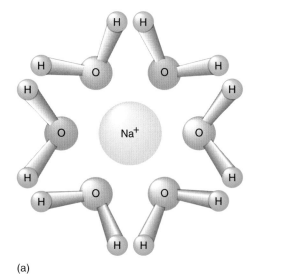

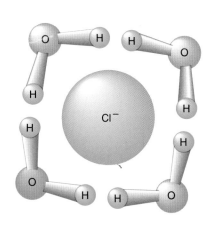

Figure 5.9 Salts dissolve in water because the polarity of the water molecule keeps positive ions separated from negative ions. (a) Sodium ions are surrounded by water molecules with their negatively charged portion attracted to the positive ion. (b) Chloride ions are surrounded by water molecules with their positively charged portion attracted to the negative ion.

(a)   (b)

As the warmed water rises, it is replaced by colder water from the surface, which is warmed in turn. Convection is a much more rapid and efficient method of distributing heat in water than are radiation and conduction.

The oceans are heated from above by solar radiation absorbed in the upper surface layers of water. The heat gained at the surface is transmitted slowly downward by the natural turbulent stirring action of the wind and the currents, as well as by the slower molecular conduction. In addition, the ocean's surface water may lose heat to the overlying atmosphere as a result of two processes: (1) direct transfer of heat from warm water to colder air by conduction and convection processes, and (2) the transfer of water vapor to the atmosphere by evaporation and atmospheric condensation. Heat transfers that warm the atmosphere from below create atmospheric convection that enhances this heat transfer. The role of these processes in oceanic and atmospheric circulation is discussed in chapters 7 and 8.

Light

The light striking the ocean surface is one of the many forms of **electromagnetic radiation** Earth receives from the Sun. The full range of this radiation may be seen in the **electromagnetic spectrum,** shown in figure 5.10. Note that visible light occupies a very narrow segment of the spectrum at wavelengths from about 390–760 nm (390–760 $\times$ 10^{-7} cm). Wavelengths shorter than visible light include ultraviolet light, X rays, and gamma rays, whereas infrared (heat) waves, microwaves, TV, and FM and AM radio waves all have wavelengths longer than visible light. Visible light may be broken down into the familiar spectrum of the rainbow: red, orange, yellow, green, blue, and violet. Each color represents a range of wavelengths; the longest wavelengths are at the red end of the spectrum, and the shortest are at the blue-violet end. When combined, these wavelengths produce white light.

As light passes through the water, it is **absorbed** and **scattered** by water molecules, ions, and suspended particles, including silt and micro-organisms. It is also absorbed by organisms for photosynthesis, to be used in their life processes. This decrease in the intensity of light over distance is known as **attenuation.**

Seawater transmits only a small portion of the electromagnetic spectrum, primarily in the visible band (fig. 5.11). Light energy is attenuated very rapidly with depth, particularly the longer infrared wavelengths. The intensity of light at any depth can be calculated using the following equation (known as Beer's Law):

$$I_z = I_0 e^{-kz}$$

where I_0 is the intensity of the light at the surface, I_z is the intensity of the light at a depth of z meters, and k is the attenuation coefficient. Beer's Law can be used to calculate the depth at which a given light-intensity ratio (I_z/I_0) occurs in the water by rearranging terms:

$$z = -\ln(I_z/I_0)/k$$

The attenuation coefficient k varies with the clarity of the water. The clearer the water, the smaller the attenuation coefficient and the greater the light penetration (fig 5.12 and table 5.4). In typical open ocean water, about 50% of the entering light energy is attenuated in the first 10 m (33 ft) ($I_z/I_0 = 0.5$), and almost all of the light is attenuated 100 m (330 ft) beneath the surface ($I_z/I_0 = 0.001$). In coastal water, where the water is not as clear, about 96% of the light energy is attenuated in the first 10 m (33 ft), and almost all the light is attenuated only 20 m (66 ft) beneath the surface.

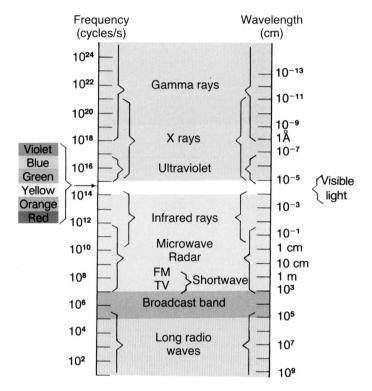

Figure 5.10   The electromagnetic spectrum.

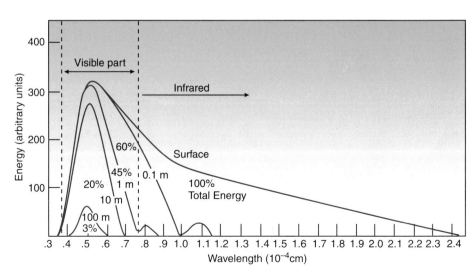

Figure 5.11   The percentage of total solar energy in the sea decreases as depth increases. The long, red wavelengths are absorbed first, and the color peak shifts toward the shorter, blue wavelengths.

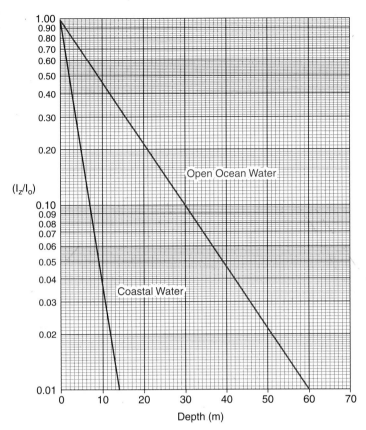

Figure 5.12 Ratio of the intensity of visible light at a depth of "z" meters (I_z) to the intensity at the surface (I_0) in open ocean water (*red*) and coastal water (*blue*). The slope of each line is equal to the negative of the attenuation coefficient for blue light (see tabular representation of this information in table 5.4).

The rate of light attenuation also depends heavily on the wavelength of the light (table 5.5). The short wavelengths in the ultraviolet regions and the long wavelengths in the red region and beyond are attenuated very rapidly (fig. 5.13). Light in the

Table 5.4	Attenuation of Visible Light in Open Ocean and Coastal Water[*]	
Depth (m)	% of Light Attenuated Open Ocean Water	% of Light Attenuated Coastal Water
0	0	0
0.1	0.8	3.3
1	7.3	28.4
10	53.2	96.5
20	78.1	99.9
30	89.8	~100
40	95.2	~100
50	97.8	~100
100	99.9	~100
150	~100	~100

[*] Values used for attenuation coefficient (k) are average values for blue light (480 nm); $k = 0.076$ m^{-1} in the open ocean and $k = 0.334$ m^{-1} in coastal water.

blue-green to blue-violet range is attenuated less rapidly and penetrates to the greatest depth (fig. 5.14).

Color perception is due to the reflection back to our eyes of wavelengths of a particular color. Because all wavelengths, or colors, are present to illuminate objects in shallow water, objects just below the surface are seen in their natural colors. Objects in deeper waters usually appear dark because they are illuminated mainly by blue light. Ocean water usually appears blue-green because wavelengths of this color, being attenuated the least, are most available to be reflected and scattered back to an observer. Coastal waters vary in color, appearing green, yellow, brown, or red; these waters usually contain silt from rivers as well as large numbers of microscopic organisms and organic substances. The presence of these inclusions is revealed by the light reflecting from depth in the water. Open-ocean water beyond the influence of land contains little suspended matter. Light penetrates deeper in water that contains less particles, and open-ocean water appears bluer than coastal waters.

Table 5.5	Attenuation of Light at Different Wavelengths[*]			
Color	Wavelength (nm)	Attenuation k (m^{-1})	% Attenuated in Upper 1 m of Water	Depth at Which 99% Is Attenuated (m)
	200	3.14	95.7	1
	250	0.588	44.5	8
Ultraviolet (10–390 nm)	300	0.154	14.3	30
	350	0.0530	5.2	87
	400	0.0209	2.1	220
Violet (390–455 nm)	450	0.0168	1.7	274
Blue (455–492 nm)	500	0.0271	2.7	170
Green (492–577 nm)	550	0.0648	6.3	71
Yellow (577–597 nm)	600	0.245	21.7	19
Orange (597–622 nm)	650	0.350	29.5	13
	700	0.650	47.8	7
Red (622–760 nm)	750	2.47	91.5	2
Infrared (760 nm–1 mm)	800	2.07	87.4	2

[*] Attenuation coefficients (k) are values for clearest ocean water.

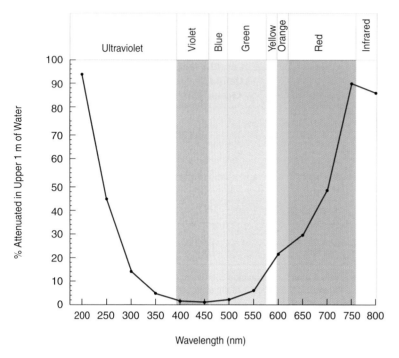

Figure 5.13 Percent of light attenuated in the upper 1 meter of clear ocean water as a function of wavelength. Light outside of the visible band is attenuated very rapidly. Within the visible band, light at blue and violet wavelengths experiences the least attenuation.

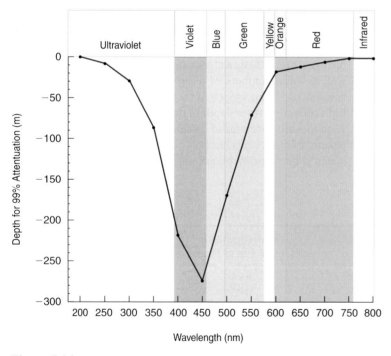

Figure 5.14 Depth (m) at which 99% of the incident light at the surface has been attenuated in the clearest ocean water. Light in the blue and blue-violet wavelengths penetrates to the greatest depth.

When light passes from air into water it is bent, or **refracted,** because the speed of light is faster in less dense air than in denser water. Because of the light refraction, objects seen through the water's surface are not where they appear to

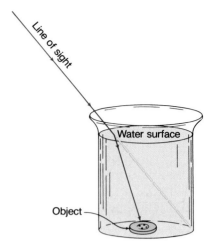

Figure 5.15 Objects that are not directly in the individual's line of sight can be seen in water because of the refraction of the light rays. The refraction is caused by the decreased speed of light in water.

be (fig. 5.15). Refraction is affected slightly by changes in salinity, temperature, and pressure.

The simplest way of measuring light attenuation in surface water is to use a **Secchi disk** (fig. 5.16). This is a disk, about 30 cm (12 in) in diameter, that is lowered on a line to the depth at which it just disappears from view. This depth can be used to estimate the average attenuation coefficient for light using the following relationship:

$$k = (1.44/z)$$

where k is the average attenuation coefficient and z is the depth in meters at which the disk can no longer be seen. Secchi disks are either entirely white, as shown in figure 5.16, or divided into alternating black and white quadrants. In water rich with living organisms or suspended silt, the Secchi disk may disappear from view at depths of 1–3 m (3–10 ft); in the open ocean, visibility usually extends down to 20–30 m (65–100 ft), although a Secchi disk reading of 79 m (260 ft) was reported from Antarctica's Weddell Sea in 1986. Although crude, a Secchi disk has the advantages of low cost, never needing adjustment, never leaking, and never requiring replacement of electronic components. However, this technique explains little about how seawater affects light.

Changes in light attenuation are measured by devices that project a beam of light over a fixed distance to an electronic photoreceptor. Such devices are lowered meter by meter into the sea, and measurements are taken at chosen intervals. The cumulative effect of attenuation may also be calculated to any depth. Different wavelengths of light are used to separate the effects of absorption and scattering by inorganic and organic particles as well as absorption by dissolved substances and phytoplankton in the water. The role of natural fluorescence of material exposed to sunlight is also under investigation. Most light is produced when atoms become excited and give off radiation in the form of light and heat. In fluorescence, only the atom's electrons are agitated; they give off radiation in the form of light but practically no heat. Fluorescence is sometimes described as "cold light." Multichannel color sensors are used to observe both the changing properties of solar radiation as it penetrates the ocean and the radiance emitted by the fluorescence of dissolved compounds. Light transmission data are used to adjust color quality and clarity of underwater photographs and video.

The Marine Optical Buoy (MOBY; fig. 5.17) was launched in 1997; it carries instruments to measure light and color at the surface and at three different depths. Measurements are transmitted by fiber optics from the buoy to instruments onboard ship. These measurements are used to improve satellites' sensing of sea surface color. The satellites monitor sea surface color to determine the abundance of phytoplankton at the sea surface, which are used as indicators of the biological productivity of the oceans.

Sound

The sea is a noisy place; waves break, fish grunt and blow bubbles, crabs snap their claws, and whales whistle and sing. Sound travels farther and faster in seawater than it does in air; the

Figure 5.16 A Secchi disk measures the transparency of water.

Figure 5.17 The Marine Optical Buoy (MOBY) is lowered into the sea. MOBY measures radiant energy entering and exiting the sea surface at three different depths. The data are used to adjust ocean color in satellite images.

average velocity of sound in seawater is 1500 m/s (5000 ft/s) compared with 334 m/s (1100 ft/s) in dry air at 20°C.

The speed of sound in seawater is a function of the axial modulus and density of the water:

$$\text{speed of sound} = \sqrt{\frac{\text{axial modulus}}{\text{density}}}$$

The axial modulus of a material is a measure of how it compresses. A material with a high axial modulus is more difficult to compress than a material with a low axial modulus; a golf ball has a higher axial modulus than a tennis ball. Both the axial modulus and the density of seawater depend on temperature, salinity, and pressure (or depth). The speed of sound in seawater increases with increasing temperature, salinity, or depth and decreases with decreasing temperature, salinity, or depth. Sound speed increases with increasing temperature because the density decreases faster than the axial modulus as the water becomes warmer. Sound speed increases with increasing salinity and depth, despite the fact that both these changes increase seawater's density, because the axial modulus of the water increases faster than its density.

Water dissipates the energy of high-frequency sounds faster than that of low-frequency sounds. Therefore, high-frequency sounds do not travel as far as the lower-frequency sounds.

Sound is reflected after striking an object; therefore, sound can be used to find objects, sense their shape, and determine their distance from the sound's source. If a sound signal is sent into the water and the time required for the return of the reflected sound, or echo, is measured accurately, the distance to the object may be determined. For example, if six seconds elapse between the outgoing sound pulse and its return, the sound has taken three seconds to travel to the object and three seconds to return. Because sound travels at an average speed of 1500 m/s in water, the object is 4500 m away, as shown in figure 5.18.

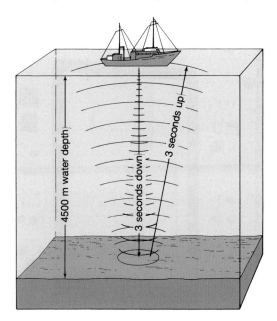

Figure 5.18 Traveling at an average speed of 1500 m per second, a sound pulse leaves the ship, travels downward, strikes the bottom, and returns. In 4500 m of water, the sound requires three seconds to reach the bottom and three seconds to return.

Water depth is measured by directing a narrow sound beam vertically to the sea floor. The sound beam passes through the nearly horizontal layers of water, in which salt content, temperature, and pressure vary. The sound speed continuously changes as it passes from layer to layer, until it reaches the sea floor and is reflected back to the ship. Little refraction, or bending, of the sound beam occurs, because its path is perpendicular to the water layers. Echo sounders, or depth recorders, are used by all modern vessels to measure the depth of water beneath the ship. Oceanographic vessels record depths on a chart, producing a continuous profile of depth as the vessel moves along its course. The precision depth recorder (PDR) on an oceanographic research vessel uses a very narrow sound beam to give detailed and continuous traces of the bottom (fig. 5.19).

Geologists can detect the properties of the sea floor by studying echo charts because some seafloor materials reflect stronger signals than others. In general, basalt reflects a stronger signal than sediment. The frequency of the sound will also determine the depth of penetration of the sound energy beneath the sea floor. Sound frequencies used by PDRs are generally 5–30khz (1khz = 1000 cycles per second). High-frequency signals do not penetrate far beneath the sea floor and are used simply to measure water depth. Lower-frequency sound energy can penetrate the seafloor sediments and reflect from boundaries between sediment layers. It is used to study sediment thickness and layering. An example of a PDR record illustrating sediment thickness and layering is seen in figure 5.19 and figure 4.1.

Depth recorders can also record large numbers of small organisms, including fish, that move toward the surface during the night and sink to greater depths during the day. These organisms form a layer known as the **deep scattering layer (DSL).** This layer reflects a portion of the sound beam energy and creates the image of a false bottom on the depth recorder trace.

Other echoes are returned from mid-depths by fish swimming in schools or by large individual fish. Echo sounders are designed and marketed as "fish finders," and persons who fish often learn to recognize fish school reflections. The echoes provide information on the depth to set nets, while the location of the school and the appearance of the echo pattern give information on the fish species.

Porpoises and whales use sound in water in the same way that bats use sound in air. The animal produces a sound, which travels outward until it reaches an object, from which the sound is reflected. The animal is able to judge the direction from which the sound returns, the distance to the reflecting object, and the properties of that object (see chapter 17). Humans have produced an underwater location technology called **sonar** (sound navigation and ranging) that uses sound in a similar way. Sonar technicians send directional pulses through the water, searching for targets that return echoes. They are then able to determine the distance and direction of the target. An electronic screen is used to display the direction of sound pulses relative to the sending vessel. If a reflective target is found, the screen also portrays the distance to the target, using the time difference between the sending of the sound pulse and the return of the echo. After much training and practice, technicians are able to distinguish among the echoes produced by a whale, a school of fish, and a submarine. It is even possible to determine the type and class of a vessel merely by listening to the sounds produced by its propellers and engines. (See the box titled "Bathymetrics" in chapter 4 for a discussion of multiple sonar devices being used in seafloor mapping.)

The target, however, may not be at the depth, distance, and angle indicated, because the sound beam may change speed as it passes through water layers of differing densities (fig. 5.20*a* and *c*). Figure 5.20*b* and *d* illustrate the refraction of sound beams and the formation of **sound shadow zones,** areas of the ocean into which sound does not penetrate. Sound beams bend toward regions in which sound travels more slowly and away from regions in which sound waves travel more rapidly.

To interpret the returning echo and to determine distance and depth correctly, the sonar operator must have information about the properties of the water through which the sound passes. Governments and their navies have conducted intensive research in the field of underwater sound. Survival for a surface vessel depends on its accuracy in locating a submarine's position, while the submarine must remain at the correct depth and distance from the sonar detector to remain invisible in the shadow zone.

At about 1000 m (3280 ft), the combination of salt content, temperature, and pressure creates a zone of minimum velocity for sound, the **sofar** (sound fixing and ranging) **channel**

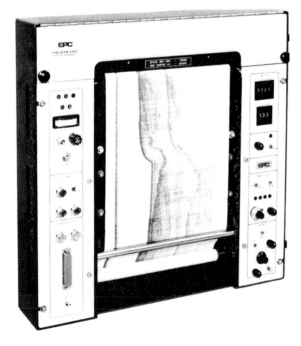

Figure 5.19   A precision depth recorder displaying bottom and subbottom profiles.

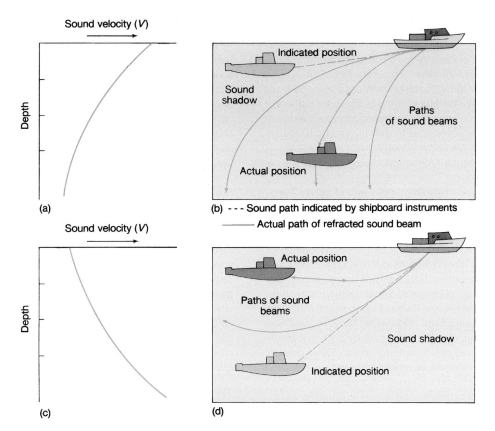

Figure 5.20 Sound waves change velocity and refract as they travel at an angle through water layers of different densities (a and c). The angle at which the sound beam leaves the ship indicates a target in the indicated, or ghost, position (b and d).

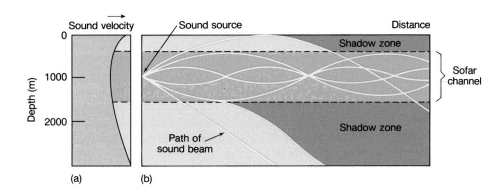

Figure 5.21 (a) The temperature, salinity, and pressure variation with depth combine to produce a minimum sound velocity at about 1000 m (3280 ft). (b) Sound generated at this depth is trapped in a layer known as the sofar channel.

(fig. 5.21*a*). Sound waves produced in the sofar channel do not escape from it unless they are directed outward at a sharp angle. Instead, the majority of the sound energy is refracted back and forth along the channel for great distances (fig. 5.21*b*). Test explosions in the channel near Australia have produced sound heard as far off as Bermuda. This channel is being used by a project known as ATOC (acoustic thermometry of ocean climate) to send sound pulses over long distances to look for long-term changes in the temperature of ocean waters. (See the box titled "Acoustic Thermometry of Ocean Climate.")

In the 1950s and 1960s, the U.S. Navy installed large arrays of hydrophones on the sea floor in areas that were potential sailing routes for foreign submarines. These acoustical nets, known as sound surveillance systems (SOSUS), were used to track and identify vessels at and below the sea surface. Oceanographers and other scientists were given access to this Cold War legacy for acoustic monitoring of marine mammals and seafloor seismic events.

Recently, a French oceanographic team studied the pure low-frequency tones of long duration received by undersea seismic stations in French Polynesia; SOSUS archives provided similar recordings. The sound source was tracked to a region of seamounts at shallow depths. It appears that erupting seamounts produce clouds of bubbles that act as a resonating chamber that allows sound waves to oscillate vertically between the volcanic source and the sea surface, producing the low-frequency tones.

Based on the principle that sound's speed in seawater is determined primarily by the temperature of the water, oceanographers Carl Wunsch of the Massachusetts Institute of Technology and Walter Munk of the Scripps Institution of Oceanography in California envisioned a transoceanic experiment to follow ocean temperature response to global warming. Because sound waves travel faster in warmer water, if the oceans are warming, the amount of time required for sound to travel from one location to another will decrease.

In 1991, the travel time of low-frequency sound pulses transmitted through the sofar channel from a site near Heard Island in the southern Indian Ocean was repeatedly measured at special listening stations around the world. The precision of sound travel time measurements from source to receiver was about 1 millisecond over a path 1000 km (660 mi) long, so temperature changes of a few thousandths of a degree could be detected. Test results were promising, and a new series of tests called acoustic thermometry of ocean climate (ATOC) began in the Pacific Ocean in 1995.

ATOC broadcasts sound from underwater sources near San Francisco and Hawaii. The sound is picked up by arrays of sensitive hydrophones as far away as Christmas Island and New Zealand (box fig. 1). After eighteen months of testing, temperature readings of the Pacific Ocean were found to be even more precise than had been projected. Scientists can detect variations as small as 20 milliseconds in the hour-long travel time of the pulses. This precision enables researchers to calculate the average ocean temperature along the sound pulse path to within 0.006°C. Repeating the measurements will allow long-term temperature changes at mid-ocean depths to be measured before they can be deduced by any other method.

Questions concerning the effect of sound signals on marine mammals began with the Heard Island tests and delayed the start of the ATOC experiment. Marine mammalogists have monitored the behavior of whale and elephant seal populations near the sound source off California. They report having seen no changes in the animals' swimming activity or distribution. Mammal researchers have also released deep-diving elephant seals farther out to sea and used satellite tags to track the paths of the animals as they returned to shore. The seals made no attempt to avoid the sound source. Additional experiments are being conducted to see whether whale vocalizations are affected.

A similar experiment that measured water temperatures in the Arctic Ocean was performed in the spring of 1994 by a joint U.S.-Russian-Canadian project, the Transarctic Acoustic Propagation Experiment. Sound signals were sent from an ice camp north of Spitsbergen to a camp 900 km (540 mi) in the Lincoln Sea and to another camp 2600 km (1600 mi) away in the Beaufort Sea. Travel times were predicted by using water temperatures from earlier research, but the measured travel times were shorter, implying that the mid-depth Atlantic water that penetrates the Arctic Ocean had warmed by 0.2°–0.4°C since the mid-1980s. The U.S.-Canada Arctic Ocean Section cruise had also measured such a warming trend, but scientists emphasize that it is too early to say whether the change is due to global warming or whether it is a part of some other natural cycle.

To Learn More About Acoustic Thermometry of Ocean Climate

Forbes, A. 1994. Acoustic Monitoring of Global Ocean Climate. *Sea Technology* 35 (5): 65–67.

Georges, T. M. 1992. Taking the Ocean's Temperature with Sound. *The World & I* (July): 282–89.

The Heard Island Experiment. 1991. *Oceanus* 34 (1): 6–8.

Mikhalevsky, P., A. Braggeroer, A. Gavrilov, and M. Slavinsky. 1995. Experiment Tests Use of Acoustics to Monitor Temperature and Ice in Arctic Ocean. *EOS* 76 (27): 265, 268–69.

5.9 Ice and Fog

Ice and fog are forms of water that are extremely hazardous to ocean navigation and transportation. At polar latitudes in the Northern Hemisphere, ice prevents the safe use of shipping routes for much of the year, and ice driven by winds and currents hampers polar research in both Arctic and Antarctic regions. Although radar helps navigation in foggy seas, the restricted visibility is a danger to all vessels, from the largest to the smallest. Processes producing ice and fog at sea are discussed in this section.

Sea Ice

Ice is present year-round at all latitudes if the elevation is high enough to keep the average temperature below the freezing point of water. This elevation varies between sea level in polar regions and approximately 5000 m (16,400 ft) at the equator. On land, ice and snow are the result of low temperatures and precipitation. In the sea and in lakes, precipitation is not necessary; the ice is formed when the temperature of the air drops below the freezing point of water.

Sea ice is formed at polar latitudes because of the low air and water temperatures. As the seawater begins to freeze, the water becomes turbid, and clouds of ice crystals are produced. As the number of ice crystals increases, a thin sheet of slush forms. Sheets of new sea ice are broken into "pancakes" by waves and wind (fig. 5.22a). As the freezing continues, the pancakes move about, unite, and form floes. Ice floes move with the currents and the wind (fig. 5.22b), collide with each other, and form ridges and hummocks, or low hills. Some floes shift constantly, breaking apart and freezing together in response to winds and water motion. Others remain anchored to the landmass. These stationary floes are called **fast ice.** As freezing continues, the ice increases in thickness. Snow and water on the ice also freeze and contribute to the flat, floating ice masses.

The dissolved salts of the seawater do not fit into the ice crystal structure and are left behind in the surface water. This very cold and dense saltier water sinks to be replaced by less-salty water, which in turn cools to its freezing point. As the ice is formed, some seawater is trapped in voids. If the ice forms slowly, most of the trapped seawater drains out and escapes; if

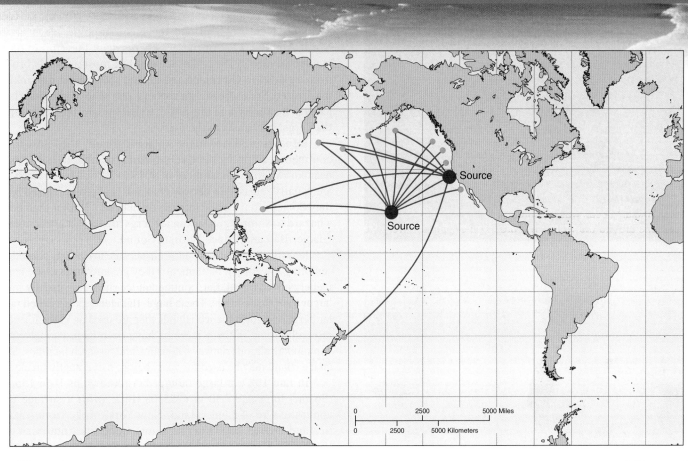

Box Figure 1 Acoustic thermometry of ocean climate (ATOC) takes the temperature of the Pacific Ocean using two sound sources, California and Hawaii, and twelve receivers.

the ice forms quickly at subfreezing temperatures, more salt water is trapped. As time passes and the ice ages, the salt water slowly escapes through the ice, and eventually the sea ice becomes fresh enough to drink when melted.

In one season, about 2 m (6 ft) of ice forms at the ocean's surface at polar latitudes. The thickness of the ice is limited by the necessity of extracting sufficient heat through the ice to cool the underlying water, allowing new ice to form beneath the existing layer. The latent heat of fusion must be removed from the water beneath the ice, and it can be extracted only by conduction of the heat through the ice. This is a slow process at polar temperatures, made slower by snow accumulating on the ice surface and acting as an insulator to the water below.

Sea ice is seasonal around the bays and shores of parts of the northeastern United States, Canada, Russia, Scandinavia, and Alaska. The ice exists year-round in the central Arctic and around Antarctica. Sea ice covers the entire Arctic Ocean in the northern winter and pushes far out to sea around Antarctica during the Southern Hemisphere's winter (fig. 5.23). Waves, currents, winds, and tides fracture the edges of the ice and cause floes to collide and pile into irregular masses to form pressure ridges (see fig. 5.22b). Ice ridges formed at the collision boundaries between ice floes may reach 10 m (30 ft) in thickness. In summer, the ice melts back along its edges and at its surface (fig. 5.22c). In areas where the ice does not melt entirely during the year, new ice is added each winter, and a thickness of 3–5 m (10–16 ft) can accumulate.

Remember that sea ice is a product of salt water; it should not be confused with freshwater glacial ice, which covers Greenland and the Antarctic continent. When pieces of land ice break off and fall into the sea, these pieces of freshwater ice are called **icebergs.**

Icebergs

Icebergs are massive and irregular in shape; they float with only about 12% of their mass above the sea surface (fig. 5.24). They are formed by glaciers, large rivers of ice that begin inland in central Greenland, Antarctica, and Alaska and inch their way toward the sea. The forward movement, the melting at the base of the glacier where it meets the ocean, and waves and tidal action cause blocks of ice to break off and float out to sea. The production of icebergs, called calving, occurs mainly during the summer.

(a)

(b)

(c)

Figure 5.22   Sea ice. (a) Pancake ice, an early stage of ice formation. (b) A pressure ridge formed by colliding floes. (c) Meltwater forms ponds on the Arctic sea ice in summer.

Icebergs produced in the Arctic drift south with the currents as far as New England and the busy shipping lanes of the North Atlantic. One of these icebergs sank the *Titanic* on its maiden voyage in 1912. This iceberg had most probably broken from a glacier in Greenland and floated south. After the sinking, the United States began an iceberg patrol to warn vessels of iceberg locations, and this effort, aided by satellite detection of the bergs, continues to the present.

Icebergs produced in the Antarctic usually stay close to the continent, caught by the circling currents, although they have been known to reach latitudes of 40°S to 50°S. Alaskan icebergs are usually released in narrow channels and semienclosed bays; they do not often escape into the open ocean. Rather than being a hazard to shipping, Alaskan icebergs are tourist attractions in Glacier Bay and Prince William Sound.

Icebergs from narrow valley glaciers are irregularly shaped, resembling towers and battlements; they are known as castle bergs. Icebergs from broad, flat, continental ice sheets are usually much larger than castle bergs. These huge, flat icebergs are called tabular bergs and form the ice islands that are used as bases for polar research. Some of these icebergs have been large enough to accommodate aircraft runways, housing, and research facilities. Such an ice island may be used for years before it eventually breaks up.

In late 1987, a large tabular berg known as B-9, 155 km (96 mi) long and 230 m (755 ft) thick, with a surface area about the size of Long Island, New York, broke from Antarctica and drifted 2000 km (1250 mi) along the Antarctic coast (fig. 5.25). This massive berg floated so deeply that its drift was the product of subsurface currents; conventionally sized tabular bergs drift with the wind. B-9 drifted clockwise and to the west in the Ross Sea for two years before it grounded and broke into three pieces. It was tracked by plane and satellite; its drift path helped increase our knowledge of currents in the Ross Sea.

The calving of a spectacularly large tabular berg is infrequent, for the seaward movement of glacial ice is only about 1 m (3 ft) per day, and a berg of B-9's size contains about seventy years of glacial advance. Formation of a berg of this size is also controlled by the spacing of the major crevasses that produce weak spots along which a berg separates from the glacier. The developing rift along which B-9 broke away was first identified in aerial photos between 1965 and 1971.

In March 2000, another massive tabular iceberg, known as B-15, broke away, or calved, from the Ross Ice Shelf in Antarctica. B-15 was one of the largest icebergs ever seen at the time of its formation. It was nearly 300 km (186 mi) in length and 40 km (25 mi) in width with a thickness of about 400 m (1312 ft) below water and 30 m (98 ft) above. Since forming, B-15 initially broke into two main icebergs and then other smaller ones, including B-17 and B-18. The calving of B-15 resulted in the retreat of the edge of the Ross Ice Shelf to a position it was in about fifty years previously.

Ongoing studies in the Antarctic indicate that the glaciers that reach the sea and form the large expanses of permanent shelf ice that surround Antarctica are receding. At the same time, the shelf ice is calving large bergs at an increasing rate. This activity may be due to global warming driving an increased rate of melting.

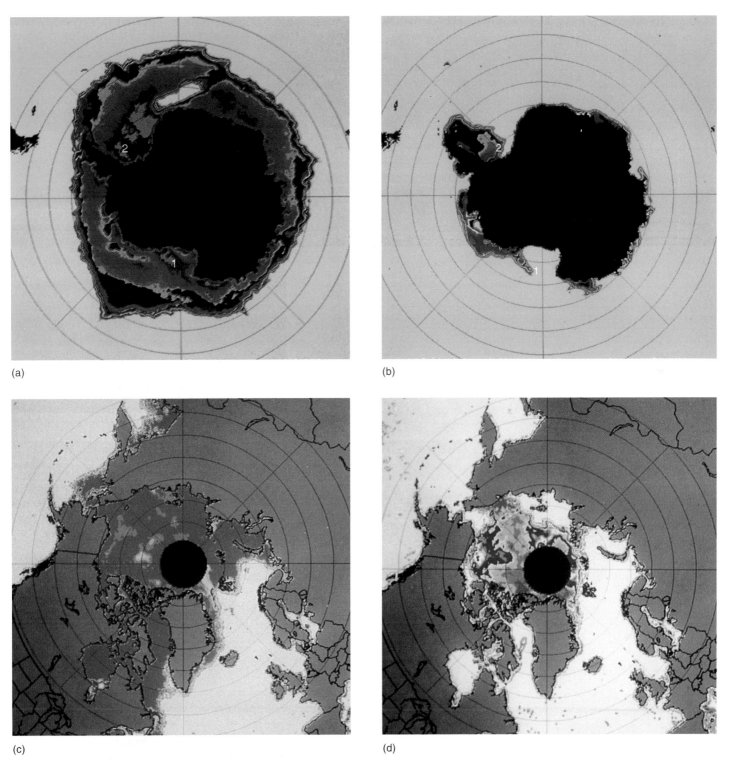

(a)

(b)

(c)

(d)

Figure 5.23 Sea ice around the Arctic and Antarctic. Data from NASA's *NIMBUS-5* satellite were used to construct seasonal maps of sea ice coverage at high latitudes. This information is used to help develop a better understanding of global climate. The Antarctic ice extends more than 100 km (60 mi) from the land into the Ross (*1*) and Weddell (*2*) Seas during the winter (a) and is much reduced during the summer (b). The Arctic Ocean is filled with ice in the winter (c), but by the next fall, the ice cover is much less (d). Short-term motions of the ice boundaries can be calculated from repeated satellite images. These show that the ice floes move up to 50 km (30 mi) per day. *Purple* indicates high ice concentration, and *blue* indicates open water in (a) and (b). *Yellow* areas are open water in (c) and (d).

(a)

(b)

Figure 5.24 (a) A castle berg drifting in the Atlantic Ocean just north of the Grand Banks. (b) A tabular berg in the South Atlantic Ocean. It is approximately 500 m (1640 ft) long.

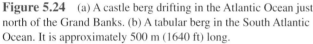

Figure 5.25 Iceberg B-9 as observed by *Landsat 4*, November 28, 1987, about four to six weeks after detachment from the Ross Ice Shelf, Antarctica.

Figure 5.26   Advective fog obscures the Golden Gate Bridge at the entrance to San Francisco Bay.

The water stored in ice caps and glaciers comprises roughly 2.11% of Earth's water supply (see table 2.3). If all of this ice melted, sea level would rise by about 79 m (259 ft), an increase of 2.11% in the average depth of the oceans of 3729 m as given in table 2.4. During the last glacial maximum the world's ice caps and glaciers contained roughly 3.5% more ice than they do today and sea level was approximately 130 m (427 ft) lower. See chapter 13 for a discussion of rising sea level over the past 120 years due to global warming.

Fog

The ability of water to remain as water vapor is a function of air temperature. Warm air can hold more moisture than cold air, and when air is cooled, its water vapor condenses around any small particles in the air, forming liquid droplets. Large accumulations of water droplets form clouds, and clouds that form close to the ground are called **fog.** Fog at sea is a navigation hazard, but the ocean scientist is interested in fog and in predicting its occurrence because of the role fog plays in transferring water and heat between the atmosphere and the ocean surface.

There are three basic types of fog, each formed by a different process. The most common fog at sea is advective fog, forming when warm air saturated with water vapor moves over colder water. This advective fog hugs the sea surface like a blanket. For example, when the warmer, moist air over the Gulf Stream flows over the colder water of the Labrador Current, the famous fogs of the Grand Banks result. In northern California, Oregon, and Washington, when offshore warm, moist air moves over the cold coastal waters, particularly in summer, a coastal fog results, giving moisture to the redwood forests and grassy cliffs and cooling the city of San Francisco (fig. 5.26).

Streamers of **sea smoke** rising from the sea surface on a cold winter day are a spectacular sight. Dry, cold air from the land or from the polar ice pack moves out over the warmer water, and the water warms the air above it. This warmed air picks up water vapor from the sea surface and rises. The rising air is cooled below the dew point, and rising ribbons of fog are formed. When sea smoke is formed, the vertical convection results in high evaporation that rapidly cools the water's surface, while advective fog's condensation returns moisture and heat to Earth's surface.

Radiative fog is the result of warm days and cold nights. Earth's surface warms and cools, as does the air above it. If the air holds enough moisture during the day, water vapor condenses as Earth cools at night, forming the low-lying, thick white fogs that are often found in river valleys and occasionally in bays and inlets along the coast. These fogs usually disappear during the morning as the Sun gradually warms the air, changing the water droplets to water vapor.

Earth's water—whether found as a liquid filling lakes, streams, and oceans; as a solid in the glaciers, snowpacks, and sea ice; or as a gas in the atmosphere—is a remarkable substance. The properties and abundance of water give Earth its habitable characteristics and make it a unique planet in our solar system.

Summary

A water molecule is made up of two positively charged atoms of hydrogen and one negatively charged atom of oxygen. The molecule has a specific shape with oppositely charged sides; it is a polar molecule. Because of this distribution of electrostatic charges, water molecules interact with each other by forming hydrogen bonds between molecules. The water molecule is very stable. The structure of the water molecule is responsible for the properties of water.

Water exists as a solid, a liquid, and a gas. Changes from one state to another require the addition or extraction of heat energy. Water has a high heat capacity; it is able to take in or give up large quantities of heat with a small change in temperature. The surface tension of water, which is related to the cohesion between water molecules at the surface, is high. The viscosity of water is a measure of its internal friction; it is primarily affected by temperature. Water is nearly incompressible. Pressure increases 1 atm for every 10 m of depth in the oceans.

The density, or mass per unit volume, of water increases with a decrease in temperature and an increase in salt content. The effect of pressure on density is small. Less-dense water floats on denser water. Pure water reaches its maximum density at 4°C. Open-ocean water does not reach its maximum density before freezing. Water vapor is less dense than air; a mixture of water vapor and air is less dense than dry air.

The ability of water to dissolve substances is exceptionally good. River water dissolves salts from the land and carries them to the sea.

Seawater transmits energy as heat, light, and sound. The sea surface layer is heated by solar radiation, and heat is transmitted downward by conduction. This is an inefficient process compared to convection.

The long red wavelengths of light are lost primarily in the first 10 m (33 ft) of seawater; only the shorter wavelengths of blue-green to blue-violet light penetrate to depths of 150 m (500 ft) or more. Light passing into water is refracted.

Attenuation, or the decrease in light over distance, is the result of absorption and scattering by the water and particles suspended in the water. Light attenuation is measured by a Secchi disk and photoreceptors.

Sound travels farther and faster in water than in air. Its speed is affected by the temperature, pressure, and salt content of the water. Echo sounders are used to measure the depth of water, and sonar is used to locate objects. Sound is refracted as it passes at an angle through water of different densities, and sound shadows are formed. The sofar channel, in which sound travels for long distances, is the result of salt content, temperature, and pressure in the oceans. The deep scattering layer, formed by small animals moving toward the surface at night and away from it during the day, reflects portions of a sound beam and creates a false bottom on a depth recorder trace.

Sea ice is formed in the extreme cold of polar latitudes. The process is self-insulating, so large thicknesses of ice do not form each winter. Icebergs are formed from glaciers that break off into the sea.

Fog occurs when water vapor condenses to form liquid droplets. Three types of fog occur: advective fog occurs when warm, water-saturated air passes over cold water; sea smoke occurs when dry, cold air moves over warm water; and radiative fog occurs when warm, moist air is cooled at night.

Key Terms

All key terms from this chapter can be viewed by term or definition when studied as flashcards on this book's website at www.mhhe.com/sverdrup10e.

covalent bond, 125
polar molecule, 125
hydrogen bond, 125
temperature, 126
degree, 126

heat, 127
calorie, 127
dew point, 127
latent heat of fusion, 128
latent heat of vaporization, 128
sublimation, 129
specific heat, 129
heat capacity, 129
cohesion, 129
surface tension, 130

viscosity, 130
ion, 132
conduction, 132
convection, 132
radiation, 132
electromagnetic radiation, 134
electromagnetic spectrum, 134
absorption, 134
scattering, 134
attenuation, 134

refraction, 136
Secchi disk, 136
deep scattering layer (DSL), 138
sonar, 138
sound shadow zone, 138
sofar channel, 138
fast ice, 140
iceberg, 141
fog, 145
sea smoke, 145

Study Questions

1. Discuss the structure of the water molecule.
2. What happens to the arrangement of water molecules when water freezes?
3. What would happen to the sea level if all the sea ice melted? Assume that the ocean area does not change and that ocean temperature stays the same.
4. Compare the heat capacity of the oceans with that of the land and the atmosphere. Compare the climate of the west and east coasts of the United States with that of the central and midwestern states. How is climate related to heat capacity, solar heating and cooling, and the direction of the prevailing wind?
5. How do the properties of the water molecule affect (a) surface tension and (b) the dissolving ability of water?
6. Explain why heating the atmosphere from below is an efficient way to distribute heat in the atmosphere, whereas heating the oceans from above is an inefficient way to distribute heat downward in the oceans.
7. If there were no scattering and only absorption of light by seawater, what color would the oceans appear? Why?
8. Both advective fog and sea smoke transfer heat between the atmosphere and the oceans. How do the rate and direction of heat transfer differ between the two? Why?

9. How are the amounts of heat energy lost and gained when water changes its state related to the bonding and motion of the water molecules?
10. If the substances in table 5.2 lose the same quantity of heat energy, which one cools to the lowest temperature?
11. Explain the effect of increasing and decreasing the salinity, temperature, and pressure on the density of seawater.

12. How are light and sound in seawater affected by changes in the water's density?
13. Where do icebergs originate? How are they formed?
14. Explain differences between *sonar* and *sofar.*
15. Why might one expect to find that traces of every known naturally occurring substance are dissolved in seawater?

Study Problems

1. If cooling and evaporation remove heat from the surface of a deep lake at the rate of 800×10^3 calories per hour and if the surface temperature of the lake is 0°C, at what rate must deep water be pumped to the surface of the lake to prevent the formation of ice?
2. Determine the freezing point of seawater at 32 g/kg salt content. Use figure 5.7.
3. If an echo sounder measures the depth of the water as 3500 m, but the instrument is shown to have a timing error of ±0.001

second over the total time period of the measurement, what is the error in the depth measurement?
4. If water from a given area of sea surface evaporates at the rate of 1 metric ton per day at 20°C, at what rate is heat being transferred to the atmosphere above this surface area? See figure 5.3.
5. If heat is removed from a kilogram of water at a constant rate that lowers the temperature of the water from 25°C to 0°C in thirty minutes, how long will it take to convert 0°C water to ice?

An evaporate salt pond in Bonaire, Netherlands Antilles.

Chapter 6

The Chemistry of Seawater

Chapter Outline

Learning Outcomes

After studying the information in this chapter students should be able to:

1. *sketch* the pattern of high and low sea surface salinity on a map of the world's oceans,
2. *explain* how sea surface salinity is modified by evaporation, precipitation, and runoff from the continents,
3. *review* the sources of major constituent ions in seawater,
4. *rank* the six most abundant constituent ions in seawater in order of their concentration,
5. *calculate* the residence time of an ion given its concentration and rate of supply,
6. *diagram* the distribution of oxygen and carbon dioxide with depth,
7. *describe* the pH scale and *explain* the role of carbon dioxide in buffering seawater pH,
8. *identify* the three ions considered important marine nutrients, and
9. *compare* and *contrast* two different methods of desalination.

Seawater is salt water, and historically, seawater has been valued for its salt. Until recently, salt was enormously important as a food preservative, and at one time, salt formed the basis for a major commercial trade. Today, although salt is still extracted from seawater, it is the water that has become increasingly valuable in many areas of the world. Seawater is also much more than salt water. Seawater is a complex solution containing dissolved gases, nutrient substances, and organic molecules as well as salts.

In this chapter, we investigate seawater, and we explore physical, chemical, and biological processes that regulate its composition. We also review the commercial extraction of salts from seawater and the possibilities of increasing our supply of fresh water by desalination.

6.1 Salts

Units of Concentration

Compounds can break apart into individual atoms or groups of atoms that have opposite electrical charges. A charged atom or group of atoms is an ion. An ion with a positive charge is a **cation;** an atom with a negative charge is an **anion.** The salts in seawater are present in dissolved form as cations and anions. For example, in a glass of seawater, you would find individual sodium cations (Na^+) and chloride anions (Cl^-). If the water was then left out in the sun to evaporate, these ions would combine to form a solid precipitate, the salt compound sodium chloride ($NaCl$).

The concentration of dissolved constituents in seawater can be expressed by weight, by volume, or in molar terms. Concentrations of different constituents vary by several orders of magnitude. When expressed by weight, concentrations are given as g/kg (parts per thousand), mg/kg (parts per million), or even μg/kg (parts per billion), depending on the abundance of the constituent. Thus, 1 g/kg (1 part per thousand) = 1000 mg/kg (10^3 parts per million) = 1 million μg/kg (10^6 parts per billion), or 1 μg/kg (1 part per billion) = 0.001 mg/kg (10^{-3} parts per million) = 0.000001 g/kg (10^{-6} parts per thousand). Similarly, measurements of concentration by volume are expressed as g/l, mg/l, or μg/l. Since 1 liter of seawater weighs very nearly 1 kilogram (approximately 1.027 kilograms), concentrations measured by weight and by volume are numerically similar (the concentration of chloride in seawater is approximately 19.35 g/kg or 19.87 g/l).

For some purposes, it is useful to express the concentration of seawater constituents in molar terms. One mole of an element, or compound, has a mass in grams equal to the atomic (or ionic, or molecular) mass of the element or compound. A mole of sodium (Na^+) contains 23 g of sodium; a mole of sulfate (SO_4^{2-}) contains $32 + (16 \times 4) = 96$ g of sulfate; and a mole of bicarbonate (HCO_3^-) contains $1 + 12 + (16 \times 3) = 61$ g of bicarbonate. Once again, because 1 liter of seawater weighs nearly 1 kilogram, molar concentrations of a given dissolved constituent expressed as moles/kg and moles/l are numerically similar.

Ocean Salinities

In the major ocean basins, 3.5% of the weight of seawater is, on the average, dissolved salt and 96.5% is water, so a typical 1000 g or 1 kg sample of seawater is made up of 965 g of water and 35 g of salt. Oceanographers measure the salt content of ocean water in grams of salt per kilogram of seawater (g/kg), or parts per thousand (‰). The total quantity of dissolved salt in seawater is known as **salinity,** and the average ocean salinity is approximately 35‰. There are about 1338.5 million km³ ($\sim 1.34 \times 10^9$ km³) of seawater in the oceans, with an average density of approximately 1.03 g/cm³ (1.03×10^{12} kg/km³) (see tables 2.4 and 5.3). The total weight of this seawater is about 1378.7 million trillion kg ($\sim 1.4 \times 10^{21}$ kg); the weight of the salt in the oceans is about 3.5% of the total weight, or approximately 48 million trillion kg ($\sim 1.1 \times 10^{20}$ lb, or $\sim 5.3 \times 10^{16}$ tons). If all the water in the oceans evaporated, this amount of salt would form a layer roughly 45.5 m (150 ft) thick over the entire surface of Earth.

The salinity of ocean surface water is associated with latitude. Latitudinal variations in evaporation and precipitation, as well as freezing, thawing, and freshwater runoff from the land, affect the amount of salt in seawater. The relationship among evaporation, precipitation, and mid-ocean surface salinity with latitude is shown in figure 6.1. Notice the low surface salinities in the cool and rainy 40°–50°N and S latitude belts, high evaporation rates and high surface salinities in the desert belts centered on 25°N and S, and low surface salinities again in the warm but rainy tropics centered at 5°N. Sea surface salinities during the Northern Hemisphere summer are shown in figure 6.2.

In coastal areas of high precipitation and river inflow, surface salinities fall below the average. For example, during periods of high flow, the water of the Columbia River lowers the Pacific Ocean's surface salinity to less than 25‰ as far as 35 km (20 mi) at sea. Also, sailors have dipped up water fresh enough to drink from the ocean surface 85 km (50 mi) from the mouth of the Amazon River. In subtropic regions of high evaporation and low freshwater input, the surface salinities of nearly landlocked seas are well above the average: 40–42‰ in the Red Sea and the Persian Gulf and 38–39‰ in the Mediterranean Sea. In the open ocean at these same latitudes, the surface salinity is closer to 36.5‰. Surface salinities change seasonally in polar areas, where the surface water forms sea ice in winter, leaving behind the salt and raising the salinity of the water under the ice. In summer, a freshwater surface layer forms when the sea ice melts. Deep-water samples from the mid-latitudes are usually slightly less salty than the surface waters in part because the deep water is formed at the surface in high latitudes with high precipitation. The formation of these deep-water types is discussed in chapter 8.

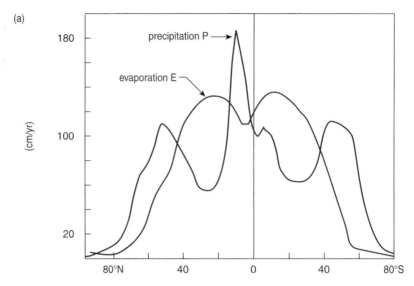

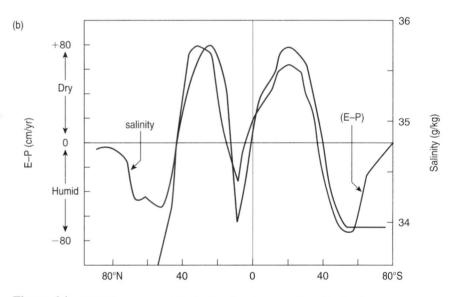

Figure 6.1   (a) Mid-ocean precipitation (*red*) and evaporation (*blue*) values as a function of latitude. (b) Mid-ocean average surface salinity values (*red*) match the average variation in [evaporation minus precipitation] values (*blue*) that occur with latitude.

Dissolved Salts

The atoms or groups of atoms that combine to form compounds can be held together by different kinds of bonds. The two hydrogen atoms and one oxygen atom in the water molecule are held together by covalent bonds in which electrons are shared between the atoms (see section 5.1). Other compounds, such as sodium chloride (NaCl), are held together by **ionic bonds** in which electrons are transferred from a metal atom (in this case, the metal sodium) to a nonmetal atom (in this case, the chlorine atom), creating ions of opposite charge that attract each other. As stated earlier, an ion with a positive charge is a cation; an ion with a negative charge is an anion. Ionic bonds are easily broken in water because of the polar nature of the water molecule. Thus, when salts are added to water, the salts dissolve, or dissociate (break apart), into ions. The dissolved salts in seawater are mostly present as cations and anions (see section 5.7).

When solid sodium chloride is added to water, the bonds between the ions break, and positively charged sodium ions and

negatively charged chloride ions are released into the water. This reaction can be written as

$$\text{sodium chloride} \rightarrow \text{sodium ion} + \text{chloride ion}$$

or it can be written with chemical abbreviations as

$$\text{NaCl} \rightarrow \text{Na}^+ + \text{Cl}^-$$

Six ions make up more than 99% of the salts dissolved in seawater. Four of these are cations: sodium (Na^+), magnesium (Mg^{2+}), calcium (Ca^{2+}), and potassium (K^+); two are anions: chloride (Cl^-) and sulfate (SO_4^{2-}). Table 6.1 lists these six ions and five more, arranged in order of their abundance in seawater. The ions listed in table 6.1 are known as the **major constituents** of seawater. Note that sodium and chloride ions account for 86% of the salt ions in seawater.

All the other elements dissolved in seawater are present in concentrations of less than one part per million and are called

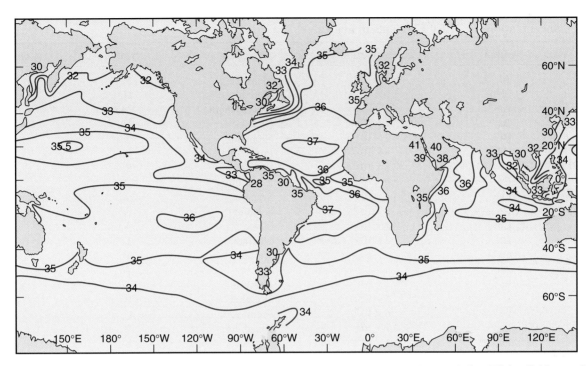

Figure 6.2   Average sea surface salinities in the Northern Hemisphere summer, given in parts per thousand (‰). (High salinities are found in areas of high evaporation; low salinities are common in coastal areas and regions of high precipitation.)

Table 6.1	**Major Constituents of Seawater[1]**					
			Concentration in Seawater			**Percentage by**
Constituent		**Symbol**	**g/kg**	**g/l**	**mole/l**	**Weight**
Chloride		Cl^-	19.35	19.87	0.560	55.07
Sodium		Na^+	10.76	11.05	0.481	30.62
Sulfate	The six most	SO_4^{2-}	2.71	2.78	0.029	7.72
Magnesium	abundant	Mg^{2+}	1.29 } 34.91	1.32 } 35.84	0.054 } 1.145	3.68 } 99.36
Calcium	ions	Ca^{2+}	0.41	0.42	0.0105	1.17
Potassium		K^+	0.39	0.40	0.0102	1.10
Bicarbonate		HCO_3^-	0.14	0.144	0.0024	0.40
Bromide		Br^-	0.067	0.069	0.00086	0.19
Strontium		Sr^{2+}	0.008	0.008	0.00009	0.02
Boron		B^{3+}	0.004	0.004	0.00037	0.01
Fluoride		F^-	0.001	0.001	0.00005	0.02
Total			35.13	36.07	1.1485	99.99

1. Nutrients and dissolved gases are not included.

trace elements (table 6.2). Most trace elements are present in such small concentrations that it is common to report their concentrations at the parts per billion level. Some of these elements are important to organisms that are able to concentrate the ions. For example, long before iodine could be determined chemically as a trace element of seawater, it was known that shellfish and seaweeds were rich sources for this element, and seaweed was harvested for commercial iodine extraction.

Because the ratios of the major constituents of seawater do not change with changes in total salt content and because these constituents are not generally removed or added by living organisms, the major constituents are termed **conservative**

constituents. Certain of the ions present in much smaller quantities, some dissolved gases, and assorted organic molecules and complexes do change in concentrations because of biological and chemical processes that occur in some areas of the oceans; these are known as **nonconservative constituents.**

Sources of Salt

The original sources of sea salts include the crust and the interior of Earth. The chemical composition of Earth's rocky crust can account for most of the positively charged ions found in seawater. Large quantities of cations are present in rocks that

Table 6.2		Concentrations of Trace Elements in Seawater[1]			
Element	Symbol	Concentration[2]	Element	Symbol	Concentration[2]
Aluminum	Al	5.4×10^{-1}	Manganese	Mn	3×10^{-2}
Antimony	Sb	1.5×10^{-1}	Mercury	Hg	1×10^{-3}
Arsenic	As	1.7	Molybdenum	Mo	1.1×10^{1}
Barium	Ba	1.37×10^{1}	Nickel	Ni	5×10^{-1}
Bismuth	Bi	$\leq 4.2 \times 10^{-5}$	Niobium	Nb	$\leq (4.6 \times 10^{-3})$
Cadmium	Cd	8×10^{-2}	Protactinium	Pa	5×10^{-8}
Cerium	Ce	2.8×10^{-3}	Radium	Ra	7×10^{-8}
Cesium	Cs	2.9×10^{-1}	Rubidium	Rb	1.2×10^{2}
Chromium	Cr	2×10^{-1}	Scandium	Sc	6.7×10^{-4}
Cobalt	Co	1×10^{-3}	Selenium	Se	1.3×10^{-1}
Copper	Cu	2.5×10^{-1}	Silver	Ag	2.7×10^{-3}
Gallium	Ga	2×10^{-2}	Thallium	Tl	1.2×10^{-2}
Germanium	Ge	5.1×10^{-3}	Thorium	Th	(1×10^{-2})
Gold	Au	4.9×10^{-3}	Tin	Sn	5×10^{-4}
Indium	In	1×10^{-4}	Titanium	Ti	$< (9.6 \times 10^{-1})$
Iodine	I	5×10^{1}	Tungsten	W	9×10^{-2}
Iron	Fe	6×10^{-2}	Uranium	U	3.2
Lanthanum	La	4.2×10^{-3}	Vanadium	V	1.58
Lead	Pb	2.1×10^{-3}	Yttrium	Y	1.3×10^{-2}
Lithium	Li	1.7×10^{2}	Zinc	Zn	4×10^{-1}
			Rare earths		$(0.5–3.0) \times 10^{-3}$

1. Nutrients and dissolved gases are not included.

2. Parts per billion, or μ g/kg.

Note: Parentheses indicate uncertainty about concentration.

are formed by the crystallization of molten magma from volcanic processes. The physical and chemical weathering of rock over time breaks it into small pieces and the rain dissolves out ions, which are carried to the sea by rivers. Anions are present in Earth's interior and may have been present in Earth's early atmosphere. Some anions may have been washed from the atmosphere by long periods of rainfall, but the more likely source of most of the anions is thought to have been Earth's mantle. During the formation of Earth, gases from the mantle are believed to have supplied anions to the newly forming oceans.

Acidic gases released during volcanic eruptions (for example, hydrogen sulfide, sulfur dioxide, and chlorine) dissolve in rainwater or river water and are carried to the oceans as Cl^- (chloride) and SO_4^{2-} (sulfate).

Tests show that the most abundant ions in today's rivers (table 6.3) are the least abundant ions in ocean water because the rivers have previously removed the most easily dissolved land salts and are now carrying the less soluble salts. Exceptions to this pattern are found in rivers used for irrigation. These rivers are flowing through arid soils that have not lost much of their salt content. The water is frequently used several times and passes through a number of irrigation projects on its way downriver, causing the water to become increasingly salty and unfit for irrigation purposes. This situation has produced years of continuous conflict between the United States and Mexico over the waters of the Colorado River and Rio Grande, which irrigate much of the agricultural land of the U.S. desert Southwest before becoming available to the farmlands of Mexico.

Table 6.3	Dissolved Salts in River Water		
Ion		Symbol	Percentage by Weight
Bicarbonate		HCO_3^-	48.7
Calcium		Ca^{2+}	12.5
Silicon dioxide (nonionic)		SiO_2	10.9
Sulfate		SO_4^{2-}	9.3
Chloride		Cl^-	6.5
Sodium		Na^+	5.2
Magnesium		Mg^{2+}	3.4
Potassium		K^+	1.9
Oxides (nonionic)		$(Fe, Al)_2O_3$	0.8
Nitrate		NO_3^-	0.8
Total			100.00

Note: Average river salt concentration is 0.120‰.

In addition, we know that hot-water vents located on the sea floor supply chemicals to the ocean water and also remove them. Hydrothermal activity found at the mid-ocean ridges and associated with hot spots and ridge formation may play an important role in stabilizing the ocean's salt composition. When hot magma is introduced, the cold crust cracks and becomes permeable. Pressure from the water above the sea floor is high (1 atm for every 10 m of water depth), and it forces water into cracks and voids, where it is heated to extremely high temperatures. The salinity of seawater entering the hydrothermal system is relatively constant worldwide, but the water emerging from hydrothermal vents has a salinity that may be more than double its original value. The processes controlling this salinity change and their role in the ocean's total salt balance are unclear at this time.

Regulating the Salt Balance

We know from the age of older marine sedimentary rocks that the oceans have been present on Earth for about 3.5 billion years. Chemical and geologic evidence from rocks and salt deposits leads researchers to believe that the salt composition of the oceans has been the same for about the last 1.5 billion years. The total amount of dissolved material in the world's oceans is calculated to be 5×10^{22} g for ocean water of 36‰ salinity. Each year, the runoff from the land adds another 2.5×10^{15} g, or 0.000005% of total ocean salt. If we assume that the rivers have been flowing to the sea at the same rate over the last 3.5 billion years, more dissolved material has been added by this process than is presently found in the sea. For the oceans to remain at the same salinity, the rate of addition of salt by rivers must be balanced by the removal of salt; input must balance output.

Salt ions are removed from seawater in a number of ways; some are shown in figure 6.3. Sea spray from the waves is blown ashore, depositing a film of salt on land. This salt is later returned to the oceans by runoff from the land. Over geologic time, shallow arms of the sea have become isolated, the water has evaporated, and the salts have been left behind as sedimentary deposits called **evaporites.** Salt ions can also react with each other to form insoluble products that precipitate on the ocean floor. Biological processes concentrate salts, which are removed from the water if the organisms are harvested or if the

organisms become part of the sediments. Organisms' excretion products trap ions, which are transferred to the sediments or returned to the seawater. Other biological processes remove Ca^{2+} (calcium) by incorporating it into shells, and silica is used to form the hard parts of diatoms and radiolarians. These accumulate in the sediments when the organisms die. Chapter 4 discussed the sediments produced by biological processes.

A chemical process known as **adsorption,** the adherence of ions and molecules onto a particle's surface, removes other ions and molecules from seawater. In this process, tiny clay mineral particles, weathered from rock and brought to the oceans by winds and rivers, bind ions such as K^+ (potassium) and trace metals to their surfaces. The ions sink with the clay particles and are eventually incorporated into the sediments. Strongly adsorbable ions replace weakly adsorbable ions in a process known as **ion exchange.** If the clay minerals and sediments adsorb and exchange one ion more easily than another, there is a lower concentration of the more easily adsorbed ion in the seawater. For example, potassium is more easily adsorbed than sodium, so K^+ is less concentrated than Na^+ in seawater. Nickel, cobalt, zinc, and copper are adsorbed on nodules that form on the ocean floor. The fecal pellets and skeletal remains of small organisms also act as adsorption surfaces. The settling of this organic debris transports metallic ions to the sediment, where they can be adsorbed on nodules. In these cases, the ions are removed from the water and are transferred to the sediments.

The process of forming the new crust at the ridge system of the deep-ocean floor (see chapter 3) participates in the input and output of the ions in seawater. Where molten rock rises from the mantle into the crust, magma chambers are formed. These

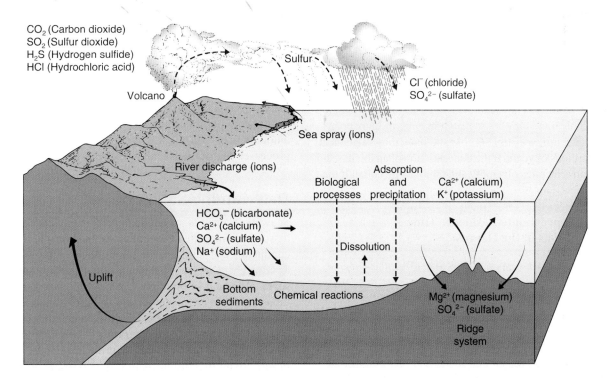

Figure 6.3 Processes that distribute and regulate the major constituents in seawater. Salt ions are added to seawater from rivers, volcanic events, ridge systems, and decay processes. Salt ions are removed from seawater by adsorption and ion exchange, spray, chemical precipitation, biological uptake, and addition to crustal rocks at ridge systems.

chambers are found mainly along plate boundaries but also above volcanic hot spots in the middle of plates. Cold water seeps down several kilometers through the fractured crust at a spreading center. The seawater becomes heated by flowing near the magma chamber. By convection, the heated water rises through the crust and reacts chemically with the rocks. Magnesium ions (Mg^{2+}) are transferred from the water to form minerals in the crust. At the same time, hydrogen ions (H^+) are released and the seawater solution becomes more acidic. Chemical reactions change sulfate (SO_4^{2-}) to sulfur and then to hydrogen sulfide (H_2S). The hot saline water dissolves metals from the crust, including copper, iron, manganese, and zinc, and releases potassium and calcium. It has been estimated that a volume of water equivalent to the entire mass of the oceans circulates through the crust at oceanic ridges every 10 million years.

The most important process for the removal of most elements from seawater is still adsorption of ions onto fine particles and their removal to the sediments. Ions deposited in the sediments are trapped there. They are not soon redissolved in the seawater, but geologic uplift elevates some marine sediments to positions above sea level. Erosion then works to dissolve and wash these deposits back to the sea.

Residence Time

The relative abundances of the salts in the sea are due in part to the ease with which they are introduced from Earth's crust and in part to the rate at which they are removed from the seawater. Sodium is moderately abundant in fresh water but is so highly soluble in seawater that it remains dissolved in the ocean. Calcium is removed rapidly from seawater to form limestone and the shells of marine organisms, but it is also replaced rapidly by calcium ions moving down rivers and in hot springs of the ridge systems on the sea floor (fig. 6.3).

The average, or mean, time that a substance remains in solution in the ocean is called its **residence time** (table 6.4). Aluminum, iron, and chromium ions have short residence times, in the hundreds of years. They react with other substances quickly and form insoluble mineral solids in the sediments. Sodium, potassium, and magnesium are very soluble and have long residence times, in the millions of years. If the concentration of an ion in seawater is constant, the rates of supply and removal are equal. If the

total amount of an ion present in the ocean is divided by either its rate of supply or its rate of removal, the residence time for that ion is known. For example, there are roughly 5.74×10^{20} g of Ca^{2+} in the oceans. It is estimated that Ca^{2+} is added to the oceans at a rate of about 5.4×10^{14} g per year. Therefore

$$\text{residence time } (Ca^{2+}) = \left(\frac{5.74 \times 10^{20} \text{ g}}{5.55 \times 10^{14} \text{ g/yr}} \right)$$
$$= 1.06 \times 10^6 \text{ years}$$

Constant Proportions

Seawater is a well-mixed solution; currents and eddies in surface and deep water, vertical mixing processes, and wave and tidal action have all helped to stir the oceans through geologic time. Because of this thorough mixing, the ionic composition of open-ocean seawater is the same from place to place and from depth to depth. That is, the ratio of one major ion or seawater constituent to another remains the same. Whether the salinity is 40‰ or 28‰, the major ions exist in the same proportions. This concept was first suggested by the chemist Alexander Marcet in 1819. In 1865, another chemist, Georg Forchhammer, analyzed several hundred seawater samples and found that these constant proportions did hold true. During the world cruise of the *Challenger* expedition (1872–76), seventy-seven water samples were collected from different depths and locations. When chemist William Dittmar analyzed these samples, he verified Forchhammer's findings and Marcet's suggestion. These analyses led to the **principle of constant proportion** (or **constant composition**) of seawater, which states that regardless of variations in salinity, the ratios between the amounts of major ions in open-ocean water are constant. Note that the principle applies to major conservative ions in the open ocean; it does not apply along the shores, where rivers may bring in large quantities of dissolved substances or may reduce the salinity to very low values. The ratios of abundance of minor nonconservative constituents vary because some of these are closely related to the life cycles of living organisms. Populations of organisms remove ions during periods of growth and reproduction, reducing the amounts in solution. Later, the population declines, and decay processes return these ions to the seawater.

Determining Salinity

Seawater transmits, or conducts, electricity because it contains dissolved ionic salts; the more ions in solution, the greater the conductance. Therefore, the salt content, or salinity, can be determined by using an instrument, called a **salinometer,** that measures electrical conductivity. Salinity readings in parts per thousand (S‰) are made quickly and directly on the water sample with an electrical probe. Because conductivity of seawater is affected by salinity and temperature, a conductivity instrument must correct for the temperature if the instrument reads directly in salinity units. This correction is also necessary if the instrument reads conductance only. If conductance and temperature are measured separately, a computer program calculates the salinity. Other direct measurement techniques are discussed in chapter 8.

Table 6.4	**Approximate Residence Time of Ions in the Oceans**
Ion	**Time in Years**
Chloride	100 million–infinite
Sodium	210 million
Magnesium	22 million
Potassium	11 million
Sulfate	11 million
Calcium	1 million
Manganese	0.0014 million (1400)
Iron	0.00014 million (140)
Aluminum	0.00010 million (100)

To be sure that all salinity determinations are comparable, the world's oceanographic laboratories use a standard method of analysis and a standard seawater reference. At present, the Institute of Oceanographic Services in Wormly, England, is responsible for the production of standard seawater adjusted to both constant chloride content and electrical conductance to ensure standard calibration of laboratory instruments.

Historically, the quantity of chloride ions in a water sample was measured to establish the salinity of the sample. To do so, silver nitrate was added because the silver combines with the chloride ions. If the amount of silver required to react with all the chloride ions in a sample is known, the amount of chloride is known. The chloride concentration measured in this way is termed **chlorinity (Cl‰)** and is measured in parts per thousand or grams of chloride per kilogram of seawater. When the chlorinity of a sample is known, the concentration of any other major constituent can be calculated by using the principle of constant proportion.

Chlorinity and salinity are related by the equation

$$\text{salinity (‰)} = 1.80655 \times \text{chlorinity (‰)}$$

or

$$S‰ = 1.80655 \times Cl‰$$

6.2 Gases

Gases move between the sea and the atmosphere at the sea surface. Atmospheric gases dissolve in seawater and are distributed to all depths by mixing processes and currents. Abundant gases in the atmosphere and in the oceans are nitrogen (N_2), oxygen (O_2), and carbon dioxide (CO_2). The percentages of each of these gases in the atmosphere and in seawater are given in table 6.5. Oxygen and carbon dioxide play important roles in the ocean because they are necessary to all life, and biological activities modify their concentrations at various depths. Although nitrogen gas is used directly only by bacteria, its use also plays an important role in ocean processes. Gases such as argon, helium, and neon are present in small amounts, but they are chemically inert and do not interact with the ocean water or its inhabitants.

The maximum amount of any gas that can be held in solution is the **saturation concentration.** The saturation concentration changes because it depends on the temperature, salinity, and pressure of the water. If the temperature or salinity decreases, the saturation concentration for the gas increases.

If the pressure decreases, the saturation concentration decreases. In other words, colder water holds more dissolved gas than warmer water, less-salty water holds more gas than more-salty water, and water under more pressure holds more gas than water under less pressure.

Distribution with Depth

During **photosynthesis,** plants, seaweeds, and phytoplankton use carbon dioxide, sunlight, and inorganic nutrients (see section 6.4) to produce organic matter. In the process of photosynthesis, oxygen is generated. Phytoplankton and seaweeds grow in surface waters where there is sufficient sunlight to carry out photosynthesis. This lighted portion of the ocean is referred to as the **euphotic zone** (euphotic means "well lit" in Greek). In coastal waters, the euphotic zone is relatively shallow and may extend to only about 15 to 20 m (49–66 ft). In the open ocean where there are less suspended particles, the euphotic zone extends much deeper to about 150 to 200 m (492–656 ft). Photosynthetic organisms produce oxygen and use carbon dioxide in surface waters. In contrast, heterotrophic organisms consume organic compounds as food and use respiration to derive energy from the consumed materials. During respiration, the organic matter is **oxidized** using oxygen to produce carbon dioxide. Thus, heterotrophic organisms consume oxygen and produce carbon dioxide. All living organisms, regardless of whether they are photosynthetic or heterotrophic, carry out respiration. As a consequence, respiration occurs at all depths within the ocean. Photosynthesis and respiration will be discussed in more detail in chapter 15. Bacteria also respire (although they, of course, do not breathe like humans). Bacterial respiration of organic matter becomes the most important factor in the removal of oxygen from seawater at depth because bacteria are the most abundant organisms deep in the water column.

The depth at which the rate of photosynthesis balances the rate of respiration is called the **compensation depth.** Above the compensation depth, photosynthetic organisms produce oxygen at the expense of carbon dioxide; below it, carbon dioxide is produced at the expense of oxygen. Oxygen can be added to the oceans only at the surface, from exchange with the atmosphere or as a waste product of photosynthesis. Carbon dioxide also enters from the atmosphere at the surface, but it is also produced at all depths from respiration.

Dissolved oxygen concentrations vary from 0–10 ml/l of seawater. Very low or zero concentrations occur in the bottom

Table 6.5	**Abundance of Gases in Air and Seawater**			
Gas	Symbol	Percentage by Volume in Atmosphere	Percentage by Volume in Surface Seawater[1]	Percentage by Volume in Total Oceans
Nitrogen	N_2	78.03	48	11
Oxygen	O_2	20.99	36	6
Carbon dioxide	CO_2	0.03	15	83
Argon, helium, neon, etc.	Ar, He, Ne	0.95	1	
Totals		100.00	100	100

1. Salinity = 36%, temperature = 20°C.

Figure 6.4 High-density ocean water is trapped behind the sill. The trapped water becomes anoxic at depth owing to continual respiration and decomposition. At the surface, sunlight enhances photosynthesis and the production of oxygen.

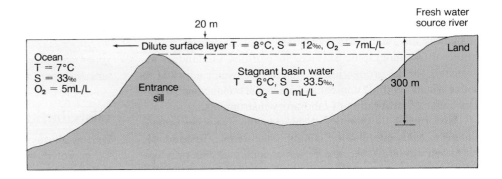

waters of isolated deep basins, which have little or no mixing with surface water. Such an area can occur at the bottom of a trench, in a deep basin behind a shallow entrance sill (as in the Black Sea), or at the bottom of a deep fjord (300–400 m; 900–1300 ft). If the deep water is only slowly flushed, respiration can use up the oxygen faster than the slow circulation to this depth is able to replace it. The bottom water becomes **anoxic,** or stripped of dissolved oxygen; **anaerobic** (or nonoxygen-using) bacteria live in such water. This condition is shown in figure 6.4. Because oxygen is more soluble in cold water than in warm water, more oxygen is found in surface waters at high latitudes than at lower latitudes. If the water is quiet, the nutrients and sunlight are abundant, and a large population of photosynthetic organisms is present, oxygen values at the surface can rise above the equilibrium (or saturation) value to 150% or more. This water is **supersaturated.** Wave action tends to liberate oxygen to the atmosphere and return the water to its 100% saturation state.

Figure 6.5 shows typical oxygen and carbon dioxide concentrations with depth. The concentration of both oxygen and carbon dioxide is influenced by biology. The concentration of oxygen is high and the concentration of carbon dioxide is low in surface waters because of photosynthesis. Below the euphotic zone, oxygen decreases as respiration of organic material removes the oxygen. The oxygen minimum occurs at about 800 m (2600 ft). Below this depth, the rate of removal of oxygen decreases because the population density of animals and the abundance of organic matter have decreased. The slow supply of oxygen to greater depths by water sinking from the surface gradually increases the concentration above that found at the oxygen minimum.

Carbon dioxide levels range between 45 and 54 ml/l throughout the oceans. The carbon dioxide concentration at the surface is low because it is used in photosynthesis. Below the surface layer, the concentration increases with depth as respiration continually produces carbon dioxide and adds it to the water. The deep water is able to hold high concentrations of CO_2 because the saturation value is high at low temperatures and high pressures. This is why calcareous oozes are preserved in the warmer, shallower water above the CCD and dissolved below it (review the discussion of biogenous sediments in chapter 4). At shallow depths, the concentration of CO_2 in the oceans is relatively low, in part because of high rates of photosynthesis and in part because the warm, shallow water has a low saturation value.

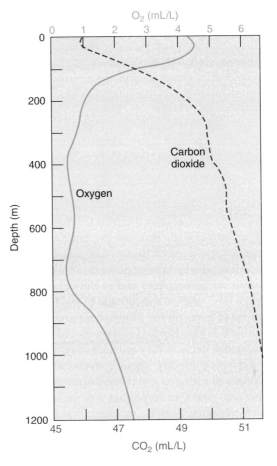

Figure 6.5 The distribution of O_2 and CO_2 with depth. Changes in O_2 concentration may exceed 400% from the surface to depth, whereas CO_2 concentrations change by less than 15%.

The Carbon Dioxide Cycle

At present, the net annual ocean uptake of carbon as carbon dioxide (CO_2) from the atmosphere is estimated to be 2 billion–3 billion metric tons per year. The rate at which the oceans absorb CO_2 is controlled by water temperature, pH (discussed in section 6.3), salinity, the chemistry of the ions (presence of calcium and carbonate ions), and biological processes, as well as mixing and circulation patterns (fig. 6.6).

The transfer of carbon from CO_2 to organic molecules by photosynthesis results in the addition of CO_2 to the intermediate and deep-ocean water when the organic material sinks and decays.

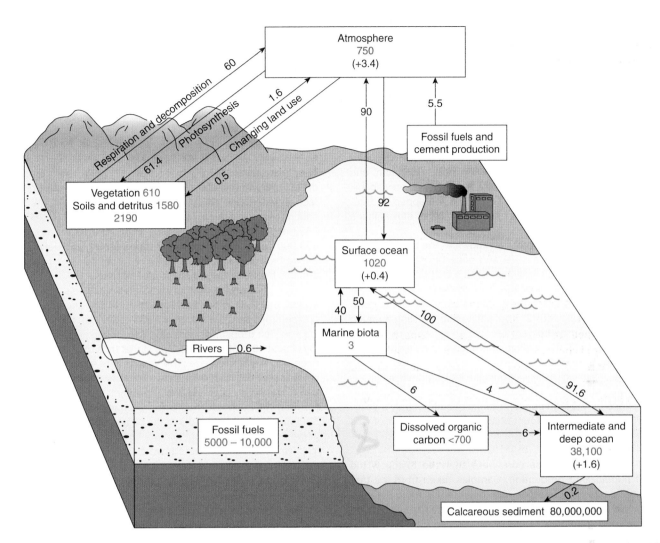

Figure 6.6 Major carbon dioxide pathways through Earth's environment. All numbers are given in billions of metric tons (1 metric ton = 10^6 g, 1 billion metric tons = 10^{15} g) of carbon. Numbers in *black* are rates of exchange per year between reservoirs. Numbers in *green* are total amounts stored in reservoirs. Numbers in *parentheses* are net annual changes in the total amount stored.

This process is often called the **biological pump.** Phytoplankton (see chapter 15) are responsible for about 40% of Earth's total production of organic material by photosynthesis. These organisms inhabit the shallow surface water, where sufficient sunlight is available for photosynthesis (see sections 5.8 and 15.2). About 90% of the phytoplankton organic matter is recycled in the euphotic zone as a result of consumption and respiration. Most of the remaining 10% (between 70% and 90%) is recycled before it reaches the sea floor, where the remainder is consumed by bottom-dwelling animals, decomposed by bacteria, or preserved in the sediment. This biochemical pump works along with the chemical solubility pump discussed previously to concentrate carbon in the deep ocean. The storage of CO_2 in Earth's reservoirs and the effect of increased production of CO_2 by the burning of fossil fuels are discussed further in chapter 7.

The Oxygen Balance

The oceans also play a large role in regulating the oxygen balance in Earth's atmosphere. Photosynthesis in the oceans releases oxygen, which is consumed by respiration and decay processes in

the same way as on land. However, some organic matter is incorporated into the seafloor sediments, preventing decay and decomposition. Therefore, oxygen is not consumed to balance the oxygen produced in photosynthesis, and 300 million metric tons of excess oxygen are produced each year. This excess amount of oxygen is not released into the atmosphere because the marine sediments are formed into rocks by Earth's geologic processes. Some of these rocks are uplifted onto land, and oxygen is eventually consumed in the weathering and oxidation of the materials in the rocks. This process balances the atmosphere's oxygen budget. The mechanisms that link and control the process are not well understood.

Measuring the Gases

The amount of dissolved oxygen in seawater samples can be measured by traditional chemical techniques in the laboratory. It is also possible to directly measure oxygen in the oceans by using specialized probes that send an electronic signal back to the ship or store the information in the testing unit. The concentration of dissolved carbon dioxide in seawater is very small. Nearly all of

Messages in Polar Ice

The chemical history of Earth's climate for the last 200,000 years is trapped in the polar ice sheets. Scientists are going back in time by drilling ice cores in Greenland and Antarctica to discover connections between climate and atmospheric chemistry and the impact of human activities on climate and atmospheric composition (box fig. 1). The chemical composition of the ice preserves, layer by layer, a record of soluble substances (sodium, chloride, sulfate, and nitrate) and heavy metals (lead, cadmium, and zinc). Dust tells how stormy the atmosphere was and when volcanic eruptions occurred, and bubbles in the ice contain trapped air, tracing the natural changes of carbon dioxide and other atmospheric gases through time.

A collaborative European program known as the Greenland Ice Core Project (GRIP) began in 1989 and ended in 1995. In 1992, GRIP reached 3029 m (9930 ft), where the ice is 200,000 years old or more. A new hole, NGRIP, is being drilled by Denmark north of the GRIP position. Thirty-five kilometers (20 mi) away, the U.S. Greenland Ice Sheet Project (GISP2) spent five years drilling through the ice sheet and into bedrock to recover an ice core of 3053 m (10,014 ft) in July 1993. The drilling operation has ceased, but research continues at the site.

Scientists analyzing these ice cores are looking back in history. The GISP2 ice core has identified some of the world's major volcanic eruptions by the high concentrations of sulfate: Italy's Vesuvius in 6955 B.C.; Oregon's Mount Mazama when Crater Lake was formed in 5676 B.C.; Iceland's Hekla in 2310 B.C.; and Alaska's Aniakchak in 1623 and 1629 B.C.

Sections of the GRIP ice core were analyzed to follow the history of human lead production. Samples taken from ice that formed 7760 years ago (depth of 1286 m [4000 ft]) in the prelead-production period were compared to samples from the times of the rise of Greek civilization, the Roman Republic and Empire, and the Medieval and Renaissance periods. Large-scale pollution from lead smelters during those periods contaminated even the Arctic, leaving the oldest documented record of human pollution.

Comparison of ice layers deposited before and after the Industrial Revolution shows that the concentration of sulfate from sulfuric acid tripled and that the concentration of nitrate from nitric acid doubled during the twentieth century. Lead continued to be a significant contaminant and increased dramatically as a byproduct of leaded-fuel combustion tied to the world's increased dependence on gasoline.

These ice cores tell us that climate changes have occurred more rapidly and more frequently than scientists had thought. Climate changes are recorded in the ice every few thousand years until the end of the last ice age, about 15,000 years ago. The climate warmed for the next 2000 years until a short period when Greenland temperatures decreased by 7°C, followed by very rapid warming, possibly over a period as short as three years, to the present interglacial stage that Earth enjoys today. These rapid, recent shifts in climate are thought to have been coupled with large shifts in the circulation of both the oceans and the atmosphere.

GISP2 core samples show massive changes in the quantities of dust and sea salt from aerosols that occur with the onset of glacial periods. When the ocean ice cover expanded, the dusts and salts originated in more southern latitudes, requiring contemporaneous, large-scale changes in atmospheric circulation to deliver the particles to the Arctic. Such events may have contributed to blocking incoming solar radiation and may partly explain the duration of the glacial periods.

In Antarctica, Russian scientists set up at the Vostok Antarctic Research Station in 1957 and began drilling ice cores in 1972. They were joined by a French team in 1984, and together, they drilled more than 2000 m (6600 ft) into the ice, capturing an ice record going back 160,000 years. In February 1998, the Russians shut down their station, having drilled the world's deepest ice core, 3623 m (12,000 ft). The bottom ice of the core dates back about 450,000 years. The Russians stopped drilling just before reaching the surface of Lake Vostok, which is twice the size of Lake Ontario and lies between the ice sheet and bedrock.

Their findings show that carbon dioxide levels increased at the end of the next to the last ice age, 140,000 years ago, and decreased rapidly again about 100,000 years ago, the start of the last ice age. Ten thousand years ago, carbon dioxide levels rose again (by almost 40%); the rising levels at that time may have played a role in ending the last ice age. The Vostok core shows a rise in carbon dioxide over the last 2000 years. Methods that compare Greenland and Antarctic ice cores against a common timeline suggest that the Northern Hemisphere has a more changeable climate than the Southern Hemisphere and that the beginning and ending of ice ages in the north trigger changes in the south.

New ice core projects have begun in West Antarctica, conducted by the United States at Siple Dome (using the drill from the Greenland GISP2 hole) and by the European Union at Dome C. The snowfall at these sites is much heavier than at the Vostok site and produces a more detailed climate record of the last 100,000 years. Data from these new cores will help researchers interpret the Vostok data.

To Learn More About Messages in Polar Ice

Fiedel, S., J. Southon, and T. Brown. 1995. The GISP Ice Core Record of Volcanism Since 7000 B.C. *Science* 267 (5195): 256–58.

Hong, S., P. Candelone, C. Patterson, and C. Boutron. 1994. Greenland Ice Evidence of Hemispheric Lead Pollution Two Millennia Ago by Greek and Roman Civilizations. *Science* 265 (5180): 1841–43.

Mayewski, P. A., et al. 1994. Changes in Atmospheric Circulation and Ocean Ice Cover over the North Atlantic During the Last 41,000 Years. *Science* 263 (5154): 1747–51.

Mayewski, P. A., et al. 1994. Record Drilling Depth Struck in Greenland. *EOS* 75 (10): 113, 119, 124.

Petit, J. R., et al. 1999. Climate and Atmospheric History of the Past 420,000 Years from the Vostok Ice Core, Antarctica. *Nature* 399: 429–36.

Stauffer, B. 1993. The Greenland Ice Core Project. *Science* 260 (5115): 1766–67.

Stone, R. 1998. Russian Outpost Readies for Otherworldly Quest. *Science* 279 (5351): 658–61.

Weiner, J. 1989. Glacier Bubbles Are Telling Us What Was in Ice Age Air. *Smithsonian* 20 (2): 78–87.

(a)

(b)

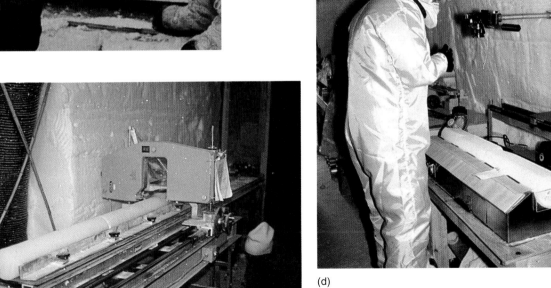

(c)

(d)

Box Figure 1 (a) Scott-Amundsen Base, 90°S, November–December 1982. Cores 10 cm (4 in) in diameter were drilled to 227 m (745 ft) with a mechanical drill. The ice at the bottom of the core is approximately 2200 years old. (b) Greenland Ice Sheet Project 2 (GISP2) drilling station, summer 1990. The tunnel houses laboratories and leads to the drilling site. The boxes at the tunnel entrance hold core sections ready for shipment to cold storage at the scientists' home laboratories. (c) A lengthwise slab is removed from a core section for analysis in the tunnel lab. (d) GISP2, July 1989. The layering of this ice core was inspected and photographed before it was analyzed for gas and particle content, chemistry, and conductivity. The core was 13 cm (5 in) in diameter.

the carbon dioxide in seawater reacts with water to form carbonic acid and its dissociation products (see section 6.3). Consequently, carbon dioxide concentrations can either be measured directly or determined by measuring the pH of the water.

6.3 The pH of Seawater

The water molecule, H_2O, can dissociate (break apart) to form a hydrogen cation, H^+, and a hydroxide anion, OH^-. Consequently, in any water solution, there will always be a combination of H_2O molecules, H^+ ions, and OH^- ions. The concentration of H_2O molecules always greatly exceeds the concentrations of the two ions. In a pure water solution (one in which there is only water molecules) at 25°C, a very small fraction of the water molecules, about 10^{-7}, will spontaneously dissociate into H^+ and OH^- ions. In other words, the concentration of both H^+ and OH^- ions will be 10^{-7}, as one in every 10 million (10^7) water molecules breaks apart. Solutions in which the concentrations of these two ions are equal are called neutral solutions.

In solutions that are not pure water, chemical reactions can remove or release hydrogen ions, making the concentrations of H^+ and OH^- unequal. The concentrations of H^+ and OH^- in a water solution are inversely proportional to each other. In other words, a tenfold increase in the concentration of one ion results in a tenfold decrease in the concentration of the other. An imbalance in the relative concentrations of these ions results in either an acidic solution (if there are more H^+ cations than OH^- anions) or an alkaline, also called basic, solution (if there are more OH^- anions than H^+ cations).

The acidity or alkalinity of a solution is measured using the **pH** scale, which ranges from a low of 0 to high of 14 (fig. 6.7). The pH scale is a logarithmic scale that measures the concentration of the hydrogen ion (written $[H^+]$) in a solution. The formal definition of pH is:

$$pH = -\log_{10}[H^+]$$

In pure water, where the concentrations of H^+ and OH^- are both 10^{-7}, the pH is equal to 7,

$$pH = -\log_{10}[10^{-7}] = -(-7) = 7,$$

and the solution is neutral. If the concentration of the hydrogen ion is increased by a factor of ten to 10^{-6}, or one part in 1 million instead of one part in 10 million, the pH drops to 6 and the solution is slightly acidic. Solutions with pH less than 7 (high H^+ concentrations) are acidic and those with pH greater than 7 (low H^+ concentrations) are alkaline, or basic.

Solutions with lower pH are more acidic than solutions with higher pH. For example, the following statements are all equivalent:

pH = 1 is more acidic than pH = 3, or

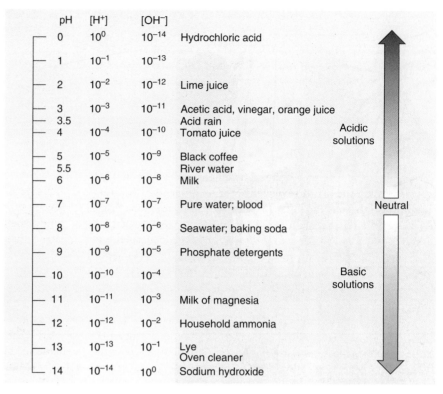

pH	$[H^+]$	$[OH^-]$	
0	10^0	10^{-14}	Hydrochloric acid
1	10^{-1}	10^{-13}	
2	10^{-2}	10^{-12}	Lime juice
3	10^{-3}	10^{-11}	Acetic acid, vinegar, orange juice
3.5			Acid rain
4	10^{-4}	10^{-10}	Tomato juice
5	10^{-5}	10^{-9}	Black coffee
5.5			River water
6	10^{-6}	10^{-8}	Milk
7	10^{-7}	10^{-7}	Pure water; blood
8	10^{-8}	10^{-6}	Seawater; baking soda
9	10^{-9}	10^{-5}	Phosphate detergents
10	10^{-10}	10^{-4}	
11	10^{-11}	10^{-3}	Milk of magnesia
12	10^{-12}	10^{-2}	Household ammonia
13	10^{-13}	10^{-1}	Lye / Oven cleaner
14	10^{-14}	10^0	Sodium hydroxide

Figure 6.7   The pH scale is a measure of the concentration of hydrogen ions in a solution. A neutral solution has a pH of 7. Acidic solutions have a pH less than 7, and alkaline, or basic, solutions have a pH greater than 7.

$[H^+] = 10^{-1}$ is more acidic than $[H^+] = 10^{-3}$, or

$[H^+] = 0.1$ is more acidic than $[H^+] = 0.001$

Similarly, solutions with higher pH are more alkaline than solutions with lower pH.

Because the hydrogen ion is very reactive, acidic water (water with a relatively high concentration of H^+ and pH less than 7) is an effective chemical weathering agent capable of decomposing and dissolving rock. The pH of rainwater is normally about 5.0–5.6 (slightly acidic), but in some heavily industrialized regions where emissions combine with water droplets to from acid rain, it can be much lower. Typical rain in the eastern United States has a pH of about 4.3, roughly ten times more acidic than normal, and can in some cases drop as low as 3, or 100 times more acidic than normal.

Seawater is slightly alkaline with a pH between 7.5 and 8.5. The pH of the world's oceans averaged over all depths is approximately 7.8. Surface water currently has an average pH of about 8.2. The pH of seawater remains relatively constant because of the buffering action of carbon dioxide in the water. A **buffer** is a substance that prevents sudden, or large, changes in the acidity or alkalinity of a solution. If some process changes the concentration of hydrogen ions in seawater, causing the pH to rise above or fall below its average, or mean value, the buffer becomes involved in chemical reactions that release or capture hydrogen ions, returning the pH to normal. When carbon dioxide dissolves in seawater, the CO_2 combines with the water to form carbonic acid (H_2CO_3). The carbonic acid rapidly dissociates into bicarbonate (HCO_3^-) and a hydrogen ion (H^+), or carbonate (CO_3^{2-}) and two hydrogen ions ($2\,H^+$). The CO_2, H_2CO_3, HCO_3^-, and CO_3^{2-} exist in equilibrium with each other and with H^+, as shown in the following equation. The double arrows indicate that the reactions can move in either direction, either producing or removing hydrogen ions as is necessary to maintain a relatively constant pH.

$$CO_2 + H_2O \Leftrightarrow H_2CO_3 \Leftrightarrow HCO_3^- + H^+ \text{ or } CO_3^{2-} + 2H^+$$

If seawater becomes too alkaline, or basic, then the reactions in this equation progress to the right, releasing hydrogen ions and decreasing the pH. If seawater becomes too acidic, then the reactions progress to the left, removing free hydrogen ions from the water and increasing the pH. This buffering capacity of carbon dioxide in seawater is important to organisms requiring a relatively constant pH for their life processes and to the chemistry of seawater, which is controlled, in part, by its pH.

From the reactions described, it is clear that the pH of seawater strongly depends on the concentration of CO_2 in the water. The higher the concentration of CO_2 in the water, the lower its pH and the more acidic it becomes. While the average pH of seawater tends to be fairly constant, it does vary with changes in the concentration of CO_2 in the water, as discussed in section 6.2.

Because the pH of seawater is inversely proportional to the concentration of CO_2, the pH of the surface water tends to be slightly higher (about 8.2 on average), or more alkaline, than average (refer back to fig. 6.5). The pH at the ocean surface may be as high as 8.5 if the water is warm and if the rate of primary production, or photosynthesis, is high. Raising the pH of the water releases carbonate ions, CO_3^{2-}, that bond with the abundant Ca^{2+} in solution to form calcium carbonate $CaCO_3$. In cold, deep water, where the concentration of CO_2 is high, the pH drops, making the water more acidic and dissolving calcium carbonate shells.

The increase in CO_2 in the atmosphere has resulted in a corresponding increase in the concentration of the gas in the oceans as CO_2 is absorbed by seawater. The increasing concentration of CO_2 in the water is producing a decrease in the pH of the water, an effect that is referred to as "ocean acidification." The average pH of the oceans is predicted to fall by up to 0.5 units by 2100 if global emissions of CO_2 continue to rise at present rates. This increase in ocean acidity could have a major impact on shallow-water marine organisms that build shells of calcium carbonate, which could dissolve rapidly in more acidic water.

The most important sources of CO_2 in seawater are direct transfer of the gas from the atmosphere, the respiration of marine organisms, and the oxidation of organic matter.

6.4 Other Substances

Nutrients

Ions required for plant or phytoplankton growth are known as nutrients; these are the fertilizers of the oceans. As on land, phytoplankton require nitrogen and phosphorus in the form of nitrate (NO_3^-) and phosphate (PO_4^{3-}) ions. A third nutrient

Table 6.6	Nutrients in Seawater	
Element	**Concentration $\mu g/kg^1$**	**Relative Molar Abundance**
Nitrogen (N)	500	16
Phosphorus (P)	70	1
Silicon (Si)	3000	40

1. Parts per billion.

required in the oceans is the silicate ion (SiO_4^{4-}), which is needed to form silica (SiO_2), the hard outer wall of the single-celled diatoms and the skeletal parts of some protozoans. These three nutrients are among the dissolved substances brought to the sea by the rivers and land runoff. Despite their importance, they are present in very low concentrations (table 6.6).

The concentrations of nutrient ions vary because some of these ions are closely related to the life cycles of organisms. The relative molar abundance of carbon, nitrogen, and phosphorus in marine phytoplankton is C:N:P = 106:16:1. This relationship is called the **Redfield Ratio.** Analysis of the composition of siliceous marine organisms makes it possible to calculate a Redfield Ratio for silicon as well: C:Si:N:P = 106:40:16:1. The consumption, decomposition, and recycling of organic matter as it sinks through the water column result in an increase in carbon:nutrient ratios as nutrients are released to the water. Nutrients are removed from the water as the plant populations grow and reproduce, temporarily reducing the amounts in solution. Later, when the populations decline, death and decay return the ions to the seawater. Nutrients are cycled to different consumers as zooplankton feed on phytoplankton. The zooplankton are in turn eaten by other consumers, and eventually the nutrients are returned to the oceans by death and bacterial decomposition. Excretory products from zooplankton and larger animals are also added to the seawater, broken down, and used by a new generation of zooplankton and phytoplankton. Nutrients are nonconservative; they do not maintain constant ratios in the way most major salt ions do. The cyclic nutrient pathways for nitrogen and phosphorus are discussed in chapter 15.

Organics

A wide variety of organic substances are present in seawater. Proteins, carbohydrates, lipids (or fats), vitamins, hormones, and their breakdown products are all present. Some are eventually oxidized or broken down into smaller molecules; others are used directly by organisms and are incorporated into their systems. Another portion of the organic matter accumulates in the sediments, where over geologic time it may slowly provide hydrocarbon molecules to form deposits of oil and gas. In the areas of the ocean that are high in plant and animal life, the surface layer may take on a green-yellow color owing to the presence of organic decay products. The incorporation of soluble organics into glacial ice at the Antarctic ice shelves is related to the formation of green ice.

6.5 Practical Considerations: Salt and Water

Chemical Resources

About 30% of the world's table salt is extracted from seawater. The industrially produced energy required to remove the water is kept to a minimum to keep extraction costs low. In warm, dry climates, seawater is allowed to flow into shallow ponds and evaporate down to a concentrated brine solution. More seawater is added, and the process is repeated several times, until a dense brine is produced. Evaporation continues until a thick, white salt deposit is left on the bottom of the pond. Several different salts are in the deposit. These salts form in the order listed in table 6.7. The salt deposit is collected and refined to separate out sodium chloride (halite), or table salt. This technique is used in southern France, Puerto Rico, and California (fig. 6.8).

In cold climate areas, salt has been recovered by freezing the seawater in similar ponds. The ice that forms is nearly fresh; the salts are concentrated in the brine beneath the ice. The brine is removed and heated to remove the last of the water.

Of the world's supply of magnesium, 60% comes from the sea, and so does 70% of the bromine. There are vast amounts of dissolved constituents in the world's seawater, including 10 million tons of gold and 4 billion tons of uranium, but the concentration is very low (one part per billion or less). In the case of gold, no method has been devised in which the cost of extraction does not exceed the value of the recovered element. At one time or another, Japan, Germany, and the United States have expressed an interest in extracting uranium. Japan set up a land-based test plant in 1986 to produce 10 kg (26 lb) of uranium from seawater each year, but the operation proved to be too expensive.

Desalination

Desalination is the process of obtaining fresh water from salt water. There are three main desalination methods:

1. processes involving a change of state of the water—liquid to solid or liquid to vapor;
2. processes requiring ion exchange columns; and
3. processes using a semipermeable membrane—electrodialysis and reverse osmosis.

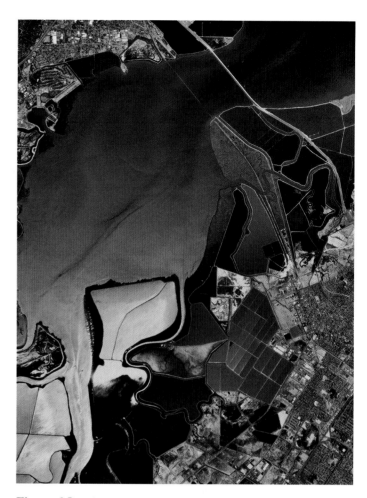

Figure 6.8   The southern end of San Francisco Bay was diked into shallow ponds, where seawater is evaporated to obtain salt. Most of these ponds are being converted back to "natural" conditions.

The simplest process involving a change of state is a solar still (fig. 6.9). In this process, a pond of seawater is capped by a low plastic dome. Solar radiation penetrates the dome and evaporates the seawater. The evaporated water condenses on the undersurface of the dome and trickles down to be caught in a trough, where it accumulates and flows to a freshwater reservoir. The rate of production is slow, and a very large system is needed to supply the water requirements of even a small community.

When water is distilled by boiling, evaporation proceeds at a rapid rate and large quantities of fresh water are produced, but the energy requirement is high. If water is introduced to a chamber with a reduced air pressure, the boiling occurs at a much lower temperature and therefore uses less energy. Change of state by freezing can also be used to recover fresh water from seawater. The energy requirement is approximately one-sixth of that needed for evaporation, but the mechanical separation of the freshwater ice from the salt brine remains difficult.

Columns containing ion exchange resins that extract ions from salt water work well with water of low salinity, but the resins need to be replaced periodically (fig. 6.10*b*). Small ion exchange units are manufactured for household use to improve drinking water quality.

Table 6.7	Sequence of Salts Formed from Evaporation of Seawater	
Order of Precipitation	Solid	% of Total Solid
1	CaCO$_3$ + MgCO$_3$	1
2	CaSO$_4$ (gypsum)	3
3	NaCl (halite)	70
4	Na-Mg-K-SO$_4$ and KCl, MgCl$_2$	26

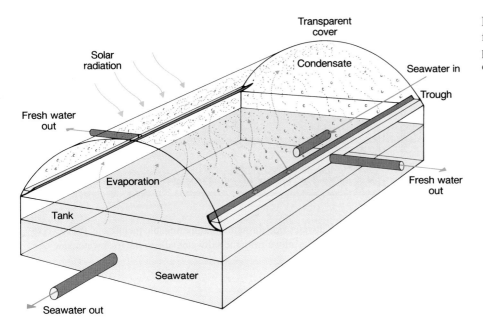

Figure 6.9 Solar energy is used to evaporate fresh water from seawater. Solar radiation penetrates the transparent cover over the seawater contained in the tank.

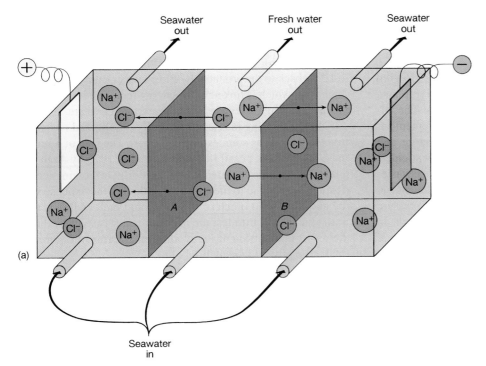

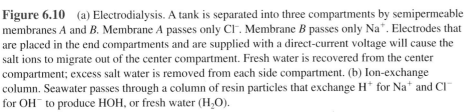

Figure 6.10 (a) Electrodialysis. A tank is separated into three compartments by semipermeable membranes *A* and *B*. Membrane *A* passes only Cl$^-$. Membrane *B* passes only Na$^+$. Electrodes that are placed in the end compartments and are supplied with a direct-current voltage will cause the salt ions to migrate out of the center compartment. Fresh water is recovered from the center compartment; excess salt water is removed from each side compartment. (b) Ion-exchange column. Seawater passes through a column of resin particles that exchange H$^+$ for Na$^+$ and Cl$^-$ for OH$^-$ to produce HOH, or fresh water (H$_2$O).

Electrodialysis uses an electrical field to transport ions out of solution and through **semipermeable membranes;** this technique also works best in low-salinity (or brackish) water (fig. 6.10*a*).

Osmosis is the movement of water across a semipermeable membrane; the water moves from the side with the higher con-

centration of water molecules (or low salinity) to the side with the lower concentration of water molecules (or high salinity); this movement creates a higher pressure on the low-water concentration (or high-salinity) side of the membrane (fig. 6.11*a*). **Reverse osmosis** produces fresh water from seawater by applying pressure

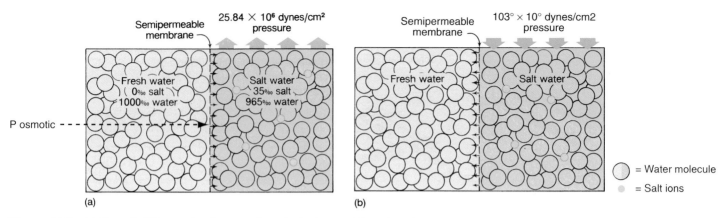

Figure 6.11 (a) Osmosis. Water molecules move from the freshwater side to the saltwater side through a semipermeable membrane. (b) Reverse osmosis. When pressure on the salt water exceeds 25.84×10^6 dynes/cm², water molecules move from the saltwater side to the freshwater side of a semipermeable membrane.

to seawater and forcing the water molecules through a semipermeable membrane, leaving behind the salt ions and other impurities (fig. 6.11b). The pressure applied to the seawater must exceed 24.5 atm, or 25.84×10^6 dynes/cm². A pressure of about 101.5 atm, or $10^3 \times 10^6$ dynes/cm², is required to achieve a reasonable rate of freshwater production. The energy requirement is about one-half that needed for the evaporative process.

Reverse osmosis is the most popular and rapidly growing form of desalination technology. As older evaporative plants wear out, reverse osmosis plants are replacing them. The advantages of reverse osmosis include no energy requirement for heating the water, no thermal pollution from the discharge, and removal of unwanted contaminants—including pesticides, bacteria, and some chemical compounds. High-salinity wastewater returning to the coastal environment can be a disadvantage. The cost of the energy required to pump the water under pressure makes the cost of this type of desalinated water very high. In southern California, desalinated water is about fifteen times more expensive than local groundwater and five times more expensive than water imported from the northern part of the state.

Because of the high price involved, desalination plants in southern California are typically operated only during periods of drought, when reservoirs and groundwater levels are low.

During years when normal or high rainfall provides sufficient fresh water, the plants are shut down and only maintenance-level work is done to ensure that they will be operational when needed.

In areas such as Kuwait, Saudi Arabia, Morocco, Malta, Israel, the West Indies, California, and the Florida Keys, water is a limiting factor for population and industrial growth. The greatest drawback to the production of fresh water from seawater is the high cost, which is linked to the energy required. In Persian Gulf countries, water costs are low because fuel costs are low.

Large floating masses of naturally desalinated ice have attracted interest as sources of fresh water. The first of a series of meetings on this possibility was hosted in 1978 by Arab interests. Discussions centered on the feasibility of towing icebergs from the Antarctic to the Red Sea. Although the expense would be enormous and much of the ice would melt as it was towed into equatorial latitudes, the need for water is so great that it is believed that sufficient ice would remain after the journey to make the project worthwhile.

Harvesting ice from Alaskan glaciers is done under permits from the Alaskan Department of Natural Resources; quantity and places of harvesting are limited. Most of the ice has been sold to Asia for processing into gourmet ice cubes.

Summary

Seawater is slightly alkaline with a pH between 7.5 and 8.5. The average pH value for all of the oceans over all depths is about 7.8. Seawater pH remains fairly constant because of the buffering action of carbon dioxide in the water.

The average salinity of ocean water is 35‰. The salinity of the surface water changes with latitude and is affected by evaporation, precipitation, and the freezing and thawing of sea ice. Soluble salts are present as ions in seawater. Positive ions are cations; negative ions are anions. Six major constituent ions make up 99% of the salt in seawater. Trace elements that are present in very small quantities are particularly important to living organisms.

Most of the positively charged ions come from the weathering and erosion of Earth's crust and are added to the sea by rivers. Gases from volcanic eruptions are dissolved in river water as anions. Because the average salinity of the oceans remains constant, the salt gain must be balanced by the removal of salt; input must equal output. Salts are removed as sea spray, evaporites, and insoluble precipitates, as well as by biological reactions, adsorption, chemical reactions, and uplift processes.

Seawater circulating near magma chambers in Earth's crust deposits dissolved metals and releases other chemicals in solution. The time that salts remain in solution, known as residence time, depends on their reactivity.

The proportion of one major ion to another remains the same for all open-ocean salinities. Ratios may vary in coastal areas and in association with biological processes.

Salinity is determined by measuring a sample's electrical conductivity. Historically, salinity has been determined chemically by measuring the quantity of chloride ions in a sample.

The saturation value of gases dissolved in seawater varies with salinity, temperature, and pressure. Carbon dioxide is added to seawater from the atmosphere and by respiration and decay processes at all depths; it is removed at the surface by photosynthesis. Oxygen is added only at the surface from the atmosphere and the photosynthetic process; it is depleted at all depths by respiration and decay. Seawater may become supersaturated with oxygen, or it may become anoxic. Carbon dioxide levels tend to change little over depth. Carbon dioxide has the additional role of buffer in keeping the pH range of ocean water between 7.5 and 8.5. Large quantities of CO_2 are absorbed by the oceans. Biological processes pump carbon as carbon dioxide into deep water, where it is fixed in the marine sediments as calcium carbonate. Atmospheric oxygen is regulated by oceanic processes. The amount of oxygen present in seawater is measured chemically and electronically. Carbon dioxide content is determined from the pH of the water.

Nutrients include the nitrates, phosphates, and silicates required for plant growth. A wide variety of organic products are also present.

Salt, magnesium, and bromine are currently being commercially extracted from seawater. Direct extraction of other chemicals is neither economic nor practical at present. Fresh water is an important product of seawater. Desalination methods include change-of-state processes, movement of ions across semipermeable membranes, and ion exchange. The practicality of desalination is determined by cost and need. Reverse osmosis has become the most popular option; it is nonpolluting but costly because of its energy requirements.

Key Terms

All key terms from this chapter can be viewed by term or definition when studied as flashcards on this book's website at **www.mhhe.com/sverdrup10e**.

cation, 149
anion, 149
salinity, 149
ionic bond, 150
major constituent, 150

trace element, 151
conservative constituent, 151
nonconservative constituent, 151
evaporite, 153
adsorption, 153
ion exchange, 153
residence time, 154
principle of constant proportion (constant composition), 154

salinometer, 154
chlorinity (Cl‰), 155
saturation concentration, 155
photosynthesis, 155
euphotic zone, 155
oxidized, 155
compensation depth, 155
anoxic, 156
anaerobic, 156
supersaturation, 156

biological pump, 157
pH, 160
buffer, 160
Redfield Ratio, 161
desalination, 162
electrodialysis, 163
semipermeable membrane, 163
osmosis, 163
reverse osmosis, 163

Study Questions

1. How is the ocean's salt balance regulated? Make a diagram showing inputs and outputs.
2. Explain the concept of the constant proportions of the major ions in seawater.
3. What is the least expensive method of desalination? Where is it most likely to be used?
4. How does the salinity of mid-ocean surface water change with latitude? What atmospheric processes produce these changes?
5. List the sources of the salts found in seawater. How is their input regulated?
6. Silicate is a nonconservative constituent of seawater and does not obey the principle of constant proportions. Explain why.
7. Compare the distributions of oxygen and carbon dioxide to depth in seawater. What processes are responsible for the distribution of each gas?
8. Explain the relationship between carbon dioxide and the pH of the oceans. Why is the buffering of seawater important?
9. Suggest several pathways a carbon dioxide molecule might follow as it is transferred from the atmosphere to the seafloor sediments.
10. Why does the addition of rain and river water to the oceans not decrease the ocean's overall salinity?
11. Why is the residence time in seawater different for different salts?
12. Explain reverse osmosis and how it produces fresh water from seawater.
13. What is the significance of the compensation depth to photosynthetic organisms?
14. Why does the concentration of dissolved substances in river water differ from the concentration of the same substances in seawater?
15. Why are nutrients considered to be nonconservative materials?

Study Problems

1. If there is 1.4×10^{21} kg of water in the oceans, what is the potential mass of NaCl (sodium chloride) in the oceans? Use table 6.1.

2. If the chloride ion (Cl^-) content of a seawater sample is 18.5‰ what is the concentration of magnesium in the same sample? Express your answer in g/kg.

3. Determine the residence time of calcium using the following information:

   Calcium ion present in seawater = 0.41 g/kg
   Seawater in the oceans = 1.4×10^{21} kg
   Calcium ion = 12.5% by weight of the average dissolved substances in river water
   Average salt content of river water = 0.12‰
   Annual river runoff = 3.6×10^{16} kg/yr

4. How many kilograms of seawater would have to be processed to obtain 1 kg of gold? Use table 6.2.

Hurricane Ele can be seen in the Central Pacific in this true-color
image taken on September 4, 2002.

Chapter 7

The Structure and Motion
of the Atmosphere

Chapter Outline

Learning Outcomes

After studying the information in this chapter students should be able to:

1. *outline* and *discuss* Earth's heat budget,

2. *distinguish* between specific heat and heat capacity,

3. *list* the layers of the atmosphere in order of ascending height and *sketch* a plot of temperature vs. elevation in each layer,

4. *argue* that global warming is enhanced by the accumulation of greenhouse gases, such as carbon dioxide, in the atmosphere,

5. *explain* the characteristics of the Antarctic ozone hole and *relate* its size to cloud formation and temperature,

6. *explain* the Coriolis effect and *describe* its direction and magnitude as a function of latitude,

7. *sketch* the pattern of major wind systems and regions of vertical motion in the atmosphere,

8. *list* three names for large, intense low pressure systems,

9. *relate* ENSO events to global weather patterns and variations in sea surface temperature, and

10. *review* the major historical and physical details of Hurricane Katrina.

---

*T*he Sun's energy reaches Earth's surface through the atmosphere, a thin shell of mixed gases we call air. The atmosphere and the ocean are in contact over 71% of Earth's surface; their interaction is continuous and dynamic. Processes that occur in the atmosphere are closely related to processes that occur in the oceans, and together they form much of what we call weather and climate. Clouds, winds, storms, rain, and fog are all the result of interplay among the Sun's energy, the atmosphere, and the oceans. This complex of interactions provides Earth's average climate and its daily weather, sometimes pleasant and stable, at other times severe and turbulent. Some of these interactions and processes are more predictable than others; some are better understood than others. Understanding the oceans requires an understanding of the atmosphere's influence on them. This chapter presents an overview of these self-adjusting relationships as well as specific examples of their combined effects.

7.1 Heating and Cooling Earth's Surface

The heating and cooling of Earth's surface are accomplished by energy exchanges that act to alter the densities of two fluid envelopes surrounding Earth, the atmosphere and the oceans. The initial source of the energy is solar radiation, which varies in both time and location on Earth's surface. This energy is absorbed, reflected, reradiated, converted into other forms of energy, and redistributed over Earth to create not only the structure but also the dynamics of the atmosphere and oceans.

Distribution of Solar Radiation

Instantaneous solar radiation per unit area of Earth surface has its greatest intensity at the equator, moderate intensity in the middle latitudes, and least intensity at the poles. If Earth had no atmosphere, the intensity of solar radiation available on a surface at right angles to the Sun's rays would be 2 calories per square centimeter per minute ($cal/cm^2/min$); this value is called the

solar constant. The solar constant is approached only at latitudes between 23½°N (the Tropic of Cancer) and 23½°S (the Tropic of Capricorn), because only between those latitudes does sunlight strike Earth at a right angle. Because of Earth's spherical shape, at all other latitudes, Earth's surface is inclined to the Sun's rays at an angle other than 90°. Compare the angles at which the Sun's rays strike Earth in figure 7.1. Because the Sun is so far away from Earth, its rays are parallel when they reach Earth. Where the rays strike Earth at right angles, the same amount of radiant energy strikes each unit area. But as the angle the Sun's rays make with Earth decreases, the rays and the surface become nearly parallel, and the unit areas receive less energy. This difference is also shown in figure 7.1.

When the Sun stands directly above the equator at noon, during either the vernal or the autumnal equinox, the radiation value is about 1.6 $cal/cm^2/min$. This value is less than the solar constant because Earth's atmosphere stands between Earth's surface and the incoming solar radiation, and the atmosphere both absorbs and reflects portions of the Sun's energy. Atmospheric interference causes the solar radiation per unit of surface area to decrease with increasing latitude in each hemisphere, because the greater the latitude, the longer the distance through the atmosphere the Sun's rays must travel (fig. 7.1). The combined effects of atmospheric path length and inclination of Earth's axis cause Earth to receive more solar heat in the tropics, less at the temperate latitudes, and the least at the poles. The intensity of solar radiation also varies as points on the turning Earth move from darkness to light and return to darkness and as the distance between the Sun and Earth changes seasonally.

Heat Budget

To maintain its long-term mean surface temperature of 16°C, Earth must lose heat as well as gain it. To maintain a constant average temperature, Earth and its atmosphere must reradiate as much heat back to space as they receive from the Sun. These gains and losses in heat are represented by a **heat budget.** If less heat was returned to space than is gained, Earth would become

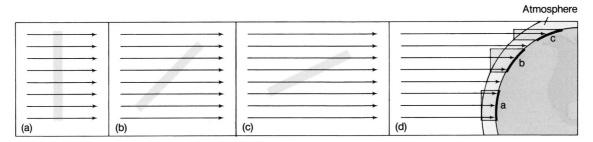

Figure 7.1 Areas of Earth's surface that are equal in size receive different levels of solar radiation as they become more oblique to the Sun's rays (see a, b, and c). As latitude increases, the Sun's rays and Earth's surface become more nearly parallel, and the solar radiation received on equal surface areas decreases (see d). Note also that the Sun's rays must travel an increasing distance through the atmosphere as latitude increases.

hotter, and if more heat was returned than is gained, Earth would become colder. In both cases, the planet would change dramatically.

To follow the long-term average deposits and with-drawals in Earth's heat budget, assume that 100 units, or 100%, of solar energy are incoming to Earth's atmosphere. Refer to figure 7.2 as you read this discussion. Thirty-one of these units are reflected back to space by Earth's atmosphere, and 4 units are reflected by Earth's surface through the atmosphere to outer space; they take no part in heating Earth or its atmosphere. Of the remaining 65 units, 47.5 are absorbed by Earth's surface, and 17.5 units are absorbed by Earth's atmosphere. Balancing the budget requires the following: these 65 units must be lost by **reradiation** of long-wave radiation to outer space; 5.5 units are reradiated from Earth's surface, and 59.5 units are reradiated from the atmosphere. The budget is balanced; incoming radia-tion balances outgoing radiation.

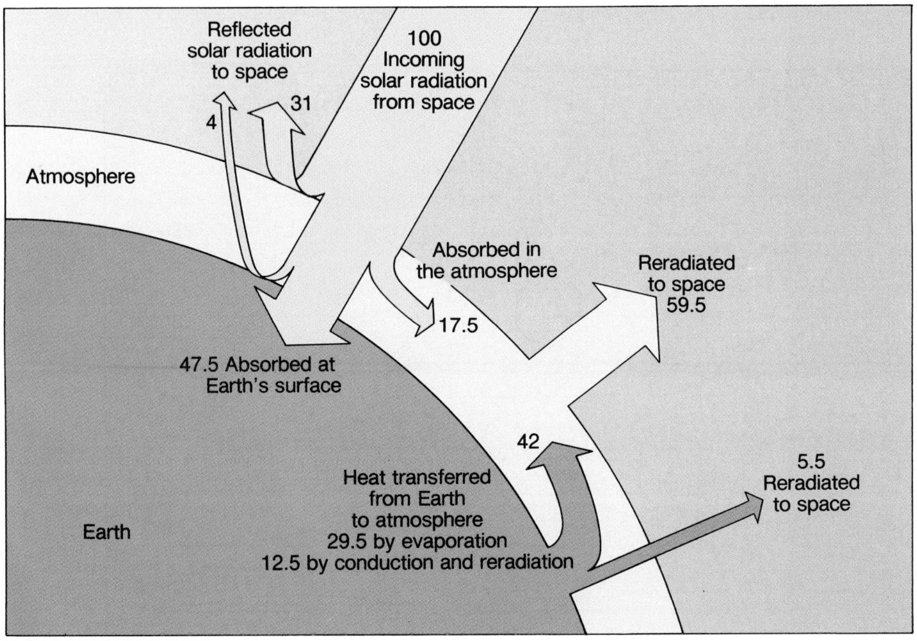

Closer inspection shows that, although Earth absorbs 47.5 incoming units, it loses only 5.5 units to space, for a gain of 42 heat units. The atmosphere, in contrast, absorbs 17.5 incoming units and loses 59.5 units to space, for a loss of 42 units. In these circumstances, it appears that the at-mosphere is cooled and Earth is heated. For Earth's surface and atmosphere to maintain their near-constant average tem-peratures, the 42 units gained by the surface must be trans-ferred to the atmosphere. First, 29.5 of the units are transferred by evaporation processes, which cool Earth's sur-face and liberate heat when water vapor condensation occurs in the atmosphere. The remaining 12.5 units are transferred to the atmosphere by conduction and reradiation and are ab-sorbed as heat, raising the temperature of the air. On the world average, the atmosphere is primarily heated from be-low by heat given off from Earth. This means the atmosphere is set into convective motion.

Earth's surface, including the oceans, is heated from above. If we consider the short-term heat budget of only a small portion of the oceans rather than the total, the fol-lowing must be taken into consideration: to-tal energy absorbed at the sea surface in that area, loss of energy due to evaporation, transfer of heat into and out of the area by currents, vertical redistribution, warming or cooling of the overlying atmosphere by heat from the sea surface, and heat rerad-iated to space from the sea surface. These processes vary with time, showing daily and seasonal variations.

Measurements made over Earth's sur-face show that, on the average, more heat over the annual cycle is gained than lost at the equatorial latitudes, while more heat is lost than gained at the higher latitudes (fig. 7.3). Winds and ocean currents remove the excess heat accumulated in the tropics and release it at high latitudes to maintain Earth's present surface temperature patterns. Figure 7.4 shows Northern Hemisphere sum-mer ocean surface temperatures. North-south

Figure 7.2 Earth's heat budget. Incoming solar energy is balanced by reflected and reradiated energy. The atmosphere's loss of heat is balanced by heat transferred from Earth to the atmosphere by evaporation, conduction, and reradiation. The figure is based on 100 units of incoming radiation, or 100% of available radiation.

deflections in the lines of constant temperature indicate the displacement of surface water by currents carrying warm water from low to high latitudes and cold water from high to low latitudes, redistributing heat energy over Earth's surface.

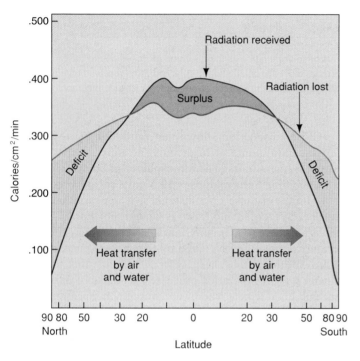

Figure 7.3 Comparison of incoming solar radiation and outgoing long-wave radiation with latitude. A transfer of energy is required to maintain a balance.

Annual Cycles of Solar Radiation

The total Earth long-term heat budget and the average distribution of incoming and outgoing radiant energy with latitude do not include the annual cycle of radiant energy changes related to the seasonal north-south migration of the Sun. When these variations are included, the annual cycle of seasonal variation in average daily solar radiation is most pronounced at the middle and higher latitudes. Here, the angle at which the Sun's rays strike Earth and the length of daylight change dramatically from summer to winter. The seasonal variation is illustrated in figure 7.5.

Changes in incident radiation produce seasonal variations in land and sea surface temperatures because of heat losses or gains. The intensity of solar radiation remains fairly constant at tropical latitudes (between 23½°N and 23½°S) over the year, because the Sun's noontime rays are always received at an angle approaching 90° and the length of the daylight period is nearly constant. During the Sun's annual migration between 23½°N and 23½°S, its rays perpendicular to Earth's surface cross the intervening latitudes twice. This twice-a-year crossing produces a small-amplitude, semiannual variation in the intensity of tropical solar radiation. The effect is most evident at the equator and can be seen in figure 7.5. Between 23½°N and 90°N and between 23½°S and 90°S, the Sun's noontime rays always strike Earth at an oblique angle. The long duration of daylight hours in summer at polar latitudes produces a high level of incident solar radiation averaged over the twenty-four-hour day. However, the intensity of radiation per unit surface area per minute of daylight and the annual average radiation level are much lower than those found at lower latitudes (see fig. 7.1). Refer back

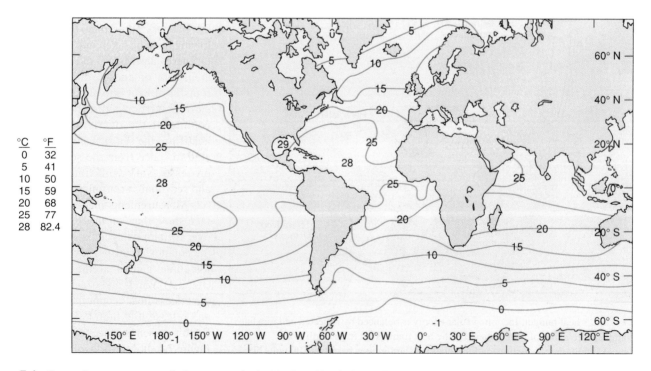

Figure 7.4 Sea surface temperatures during summer in the Northern Hemisphere, given in degrees Celsius.

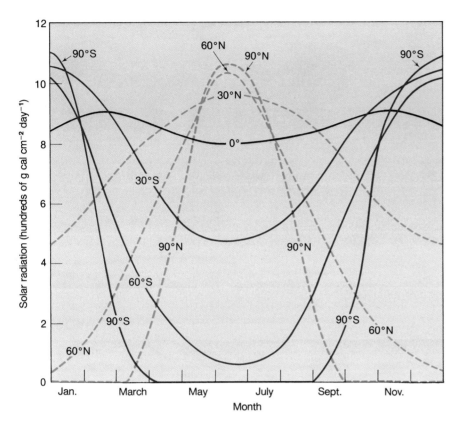

Figure 7.5 Average daily solar radiation values at different latitudes during the year. When it is summer in the Northern Hemisphere, it is winter in the Southern Hemisphere; therefore, the higher values in the Northern Hemisphere coincide with the lower values in the Southern Hemisphere. Peak values of solar radiation at the higher latitudes occur during summer in each hemisphere as a result of more hours of sunlight.

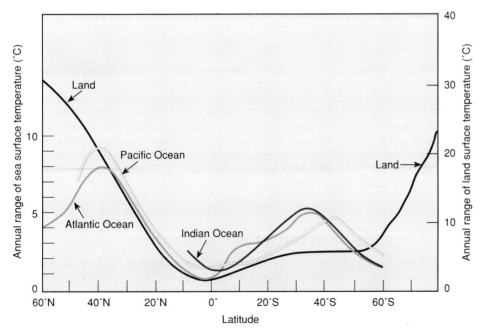

Figure 7.6 The annual range of mid-ocean sea surface temperatures is considerably less than the annual range of land surface temperatures at the higher latitudes. The maximum annual range of sea surface temperatures occurs at the middle latitudes.

to chapter 2 to review the relationship of Earth, the Sun, and Earth's seasons.

Specific Heat and Heat Capacity

Land and sea respond differently to solar radiation because of the difference in the specific heat of land materials and water (review table 5.2). The land has relatively low heat capacity because of the low specific heat of rock and soil. Consequently, the land can undergo large temperature changes as heat is gained or lost between day and night or summer and winter. The oceans have a very high heat capacity due to the high specific heat of water. As a result, the oceans can absorb and release large amounts of heat with little change in temperature.

The average annual range of surface temperatures for land and ocean are given in figure 7.6. Note that the seasonal temperature ranges at high latitudes are greater for land than for the oceans and that the annual range of ocean surface temperatures is greatest at the middle latitudes. Because of the unequal distribution of land between the two hemispheres, the summer-to-winter variation in temperature for land at the middle latitudes is much greater in the Northern Hemisphere than in the Southern Hemisphere. Because of the lack of land in the Southern Hemisphere, the oceans, with their high heat capacity, control the annual temperature range in southern middle latitudes. These differences between land and ocean annual surface temperatures can be seen in figure 7.7. Compare the summer (fig. 7.7a) and winter (fig. 7.7b) temperatures of landmasses and water areas at about 60°N.

Heat that is absorbed at the ocean surface in summer is transferred downward by winds, waves, and currents. In winter, heat is transferred upward toward the cooling surface. Heat is also transferred to and from the atmosphere at the sea surface. The net effect of these processes is the small annual change in mid-ocean surface temperatures: 0°–2°C in the tropics and 5°–8°C at the middle latitudes. The smaller 2°–4°C change in temperature at polar latitudes results from the heat transferred locally in the formation and melting of sea ice.

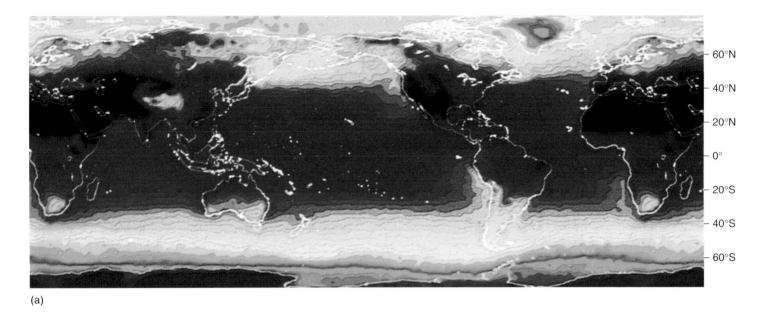

(a)

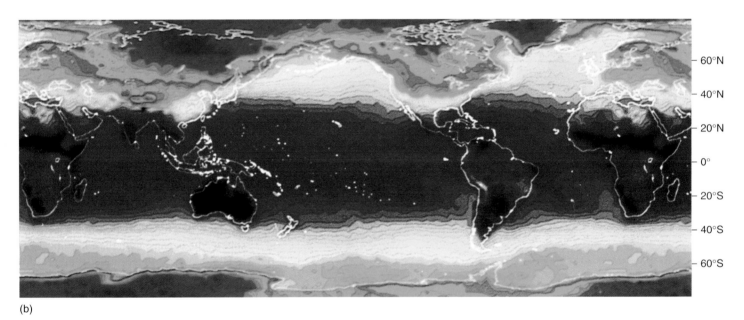

(b)

Figure 7.7   Meteorological satellites, such as NOAA's *TIROS,* carry high-resolution infrared sensors that measure long-wave radiation emitted from Earth's surface and atmosphere. This radiation is related to Earth's surface temperatures. *Green* and *blue* indicate temperatures below 0°C; warmer temperatures are shown in *red* and *black.* (a) July data show the Northern Hemisphere landmasses with considerably warmer temperatures, but the ocean waters do not change dramatically from winter to summer. The oceans' surface temperature distribution moves north and south with the change in seasons. North-south currents along the coasts of continents are also visible. (b) In January, Siberia and Canada show surface temperatures near −30°C; at the same time, latitudes between 30°S and 50°S show warm summer temperatures.

7.2 The Atmosphere

The absorption, reflection, and transmission of solar energy in the atmosphere depend on the gaseous composition of the atmosphere, suspended particles, and the abundance and types of clouds. The transfer of heat energy from Earth's surface acts to heat the atmosphere from below and set this gaseous fluid into convective motion. This convective motion produces the winds, which redistribute heat over Earth's surface and produce waves and currents in the oceans.

Structure of the Atmosphere

The atmosphere is a nearly homogeneous mixture of gases extending 90 km (54 mi) above Earth. Ninety-nine percent of the mass of atmospheric gases is contained in a layer extending upward 30 km (18 mi), and 90% is within a layer extending only 15 km (9 mi) above Earth's surface. The lowest layer of the atmosphere is the **troposphere;** here, the temperature decreases with altitude, changing from a mean Earth surface value of 16°C to −60°C at an altitude of 12 km (7 mi). The tropopause marks the minimum temperature zone between the troposphere and the layer above it, the **stratosphere.** In the stratosphere, the temperature increases with increasing altitude until the stratopause is reached at 50 km (30 mi) (fig. 7.8).

The troposphere is warmed from below by heat energy reradiated and conducted from Earth's surface and by condensation of water vapor in the upper troposphere (see fig. 7.2). Precipitation, evaporation, convective circulation, wind systems, and clouds are all found within the troposphere. **Ozone,** a highly reactive form of oxygen, occurs principally in the stratosphere. Each ozone molecule, O_3, is made up of three atoms of oxygen instead of two, as found in oxygen, O_2. Ozone absorbs ultraviolet radiation from sunlight and therefore raises the temperature of the stratosphere. By absorbing ultraviolet radiation, the ozone lowers the incidence of ultraviolet light at Earth's surface, protecting living organisms from harmful high-intensity ultraviolet radiation.

At altitudes higher than 50 km (30 mi), there is little absorption of solar radiation, so the temperature again decreases with height in the layer known as the **mesosphere.** Here, the number of molecules per cubic centimeter is reduced by 1000,

and the pressure is only 1/1000 of the atmospheric pressure at Earth's surface. The mesosphere extends upward to 90 km (54 mi), and above the mesosphere, the **thermosphere** extends out into space. (See again fig. 7.8.)

The emphasis here is on the troposphere, for within this layer, heat and water move between Earth's surface and the atmosphere, causing the motions that produce the winds, the weather, the ocean's waves, and ocean surface currents. The tropopause is also of interest because this is the region of the high-altitude winds called jet streams that play a role in wind systems and storm tracks.

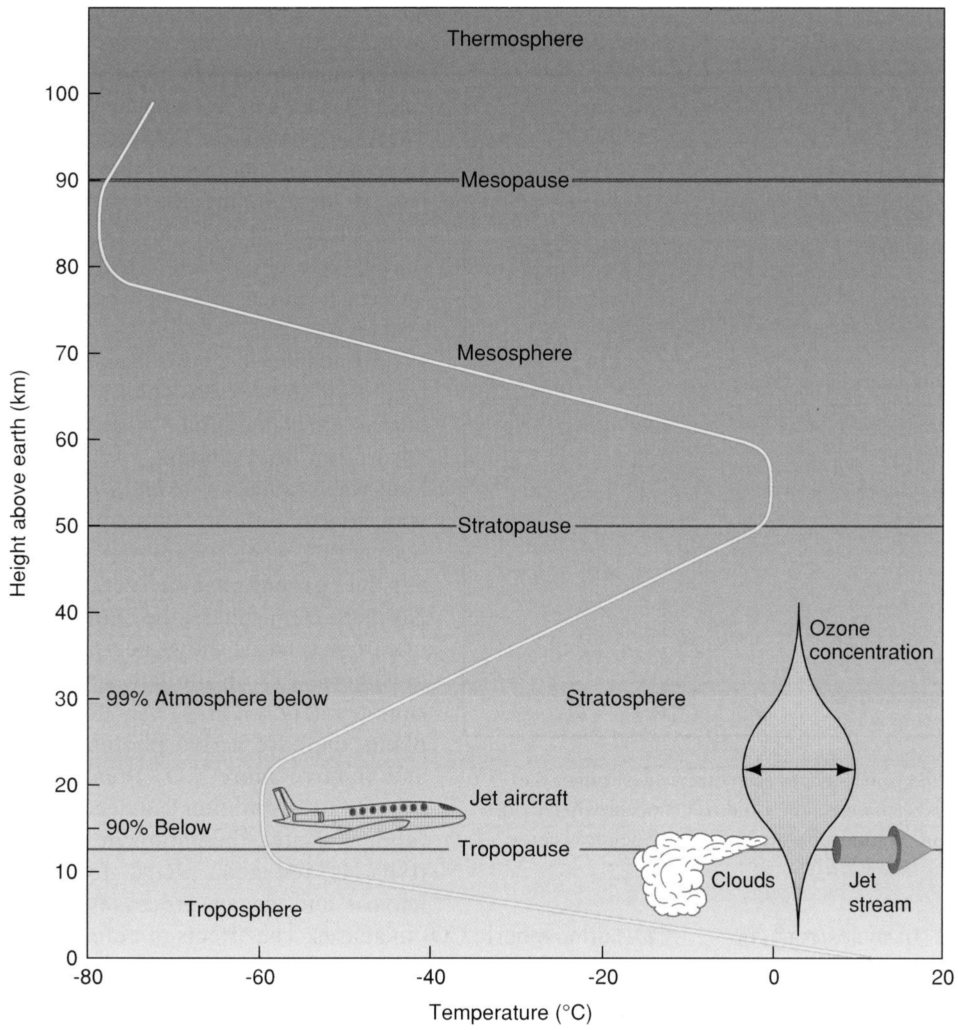

Composition of Air

The atmosphere is composed of gases, suspended microscopic particles, and water droplets. This mixture is commonly and simply called air. Atmospheric gases are typically categorized as being permanent or variable. The permanent gases are present in a constant relative percentage of the atmosphere's total volume, while the concentration of the variable gases changes with time and location (table 7.1).

The density of air is controlled by three variables: temperature, the amount of water vapor in the air, and altitude. The density of air decreases with increasing temperature and increases with decreasing temperature. In other words, warm air is generally lighter than cold air. Air density decreases if its humidity increases or the concentration of water vapor increases, and it increases if the humidity decreases. In general, moist air is lighter than dry air. When water vapor is added to the atmosphere, the relatively low molecular-weight water molecules replace higher molecular-weight permanent gases (compare the molecular weight of water to the molecular weights of the permanent gases listed in table 7.1). Thus, cold, dry air is more dense than warm, moist air. Finally, the density of air decreases with increasing altitude. The air at any given altitude is compressed by the weight of the column of air above it. The greater the compression, the greater the density. Changes in the density allow air to move vertically and cause atmospheric convective motion.

Atmospheric Pressure

Atmospheric pressure is the force with which a column of overlying air presses on an area of Earth's surface. The average atmospheric pressure at

Figure 7.8 The structure of the atmosphere. The distribution of air temperature with height is shown by the *yellow line.*

Table 7.1	Composition of the Atmosphere		
Permanent Gases			
Gas	Formula	Percent by Volume	Molecular Weight
Nitrogen	N_2	78.08	28.01
Oxygen	O_2	20.95	32.00
Argon	Ar	0.93	39.95
Neon	Ne	1.8×10^{-3}	20.18
Helium	He	5.0×10^{-4}	4.00
Hydrogen	H_2	5.0×10^{-5}	2.02
Xenon	Xe	9.0×10^{-6}	131.30
Variable Gases			
Gas	Formula	Percent by Volume	Molecular Weight
Water vapor	H_2O	0 to 4	18.02
Carbon dioxide	CO_2	3.5×10^{-2}	44.01
Methane	CH_4	1.7×10^{-4}	16.04
Nitrous oxide	N_2O	3.0×10^{-5}	44.01
Ozone	O_3	4.0×10^{-6}	48.00

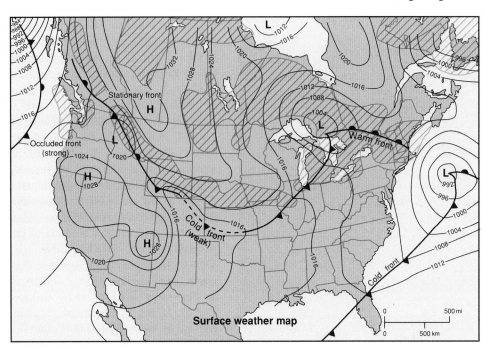

Figure 7.9 Isobars are in millibars. The closer the isobars are to each other, the stronger the winds. *H* is high pressure; *L* is low pressure; *cross-hatching* is rain. Cold fronts occur when cold, dense air wedges itself under less dense, warmer air. Warm fronts are produced when less dense, warm air moves over denser, colder air.

sea level is 1013.25 millibars (1 bar = 1×10^6 dynes/cm^2), or 14.7 lb/in^2. This standard atmospheric pressure is also equal to the pressure produced by a column of mercury standing 760 mm (29.92 in) high. Barometers may measure atmospheric pressure in millibars or in millimeters or inches of mercury. Pressure can also be recorded in torrs, where 1 torr equals the pressure of a column of mercury 1 mm high. Atmospheric pressure distribution is shown on weather charts by lines of constant pressure, or **isobars** (fig. 7.9). Where the density of the air is less than average, atmospheric pressure is below average, and a **low-pressure zone** of rising air is formed. Regions of air with a density greater than average are known as **high-pressure zones** of descending air.

7.3 Greenhouse Gases

The atmosphere exerts primary control over Earth's climate, and during the last fifty years, it has become increasingly evident that the balance of the gases present in Earth's atmosphere is changing. Human activities appear to play a significant role in these changes, and there is particular concern about the changes in concentration of greenhouse gases, which include, most importantly, water vapor and carbon dioxide, as well as nitrous oxide and others.

Carbon Dioxide and the Greenhouse Effect

There are three active reservoirs for carbon dioxide (CO_2): the atmosphere, the oceans, and the terrestrial system; in addition, there is the geologic reservoir of Earth's crust. The oceans store the largest amount of CO_2, and the atmosphere has the smallest amount (fig. 7.10). The atmosphere is the link with the other reservoirs, and the ocean plays a major part in determining the atmosphere's concentration of CO_2 through physical (mixing and circulation), chemical, and biological processes (see section 6.3, chapter 6).

Atmospheric CO_2 is transparent to incoming short-wave solar radiation, but at the same time, it reduces by absorption the amount of outgoing long-wave radiation from Earth. In this way, Earth and its atmosphere are warmed by what is commonly known as the **greenhouse effect.** In the Northern Hemisphere, the natural cycle of carbon dioxide shows decreasing atmospheric CO_2 in the late spring and summer (fig. 7.11a). At this time, plants increase active photosynthesis and remove more CO_2 than is contributed by respiration and decay. In the fall and winter, the photosynthetic activity is reduced, plants lose their leaves, and decay processes release CO_2; atmospheric CO_2 increases. The effects of deforestation and conversion of forest land to agriculture, the burning of fossil fuels, and the growth of human populations have been superimposed on this natural cycle. In preindustrial times, the human impact on this seasonal cycle was small and the cycle was in balance.

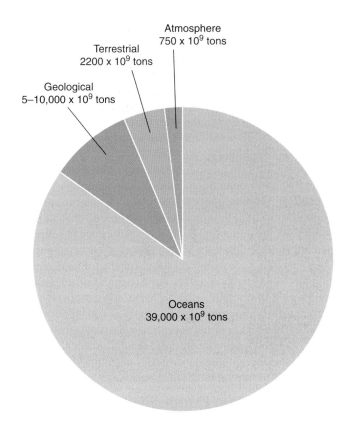

Atmosphere
750 x 10^9 tons

Terrestrial
2200 x 10^9 tons

Geological
5–10,000 x 10^9 tons

Oceans
39,000 x 10^9 tons

Figure 7.10 World carbon dioxide distribution.

Since 1850 and the start of the Industrial Revolution, however, the concentration of CO_2 in the atmosphere has increased from 280 parts per million (ppm) to 380 ppm. These are the highest values observed for the last 420,000 years based on measurements of CO_2 concentrations in air bubbles trapped in the Vostok ice core (see box "Messages in Polar Ice" in chapter 6). Recently, the average rate of increase has been 1.5–2 ppm per year. For nearly fifty years, scientists have been recording a steady increase in the CO_2 concentration in the atmosphere due to the burning of coal, oil, and other fossil fuels (fig. 7.11*b*). If this trend continues, the concentration of CO_2 will double its 1850 value sometime before the end of this century. This increase will warm Earth and alter its average heat budget by reducing the surface heat loss to space by long-wave radiation. Instead, more long-wave radiation will be absorbed into the atmosphere, increasing its temperature and forcing it to lose more long-wave radiation to space in order to maintain the heat budget of Earth and atmosphere systems.

Based on this trend of increasing CO_2, climate researchers have predicted global warming of 2°–4°C in the next hundred years. Accurate measurements of Earth's warming periods are relatively recent. A twenty-year warming period began in 1920; temperatures increased again in 1977 and remained high through the 1980s. Neither of these periods achieved the warming rate of approximately 0.25°C per decade predicted by the present CO_2 concentrations.

Because different researchers assign different levels of importance to the many factors that affect Earth's global temperature, results of research studies vary, and debate is spirited over the causes and the rate of global warming. Some researchers point to changes in Sun spot cycles; others consider atmospheric aerosols and dust from human and natural sources significant because they could govern cloud formation. Natural events promote global cooling as well as global warming. For example, the 1991 discharge of large amounts of debris and gas by the eruption of the Philippine volcano Mount Pinatubo caused cooling of about 0.5°C and made 1992 the coolest year since 1986. This cool period has been followed by continued warming, each succeeding year warmer than the last. In time, reliable global models may show us how much and by what means changes in Earth and atmospheric temperatures are related to increasing levels of CO_2.

The annual world production of CO_2 from fossil fuels and the burning of tropical forests is estimated at approximately 7 billion tons. The increase in atmospheric CO_2 accounts for about 3 billion tons annually; however, this rate is not constant but varies with events such as El Niño (see section 7.8). The most recent research on storage of CO_2 in the oceans estimates that about 2 billion tons enter the oceans and ocean sediments annually; refer to the discussion of carbonate sediments in chapter 4 and of dissolved CO_2 in seawater in chapter 6. At least 1 billion tons are thought to be taken up by the temperate-zone forests and the intense growth of the plants of tropical regions. The exact rates at which this CO_2 is transferred from one reservoir to another are difficult to measure, and the pathways followed by the CO_2 are difficult to trace. During the 1990s, much was learned about the global CO_2 cycle, but predicting specific Earth reactions to increases of greenhouse gases requires further study.

In 1997, the Third Conference of Parties to the Framework Convention on Climate Change met in Kyoto, Japan. The Kyoto pact calls for thirty-eight industrial nations to cut their emissions of greenhouse gases by an average of 5.2% from 1990 levels by the year 2012. If achieved, this would reduce projected 2012 emissions by two-thirds, but they will not prevent total greenhouse emissions from rising because developing nations that are not bound by the treaty, such as China and India, are expected to account for a significant amount of the world's total greenhouse emissions by the year 2012. The United States supports a plan that would allow developed nations to count CO_2 absorbed by forests, so-called carbon sinks, against emissions reduction targets. The European Union is concerned that countries may claim that all their greenhouse gases are being absorbed by sinks and that they therefore do not need to make any actual reductions in the CO_2 they emit from vehicle exhausts and other sources. Representatives of the countries involved in the treaty failed to reach agreement on this issue at a meeting at The Hague, Netherlands, in the fall of 2000. The United States chose not to attend a summer 2001 meeting in Bonn, Germany, which was attended by representatives of 178 countries. The United States felt that limits on carbon dioxide emissions could seriously harm the economy. This meeting also ended without a final agreement on limiting emissions. After years of opposition, Russia ratified the treaty in 2004. Russia's ratification was crucial in achieving the minimum necessary conditions for the treaty to go into effect,

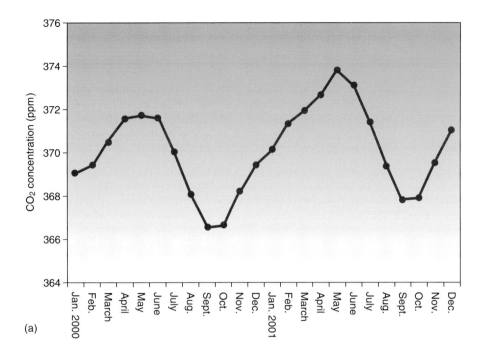

(a)

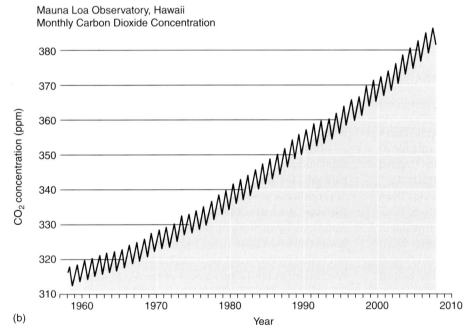

(b)

Figure 7.11 Concentration of atmospheric carbon dioxide in parts per million (ppm) of dry air versus time observed at Mauna Loa Observatory, Hawaii. (a) Two years of monthly observations showing the seasonal variation due to Northern Hemisphere photosynthesis, respiration, and decay. (b) Fifty years of almost continuous monthly observations showing the increase in average annual CO_2 concentration.

1 m (3 ft) by the year 2100. A decrease in sea ice could provide more open water for marine phytoplankton populations with a corresponding increase of photosynthesis, leading to increased carbon storage in the ocean reservoir. Other possibilities are related to the interaction between clouds, the oceans, and Earth's surface (see the box titled "Clouds and Climate"). Changes in sea surface temperature could also affect the oceanic circulation (see chapters 8 and 9) and the surface wind systems (covered in "The Atmosphere in Motion" section in this chapter) that drive the ocean currents. Changes in the ocean current patterns would modify the transfer of heat and water vapor from low to high latitudes and alter Earth's climate patterns. Also, increases in available atmospheric CO_2 have the potential to stimulate photosynthesis both on the land and in the oceans with unknown effects.

The Ozone Problem

Depletion of the stratospheric ozone layer that screens Earth from much of the Sun's ultraviolet radiation was first reported in 1985 by members of the British Antarctic Survey, who discovered that significant ozone loss had been occurring over Antarctica since the late 1970s. The Antarctic "ozone hole" is the result of a large-scale destruction of the ozone layer over Antarctica that occurs when temperatures in the ozone layer drop low enough for clouds to form. These clouds form in the stratosphere, at heights of between 10 and 30 km, when the temperature there falls below $-80°C$. In these clouds, chemical reactions take place that lead to the destruction of ozone. Without the clouds, there is little or no ozone destruction. Only during the Antarctic winter does the atmosphere get cold enough for these clouds to form widely through the center of the ozone layer. The Antarctic ozone hole is usually largest in early September (spring season in the Southern Hemisphere) after the ozone has been decreasing through the Antarctic winter, and deepest in late September to early October (fig. 7.12). The size and shape of the hole vary from year to year with natural variations in the temperature of the stratosphere (fig. 7.13). The word *hole* is a misnomer; the hole is really an area over the pole that experiences a significant reduction (up to 70%) in the ozone concentrations normally found over Antarctica. Satellites carrying total ozone mapping spectrometers have monitored the ozone situation since 1978 with increasing accuracy. Currently,

which happened early in 2005. Under the treaty, the European Union committed to reducing its emissions 8% below 1990 levels; Japan and Canada committed to a 6% cut; and Russia committed to limit emissions right at 1990 levels. Australia and the United States have chosen not to ratify the treaty.

With global warming, several scenarios are possible. It is expected that such warming would affect the higher latitudes, causing melting of polar land ice and raising sea level about

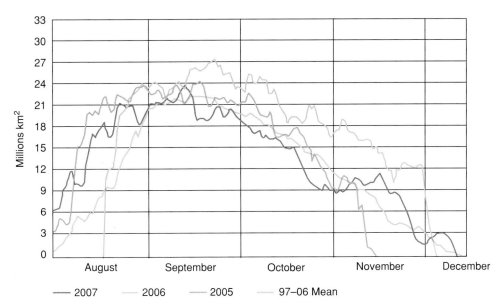

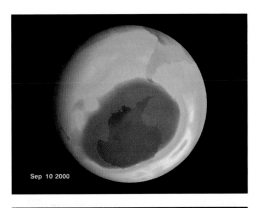

Sep 10 2000

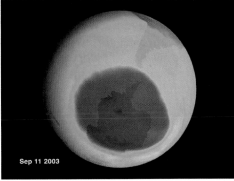

Sep 11 2003

Figure 7.12 Variation in the area of the Antarctic ozone hole from late winter to early spring in the Southern Hemisphere. Areas in millions of square kilometers for 2007 (*red*), 2006 (*blue*), 2005 (*green*), and the mean area during the period 1997–2006 (*yellow*). Areas of four continents are included for comparison. (For additional information see: *http://www.cpc.ncep.noaa.gov/products/ stratosphere/sbuv2to/ozone_hole.shtml*).

Figure 7.13 Comparison of the Antarctic ozone hole on September 10, 2000 (*top*) and September 11, 2003 (*bottom*). The maximum area in 2000 was 29.8×10^6 km^2 and in 2003 was 28.3×10^6 km^2.

NASA's *Earth Probe* satellite is monitoring both the Antarctic and Arctic ozone holes.

The Northern and Southern Hemispheres have different temperature conditions in the ozone layer. The temperature of the

Arctic ozone layer during winter is normally some 10 degrees warmer than that of the Antarctic. This means that clouds in the stratosphere over the Arctic are rare, but sometimes the temperature is lower than normal and they do form. Under these circumstances significant ozone depletion can take place over the Arctic, but it is usually for a much shorter period of time and covers a smaller area than in the Antarctic. The Arctic ozone hole covered approximately 6×10^6 km^2 in March 1997. The Arctic winter of 1996–97 was the fifth winter in a row in which the stratosphere over the Arctic had been particularly cold. Measurements taken early in 2000 found cumulative ozone losses of more than 60% over the Arctic. There are three possible explanations for the colder conditions in the Arctic stratosphere between 1992 and 1997. Under conditions of decreasing ozone, the stratosphere is less able to absorb solar radiation and warm itself; the result is a destructive system of increasing cold that leads to lower ozone levels. Also, the increasing greenhouse gases trap heat in the lower atmosphere, and, because this heat does not reach the upper atmosphere, the stratosphere gets cooler. Last, some other variations in Earth's climate or changes in the weather may be at work.

The most widely accepted theory of ozone destruction is related to the release of chlorine into the atmosphere. Chlorine is commonly released as a component of chlorofluorocarbons (CFCs). CFCs are used as coolants for refrigeration and air conditioning, as solvents, and in the production of insulating foams. CFCs are distributed throughout the troposphere by the winds and gradually leak into the stratosphere, where the ultraviolet light breaks them apart. Gases in the atmosphere react with the chlorine and trap it as inert molecules in stratospheric clouds, which are common during the polar winter. In the presence of sunlight, chlorine is liberated and free to attack ozone molecules.

CFCs appear to have reached their maximum level in the troposphere and are expected to decline gradually as efforts to reduce their production continue. Because it takes many years for CFCs in the troposphere to work their way into the stratosphere, they will continue to destroy ozone for many years to come.

Recently, methyl bromide has been identified as an even more efficient agent for breaking down ozone. It is thought to be responsible for as much as 20% of the Antarctic ozone depletion and 5%–10% of Earth's global ozone loss. Estimates place the input of methyl bromide to the atmosphere at 137 gigagrams (Gg) per year, of which 86 Gg/yr act directly to destroy ozone. Sources include microscopic single-celled organisms at the sea surface, pesticides, industry, and slow-smoldering burning of vegetation.

At ground level, ozone is a pollutant and a health hazard; in the stratosphere, it absorbs most of the ultraviolet radiation from the Sun, protecting life forms on land and at the sea surface. The loss of ozone is a significant concern because a 50% decrease in ozone is estimated to cause a 350% increase in ultraviolet radiation reaching Earth's surface. Increased ultraviolet radiation is responsible for increases in skin cancers and has been shown to affect the growth and reproduction of some organisms. The effects of increased ultraviolet radiation on the photosynthetic organisms of the surface layers of the Southern Ocean are discussed in chapter 15.

7.4 The Role of Sulfur Compounds

It is currently estimated that 20–40 million tons of sulfur from dimethyl sulfide (DMS) is added to the atmosphere each year. Dimethyl sulfide is a gas produced, both directly and indirectly, by phytoplankton at the ocean's surface. It is in part responsible for the characteristic odor of the sea that one notices

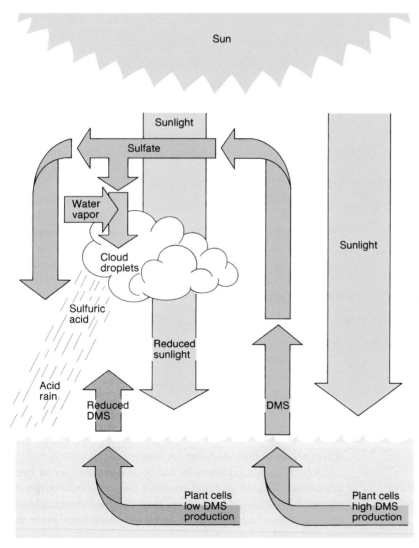

Figure 7.14 A proposed self-supporting thermal control system linking dimethyl sulfide (DMS) to the sea surface and the atmosphere, a cycle among plants, the Sun, and clouds. Acid rain, pH = 3.5, has little impact on seawater because of seawater's carbon dioxide buffer.

when approaching the coast from the land. Once in the atmosphere, DMS changes rapidly. One of its main products is sulfur in the form of sulfate, which combines with atmospheric water to form sulfuric acid, which is returned to Earth's surface with the rain (fig. 7.14). The quantities of sulfuric acid returned to Earth's surface by this process are far below the values associated with the problem of "acid rain." Because DMS forms airborne particles, it plays a role in controlling the density of clouds that appear over the ocean. Dimethyl sulfide changes the clouds' reflective properties, reduces incoming radiation, and decreases the heating of the ocean's surface. If a DMS buildup results in an excess of clouds, less light strikes the sea surface, surface temperatures drop, and the phytoplankton production of DMS decreases. If less DMS is produced, there is less cloud cover and less reflection of light. In this way, DMS may act as a feedback mechanism, or self-regulating thermostat, to help control ocean surface temperatures, another link in the complexity of interactions between the atmosphere and the oceans.

In industrialized areas of the Northern Hemisphere, the burning of fossil fuels creates another source of sulfates. Like the sulfates from DMS, these industrially produced sulfates contribute to acid rain; they also form clouds that block solar radiation and cool Earth's surface.

Unlike its effect on land, acid rain does not have a great impact on the ocean. When acid rain falls on the sea surface, the buffering capacity of seawater neutralizes its impact (review section 6.3 in chapter 6).

7.5 The Atmosphere in Motion

The air moves because at one place less dense air rises, while in another place denser air sinks toward Earth. Between these areas, the air that flows horizontally along Earth's surface is the wind. This process is shown in figure 7.15; note that there are really two horizontal airflows, or wind levels, moving in opposite directions: one at Earth's surface and one aloft. Air circulating in this manner forms a convection cell based on vertical air movements due to changes in the air's density. Less-dense, warmer air rises, and cooler, denser air sinks.

Winds on a Nonrotating Earth

The heating and cooling of air and the gains and losses of water vapor in air are related to the unequal distribution of the Sun's energy over Earth's surface, the presence or absence of water, and the variation in temperature of Earth's surface materials in response to heating. These act to affect the air's density. Imagine a model Earth with no continents and with no rotation but heated like the real Earth. On this stationary model covered with uniform layers of atmosphere and water,

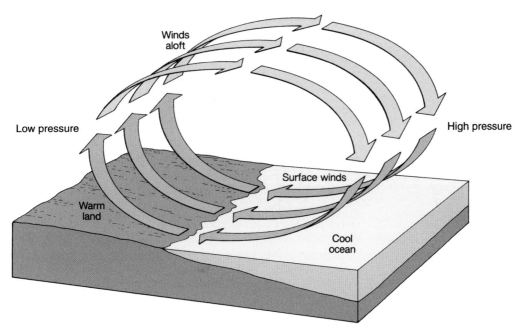

Figure 7.15 A convection cell is formed in the atmosphere when air is warmed at one location and cooled at another.

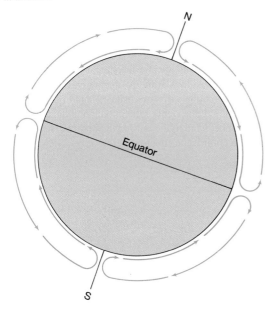

Figure 7.16 Heating at the equator and cooling at the poles produce a single large convection cell in each hemisphere on a nonrotating, water-covered model Earth.

Table 7.2	Eastward Speed of Earth's Surface with Latitude	
Latitude	**Speed (km/h)**	**Speed (mi/h)**
90°N and S	0	0
75°N and S	433	269
60°N and S	837	520
45°N and S	1184	735
30°N and S	1450	900
15°N and S	1617	1004
0°	1674	1040

the wind pattern is very simple. Around the equator, the air, warmed from below, rises. Once aloft, the air flows toward the poles, where it is cooled and sinks to flow back toward the equator. Because of the unequal distribution of the Sun's heat over the model's surface, large amounts of heat and water vapor are transferred to the atmosphere around the equator. This less-dense air rises, and as it rises, it cools; condensation exceeds evaporation, clouds form, and it rains. Equatorial regions are known for their warm, wet climate. The cool, dry air remains aloft and flows toward the poles, where it sinks, producing high evaporation, dense air, and a zone of high atmospheric pressure. Such air movement is shown in figure 7.16; note each hemisphere's two large convective circulation cells, each extending from a pole to the equator. In this model, the Northern Hemisphere surface winds blow from north to south, and the upper winds blow from south to north; in the Southern Hemisphere, the reverse is true, with the surface winds blowing from south to north and the upper winds blowing from north to south.

It is important to remember that winds are named for the direction *from which they blow.* A north wind blows from north to south; a south wind blows from south to north. In this model, the Northern Hemisphere surface winds are north winds, and the Southern Hemisphere surface winds are south winds.

The Effects of Rotation

Consider next the same model of a water-covered Earth but with rotation added. In this case, observations of moving parcels are judged relative to coordinates, latitude and longitude, that are fixed to and rotate with Earth. Referencing motion to this rotating coordinate system creates apparent forces that appear to deflect moving bodies. Gravity holds Earth's atmosphere captive, but the atmosphere is not rigidly attached to Earth's surface. Little or low friction exists between Earth's surface and the atmosphere, and the atmosphere moves somewhat independently of Earth's surface. For example, a parcel of air that appears to be stationary above a point on the equator is actually turning with Earth and moving eastward at a speed of about 1674 km (1040 mi) per hour. If a south wind blows this parcel of air due north, it is moved across circles of latitude with progressively smaller circumferences. At these higher latitudes, points on Earth's surface move eastward more slowly (table 7.2). At 60°N, the eastward speed of a point on Earth's surface due to rotation is only half the speed at the equator. For this reason, a parcel of air that was originally moving only northward relative to Earth at the equator carries with it its equatorial eastward speed so that it is now moving

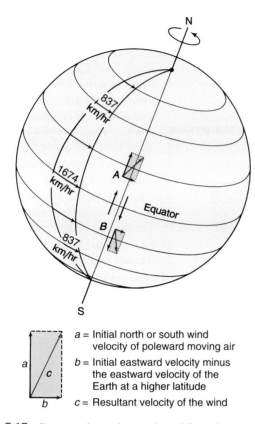

a = Initial north or south wind velocity of poleward moving air

b = Initial eastward velocity minus the eastward velocity of the Earth at a higher latitude

c = Resultant velocity of the wind

Figure 7.17   Because air moving northward from the equator to point *A* carries with it its higher equatorial eastward velocity, the air is deflected to the right of the initial northward wind direction in the Northern Hemisphere. In the Southern Hemisphere, air moving southward to point *B* is deflected to its left.

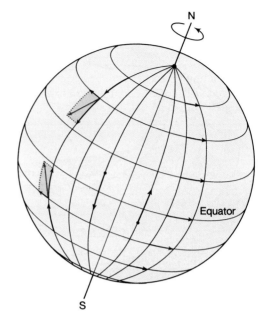

a = Initial north or south wind velocity moving toward the equator

b = Initial eastward velocity minus the eastward velocity of the Earth at a lower latitude

c = Resultant velocity of the wind

Figure 7.18   Air moving toward the equator passes from a latitude of lower eastward speed to a latitude of higher eastward speed. The result is it falls behind the eastward displacement of Earth's surface, causing a deflection to the right of the initial wind direction in the Northern Hemisphere and to the left in the Southern Hemisphere. There is no deflection at the equator.

eastward at a speed that is greater than the eastward speed of Earth's surface at this higher latitude. Therefore, the air parcel is displaced to the east, relative to Earth's surface, as it moves from low to high latitudes; in the Northern Hemisphere, this deflection is *to the right of the direction of the air motion.* This relationship is illustrated in figure 7.17; follow the arrows at *A*.

If the same situation occurs in the Southern Hemisphere, with a parcel of air moving southward from the equator due to a north wind, the deflection is still to the east relative to Earth's surface. However, this deflection is now *to the left of the direction of the air motion.* See again figure 7.17 and follow the arrows at *B*.

If an air parcel moves southward toward the equator in the Northern Hemisphere, then it moves from a latitude where a position on Earth has a low eastward speed to a latitude where a position on Earth has a higher eastward speed. These positions move to the east, relative to the air, when the air moves from a higher to a lower latitude. This relationship causes the air to fall behind the position on Earth, so that the air appears to move westward relative to Earth. This pattern is illustrated in figure 7.18. The deflection is still to the right of the direction of motion in the Northern Hemisphere. A similar pattern in the Southern Hemisphere shows that the deflection is to the left of the direction of motion.

Deflection due to air moving to the east or to the west is shown in figure 7.19. Air moving eastward is moving eastward

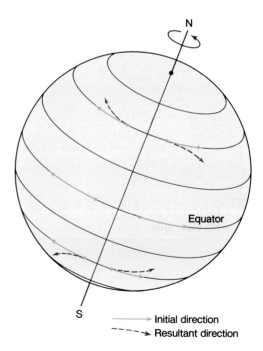

Initial direction
Resultant direction

Figure 7.19   The deflection of eastward-moving and westward-moving air. There is no deflection at the equator.

faster than Earth beneath it, with reference to Earth's axis. The air is affected by a centrifugal force acting outward from Earth's axis of rotation that is stronger than the centrifugal force that is acting at Earth's surface. This small excess force acting on the air is in part acting against Earth's gravity and in part acting parallel to Earth's surface directed toward lower latitudes. That part of the centrifugal force acting against Earth's gravity is so small that it has very little effect. That part acting parallel to Earth's surface is unopposed and causes a deflection of the moving air to lower latitudes, or to the right of its motion in the Northern Hemisphere. Westward-moving air is moving eastward more slowly, relative to Earth's surface. This air is affected by an outward-acting centrifugal force that is weaker than the centrifugal force that is acting at Earth's surface. Because the centrifugal force acting at Earth's surface is greater than the centrifugal force acting on the air, there is a weak force acting toward Earth's axis of rotation. This force is acting in part in the direction of gravity and in part toward higher latitudes. Note that the deflection continues to be to the right of the initial wind motion in the Northern Hemisphere. In the Southern Hemisphere, the deflection of eastward-moving air toward lower latitudes and of westward-moving air toward higher latitudes results in a deflection of the air to the left of its original direction of motion.

If we were to move from Earth's surface out into space and were able to watch the motions of Earth and the atmosphere, we would see the atmosphere's independent motion and we would also see Earth turning out from under the moving parcels of air. But we live on a moving Earth's surface, and we make our measurements of air motion relative to a coordinate system fixed to this surface. Therefore, from Earth's surface, we see the moving air parcels deflected from their paths—deflected to the right in the Northern Hemisphere and to the left in the Southern Hemisphere.

The apparent deflection of the moving air relative to Earth's surface is called the **Coriolis effect,** after Gaspard Gustave de Coriolis (1792–1843), who mathematically solved the problem of deflection in frictionless motion when the motion is referred to a rotating body and its coordinate system. Because one of the laws of motion in physics tells us that a body set in motion along a straight line continues to move along that straight line unless it is acted upon by a force, such as a push or a pull, the Coriolis effect is often called the Coriolis force. If the term *force* is used, remember that it is an *apparent* force and that a deflection appears only when the motion of an object that is subject to little or no friction is judged against a rotating frame of reference. The magnitude of the Coriolis force increases with increasing latitude, increases with the speed of the moving air, and is dependent on the rotation rate of Earth on its axis.

Wind Bands

Applying rotation, and therefore the Coriolis effect, to the non-rotating model Earth modifies the winds considerably. Refer to figure 7.20 as you read the following description. The air rises at the equator and flows aloft to the north and the south, but it cannot continue to move northward and southward without being deflected to the right in the Northern Hemisphere and to the left in the Southern Hemisphere. This deflection short-circuits the large, hemispheric, atmospheric convection cells of the stationary model Earth. The deflected air aloft sinks at 30°N and 30°S; it moves along the water-covered surface, either back toward the equator or toward 60°N and 60°S. The upper-level air that reaches the poles cools, sinks, and moves to lower latitudes, warming and picking up water vapor; at 60°N and 60°S, it rises again. The result is three convection cells in each hemisphere wrapped around the rotating Earth.

Consider the flow of surface air in the three-cell system. Between 0° and 30°N and 30°S, the surface winds are deflected relative to Earth, blowing from the north and east in the Northern Hemisphere and from the south and east in the Southern Hemisphere. This deflection creates bands of moving air known as the **trade winds:** the northeast trade winds, north of the equator, and the southeast trade winds, south of the equator. Between 30°N and 60°N, the deflected surface flow produces winds that blow from the south and west, while between 30°S and 60°S, they blow from the north and west. In both hemispheres, these winds are called the **westerlies.** Between 60°N and the North Pole, the winds blow from the north and east, while between 60°S and the South Pole, they blow from the south and east. In both cases, they are called the **polar easterlies.** The six surface wind bands are shown in figures 7.20 and 7.21.

At 0° and 60°N and 60°S, moist, low-density air rises in areas of low atmospheric pressure; these are zones of clouds and rain. Zones of high-density descending air at 30° and 90°N and 90°S are areas of high atmospheric pressure, dry air with low precipitation, and clear skies.

The salinity of mid-ocean surface water is controlled by the distribution of the world's evaporation and precipitation zones (see figs. 6.1 and 6.2). Salinity of ocean water is measured in grams of salt per kilogram of seawater and is expressed in parts per thousand (‰). The average ocean salinity is appoximately 35‰. In the tropics, the precipitation is heavy on land and at sea. On land, this produces tropical rain forests; at sea, the surface water has a low salinity, around 34.5‰. At approximately 30°N and 30°S, the evaporation rates are high. These are the latitudes of the world's deserts and of increased salinity in the surface waters, around 36.7‰. Farther north and south, from 50°–60°N and from 50°–60°S, precipitation is again heavy, producing cool but less-salty surface water, around 34.0‰, and the heavily forested areas of the Northern Hemisphere. At the polar latitudes, sea ice is formed in the winter. During the freezing process, the salinity of the water beneath the ice increases, and in the summer, the thawing of the ice reduces the surface salinity once more (see fig. 7.20).

Wind belts are formed when air flows over Earth's surface from regions of higher atmospheric pressure to areas of lower atmospheric pressure. In the zones of vertical motion, between the wind belts, the surface winds are unsteady. Such areas were troublesome to the early sailors, who depended on steady winds for propulsion. The area of rising air at the equator is known as the **doldrums,** and the high-pressure areas at 30°N and 30°S are known as the **horse latitudes.** In all these areas, sailing ships

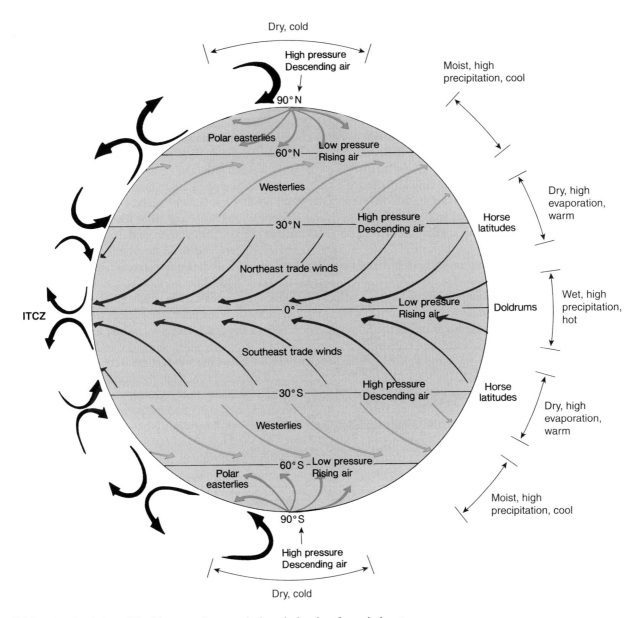

Figure 7.20 The circulation of Earth's atmosphere results in a six-band surface wind system.

could find themselves becalmed for days. The origin of the word *doldrum* is obscure, but it is likely related to its meaning of a period of low spirits, listlessness, or despondency, feelings often felt by sailors becalmed in this region. The formal name for the zone of low pressure and rising air near the equator is the **Intertropical Convergence Zone (ITCZ)** (fig. 7.20). It is along the ITCZ that the wind systems of the Northern and Southern Hemispheres converge to produce rising air, low pressure, and high precipitation. The horse latitudes are said to have gotten their name from the stories of ships' carrying horses that were thrown overboard when the ships were becalmed and the freshwater supply became too low to support both the sailors and the animals.

Every day for decades, thousands of measurements of air pressure, wind speed, and wind direction have been taken by meteorological stations on land and by ships and buoys at sea. These data are converted into atmospheric wind and pressure

charts that are used to make the world's weather forecasts. Satellites can provide another way to report wind velocities at Earth's surface. Microwave radar signals sent from the satellite are scattered by the waves on the sea's surface and returned to the satellite. These returned signals are then interpreted as measurements related to the speed and direction of the surface winds.

The radar signals are sent and received by an instrument called a scatterometer, which is carried by the satellite. The first scatterometer was sent into space in the late 1970s aboard the satellite *SEASAT*. The most recent model, *NASA Scatterometer (NSCAT),* went into space on Japan's *Advanced Environmental Orbiting Satellite (ADEOS)* in August 1996, but *ADEOS* remained active only until June 1997. For eight-and-a-half months, *NSCAT* measured surface wind data that were verified by data from ships and buoys (see fig. 7.21). Such data are not only valuable for weather forecasting but also yield data that can be used

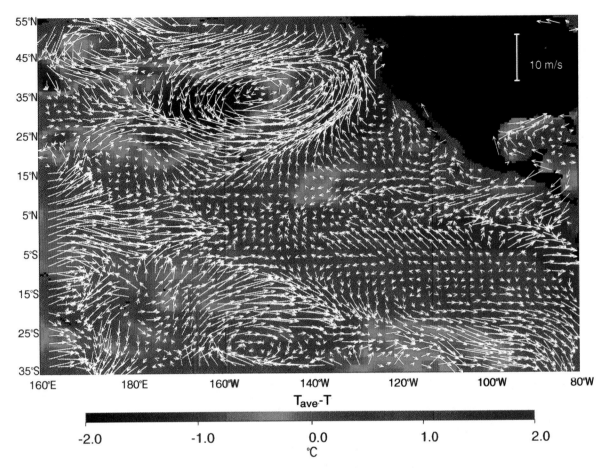

$T_{ave}-T$

-2.0   -1.0   0.0   1.0   2.0

°C

Figure 7.21 Surface wind anomalies *(white arrows)* measured by NASA's high-resolution scatterometer (NSCAT) are superimposed on the map of sea surface temperature anomalies *(colors)* derived from advanced very high-resolution radiometer (AVHRR) measurements. The *white arrows* and *colors* indicate the deviation of winds and sea surface temperatures from the average conditions present for the period May 22–31, 1997. Notice the elevated temperature *(red)* in the eastern Pacific due to El Niño conditions and the weak and reversed trade winds in the same area.

in computer models that couple the ocean and atmosphere, enabling meteorologists, climatologists, and oceanographers to investigate and better understand the couplings of the ocean-atmosphere system.

7.6 Modifying the Wind Bands

Moving from the rotating, water-covered model of Earth to the real Earth requires the consideration of three factors: (1) seasonal changes in Earth's surface temperature due to solar heating, (2) the addition of the large continental land blocks, and (3) the difference in heat capacity of land and water. Both ocean and land surfaces remain warm at the equatorial latitudes and cold at the polar latitudes all through the year, but the middle latitudes have seasonal temperature changes, ranging from warm in summer to cold in winter. Keep in mind that land surface temperatures have a greater seasonal fluctuation than ocean surface temperatures because the heat capacity of the ocean water is greater than the heat capacity of land and because ocean water has the ability to transfer heat from the surface to depth in summer and from depth to the surface in winter. Land does not transfer heat in this way.

Seasonal Changes

The presence of land and water in near-equal amounts at the middle latitudes in the Northern Hemisphere produces average seasonal patterns of atmospheric pressure. During the warm summer months, the land is warmer than the ocean (see fig. 7.7). The air over the land is heated from below and rises, creating a low-pressure area, while the air over the sea cools and sinks, producing a high-pressure area over the water. In the summer, low-pressure zones at 60°N and 0° tend to combine over the land, cutting through the high-pressure belt along 30°N latitude. This breaks the high-pressure belt into several high-pressure cells over the oceans rather than maintaining the latitudinal zones of pressure stretching continuously around Earth. In the winter, the reverse is true at the middle latitudes (see again fig. 7.7). The land becomes colder than the water. The air over the land is cooled, increasing its density and producing a region of descending air and high atmospheric pressure over the land. The air over the relatively warm water is heated from below, decreasing its density and producing a region of rising air and low atmospheric pressure over the water.

Over the land, the polar high-pressure zone spreads toward the high-pressure zone at 30°N, breaking the low-pressure belt centered about 60°N into discrete low-pressure cells that are centered over the warmer ocean water. This middle-latitude, seasonal alternation of high- and low-pressure cells breaks up the latitudinal pressure zones and wind belts that are seen in the water-covered model to produce the air-pressure distribution shown in figure 7.22.

During the Northern Hemisphere's summer, the air in the high-pressure cells over the central portion of the North Atlantic and North Pacific Oceans descends and flows outward toward the continental low-pressure areas. As the descending air moves outward, it is deflected to its right, producing winds that spiral in a clockwise direction about the high-pressure cells. The wind

circulation about a high-pressure cell in the Northern Hemisphere is always clockwise. On the northern side of these high-pressure cells are the westerlies, and on the southern side are the northeasterlies; the eastern side of the cell has northerly winds, and the western side has southerly ones. In the winter, the air circulates counterclockwise about the low-pressure cell over the northern oceans, and the prevailing wind directions reverse. The wind circulation pattern about a low-pressure cell in the Northern Hemisphere is always counterclockwise. The wind directions related to these pressure cells are shown in figure 7.23. Along the Pacific coast of the United States, the northerly winds cool the coastal areas in the summer, and the southerly winds warm them in the

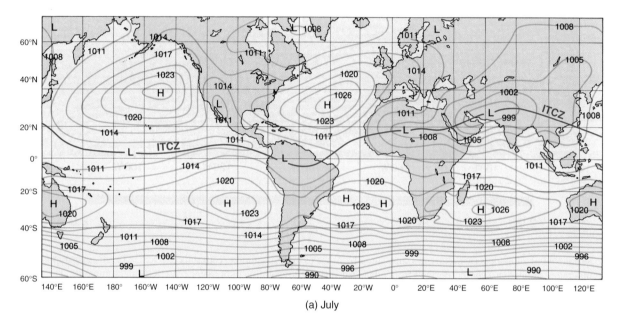

(a) July

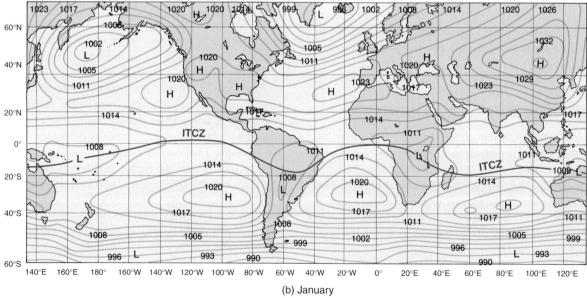

(b) January

Figure 7.22   Average sea-level atmospheric pressures expressed in millibars for (a) July and (b) January. Large landmasses at mid-latitudes in the Northern Hemisphere cause high and low atmospheric pressure cells to change their positions with the seasons. In the Southern Hemisphere, large landmasses do not exist at mid-latitudes, and average air-pressure distribution changes little with the seasons. The location of the Intertropical Convergence Zone (ITCZ) shifts with the seasons.

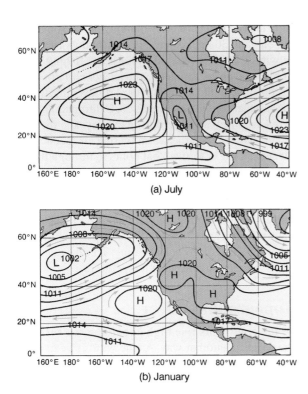

(a) July

(b) January

Figure 7.23 Atmospheric pressure cells control the direction of the prevailing winds. Atmospheric pressures are expressed in millibars. In the Northern Hemisphere, air flows outward and clockwise around a region of high pressure and inward and counterclockwise around a low-pressure area. (a) Average wind conditions for summer. (b) Average wind conditions for winter.

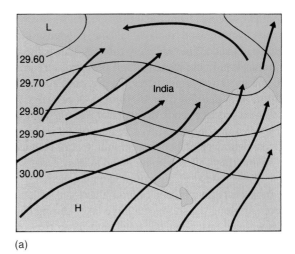

(a)

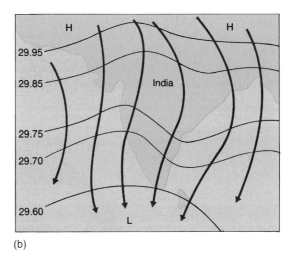

(b)

Figure 7.24 The seasonal reversal in wind patterns associated with (a) the summer (wet) monsoon and (b) the winter (dry) monsoon. The isobars of pressure are given in inches of mercury. Land topography increases friction between winds and land, thereby decreasing the Coriolis effect. The winds blow more directly from high- to low-pressure areas across the isobars.

winter. The eastern United States receives warm, moist air from the low latitudes in the summer, and cold air moves down from the high latitudes in the winter. Although these seasonal changes modify the wind and pressure belts that were developed on the water-covered model, the generalized wind and pressure belts are still identifiable over Earth in the northern latitudes when the atmospheric pressures are averaged over the annual cycle.

Rotation of the airflow around high- and low-pressure cells is reversed in the Southern Hemisphere because the Coriolis effect is opposite to that in the Northern Hemisphere. At the middle latitudes in the Southern Hemisphere, there is little land, and the water temperature predominates. There is little seasonal effect; the atmospheric pressure and wind patterns created by surface temperatures change little over the annual cycle and are very similar to those developed for the water-covered model (see figs. 7.20 and 7.22).

Seasonal changes in the surface temperature of the oceans and continents also influence the position of the ITCZ (fig. 7.22). Its actual location is generally associated with the zone of highest surface temperature. During the Northern Hemisphere summer, it is deflected north of the equator over the continents as the land heats up faster than the water. The ITCZ may be deflected as far as 10°–20°N over Africa and southern Asia. During the Southern Hemisphere summer, the ITCZ will be deflected as far as 10°–20°S over Africa and South America. The ITCZ is the

rainiest latitude zone in the world, with many locations experiencing rainfall more than 200 days each year. The weather at many areas along the equator is dominated by the ITCZ year-round with almost no dry season. Areas near the extreme northern and southern swings of the ITCZ, however, are subject to brief dry seasons when the zone shifts away. For example, Iquitos, Peru (3°S), is close enough to the equator that it is always influenced by the ITCZ, while San José, Costa Rica (10°N), has a relatively dry season from January to March, when the ITCZ is displaced to the south.

The Monsoon Effect

The differences in temperature between land and water produce large-scale and small-scale effects in coastal areas. In the summer along the west coast of India and in Southeast Asia, the air rises over the hot land, creating a low-pressure system (fig. 7.24a). The rising air is replaced by warm, moist air carried on the southwest winds from the Indian Ocean. As this onshore airflow rises over the land, it cools, causing condensation. This produces clouds and a time of heavy rainfall. This is the wet, or summer, **monsoon.** In the winter, a high-pressure cell forms over the land, and northeast winds carry the dry, cool air southward from central Asia and out over the Indian Ocean (fig. 7.24*b*). This movement produces cool, dry weather over the land, known as the dry, or winter, monsoon. For years, coastal traders in the Indian Ocean who were dependent on sailing craft planned their voyages so that they sailed with the wind to their destination on one phase of the monsoon and returned on the next.

The monsoon effect is seen on a smaller local scale along a coastal area or along the shore of a large lake. During daylight, the land is warmed faster than the water, and the air rises over the land. The air over the water moves inland to replace it, creating an **onshore** breeze. At night, the land cools rapidly, and the water becomes warmer than the land. Air rises over the water, and the air from the land replaces it, this time creating an **offshore** breeze. Such a local diurnal (once-a-day) wind shift is referred to as a land-sea breeze (fig. 7.25). The onshore breeze reaches its peak in the afternoon, when the temperature difference between the land and the water is at its maximum. The offshore breeze is strongest in the late night and early morning hours; sometimes one can smell the land 30 km (20 mi) at sea, its odors carried by the offshore winds. This daily wind cycle helps fishing boats that depend only on their sails to leave the harbor early in the morning and return late in the afternoon or early evening. This effect brings the summer fogs to San Francisco, California. The area east of San Francisco Bay heats up during the day, and the warm air over the land rises. When the warm marine air reaches the cold coastal waters, a fog is formed that pours in through the Golden Gate, obscuring first the Golden Gate Bridge and then the city (see fig. 5.25). When the air reaches the east side of the bay, it warms, and the fog dissipates. At night, the flow is reversed, and the city may be swept free of the fog. Sometimes San Francisco will remain foggy all night and the fog will not disperse until the morning, when it is burned off by solar heating. This happens when the land and air are close to the same temperature.

The Topographic Effect

Because the continents rise high above the sea surface, they affect the winds in another way. As the winds sweep across the ocean, they reach the land and are forced to rise as they

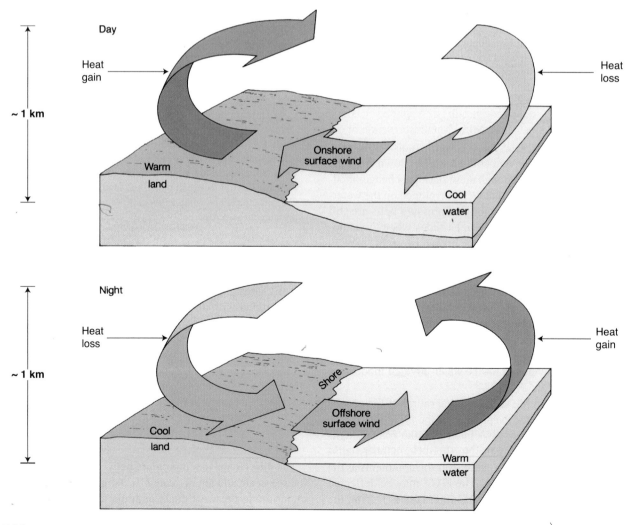

Figure 7.25   Differences between day and night land-sea temperatures produce an onshore breeze during the day and an offshore breeze at night.

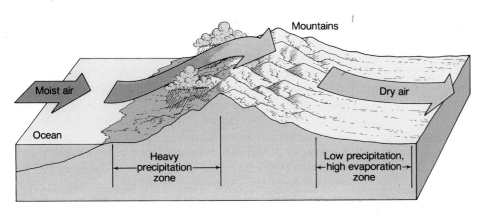

Figure 7.26 Moist air rising over the land expands, cools, and loses its moisture on the mountains' windward side. Descending air compresses, warms, and becomes drier, creating a rain shadow on the leeward side.

encounter the land, as shown in figure 7.26. The upward deflection cools the air, causing rain on the windward side of the islands and mountains; on the leeward (or sheltered) side, there is a low-precipitation zone, sometimes called a **rain shadow.** For example, the windward sides of the Hawaiian mountains have high precipitation and lush vegetation, whereas the leeward sides are much drier and require irrigation. On the west coast of Washington, the westerlies moving across the North Pacific produce the Olympic Rain Forest as the air rises to clear the Olympic Mountains. On the western side of the Olympic Mountains, the rainfall is as much as 5 m (200 in) per year; 100 km (60 mi) away, in the rain shadow on the leeward side of the mountains, it is 40–50 cm (16–20 in) per year. The west side of the mountains of Vancouver Island in British Columbia has a high rainfall; the eastern side of the island is known for its sunshine and scenic cruising. The southeast trade winds on the east coast of South America sweep up and across the lowlands and then rise to cross the Andes Mountains, producing rainfall, large river systems, and lush vegetation on the eastern side of the Andes and a desert on their western slope. Over the Indian subcontinent in summer, the air approaching the Himalayas rises, intensifying the wet monsoon; air flows down from these mountains during the winter, the time of the dry monsoon. The control of precipitation patterns due to elevation changes is called the **orographic effect.**

Jet Streams

Centered over zones of sinking and rising air near 50°–60°N and S and 30°N and 30°S are the **jet streams,** high-speed winds of the upper troposphere. The polar jet streams (50°–60°N and S) are westerlies, lying above the boundary between the polar easterlies and the westerlies; the subtropical jet streams (30°N and 30°S) are easterlies, found above the zone between the trades and the westerlies (fig. 7.27). The Northern Hemisphere polar jet stream is discussed in the following example.

In the winter, the polar jet stream flows rapidly, reaching 250 km (150 m) per hour, and oscillates more than 2000 km (1250 mi) north to south. Wind speeds and displacement are reduced in the summer. The westerlies move the wave form of the stream and its associated high- and low-pressure systems eastward around Earth. Oscillations in jet stream flow are caused by strong high-pressure systems to the south and intense low-pressure systems to the north. The location and movement of the polar jet stream are controlled by the location and strength of the boundary between subtropical and polar air and the shape of the temperate zone's alternating high- and low-pressure systems.

The position of the polar jet stream and the boundary between subtropical and polar air are important in determining the weather of the temperate zone. Circulation of air around these pressure systems transports warm subtropical air to higher latitudes and cold polar air to lower latitudes; refer to the discussion of the heat budget in "Heating and Cooling Earth's Surface" in this chapter and to figure 7.2. The most important source of heat in this system is the water vapor gained by evaporation in the subtropics and condensation as the air cools at higher latitudes. In the equatorial regions, great cumulus cloud systems form in the rising air of the doldrums and are moved westward by the subtropical jets.

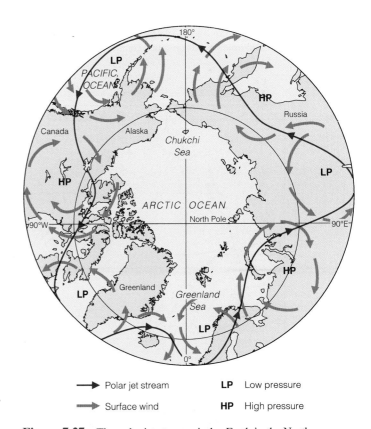

Figure 7.27 The polar jet stream circles Earth in the Northern Hemisphere above the boundary between the polar easterly winds and the westerly winds. It is deflected north and south by the alternating air-pressure cells of the northern temperate zone.

7.7 Hurricanes

Although the trade winds of the tropics blow steadily, they may develop variations in speed and direction when the winds move over waters of changing temperature. These variations cause moving air to converge and then, as the air oscillates, to diverge; the resulting pressure disturbance is known as an easterly wave. It is visible as a sharp wrinkle in the isobars over the tropical oceans.

If the sea surface temperature is above 27°C, the atmospheric pressure decreases and the evaporation rate increases; the easterly wave develops into an intense, isolated, low-pressure cell, and a tropical depression is formed. The winds circle this depression counterclockwise in the Northern Hemisphere and clockwise in the Southern Hemisphere, they build in strength, and the weather disturbance becomes a tropical storm and then a **hurricane** (fig. 7.28). The strong winds, in excess of 130 km (70 knots) per hour, that rotate about the center of this low-pressure system extract water vapor, and therefore heat, from the sea surface. The large amounts of heat energy liberated from the condensing water vapor fuel the storm winds, raising them to destructive levels, up to 300 km per hour (160 knots). A major hurricane contains energy exceeding that of a large nuclear explosion; fortunately, the energy is released much more slowly. The energy generated by a hurricane is about 3×10^{12} watt-hours per day, the equivalent of the energy in 1 million tons of TNT. When these storms move over colder water or over land, the hurricane is robbed of its energy source and begins to dissipate. These storms bring strong winds and high precipitation because of the very high rate of condensation associated with the rising warm, moist tropical air.

Hurricanes can form on either side of the equator but not at the equator because the Coriolis effect is zero. When a storm of this type is formed in the western Pacific Ocean, it is called a **typhoon,** or **cyclone,** instead of a hurricane. Areas that give rise to hurricanes and typhoons and their typical storm tracks are shown in figure 7.29. The storm surge, a mound of high water that accompanies these storms, is discussed in section 7.9.

Figure 7.28   Hurricane Allen in the western Caribbean, photographed by a NOAA satellite in 1980. The center, or eye, of the storm is the clear, calm hole. Winds circulate counterclockwise about the eye.

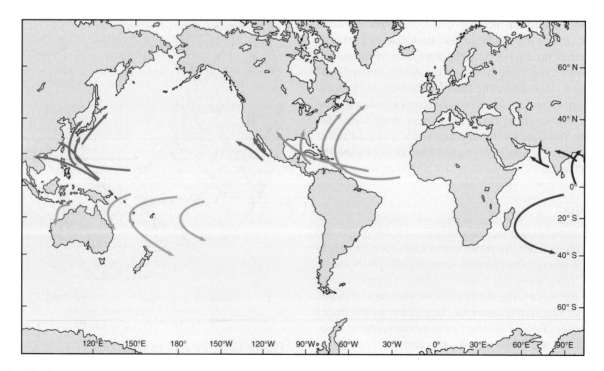

Figure 7.29   Hurricanes, typhoons, and cyclones form on either side of the equator in tropical seas. These storms follow preferred paths in different areas of the oceans.

Forecasting the path, strength, and destination of hurricanes is of enormous importance. Models that incorporate satellite data, weather observations, and atmospheric data are used to make these forecasts. The relatively new geostationary operational environmental satellites (GOES) continuously monitor the position, speed, and path of these intense storms. Land-based Doppler radar can see into a hurricane as the storm approaches land and can measure its wind speed and wind distribution in great detail.

Combining these data, scientists are attempting to create other models to predict the role that global warming may play in the frequency or severity of hurricanes. Early results indicate that increasing sea surface temperature in the Pacific Ocean could increase storm intensity. However, researchers studying the records of past storms in the Atlantic find that the frequency of intense storms is decreasing and that there are no data pointing to worsening conditions with a warmer climate. What is evident is that hurricane damage is growing because of greater population concentrations and property development in hurricane-prone areas.

7.8 El Niño–Southern Oscillation

An El Niño is a period of anomalous climatic conditions centered in the tropical Pacific. El Niño events occur approximately every three to seven years. Once developed, they tend to last for roughly a year, although occasionally they may persist for eighteen months or more. El Niño events are perturbations of the coupled ocean-atmosphere system. We do not yet know whether the perturbations begin in the atmosphere or the ocean. We will consider what happens in the atmosphere first and then look at what happens in the ocean.

Under normal conditions, a low-pressure area in the atmosphere, the Indonesian Low, is located north of Australia over Indonesia in the western equatorial Pacific Ocean, and a corresponding high-pressure area, the South Pacific High, is located between Tahiti and Easter Island in the southeastern Pacific Ocean (fig. 7.30a). The prevailing surface winds over the equatorial Pacific from the South Pacific High to the Indonesian Low are the southeast trade winds. The strength of the trade winds depends on the difference in surface atmospheric pressure between the South Pacific High, typically measured at Tahiti, and the Indonesian Low, typically measured at Darwin, Australia. There is a corresponding return flow of air from the Indonesian Low back to the South Pacific High in the upper atmosphere that completes an atmospheric circulation cell known as the Walker circulation (fig. 7.30b). The resulting trade winds drive water away from the west coast of Central and South America. This causes a thick accumulation of warm water with high sea-surface temperatures and elevated sea surface in the western equatorial Pacific Ocean (fig. 7.31a). Sea-surface temperatures are about 8°C greater and sea-surface elevation is about 50 cm (20 in) higher in the western Pacific compared to the eastern Pacific. As water is driven away from the west coast of Central and South America, upwelling brings cold, deep water to the surface along the west coast of Peru. This results in relatively low sea-surface temperatures in the eastern equatorial Pacific Ocean.

During an El Niño event, the surface pressure in the area of the Indonesian Low is unusually high, while the surface pressure in the area of the South Pacific High is unusually low. As a result, the Indonesian Low moves eastward into the central tropical Pacific. A region of high pressure moves in to replace it, thus reversing the normal surface pressure gradient across the Pacific. As a result, the trade winds weaken or even reverse direction. This periodic reversal in position of the high- and low-pressure regions on either side of the tropical Pacific Ocean is called the **Southern Oscillation.** As the Indonesian Low moves into the central Pacific, the ITCZ also changes position, moving southward and eastward. The collapse of the trade winds occurs abruptly. As a result of the Southern Oscillation, the deep, warm pool of surface water that normally occupies the western equatorial Pacific moves toward the eastern equatorial Pacific over a period of two to three months to accumulate along the coast of the Americas (fig. 7.31b). This effectively shuts down the upwelling of deep water off the Peruvian coast and raises the sea-surface temperature by several degrees centigrade. The flow of water from the west also raises the sea-surface elevation by 30 cm (12 in) or more (the change in elevation in the eastern equatorial Pacific between January 1997 and November 1997 shown in figure 7.31 was approximately 34 cm). This sequence of events is known as **El Niño,** or Christ Child, named for its frequent occurrence around the Christmas season. Eventually, the normal atmospheric pressure distribution is reestablished, the trade winds again blow to the west, and the equatorial Pacific Ocean again has warm water in the west and cool water in the east, signaling the end of the El Niño. Figure 7.31c illustrates the transition to normal conditions following the end of the 1997–98 El Niño. Because of the relationship between El Niño and the Southern Oscillation, the two events together are known as ENSO (an acronym for El Niño–Southern Oscillation).

The climatic effects of El Niño are highly variable and appear to depend on the size of the warm pool of water and its temperature. The same region may experience higher-than-normal rainfall and flooding during one event and drought conditions during the next event. However, there do appear to be some very regular consequences of El Niño events. The northern United States and Canada generally experience warmer-than-normal winters. The eastern United States and normally dry regions of Peru and Ecuador typically have high rainfall, while Indonesia, Australia, and the Philippines experience drought. In addition, El Niño years are associated with less-intense hurricane seasons in the Atlantic Ocean.

During the severe El Niño of 1982–83, ocean surface temperatures off Peru rose 7°C above normal and tropical species were displaced as far north as the Gulf of Alaska. In 1982–83, the polar jet stream was displaced far southward over the Pacific Ocean, bringing unusually dry conditions to Hawaii but high winds and high precipitation to areas along the west coast of the United States, while the eastern United States had its mildest winter in twenty-five years. Heavy rains occurred in Ecuador, Peru, and Polynesia, but drought conditions were experienced in Central America, Africa, Indonesia, Australia, India, and China. At the same time, lower sea-surface temperatures in the North Atlantic made the hurricane season the quietest in over

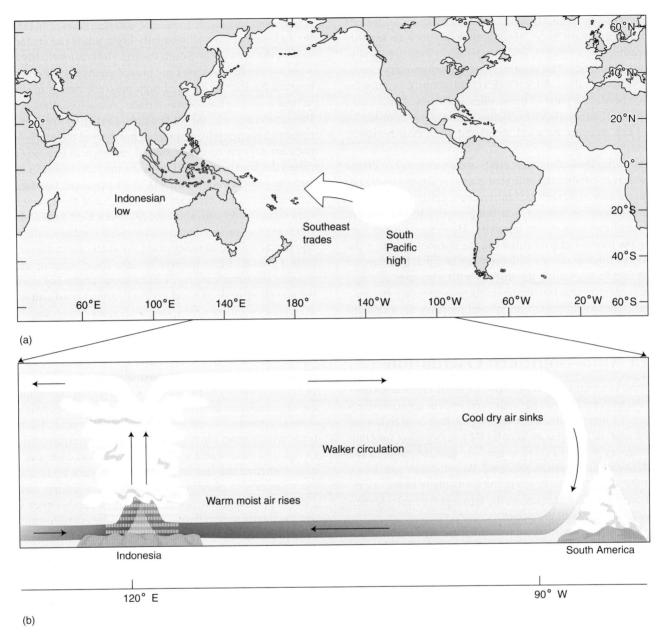

(a)

(b)

Figure 7.30 (a) Schematic representation of normal (not an El Niño event) atmospheric conditions in the tropical Pacific Ocean. (b) Walker circulation between the South Pacific High and the Indonesian Low. Surface winds are the southeast trade winds.

fifty years. Damages from this El Niño event were estimated at $8 billion worldwide ($2 billion in the United States). The severity of this El Niño and its worldwide effects led to the initiation in 1985 of a decade-long study called the Tropical Ocean Global Atmosphere (TOGA) program. TOGA was designed to monitor ocean and atmospheric conditions in the equatorial South Pacific Ocean for the purpose of predicting El Niño events. The ability to forecast these events and so regulate fisheries, predict agricultural droughts, and prepare for severe weather conditions has great economic and commercial importance around the world. Continued monitoring is being done using seventy stationary buoys deployed in the international Tropical Atmosphere and Ocean (TAO) project. Data collected during the TOGA and TAO projects have led to improved ocean-atmosphere models that

have been used with some success to predict the onset of El Niño conditions, but further improvement is needed to accurately predict the severity and length of El Niños.

During 1991–92, the southern displacement of the polar jet stream over the Pacific brought heavy winter rains to southern California and the U.S. Gulf Coast, and a mild, low-precipitation winter to the coastal regions of Oregon and Washington. The jet stream's extreme oscillation, north over the central United States and south over the eastern United States, gave New England and the maritime provinces of Canada extreme cold and heavy snow.

The strongest El Niño on record occurred in 1997–98. The greatest warming occurred in the eastern tropical Pacific, where surface water temperatures were as much as 8°C above normal near the Galápagos Islands and off the coast of Peru. Normally

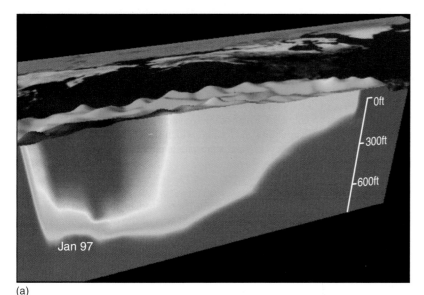

(a)

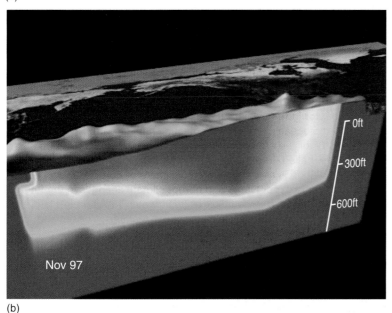

(b)

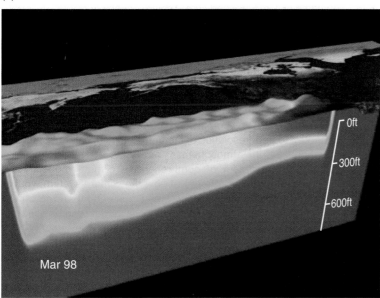

(c)

Figure 7.31 These images show sea-surface topography from NASA's *TOPEX* satellite, sea-surface temperatures from NOAA's *AVHRR* satellite sensor, and sea temperature below the surface as measured by NOAA's network of TAO moored buoys in the equatorial Pacific Ocean. *Red* is 30°C, and *blue* is 8°C. (a) Normal conditions as seen in January 1997, (b) El Niño conditions during November 1997, (c) the end of El Niño conditions and the beginning of a return to normal conditions in March 1998.

dry regions in Ecuador and Peru that usually receive only 10–13 cm (4–5 in) of rain annually had as much as 350 cm (138 in, or 11.5 ft). Severe drought struck areas of the west Pacific Ocean in the Philippines, Indonesia, and Australia. Because El Niño tends to reduce precipitation in the wet monsoon areas of Asia, a weak rainy season was predicted, but India's rains were 2% above normal. A series of strong storms caused severe beach erosion, flooding, and landslides along the California coast. At the same time, the Pacific Northwest and Midwest winters were mild and the southern portion of the United States experienced wetter-than-normal conditions. The jet stream was diverted far to the south over North America, inhibiting the growth of hurricanes in the Atlantic Ocean. By the spring of 1998, Pacific Ocean sea-surface temperatures were returning to normal and signaling the end of this El Niño event.

Between El Niño events, surface temperatures off Peru may drop below normal; an event of this type is known as **La Niña,** "the girl child." These colder-than-normal years also produce wide-scale meteorological effects. The trade winds strengthen, and surface-water temperatures of the eastern tropical Pacific are colder, whereas those to the west are warmer than normal. These changes help establish very dry conditions over the coastal areas of Peru and Chile, while rainfall and flooding increase in India, Myanmar, and Thailand.

Computerized ENSO models attempt to forecast El Niño events by means of the ENSO index (fig. 7.32). The ENSO index is calculated from measurements of sea-surface temperature, sea-level pressure, surface-air temperature, the east-west and north-south velocity components of the trade winds, and the total amount of cloudiness. Positive values of the ENSO index, also known as the ENSO warm phase, are associated with conditions occurring during El Niño events, while negative values of the index, known as the ENSO cool phase, are associated with conditions at the onset of La Niña events. Normal conditions correspond to an ENSO index at or near zero. Computer models successfully predicted the onset of the 1991–92 El Niño event, but the polar jet stream split over the eastern Pacific, and not all predictions came to pass. The 1997–98 episode was predicted six months in advance, although distinguishing the signs of the approaching El Niño from the unusually warm conditions that persisted throughout the 1990s was difficult. The advance warnings of the

Field Notes

The Oceans and Climate Change
by Dr. LuAnne Thompson

LuAnne Thompson is an Associate Professor of Oceanography, Adjunct Associate Professor of Atmospheric Sciences, and Interim Director of the Program on Climate Change at the University of Washington. Her research interest is the role of the ocean in climate variability and change, which she explores using numerical models of ocean circulation, biogeochemical cycles, and atmospheric circulation.

The ocean serves as the memory of the climate system, storing heat, fresh water, and chemicals over time spans from decades to millennia. This memory results from the chemical and physical properties of seawater. First, water is heavier than air. A 10 meter column of water is heavier than a column of air that extends to the top of the Earth's atmosphere. Second, water has four-fold higher specific heat than air and five-fold higher specific heat than soil. Because it takes vastly more energy to heat up the ocean, ocean temperature is much more resistant to change than air or land temperature. Although the temperature of surface waters of the ocean varies according to latitude, the temperature of deep water in all oceans is almost always close to freezing. Finally, gases such as carbon dioxide dissolve in water making the ocean a major storage depot. In fact, fifty-fold more carbon dioxide is stored in the ocean than in the atmosphere. Without absorption of carbon dioxide by the ocean, atmospheric concentrations of this greenhouse gas would be even higher. Thus the oceans provide a damper that keeps the Earth's climate relatively constant and benign.

Over the past fifty years, large research programs have greatly informed how this ocean damper works. In 1957–58, under the auspices of the International Geophysical Year, detailed profiles of temperature and salinity were done in the Atlantic Ocean. These surveys allowed oceanographers to see clearly that the sources of the deep (below 1500 m depth) and intermediate water (between 500 and 1500 m depth) of the world's oceans come from geographically isolated regions, with deep water forming in the northern North Atlantic and near Antarctica, and intermediate water originating at high latitudes and in marginal seas.

In the 1970s, data collected during the Geosecs (Geochemical Ocean Sections Study) Program greatly added to our knowledge of the role of the ocean in global cycles of nutrients and gases such as carbon dioxide and oxygen. By this time, a fairly complete picture of the full three-dimensional structure of the ocean water properties was developed and there was general acceptance that the modern ocean system was relatively stable. In the 1980s a simple cartoon of the global ocean circulation called the great ocean conveyor belt (see Figure 9.14) was popularized. This cartoon suggests that cold water created in the high northern latitudes of the North Atlantic sinks to the abyss and moves slowly toward the North Pacific and Indian Oceans where it rises to the surface, warms, and then makes its way back to the Nordic Seas via surface currents and ultimately the Gulf Stream.

In the mid-1980s through the 1990s, extensive and detailed ocean surveys were conducted as part of the World Ocean Circulation Experiment (WOCE). These studies generated the most comprehensive observational analyses of the ocean general circulation and indicated that the conveyer belt circulation model is an oversimplification. Analyses of WOCE data clearly emphasize the significance of deep water formed near Antarctica, and suggest that much of the upwelling of deep water may occur in the Southern Ocean, and that there are multiple routes by which water is exchanged among the various ocean basins.

The role the ocean plays in climate variability and change is becoming clearer. Geochemical tracers show that deep-water cycles in the ocean take several centuries, while water cycles through waters above the thermocline much more quickly, on the order of decades. Analyses of satellite measurements suggest that the oceans carry the bulk of the excess heat away from the equator in the tropics. However, the amount of heat transferred by the oceans' amount falls off at higher latitudes, and poleward of 40° of the atmosphere carries the lion's share. These results have called into question how important the conveyor belt circulation is to maintaining the climate of Europe, and whether we would expect large and abrupt climate changes in the future if there were a large influx of fresh water into the high latitude North Atlantic Ocean from the rapid melting of Greenland ice, for example.

At the same time that more comprehensive observations of the oceans were being made, the first general circulation models of the ocean and the climate system were constructed. In the early 1970s numerical modeling of ocean circulation was born with the construction of the first general circulation model of the ocean that qualitatively reproduced the major ocean currents. Shortly thereafter, this model was coupled to a model atmosphere and exchange of heat and water across the air-sea interface was modeled. These models are the predecessors of the models used today. Written in Fortran, a scientific programming language, they divided the ocean up into boxes, with the sides of each box about 200–300 km, and the depth about 100–500 m. The original models were written on paper computer cards. Each model run took weeks or more of computer time even though necessarily restricted to poor resolution of features and physical processes in the ocean and the atmosphere.

The development of comprehensive climate models has allowed testing of hypotheses of how the climate system would respond to changes in both the greenhouse effect and to other climatic perturbations. For example, these models can be used to ask if Europe would rapidly cool if a significant fraction of

Greenland were to melt. In experiments where a large influx of fresh water caps off the North Atlantic, there is significant cooling over Scandinavia, but not to the south, suggesting that the effect of the heat transport by the ocean on climate for areas surrounding the North Atlantic Ocean may be much smaller than previously thought. In fact, the most likely scenario is that Scandinavians may be saved from some effects of global warming by the changes wrought by a melting Greenland, but that they will not experience rapid or extreme cooling. Unlike the scenario suggested in the movie *The Day After Tomorrow*, the picture emerging is that ocean currents with relatively small spatial scales play an important role in controlling the redistribution of both fresh water and heat around the globe, and that the simple schematic ideas about the large-scale ocean circulation are not sufficient for understanding the role the oceans will play in a changing climate. Also, while the lessons learned about the climate system from data originating during glacial times are useful, they are probably not the best indicators for what may happen in the future.

One aspect of how the oceans will respond as the climate warms is certain—sea level will rise. The heat-trapping capacity of the planet is increased with the addition of greenhouse gases, such as carbon dioxide, to the atmosphere by human activities. Some 80 to 90% of this heat has been transferred to the oceans, resulting in thermal expansion of the seawater. Over the last decade, global warming has caused an increase in sea level, about half of which is due to thermal expansion of the ocean (warmer water expands), and half due to melting of land ice outside of the polar regions. The recent (2007) dramatic reduction in summer sea-ice extent in the Arctic is not associated with sea-level rise because sea ice floats on the top of the ocean. Because the ocean and atmosphere equilibrate slowly, even if we were to maintain the amount of human-induced greenhouse gases in the atmosphere at 2007 levels, the climate system would warm by an additional 0.5°C by the end of the twenty-first century and sea level would rise through thermal expansion by about 10 cm.

Modeling studies suggest that, as the upper ocean warms, ocean stratification will increase, decreasing the exchange of surface waters with those at depth. This will decrease the supply of nutrients to the surface ocean, thus decreasing primary productivity and likely the activity of the biological pump. Most climate models do not take these effects into account when predicting the future climate.

As the climate changes, ocean chemistry will also change. As of 2007, humans are releasing carbon into the atmosphere at a rate of around 7 gigatons (15,000,000,000,000 pounds) per year. Historically, about one-third to one-half of human (or anthropogenic) carbon releases has been absorbed by the ocean. Much of this absorption occurs at high latitudes where the ocean is cold and can hold more carbon dioxide than at lower latitudes (think of how flat a warm can of soda is compared to a cold one). The surface ocean carbon dioxide concentrations can come into equilibrium with the atmosphere in less than a century, which tells us that, if we stop emitting carbon dioxide into the atmosphere now, it will take about a century for the amount in the atmosphere to approach the levels last seen in the nineteenth century. It will take

several millennia before the carbon dioxide stored in the water column is buried in sediments at the bottom of the ocean.

The addition of carbon dioxide to the water column is not benign. Carbon dioxide dissolved in seawater forms carbonic acid, lowering the slightly alkaline pH level of the ocean. The shift toward increasing acidity and associated changes in ocean chemistry due to increasing carbon dioxide levels in the overlying atmosphere will likely impact marine organisms that build shells out of calcium carbonate. In particular, corals are in danger as "ocean acidification" combines with other environmental stresses. The level of acidification will vary geographically, with high latitudes (colder water) being more affected. In the Southern Ocean, the shells of pterapods, an important component of the food web, will be in danger of acid corrosion by the end of the twenty-first century.

There are several tools to bring to bear on the challenges of understanding the role of the oceans in climate and climate change. Numerical models of the climate system are improving constantly, because of incorporation of improved understanding of fundamental climate processes, but also because computers keep getting more powerful. For instance, the increase in computing power alone allows improved modeling of the strong ocean currents that play such an important role in moving heat, salt, and chemicals around in the ocean. Scientists are also advancing the inclusion of ocean biogeochemistry in these models.

Continued ocean observations are also vital to increasing our understanding of the climate system. The Global Ocean Observing System, which is currently being implemented, is already providing insights into the changing ocean. Satellites that measure sea surface height variations provide scientists with data to monitor sea-level and surface ocean currents. Satellite sea surface temperature, sea surface salinity, and surface winds provide global ground truths for models of the climate system. In addition, the subsurface ocean is observed by over 3000 autonomous floats that drift freely at 1000-m depth and every ten days sink to 2000 m before rising to the surface, measuring temperature and salinity (and recently dissolved oxygen and other chemical parameters), which is then reported back via satellite before the float sinks back to 1000 m again. Repeat measurements of water properties and currents in the deepest parts of the oceans are also being made, although currently, these measurements are only feasible from ships, so they are less frequent. These measurements are revealing significant warming in many parts of the deep ocean.

Humans are changing the climate system in ways that we are only now beginning to understand. Scientists are just now beginning to be able to accurately and globally measure ocean variations through now technologies, including satellites and autonomous platforms, and to model the oceans, a very important aspect of the climate system. Changes are evident in the most remote areas of the ocean, with melting ice in the high latitudes and warming all the way to the ocean floor. These observations are evidence of the magnitude of the changes that we are making in the Earth's climate.

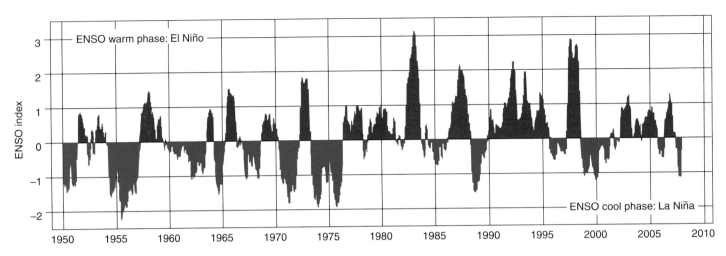

Figure 7.32   The ENSO index from 1950 to 2007. ENSO warm phases *(red)* correspond to El Niño events and cool phases *(blue)* correspond to La Niña events. The larger the positive value of the index, the stronger the El Niño, while the larger the negative value, the stronger the La Niña.

1997–98 El Niño are estimated to have saved $1 billion to $2 billion in property damage in the United States.

The cyclic alternation between these two events has been quite regular for the last hundred years, except for the periods between 1880 and 1900, when La Niña conditions prevailed, and the 1975–97 period of El Niños. In 2000–2001, a La Niña prevailed. This led to decreased rainfall in the Pacific Northwest and increased rain in California. The reduced rain and snowpack in Washington caused problems with irrigation and power production. El Niño conditions returned in the winter of 2002 and continued generally through late summer 2007 with a brief period of La Niña conditions in 2006. In August 2007 a prolonged period of La Niña conditions began signaled by substantially below-average sea surface temperatures in the central and eastern equatorial Pacific, and enhanced Trade Winds. Climate models predict La Niña conditions continuing at least through the spring of 2008 with gradual weakening and a 50% probability of a return to ENSO-neutral conditions by mid-2008.

In addition to ENSO events, another natural cycle in the ocean-atmosphere system in the Pacific has been identified with a period of roughly seventeen to twenty-six years. This cycle is known as the Pacific Decadal Oscillation (PDO) (see chapter 9, section 9.6). The PDO appears to have a direct influence on sea-surface temperatures. Satellite data indicate that the Pacific Ocean was in the warm phase of the PDO from 1977–99 and that it has now entered the cool phase. The dominance of El Niño conditions from 1975–97 suggests that El Niño events may occur more frequently during the PDO warm phase and they may be suppressed during the PDO cool phase.

7.9 Practical Considerations: Storm Tides and Storm Surges

Periods of excessive high water along a coast associated with changes in atmospheric pressure and the wind's action on the sea surface are known as **storm surges** or **storm tides.** These storm surges, combined with normal high-tide conditions, can spell disaster for low-lying coastal areas.

Strong storms at sea, such as hurricanes or typhoons, are centered about intense low-pressure systems in the atmosphere. Under the low-pressure area in the center of the storm, the sea surface rises up into a dome, or hill, while the surface is lower farther away from the center, where the atmospheric pressure is greater. The atmospheric pressure change between the outside of a hurricane and its center can be as large as 7.5 cm (3 in) of mercury. Because mercury is 13.6 times denser than water, this pressure change produces a 97 cm (38 in) change in elevation of the water between regions outside the storm and those at the storm's center. In addition, the surface winds spiraling toward the center of the storm add to the sea's elevation under the storm center.

The elevated dome of water travels across the sea under the storm center and raises the water level at the shore when the storm reaches the coast. During the storm, the drag of the wind on the sea surface pushes surface water in the approximate direction of the wind. On the side of the storm where the winds are toward the shore, the water moving toward the shore is piled up against the coast, increasing the height of the sea surface and producing the storm surge. The water continues to pile against the shore, steepening the slope of the water until the tendency of the water to flow downhill and back to sea is equal to the force of the wind driving the water ashore. If the water along the shore is deep, some of the water moving landward downwells and returns seaward; this process decreases the height of the water being driven ashore. Along a shallow shore, the landward-moving layer of water extends to the sea floor; there is little downwelling or seaward return flow, and the height of the water along the shore is much greater. A storm surge may sustain a high water level for many hours until the storm winds diminish. Storm surges are often confused with tsunamis; these great sea waves are discussed in chapter 10.

Along shallow areas of the East and Gulf Coasts of the United States, storm surges have caused considerable damage. In 1900, a storm surge caused "the great Galveston flood." It produced water depths of 4 m (13 ft) over the island of

The costliest and most destructive natural disaster in United States history was Hurricane Katrina with damage estimated to be in excess of $200 billion, over a million people displaced, roughly 5 million people without power, and over 1200 deaths. Katrina first made landfall just north of Miami, Florida, on August 25, 2005. As it moved across Florida it weakened but then wind speeds increased as it moved into the Gulf of Mexico over warm water (fig. 7.33). Warm water enters the Gulf through the Yucatan Strait as the Yucatan Current. It then flows clockwise in the Loop Current before exiting the Gulf as the Florida Current. The Loop Current occasionally pinches off to form clockwise rotating eddies or "gyres" of warm water that can slowly drift westward at speeds of 2–5 km/day (1.2–3 mi/day) (the formation of eddies is discussed in detail in Chapter 9). As Katrina moved into the Gulf it passed over the warm (roughly 0.5°C or 1.0°F warmer than average) Loop Current and quickly strengthened, having maximum sustained winds of 280 km/hr (175 mi/hr) with gusts as high as 344 km/hr (215 mi/hr) (fig. 7.34). After moving into the Gulf of Mexico, it made landfall again on August 29 along the central Gulf Coast near New Orleans, Louisiana, with sustained winds of 235 km/hr (145 mi/hr). It then moved east along the Louisiana coastline and a few hours later made landfall a third time near the Louisiana/Mississippi border. Because of the size and strength of the storm, record storm surges inundated the entire Mississippi Gulf coast. A region roughly the size of the United Kingdom covering 233,000 km² (90,000 mi²) was declared a Federal Disaster Area.

New Orleans was spared the worst of the winds when Katrina turned slightly to the east just before landfall. However, torrential rain and storm surge raised the level of Lake Pontchartrain, and on August 30 at 1:30 A.M. the 17th Street Canal barriers failed, and an estimated 224 billion gallons of water poured in, flooding roughly 80% of New Orleans (fig. 7.35).

Hurricane Katrina produced a 3 to 10 m (10 to 30 ft) storm surge along over 200 continuous miles of coastline from southeast Louisiana through the Florida panhandle. The 11.3 m (37 ft) storm surge that struck Pass Christian, Mississippi, is the highest ever recorded in the United States (figs. 7.36 and 7.37). Record storm surges that had not occurred in at least the last 150 years flooded the entire Mississippi coast.

Following the devastation of Katrina, Hurricane Rita moved into the Gulf of Mexico as the strongest measured hurricane to

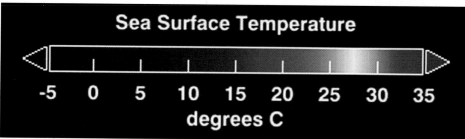

Figure 7.33 Hurricane Katrina gained strength as it moved over warm ocean water in the Gulf of Mexico. This image depicts a three-day average of actual sea surface temperatures (SSTs) for the Caribbean Sea and the Atlantic Ocean, from August 25–27, 2005. Areas in yellow, orange, or red represent 27.8°C (82°F) or above. Hurricanes need SSTs of 27.8°C (82°F) or warmer to strengthen.

Galveston, Texas, destroying the city and killing 5000 people. Hurricane Camille, in 1969, was one of the strongest storms ever to hit the Gulf Coast. The high water levels caused severe property damage, and several hundred people were killed. In 1989, a storm surge of 5 m (16.5 ft) came ashore with Hurricane Hugo in South Carolina, destroying much property along the coast and in the city of Charleston. Great loss of life was averted by early warnings and mass evacuation of low-lying coastal areas. In August 1992, Hurricane Andrew struck Florida with the same intensity that Hurricane Hugo brought to the Carolina coast, but the coastal water damage in Florida was much less. Although the winds were high (220 km/h, or 138 mi/h), the storm surge was only 2.4 m (8 ft). The storm traveled a short distance between the Bahamas and Florida at a fast rate of speed; it did not have enough time to build up a large mass of onshore moving water. In addition, some of Andrew's wave energy was absorbed by offshore reefs and coastal mangrove swamps. On shore, however, the wind damage was very severe.

ever enter the Gulf. Moving into the northern Gulf, it traveled over cooler water and lost energy as it made landfall near the Texas/Louisiana border on September 24, 2005, with sustained wind speeds of 190 km/hr (120 mi/hr). The effects of Hurricane Rita were not as severe as those of Katrina. Storm surges were generally less than 3 m (10 ft), rising to 4.5 to 6 m (15 to 20 ft) in some areas of southwestern Louisiana. Over 2 million people were left without power, and property damages were estimated to be $8 to 11 billion.

Severe storms over the shallow Bay of Bengal spawned storm surges that took an estimated 300,000 lives in 1970. Another 10,000 people were killed in 1985, and in 1991, yet another storm surge struck this same area of Bangladesh, killing an estimated 139,000 people. The expanding population of this area needs land

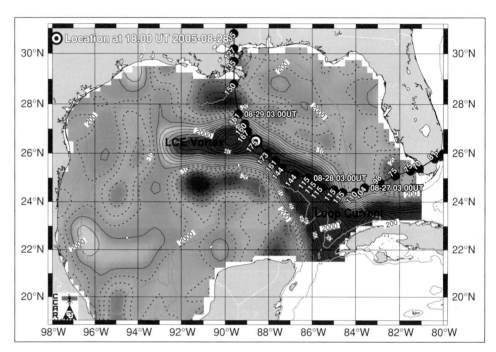

Figure 7.34   An overlay of Hurricane Katrina's track and maximum sustained wind speeds (mph) on the August 28, 2005, sea surface height (SSH) map. The units for SSH are dynamic cm and the contour interval is 5 cm. The SSH map approximates the total sea surface height signal associated with the general ocean circulation in the Gulf and is *false* colored over the range from –30 to +30 cm to highlight cold (*blue*) and warm (*red*) circulation features. At the time of the passage of Katrina, the Loop Current was extended far into the northern and western Gulf and was in the process of shedding a large, warm Loop Current Eddy (LCE) that was named "Vortex." The warm waters associated with the Loop Current and LCE Vortex contributed to the remarkable size and intensity of Hurricane Katrina.

Figure 7.35   A Texas Army National Guard Blackhawk helicopter deposits a 6000 pound-plus bag of sand and gravel on-target, Sunday, September 4, as work progresses to close the breach in the 17th Street Canal. *U.S. Army Corp of Engineers photo by Alan Dooley.*

Figure 7.36 What remained of the Pass Christian, Mississippi, Middle School following Hurricane Katrina. *Photograph taken on September 19, 2005, by Mark Wolfe/FEMA.*

Figure 7.37 Aerial photograph of Mississippi gulf coast Highway I-90 destroyed as a result of winds and tidal surge from Hurricane Katrina. This section of bridge connects Pass Christian, near Gulfport, to Bay St. Louis. *Photograph taken on October 4, 2005, by John Fleck/FEMA.*

for homes and farms, and as fast as new land appears in the delta of the Ganges River, the people move seaward. Although our ability to predict the severity and the route over which a hurricane will pass is improving, other factors work against the potential life-saving results of this ability. When a storm comes to a heavily populated, low-lying coast, the water rises rapidly and covers hundreds of square miles; such conditions make the evacuation of thousands of people extremely difficult.

The Netherlands has constructed barrier dams to protect its coast from the storm tides of the North Sea. England has installed gates across the River Thames to bar a storm tide moving up the river. The great cost of engineering projects such as these may be justified in heavily populated industrial areas, but storm tides cannot be prevented, and along many stretches of shore, there is no defense. The costs to taxpayers and their governments for emergency services and cleanup of storm tide damage are enormous. Would it be wiser to vacate areas historically prone to storm surges and high-water damage and use them only in nonpermanent ways?

Summary

The intensity of solar radiation over Earth's surface varies with latitude. Incoming radiation and outgoing radiation are equal when averaged with time over the whole Earth. Reflection, reradiation, evaporation, conduction, and absorption keep the heat budget in balance with the incoming solar radiation. In any local area of the oceans, there is daily and seasonal variation. Winds and ocean currents move heat from one ocean area to another to maintain the surface temperature patterns.

The oceans gain and lose large quantities of heat, but their temperature changes very little. The heat capacity of the oceans is high compared to that of the land and the atmosphere. Mixing and evaporative cooling work to reduce the temperature contrast between the sea surface and deeper water.

Clouds and weather occur in the troposphere, where temperature decreases with altitude. In the stratosphere, temperature increases with altitude because of the ozone and its ability to absorb ultraviolet radiation. The temperature of the higher mesosphere and thermosphere decreases with height.

The atmosphere is a mixture of gases, including water vapor. Atmospheric pressure is the force with which air presses on Earth's surface; high-density air creates high-pressure zones, and low-density air forms low-pressure zones.

Concern that Earth's climate may be changing is related to changes in the balance of the gases in the atmosphere. Increases in CO_2 concentration due to burning and use of fossil fuels are leading to the prediction of global warming because carbon dioxide traps outgoing long-wave radiation by the greenhouse effect. The ozone layer has been significantly depleted, most probably due to the release of chlorine into the atmosphere.

Dimethyl sulfide produced by plant cells at the sea surface is self-regulated by its role in cloud formation. Sulfur gases from industry and volcanic eruptions are also related to cloud formation and climate change.

The density of air is controlled by temperature, pressure, and water vapor content. The winds are the horizontal air motion in convection cells produced by heating the atmosphere from below. Winds are named for the direction from which they blow.

Because of the Coriolis effect, winds are deflected to their right in the Northern Hemisphere and to their left in the Southern Hemisphere. This action produces a three-celled wind system in each hemisphere that results in the surface wind bands of the trade winds, the westerlies, and the polar easterlies. Zones of rising air occur at 0° and at 60°N and S; these are low-pressure areas of clouds and rain. Zones of descending air at 30° and 90°N and S are high-pressure areas of clear skies and low precipitation. Surface winds are unsteady and unreliable at the zones of rising and sinking air, producing the doldrums and the horse latitudes. The doldrum belt is displaced north of Earth's geographic equator.

Seasonal atmospheric pressure changes modify these wind bands and cause coastal winds to change direction seasonally in the Northern Hemisphere. Differences in temperature between land and water produce the monsoon effect. The seasonal reversal in wind pattern causes the wet and dry monsoons of the Indian Ocean; a similar daily reversal causes the onshore and offshore winds of any coastal area. Winds from the ocean rising to cross the land also produce heavy rainfall. The jet streams are high-altitude, fast-moving winds found in both hemispheres; when greatly displaced, abnormal weather patterns result. Hurricanes develop from tropical low-pressure systems; these storms carry enormous amounts of energy.

In some years, warm tropical surface water moves eastward across the Pacific and accumulates along the west coast of the Americas, blocking the normal upwelling. This phenomenon is El Niño; it is associated with changes in atmospheric pressure and wind direction. Colder-than-normal surface-water temperatures off coastal Peru and associated weather phenomena are known as La Niña; La Niña episodes appear to alternate with El Niño events.

A storm tide or a storm surge is a storm-elevated sea surface. When the storm reaches the shore, severe coastal flooding and destruction are possible.

Key Terms

All key terms from this chapter can be viewed by term or definition when studied as flashcards on this book's website at www.mhhe.com/sverdrup10e.

solar constant, 168
heat budget, 168
reradiation, 169
troposphere, 172
stratosphere, 173

ozone, 173
mesosphere, 173
thermosphere, 173
atmospheric pressure, 173
isobar, 174
low-pressure zone, 174
high-pressure zone, 174
greenhouse effect, 174
Coriolis effect, 181
trade winds, 181

westerlies, 181
polar easterlies, 181
doldrums, 181
horse latitudes, 181
Intertropical Convergence Zone (ITCZ), 182
monsoon, 186
onshore, 186
offshore, 186
rain shadow, 187

orographic effect, 187
jet stream, 187
hurricane, 188
typhoon, 188
cyclone, 188
Southern Oscillation, 189
El Niño, 189
La Niña, 191
storm surge, 194
storm tide, 194

Study Questions

1. Write an equation for the whole Earth's heat budget. Include all the factors for incoming and outgoing energy.
2. How does the troposphere differ from the stratosphere?
3. What controls the density of air? Why is moist air less dense than dry air?
4. Why are the carbon dioxide and ozone concentrations of the atmosphere changing? Why is there concern about these changes?
5. How do airborne sulfur compounds affect cloud cover and the acid rain problem? What are natural and industrial sources of these compounds?
6. Explain why, on the world average, Earth's atmosphere is in motion. Take into consideration the distribution of Earth's surface area with latitude and the variation in heat loss and heat gain with latitude.
7. Why are regions of low barometric pressure and rising air noted for their excessive precipitation?
8. A frictionless projectile is fired from the North Pole and is aimed along the prime meridian. It takes three hours to reach its landing point, halfway to the equator. Where does it land? (Give latitude and longitude.) If the same projectile is fired from the South Pole under the same circumstances, where does it land? (Give latitude and longitude.)
9. Why does the circulation of the atmosphere depend on its transparency to solar radiation? What would happen to atmospheric circulation if the upper atmosphere absorbed most of the solar radiation?
10. Plot the six major wind belts on a map of the world. Add bands of rising air and descending air, the doldrums, and the horse latitudes.
11. Explain why the northeast and southeast trade winds are steady in strength and direction year-round but the Northern Hemisphere westerlies alternate from northwest to southwest with summer and winter along the west coast of the United States.
12. Why are the westerlies of the Southern Hemisphere more consistent than the westerlies of the Northern Hemisphere?
13. What are the early signs that alert forecasters to the onset of an El Niño event?
14. In what way does the polar jet stream influence the transfer of heat from low to high latitudes?
15. How do hurricanes produce storm surges? Why is a storm surge more severe along a coast with a wide, shallow continental shelf than along a coast with a narrow continental shelf?
16. Why do the windward sides of the Hawaiian Islands receive more rain than the leeward sides?

Storm skies over the ocean.

Chapter 8

Circulation and Ocean Structure

Chapter Outline

Learning Outcomes

After studying the information in this chapter students should be able to:

1. *estimate* the density of a mixture of two samples of seawater that have the same density but different temperatures and salinities,

2. *describe* and *sketch* changes in the seasonal thermocline at mid-latitudes through the year,

3. *plot* temperature and salinity as a function of depth and *identify* the thermocline and halocline,

4. *list* five different water masses and *describe* how they form,

5. *relate* surface convergence and divergence to downwelling and upwelling,

6. *describe* the properties of water masses in each ocean basin, and

7. *discuss* how thermal energy can be extracted from the oceans.

Hidden below the ocean's surface is its structure. If we could remove a slice of ocean water in the same way we might cut a slice of cake, we would find that, like a cake, the ocean is a layered system. The layers are invisible to us, but they can be detected by measuring the changing salt content and temperature and by calculating the density of the water from the surface to the ocean floor. This layered structure is a dynamic response to processes that occur at the surface: the gain and loss of heat, the evaporation and addition of water, the freezing and thawing of ice, and the movement of water in response to wind. These surface processes produce a series of horizontally moving layers of water, as well as local areas of vertical motion. In this chapter, we will study both the surface processes and their below-the-surface results in order to understand why the ocean is structured in this way and how the structure is maintained. We will also explore the ways in which oceanographers gather data about this layered system, and we will survey the possibilities of extracting useful energy from it.

8.1 Density Structure

Surface Processes

Absorption of solar energy and energy exchanges between the sea surface and the atmosphere determine sea-surface temperatures and the zones of precipitation and evaporation that govern surface salinities. Variations in temperature and salinity combine to control the density of ocean surface water. Many combinations of salinities and temperatures produce the same density; the result is shown in lines of constant density (fig. 8.1). Figure 8.1, called a T-S diagram, is related to the density values in figure 5.8 in chapter 5.

As the salinity increases, the density increases; as the temperature increases, the density decreases. Salinity may be increased by evaporation or by the formation of sea ice; it may be decreased by precipitation, the inflow of river water, the melting of ice, or a combination of these factors. Changes in pressure also affect density. As the pressure increases, the density increases. However, because pressure plays a minor

role in determining the density of surface water, its effects are not considered here.

Notice that in figure 8.1 the lines of constant density are curved. Therefore, two types of water having the same density but different values of salinity and temperature when mixed form a water that lies on their mixing line. This water has a density greater than either of the original water types, and it will sink. This mixing and sinking process is known as **caballing** and occurs in all oceans when surface waters converge.

Less-dense water remains at the surface (for example, the warm, low-salinity surface water of the equatorial latitudes). Although the surface water at 30°N and 30°S latitudes is warm,

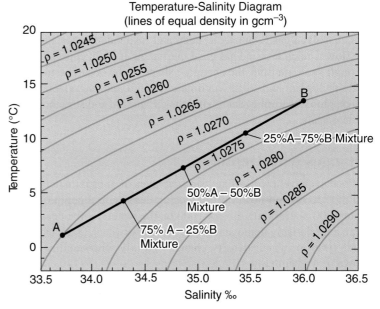

Temperature-Salinity Diagram
(lines of equal density in gcm^{-3})

Figure 8.1 The density of seawater, measured in grams per cubic centimeter, is abbreviated as ρ (rho) and varies with temperature and salinity. Many combinations of salinity and temperature produce the same density. Low densities are at the *upper left* and high densities at the *lower right*. The *straight line* is the mixing line for waters *A* and *B*, both with the same density. A mixture of *A* and *B* lies on the mixing line and is more dense than either *A* or *B*.

it has a higher salinity than surface waters at the equator. Therefore, it is denser than the warm, low-salinity equatorial water. This 30° latitude surface water sinks below the equatorial surface water; it extends from the surface at 30°N to below the less-dense equatorial layer and back to the surface at 30°S. The combined salinity and temperature in surface waters at 50°–60°N and 50°–60°S produce a water that is denser than either the equatorial or the 30° latitude surface water. The 50°–60° latitude water therefore sinks below the equatorial and 30° surface water and extends from the surface in one hemisphere below the other water types to the surface in the other hemisphere. Winter conditions in the subpolar regions lower the surface water temperature and, if sea ice forms, increase the salinity. The result

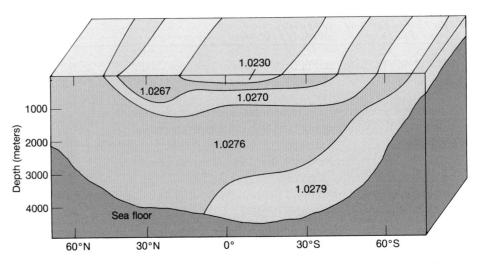

Figure 8.2   Waters of different densities form a layered ocean. The density of each layer is determined at the surface by the climate at the latitude at which it is formed. Density values are given in grams per cubic centimeter.

is a dense surface water that sinks at the subpolar latitudes. These variations in surface water properties and the resulting density changes produce a density-layered ocean. This layered system is shown in figure 8.2. The thickness and horizontal extent of each layer are related to the rate at which the water of each layer is formed and the size of the surface region over which it is formed.

Wind-driven surface currents also directly affect the density structure of the upper ocean and continually move surface waters from one area to another. Turbulence associated with tidal motion as well as wind-driven surface currents mix water across the density layers. Seawater is circulated throughout the oceans, and each volume or mass of water carries with it salt ions, dissolved gases, nutrients, heat, and suspended materials. The flow is relentless and has great inertia, but it is not constant. Through geologic time, the changing shape of the ocean basins (plate tectonics) and shifts in climate (glacial and interglacial cycles, global warming) act to alter the circulation and structure of the oceans.

Changes with Depth

The oceans have a well-mixed surface layer of approximately 100 m (330 ft) and layers of increasing density to a depth of about 1000 m (3300 ft). Below 1000 m, the waters of the deep ocean are relatively homogeneous. A region between 100 m and about 1000 m, where density changes rapidly with depth, is known as a **pycnocline** (fig. 8.3).

Below the 100 m surface layer, the temperature decreases rapidly with depth to the 1000 m level. A zone with a rapid change in temperature with depth is called a **thermocline.** Below the thermocline, the temperature is relatively uniform over depth, showing a small decrease to the ocean bottom. A similar situation occurs with salinity. Below the surface water at the middle latitudes, the salinity increases rapidly to about 1000 m; this zone of relatively large change in salinity with depth is called the **halocline.** Beneath the halocline, relatively uniform

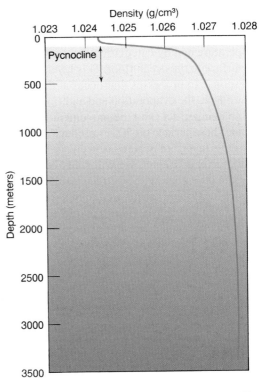

Figure 8.3   Density increases with depth in seawater. The pycnocline is the region in which density changes rapidly with depth. (Based on data from the northeastern Pacific Ocean.)

conditions extend to the ocean bottom. Both a thermocline and a halocline are shown in figure 8.4.

If the density of the water increases with depth, the water column from surface to depth is **stable.** If there is more-dense water on top of less-dense water, the water column is **unstable.** An unstable water column cannot persist; the denser surface water sinks and the less-dense water at depth rises to replace the surface water. Vertical convective **overturn** of the water takes place. If the water column has the same density over depth, it has neutral stability and is termed **isopycnal.** A neutrally stable water column is easily vertically mixed by wind, wave action, and currents. If the water temperature is unchanging over depth, the water column is **isothermal;** if the salinity is constant over depth, it is **isohaline.**

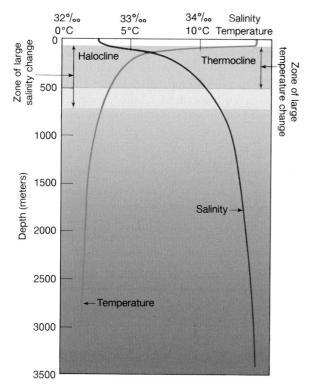

Figure 8.4 Temperature and salinity values change with depth in seawater. Rapid changes in temperature and salinity with depth produce a thermocline and a halocline, respectively. (Based on data from the northeastern Pacific Ocean.)

Density-Driven Circulation

Processes that increase the water's density at the surface cause density-driven **vertical circulation.** This overturn may reach only to shallow depths or extend to the deep-sea floor, ensuring an eventual top-to-bottom exchange of water. Because the density is normally controlled by surface changes in temperature and salinity, this vertical circulation is often called **thermohaline circulation.** An excellent example of thermohaline circulation occurs in the Weddell Sea of Antarctica, where winter cooling and freezing produce dense, cold, saltier surface water that sinks to the sea floor. This water descending along the coast of Antarctica is the densest water found in the open oceans (see fig. 8.2).

At the temperate latitudes in the open ocean, the surface water's temperature changes with the seasons. This effect is illustrated in figure 8.5. During the summer, the surface water warms and the water column is stable, but in the fall, the surface water cools, its density increases, and overturn begins. Winter storms and winter cooling continue the mixing process. The shallow thermocline formed during the previous summer is lost, and the upper portion of the water column is vertically mixed and becomes isothermal to greater depths. Spring brings warming, and the shallow thermocline begins to reestablish itself; the water column becomes stable and remains so through the summer.

Seasonal temperature changes are more important than salinity changes in altering density in the open ocean at subarctic and high temperate latitudes. For example, in the Atlantic Ocean, the surface water at 30°N and 30°S has a high salinity but is warm year-round, so it stays at the surface. In contrast, water from 50°–60°N in the North Atlantic has a lower salinity but is cold, especially during the winter. This cold water sinks and flows below the saltier but warmer surface water.

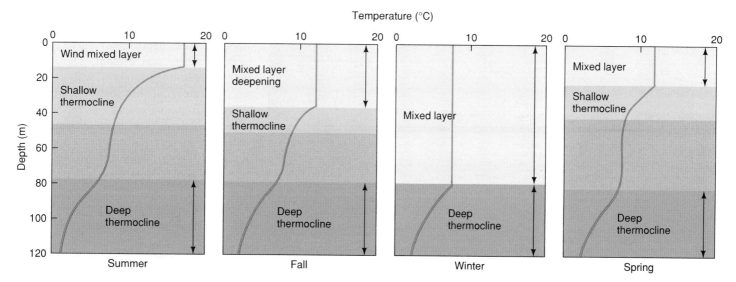

Figure 8.5 The surface-layer temperature structure varies over the year. In the absence of strong winds and wave action in the summer, solar heating produces a shallow thermocline. During the fall and winter, the surface cooling and storm conditions cause mixing and vertical overturn, which eliminate the shallow thermocline and produce a deep wind-mixed layer. In spring, the thermocline re-forms. The deep thermocline is below the influence of seasonal surface changes and remains in place year-round.

Close to shore, the salinity of seawater can become more important than the temperature in controlling density. Salinity is particularly important in semienclosed bays, sounds, and fjords that receive large amounts of freshwater runoff. Here, extremely cold (0°–1°C) fresh water from melted ice is added to the sea. In some cases, the salinity becomes so low that the diluted seawater does not sink but remains at the surface as a seaward-moving **freshwater lid.** In polar regions, when sea ice melts, a layer of fresh water forms at the surface and slowly dilutes the underlying surface water.

8.2 Upwelling and Downwelling

When dense water from the surface sinks and reaches a level at which it is denser than the water above but less dense than the water below, it spreads horizontally as more water descends. At the surface, water moves horizontally into the region where sinking is occurring. The dense water that has descended displaces deeper water upward, completing the cycle. Because water is a fixed quantity in the oceans, it cannot be accumulated at one location or removed at another location without movement of water between those locations. This concept is called **continuity of flow.** Areas of thermohaline circulation where water converges and sinks are called **downwelling zones;** areas of diverging rising waters are **upwelling zones.** Downwelling is a mechanism that transports oxygen-rich surface water to depth, where it is needed for the deep-living animals. Upwelling returns low oxygen-content water with dissolved, decay-produced nutrients to the surface, where the nutrients act as fertilizers to promote photosynthesis and the production of more oxygen in the sunlit surface waters.

Upwelling and downwelling refer to vertical motion of water upward or downward. They are present in thermohaline circulation but can also be caused by wind-driven surface currents. When the surface waters are driven together by the wind or against a coast, a surface **convergence** is formed. Water at a surface convergence sinks, or downwells. When the wind blows surface waters away from an area or coast, a surface **divergence** occurs and water upwells from below (fig. 8.6). Surface convergences and divergences created by wind-driven currents are discussed in chapter 9.

The speed of upwelling and downwelling water is about 0.1–1.5 m (5 ft) per day. Compare these flows to oceanic surface currents, which reach speeds of 1.5 m/s. Horizontal movement at mid-ocean depth due to the thermohaline flow is about 0.01 cm (0.004 in) per second. Water caught in this slow but relentlessly moving cycle may spend 1000 years at the greater ocean depths before it again reaches the surface.

8.3 The Layered Oceans

Oceanographers have taken salinity and temperature measurements with depth at many locations and for many years. Gradually, they accumulated sufficient data to identify the layers of water that make up each ocean and the surface source of the water forming each layer. The structure of an ocean is determined by the properties of these layers. Each layer received its characteristic salinity, temperature, and density at the surface. The water's density controls the depth to which the water sinks; the thickness and horizontal extent of each layer are related to the rate of its formation and the size of the surface source region. Water that sinks from the surface to spread out at depth and slowly mixes with adjacent layers eventually rises at another location. In all cases, water that sinks displaces an equivalent volume of water upward toward the surface at some other location so that the oceans' vertical circulation is continuous.

The layers of water just described are associated with depth zones: surface, intermediate, deep, and bottom. The surface zone extends to 200 m (650 ft), and the intermediate zone lies between 300 and 2000 m (1000 and 6500 ft). Deep water is found between 2000 and 4000 m (6500 and 13,000 ft), and bottom water is below 4000 m (13,000 ft).

The Atlantic Ocean

The properties of the layers of water making up the Atlantic Ocean are shown in figure 8.6. At the surface in the North Atlantic, water from high northern latitudes moves southward, while water from low latitudes moves northward along the coast of North America and then east across the North Atlantic. These waters converge in areas of

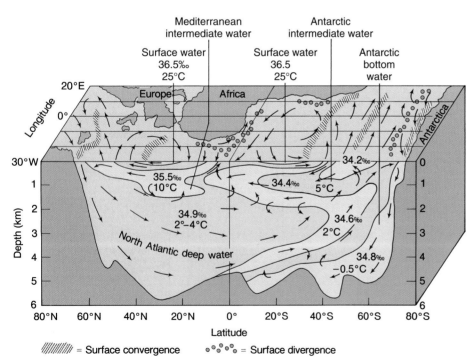

Figure 8.6 The anatomy of the Atlantic Ocean. Surface and subsurface circulation are related here. The surface currents converge to produce downwelling; water sinks to its density level and flows horizontally. Water at depth rises under zones of surface divergence.

cool temperatures and high precipitation at approximately 50°–60°N in the Norwegian Sea and at the boundaries of the Gulf Stream and the Greenland and Labrador Currents. The resulting mixed water has a salinity of about 34.9‰ and a temperature of 2°–4°C. This water, known as **North Atlantic deep water (NADW),** sinks and moves southward. North Atlantic deep water from the Norwegian Sea moves south along the east side of the Atlantic, while water formed at the boundary of the Labrador Current and the Gulf Stream flows along the western side. Above this water at 30°N, a low-density lens of very salty (36.5‰) but very warm (25°C) surface water remains trapped by the circular movement of the major oceanic surface currents. Between this surface water and the North Atlantic deep water lies water of intermediate temperature (10°C) and salinity (35.5‰). This water is a mixture of surface water and the upwelled colder, saltier water from the subtropical regions. It moves northward to reappear at the surface south of the convergence in the North Atlantic.

Near the equator, the upper boundary of the North Atlantic deep water is formed by water produced at the convergence centered about 40°S. This is **Antarctic intermediate water (AAIW).** Because it is warmer (5°C) and less salty (34.4‰) than the North Atlantic deep water, it is less dense and remains above the denser and saltier water below. Along the edge of Antarctica, very cold (−0.5°C), salty (34.8‰), and dense water is produced at the surface by sea ice formation during the Southern Hemisphere's winter. This is **Antarctic bottom water (AABW),** the densest water in the oceans. This water sinks to the ocean floor and flows slowly northward, creeping beneath North Atlantic deep water, as it continues on through the deep South Atlantic ocean basins west of the Mid-Atlantic Ridge. Antarctic bottom water does not accumulate enough thickness to be able to flow over the mid-ocean ridge system into the basins on the African side of the ridge; it is confined to the deep basins on the west side of the South Atlantic and has been found as far north as the equator.

At the same time, the North Atlantic deep water between the Antarctic bottom and intermediate waters rises to the ocean's surface in the area of the 60°S divergence. As it reaches the surface, it splits; part moves northward as **South Atlantic surface water** and Antarctic intermediate water; part moves southward toward Antarctica, to be cooled and modified to form Antarctic bottom water. A mixture of North Atlantic deep water and Antarctic bottom water becomes the circumpolar water for the Southern Ocean that flows around Antarctica. The Antarctic circumpolar water becomes the source of the deep water found in the Indian and Pacific Oceans. In this way, the Atlantic Ocean and its circulation play a defining role in the structure and circulation of all the oceans.

Warm (25°C), salty (36.5‰) surface water in the South Atlantic is also caught by the circular current pattern at the surface and is centered about 30°S. South of the southern tips of South America and Africa, the water flows eastward, driven by the prevailing westerly winds, which move the water around and around Antarctica.

Because the Atlantic Ocean is a narrow, confined ocean of relatively small volume but great north-south extent, the water types are readily identifiable and the movement of the layers can be followed quite easily. In addition, the bordering nations of the Atlantic have had a long-standing interest in oceanography, so the vertical circulation and layering of the Atlantic are the most studied and the best understood of all the oceans.

The Pacific Ocean

In the vast Pacific Ocean, waters that sink from relatively small areas of surface convergences lose their identity rapidly, making the layers difficult to distinguish. Antarctic bottom water forms in small amounts along the Pacific rim of Antarctica, but it is quickly lost in the great volume of the Pacific Ocean. The deeper water of the South Pacific Ocean is the water of the Antarctic circumpolar flow. Because the North Pacific is isolated from the Arctic Ocean, only a small amount of water comparable to North Atlantic deep water can be formed. In the extreme western North Pacific, convergence of the southward-flowing cold water from the Bering Sea and the Sea of Okhotsk and the northward-moving water from the lower latitudes produces only a small volume of water that sinks to mid-depths. There is no large source of deep water similar to that found in the North Atlantic. Warm, salty surface water occurs at subtropical latitudes (30°N and 30°S) in each hemisphere, and Antarctic intermediate water is produced in small quantities, but its influence is small. Deep-water flows in the Pacific are sluggish, and conditions are very uniform below 2000 m (6600 ft). The slow circulation of the Pacific means that it has the oldest water at depths where age is measured as time from the water's last contact with the surface. Residence time for deep water in the Pacific is about twice that of deep water in the Atlantic.

The Indian Ocean

The Indian Ocean is principally an ocean of the Southern Hemisphere and has no counterpart of the North Atlantic deep water. Small amounts of Antarctic bottom water are soon mixed with the deeper waters to form a fairly uniform mixture of Antarctic circumpolar water brought into the Indian Ocean by the Antarctic circumpolar current. There is a small amount of Antarctic intermediate water, and in the subtropics, a lens of warm, salty water occurs at the surface.

Comparing the Major Oceans

Temperature and salinity distributions as functions of depth are shown in figure 8.7 for each of the oceans. The Atlantic Ocean salinity and temperature values form patterns with depth (fig. 8.7a and b) that are clearly identifiable and can be related to figure 8.6. Salinity and temperature values in the Pacific (fig. 8.7c and d) identify the high-salinity surface water lenses and the mixture of Antarctic bottom water and circumpolar deep water, with a 34.7‰ salinity and a 1°C temperature. A minor intrusion of low-salinity water in the north is the result of the surface convergence in the far northwestern Pacific. Note the large volume of deep water that shows a uniform salinity in the Pacific Ocean. Temperature values also follow a less-definite pattern than in the Atlantic, emphasizing the uniformity of most of this deep water. The Indian Ocean values (fig. 8.7*e* and *f*) resemble those for the South Atlantic, without the presence of water that is comparable to North Atlantic deep water. The Arctic Ocean is an extension of the North Atlantic; see figure 8.8 and the following discussion.

The Arctic Ocean

Some oceanographers consider the Arctic Ocean to be an extension of the North Atlantic, but it differs from the Atlantic Ocean in many significant ways. About one-third of its area, 8×10^6 km^2 (3.1×10^6 mi^2), is covered by extensive continental shelves, the widest of any ocean. Two basins, the Eurasian to the east and the Canadian to the west, occupy the central portion of the ocean; they are separated by the Lomonosov Ridge extending due north from Greenland (fig. 8.8). The Eurasian basin is the deeper basin, 5000 m (16,400 ft); it is connected to the North Atlantic through a gap in the continental shelf between Spitsbergen and Greenland. The larger, shallower Canadian basin is about 3800 m (12,450 ft) deep. Both basins contain spreading centers that are northern extensions of the Mid-Atlantic Ridge system.

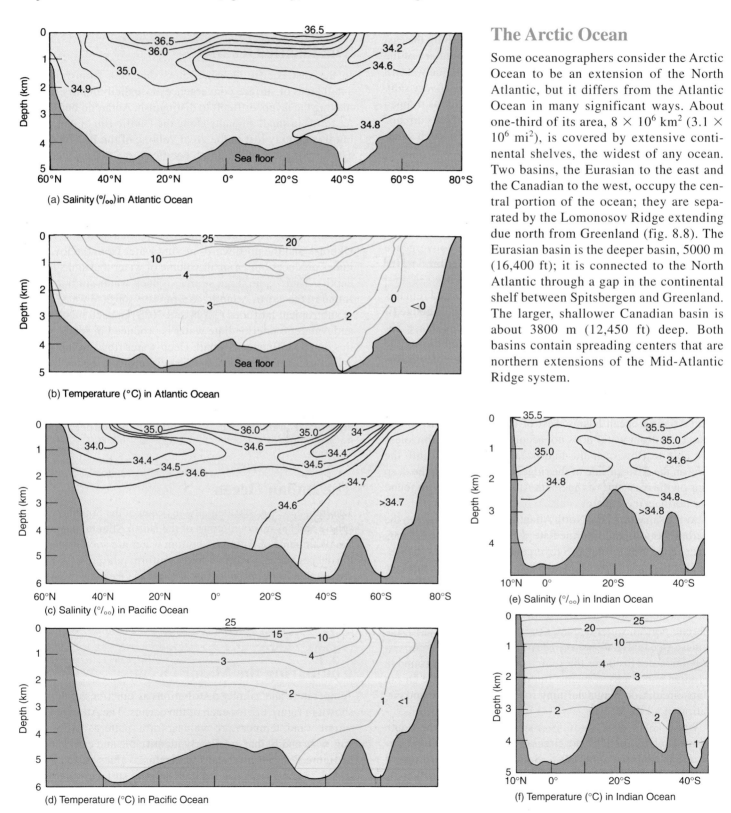

Figure 8.7 Mid-ocean salinity and temperature profiles of the Atlantic, Pacific, and Indian Oceans. Temperature and salinity are shown as functions of depth and latitude.

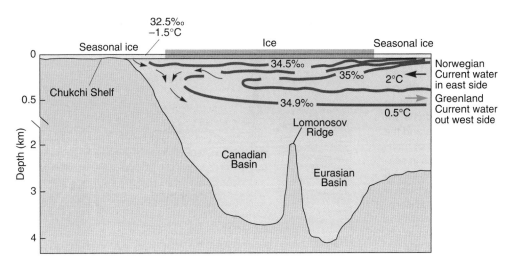

Figure 8.8   The Arctic Ocean. Water is supplied to a shallow surface layer by the Atlantic Ocean's Norwegian Current. Water exits the Arctic Ocean into the Atlantic Ocean by the Greenland Current. The circulation is confined almost entirely to the upper 500 m (1650 ft) of the Arctic Ocean.

survived the summer of 2002 was also found to be thinner than usual. Ice reflects solar radiation, but exposed water absorbs solar radiation and so accelerates the melting process. If the present situation continues, an estimated 20% of ice cover will be lost by 2050, which could lead to problems for polar bears, seals, and natives of the high northern latitudes. Under such conditions, Greenland's glaciers would also melt and release large amounts of fresh water to the North Atlantic. Such water would both cool and reduce the density of North Atlantic surface water. In these circumstances, the production of North Atlantic deep water and the circulation of the Atlantic would decrease.

The density of the Arctic Ocean water is controlled more by salinity than by temperature. Its surface layer is formed from low-salinity water entering from the Bering Sea, fresh water from Siberian and Canadian rivers, and seasonal melting of sea ice. The surface layer from these combined sources is about 80 m (250 ft) deep and has a low salinity (32.5‰) and a low temperature (−1.5°C). Below the surface layer, salinity increases with depth in the halocline layer, 200 m (650 ft) thick, to reach 34.5‰ at its base. The cold, salty water of the halocline layer is produced by the annual freezing and formation of sea ice over the continental shelves. This water sinks and moves across the shelves to spread out in the central ocean basins. West of Spitsbergen, North Atlantic water (2°C and 35‰) enters the Arctic Ocean and is cooled as it flows under the halocline and fills the Arctic Ocean basins. This water upwells along the edge of the continental shelves, mixing with the water formed during freezing, and exits the Arctic as water of 0.5°C temperature and 34.9‰ salinity along the edge of the shelf adjacent to Greenland. This exiting water moves south along the coast of Greenland and enters the North Atlantic south of Greenland and Iceland, where it combines with Gulf Stream water to form North Atlantic deep water.

Many aspects of the structure of the Arctic Ocean are still a puzzle. Approximately 70% of the Arctic Ocean observations to date were made by the former Soviet Union. Since the cessation of the Cold War, detailed bathymetric and water column data have been declassified and made available to the Arctic research community. One data set contained 1.3 million salinity and temperature measurements taken over a period of many years from icebreakers, drifting ice camps, and buoys. The data are being distributed in a four-volume *U.S.-Russian Arctic Atlas.* The first volume was released in January 1997.

The average minimum Arctic Sea ice cover since 1978 has been 6.22×10^6 km^2 (2.4×10^6 mi^2). In 2002, measurements of Arctic Ocean ice cover showed the summer minimum to have decreased to 5.18×10^6 km^2 (2×10^6 mi^2). The sea ice that

Bordering Seas

Two small but specific water types from bordering seas are readily identifiable, one in the North Atlantic and one in the Indian Ocean. The water from the Mediterranean Sea has a temperature of about 13°C and a salinity of 37.3‰ as it leaves the Strait of Gibraltar. This water, mixing with Atlantic Ocean water, forms an intermediate density water, **Mediterranean Intermediate Water (MIW),** that sinks in the North Atlantic to a depth of approximately 1000 m (3300 ft) (fig. 8.6). The influence of Mediterranean water can be traced 2500 km (1500 mi) from the Strait of Gibraltar before it is lost through modification and mixing. In the Indian Ocean, the initially very salty (40–41‰) water from the Red Sea has been found in a spreading layer at 3000 m (10,000 ft) depth more than 200 km (124 mi) south of its source.

Studies to determine where and how the deep water of the Mediterranean is formed were made in the 1980s and again in 1995. The Mediterranean Sea is divided into east and west basins at the Strait of Sicily (fig. 8.9). Deep and intermediate Mediterranean waters formed in the east basin move through the Strait of Sicily into the west basin and then through the Strait of Gibraltar into the Atlantic Ocean. The 1980 studies showed that high evaporation rates and winter cooling in the southern Adriatic Sea caused surface water to sink, and this water then moved into the east basin and from there into the west basin on its way to the Atlantic. The 1995 studies indicate that the source of the east basin water is no longer the Adriatic Sea; instead, the water is being formed in the Aegean Sea. The Aegean Sea water is saltier and denser than that produced in the Adriatic Sea, and the Aegean water is now displacing older Adriatic water upward and westward into the Atlantic Ocean. It appears that small climate changes in the eastern Mediterranean Sea affect winter cooling and evaporation rates and play a significant role in the location and strength of the processes that produce the deep waters of the Mediterranean. Recent studies of surface currents show that Atlantic Ocean water flows

*N*ot all oceanographic research involves ships or is conducted in warm, tropical regions. More and more interest is being shown in the Arctic Ocean, the world's northernmost but least-studied ocean. Understanding Northern Hemisphere climate changes and making long-range weather forecasts for northern Europe and North America require that we understand the effects of periodic increases in ice and the large volumes of meltwater that enter the North Atlantic from the Arctic. Changes in sea-ice cover directly affect Earth's heat budget. When the Arctic Ocean is ice-covered, reflectance of solar energy is high and absorption is low, but when little ice is exposed, reflectance is low and absorption is high. New concerns have arisen that circulation within the Arctic Ocean may distribute persistent organic pollutants from adjacent land sources, contaminating marine life and placing the native peoples dependent on this marine life at risk; again, we need more information on the circulation of the Arctic Ocean to evaluate this problem.

In 1994, a joint program between Canada and the United States used icebreakers to traverse 1600 nautical miles of the Arctic Ocean from the Bering Sea to the pole. The main goal was to establish the Arctic Ocean's role in global climate change. Studies were made of the water, sea ice, sea floor, marine organisms, distribution of pollutants, and polar bear populations (box figs. 1 and 2). The overlying theme was to establish the Arctic Ocean's role in global climate change by studying the area from the atmosphere through the water and ice down to the sea floor.

On January 1, 1994, the Arctic Climate System Study (ACSYS) was organized. The observational phase of this program continued for ten years. Its primary goal was to understand the role of the Arctic in producing global climate. ACSYS encouraged and coordinated national and international research activities concerning ocean circulation, ice cover, exchanges of water in the Arctic Ocean, long-term climatic research, and monitoring programs. The Surface Heat Budget of the Arctic (SHEBA) was a subprogram of ACSYS, as well as a U.S.-Canadian cooperative research program. This program froze a Canadian Coast Guard icebreaker into the Arctic ice in October 1997, 248 km (400 mi) north of Prudhoe Bay, Alaska. For the next year, the ship acted as hotel and supply base for the SHEBA program (box fig. 3). The data from studies of air-sea exchange processes affecting Arctic Ocean heat budgets helped scientists better understand the role of Arctic processes in global climate models.

In the spring of 2000, researchers began a new program known as the National Science Foundation's North Pole Environmental Observatory. To learn more about how the northernmost ocean regulates global climate, buoys were placed on the polar ice to drift with the ice pack. These buoys are equipped with sensors that extend through the ice and gather data on changes in the thickness of the ice and the properties of the upper ocean. Information is relayed via satellites. Another installation of drifting buoys was completed in the spring of 2001. One buoy was developed by the Japanese to measure ocean temperature and salinity, current profiles, and atmospheric temperature and pressure, as well as wind velocity. Other buoys include a meteorological buoy that records wind speed, temperature, and

Box Figure 1 Scientists and their gear are offloaded to the ice from the U.S. Coast Guard icebreaker *Polar Sea* during its 1994 Arctic Ocean Section expedition.

atmospheric pressure; two radiometer buoys that measure radiation from the sun and reflected radiation from the ice; and two ice-mass balance buoys that record ice temperature and snow thickness.

Another installation in 2001 was a 4250 m (14,000 ft) long instrumented cable with 4000 kg (9000 lb) of gear that was moored to the sea floor under the ice (box fig. 4). The mooring, supported by submerged floats, carries conductivity-temperature recorders to monitor salinity and temperature changes, current meters to record speed and direction of flow, a current profiler to detail ice drift and vertical structure of currents, and sonar to measure ice thickness. Each instrument must record its data internally because a satellite cannot retrieve data from under the ice. The data were recovered in

Box Figure 2 A polar bear is tranquilized and tested for contaminants during the 1994 Arctic Ocean Section expedition.

Box Figure 4 Deploying a 4250 m mooring at the North Pole. The mooring is outfitted with a suite of oceanographic instruments. An acoustic Doppler current profiler is located near the top of the mooring.

Box Figure 3 The Canadian Coast Guard icebreaker *Des Groseilliers* serves as the support base for SHEBA research huts on the Arctic ice.

Box Figure 5 The building on the ice combines living quarters and a scientific laboratory. Equipment includes an A-frame and buoys ready for deployment through the ice.

the following year when the mooring was retrieved. Another part of the 2001 program includes using ski-equipped aircraft to establish five camps over a 485 km (300 mi) route from the pole to Alaska. At each camp (box fig. 5), members of the research team make measurements through the ice and collect samples for chemical tests. These data continue the effort of 2000, when similar data were obtained over a 565 km (350 mi) route between the pole and Canada. The data that are being collected will be combined with satellite and meteorological data to study changes in the Arctic Ocean and to determine how fast the changes are occurring.

Figure 8.9 Surface currents in the Mediterranean Sea. A *line* through the Strait of Sicily separates the Mediterranean Sea into the western Mediterranean basin and the eastern Mediterranean basin. Deep and intermediate waters are formed in the eastern basin, move through the Strait of Sicily into the western basin, continue westward, and exit through the Strait of Gibraltar into the Atlantic Ocean. Before 1987, cool, high-salinity, deep water was formed principally in the Adriatic Sea; by the winter of 1995, the source of this water had changed to the Aegean Sea.

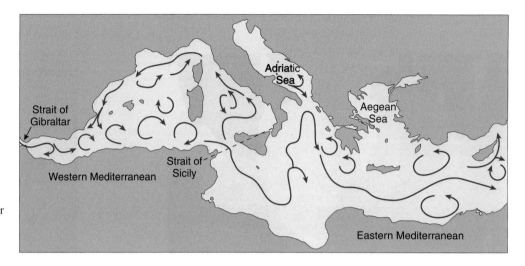

into the Mediterranean Sea to replace seawater that exits the Mediterranean at depth (fig. 8.9).

Oceanographers are interested in what happens in the Mediterranean Sea because the forces that drive the Mediterranean system are comparable to the larger-scale processes governing the open ocean; thus, the Mediterranean Sea can be used as a model for the larger ocean system.

Internal Mixing

Mixing between waters in the ocean is most active when turbulence and energy of motion are available to stir the waters and blend their properties. At the sea surface, wind-driven waves and currents supply energy for mixing, and the tides create currents at all depths. The large eddies that may form at the boundaries of currents also stir together dissimilar waters, acting to homogenize them. When surface currents coverage, mixing at current boundaries may produce caballing of the mixed water. When currents and their associated turbulence are weak, mixing is reduced. Mixing by diffusion occurs continually at the molecular level, but diffusion is much weaker than mixing by turbulent processes.

If a parcel of water is displaced vertically by turbulence, buoyancy forces tend to return the parcel to its original density level. Therefore, vertical mixing between the water types that form the oceans' internal layers is weak. Horizontal mixing is more efficient because it requires less energy than vertical mixing. A parcel of water displaced horizontally along a surface of constant density remains at its new position and shares its properties with the surrounding water.

In areas under warm, high-salinity surface water with an appreciable salinity and temperature decrease with depth, internal vertical mixing processes occur despite the stability of the water column. Vertical columnar flows, approximately 3 cm (5 in) in diameter, are called salt fingers; they develop and mix the water vertically, causing a stair-step salinity and temperature change with depth. This phenomenon is caused by the ability of seawater to gain or lose heat faster by conduction than it gains or loses salt by diffusion. This causes the density of the vertically moving water to change relative to that of

the surrounding water, and the moving water is propelled either up or down. Salt fingers mix water over limited depths, creating homogeneous layers 30 m (100 ft) thick. These layers exist from about 150–700 m (500–2300 ft) deep and are estimated to occur over large areas of the oceans when the required conditions are present.

8.4 Measurement Techniques

Measuring the salinity, temperature, dissolved gases, nutrients, suspended matter, and other characteristics of seawater in situ requires that the oceanographer devise specialized sampling and measuring equipment and a platform from which the equipment can be used or deployed. The research vessel is the traditional platform from which samples are taken and measurements are made (see the photo essay "Going to Sea," pages 266–269). These vessels are equipped with winches and cables that launch, lower, and retrieve instruments; laboratories for specialized onboard research; and a large array of measuring devices, including depth recorders, sonar, speed and direction sensors, atmospheric and solar radiation sensors, and many more. However, research time at sea is expensive, for vessels require fuel, living amenities, and professional merchant crews as well as oceanographers. Total vessel and crew costs are high: $25,000 and more per operating day, not including the scientific party and equipment. Therefore, researchers need to be able to gather accurate information in the least amount of time.

The oceanography of the first fifty years of the twentieth century depended on robust mechanical devices that were lowered into the sea, took their samples at preset depths, and returned those samples to a ship for later analysis, but their use is decreasing rapidly.

Although early water bottle data were sparse, they were sufficient to indicate the depth structure of the oceans. Estimated flow rates of water in deep-ocean circulation were made by combining dissolved oxygen measurements and estimated average values of oxygen utilization. Using these values, researchers calculated the time required for a water parcel to leave the ocean surface and arrive at depth. The calculated times coupled with water parcel positions yielded water transport rates.

Direct measurements of transport rates were later made by suspending neutrally buoyant floats in the moving water. The floats emitted periodic sound pulses (pings) that were used to locate them in the water flow. Indirect measurements of ocean circulation also involved isotope tracers that change their properties with time. Tritium, a radioactive isotope resulting from nuclear testing in the 1950s and 1960s, has a half-life of 12.45 years. It was used in large ocean-circulation studies concerned with short-term displacements of water at depth. Longer-term changes in the circulation used isotopes such as C^{14}, which has a much longer half-life, 5570 years.

The conductivity-temperature-depth sensor, or CTD (fig. 8.10), is today's workhorse for determining water properties and providing a depth profile of salinity by measuring electrical conductivity of the seawater. A temperature profile is obtained with an electrical resistance thermometer, and depth readings are made with a pressure sensor. The CTD is lowered in a protective cage that contains a series of water bottles and may also carry other sensors such as pH probes, chlorophyll sensors, optical scanners, and dissolved oxygen sensors. Conductivity and temperature are monitored continuously as the CTD is lowered, and the data are returned to the ship as electronic signals through the suspending cable. Onboard ship, the CTD data may be fed directly into a computer, recorded on a chart, or made available as numerical data. Continuous profiles of data to several thousand meters can be acquired in this way, whereas the old water bottles collected only one sample at each depth for later analysis.

CTD systems still require a research vessel to stop and remain on station while the instrument is lowered, measurements are made, and the CTD is recovered. However, the CTD can also be left in place, suspended from a surface vessel or buoy; the collected data can be stored in digital form for retrieval at another time.

Seasoar (fig. 8.11) is a modified CTD device that allows sampling while the vessel is underway. The Seasoar is a winged vehicle that is towed behind a ship. The device controls its depth by diving planes and can be compared to a kite flying upside down underwater. Its vertical range extends from the surface to about 350 m (1100 ft), and its dive cycle takes about ten minutes to complete, compared with the standard CTD, which, with stop and start time for the vessel, takes about forty minutes. The Seasoar's sensors record continuously, both horizontally and vertically, while it dives.

What is required to sample an ocean and determine its structure and circulation as well as any changes that may be occurring? In the 1980s, most oceanographers would have said more time, more effort, more equipment, and more money than are available. But in the 1990s, a program was developed to do just that. The very large and very successful World Ocean Circulation Experiment (WOCE), a multi-institutional, multifaceted program, combined satellite data, buoys, direct measurement of currents and water properties, and the Global Positioning System. Data from this program greatly improved oceanographers' abilities to model the oceans and understand ocean-atmosphere interactions.

Figure 8.10 Retrieving a conductivity-temperature-depth sensor, or CTD. The CTD is attached below a rosette of water bottles. Data are relayed to the ship's electronic processing and data center. Water samples are taken when an interesting water structure is found or when water samples are required to calibrate the CTD.

In the spring of 2003, the National Oceanographic and Atmospheric Administration (NOAA) sponsored a cruise from Iceland to Madeira in the Atlantic Ocean and then on to Fortaleza, Brazil. This two-month trip repeated one of the sections of the WOCE program, allowing researchers to look for patterns of natural events that help to explain observed changes in oceanic structure and circulation since the WOCE cruise. Researchers also looked for patterns of natural events to guide them in planning future cruises.

To sample larger ocean areas than can be served by a single research ship and for continuous sampling over long periods of time, large, instrumented buoys are moored at sea and their data transferred by satellite or radio link to a research ship or laboratory for analysis. Other surface buoy systems are released to drift with the currents, monitoring water properties at changing locations and reporting back to ship or shore via satellite. Small buoys also drift with the surface currents, sink

Figure 8.11 Seasoar is a conductivity-temperature-depth sensor that is towed by a ship. The wings allow Seasoar to move vertically while it is towed horizontally.

Figure 8.12 The SLOCUM glider has multiple sensors designed to monitor water properties independently for up to five years. While at the surface it is able to send data and update its navigation by satellite.

to predetermined depths, and then return to the surface to drift again. These instruments monitor water properties as they rise to the surface and send this information back to a research vessel or a satellite.

Continual improvements in electronics, computers, and sensor systems have resulted in oceanographers devising independent and relatively inexpensive devices that are placed in the sea in large numbers. These independent, untethered instruments sink to predetermined depths, making measurements as they rise through the water column on their return to the surface. These data-gathering devices are known as profilers and carry a wide variety of sensors. The availability of these tools is beginning to provide the density of data needed to determine both long- and short-term trends in the oceans.

One type of profiler functions as a winged glider (fig. 8.12; see the box titled "Ocean Gliders"). When this profiler sinks and rises, it flies down and up at an angle, measuring properties along a diagonal over its operating depth. This flying ability allows the profiler to wander partially independent of the currents. The Argo profiler (see fig. 1.19) does not fly but sinks and rises vertically only. Lateral displacement is caused by the currents encountered over its vertical path. Once at the surface, both types of profilers broadcast their recorded data and positions to a satellite, then sink and repeat the process.

Specialized satellites measure a large variety of sea surface and atmospheric conditions, sea topography, wind speeds, plankton abundance, air and water temperatures, waves, and more. Satellites are very expensive monitoring devices, but their ability to survey the global oceans repeatedly and collect huge amounts of data makes them cost-effective as they collect data that cannot be obtained by any other means.

Not all marine studies require a large research vessel with its sophisticated and expensive equipment. Inshore, shallow-water studies usually use smaller vessels with less-sophisticated equipment. Small winches driven by power or even by hand are used to lower mechanical water bottles with their thermometers to the required depth. Lightweight electronic instruments are also available and include devices to measure conductivity, temperature, dissolved oxygen, optical properties, pH, and fluorescence of chlorophyll.

8.5 Practical Considerations: Ocean Thermal Energy Conversion

An indirect form of solar energy and an alternative to fossil fuel production of energy is found in the sea. The transparency and heat capacity of water allow large amounts of solar energy to be stored in the ocean, and this heat can be extracted independent of daily and seasonal changes in the available solar radiation.

Ocean thermal energy conversion (OTEC) depends on the difference in temperature between surface water and water at 600–1000 m (2000–3300 ft) depth. There are two types of OTEC systems: (1) closed cycle (fig. 8.13*a*), which uses a contained working fluid with a low boiling point, such as ammonia, and (2) open cycle (fig. 8.13*b*), which directly converts seawater to steam.

In a closed system, the warm surface water is passed over the evaporator chamber containing the ammonia, and the ammonia is vaporized by the heat derived from the warm seawater. The vapor builds up pressure in a closed system, and this gas under pressure is used to spin a turbine, which generates power. After the pressure has been released, the ammonia

Ocean Gliders

*A*n oceanographic research vessel can monitor only a small portion of an ocean during any one cruise, and the cost is very high (see the photo essay "Going to Sea"). To increase the size of the area that can be surveyed, as well as lower the cost, oceanographers have been experimenting with independent, unstaffed vehicles (see figs. 8.13 and 8.14). A new type of survey vehicle is an ocean glider, which functions in the water much as a conventional sailplane does in the atmosphere. Unlike a sailplane, which has to be towed aloft and ultimately falls back to the ground, an ocean glider has the ability to change its buoyancy so that it can glide up as well as down.

An oceanographic glider, like an airborne glider, has no propeller but is powered by its buoyancy. Buoyancy is changed by adjusting glider volume in order to sink or rise. The presence of wings translates this vertical force into motion along a slanting glide path. The limited onboard battery energy is devoted to buoyancy control, sampling sensors, navigation, and communication systems. Through energy efficiency, the glider remains operational for long periods, up to a year, and collects and reports large quantities of data.

At present, research groups at the Scripps Institution of Oceanography, the Woods Hole Oceanographic Institution, the Webb Research Corporation, and the University of Washington are developing gliders with the support of the U.S. Navy, the U.S. National Science Foundation, and the National Oceanic and Atmospheric Administration. Different glider designs are being used to address different ocean sampling needs, such as performing long open-ocean transects, maintaining geographic position while profiling vertically, or operating in shallow regions. One under development by Webb Research, known as the SLOCUM Glider, extracts energy from ocean thermal stratification to change buoyancy (see fig. 8.12). Others use battery energy for propulsion.

At the University of Washington, Seagliders are being developed and evaluated (box fig. 1). Seagliders are small (1.8 m [7 ft] long), light (52 kg [115 lb]) battery-powered gliders that can be easily launched and retrieved from small vessels rather than relying on research ships. They alternatively dive and climb through the water column as oil is transferred between an internal reservoir and an external bladder. When the glider is launched at the surface, its density is just slightly less than the density of seawater. A small amount of oil is bled from the external bladder to the inside of the pressure case, decreasing the bladder volume and increasing the density of the Seaglider, causing it to sink. When sinking motion starts, the wings on the Seaglider provide lift, and the glider moves forward as it sinks. When the vehicle reaches its programmed depth, an electric pump transfers oil from inside the pressure case to the external bladder, decreasing the vehicle density. As the less-dense glider rises toward the surface, the wings act to hold it down so that, it moves forward. Its forward speed is only about 0.25 m/s (0.5 knots),

Box Figure 1 A Seaglider being launched for a trial ocean run.

but because drag increases quadratically with speed, the glider has a range and endurance much greater than conventional propeller-driven autonomous underwater vehicles.

Seagliders carry sensors and recorders for temperature, salinity, dissolved oxygen, fluorescence, and optical backscatter. GPS navigation and telemetry equipment are also onboard; antennae are housed in the trailing wand (box fig. 2). At the surface, the glider determines its position using GPS and sends data via a global satellite phone link. Researchers are able to communicate with the glider and guide it under remote control.

To measure surface currents, the drifting glider can be tracked by GPS. When the glider surfaces from a dive, the difference between the glider's actual position and its intended position is the result of displacement by the average of the currents over both its sinking and rising paths.

Seagliders are designed to dive as deep as 1000 m (3000 ft) and cover as much as 10 km (5.3 nautical mi) horizontally in one dive cycle. In a six-month mission, they can travel as much as 5000 km (2700 nautical mi) through the ocean.

(continued)

---

is passed to a condenser, where it is cooled by cold water pumped up from depth. Cooling condenses the ammonia to the liquid state, and the liquid is pumped back to the evaporator to repeat the cycle.

In an open system, large quantities of warm seawater are converted to steam in a low-pressure vacuum chamber, and the steam is used as the working fluid. Because less than 0.5% of

the incoming water is turned into steam, large quantities of warm water must be used. The steam passes through a turbine and into a condenser cooled by cold water from depth and condenses to desalinated water.

OTEC plants using either system can be located onshore, offshore, or on a ship that moves from place to place. Figure 8.13*c* is an engineering concept design for an

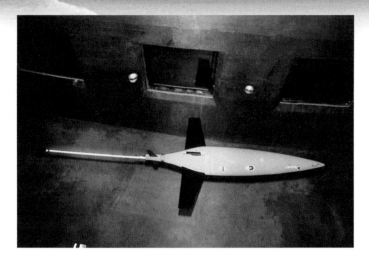

To Learn More About Ocean Gliders

Eriksen, C. C., T. J. Osse, R. D. Light, T. Wen, T. W. Lehman, P. L. Sabin, J. W. Ballard, and A. M. Chiodi. 2001. Seaglider: A Long Range Autonomous Underwater Vehicle for Oceanographic Research. *I.E.E.E. Journal of Oceanic Engineering,* 2001 26(4):421–23.

Box Figure 2   A Seaglider moving upward in a testing tank.

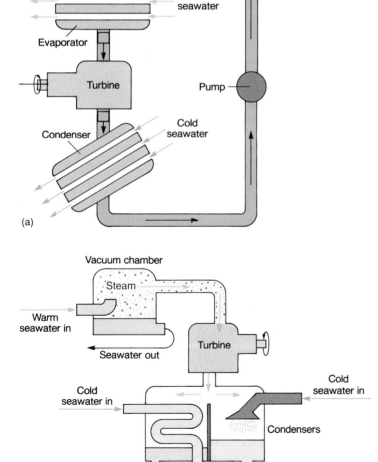

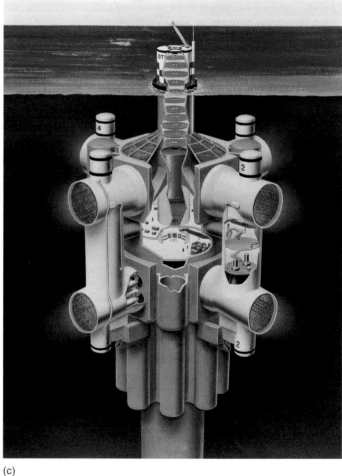

Figure 8.13   (a) The simplified working system of an ocean thermal energy conversion (OTEC) closed-system electrical generator. (b) The simplified working system of an OTEC open-system electrical generator. (c) The concept drawing for the Lockheed OTEC Spar.

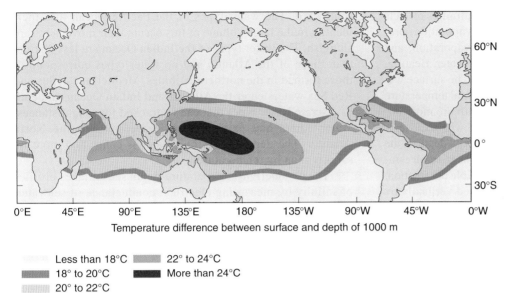

Temperature difference between surface and depth of 1000 m

Less than 18°C	22° to 24°C
18° to 20°C	More than 24°C
20° to 22°C	

Figure 8.14 OTEC plants are ideally suited to areas with a large temperature difference between the surface and deep waters (at least 20°C). The conditions are generally found in tropical and equatorial latitudes. Image from the National Renewable Energy Laboratory.

nutrient-poor tropical surface waters, changing the productivity of the surface water.

The Natural Energy Laboratory of Hawaii near Keahole Point operated a land-based, open-system OTEC plant for six years (1993–98). The project produced a net of 100 kW of power per day using the temperature difference between the water at the ocean surface and the water about 800 m (2500 ft) below the surface. The plant also produced 26,000 l (7000 gal) of desalinated water per day. Aquaculture projects using the nutrient-rich water from depth included production of oysters, shrimp, fish, black pearls, and various seaweeds.

Work on OTEC technology slowed in the late 1990s because it was too costly compared to generating energy using relatively inexpensive oil. The dramatic increase in the price of oil in recent years has resulted in a renewed interest in OTEC. In 2006 plans for building the world's two largest OTEC power plants in Hawaii were announced. One power plant is planned to be operational at the Natural Energy Laboratory of Hawaii Authority in Kona in 2008. This power plant will produce up to 1.2 megawatts, of which about one-third will be used to run pumps and keep the system going. That leaves a net production of 800 kilowatts. The second power plant is planned to be located at an undisclosed ocean location for military use. This plant will produce a net of 8.2 megawatts plus 1.25 million gallons of fresh water a day.

independent, free-floating OTEC plant. The Lockheed Spar has not been built, and there are no present plans to do so. OTEC requires at least a 20°C difference in temperature between surface and depth to generate useful amounts of energy. These power plants would be located at latitudes between approximately 25°N and 25°S where both warm surface water and cold deep water are available; seawater temperature difference over depth must average 22°C (fig. 8.14). W. H. Avery and C. Wu (*Renewable Energy from the Ocean, a Guide to OTEC,* Oxford University Press, New York, 1994) estimated that the area of the world's oceans meeting OTEC requirements is 60×10^6 km^2 (23×10^6 mi^2). The total power generated within this area would exceed 10×10^6 megawatts (MW); the total U.S. electricity generating capacity is about 1.7×10^6 MW.

Engineers studying the potential and feasibility of OTEC have estimated that about 0.2 MW of usable power can be extracted per square kilometer of tropical ocean surface; that is about 0.07% of the average absorbed solar energy. The process is considered safe and environmentally benign; however, the pumping of cold water from depth to the sea surface brings nutrients from depth and liberates them into the

Other ideas for using cold deep-ocean water, or DOW, pumped from 630 m (2000 ft) include air conditioning and industrial cooling systems. The condensate that forms on pipes carrying DOW is fresh water, formed at a rate of about 5% of the flow of the cold water. A flow of 76,000 l (20,000 gal) per minute generates an estimated 3800 l (1000 gal) per minute of fresh water. Lobsters, flat fish, shrimp, abalone, oyster, and other organisms have been raised in the tropics using DOW. Tropical agricultural experiments have also used DOW, pumping it through pipes placed at root depth in the soil. This system chills the soil to 10°C and produces freshwater condensate on the pipes and on the soil.

Summary

The absorption and exchange of energy at the sea surface govern the properties of surface seawater. Sea-surface exchanges of heat, radiant energy, and water alter the temperature and salinity of the surface water and affect the density of the water. Many combinations of salinity and temperature can produce seawater of the same density. When waters with different properties but the same density are mixed, the resulting water has a greater density than either of its components. The density-driven vertical circulation that results is known as caballing. The waters of different densities produced at the sea surface and the resulting vertical circulation create a layered ocean that is primarily stably stratified.

The geographical distribution of surface salinities reflects Earth's latitudinal and seasonal patterns of evaporation, precipitation, and sea-ice formation. Cooling, evaporation, and freezing increase the density of the sea-surface water. Heating, precipitation, and ice melt decrease its density. The surface water changes its density with changes in salinity and temperature that are keyed to latitude. The densities at depth are more homogeneous. Thermoclines and haloclines form where the temperature and salt concentrations change rapidly with depth. If the density increases with depth, the water column is stable; unstable water columns overturn and return to a stable distribution. Neutrally stable water columns are easily mixed vertically by winds and waves.

Vertical circulation driven by changes in surface density is known as thermohaline circulation. In the open ocean, temperature is generally more important than salinity in determining the surface density. Salinity is the more important factor close to shore and in areas of large seasonal ice melt.

Water sinks at downwellings and rises at upwellings. A downwelling occurs at the convergence of surface currents and transfers oxygen to depth. Upwellings bring nutrients to the surface and occur at zones of surface current divergence.

The oceans are layered systems. The layers (or water types) are identified by specific ranges of temperature and salinity. The water types of the Atlantic Ocean are formed at the surface at different latitudes. They sink and flow northward or southward. The water types of the Pacific Ocean lose their identity in the large volume of this ocean; their movements are sluggish. The water types of the Indian Ocean are less distinct than those of the Atlantic and no water types correspond to those formed in the northern latitudes. Mediterranean Sea and Red Sea waters enter the Atlantic and Indian Oceans at depth as discrete water types that can be tracked for long distances. Water enters and exits the Arctic Ocean from the North Atlantic. The density of Arctic water is controlled more by salinity than by temperature.

The conductivity-temperature-depth sensor takes profiles of salinity by measuring electrical conductance of seawater. Temperature is obtained with an electrical resistance thermometer, and depth is obtained with pressure sensors. Seasoar is a towed CTD modified to take samples when a ship is underway. To sample large areas, moored buoys, buoys that drift with the currents, and drifting instruments are used. Profilers sink to predetermined depths, making measurements as they rise. All are able to send their positions and data via satellite. Satellites also monitor the oceans directly. Shallow-water measurements can be made with mechanical water bottles and thermometers.

Ocean thermal energy conversion generates energy by using the difference in temperature between ocean surface water and deeper water. A land-based, open-system OTEC plant is operated in Hawaii.

Key Terms

All key terms from this chapter can be viewed by term or definition when studied as flashcards on this book's website at www.mhhe.com/sverdrup10e.

caballing, 201
pycnocline, 202
thermocline, 202
halocline, 202

stable water column, 202
unstable water column, 202
overturn, 202
isopycnal, 203
isothermal, 203
isohaline, 203
vertical circulation, 203
thermohaline circulation, 203
freshwater lid, 204

continuity of flow, 204
downwelling zones, 204
upwelling zones, 204
convergence, 204
divergence, 204
North Atlantic deep water (NADW), 205
Antarctic intermediate water (AAIW), 205

Antarctic bottom water (AABW), 205
South Atlantic surface water, 205
Mediterranean Intermediate Water (MIW), 207
ocean thermal energy conversion (OTEC), 212

Study Questions

1. Why does the mixing of two water types at the sea surface result in a mixed water type that sinks to a deeper depth?
2. What natural processes alter the surface salinity of the oceans? How do these processes vary with latitude?
3. Describe the changes in water density in the upper ocean layer over the annual cycles at tropical, polar, and temperate latitudes. Indicate when periods of stable and unstable conditions exist in the upper water column.
4. Why are water layers more prominent in the Atlantic Ocean than in the Pacific Ocean?
5. How do salt fingers form, and how do they contribute to vertical mixing?
6. How can heat be transferred from place to place in the oceans? Why is this heat transfer ignored when considering the world's total heat budget?
7. If the upper layers of an ocean area are homogeneous in salinity, explain why the thermocline coincides with the pycnocline.
8. How does Mediterranean Sea water alter the salinity and temperature of the North Atlantic? Use figures 8.6 and 8.7a and b. At what depth does this change occur?
9. A water sample taken at 4000 m in the Atlantic Ocean has a salinity of 34.8‰ and a temperature of 3°C. At approximately what latitude was this water last at the surface?

10. The following data were taken from a sampling station located at 79°N, 145°W:

Depth (meters)	Temp. (°C)	Salinity (‰)	Density g/cm³
0	1.28	33.29	1.02659
50	1.29	33.30	1.02659
100	1.36	33.35	1.02669
150	1.39	33.55	1.02694
200	2.73	33.76	1.02701
300	3.07	33.87	1.02708
400	3.12	34.03	1.02713
500	3.14	34.13	1.02721

a. In what ocean region is this station?

b. Is the water column stable or unstable?

c. Does the temperature or the salt content control the density?

d. How deep is the wind-mixed layer?

e. At what time of year were these data obtained?

Explain your answers.

11. Why is the surface of Earth not heated equally all over by the Sun's radiation?

12. What does a CTD measure, and why is the CTD the equipment of choice for routine oceanographic measurements?

13. Where do the deeper waters in the bottom of the Atlantic Ocean have their origins?

14. Distinguish between the terms in each pair:

a. halocline—thermocline

b. upwelling—downwelling

c. water mass—water type

15. Why is convective overturn in the sea's upper layers more likely to occur at temperate and high latitudes than at low latitudes?

Between Yemen and Somalia, the waters of the Gulf of Aden swirl in topographically squared off eddies which, in this SeaWIFS image, are made visible by the chlorophyll-bearing phytoplankton that they carry. This image was collected on November 1, 2003.

Chapter 9

The Surface Currents

Chapter Outline

Learning Outcomes

After studying the information in this chapter students should be able to:

1. *describe* and *sketch* the motion of water in the Ekman layer,
2. *diagram* the formation of surface current gyres,
3. *locate* the major surface currents on a map of the oceans,
4. *explain* the process of western intensification,
5. *relate* patterns of surface convergence and divergence with downwelling and upwelling, and
6. *sketch* the gross structure of combined wind-driven and thermohaline circulation in the oceans.

Earth is surrounded by two great oceans: an ocean of air and an ocean of water. Both are in constant motion, driven by the energy of the Sun and the gravity of Earth. Their motions are linked; the winds give energy to the sea surface and the currents are the result. The currents carry heat from one location to another, altering Earth's surface temperature patterns and modifying the air above. The interaction between the atmosphere and the ocean is dynamic; as one system drives the other, the driven system acts to alter the properties of the driving system.

In this chapter, we explore the formation of the ocean's surface currents. We follow these currents as they flow, merge, and move away from each other. We examine both horizontal and vertical circulation, inspect the coupling of these water motions, and consider the ways in which they are linked to the overall interaction between the atmosphere and the ocean.

9.1 Surface Currents

When the winds blow over the oceans, they set the surface water in motion, driving the large-scale surface currents in nearly constant patterns. The density of water is about 1000 times greater than the density of air, and once in motion, the mass of the moving water is so great that its inertia keeps it flowing. The currents flow more in response to the average atmospheric circulation than to the daily weather and its short-term changes; however, the major currents do shift slightly in response to seasonal changes in the winds. The currents are further modified by interactions between the currents and along zones of converging and diverging water. The major surface currents have been called the rivers of the sea; they have no banks to contain them, but they maintain their average course.

Because the frictional coupling between the ocean water and Earth's surface is small, the moving water is deflected by the Coriolis effect in the same way that moving air is deflected (see chapter 7). But because water moves more slowly than air, it takes longer to move water the same distance as air. During this longer time period, Earth rotates farther out from under the water than from under the wind. Therefore, the slower-moving water appears to be deflected to a greater degree than the overlying air. The surface-current acted upon by the Coriolis effect is deflected to the right of the driving wind direction in the Northern Hemisphere and to the left in

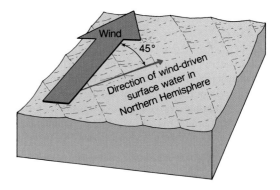

Figure 9.1 A wind-driven surface current moves at an angle of 45° to the direction of the wind; this angle is to the right in the Northern Hemisphere and to the left in the Southern Hemisphere.

the Southern Hemisphere. In the open sea, the surface flow is deflected at a 45° angle from the wind direction, as shown in figure 9.1.

The Ekman Spiral and Ekman Transport

Wind-driven surface water sets the water immediately below it in motion. But because of low-friction coupling in the water, this next deeper layer moves more slowly than the surface layer and is deflected to the right (Northern Hemisphere) or left (Southern Hemisphere) of the surface-layer direction. The same is true for the next layer down and the next. The result is a spiral in which each deeper layer moves more slowly and with a greater angle of deflection to the surface flow. This current spiral is called the **Ekman spiral,** after the physicist V. Walfrid Ekman, who developed its mathematical relationship. The spiral extends to a depth of approximately 100–150 m (330–500 ft), where the much reduced current will be moving in the opposite direction to the surface current. Over the depth of the spiral, the average flow of the water set in motion by the wind, or the net flow (**Ekman transport),** moves 90° to the right or left of the surface wind, depending on the hemisphere (fig. 9.2). This relationship is in contrast to the surface water, which moves at an angle of 45° to the wind direction.

Ocean Gyres

Refer to figure 9.3 as you read the description of surface currents in the major oceans. In the Northern Hemisphere, the wind-driven

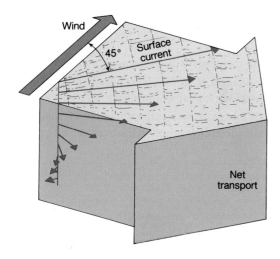

Figure 9.2 Water is set in motion by the wind. The direction and speed of flow change with depth to form the Ekman spiral. This change with depth is a result of Earth's rotation and the inability of water, due to low friction, to transmit a driving force downward with 100% efficiency. The net transport over the wind-driven column is 90° to the right of the wind in the Northern Hemisphere and 90° to the left in the Southern Hemisphere.

Ekman transport under the westerlies moves 90° to the right of the westerlies, away from an ocean's western shore or boundary, and along the 40°–50°N latitudes until it reaches its eastern shore or boundary. The Ekman transport under the trade winds moves 90° to the right of the trades, away from an ocean's eastern boundary, and along the 10°–20° latitudes until the current reaches the western boundary of the ocean. When Northern Hemisphere water moving with the trade winds accumulates at the land boundary on the west side of an ocean, the water flows north to the latitude of the westerlies and then eastward across the ocean. Water accumulating on the east side of a northern ocean flows south, toward the region from which the water moves westward under the trade winds. In Southern Hemisphere oceans, the east-west wind-driven Ekman transport is deflected 90° to the left of the trade winds and the westerlies. Water accumulating on the eastern side of a Southern Hemisphere ocean moves north, and water on the western side of a southern ocean moves south.

The Ekman transport causes an accumulation of water at the center of the circular flow pattern that results in an elevated convergence. As the elevation builds, the wind-driven surface flow moves more closely in line with the driving winds, and the surface current circulates around the convergence zone. In each hemisphere, this pattern produces continuous flow and a series of interconnecting surface currents moving in a circular path centered on 30° latitude. The rotation is clockwise in the Northern Hemisphere and counterclockwise in the Southern Hemisphere. These large, circular-motion, wind-driven current systems are known as **gyres.** In the more southern latitudes, there is no land between the Atlantic, Pacific, and Indian Oceans; here, the surface currents, driven by the westerlies, continue around Earth in a circumpolar flow around Antarctica.

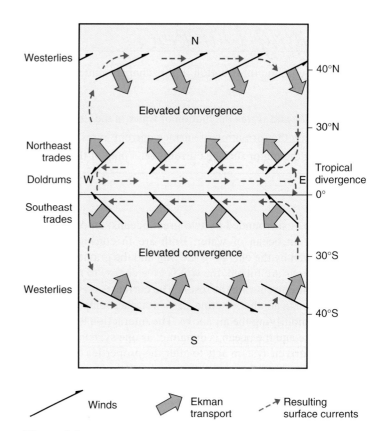

Figure 9.3 Wind-driven transport and resulting surface currents in an ocean bounded by land to the east and to the west. The currents form large oceanic gyres that rotate clockwise in the Northern Hemisphere and counterclockwise in the Southern Hemisphere.

Geostrophic Flow

If Ekman transport is applied to oceans with eastern and western land boundaries, a portion of the wind-driven surface water is deflected toward the center of each of the large, circular current gyres just described (fig. 9.3). A convergent lens of surface water is elevated more than 1 m (3 ft) above the equilibrium sea level, and this lens depresses the underlying denser water.

The thickness of the surface lens is about 1000 times greater than the elevation of the lens above sea level. This is because the difference in density between the surface water and the deeper water is only about 1/1000 of the density difference between air and water at the sea surface. The surface slope of the mound increases as deflected water moves inward until the outward pressure driving the water away from the gyre center equals the Coriolis effect, acting to deflect the moving water into the raised central mound. At this balance point, **geostrophic flow** is said to exist, and no further deflection of the moving water occurs. Instead, the currents flow smoothly around the gyre parallel to its elevation contours. See figure 9.4 for a diagram of this process. Using the subsurface water-density distribution to describe the extent of the depression of the deeper water, oceanographers are able to calculate the elevation and slope of the sea surface and so calculate the velocity, volume transport, and depth of the currents present in the geostrophic flow around the mound. It is also possible to measure the topography of the sea surface using satellites

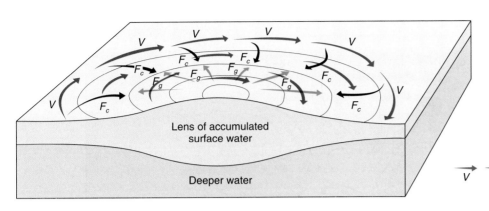

Figure 9.4 Geostrophic flow (*V*) exists around a gyre when *Fc,* the inward deflection force due to the Coriolis effect, is balanced by *Fg,* the outward-acting pressure force created by the elevated water and gravity. This example is of a clockwise gyre in the Northern Hemisphere.

and to calculate the geostrophic flow that maintains the topography. The region of the Sargasso Sea in the North Atlantic Ocean is the classic example of a gyre in geostrophic balance; it is discussed following the Atlantic currents in section 9.2.

9.2 Wind-Driven Ocean Currents

The currents that make up the large oceanic gyre systems and other major currents have been given names and descriptions based on their average positions. These are presented here ocean by ocean and can be followed on figure 9.5. As you follow these current paths, review their associations with the large gyre systems and their overlying wind belts.

Pacific Ocean Currents

In the North Pacific Ocean, the northeast trade winds push the water toward the west and northwest; this is the **North Equatorial Current.** The westerlies create the **North Pacific Current,** or **North Pacific Drift,** moving from west to east. Note that the trade winds move the water away from Central and South America and pile it up against Asia, while the westerlies move the water away from Asia and push it against the west coast of North America. The water that accumulates in one area must flow toward areas from which the water has been removed. This movement forms two currents: the **California Current,** moving from north to south along the western coast of North America, and the **Kuroshio Current,** moving from south to north along the east coast of Japan. The Kuroshio and California Currents are not wind-driven currents; they provide continuity of flow and complete a circular motion centered around 30°N latitude. This circular, clockwise flow of water is called the North Pacific gyre. Other major North Pacific currents include the **Oyashio Current,** driven by the polar easterlies, and the **Alaska Current,** fed by water from the North Pacific Current and moving in a counterclockwise gyre in the Gulf of Alaska. Little exchange of water occurs through the Bering Strait between the North Pacific and the Arctic Ocean; no current exists that is comparable to the Atlantic Ocean's Norwegian Current, which moves warm water to the Arctic Ocean.

In the South Pacific Ocean, the southeast trade winds move the water to the left of the wind and westward, forming the **South Equatorial Current.** The westerly winds push the water to the east; at these southern latitudes, the surface current so formed can move almost continuously around Earth. This current is the **West Wind Drift.** The tips of South America and Africa deflect a portion of this flow northward on the east sides of the South Pacific and South Atlantic Oceans. As in the North Pacific, continuity currents form between the South Equatorial Current and the West Wind Drift. The **Peru Current,** or **Humbolt Current,** flows from south to north along the coast of South America, while the **East Australia Current** can be seen moving weakly from north to south on the west side of the ocean. These four currents form the counterclockwise South Pacific gyre.

The North Pacific and South Pacific gyres form on either side of 5°N because the meteorological equator or doldrums belt is displaced northward from the geographic equator (0°), owing to the unequal heating of the Northern and Southern Hemispheres. Also between the North and South Equatorial Currents, in the zone of the doldrums is a current moving in the opposite direction, from west to east. This is a continuity current known as the **Equatorial Countercurrent,** which helps to return accumulated surface water eastward across the Pacific. Under the South Equatorial Current is a subsurface current flowing from west to east called the **Cromwell Current.** This cold-water continuity current also returns water accumulated in the western Pacific.

Atlantic Ocean Currents

The North Atlantic westerly winds move the water eastward as the **North Atlantic Current,** or **North Atlantic Drift.** The northeast trade winds push the water to the west, forming the **North Equatorial Current.** The north-south continuity currents are the **Gulf Stream,** flowing northward along the coast of North America, and the **Canary Current,** moving to the south on the eastern side of the North Atlantic. The Gulf Stream is fed by the **Florida Current** and the North Equatorial Current. The North Atlantic gyre rotates clockwise. The polar easterlies provide the driving force for the **Labrador** and **East Greenland Currents,** which balance water flowing into the Arctic Ocean from the **Norwegian Current.**

In the South Atlantic, the westerlies continue the West Wind Drift. The southeast trade winds move the water to the west, but the bulge of Brazil deflects part of the **South Equatorial Current** northward into the Caribbean Sea and eventually into the Gulf of Mexico, where it exits as the Florida Current and joins the Gulf Stream. A portion of the South Equatorial Current moves south of the Brazilian bulge along the western side of the South Atlantic

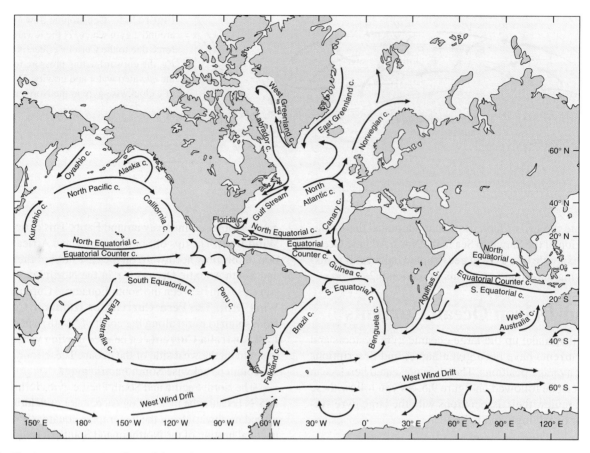

Figure 9.5   The long-term average flow of the major surface currents of the oceans.

to form the **Brazil Current.** The **Benguela Current** moves northward along the African coast. The South Atlantic gyre is complete, and it rotates counterclockwise.

Because much of the South Equatorial Current is deflected across the equator, the Equatorial Countercurrent appears only weakly in the eastern portion of the mid-Atlantic. The northward movement of South Atlantic surface water across the equator results in a net flow of surface water from the Southern Hemisphere to the Northern Hemisphere. This flow is balanced by a flow of water at depth from the Northern Hemisphere to the Southern Hemisphere. This deep-water return flow is the North Atlantic deep water, discussed in chapter 8. Again, the equatorial currents are displaced northward, although not as markedly as in the Pacific Ocean.

The Sargasso Sea marks the middle of an ocean gyre. It is located in the central North Atlantic Ocean, and its boundaries are the Gulf Stream on the west, the North Atlantic Current to the north, the Canary Current on the east, and the North Equatorial Current to the south. The circular motion of the gyre currents isolates a lens of clear, warm, downwelling water 1000 m (3000 ft) deep. The region is famous for the floating mats of *Sargassum,* a brown seaweed, stretching across its surface. The extent of the floating seaweed frightened early sailors, who told stories of ships imprisoned by the weed and sea monsters lurking below the surface. Except for the floating *Sargassum,* with its rich and specialized ecological community, the clear water is nearly a biological desert.

Indian Ocean Currents

The Indian Ocean is mainly a Southern Hemisphere ocean. The southeast trade winds push the water to the west, creating the **South Equatorial Current.** The Southern Hemisphere westerlies still move the water eastward in the West Wind Drift. The gyre is completed by the **West Australia Current** moving northward and the **Agulhas Current** moving southward along the east coast of Africa. Because this is a Southern Hemisphere ocean, the currents are deflected left of the wind direction, and the gyre rotates counterclockwise. The northeast trade winds in winter drive the **North Equatorial Current** to the west, and the **Equatorial Countercurrent** returns water eastward toward Australia. Again, these equatorial currents are displaced approximately 5°N. With the coming of the wet monsoon season and its west winds, these currents are reduced. The strong seasonal monsoon effect controls the surface flow of the Northern Hemisphere portion of the Indian Ocean. In the summer, the winds blow the surface water eastward, and in the winter, they blow it westward. This strong seasonal shift is unlike anything found in the Atlantic or the Pacific Ocean.

Arctic Ocean Currents

The relentless drift of water and ice in the Arctic Ocean moves in a large clockwise gyre driven by the polar easterly winds. This gyre is centered not on the North Pole, as early explorers

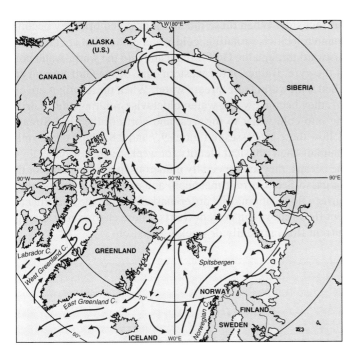

Figure 9.6   The circulation in the Arctic Ocean is driven by the polar easterlies, which produce a large, clockwise gyre. Water enters the Arctic Ocean from the North Atlantic by way of the Norwegian Current and exits to the Atlantic by the East Greenland Current and the Labrador Current.

expected, but is offset over the Canadian basin at 150°W and 80°N (fig. 9.6). Although the currents and the winds move the ice slowly at 0.1 knot (2 mi/day), Arctic explorers trying to reach the North Pole found that they traveled south with the drifting ice and water at speeds almost equal to their difficult progress north.

The Arctic Ocean is supplied from the North Atlantic by the Norwegian Current; some of this flow enters west of Spitsbergen, but most flows along the coast of Norway and moves eastward along the Siberian coast into the Chukchi Sea. A small inflow of water entering the Arctic through the Bering Strait brings water from the Bering Sea to join the eastward flow along Siberia and the large Arctic gyre. The western side of the gyre crosses the center of the Arctic Ocean to split north of Greenland. Here, the larger flow forms the East Greenland Current flowing south and taking Arctic Ocean water into the North Atlantic. The lesser flow moves along the west side of Greenland to join the Labrador Current and move south along the Canadian coast.

Outflow from Siberian rivers is caught in the eastward flow of water and ice along Siberia. Eventually, this discharge joins the gyre, distributing sediments and pollutants throughout the Arctic (see the box in chapter 8 titled "Arctic Ocean Studies").

9.3 Current Flow

Current Speed

Wind-driven open-ocean surface currents move at speeds that are about 1/100 of the wind speed measured 10 m (30 ft) above the sea surface. The water moves between 0.25 and 1.0 knot, or

0.1–0.5 m (0.3–1.5 ft) per second. Currents flow faster when a large volume of water is forced to flow through a narrow gap. For example, the North and part of the South Atlantic Equatorial Currents flow into the Caribbean Sea, then into the Gulf of Mexico, and finally exit to the North Atlantic as the Florida Current through the narrow gap between Florida and Cuba. The Florida Current's speed may exceed 3 knots, or 1.5 m (5 ft) per second. Once into the Atlantic Ocean this current turns north and becomes the Gulf Stream.

The flow is distributed over the width and depth of the current. When the cross-sectional area of the current expands, the current slows down; when the cross-sectional area decreases, the current speeds up. Speed of flow may not be directly related to surface wind speed but can be affected by the depth and width of the current as determined by land barriers, by the presence of another current, or by the rotation of Earth, as explained in the "Western Intensification" section of this chapter.

Current Volume Transport

Major ocean currents transport enormous volumes of water. A convenient unit to report transport volume is the Sverdrup (Sv) (named after Harald Sverdrup, a leading oceanographer of the last century and former Director of the Scripps Institution of Oceanography). A Sverdrup equals 1 million cubic meters ($\sim 3.5 \times 10^7$ ft^3) per second. The transport rate of fresh water in all of the world's rivers into the ocean is about 1 Sv. Transport rates of ocean currents are difficult to measure accurately and can vary by both location in the current and time of the year. The Gulf Stream transports about 30 Sv passing through the Strait of Florida as the Florida Current. This increases steadily as it moves north along the coast until it transports about 80 Sv near Cape Hatteras. The transport of the Gulf Stream continues to increase downstream of Cape Hatteras at a rate of 8 Sv every 100 km, reaching a maximum transport of about 150 Sv at 55°W. The downstream increase in transport between Cape Hatteras and 55°W is thought to be caused by increased velocities in the deep waters of the Gulf Stream. The current transports a maximum amount of water in the fall and a minimum in the spring.

Western Intensification

In the North Atlantic and North Pacific, the currents flowing on the western side of each ocean tend to be much stronger, deeper, and narrower in cross section than the currents on the eastern side. This phenomenon is known as the **western intensification** of currents. The Gulf Stream and Kuroshio Currents are faster and narrower than the Canary and California Currents, although both the eastern and western boundary currents transport about the same amount of water to preserve continuity of flow around the gyres. Western intensification of currents traveling from low to high latitudes is related to (1) the eastward turning of Earth, (2) the increase in the Coriolis effect with increasing latitude, (3) the changing strength and direction of the east-west wind field (trade winds and westerlies) with latitude, and (4) the friction between land masses and ocean water currents. These factors cause a compression of the currents on the western side of the oceans, where water is moving from lower to higher latitudes. This compression requires that the current speed increase to transport the water circulating about the gyre. On the eastern side of the gyre, where currents are moving from higher to lower latitudes, the currents are stretched in the east-west direction. Here, the current's speed is reduced, but it still transports the required volume of water. The changing speed of flow around the gyre causes the Coriolis effect to vary. Where the current speed is high, the Coriolis effect is large, and a steeper surface slope is required to create a geostrophic flow balance.

Fast-flowing, western-boundary currents move warm equatorial surface water to higher latitudes. Both the Gulf Stream and the Kuroshio Current bring heat from equatorial latitudes to moderate the climates of Japan and northern Asia (in the case of the Kuroshio) and the British Isles and northern Europe (in the case of the Gulf Stream, via the North Atlantic and Norwegian Currents). Western intensification is obscured in the South Pacific and South Atlantic, because both Africa and South America deflect portions of the West Wind Drift and create strong currents on the eastern side of these oceans. The deflection of water from the Atlantic's South Equatorial Current to the Northern Hemisphere removes water from the South Atlantic gyre and strengthens the Gulf Stream. The flow of surface water from the Pacific to the Indian Ocean through the islands of Indonesia also helps to prevent the development of strongly flowing currents on the west side of Southern Hemisphere oceans.

9.4 Eddies

When a narrow, fast-moving current moves into or through slower-moving water, the force of its flow displaces the quieter water and captures additional water as it does so. The current oscillates and develops waves along its boundary that are known as meanders. These meanders break off to form **eddies,** or pockets of water moving with a circular motion; eddies take with them energy of motion from the main flow and gradually dissipate this energy through friction. Eddies also act to mix and blend water.

As the Gulf Stream moves away from the North American coast, it is likely to develop a meandering path. The western edge of the Gulf Stream develops oscillations, and the indentations are filled by cold water from the Labrador

Current side. When these indentations pinch off, they become counterclockwise-rotating, cold-water eddies that are displaced eastward through the Gulf Stream and into the warm-water core of the gyre. Bulges at the western edge of the Gulf Stream are filled with warm Gulf Stream water. When these bulges are cut off, they become warm-water, clockwise-rotating eddies drifting into cold water to the west and north of the Gulf Stream. Follow these processes in figures 9.7 and 9.8. These eddies may maintain their physical identity for weeks as they wander about the oceans; they are especially numerous in the area north of the Sargasso Sea. Current meanders and eddy formation produce surface flow patterns that differ markedly from the uniform current flows shown on current charts. These charts show average current flow, not daily or weekly variations.

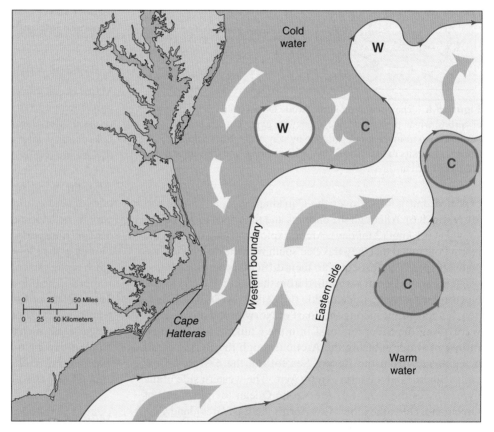

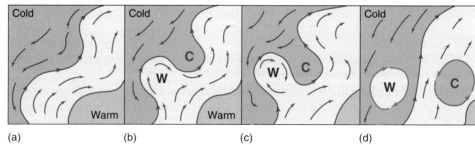

Figure 9.7   The western boundary of the Gulf Stream is defined by sharp changes in current velocity and direction. Meanders form at this boundary after the Gulf Stream leaves the U.S. coast at Cape Hatteras. The amplitude of the meanders increases as they move downstream (a and b). In time, the current flow pinches off the meander (c). The current boundary re-forms, and isolated rotating cells of warm water (*W*) wander into the cold water, while cells of cold water (*C*) drift through the Gulf Stream into the warm water (d).

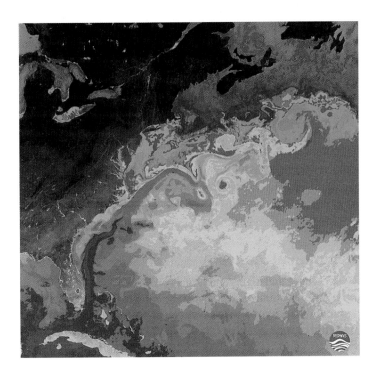

Figure 9.8 A composite satellite image of the sea surface reveals the warm (*orange* and *yellow*) and cold (*green* and *blue*) eddies that form along the Gulf Stream. (*Reddish blue* areas at the top are the coldest waters.) These eddies may stir the water column right down to the ocean floor, kicking up blizzards of sediment.

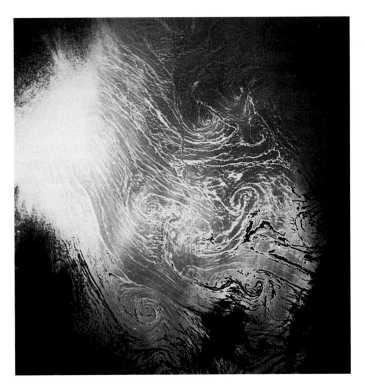

Figure 9.9 Space shuttle view. Sunlight reflected off the Mediterranean reveals spiral eddies; their effects on climate are being monitored. Dimensions of this image are 500 km × 500 km (310 mi × 310 mi).

Large and small eddies generated by horizontal flows or currents exist in all parts of the oceans; these eddies are of varying sizes, ranging from 10 to several hundred kilometers in diameter. Each eddy contains water with specific chemical and physical properties and maintains its identity and rotational inertia as it wanders through the oceans. Eddies may appear at the sea surface or be embedded in waters at any depth.

Eddies rotate in a clockwise or counterclockwise direction. They stir the ocean until they gradually dissipate because of fluid friction, losing their chemical and thermal identity and their energy of motion. By testing the water properties of an eddy, oceanographers are able to determine the eddy's place of origin. Small surface eddies encountered 800 km (500 mi) southeast of Cape Hatteras in the North Atlantic have been found with water properties of the eastern Atlantic near Gibraltar, more than 4000 km (2500 mi) away. Eddies from the Strait of Gibraltar are formed in the salty water of the Mediterranean as it sinks and spreads out into the Atlantic 500–1000 m (1600–3300 ft) down; these eddies have been nicknamed "Meddies." Deep-water eddies near Cape Hatteras may come from the eastern and western Atlantic, the Caribbean, or Iceland. Researchers estimate that some of these eddies are several years old; age determination is based on drift rates, distance from source, and biological consumption of oxygen.

The rotational water speed in the large eddies that form at the western boundary of the Gulf Stream is about 0.51 m/s (1 knot), but because of the water's density, the force of the flow is similar to that generated by a 35-knot wind. The diameter of the eddies may be as much as 325 km (200 mi), and their effect may reach to the sea floor. At the sea floor, the rotation rate is zero; therefore, a few meters above the bottom the speed of rotation diminishes very rapidly, and considerable turbulence is generated as the energy of the eddy is dissipated. These eddies are similar in many ways to the winds rotating about atmospheric pressure cells, and they are sometimes called abyssal storms. As the eddies wander through the oceans, they stir up bottom sediments, producing ripples and sand waves in their wakes; they also mix the water, creating homogeneous water properties over large areas. Eventually, the eddies lose their energy to turbulence and blend into the surrounding water.

Eddies constantly form, migrate, and dissipate at all depths. Eddy motion is superimposed on the mean flow of the oceans. To understand the role of eddies in mixing the oceans, we need more data and better tracking of eddy size, position, and rate of dissipation. Satellites are important tools for detecting surface eddies because they can precisely measure temperature, increased elevation, and light reflection of the sea surface (fig. 9.9). Deep-water eddies are monitored by using instruments designed to float at a mid-depth density layer. The instruments are caught up in the eddies, moving with them and sending out acoustic signals that are monitored through the sofar channel. In this way, the rate of deep-water eddy formation, the numbers of major eddies, their movements, and their life spans can be observed.

9.5 Convergence and Divergence

Changes in density and the accompanying concepts of up-welling, downwelling, convergence, divergence, and continuity of flow were introduced in section 8.2 of chapter 8. The convergence and divergence zones discussed in this chapter are the product of wind-driven surface currents that produce large-scale areas of convergence and divergence at the sea surface. For example, convergence zones are at the centers of the large oceanic gyres, and when wind-driven surface currents collide or are forced against landmasses, they produce convergences. When surface currents move away from each other or away from a landmass, they produce a surface divergence. Upwellings and downwellings of this type are nearly permanent but do react to seasonal changes in Earth's surface winds.

Langmuir Cells

A strong wind blowing across the sea surface often causes streaks of foam and surface debris that are seen trailing off in the direction the wind is blowing. These streaks are called windrows and may be 100 m (330 ft) in length. Windrows mark the convergence zones of shallow circulation cells known as **Langmuir cells.** These cells are composed of paired right- and left-handed helixes (fig. 9.10). The spacing between the windrows varies between 5 and 50 m (16 and 160 ft), and the rows are closer together if the thermocline depth is shallow. The vertical extent of these Langmuir cells is 4–10 m (12–30 ft). The distance between rows becomes larger at higher wind speeds. Langmuir cells are not long-lasting but help to mix the surface waters and organize the distribution of suspended organic matter in convergence and divergence zones. Sinking particles and organisms tend to congregate at depth in the regions of rising currents, while floating particles and organisms accumulate at the surface where currents converge and descend.

Permanent Zones

Convergence and divergence zones on an ocean scale are shown in figures 8.6 and 9.11. There are five major zones of convergence: the **tropical convergence** at the equator and the two **subtropical convergences** at approximately 30°–40°N and S. These convergences mark the centers of the large ocean gyres. The **Arctic** and **Antarctic convergences** are found at about 50°N and S. Surface convergence zones are regions of downwelling. These areas are low in nutrients and biological productivity. There are three major divergence zones: the two **tropical divergences** and the **Antarctic divergence.** Upwelling associated with divergences delivers nutrients to the surface waters to supply the food chains that support the anchovy and tropical tuna fisheries and the richly productive waters of Antarctica.

In coastal areas where trade winds move the surface waters away from the western side of continents, upwelling occurs nearly continuously throughout the year. For example, the trade winds drive upwellings off the west coasts of Africa and South America that are very productive and yield large fish catches.

Seasonal Zones

Off the west coast of North America, downwelling and upwelling occur seasonally as the Northern Hemisphere temperate wind pattern changes from southerly in winter to northerly in summer. The downwelling and upwelling occur because of the change in direction of Ekman transport (fig. 9.12). Remember that the wind-driven Ekman transport moves at an angle of 90° to the right or left of the wind direction, depending on the hemisphere.

Along this coast, the average wind blows from the north in the summer, and

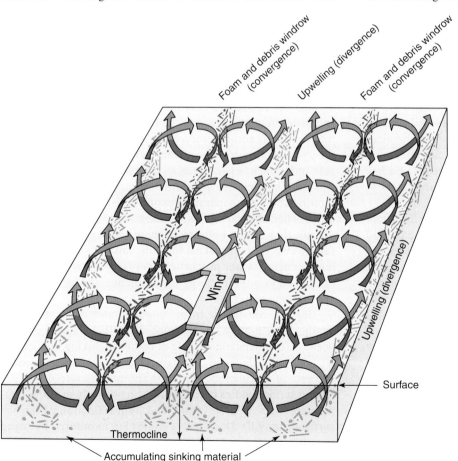

Figure 9.10 Langmuir cells are helixes of near-surface, wind-driven water motion. The cells extend downwind to form windrows of surface debris along convergences. At the same time, sinking materials are swept into zones of upwelling water.

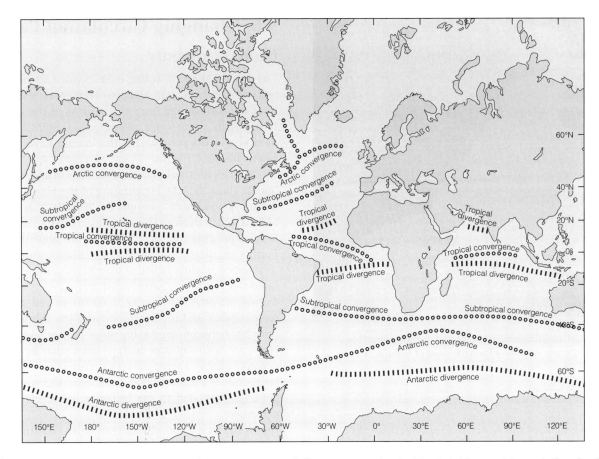

Figure 9.11 The principal zones of open-ocean surface convergence and divergence associated with wind-driven and thermohaline circulation.

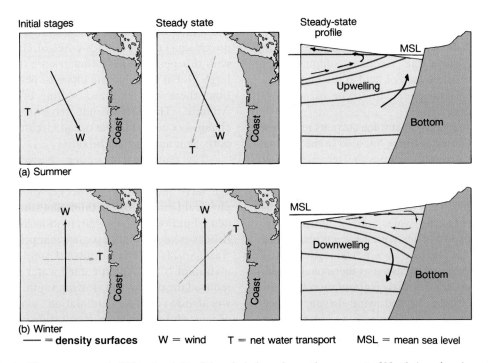

Figure 9.12 Initially, the Ekman transport is 90° to the right of the wind along the northwest coast of North America. As water is transported (a) away from the shore in summer or (b) toward the shore in winter, a sea surface slope is produced. This slope creates a gravitational force that alters the direction of the Ekman transport to produce a geostrophic balance and a new steady-state direction.

Figure 9.13 Coastal upwelling (shades of *purple*) appears as patches of cold water extending from the coast of California.

the net movement of the water is to the west, or 90° to the right of the wind. This action results in the offshore transport of the surface water and the upwelling of deeper water along the coast to replace it. The upwelling zone is evident from central California to Vancouver Island (fig. 9.13). In winter, the winds blow from the south; the wind-driven surface waters move to the east, onshore against the coast; and downwelling occurs (fig. 9.14; see also fig. 7.23). The summer upwelling pattern is what produces the band of cold coastal water at San Francisco that helps cause the frequent summer fogs (see chapter 5). Because of the lack of land at the middle latitudes in the Southern Hemisphere, this type of seasonal upwelling is less common.

Convergence and divergence of surface currents result not only in upwellings and downwellings but also in the mixing of water from different geographic areas. Waters carried by the surface currents converge, share properties due to mixing, and form new water mixtures with specific ranges of temperature and salinity, which then sink to their appropriate density levels. After sinking, the water moves horizontally, blending and sharing its properties with adjacent water, and eventually, it rises to the surface at a new location. This process, known as caballing, is discussed as thermohaline circulation in the section on density-driven circulation in chapter 8. Thermohaline circulation and wind-driven surface currents are closely related. The connections are so close among formation of water mixtures, upwelling and downwelling, and convergence and divergence that it is difficult to assign a priority of importance to one process over another. Changes in these flows and events that might cause such changes are discussed in section 9.6.

9.6 Changing Circulation Patterns

Global Currents

The interchange of water over depth and between oceans redistributes heat, salt, and dissolved gases; use figure 9.15 to trace the surface and deep-water flows of the world's oceans. Is it possible that some phenomenon could alter this water motion and trigger changes that would have major climatic consequences?

Studies of fossil marine organisms in cores of marine sediments show that the temperature of the North Atlantic Ocean during glacial and interglacial periods can be related to Milankovitch cycles. These cycles are associated with regular long-term changes in Earth's orbit and the relation of Earth's axis to the Sun, producing variations in the amount and distribution of solar radiation over Earth's surface. Twenty-three thousand years ago, Earth had its period of minimum solar radiation. About 6000 years later, Earth endured its coldest ocean temperatures and its maximum land ice volumes. Notice the time lag between decreasing solar energy and Earth cooling. However, sediment-core analysis shows that the warming of the oceans to today's temperatures occurred rapidly with high solar radiation values 12,000 years ago. The warming, like the cooling, is related to the Milankovitch cycles of Earth-Sun placement.

Water is a good absorber of solar radiation, but ice is a poor absorber and a good reflector, so the land ice melted slowly as the oceans warmed rapidly. Eleven thousand years ago, the surface temperatures of the water in the North Atlantic and the air temperatures in northern Europe decreased suddenly; the sudden decrease was unrelated to the long-term solar cycles. The temperatures remained low for about 700 years in a mini-ice age; large changes (20%) in atmospheric carbon dioxide gas trapped in ice cores of Greenland are associated with this period. These changes are interpreted to mean that large global atmospheric changes occurred, and at the same time, there was localized cooling of Europe and the North Atlantic. This could not have happened unless major changes occurred in the ocean circulation system that transports heat and carbon dioxide.

The required connections between global atmospheric changes and ocean circulation during the time of this mini-ice age have led Wallace Broecker of Lamont Doherty Geophysical Observatory to think that something triggered a sudden shutdown of the circulation of the North Atlantic. Broecker suggests that some event prevented the transport of warm surface water northward and interfered with the formation of North Atlantic deep water and in so doing greatly reduced the deep water flowing south. Under such conditions, worldwide oceanic circulation would change, and the oceans' ability to transport and store carbon dioxide would be altered. A change in the carbon dioxide storage capacity of the oceans requires a corresponding change in atmospheric carbon dioxide levels.

Broecker has proposed that the trigger for such an episode was a sudden change in the location where meltwater from the

January Global Wind-induced Upwelling (cm/day)

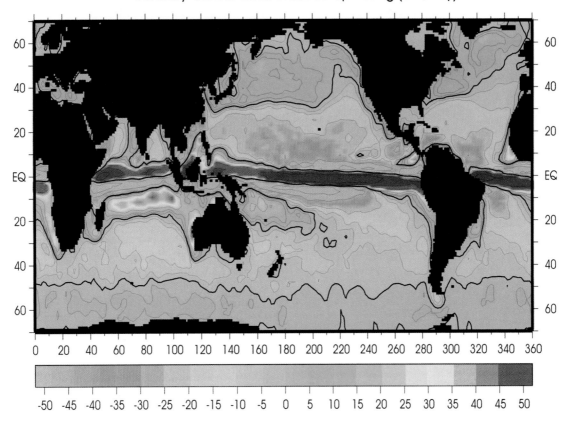

July Global Wind-induced Upwelling (cm/day)

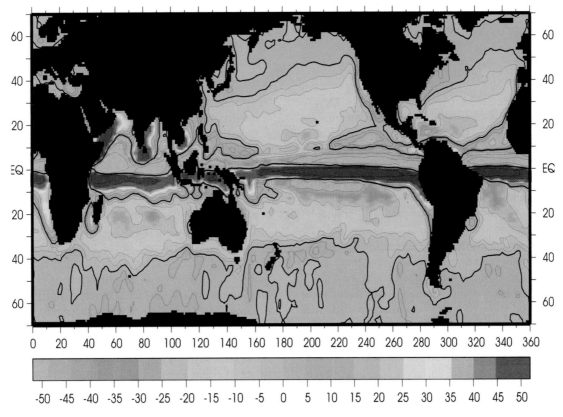

Figure 9.14 Surface winds produce surface divergences and convergences that create areas of upwelling and downwelling. Average surface wind stress values were used to produce this model of global upwelling and downwelling for January (a) and July (b). Negative values indicate downwelling. Speed of vertical motion is given in centimeters per day.

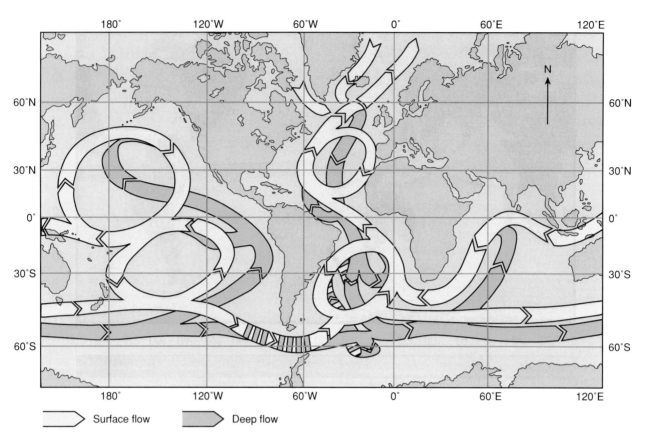

Figure 9.15 The combined wind-driven and thermohaline flow of ocean water within and between oceans circulates at the surface (*yellow arrows*) and at depth (*green arrows*). Cold surface water sinks to the sea floor in the North Atlantic Ocean, then flows south to be further cooled by the Antarctic Bottom water formed in the Weddell Sea. This deep water moves eastward around Antarctica, feeding into the surface layer of the Indian Ocean and into the deep basins of the Pacific Ocean. A return flow of surface water from the Pacific and Indian Oceans moves north to replace the surface water in the North Atlantic Ocean.

receding North American ice sheet entered the Atlantic Ocean. Large quantities of cold, low-density, low-salinity water entering the oceans would significantly affect the salinity and density of the North Atlantic's surface waters. Geologic studies support Broecker's ideas. We know that the majority of the meltwater from the early stages of the receding glaciers in North America converged on the Mississippi River system and drained into the Gulf of Mexico, not directly into the Atlantic. But gradually, as the ice margin receded, meltwater was channeled through the Hudson River drainage system, the Great Lakes and the St. Lawrence River valley, and then into the North Atlantic. The result was a diluting and cooling of the Atlantic surface water.

At the same time, the low-salinity surface water did not sink, and, therefore, the production of North Atlantic deep water was slowed, reducing the northward flow of warm Gulf Stream water and affecting the circulation in all oceans. This event ceased as rapidly as it had started. As the melting of the glaciers slowed, the influx of meltwater was reduced. Surface warming and evaporation increased surface water density, and the formation of North Atlantic deep water resumed. The circulation of the North Atlantic began a rapid recovery.

Sediment cores obtained by the Ocean Drilling Program have allowed measurements of (1) oxygen isotopes to determine the water temperature and (2) the salt content of water found between the grains of age-dated sediments. This research indicates that during the last glacial maximum (LGM), the deep waters of the Pacific, Southern, and Atlantic Oceans were about the same temperature but that variations in salinity existed over time between the oceans. This data indicate that during the LGM, the density stratification in the deep ocean was salinity-controlled and not temperature-controlled, as it is today. During the LGM, the deep water with the greatest salinity was formed in the Southern Ocean; this salinity control of deep-ocean density suggests the freshwater budget at the poles was also different at that time. It is possible that a "salt switch" exists that abruptly changes the mode of deep-water circulation when sufficient fresh water is suddenly added or withdrawn from the sea at high latitudes. Understanding what happens to cause such changes in deep-water circulation is a key to understanding how climate changes relate to ocean conditions.

North Pacific Oscillations

W. James Ingraham, Jr., an oceanographer at NOAA's Seattle laboratory, has developed a North Pacific Current model, or

Ocean Drifters

*I*n May 1990, a severe storm in the North Pacific caused the container ship *Hansa Carrier,* en route from Korea to the United States, to lose overboard twenty-one deck-cargo containers, each approximately 40 ft long (box fig. 1). Among the items lost were 39,466 pairs of Nike brand athletic shoes, and these began washing up along the beaches of Washington, Oregon, and British Columbia six months to a year later. Although the shoes had been drifting in the ocean for almost a year, they were wearable after washing and having the barnacles and the oil removed. However, the shoes of a pair had not been tied together for shipping, and pairs did not come ashore together. As beach residents recovered the shoes (some with a retail value of $100 a pair), swap meets were held in coastal communities to match the pairs.

In May 1991, I was having lunch with my parents, Gene and Paul, and Gene pointed out the news of the beached Nikes. I was intrigued and realized that 78,932 shoes was a very large number of drifting objects, compared with the 33,869 drift bottles used in a 1956–59 study of North Pacific currents. I contacted Steve McLeod, an Oregon artist and shoe collector who had information on locations and dates for some 1600 shoes that had been found between northern California and the Queen Charlotte Islands in British Columbia. I contacted additional beachcombers and constructed a map showing the times and locations where batches of 100 or more shoes had been found (box fig. 2). Next, I visited Jim Ingraham at NOAA's National Marine Fisheries Service's offices in Seattle to study his computer model of Pacific Ocean currents and wind systems north of 30°N latitude. Using the spill date (May 27, 1990), the spill location (161°W; 48°N), and the dates of the first shoe landings

Box Figure 1   After the storm, the *Hansa Carrier* docked in Seattle.

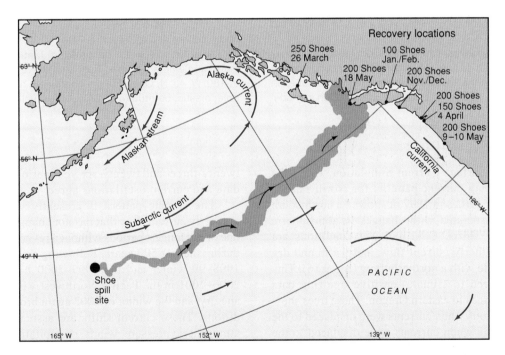

Box Figure 2   Site where 80,000 Nike shoes washed overboard on May 27, 1990, and dates and locations where 1300 shoes were discovered by beachcombers (*dots at upper right*). Drift of the shoes was simulated with a computer model (*colored plume*).

(continued)

on Vancouver Island and Washington beaches between Thanksgiving and Christmas 1990, I found that the shoe drift rates agreed with the computer model's predicted currents.

News of Jim's and my interest in the shoe spill reached an Oregon news reporter, and the story was then picked up by the Associated Press, resulting in a quick dissemination of the news nationwide. Readers sent letters describing their own shoe finds, and even reports of single shoes were valuable because we found that each shoe had within it a Nike purchase order number that could be traced to a specific cargo container. We were able to determine from these numbers that only four of the five containers broke open so that only 61,820 shoes were left afloat.

The computer model and previous experiments with satellite-tracked drifters showed that there would have been little scattering of the shoes as the ocean currents carried them eastward and approximately 1500 miles from the spill site to shore, but the shoes were found scattered from California to northern British Columbia. The north-south scattering is related to coastal currents that flowed northward in winter, carrying the shoes to the Queen Charlotte Islands, and southward in spring and summer, bringing the shoes to Oregon and California.

I was interested to see where the shoes might have gone if they had been lost on the same date but under different conditions in other years. The computer allowed simulations for May 27 of each year from 1946–91. Box figure 3 shows the wide variation in model

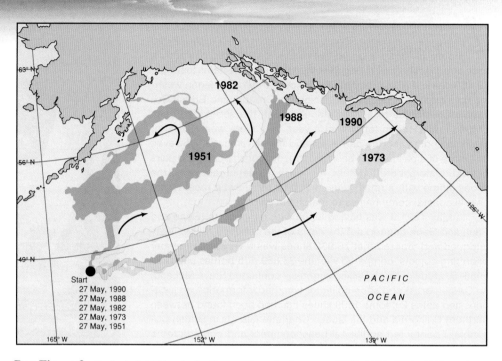

Box Figure 3 Projected drift tracks for the sneakers for the years 1951, 1973, 1982, 1988, and 1990, based on computer modeling of ocean currents and weather.

predicted drift routes. If the shoes had been lost in 1951, they would have traveled in the loop of the Alaska Current. If they had been lost in 1982, they would have been carried far to the north during the very strong El Niño of 1982–83, and if lost in 1973, they would have come ashore at the Columbia River.

Another spill event occurred in January 1992, when twelve cargo containers were lost from another vessel in the North Pacific at 180°W, 45°N. One of these containers held 29,000 small, floatable, plastic, bathtub toys. Blue turtles, yellow ducks, red beavers,

---

North Pacific Ocean Surface Current Simulation. When the model was used to map current patterns for North Pacific surface water between 1902 and 1997, it showed a north-south current oscillation associated with changes in atmospheric pressure and climate shifts. Cold and wet conditions are associated with a southerly current flow, and warm and dry conditions predominate with a northerly flow (fig. 9.16). This evidence was compared to climate-sensitive tree-ring data from western juniper trees in eastern Oregon. These trees show wide rings during periods when currents were displaced to the north and narrow rings when currents were displaced to the south. Tree-ring data cover many more years than oceanographic and meteorological data, and from the tree rings, it is calculated that thirty-four north-south oscillations have occurred since the time of Columbus. The most common time period between fast and slow growth is seventeen years, with

twenty-three- and twenty-six-year periods common. Tree-ring data and current oscillations agree.

The climate pattern is presently warm and dry with a northerly current flow that has not changed since 1967. This is one of the longest periods without a reversal that has been found during the last 500 years. Following the wet conditions of the 1998–99 winter and the 1999–2000 La Niña, the winter of 2000–2001 in the Pacific Northwest was exceptionally warm and dry, and the winter of 2002 again brought wet El Niño conditions. These current shifts are associated with changes in atmospheric pressure, winds, precipitation, and water temperature; together, they are known as the Pacific Decadal Oscillation, or PDO. Because the PDO affects coastal surface temperatures from California to Alaska, it is thought to affect the survival of fish stocks of the area, especially salmon. Commercial fishing is discussed further in chapter 17.

Box Figure 4 Shoes and toys after their rescue from Pacific cargo spills.

and green frogs began arriving on beaches near Sitka, Alaska, in November 1992. Advertisements asking for news of toy strandings were placed in local newspapers and the Canadian lighthouse keepers' newsletter. Beachcombers reported a total of about 400 of these toys (box fig. 4).

None of the toys was found south of 55°N latitude, suggesting that once the toys reached the vicinity of Sitka, they drifted to the north. This time, the computer model showed that if the toys had continued to float with the Alaska Current, they would have moved with the Alaska stream through the Aleutian Islands into the Bering Sea. Some may have continued across the Bering Sea into the Arctic Ocean to the vicinity of Point Barrow and from there would have drifted north of Siberia with the Arctic pack ice. Eventually, some plastic turtles, ducks, beavers, or frogs may have come to rest on the coast of western Europe. If the toys had turned south, they might have merged with the Kuroshio Current and been carried past the location where they were spilled.

Two thousand five hundred cases of hockey gloves (34,300 gloves) were lost from a container ship in the North Pacific after a December 1994 fire. In August 1995, a fishing vessel found seven gloves 800 miles west of the Oregon coast, and by January 1996, the barnacle-covered gloves began to arrive on Washington beaches. The most northerly glove sighting came from Prince William Sound, Alaska, in August 1996. The gloves are expected to follow the tub toys along the coast of Alaska and into the Arctic. More than 4 million Lego pieces lost off the English coast in 1997 are expected to be distributed along Northern Hemisphere shores by 2020.

And one more time, as it is said, what goes around comes around. On December 15, 2002, a ship carrying cargo containers on its way from Los Angeles to Tacoma, Washington, ran into 25-ft seas off Cape Mendocino in northern California. The vessel rolled and several cargo containers fell overboard. A cargo container of 33,000 Nike athletic shoes was among those lost, and by mid-January 2003, the shoes were coming ashore along the Washington coast. Riding the Davidson Current northward, the shoes had traveled about 833 km (450 nautical miles). Once again, the pairs of shoes were not tied together, and notices were posted in the search for mates.

Curtis C. Ebbesmeyer, Ph.D.
Seattle, Washington

To Learn More About Ocean Drifters

Krajick, K. 2001. Message in a Bottle. *Smithsonian* 32 (4): 36–47.

North Atlantic Oscillations

A large pool of cold, low-salinity surface water appeared off Greenland, north of Iceland, in 1968. It was about 0.5‰ less salty and 1° and 2°C colder than usual. Within two years, this cool pool had moved west into the Labrador Sea off eastern Canada; then it crossed the Atlantic and, in the mid-1970s, moved north into the Norwegian Sea. It had returned to its place of origin by the early 1980s. Follow the path of this pool in figure 9.17. During this period, harsh winters plagued Europe, and the entire Northern Hemisphere had cooler-than-average temperatures for more than ten years.

Recent studies have further investigated the North Atlantic, seeking to improve our ability to predict climate. The new analysis shows pools of warm and cool surface water that circle the North Atlantic; the pools seem to have life spans of four to ten years. Because of the alternation of climatic conditions, this is referred to as the North Atlantic Oscillation (NAO). Investigators are finding pieces of the puzzle that are associated with NAO but have not yet found the factors that control the system. A counterclockwise wind circulation centered over Iceland and a high-pressure clockwise circulation residing near the Azores are the usual situation. If the air-pressure differences between these two locations are large, then strong westerly winds supply Europe with heat from the North Atlantic Current. If the air-pressure difference decreases, then weaker-than-normal westerlies drive less warm water into the Norwegian Current and less heat is delivered to Europe. The periods of 1950–71 and 1976–80 were recognized as prolonged cool periods for Europe. The winds may also drive cold, low-salinity water from the Arctic into the area where North Atlantic deep water is formed. The influx of this water may cause a small-scale reduction in the

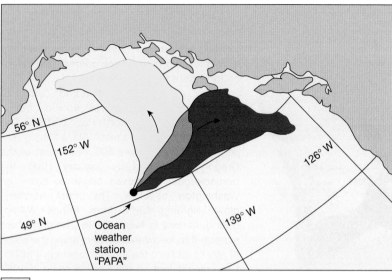

Figure 9.16 North-south current shifts in the Northeast Pacific are associated with changes in atmospheric pressure, winds, precipitation, and water temperature. This phenomenon is known as the Pacific Decadal Oscillation (PDO).

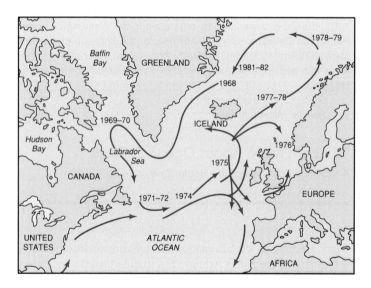

Figure 9.17 *Purple arrows* follow the path of the North Atlantic cool pool. *Red arrows* show warm-water flow from the Gulf Stream.

formation of North Atlantic deep water and an accompanying reduction of warm surface water moving northward. How much freshening is required to change ocean circulation? This question is unanswered, but if the current melting of Arctic sea ice continues, more information may soon be available.

Natural cycles operate at a variety of time scales and magnitudes, and they interact and react in ways we do not fully comprehend. In the enormously complex interactions that they are observing, oceanographers and meteorologists are beginning to identify the strands that united the ocean-atmosphere and the world current systems. Scientists use mathematical equations

assembled into a model to describe interactions between the oceans and the atmosphere. Supercomputers are needed to make the millions of calculations that a model requires to predict changes in environmental conditions over a given time period. This process is repeated many times to predict conditions. Models are verified and tuned by adjusting the equations so that the predicted model changes closely agree with changes that have been observed in the past. Modeling is nearly as much an art as a science. Sometimes predictions work; sometimes they don't. It depends on how well the model has been conceived, how much data are available, and the capability of the computers to process the data. The model may be constructed to approximate the whole ocean-atmosphere system or to apply only to part of Earth. Eventually, long-term predictions concerning the oceans and the atmosphere will become possible; how we use such information and whether it will benefit ocean resources are not yet clear.

9.7 Measuring the Currents

Direct measurements of currents fall into two groups: (1) those that follow a parcel of the moving water and (2) those that measure the speed and direction of the water as it passes a fixed point. Moving waters may be followed with buoys designed to float at predetermined depths. These buoys signal their positions acoustically to a research vessel or shore station; their paths are followed and their speed and displacement due to the current are calculated. Autonomous profilers such as the Argo drifter (see fig. 8.14) also measure currents. Surface water may be labeled with buoys or with dye that can be photographed from the air. Buoy positions may also be tracked by satellites using GPS. A series of pictures or position fixes may be used to calculate the speed and direction of a current from the buoys' drift rates. Buoys can be instrumented to measure other water properties such as temperature and salinity (fig. 9.18).

Figure 9.18 Retrieving an oceanographic research buoy in the Arabian Sea (northwest Indian Ocean). This buoy is equipped with weather, radiation, and water property sensors.

(a)

(b)

Figure 9.19 (a) An internally recording Aanderaa current meter. The vane orients the meter to the current while the rotor determines current speed. (b) A Doppler current meter sends out sound pulses in four directions. The frequency shift of the returning echoes allows the detection of the current. These meters are also equipped with salinity-temperature-depth sensors as well as instruments for measuring water turbidity and oxygen. The data may be stored internally and collected at another time.

A variation of this technique uses **drift bottles.** Thousands of sealed bottles, each containing a postcard, are released at a known position. When the bottles are washed ashore, the finders are requested to record the time and location of the find and return the card. In this case, only the release point, the recovery points, and the elapsed time are known; the actual path of motion is assumed. See the box titled "Ocean Drifters."

Sensors used to measure current speed and direction at fixed locations are called **current meters.** The current meters used by oceanographers over the last twenty years include a rotor to measure speed and a vane to measure direction of flow (fig. 9.19*a*). If the current meter is lowered from a stationary vessel, the measurements can be returned to the ship by a cable or stored by the meter for reading upon retrieval. If the current meter is attached to an independent, bottom-moored buoy system, the signals can be transmitted to the ship as radio signals or stored on tape in the meter, to be removed when the buoy and meter are retrieved.

To measure a current at a location, a current meter must not move. Although a vessel can be moored in shallow water so that it does not move, it is very difficult, if not impossible, to moor a ship or surface platform in the open sea so that it will not move and thus move the current meter. The solution is to attach the current meter to

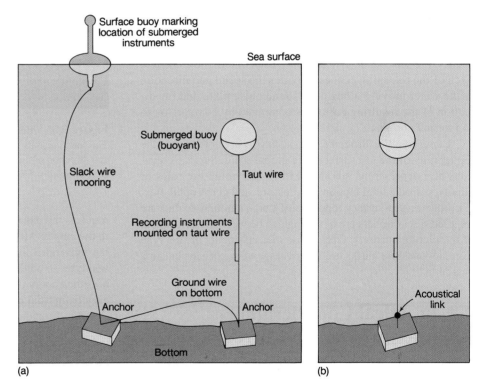
(a)   (b)

Figure 9.20 Taut-wire moorage. (a) Recovery is accomplished by retrieving the surface buoy and hauling in the wire. If the surface buoy is lost, it is possible to grapple for the ground wire. (b) In this system, a sound signal disconnects the anchor, and the equipment floats to the surface.

a buoy system that is entirely submerged and not affected by winds or waves. See figure 9.20 for a diagram of this taut-wire moorage system. The string of anchors, current meters, wire, and floats is preassembled on deck and is launched, surface float first, over the stern of the slowly moving ship. As the ship moves away, the float, meters, and cable are stretched out on the surface. When the vessel reaches the sampling site, the anchor is pushed overboard to pull the entire string down into the water. The floats and meters are retrieved by grappling from the surface for the ground wire or by sending a sound signal to a special acoustical link (fig. 9.20*b*), which detaches the wire from the anchor. The anchor is discarded, and the buoyed equipment returns to the surface. The technique is straightforward, but many problems can occur in launching, finding, and retrieving instruments from the heaving deck of a ship at sea. Whenever oceanographers send their increasingly sophisticated equipment over the side, they must cross their fingers and hope to see it again.

A new technique for measuring currents does not need the energy of the moving water to run the rotor of a current meter. This technique uses sound pulses and takes advantage of the change in the pitch, or frequency, of sound as it is reflected from particles suspended in the moving water. When sound is reflected off particles moving toward the meter, the pitch increases; when the sound is reflected off particles moving away from the meter, the pitch decreases. This is the **Doppler effect;** the same effect increases the pitch of the horn or siren of an approaching vehicle and decreases the pitch as the vehicle passes and moves away. To use this effect, the sound source is mounted on a vessel or a buoy system, or placed on the sea floor. A seafloor-mounted Doppler current meter is shown in figure 9.21. Four beams of sound pulses, set at a precisely known frequency, are sent out at right angles to each other. The change in pitch of the returning echoes provides the speed and direction of the water moving along each sound pulse path, and the direction of the resulting current is computed by comparison to an internal compass.

Using satellite altimeter data, scientists are able to map the topography of the sea surface on a global scale. Surface elevations and depressions are analyzed to determine the roles of gravity, periodic tidal motion, air pressure, and geostrophic flow in producing sea-surface topography. Large amounts of data are required if researchers are to distinguish between the assortment of interacting currents. Oceanwide measurements of this type were not possible until satellite coverage of the oceans became available.

9.8 Practical Considerations: Energy from the Currents

The massive oceanic surface currents of the world are untapped reservoirs of energy. Their total energy flux has been estimated at 2.8×10^{14} (280 trillion) watt-hours. Because of their link to winds and surface heating processes, the ocean currents are considered as indirect sources of solar energy. If the total energy of a current was removed by conversion to electrical power, that current would cease to exist; but only a small portion of any ocean current's energy can be harnessed, owing to the current's size. Harnessing the energy from these open-ocean currents

Figure 9.21 Retrieving a Doppler current meter from the sea floor. The sound signals from the meter are analyzed to determine the speed and direction of the current. *Courtesy of Mark Holmes, School of Oceanography, University of Washington.*

requires the use of turbine-driven generators anchored in place in the current stream. Large turbine blades would be driven by the moving water, just as windmill blades are moved by the wind; these blades could be used to turn generators and to harness the energy of the water flow. (See also the discussion on energy from tidal currents in chapter 11.)

The Florida Current and the Gulf Stream are reasonably swift and continuous currents moving close to shore in areas where there is a demand for power. If ocean currents are developed as energy sources, these currents are among the most likely. But most of the wind-driven oceanic currents generally move too slowly and are found too far from where the power is needed. In addition, the impact on other uses of the sea—transport, fishing, recreation—needs to be considered. The cost of constructing, mooring, and maintaining current-driven power-generating devices in the open sea makes them noncompetitive with other sources of power at this time.

Summary

Winds push the surface water 45° to the right of their direction in the Northern Hemisphere and 45° to the left in the Southern Hemisphere. Wind moves the water in layers that are deflected by the Coriolis effect to form the Ekman spiral; net flow over the depth of the spiral is deflected 90°. Geostrophic flow is produced when the force of gravity balances the Coriolis effect. Large surface gyres are observed in each ocean. Northern Hemisphere gyres rotate clockwise, and Southern Hemisphere gyres rotate counterclockwise. The currents of the northern Indian Ocean change with the seasonal monsoons.

Large oceanic current systems have names and descriptions based on their average locations. The water transport and speed of a current are affected by the current's cross-sectional area, by other currents, by westward intensification, and by wind speed. Eddies are formed at the surface when a fast-moving current develops waves along its boundary that break off from the parent current. Eddies occur at all depths, wander long distances, and gradually lose their identity.

Downwelling is produced by converging surface currents, and upwelling is produced by diverging surface currents.

Upwelling and downwelling may be shallow and short-lived, as in Langmuir cells, or these processes may involve large volumes and large areas of the oceans. Upwelling occurs nearly continuously along the western sides of the continents in the trade-wind belts, where surface water diverges from the coast. Seasonal upwellings and downwellings occur in coastal areas that have changing wind patterns and an alternating coastal flow of water onshore and offshore due to the Ekman transport.

Cyclic global circulation changes are part of Earth's normal dynamic system. Sudden changes in global ocean circulation may lead to major climate changes and are thought to be triggered by localized events in the North Atlantic. Both the North Pacific and the North Atlantic show decadal oscillations in current flow and climate.

A variety of techniques are available to measure currents: by following the water, by measuring the water's speed and direction as it moves past a fixed point, or by using changes in the frequency of sound.

Deriving energy from the oceanic current flows is not practical at the present time.

Key Terms

All key terms from this chapter can be viewed by term, or by definition when studied as flashcards on this book's website at www.mhhe.com/sverdrup10e.

Ekman spiral, 219
Ekman transport, 219
gyre, 220
geostrophic flow, 220
western intensification, 223
eddy, 224

Langmuir cell, 226
tropical convergence, 226
subtropical convergence, 226
Arctic convergence, 226
Antarctic convergence, 226

tropical divergence, 226
Antarctic divergence, 226
drift bottle, 235
current meter, 235
Doppler effect, 236

Study Questions

1. According to the net transport of the Ekman spiral, wind-driven water is directed toward the center of a large oceanic current gyre. Why does the current not flow to the gyre's center but instead flows in a clockwise circular path about a gyre in the Northern Hemisphere?
2. How is wind-driven Ekman transport related to coastal upwelling and downwelling?
3. On a map of the world, plot the oceanographic equator, the six major wind belts, the current system of each ocean, and the main areas of surface convergence and divergence.
4. Why does a flow of water that is constant in volume transport per time increase its speed when it passes through a narrow opening?

5. Explain how wind and current directions are specified.
6. What are eddies? How are they formed? Where can they be found?
7. Explain why the large Northern Hemisphere mid-ocean gyres tend to flow in a circular path, although the driving force of the winds is to the east on the higher-latitude side of the gyre and to the west on the lower-latitude side.
8. Explain why surface convergences are located at the centers of the large subtropical gyres in both the northern and southern Atlantic and Pacific Oceans.
9. Explain why upwellings on the lee side of continents at subtropical latitudes function continuously while upwelling on the west side of North America is seasonal.

10. Why is there a net northward flow of surface water across the equator in the Atlantic Ocean but not in the Pacific Ocean?

11. How have tree-ring data helped describe climate and ocean current fluctuations in the North Pacific?

12. How can the cross section of a current and its average speed be used to calculate the volume of water transport?

13. Why is there a net flow of surface water from the Pacific through the Indian to the Atlantic Ocean?

14. If the production of North Atlantic deep water were severely reduced, what changes would you expect to find in (a) Gulf Stream flow, (b) upwelling in the Pacific Ocean, (c) world sea-surface temperatures?

15. How are sound and the Doppler effect used to measure ocean currents?

A wind-blown wave breaking as it nears the shore.

Chapter

10

The Waves

Chapter Outline

Learning Outcomes

After studying the information in this chapter students should be able to:

1. *describe* the process of wave formation, including wave generating and restoring forces,
2. *label* the basic characteristics of a wave, including wave crest, trough, height, and wavelength,
3. *define* wavelength, wave period, wave frequency, and wave steepness,
4. *diagram* wave-induced water motion as a function of depth for both deep- and shallow-water waves,

5. *characterize* deep- and shallow-water waves,
6. *calculate* the speed of deep- and shallow-water waves given wave period, wavelength, wave frequency, and water depth,
7. *review* the factors controlling maximum potential wave height,
8. *diagram* the processes of wave refraction, diffraction, and reflection,
9. *discuss* the formation of tsunamis and *report* the history and characteristics of the 2004 Indian Ocean tsunami, and
10. *explain* the formation and properties of internal and standing waves.

We have all seen water waves. This up-and-down motion occurs on the surface of oceans, seas, lakes, and ponds, and we speak of waves, ripples, swells, breakers, whitecaps, and surf. Sometimes the waves are related to the wind, or to a passing ship, or even to a stone thrown into the water. Waves may swamp a small boat or, when large, smash and twist the bow of a supertanker. The storm waves of winter crash against our coasts, eating away the shore and damaging structures, such as docks and breakwaters. Heavy wave action may force commercial vessels to slow their speed and lengthen their sailing time between ports, to the inconvenience and increased cost of the shippers. Special waves associated with seismic disturbances have killed thousands of coastal inhabitants and severely damaged their cities and towns. Surfers search for the perfect wave, and ancient peoples navigated by the patterns that waves form.

In this chapter, we describe different kinds of waves, investigate their origin and behavior, and study their major characteristics. Our study is somewhat superficial and simplified, because waves are among the most complex of the ocean's phenomena and because no two waves are exactly the same. Books are written about waves; mathematical models have been devised to explain waves; wave tanks are built to study waves. However, there is much we can learn if we follow the sequence of events beginning at sea where a wave is born and ending at last in the spray and surf of a faraway shore.

10.1 How a Wave Begins

Imagine that you are standing on the beach looking out across a perfectly smooth surface of water. The production or generation of waves requires a disturbing force. You throw a stone into the water or a large, naturally occurring landslide comes down into the water. A pulse of energy is introduced, and waves are produced; this disturbing force is called a **generating force.** The waves produced by the generating force move away from the point of disturbance. Of course, the magnitude of the disturbance and the resulting wave size are very different in these two examples.

In the case of the thrown stone, the stone strikes the surface of the water and displaces, or pushes aside, the water. As the stone sinks, the displaced water flows from all sides into the space left behind, and as this water rushes back, the water at the center is forced upward. The elevated water falls back, causing a depression of the surface that is refilled, starting another cycle. This process sets up a series of waves, or oscillations, at the air-sea boundary that move outward and away from the point of disturbance until they are dissipated through friction among the water molecules.

In this example, the wave-generating force is the stone as it strikes the surface of the water. The force that causes the water to return to its undisturbed surface level is the **restoring force.** If the stone thrown into the water is small, the waves are small, and the restoring force is the surface tension of the water surface. (Surface tension, or the elastic quality of the surface due to the cohesive behavior of the water molecules, is discussed in chapter 5.) Very small water waves are affected by surface tension.

When the landslide is the generating force, much larger waves are pulled back to the undisturbed surface level of the water by the restoring force of gravity. When waves are of sufficiently large size so that the restoring force of Earth's gravity is more important than surface tension, the waves are called **gravity waves.**

The most common generating force for water waves is the moving air, or wind. As the wind blows across a water surface, the friction, or drag, between the air and the water tends to stretch the surface, resulting in wrinkles; surface tension acts on these wrinkles to restore a smooth surface. The wind and the surface tension create small waves, called **ripples,** or **capillary waves** (fig. 10.1). Patches of these very small waves are seen forming, moving, and disappearing as they are driven by pulses of wind. These patches darken the surface of the water and move quickly, keeping pace with the gusts of wind; sailors call these fast-moving patches **cat's-paws.** These waves die out rapidly while new ripples form constantly in front of each moving wind gust.

When the wind blows, energy is transferred to the water over large areas, for varying lengths of time, and at different

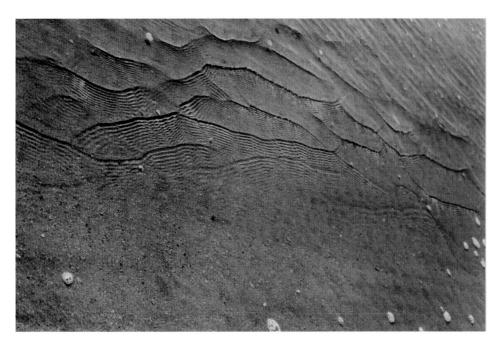

Figure 10.1 Short wavelength capillary waves in front of ordinary gravity waves on a beach. The gravity waves are the larger waves and are propagating "down" toward a region of shallow water. The capillary waves are the very short waves propagating in front of the larger waves. (Photo taken by Fabrice Neyret.)

Figure 10.2 Wind-generated gravity waves at sea.

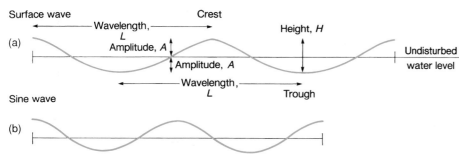

Figure 10.3 The profile of an ideal sea surface wave (a) differs from the shape of a sine wave (b).

rates. As waves form, the surface becomes rougher, and it is easier for the wind to grip the roughened water surface and add energy. As wind speeds increase or winds blow for longer times over the water, the waves become larger, and the restoring force changes from surface tension to gravity (fig. 10.2).

10.2 Anatomy of a Wave

In any discussion of waves, certain terms are used; refer to figure 10.3 for help in understanding these terms. The part of the wave that is elevated the highest above the undisturbed sea surface is called the **crest;** the part that is depressed the lowest below the surface is called the **trough.** The distance between two successive crests or two successive troughs is the length of the wave, or its **wavelength.** The **wave height** is the vertical distance from the top of the crest to the bottom of the trough. Sometimes the term **amplitude** is used. The amplitude is equal to one-half the wave height, or the distance from either the crest or the trough to the undisturbed water level, or **equilibrium surface.** The oceanographer characterizes a wave not only by its length and height (or amplitude) but also by its **period.** The period is the time required for two successive crests or troughs to pass a point in space. If you are standing on a piling and start a stopwatch as the crest of a wave passes and then stop the stopwatch as the crest of the next wave passes, the measured time is the period of the wave.

The dimensions and characteristics of waves vary greatly, but the regularity in the rise and fall of the water surface and the relationship between wavelength and wave period allow mathematical approximations to be made, which give us insight into the behavior and properties of waves. These calculations are done by relating real waves to simple model waves. Figure 10.3*a* has been drawn to show that real waves tend to have a trough that is flatter than the crest; by contrast, figure 10.3*b* is that of a symmetrical sine wave. This regular wave, which only approximates a water wave in nature, is one of the wave forms used by physical oceanographers and mathematicians to explore and explain wave motion. The relationships presented in sections 10.3 through 10.7 are based on this regular sine-wave form.

10.3 Wave Motion

As a wave form moves across the water surface, particles of water are set in motion. Seaward, beyond the surf and breaker zone, where the surface undulates quietly, the water is not moving toward the shore. Such an ocean wave does not represent a flow of water but instead represents a flow of motion or energy from its origin to its eventual dissipation at sea or against the land. To understand what is happening during the passing of a wave, let us follow the motion of the water particles as the wave moves through the water.

As the wave crest approaches, the surface water particles rise and move forward. Immediately under the crest the particles have stopped rising and are moving forward at the speed of the crest. When the crest passes, the particles begin to fall and to slow in their forward motion, reaching a maximum falling speed and a zero forward speed when the midpoint between crest and trough passes. As the trough advances, the particles slow their falling rate and start to move backward, until at the bottom of the trough they have attained their maximum backward speed and neither rise nor fall. As the remainder of the trough passes, the water particles begin to slow their backward speed and start to rise again, until the midpoint between trough and crest passes. At this point, the water particles start their forward motion again and continue to rise with the advancing crest. This motion (rising, moving forward, falling, reversing direction, and rising again) creates a circular path, or **orbit,** for the water particles as the wave passes. Follow this motion in figure 10.4.

It is the orbital motion of the water particles that causes a floating object to bob, or move up and down, forward and backward, as the waves pass. This motion affects a fishing boat, swimmer, seagull, or any other floating object on the surface seaward of the surf zone. The surface water particles trace an orbit with a diameter equal to the height of the wave. This same type of motion is transferred to the water particles below the surface, but less energy of motion is found at each succeeding depth. The diameter of the orbits becomes smaller and smaller as depth increases. At a depth equal to one-half the wavelength, the orbital motion has decreased to almost zero; notice how the orbit decreases in diameter in figure 10.4. Submarines dive during rough weather for a quiet ride because the wave motion does not extend far below the surface.

The orbits just described are based on the sine wave (see fig. 10.3b), not on real sea waves (see fig. 10.3a). The water

particles of actual sea waves move in orbits whose forward speed at the top of the orbit is slightly greater than the reverse speed at the bottom of the orbit. Therefore, each orbit made by a water particle does move the water slightly forward in the direction the waves travel. This movement is due to the shape of the real waves, whose crests are sharper than their troughs; again see figure 10.3a. This difference means that a very slow transport of water in the direction of the waves occurs in nature, but this motion is ignored in calculations based on simple wave models.

10.4 Wave Speed

A wave's speed across the sea surface is related to its wavelength and wave period. The speed of any surface wave *(C)* is equal to the length of the wave *(L)* divided by the period *(T):*

$$\text{speed} = \frac{\text{length of wave}}{\text{wave period}}$$

or

$$C = \frac{L}{T}$$

Once a wave is created, the speed at which the wave moves may change, but *its period remains the same* (period is determined by generating force).

In wave studies, *C* stands for *celerity,* a term traditionally used to identify wave speed and to distinguish wave speed from the group speed of waves. Group speed is usually identified by *V* (see section 10.5).

10.5 Deep-Water Waves

"Deep water" has a precise meaning for the oceanographer studying waves. To be a **deep-water wave,** the wave must occur in water that is deeper than one-half the wave's length. Under this condition, the water particle orbits of the wave do not reach the sea floor. For example, a wave 15 m (50 ft) long must occur in water that is deeper than 7.5 m (26 ft) to be considered a deep-water wave and to behave as the waves described next.

The wavelength *(L)* of deep-water waves is derived from the wave period *(T),* and, because wavelength *(L)* is a function of wave period *(T),* wave speed *(C)* is also derived from wave period *(T).* The oceanographer at sea determines the wave period *(T)* by direct measurement and calculates the wavelength *(L)* from its relationship with gravity and wave period *(T).* In deep water, the wavelength is equal to the Earth's acceleration due to gravity *(g)* divided by 2π times the square of the wave period *(T).* The value of Earth's gravity, *g,* is 9.81 m/s².

$$L = \frac{g}{2\pi}T^2 \qquad \text{or} \qquad L = 1.56 \ m/s^2 T^2$$

This deep-water wave equation, when combined with the general wave speed

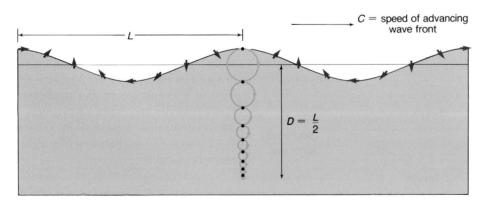

Figure 10.4 The moving wave sets the water particles in motion. *Red arrows* indicate the direction a water particle at the surface moves as the wave propagates to the right (in the direction of wave propagation at the crest and in the opposite direction at the trough). The diameter of a water particle's orbit at the surface is determined by wave height. Below the surface, the diameter decreases and orbital motion ceases at a depth *(D)* equal to one-half the wavelength.

equation $C = L/T$, is used to determine wave speed (C) from either wavelength (L) only or wave period (T) only:

$$C = \frac{L}{T} = \frac{g}{2\pi}T \quad \text{or} \quad C^2 = \frac{L^2}{T^2} = \frac{g}{2\pi}L$$

$$C = 1.56T \quad \text{or} \quad C^2 = 1.56L$$

See appendix C for a more complete wave equation and the method used to determine the equations for deep-water wavelength and celerity.

Storm Centers

Most waves observed at sea are **progressive wind waves.** They are generated by the wind, are restored by gravity, and progress in a particular direction. These waves are formed in local storm centers or by the steady winds of the trade and westerly wind belts. In an active storm area covering thousands of square kilometers, the winds are not steady but vary in strength and direction. Storm-area winds flow in a circular pattern about the low-pressure **storm center,** creating waves that move outward and away from the storm in all directions. When a storm center moves across the sea surface in a direction following the waves, wave heights are increased because the winds supply energy for a longer time and over a longer distance. You may gain some idea of the size of storms and their direction of travel by viewing the cloud patterns in the satellite photographs that are presented in televised weather forecasts.

In a storm area, the sea surface appears as a jumble and confusion of waves of all heights, lengths, and periods; there are no regular patterns. Capillary waves ride the backs of small gravity waves, which in turn are superimposed on still higher and longer gravity waves. This turmoil of mixed waves is called a "sea," and sailors use the expression "There is a sea building" to refer to the growth of these waves under storm conditions. When waves are being generated, they are forced to increase in size and speed by the continuing input of energy; these are known as **forced waves.** Because of variations in the winds of the storm area, energy at different intensities is transferred to the sea surface at different pulse rates, resulting in waves with a variety of periods and heights. Remember, wave periods are a function of the generating force; the speed at which a wave moves away from a storm may change, but its period remains the same.

Dispersion

When the waves move away from the storm, they are no longer wind-driven forced waves but become **free waves** moving at speeds due to their periods and wavelengths. Waves with long periods and long wavelengths have a greater speed than waves with short periods and

short wavelengths. The faster, longer waves gradually move through and ahead of the shorter, slower waves; this process is called **sorting,** or **dispersion.** Groups of these faster waves move as **wave trains,** or packets of similar waves with approximately the same period and speed. Because of dispersion, the distribution of observed waves from any single storm changes with time. Near the storm center, the waves are not yet sorted, while farther away, the faster, longer-period waves are out ahead of the slower, shorter-period waves. This process is shown in figure 10.5.

Away from the storm, the faster, longer-period waves appear as a regular pattern of crests and troughs moving across the sea surface. These uniform, free waves are called **swell;** they carry considerable energy, which they lose very slowly (fig. 10.6). The distribution of the waves from a given storm and the energy associated with particular wave periods change predictably with time, allowing the oceanographer to follow wave trains from a single storm over long distances. Groups of large, long-period waves created by storms between 40°S and 50°S in the Pacific Ocean have been traced across the entire length of that ocean, until they die on the shores and beaches of Alaska.

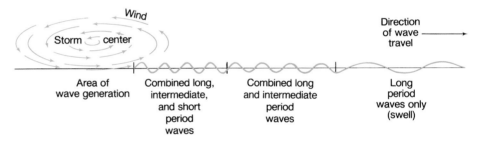

Figure 10.5 Dispersion. The longer waves travel faster than the shorter waves. Waves are shown here moving in only one direction.

Figure 10.6 Waves of uniform wavelength and period, known as swell, approaching the coast at the entrance to Grays Harbor, Washington.

Group Speed

Consider again the waves formed by a stone thrown into the water. The wave group, or train, is seen as a ring of waves moving outward from the point of disturbance. Careful observation shows that waves constantly form on the inside of the ring as it moves across the water. As each new wave joins the train on the inside of the ring, a wave is lost from the leading edge, or outside, of the ring, and the number of waves remains the same. The outside wave's energy is lost in advancing the wave form into undisturbed water. Therefore, the speed of each individual wave *(C)* in the group is greater than the speed of the leading edge of the wave train, and the wave ring moves outward at a speed one-half that of the individual waves. This speed is known as the **group speed,** the speed at which wave energy is transported away from its source under deep-water conditions:

group speed = ½ wave speed = speed of energy transport

or

$$V = \frac{C}{2}$$

Wave Interaction

Waves that escape a storm and are no longer receiving energy from the storm winds tend to flatten out slightly, and their crests become more rounded. These waves moving across the ocean surface as swell are likely to meet other trains of swell moving away from other storm centers. When the two wave trains meet, they pass through each other and continue on. Wave trains may intersect at any angle, and many possible interference patterns may result. If the two wave trains intersect each other sharply, as at a right angle, then a checkerboard pattern is formed (fig. 10.7).

If the waves have similar lengths and heights and approach from opposite directions (fig. 10.8a) and if the crests of one wave train coincide with the crests of another train, the wave trains reinforce each other by constructive interference. If the crests of a wave train coincide with the troughs of another wave train, the waves are canceled through destructive interference (fig. 10.8b). In this way, two or more similar wave trains traveling in the

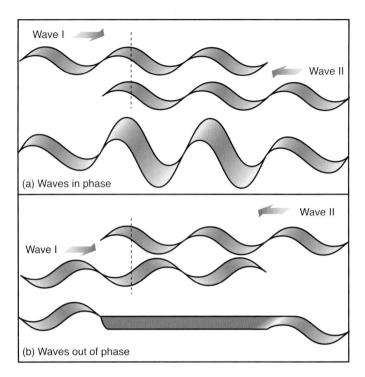

Figure 10.8   Waves approach each other. (a) If the crests or the troughs of the approaching waves coincide, the height of the combined waves increases, and the waves are in phase; this is constructive interference. (b) If the crests of one wave and the troughs of the other wave coincide, the waves cancel each other, and the waves are out of phase; this is destructive interference.

same or opposite direction and passing through each other can join together in phase and suddenly develop large-amplitude waves unrelated to local storms. If these waves become too high, they may break, lose some of their energy, and create new, smaller waves. If such high waves overtake a vessel, they can cause severe damage.

10.6 Wave Height

The height of wind waves is controlled by the interaction of several factors. The three most important factors are (1) wind speed (how fast the wind is blowing), (2) wind duration (how long the wind blows), and (3) **fetch** (the distance over water that the wind blows in the same direction). The wave height may be limited by any one of these factors. If the wind speed is very low, large waves are not produced, no matter how long the wind blows over an unlimited fetch. If the wind's speed is great but it blows for only a few minutes, no high waves are produced despite unlimited wind strength and fetch. Also, if very strong winds blow for a long period over a very short fetch, no high waves form. When no single one of these three factors is limiting, spectacular wind waves are formed at sea.

Table 10.1 lists the maximum and significant wave heights possible for certain average wind speeds when fetch and wind duration are not limiting. The significant wave height is defined as the average wave height of the highest one-third of the waves in a long record of measured wave heights. For example, if a wave-height record of 1200 successive storm waves

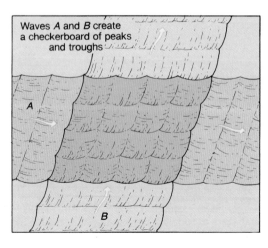

Figure 10.7   Waves meeting at right angles create a checkerboard pattern.

Table 10.1	The Relationship Between Wind Speed and Wave Height						
Average Wind Speed		Significant Wave Height	Significant Wave Period	Significant Wave Speed	Maximum Wave Height	Minimum Fetch[1]	Minimum Wind Duration[1]
(knots)	(m/s)	(m)	(s)	(m/s)	(m)	(km)	(h)
10	5.1	1.22	5.5	8.58	2.19	16	2.4
20	10.2	2.44	7.3	11.39	4.39	110	10
30	15.3	5.79	12.5	19.50	10.43	450	23
40	20.4	14.33	18.0	28.00	25.79	1136	42
50	25.5	16.77	21.0	32.76	30.19	2272	69

1. Minimum fetch and minimum wind duration are distances and times required when wind speed is the only limiting factor in wave development.

is made and the individual wave heights are arranged in order of height, the average height of the 400 highest waves defines the significant wave height. Significant wave heights are forecast from wind data, and the maximum wave heights are related to the significant wave heights by calculation.

The area of the ocean in the vicinity of 40°–50°S latitude is ideal for the production of high waves. Here, in an area noted for high-intensity storms and strong winds of long duration (what sailors have called the "roaring forties and furious fifties"), there are no landmasses to interfere and limit the fetch length. The westerly winds blow almost continuously around Earth, adding energy to the sea surface for long periods of time and over great distances, resulting in waves that move in the same direction as the wind. Although this area is ideal for the production of high waves, such waves can occur anywhere in the open sea, given the proper storm conditions.

A typical maximum fetch for a local storm over the ocean is approximately 920 km (500 nautical mi). Because storm winds circulate around a low-pressure disturbance, the winds continue to follow the waves on the side of the storm, along which wave direction is the same as the storm direction. This increases both the fetch and the duration of time over which the wind adds energy to the waves. If the waves move fast enough, their speeds exceed the speed of the moving storm center; the waves escape the generating wind, become free waves, and do not grow larger. Waves 10–15 m (33–49 ft) high are not uncommon under severe storm conditions; such waves are typically between 100 and 200 m (330 and 660 ft) long. This length is about the same as the length of some modern ships, and a vessel of this length encounters hazardous sailing conditions, because the ship may become suspended between the crests of two waves and break its back.

Giant waves over 30.5 m (100 ft) high are rare. In 1933, the USS *Ramapo,* a Navy tanker en route from Manila to San Diego, encountered a severe storm, or typhoon. As the ship was running downwind to ease the ride, it was overtaken by waves that, as measured against the ship's superstructure by the officer on watch, were 112 ft (34.2 m) high. The period of the waves was measured at 14.8 seconds; the wave speed was calculated at 27 m (90 ft) per second and the wavelength at 329 m (1100 ft). Other storm waves in this size category have been reported, but none has been as well documented. It is also probable that ships confronted with such waves do not always survive to report the incidents.

Episodic Waves

Large waves, or **episodic waves,** can suddenly appear unrelated to local sea conditions. An episodic wave is an abnormally high wave that occurs because of a combination of intersecting wave trains, changing depths, and currents. We do not know a great deal about these waves, as they do not exist for long and they can and do swamp ships, often eliminating any witnesses. They occur most frequently near the edge of the continental shelf, in water about 200 m (660 ft) deep, and in certain geographic areas with particular prevailing wind, wave, and current patterns.

The area where the Agulhas Current sweeps south along the east coast of South Africa and meets the storm waves arising in the Southern Ocean is noted for such waves. Storm waves from more than one storm may combine constructively and run into the current and against the continental shelf, producing occasional episodic waves. This area is also one of the world's busiest sea routes, as supertankers carrying oil from the Middle East ride the Agulhas Current on their trip southward to round the Cape of Good Hope en route to Europe and America. The situation is an invitation to trouble, and tankers have been damaged and lost in this area (fig. 10.9). In the North Atlantic, strong northeasterly gales send large storm waves into the edge of the northward-moving Gulf Stream near the border of the continental shelf, resulting in the formation of large waves. The shallow North Sea also seems to provide suitable conditions for extremely high episodic waves during its severe winter storms.

Researchers studying these waves describe them as having a height equal to a seven- or eight-story building (20–30 m, or 70–100 ft) and moving at a speed of 50 knots, with a wavelength approaching a half mile (0.9 km). If such a wave topples onto a vessel that has dropped its bow into the preceding trough, there is no escape from the thousands of tons of water crashing on the deck. In the North Sea, maximum potential wave height for an episodic wave is calculated as 33.8 m (111 ft), but 22.9 m (75 ft) is the highest that has been observed. In the Agulhas Current area, researchers searched the storm records for a twenty-year period and calculated a possible maximum wave height of about 57.9 m (190 ft).

There are many disappearances of vessels for which episodic waves are now suspected of being the chief cause. Many of the casualties are tankers or bulk carriers of ore, grain, and the like. These vessels are susceptible not only because so many are at sea at any one time but also because of their design. Bulk carriers comprise a bow section and an aft (or rear) section that includes the engine and the crew accommodations. These sections are separated by a series of flat-bottomed boxes, or storage tanks, which make up the majority of the vessel's length. Because these vessels are about 300 m (1000 ft) long, they are subject to great wrenching forces if a large portion of the hull is left unsupported or is only partially supported while suspended between wave crests. More-traditional and smaller hull forms are stronger, ride more easily, and are less likely to be destroyed by these severe waves.

Wave Energy

A wave's energy is present as **potential energy,** due to the change in elevation of the water surface, and as **kinetic energy,** due to the motion of the water particles in their orbits. The higher the wave, the larger the diameter of the water particle orbit and the greater the speed of the orbiting particle; therefore, the greater the kinetic and potential energy. The energy in a deep-water wave is nearly equally divided between kinetic and potential energy. The total energy of a wave is distributed over one wavelength per unit width (1 m) of crest from the sea surface to a depth of $L/2$ and is related to the square of the wave height. This relationship is demonstrated in figure 10.10. The energy density in a wave is the energy under a unit surface area of the wave. Energy density is also related to the square of the wave height.

Wave Steepness

There is a maximum possible height for any given wavelength. This maximum value is determined by the ratio of the wave's height to the wavelength, and it is the measure of the **steepness** of the wave:

$$\text{steepness} = \frac{\text{height}}{\text{length}} \quad \text{or} \quad S = \frac{H}{L}$$

If the ratio of the height to the length exceeds 1:7, the wave becomes too steep and the wave breaks. For example, if the wavelength is 70 m (230 ft), the wave will break when the wave height reaches 10 m (33 ft). The angle formed at the wave crest approaches 120°, and the wave becomes unstable. Under this condition, the wave cannot maintain its shape; it collapses and breaks (fig. 10.11).

Small, unstable breaking waves are quite common. When wind speeds reach 8–9 m/s (16–18 knots), waves known as whitecaps can be observed. These waves have short wavelengths (about 2 m), and the wind increases their height rapidly. As each wave reaches the critical steepness and crest angle, it breaks and is replaced by another wave produced by the rising wind.

Long waves at sea usually have a height well below their maximum value. Sufficient wind energy to force them to their maximum height rarely occurs. If a long wave does attain maximum height and breaks in deep water, tons of water

Figure 10.9 A giant wave breaking over the bow of the *ESSO Nederland* southbound in the Agulhas Current. The bow of this supertanker is about 25 m (80 ft) above the water.

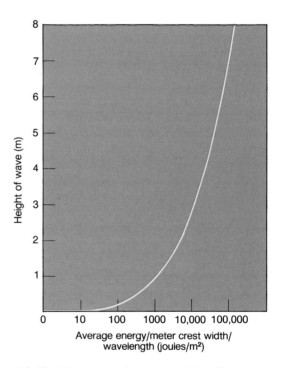

Figure 10.10 Wave energy increases rapidly with the square of the wave height. Average wave energy is calculated per unit width of the crest and averaged over the wavelength *(L)* and the depth *(L/2)*.

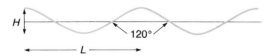

Figure 10.11 Wave steepness. When *H/L* approaches 1:7, the wave's crest angle approaches 120° and the wave breaks.

Table 10.2	Universal Sea State Code	
Sea State Code	**Description**	**Average Wave Height**
SS0	Sea like a mirror; wind less than 1 knot	0
SS1	A smooth sea; ripples, no foam; very light winds, 1–3 knots, not felt on face	0–0.3 m 0–1 ft
SS2	A slight sea; small wavelets; winds light to gentle, 4–6 knots, felt on face; light flags wave	0.3–0.6 m 1–2 ft
SS3	A moderate sea; large wavelets, crests begin to break; winds gentle to moderate, 7–10 knots; light flags fully extend	0.6–1.2 m 2–4 ft
SS4	A rough sea; moderate waves, many crests break, whitecaps, some wind-blown spray; winds moderate to strong breeze, 11–27 knots; wind whistles in the rigging	1.2–2.4 m 4–8 ft
SS5	A very rough sea; waves heap up, forming foam streaks and spindrift; winds moderate to fresh gale, 28–40 knots; wind affects walking	2.4–4.0 m 8–13 ft
SS6	A high sea; sea begins to roll, forming very definite foam streaks and considerable spray; winds at strong gale, 41–47 knots; loose gear and light canvas may be blown about or ripped	4.0–6.1 m 13–20 ft
SS7	A very high sea; very high, steep waves with wind-driven overhanging crests; sea surface whitens due to dense coverage with foam; visibility reduced due to wind-blown spray; winds at whole gale force, 48–55 knots	6.1–9.1 m 20–30 ft
SS8	Mountainous seas; very high-rolling breaking waves; sea surface foam-covered; very poor visibility; winds at storm level, 56–63 knots	9.1–13.7 m 30–45 ft
SS9	Air filled with foam; sea surface white with spray; winds 64 knots and above	13.7 m and above 45 ft and above

are sent crashing to the surface. The energy of the wave is lost in turbulence and in the production of smaller waves. Rather than breaking, under such conditions it is more likely that the top of a large wave will be torn off by the wind and cascade down the wave face. This action does not completely destroy the wave and is not considered to be the true collapse, or breaking, of such a wave.

In addition, waves break and dissipate if intersecting wave trains pass through each other in proper phase to form a combined wave with sufficient height to exceed the critical steepness. Waves sometimes run into a strong opposing current, forcing the waves to slow down. Remember that the speed of all waves equals the wavelength divided by the period ($C = L/T$) and that a wave's period does not change. If the speed of a wave is reduced by an opposing current, its wavelength must shorten. In such a case, the wave's energy is confined to a shorter length, so the wave increases in height to satisfy the height-energy relationship. If the increase in height exceeds the maximum allowable height-to-length ratio for the shorter wavelength, the wave breaks. Crossing a sandbar into a harbor or river mouth during an outgoing or falling tide is dangerous because the waves moving over the bar and against the tidal current steepen and break. Entering a harbor or river should be done at the change of the tide or on the rising tide, when the tidal current moves with the waves, stretching their wavelengths and decreasing their heights.

Universal Sea State Code

In 1806, Admiral Sir Francis Beaufort of the British navy adapted a wind-estimation system from land to sea use. On land, the clues to wind speed included smoke drift, the rustle of leaves, the flapping of flags, slates blowing from roofs, and uprooting of trees. Admiral Beaufort related observations of the sea-surface state to wind speed and designed a 0–12 (calm to hurricane) wind scale with typical wave descriptions for each level of wind speed. This Beaufort Scale was adopted by the U.S. Navy in 1838, and the scale was extended from 0–17. At present, a Universal Sea State Code of 0–9, based on the Beaufort Scale, is in international use for wind speeds and related sea-surface conditions (table 10.2).

10.7 Shallow-Water Waves

As a deep-water wave approaches the shore and moves into shallow water, the reduced depth begins to affect the shape of the orbits made by the water particles. The orbits gradually become flattened circles, or ellipses (fig. 10.12). The wave begins to "feel" the bottom, and the resulting friction and compression of the orbits reduce the forward speed of the wave.

Remember that (1) the speed of all waves is equal to the wavelength divided by period, and (2) the period of a wave does not change. Therefore, when the wave "feels bottom," it slows, and the accompanying reduction in the wavelength and speed results in increased height and steepness as the wave's energy is condensed in a smaller water volume.

When the wave enters water with a depth of less than one-twentieth the wavelength ($D < L/20$), the wave becomes a **shallow-water wave** (fig. 10.13). The group speed, V, of shallow-water waves is equal to the speed, C, of each wave in the group. While the length and speed of a deep-water wave are determined by the wave period, the shallow-water wavelength and speed are controlled only by the water depth. Here the wavelength and speed

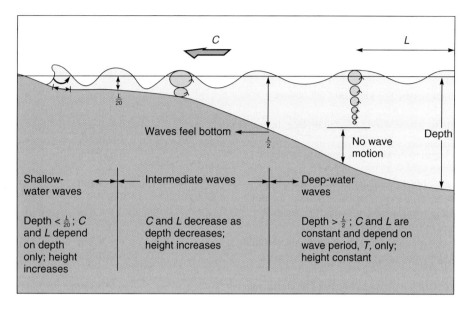

Figure 10.12   Deep-water waves become intermediate waves and then shallow-water waves as depth decreases and wave motion interacts with the sea floor.

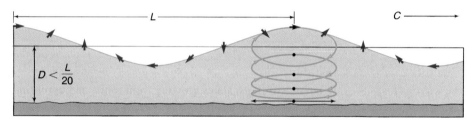

Figure 10.13   Shallow-water wave particles move in elliptical orbits. The orbits flatten with depth due to interference from the sea floor. *Red arrows* indicate the direction a water particle at the surface moves as the wave propagates to the right (in the direction of wave propagation at the crest and in the opposite direction at the trough).

are determined by the square root of the product of Earth's gravitational acceleration (g, 9.81 m/s^2) and depth *(D):*

$$C = \sqrt{gD} \quad \text{or} \quad C = 3.13\sqrt{D}$$
$$\text{or} \quad L = 3.13\ T\sqrt{D}$$

When the condition for a shallow-water wave is met, the orbits of the water particles are elliptical; they become flatter with depth until, at the sea floor, only a back-and-forth oscillatory motion remains (fig. 10.13). Note that the horizontal dimension of the orbit remains unchanged in shallow water. See appendix C for a more complete equation and the method used to calculate approximations of shallow-water wavelength and celerity.

When the water depth is between $L/2$ and $L/20$, the speed of the wave is also slowed. Waves in this depth range are called intermediate waves. No simple algebraic equation exists to determine the speed of these intermediate waves. Methods for working with intermediate waves using the full wave-speed equation are found in appendix C.

Refraction

Waves are refracted, or bent, as they move from deep to shallow water, begin to feel the bottom, and change wavelength and wave speed. When waves from a distant storm center approach the shore, they are likely to approach the beach at an angle. One end of the wave crest comes into shallow water and begins to feel bottom, while the other end is in deeper water. The shallow-water end moves more slowly than that portion of the wave in deeper water. The result is that the wave crests bend, or refract, and tend to become oriented parallel to the shore. This pattern is shown in figure 10.14. The **wave rays** drawn perpendicular to the crests show the direction of motion of the wave crests. Note that the refraction of water waves is similar to the refraction of light and sound waves, described in chapter 5.

When approaching an irregular coastline with headlands jutting into the ocean and bays set back into the land, waves may encounter a submerged ridge seaward from a headland or a depression in front of a bay. As waves approach such a coastline and feel the bottom, the portion of the wave crest over the ridge slows down more than the wave crests on either side. Therefore, the crests wrap around the headland in the pattern shown in figure 10.15. This refraction pattern focuses wave energy on the headland. Waves must gain in height as their wavelength is shortened and the total wave energy is crowded into a smaller volume of water; this increases the energy per unit of surface area. Therefore, more energy is expended on a unit length of shore at the point of the headland than on a unit length of shore elsewhere.

The central area at the mouth of the bay is usually deeper than the areas to each side, so the advancing waves slow down more on the sides than in the center. Because the wavelengths remain long in the center and shorten on each side, the wave crests bulge toward the center of the bay. The waves in the center of the bay do not shorten as much; therefore, they have less height and less energy to expend per length of shoreline. This pattern is shown in figure 10.16. The result is an environment of low wave energy, providing sheltered water in the bay. Overall, this unequal distribution of wave energy along the coast results in a wearing down of the headlands and a filling in of the bays, as the sand and mud settle out in the quieter water. If all coastal materials had the same resistance to wave erosion, this process would lead in time to a straightening of the coastline; but the rocky structure of many headlands resists wave erosion; therefore, the cliffs remain.

Reflection

A straight, smooth, vertical barrier in water deep enough to prevent waves from breaking reflects the waves (fig. 10.17). The barrier may be a cliff, steep beach, breakwater, bulkhead, or other structure. The reflected waves pass through the incoming waves to produce an interference pattern, and steep, choppy seas often result. If the waves reflect directly back on themselves, the

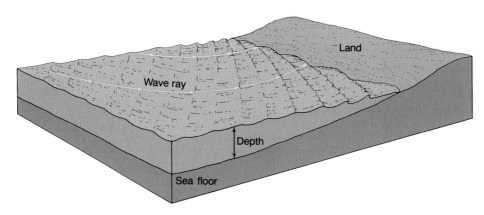

Figure 10.14 Waves moving inshore at an oblique angle to the depth contours are refracted. One end of the wave reaches a depth of *L/2* or less and slows, while the other end of the wave maintains its speed in deeper water. Wave rays drawn perpendicular to the crests show the direction of wave travel and the bending of the wave crests.

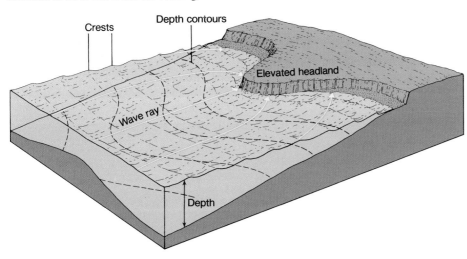

Figure 10.15 The energy of waves refracted over a shallow, submerged ridge is focused on the headland. The converging wave rays show the wave energy being crowded into a small volume of water, increasing the energy per unit length of wave crest as the height of the wave increases.

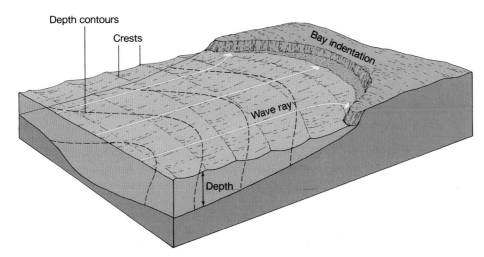

Figure 10.16 Waves refracted by the shallow depths on each side of the bay deliver lower levels of energy inside the bay. The diverging wave rays show the spreading of energy over a larger volume of water, decreasing the energy per unit length of wave crest as the wave height decreases.

resulting waves appear to stand still, rising and falling in place. The behavior of waves reflected from a curved vertical surface depends on the type of curvature. If the curvature is convex, the reflected wave rays spread and disperse the wave energy, but if the curved surface is concave, the reflected wave rays converge and the energy is focused. This situation is similar to the reflection of light from curved mirrors. Great care must be taken in designing walls and barriers to protect an area from waves to be sure that the energy of reflected waves is not focused in a way that will result in another area being damaged.

Diffraction

Another phenomenon that is associated with waves as they approach the shore or other obstacles is **diffraction.** Diffraction is caused by the spread of wave energy sideways to the direction of wave travel. If waves move toward a barrier with a small opening (fig. 10.18*a*), some wave energy passes through the small opening to the other side. Once the waves have passed through the opening, their crests decrease in height and radiate out and away from the gap. A portion of the wave energy is diffracted, transported sideways from its original direction. If more than one gap is open to the waves, the patterns produced by the spreading waves from the openings may intersect and form interference patterns as the waves move through each other (fig. 10.18*b*).

If the waves approach a barrier without an opening, diffraction can still occur. Energy will be transported at right angles to the wave crests as the waves pass the end of the barrier (fig. 10.19). Note that the pattern produced is one-half of the pattern observed when the waves pass through a narrow opening, as in figure 10.18*a*. Energy is transported behind the sheltered (or lee) side of the barrier. This effect is an important consideration in planning the construction of breakwaters and other coastal barriers intended to protect vessels in harbors from wave action and possible damage.

Navigation from Wave Direction

In areas of the world where the winds blow steadily and from one direction, as in the trade-wind belts, waves at sea are very

Figure 10.17   The wave from a ship's wake is reflected from a concrete wall. The principal wave is moving from *lower left* to *upper right* of the photograph. The reflected wave is moving from *lower right* to *upper left*. A checkerboard pattern is formed as the principal wave and reflected wave move through each other at approximately right angles.

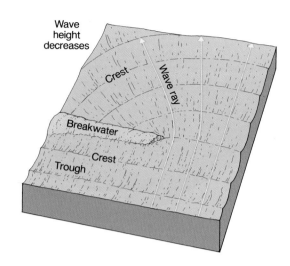

Figure 10.19   Diffraction occurs behind the breakwater.

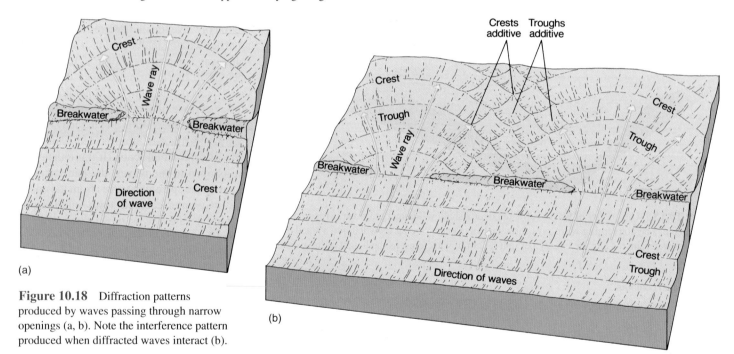

Figure 10.18   Diffraction patterns produced by waves passing through narrow openings (a, b). Note the interference pattern produced when diffracted waves interact (b).

regular in their direction of motion, and this regularity allows a vessel to maintain a constant course relative to the waves. Because waves change speed, shape, and height with water depth, and because waves change direction and pattern because of refraction, diffraction, and reflection, it is possible to deduce the presence of shoals, bars, islands, and coasts from the changes in wave patterns.

Careful direct observation of wave patterns, or the sensing of wave patterns through the motions of a small craft, even at night, allows the detection of a change in the angle between waves and wind direction or a change in the wave pattern. Although these changing patterns are subtle, they can be detected many miles downwind from islands and can be used to bring a small vessel to shore when no landmarks are visible.

The Polynesians of the past lived with the sea; it was their home, and they were acutely sensitive to its changes. They made long voyages in their small canoes using their knowledge of wave patterns. They had no theory to explain these patterns, but they understood the association with winds, shores, and islands. Combining their knowledge of star positions, cloud forms over land and sea, and bird flights with their knowledge of waves, they sailed many hundreds, even thousands, of miles across the open ocean to reach their destinations. No navigational tools were required other than charts constructed from twigs and shells, showing island positions relative to stars, wind direction, and swell. An example of such a chart is found in chapter 1. These people learned by living with and observing nature in an area where nature cooperated by behaving in regular and predictable ways.

10.8 The Surf Zone

The surf zone is the shallow area along the coast in which the waves slow rapidly, steepen, break, and disappear in the turbulence and spray of expended energy. The width of this zone is variable and is related to both the wavelength and height of the arriving waves and the changing depth pattern. Longer, higher waves, which feel the bottom before shorter waves, become unstable and break farther offshore in deeper water. If shallow depths extend offshore for some distance, the surf zone is wider than it is over a sharply sloping shore.

Breakers

Breakers form in the surf zone because the water particle motion at depth is affected by the bottom. Orbital motion is slowed and compressed vertically, but the orbit speed of water particles near the crest of the wave is not slowed as much. The particles at the wave crest move faster toward the shore than the rest of the wave form, resulting in the curling of the crest and the eventual breaking of the wave. The two most common types of breakers are **plungers** and **spillers** (fig. 10.20).

Plunging breakers form on narrow, steep beach slopes. The curling crest outruns the rest of the wave, curves over the air below it, and breaks with a sudden loss of energy and a splash. The more common spilling breaker is found over wider, flatter beaches, where the energy is extracted more gradually as the wave moves over the shallow bottom. This action results in the less dramatic wave form, consisting of turbulent water and bubbles flowing down the collapsing wave face. The spilling breakers last longer than the plungers because they lose energy more gradually. Therefore, spillers give surfers a longer ride, but plungers give them a more exciting one.

The slow curling over of the crest observed on some breakers begins at a point on the crest and then moves lengthwise along the wave crest as the wave approaches shore. This movement of the curl along the crest occurs because waves are seldom exactly parallel to a beach. The curl begins at the point on the crest that is in the shallowest water or the point at which the crest height is slightly greater, and it moves along the crest as the rest of the wave approaches the beach. The result is the "tube" so sought after by surfers (fig. 10.21).

If the waves approaching the beach are uniform in length, period, and height, they are the swells from some far-distant storm, which have had time and distance to sort into uniform groups. For example, the long surfing waves of the California

(a)

(b)

Figure 10.20 Breaking waves. A plunger (a) loses energy more quickly than a spiller (b).

Figure 10.21 A surfer rides the tube of a large curling wave in Hawaii.

beaches in summer begin in the winter storms of the South Pacific and Antarctic Oceans. If the waves are of different heights, lengths, and periods and break at varying distances from the beach, then unsorted waves have arrived and are probably the product of a nearby local storm superimposed on the swell.

Water Transport

The small net drift of water in the direction the waves are traveling (refer to the discussion in section 10.3 on the water particle orbits of real waves) is intensified in the surf zone, as the shoreward motion of the water particles at the crest becomes greater than the return particle motion at the trough. Because the crests usually approach the beach at an angle, the surf zone transport of water flows both toward the beach and along the beach. The result is that water accumulates against the beach and flows along the beach until it can flow seaward again and return to the area beyond the surf zone. This return flow generally occurs in quieter water with smaller wave heights, for example, in areas with troughs or depressions in the sea floor.

Because regions of seaward return flow may be narrow and some distance apart, the flow in these areas must be swift in order to carry enough water beyond the surf zone to balance the slower but more extensive flow toward the beach. These regions of rapid seaward flow are called **rip currents.** Rip currents can be a major hazard to surf swimmers. In the spring of 1994, five swimmers were drowned and six others were hurt in a single rip current system at American Beach, Florida, and in 1995, three more swimmers drowned in severe rip currents caused by storm waves from Hurricane Felix.

Swimmers who unknowingly venture into a rip current will find themselves carried seaward and unable to swim back to shore against the flow; they must swim parallel to the beach, or across the rip current, and then return to shore. Because of the danger associated with rip currents, swimmers should be on the lookout for indicators of their presence, including (1) turbid water and floating debris moving seaward through the surf zone, (2) areas of reduced wave heights in the surf zone, and (3) depressions in the beach running perpendicular to the shore.

Wave action on the beach stirs up the sand particles and temporarily suspends them in the water. The sand is carried along the beach parallel to the shore until the rip current is reached, and the sand is transported seaward. Viewed from a height, such as a high cliff or a low-flying airplane, rip currents are seen as streaks of discolored turbid water extending seaward through the clearer water of the outer surf zone (fig. 10.22).

Energy Release

Watching the heavy surf pounding a beach from a safe vantage point is an exciting and exhilarating experience; the trick is to determine at what point one is safe. In a narrow surf zone during a period of very large waves, the wave energy must be expended rapidly over a short distance. Under these conditions, the height of the waves and the forward motion of the water particles combine to send the water high up on the beach. The

Figure 10.22   Small rip currents carry turbid water seaward through the surf zone.

accompanying release of energy is explosive and can result in rocks and debris from the water's edge being hurled high up on the beach by the force of the water. Minot's Lighthouse on the south side of Massachusetts Bay is 30 m (100 ft) high, but it is regularly engulfed in spray. Lonely Tillamook Light off the Oregon coast had to have steel gratings installed to protect the glass that shields the light 40 m (130 ft) above sea level, after the glass had been broken several times by wave-thrown rocks. In every winter storm, waves displace boulders weighing many tons from breakwaters along the world's coasts.

Waves do not always expend their energy on the shore. Some break farther seaward on sandbars such as those associated with river mouths and estuaries. The famous bar at the mouth of the Columbia River is responsible for extremely hazardous conditions for both fishing and commercial vessels, and in winter it often produces casualties. On an ebbing (or falling) tide, waves approaching the bar against the current rise and break at heights up to 20 m (66 ft). The Coast Guard uses the Columbia River entrance to train its crews to operate roll-over, or self-righting, rescue boats, which are used during storms and heavy surf conditions.

The great driftwood logs found stranded on some beaches may be set afloat again at high tide during severe winter storms and can become lethal battering rams when hurled shoreward by the surf. Even with less severe waves, beach logs lying near the water's edge have a dangerous potential. The unsuspecting vacationer sits or plays on such a log, and when the occasional higher wave rolls it over, the person may become trapped under the log and crushed or drowned by the succeeding waves. Logs in or near the surf represent a great danger for the unwary. Where are you safe? You must be the judge.

10.9 Tsunami

Sudden movements of Earth's crust may produce a **seismic sea wave,** or **tsunami.** These waves are often incorrectly called tidal waves. Because a seismic sea wave has nothing to do with tides, oceanographers have adopted the Japanese word *tsunami,* meaning harbor wave, to replace the misleading term *tidal wave.* Tsunami is a synonym for seismic sea wave.

If a large area, maybe several hundred square kilometers, of Earth's crust below the sea surface is suddenly displaced, it may cause a sudden rise or fall in the overlying sea surface. In the case of a rise, gravity causes the suddenly elevated water to return to the equilibrium surface level; if a depression is produced, gravity causes the surrounding water to flow into it. Both events result in the production of waves with extremely long wavelengths (100–200 km, or 60–120 mi) and long periods as well (ten–twenty minutes). The average depth of the oceans is about 4000 m (4 km, or 13,000 ft); this depth is less than one-twentieth the wavelength of these waves, so tsunamis are shallow-water waves. These waves radiate from the point of the seismic disturbance at a speed determined by the ocean's depth ($C = \sqrt{gD}$) and move across the oceans at about 200 m/s (400 mph). Because they are shallow-water waves, tsunamis may be refracted, diffracted, or reflected in mid-ocean by changes in seafloor topography and by mid-ocean islands.

When a tsunami leaves its point of origin, it may have a height of several meters, but this height is distributed over its many-kilometer wavelength. It is not easily seen or felt when superimposed on the other motions of the sea's surface, and a vessel in the open ocean is in little or no danger if a tsunami passes. The danger occurs only if the vessel has the misfortune to be directly above the area of the original seismic disturbance.

The energy of a tsunami is distributed from the ocean surface to the ocean floor and over the length of the wave. When the path of the wave is blocked by a coast or an island, the wave behaves like any other shallow-water wave; it slows (to approximately 80 km [50 mi] per hour), its wavelength decreases, and its energy is compressed into a smaller water volume as the depth rapidly decreases. This sudden confinement of energy to a smaller volume increases the energy density and causes the wave height to build rapidly, and the loss of energy is equally rapid when the wave breaks. A tremendous surge of moving water races up over the land, flooding the coast for a period that lasts five to ten minutes before the water flows seaward, exposing the nearshore sea floor. The surge destroys buildings and docks; large ships may be left far up on the shore. The receding water carries the debris from the surge, battering and buffeting everything in its path. In bays and inlets, the rising and falling of the water may not be as destructive as the extreme

currents that form at the entrance of a bay or an inlet where large volumes of water oscillate in and out, producing dangerous and destructive current surges. If these oscillating flows occur at the natural oscillation period of the bay, large changes in water level are created (see section 10.11).

The leading edge of the tsunami wave group may be either a crest or a trough. If the initial crustal disturbance is an upward motion, a crest arrives first; if the crustal motion is downward, a trough precedes the crest. If a trough arrives at the shore first, the water level drops by as much as 3–4 m (10–13 ft) within two to three minutes. People who follow the receding water to inspect exposed sea life may drown, for in another four to five minutes, the water rises 6–8 m (20–26 ft), and they will not be able to outrun the wall of advancing water. In still another four to five minutes, a trough with a water-level drop of 6–8 m (20–26 ft) arrives, and the water flowing away from the beach will carry debris out to sea.

Tsunamis are most likely to occur in ocean basins that are tectonically active. The Pacific Ocean, ringed by crustal faults and volcanic activity, is the birthplace of most tsunamis. They also appear in the Indian Ocean, in the Caribbean Sea, which is bounded by an active island arc system, and in the Mediterranean Sea. The spectacular havoc and destruction caused by these waves are well recorded. In August 1883, the Indonesian island of Krakatoa erupted and was nearly destroyed in a gigantic volcanic explosion that hurled several cubic miles of material into the air. A series of tsunamis followed; these waves had unusually long periods of one to two hours. The town of Merak, 53 km (33 mi) away and on another island, was inundated by waves over 30.5 m (or 100 ft) high, and a large ship was carried nearly 3 km (2 mi) inland and left stranded 9.2 m (30 ft) above sea level. More than 35,000 people died from these enormous waves, and as the waves moved across the oceans, water-level recorders were affected as far away as Cape Horn (12,500 km, or 7800 mi) and Panama (18,200 km, or 11,400 mi). The waves were traveling with a speed calculated at about 200 m/s, or 400 mi/h.

On April 1, 1946, an earthquake in the area of the Aleutian Trench off Alaska produced a series of tsunamis. These waves heavily damaged Hilo, Hawaii, killing more than 150 people. Not only the portion of the island facing the oncoming waves was damaged; the waves bent around the island by refraction and were diffracted when passing between islands, producing high waves that struck the sheltered side of the island. The tsunami waves from the 1946 earthquake were highest near their source in the Aleutian Islands, where they destroyed a concrete lighthouse 10 m (33 ft) above sea level at Scotch Cap and killed the crew. A radio mast mounted 33 m (108 ft) above sea level in the same area was also destroyed. In 1957, Hawaii was hit again, with waves higher than the 1946 series; but no lives were lost because early warnings alerted people to evacuate. In 1964, the Alaska earthquake that severely damaged Anchorage and Alaskan coastal towns produced tsunamis that selectively hit areas on the west coast of Vancouver Island and the northern coast of California.

On September 1, 1992, a magnitude-7.0 earthquake occurred 100 km (60 mi) off the Nicaraguan coast. The tsunamis produced by this earthquake killed 170 people, injured 500, and destroyed 1500 homes. Eyewitnesses recall a single large wave, preceded by a trough that lowered the harbor depth by 7 m (23 ft). In this case, the maximum run-up was 10 m (33 ft); run-ups of 2–6 m (7–19 ft) were more common. This event was triggered by an earthquake in a previously stationary section of a fault region associated with the Cocos Plate where it is being subducted under the Caribbean Plate.

On December 12, 1992, an earthquake of magnitude 7.5 struck Flores Island in the Sunda and Banda island arc systems. The quake epicenter was only 50 km (30 mi) northwest of the town of Maumere, and five minutes after the quake, the first of a series of tsunamis struck the town. The earthquake and the tsunamis caused 2080 reported deaths and 2144 injuries. Average water run-up on the shore was 5 m (16 ft), with a maximum of 19.8 m (65 ft) at the northeast corner of the island.

After ten o'clock at night on July 12, 1993, a magnitude-7.8 earthquake occurred in the Sea of Japan at the boundary of the Eurasian and North American Plates and very close to Okushiri Island off the southwest coast of Hokkaido. Six minutes after the earthquake began, a tsunami swept over a 5 m (16 ft) seawall, destroyed buildings, and ruptured propane tanks that set fire to the town of Aonae at the southern end of Okushiri Island. In this case, the longest wave run-up was 30 m (100 ft), with more typical values at 10–15 m (33–50 ft). One hundred eighty-five people died, and the property loss was estimated at $600 million (fig. 10.23).

A massive tsunami struck the northwest coast of Papua, New Guinea, on July 17, 1998. The 7–15 m (22–50 ft) high tsunami swept across the barrier beach parallel to Sissano Lagoon, destroying four fishing villages (fig. 10.24). Three waves swept debris and bodies into the lagoon and the mangrove trees beyond. More than 2000 persons were killed or reported missing. The tsunami was triggered by a close-to-shore underwater earthquake (magnitude 7 on the Richter scale) that caused a 2 m (6 ft) vertical drop along a 40 km (24 mi) fault in the sea floor, followed by an underwater landslide. Although the island experienced tremors thirty minutes before the arrival of the first wave, their significance was not recognized in the shore area.

The most destructive tsunami in history in terms of loss of life was the December 26, 2004, Sumatra tsunami that devastated coastlines in the Indian Ocean basin. (See *Field Notes* entitled "Modeling the December 26, 2004, Sumatra Tsunami.")

Figure 10.23 Aonae at the southern tip of Okushiri Island in the Sea of Japan after the tsunami and fire of 1993. The area on the *left* extending to the tip of the island was leveled despite the protecting seawall. Debris left in the streets by the surging water prevented fire control equipment from reaching the burning buildings.

Figure 10.24 The tsunami that struck the northwest coast of Papua, New Guinea, July 17, 1998, destroyed a chain of villages along 30 km (18 mi) of beach bordering Sissano Lagoon. Buildings, trees, and people were swept away by three waves 7–15 m (22–50 ft) high. The eroded beach and the shattered palms were all that was left of many communities.

Shortly before eight o'clock in the morning, a massive earthquake ruptured the sea floor off the northwest coast of Sumatra, resulting in as much as 10 m (33 ft) of vertical displacement. This earthquake was the result of the subduction of the Indian Plate beneath the Eurasian Plate. An international team of scientists studying the effects of the tsunami in Sumatra documented wave heights of 20–30 m (65–100 ft) at the island's northwest end and found evidence suggesting that wave heights may have ranged from 15–30 m (50–100 ft) along at least a 100 km (62 mi) stretch of the northwest coast. The worst damage is thought to have occurred in the province of Aceh, roughly 100 km (62 mi) from the epicenter of the earthquake (fig. 10.25*a–c*). About one-third of the 320,000 residents of Aceh's capital, Banda Aceh, are presumed to have been killed.

The tragic loss of life caused by the Sumatra tsunami was compounded by the lack of tsunami detection instruments in the Indian Ocean basin, similar to those in the Pacific Ocean basin, and a coordinated tsunami alert plan in the region. Tsunami Prediction and Warning Centers were first located in Hawaii and Alaska in 1946, after the April 1, 1946, tsunamis that struck Hawaii. Tsunami detection in the Pacific Ocean basin is now aided by the National Oceanic and Atmospheric Administration's (NOAA) Deep-Ocean Assessment and Reporting of Tsunamis (DART) program. The DART program operates forty open-ocean tsunameters for detecting tsunamis (fig. 10.26). Additional tsunameters are operated by the Australian, Chilean, Indonesian, and Thai governments. Each instrument consists of a surface buoy (fig. 10.27*a*) for real-time data transmission connected to an anchored seafloor bottom package that includes a bottom pressure recorder (fig. 10.27*b*). Each instrument is designed to be deployed for 24 months at depths of up to 6000 m (19,700 ft). These instruments have measured tsunamis characterized by amplitudes less than 1 cm in the deep ocean. Data are transmitted from the bottom pressure recorder on the sea floor to the surface buoy and then relayed via satellite to ground stations.

The tsunameters operate in two modes, standard and event. In standard mode, they measure water pressure due to the height of the water column above them every 15 seconds to obtain an average value over a 15 minute period of time. In this mode, four measurements are transmitted each hour. When a computer on the instrument detects a possible tsunami, the instrument goes into event mode. In event mode, the instrument transmits 15-second values during the initial few minutes, followed by 1-minute averages. The system returns to standard mode after 4 hours of 1-minute transmissions if no additional events are detected.

(a)

(b)

(c)

Figure 10.25   (a) A mosque is left standing amid the rubble in Banda Aceh, Sumatra. Several mosques survived and may have been saved by the open ground floor that is part of their design. The tsunami waves reached the middle of the second floor. (b) Boat carried inland by the tsunami. (c) Typical tsunami damage in Banda Aceh. (Photos courtesy of the United States Geological Survey.)

Field Notes

Modeling the December 26, 2004, Sumatra Tsunami

by Dr. Eddie Bernard

Dr. Eddie Bernard is the Director of the National Oceanic and Atmospheric Administration's Pacific Marine Environmental Laboratory in Seattle, Washington. He directs a broad range of oceanographic research programs, including ocean climate dynamics, fisheries oceanography, El Niño forecasts, tsunamis, and seafloor spreading. Dr. Bernard is an expert in the study of tsunamis.

At 07:59 Local Time (00.59 UTC) on December 26, 2004, a magnitude-9.3 megathrust earthquake occurred along 1300 km (800 mi) of the oceanic subduction zone located 100 km (62 mi) west of Sumatra and the Nicobar and Andaman Islands in the eastern Indian Ocean. Highly destructive tsunamis were generated by up to 10 m (33 ft) vertical displacements of the sea floor associated with massive (more than 20 m [65 ft] horizontally) sudden movements of adjacent plates during this event. Although the exact numbers will never be accurately known, it is estimated that 237,000 people died and over $13 billion in damage occurred. Some economists estimate that the tsunami devastation will place about 1 million people in poverty for the rest of their lives. The tsunami was in excess of 30 m (100 ft) as it assaulted the Sumatra coastline (box fig. 1) and was recorded around the world. This tsunami is the first for which there are high-quality worldwide tide gauge measurements and for which there are multiple-satellite altimetry passes that were able to measure the tsunami wave height in the open ocean. These widespread coastal and open-ocean measurements of the tsunami height have been used to further refine a global tsunami numerical model, known as MOST (Method of Splitting Tsunami), used to predict the propagation and wave heights of tsunamis all over the world. The objective of tsunami modeling is to develop faster and more reliable forecasts of tsunamis striking coastal regions. A comparison of the actual measured tsunami heights with the predicted heights from the MOST model have revealed some factors that contributed to the propagation of the tsunami's energy thousands of kilometers throughout the world oceans.

The first instrumental tsunami measurements were available about three hours after the earthquake from the real-time reporting tide gauge at the Cocos Islands (box fig. 2) located approximately 1700 km (1056 mi) south of the earthquake source area. Data from this gauge revealed a 30 cm (11.8 in) high first wave followed by a long train of water level oscillations with maximum peak-to-trough ranges of 53 cm (21 in). Gauge data and inundation measurements from sites in India and Sri Lanka at similar distances from the epicenter yielded amplitudes almost ten times greater than the Cocos Islands values. These significant wave height differences were consistent with numerical modeling results that clearly demonstrate the highly directional nature of the Sumatra tsunami (box fig. 2).

Satellite altimetry measurements of tsunami amplitude were obtained from the *Jason-1* and *Topex/Poseidon* satellites (see chapter 1, section 1.10) as they transited the Indian Ocean about

Box Figure 1 Tsunami inundation along the northern Sumatra coastline where flooding exceeded 30 m (98 ft) and caused the most deaths and damage. The white staff in the center of the photograph is 5 m (16.5 ft). Photo courtesy of Jose Borrero, University of Southern California.

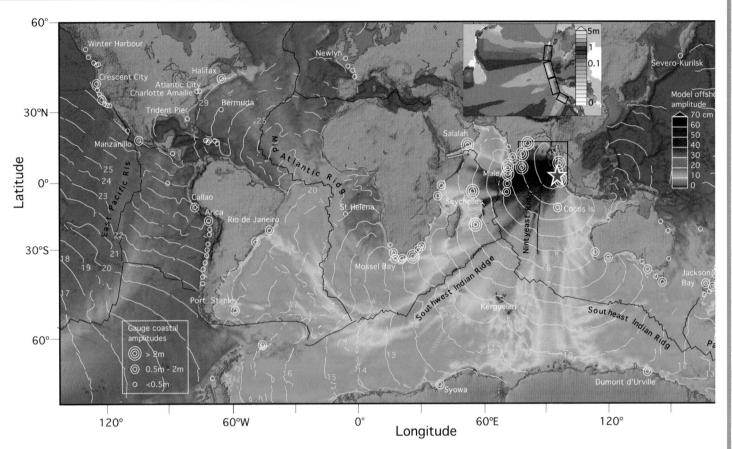

Box Figure 2 Global chart showing energy propagation of the 2004 Sumatra tsunami calculated from MOST. *Star* denotes the epicenter of the earthquake. *Filled colors* show maximum computed tsunami heights during forty-four hours of wave propagation simulation. *Contours* show computed arrival time of tsunami waves. *Circles* denote the locations and amplitudes of tsunami waves in three range categories for selected tide gauge stations. *Inset* shows fault geometry of the model source and close-up of the computed wave heights in the Bay of Bengal. Distribution of the slip among four sub-faults (from south to north: 21 m [69 ft], 13 m [43 ft], 17 m [56 ft], 2 m [6.6 ft]) provides best fit for satellite altimetry data and correlates well with seismic and geodetic data inversions (from Titov, et al., 2005).

150 km (93 mi) apart approximately two hours after the quake. The tracks crossed the spreading front of the tsunami waves in the Bay of Bengal down to about 1200 km (745 mi) southward from Sri Lanka. The measurements revealed amplitudes of about 50–70 cm (20–28 in) of the leading tsunami wave at this location in the Indian Ocean.

Box figure 2 summarizes simulation results from the tsunami numerical model MOST for a model tsunami source constrained to produce the observed open-ocean satellite wave height measurements and the known characteristics of the earthquake that generated the tsunami. Model results support suggestions that there are two main factors affecting the direction the tsunami wave travels and its height: the orientation of the earthquake source region and the effect of mid-ocean ridges to act as wave guides.

For waves far from the epicenter of the earthquake, seafloor topography is the main factor determining the directionality of energy propagation. Analysis of the global tsunami model illustrates the role of mid-ocean ridges in guiding inter-ocean tsunami propagation (box fig. 2). The Southwest Indian Ridge and the Mid-Atlantic Ridge served as wave-guides for tsunami energy propagation into the Atlantic Ocean while the Southeast Indian Ridge, Pacific-Antarctic Ridge, and the East Pacific Rise served as guides for waves entering the Pacific. Results further show that ridges act as wave guides only until their curvature exceeds critical angles at locations along the tsunami wave paths. For example, the sharp bend of the Mid-Atlantic Ridge in the South Atlantic results in the tsunami ray leaving the wave-guide near 40°S and hitting the Atlantic coast of South America with relatively high wave amplitudes. The model predicts the large (~1 m [~3 ft]) peak waves observed at Rio de Janeiro, Brazil (box fig. 2). The Ninety-East Ridge caused focusing of wave energy and increased tsunami height southward toward the coast of Antarctica. There were no gauge stations on the coast of Antarctica directly in line with the beam of

(continued)

(concluded)

tsunami energy arriving from the Ninety-East Ridge, and only moderate (60–70 cm [24–28 in]) peak-to-peak waves were recorded at the French Dumont d'Urville station and the Japanese station Syowa on the coast of Antarctica.

For most of the eastern and central Indian Ocean records, the first few waves were the largest (up to twelve hours of anomalously high wave intensity), followed by relatively rapid exponential wave attenuation. Model simulations illustrate that these records are from locations where the largest tsunami waves followed a direct route from the source following initial focusing by the source configuration. Tide gauge recordings from the western Indian and other oceans show increased tsunami duration, with maximum waves arriving later in the first wave train. This demonstrates increased input from waves that reached the gauge locations after scattering or refracting from shallow submarine features and reflecting from the coasts.

The prolonged tsunami records for the Atlantic Ocean are consistent with substantial tsunami energy propagation along the Mid-Atlantic Ridge wave-guide. In the Pacific Ocean, wave trains for the Sumatra tsunami often contained two or more distinct "packets" with different wave height and frequency characteristics. Because it is so vast, the Pacific Ocean allows for two different propagation paths for most coastal locations.

Although no direct tsunami damage has been reported for the 2004 event outside of the Indian Ocean basin, the numerical study demonstrated the ability for tsunami energy to be transported throughout the world ocean. Thus, large tsunamis can propagate substantial and damaging wave energy to distant coasts, including different oceans, through a combination of source focusing and bathymetric wave-guides.

To Learn More About the Sumatra Tsunami

Smith, W. H. F., R. Scharroo, V. V. Titov, D. Arcas, and B. K. Arbic. 2005. Satellite Altimeters Measure Tsunami, Early Model Estimates Confirmed, *Oceanography*, 18 (2), p. 10–12.

Titov, V. V., A. B. Rabinovich, H. O. Mofjeld, R. E. Thomson, and F. I. González. 2005. The global reach of the 26 December 2004 Sumatra Tsunami, *Science*, 309 (5743), pp. 2045–2048.

Internet References

http://www.pmel.noaa.gov/tsunami/
http://www.noaa.gov/tsunamis.html
http://www.tsunami.noaa.gov/
http://www.geophys.washington.edu/tsunami/general/warning/warning.html

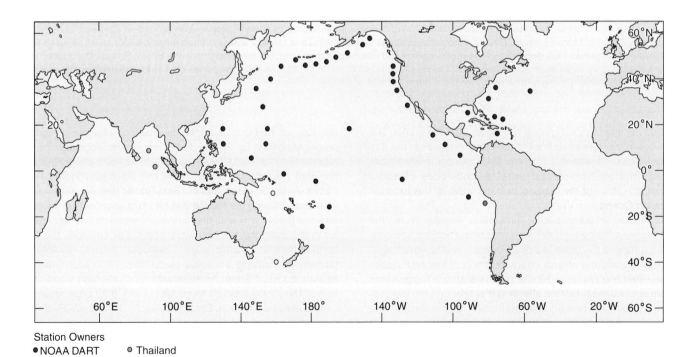

Station Owners
- NOAA DART
- Australia
- Chile
- Thailand
- Indonesia

Figure 10.26 NOAA's Project DART (Deep-ocean Assessment and Reporting of Tsunamis). Tsunameters are deployed near regions with a history of tsunami generation, to ensure measurement of the waves as they propagate toward coastal communities and to acquire data critical to real-time forecasts. Locations of the thirty-six tsunameters comprising the network (red circles) are shown on this map along with additional tsunameters operated by the Australian, Chilean, Indonesian, and Thai governments.

(a)

(b)

Figure 10.27 (a) A DART tsunameter surface buoy. (b) A bottom pressure recorder used to detect variations in sea surface elevation by measuring changes in water pressure. Data obtained by the pressure recorder are transmitted to the buoy and then on to a satellite which relays the data to tsunami warning centers.

10.10 Internal Waves

The waves discussed to this point have all formed at the interface of the atmosphere and the ocean. This interface marks the common boundary between two fluids of different densities, air and water. Another interface between two fluids lies below the ocean surface at the pycnocline that separates the shallow mixed layer from the denser underlying water. In this case, the boundary is less abrupt and the density difference is not as great as it is at the air-water boundary. The waves that form along this boundary are known as **internal waves** (fig. 10.28). These internal waves cause the boundary to oscillate as the wave form progresses between the water layers.

Internal waves are slower than surface waves. They typically have wavelengths from hundreds of meters to tens of kilometers and periods from tens of minutes to several hours. Their height often exceeds 50 m and may be limited by the thickness of the surface layer. The orbital motion generated by internal waves of water particles is illustrated in fig. 10.28a. The radius of the circular motion of the water particles is largest at the density boundary (pycnocline or thermocline) depth and decreases downward as well as upward from this depth. When wave heights are large, the crests of the internal waves may show at the sea surface as moving bands. The water over the crests of the internal waves often shows ripples. If the amplitude of the wave approaches the thickness of the surface layer,

the deeper water may also be seen breaking the sea surface. The water over the trough of the internal wave is generally smooth. Sometimes, instead of visible bands, elevations and depressions of the sea surface occur as the internal waves pass (fig. 10.28).

Many processes are responsible for internal waves. A low-pressure storm system may elevate the sea surface and depress the pycnocline. When the storm moves away, the displaced pycnocline will oscillate as it returns to its equilibrium level. If the speed of a surface current changes abruptly at the pycnocline, internal waves may be generated. Currents moving over rough bottom topography may also produce internal waves. If a thin layer of low-density surface water allows a ship's propeller to reach the pycnocline, the energy from the propeller creates internal waves; under this condition, the ship's propeller becomes inefficient because internal waves created at the pycnocline carry energy away from the vessel instead of driving the vessel forward. This results in a loss of speed that mariners call the "dead water effect."

The relationship of wavelength and depth to wave speed is similar for both internal and surface waves (see section 10.5). When the internal waves are short relative to the water depth, the density (rho = ρ) of both layers must be included in the equation. Wave speed squared (C^2) is equal to Earth's acceleration due to gravity (g) divided by 2π times the wavelength (L) times a ratio of densities, where ρ (rho) is the density of the lower layer and ρ' is the density of the upper layer:

$$C^2 = \frac{g}{2\pi} L \left[\frac{\rho - \rho'}{\rho + \rho'} \right]$$

When the internal waves are long relative to the water depth, it is necessary to include other relationships. For this equation, see appendix C.

Sea surface

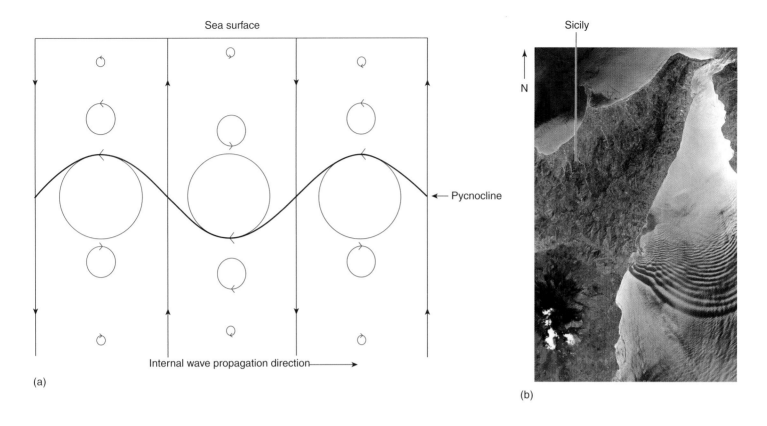

(a)

Internal wave propagation direction

Pycnocline

Sicily

N

(b)

Strait of Messina
October 25, 1995

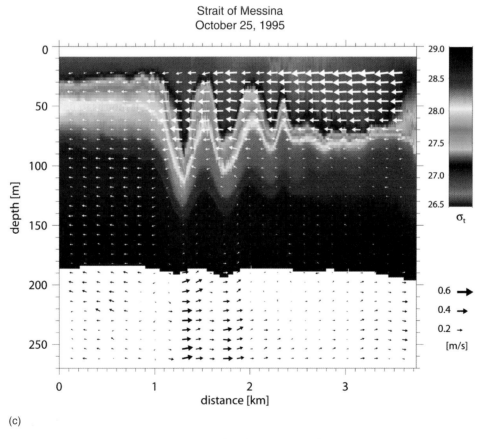

(c)

Figure 10.28 (a) Schematic illustration of an internal wave propagating along the base of the pycnocline (heavy solid line). The orbital motion of water particles is indicated by dashed lines. (b) Satellite image (58 km x 90 km, 36 mi x 56 mi) of internal waves moving through the Strait of Messina separating the island of Sicily from the Italian peninsula. (c) Color-coded density variations with depth (as measured by a CTD, see fig. 8.10) revealing the movement of internal waves in the Strait of Messina. The speed of the waves (m/s) is indicated by the arrows.

10.11 Standing Waves

Deep-water waves, shallow-water waves, and internal waves are all progressive waves; they have a speed and move in a direction. **Standing waves** do not progress; they are progressive waves reflected back on themselves and appear as an alternation between a trough and a crest at a fixed position. They occur in ocean basins, partly enclosed bays and seas, and estuaries. A standing wave can be demonstrated by slowly lifting one end of a container partially filled with water and then rapidly but gently returning it to a level position. If this is done, the surface alternately rises at one end and falls at the other end of the container. The surface oscillates about a point at the center of the container, the **node;** the alternations of low and high water at each end are the **antinodes** (fig. 10.29a). A standing wave is a progressive wave reflected back on itself; the reflection cancels out the forward motions of the initial and reflected waves. If different-sized containers are treated the same way, the period of oscillation increases as the length of the container increases or its depth decreases.

Notice that the single-node standing wave contains one-half of a wave form (fig. 10.29a). The crest is at one end of the container, and the trough is at the other end. As the wave oscillates, a trough replaces the crest, and a crest replaces the trough. One wavelength, the distance from crest to crest or from trough to trough, is twice the length of the container. By rapidly tilting the basin back and forth at the correct rate, one can produce a wave with more than one node (fig. 10.29b). In the case of two nodes, there is a crest at either end of the container and a trough in the center; this configuration alternates with a trough at each end and the crest in the center. The two nodes are one-quarter of the basin's length from each end. In this case, note that the wavelength is equal to the basin's length. The oscillation period of the wave with two nodes is one-half that of the wave with a single node.

Standing waves in bays or inlets with an open end behave somewhat differently than standing waves in closed basins. A node is usually located at the entrance to the open-ended bay, so only one-quarter of the wavelength is inside the bay.

Figure 10.29a A standing wave oscillating about a single node in a basin. The time for one oscillation is the period of the wave, *T.*

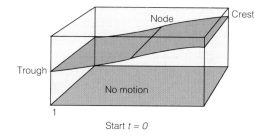
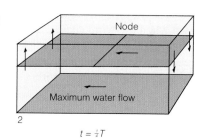
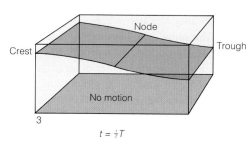
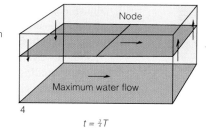
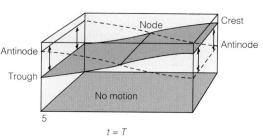

Figure 10.29b A standing wave oscillating about two nodes. The time for one oscillation is the period of the wave.

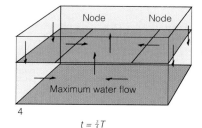
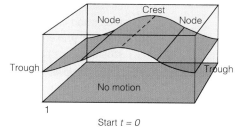
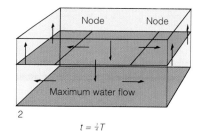
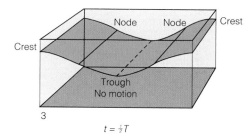
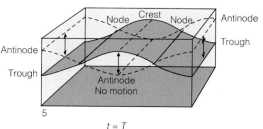

There is little or no rise and fall of the water surface at the entrance, but a large rise and fall occurs at the closed end of the bay (fig. 10.30). Multiple nodes may also be present in open-ended basins.

Standing waves that occur in natural basins are called **seiches,** and the oscillation of the surface is called seiching. In natural basins, the length dimension usually greatly exceeds the depth. Therefore, a standing wave of one node in such a basin behaves as a reflecting shallow-water wave, with the wavelength determined by the length or width of the basin. In water with distinct layers having sharp density boundaries, standing waves may occur along the fluid boundaries as well as at the air-sea boundary. The oscillation of the internal standing waves is slower than the oscillation of the sea surface.

Standing waves may be triggered by tectonic movements that suddenly shake a basin, causing the water to oscillate at a period defined by the dimensions of the basin. This phenomenon occurs during an earthquake when water sloshes back and forth in swimming pools. If storm winds create a change in surface level to produce storm surges, the surface may oscillate as a standing wave in the act of returning to its normal level when the wind ceases. The movement of an air-pressure disturbance over a lake may also cause periodic water-level changes, reaching a meter or more in height. Tidal currents moving through an area with a sharp pycnocline and

an irregular bottom topography may create internal waves that sometimes produce seiches.

If the period of the disturbing force is a multiple of the natural period of oscillation of the basin (an ocean basin or a smaller coastal basin), the height of the standing wave is greatly increased. For example, if a child is riding on a swing, a gentle push timed with each swing period forces the swing higher and higher. The push may be delivered each time the swing passes, every other time, or every third time; all are multiples of the natural period of the swing. In chapter 11, we will learn that repeating tidal forces at the entrance to a bay can produce standing waves in those basins that have natural periods of oscillation approximating the tidal period. In the open ocean, large oceanic basins sometimes have natural periods of oscillation that promote standing wave tides (chapter 11).

A standing wave in a basin is like a water pendulum. The wave's natural period of oscillation is

$$T = \left(\frac{1}{n}\right)\left(\frac{L}{\sqrt{gD}}\right)$$

where n is the number of nodes present, D is the depth of water in the basin, g is Earth's acceleration due to gravity, and L is the wavelength. L equals twice the basin length, l, in a closed basin and four times the basin length in a basin with an open end. This equation is related to the shallow-water wave equation when the number of nodes is equal to 1:

$$\frac{L}{T} = n\sqrt{gD}$$

A progressive wave directly reflected back on itself produces a standing wave, because the two waves—original and reflected—are moving at the same speed but in opposite directions. The checkerboard interference pattern produced by two matched wave systems approaching each other at an angle also creates standing waves, with crests and troughs alternating with each other in fixed positions (see fig. 10.7).

Figure 10.31 shows the relationship between the distribution of total oceanic wave energy and wave period. The energy of ordinary wind waves is high because these waves are always present and well distributed through all the oceans. Storm waves are larger and carry more energy, but they do not occur as frequently and are present over much less of the ocean area. Therefore, storm waves have less total energy than ordinary wind waves. Tsunami-type waves contain a large amount of energy, but they are infrequent and confined to fewer areas of the oceans. The tides, when considered as waves, concentrate their energy in two

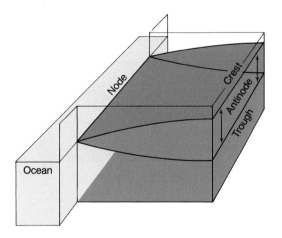

Figure 10.30   A standing wave oscillates about the node located at the opening to a basin. The antinodes produce the rise and fall of water at the closed end of the basin. This type of oscillation is produced by alternating water inflow and outflow at a period equal to the natural period of the basin.

Figure 10.31   The distribution of wave energy with wave period.

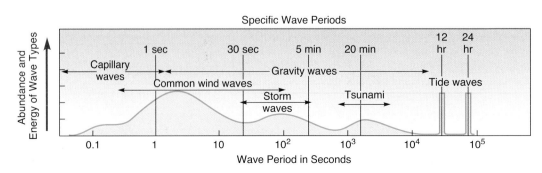

narrow bands centered on the twice-daily and once-daily tidal periods. Tide wave forms are discussed in chapter 11.

10.12 Practical Considerations: Energy from Waves

A tremendous amount of energy exists in ocean waves. The power of all waves is estimated at 2.7×10^{12} watts, which is about equal to 3000 times the power-generating capacity of Hoover Dam. Unfortunately for human needs, this energy is widely dispersed and not constant at any given location or time. It is, therefore, difficult to tap this supply to produce power, except in small quantities.

Wave energy can be harnessed in three basic ways: (1) using the changing level of the water to lift an object, which can then do useful work because of its potential energy; (2) using the orbital motion of the water particles or the changing tilt of the sea surface to rock an object to and fro; and (3) using rising water to compress air or water in a chamber. A combination of these may also be used. If the wave motion is used directly or indirectly to turn a generator, electrical energy may be produced.

Consider a large surface float with a hollow cylinder extending down into the sea (fig. 10.32). Inside the cylinder is a piston, and the up-and-down motion of the surface float causes the cylinder to move up and down over the piston, while the large drag plate restricts the motion of the piston. The system takes in water as the surface buoy rises on the crest of the waves and squirts water out as the surface buoy drops with the passing of the trough. The pumped water can be used to turn a turbine, but because wave energy is distributed over a volume of water, this mechanism does not withdraw much of the passing wave's energy. This system can be adapted to pump air rather than water. An air-compression system called *Sperboy* has been developed and tested in Great Britain. Air displaced by the oscillating water column is passed through turbine generators to produce energy. *Sperboy* is designed to be deployed in large arrays 8 to 12 miles offshore providing large-scale energy generation at a competitive cost.

Another system constructs a tapered channel perpendicular to the shore. Incoming waves force the water up 2–3 m (6–10 ft) in the narrow end of the channel, where it spills into an elevated storage tank, then down through a turbine. This system is used to generate power by a 75-kilowatt plant on Scotland's Isle of Islay, a 350-kilowatt plant at Toftestalen, Norway, and two 1500-kilowatt plants, one in Java and the other in Australia. Both the surface float and the tapered channel are examples of changing the level of an object or the water itself to create potential energy.

Wave power systems that use the orbital or rocking motion of the waves are under study in Great Britain. Long strings of mechanical power units are moored in water where waves are abundant. Each passing wave makes the power units move relative to

each other, causing pumps to move oil that passes through a turbine in a closed system.

In Western Australia, the Azores, and Japan, other systems using wave energy to compress air are being developed (fig. 10.33). Air traps can be installed along a wave-exposed

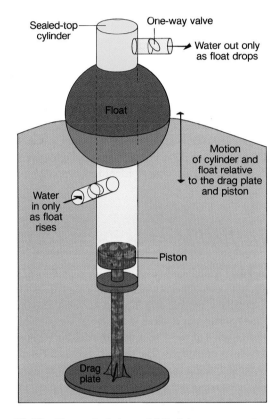

Figure 10.32 The vertical rise and fall of the waves can be used to power a pump.

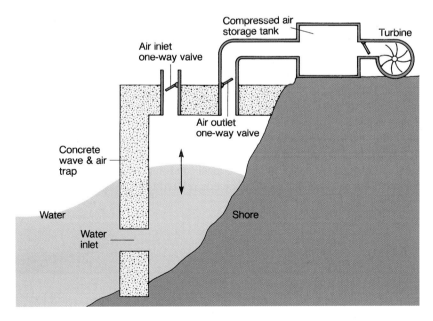

Figure 10.33 Each rise and fall of the waves pumps pulses of compressed air into a storage tank. A smooth flow of compressed air from the storage tank turns a turbine that generates electricity.

coast so that the crest of a wave moving into the trap compresses air, forcing it through a one-way valve; the air traps can also be constructed to pass air in either direction. This compressed air powers a turbine. The trough of the wave allows more air to enter the trap, readying it for compression by the following wave crest.

Shores that are continually pounded by large-amplitude waves are most likely to be developed for wave power. Great Britain has a coastline with frequent high-energy waves and an average wave power of about 5.5×10^4 watts (or 55 kilowatts) per meter of coastline. If the wave energy could be completely harnessed along 1000 km (620 mi) of coast, it would generate enough power to supply 50% of Great Britain's present power needs. Along the northern California coast, waves are estimated to expend 23×10^6 kilowatts of power annually; it is thought that 4.6×10^6 kilowatts, or 20%, could be harvested to generate electrical power. The Pacific Gas & Electric Company, a northern

California utility, has considered installing a generating device in a breakwater planned for Fort Bragg, California.

When we think about wave energy systems, thoughtful consideration needs to be given to items other than cost. If all the energy were extracted from the waves in a coastal area, what effect would this action have on the shore area? If the nearshore areas are covered with wave energy absorbers 5–10 m (15–33 ft) apart, what will the effect be on other ocean uses? Since the individual units collect energy at a slow rate, can they collect enough energy over their projected life span to exceed the energy used to fabricate and maintain them? Answers to these questions will help us understand that the harvesting of wave energy is not without an effect on the environment, that it may not be either cost- or energy-effective, and that its location may present enormous problems for installation, maintenance, and transport of energy to sites of energy use.

Summary

When the water's surface is disturbed, a wave is formed by the interaction between generating and restoring forces. The wind produces capillary waves, which grow to form gravity waves. The elevated portion of a wave is the crest; the depressed portion is the trough. The wavelength is the distance between two successive crests or troughs. The wave height is the distance between the crest and the trough. Wave period measures the time required for two successive crests or troughs to pass a location. The moving wave form causes water particles to move in orbits. The wave's speed is related to wavelength and period.

Deep-water waves occur in water deeper than one-half the wavelength. Wind waves generated in storm centers are deep-water waves. The period of a wave is a function of its generating force and does not change. Long-period waves move out from the storm center, forming long, regular waves, or swell. The faster waves move through the slower waves and form groups, or trains, of waves. The longer waves are followed by the shorter waves. This process is known as sorting, or dispersion. The speed of a group of waves is half the speed of the individual waves in deep water. Swells from different storms cross, cancel, and combine with each other as they move across the ocean.

Wave height depends on wind speed, wind duration, and fetch. Single large waves unrelated to local conditions are called episodic waves. The energy of a wave is related to its height.

When the ratio of the height to the length of a wave, or its steepness, exceeds 1:7, the wave breaks. The Universal Sea State Code relates wind speeds and sea-surface conditions.

Shallow-water waves occur when the depth is less than one-twentieth the wavelength. The speed of a shallow-water wave depends on the depth of the water. As the wave moves toward shore and decreasing depth, it slows, shortens, and increases in height. Waves coming into shore are refracted, reflected, and diffracted. The patterns produced by these processes helped people in ancient times to navigate from island to island.

In the surf zone, breaking waves produce a water movement toward the shore. Breaking waves are classified as plungers or spillers. Water moves along the beach as well as toward it; it is returned seaward through the surf zone by rip currents.

Tsunamis are seismic sea waves. They behave as shallow-water waves, producing severe coastal destruction and flooding.

Internal waves occur between water layers of different densities. Standing waves, or seiches, occur in basins as the sea surface oscillates about a node. Alternate troughs and crests occur at the antinodes.

The energy of the waves can be harnessed by using either the water-level changes or the changing surface angle associated with them. Difficulties include cost, location, environmental effects, and lack of wave regularity.

Key Terms

All key terms from this chapter can be viewed by term or definition when studied as flashcards on this book's website at www.mhhe.com/sverdrup10e.

generating force, 240	cat's-paws, 240	amplitude, 241
restoring force, 240	crest, 241	equilibrium surface, 241
gravity wave, 240	trough, 241	wave period, 241
ripple, 240	wavelength, 241	orbit, 242
capillary wave, 240	wave height, 241	deep-water wave, 242

Study Questions

1. A surfboard slides downward on the face of a wave. The steepness of the wave face is governed by the decrease of L (wavelength) and the increase of H (wave height) as the wave slows in shallow water. How must the surfer adjust the board in order to stay on the face of the wave as the wave approaches shallow water?
2. Locate a small pond or pool and drop a stone into it. Describe what happens (1) to an individual wave and (2) to the group of waves. Try to determine the group speed and the individual wave speed.
3. Drop two stones into the pond at a short distance from each other. Describe what happens when the wave rings produced by the two stones pass through each other. Do the heights of the waves change when they intersect? Do the wave trains pass through each other and continue on?
4. List the forces that act on a smooth-water surface to create deep-water wind waves.
5. If you were sailing at night in the trade-wind belt, how could you use the waves to keep you on a course of constant direction?
6. Make a sketch of an ideal progressive wind wave in deep water. Label the parts.
7. What happens to a deep-water progressive wave when it moves into shallow water and up a sloping beach?
8. Compare a tsunami and a storm surge (see chapter 7). How are they the same? How are they different?
9. Distinguish among (a) sea and swell, (b) wave height and wave steepness, (c) wave height and wave amplitude, (d) plunger and spiller, and (e) node and antinode.
10. What is the effect of sorting (dispersion) on waves moving away from a storm center?
11. How do refraction, reflection, and diffraction affect a wave?
12. How is a standing wave related to a progressive wave?
13. Explain two ways in which wave energy could be harnessed to provide useful power. What are the advantages and disadvantages of each method?
14. If a group of mixed waves is generated in a sudden storm, why does it take more time for the group to pass an island far from the storm center than to pass an island near it?
15. A depression in the sea floor at right angles to a straight coastline may be the site of a rip current. Why?

Study Problems

1. Using the equations $C = L/T$ and $L = (g/2\pi)T^2$, show that wave speed can be determined from (a) wave period only and (b) wavelength only.
2. What is the period of a wave moving in deep water at 10 m/s, if its wavelength is 64 m? When it enters shallow water, what will happen to the wave's speed, length, period, and height? How high will the wave have to be to break in deep water?
3. A submarine earthquake produces a tsunami in the Gulf of Alaska. How long will it take the tsunami to reach Hawaii if the average depth of the ocean over which the waves travel is 3.8 km and the distance is 4600 km?
4. Explain wave dispersion. How far from a storm center will waves with periods of twelve seconds, nine seconds, and six seconds have traveled after twelve hours? If the six-second waves arrive at your beach ten hours after the twelve-second waves, how far away is the storm?
5. Fill a rectangular aquarium or dishpan approximately one-third full of water. Measure the water depth (D). Carefully lift one end of the container and set it down rapidly and smoothly. Time the period between successive high waters at one end (T); this is the wave period. The wavelength (L) is twice the length of the container. Show that C (the wave speed) determined from L/T is equal to C determined from $n\sqrt{gD}$.

Going to Sea

Oceanographic research vessels are operated by the U.S. Navy, NOAA, universities, private corporations, and foundations. They are the platforms staffed by professional crews and provide accommodations, laboratory space, and the equipment required to handle deep-sea sampling gear. Each vessel is scheduled for long periods at sea, so scientists and equipment are flown to and from the nearest port as the scientific vessel changes. Research vessels are often shared by scientific programs from different institutions to create cost-effective use. While the ships operate approximately 250 days at sea each year, a specific research cruise may only last a month.

The University of Washington's Research vessel, Thomas G. Thompson, *began its research duties in 1992.*

Scientists and students cool off in a purpose-built swimming pool.

Everyone on board enjoys dragging a line when the ship is under way. Sometimes they are lucky and fresh fish is on the menu.

Filtering water for chlorophyll a and POC/PON measurements in a shipboard laboratory cold room on the CCGS Amundsen. *When working in Arctic environments, biological oceanographers must take care to process many samples at as close to in situ temperature as possible.*

Arctic Ocean recovery of a Mock-Ness net tow used to sample zooplankton communities and larval fish on the CCGS Amundsen.

Deployment of a spade corer from the CCGS Amundsen *to sample bottom sediments in the Arctic.*

Subsampling a kasten core in the wet lab on the R/V Endeavor *in the northwest Mediterranean.*

Monitoring weather conditions and winch cable payout when coring on the R/V Endeavor *in the northwest Mediterranean.*

Recovery of the HOV Alvin *on board R/V* Atlantis *after a dive to 2200 m (7263 ft) on the Endeavour Segment of the Juan de Fuca Ridge.* Alvin *has a depth capability of 4500 m (14,764 ft) and is able to reach nearly 63% of the global ocean floor with two scientists and a pilot inside. A new* Alvin, *expected to be completed in spring 2010, will have a depth capability of 6500 m (21,325 ft) and will be capable of taking three scientists to 98% of the sea floor with increased speed, battery life, and visibility to conduct their science.*

Readying a sediment trap for deployment in Glacier Bay, Alaska.

Deploying a zooplankton net in Glacier Bay, Alaska.

Deploying a CTD in Glacier Bay, Alaska.

An oceanographer's first dive in Alvin *always ends with a cold seawater bath, as well as other pranks (shoes frozen in blocks of ice or a fish in bed) to celebrate the occasion.*

Oceanographers aboard the R/V Knorr *await deployment of the Niskin bottles mounted on a CTD (conductivity, temperature, depth) rosette. Water sampling is being conducted in the Benguela Current, located off the southwestern coast of Africa.*

A small boat from the NOAA ship HI'I ALAKAI (shown in the background) collects water samples from various locations around American Samoa. These operations are part of NOAA's Coral Reef Conservation Program (www.coralreef.noaa.gov).

Shallow water (<30 m) CTD cast being conducted from a small boat in the nearshore waters off American Samoa.

Hopewell Rocks on the beach at low tide, Bay of Fundy, Nova Scotia, Canada.

Chapter 11

The Tides

Chapter Outline

Learning Outcomes

After studying the information in this chapter students should be able to:

1. *compare* and *contrast* diurnal, semidiurnal, and semidiurnal mixed tides,

2. *label* the basic characteristics of the three tidal patterns listed above,

3. *explain* why the Moon's tide-raising force is greater than the Sun's despite the much larger mass and gravitational attraction of the Sun,

4. *diagram* the Earth-Moon-Sun system during spring and neap tides,

5. *calculate* the time of the next high tide for diurnal and semidiurnal tides, given the time of the last high tide,

6. *sketch* the Earth-Moon system that leads to declinational tides,

7. *describe* the effect that distance from the amphidromic point has on tidal range,

8. *illustrate* the motion of the ocean surface in a rotary standing tide, and

9. *discuss* the prospects for capturing energy from the tides.

Best known as the rise and fall of the sea around the edge of the land, the tides are caused by the gravitational attraction between Earth and the Sun and between Earth and the Moon. Far out at sea tidal changes go unnoticed, but along the shores and beaches the tides govern many of our water-related activities, both commercial and recreational. Early sailors from the Mediterranean Sea, where the daily tidal range is less than 1 m (3 ft), ventured out into the Atlantic and sailed northward to the British Isles; to their amazement, they found a tidal range in excess of 10 m (30 ft). The movement of the tide into and out of bays and harbors has been helpful to sailors beaching their boats and to food-gatherers searching the shore for edible plants and animals, but it is also recognized as a hazard by navigators and can produce some spectacular effects when rushing through narrow channels.

In this chapter, we survey tide patterns around the world and explore the tides in two ways: one is a theoretical consideration of the tides on an Earth with no land; the other is a study of the natural situation. We also show how to use available tide data to predict water-level changes and coastal tidal currents.

11.1 Tide Patterns

Measurements of tidal movements around the world show us that the tides behave differently in different places. In some coastal areas, a regular pattern occurs of one high tide and one low tide each day; this is a **diurnal tide.** In other areas, a cyclic high water–low water sequence is repeated twice in one day; this is a **semidiurnal tide.** In a semidiurnal tidal pattern, the two high tides reach about the same height and the two low tides drop to about the same level. A tide in which the high tides regularly reach different heights and the low tides drop regularly to different levels is called a **semidiurnal mixed tide.** This type of tide has a diurnal (or daily) inequality, created by combining diurnal and semidiurnal tide patterns. The tide curves in figure 11.1 show each type of tide.

Curves for typical tides at some U.S. coastal cities are shown in figure 11.2.

11.2 Tide Levels

In a uniform diurnal or semidiurnal tidal system, the greatest height to which the tide rises on any day is known as **high water,** and the lowest point to which it drops is called **low water.** In a mixed-tide system, it is necessary to refer to **higher high water** and **lower high water,** as well as **higher low water** and **lower low water** (see fig. 11.1).

Tide measurements taken over many years are used to calculate the **average** (or **mean**) **tide** levels. Averaging all water levels over many years gives the local mean tide level. Averages are also calculated for the high-water and low-water levels, as mean high water and mean low water. For mixed tides, mean higher high water, mean lower high water, mean higher low water, and mean lower low water are calculated.

Because the depth of coastal water is important to safe navigation, an average low-water reference level is established; depths are measured from this level for navigational charts. The tide level is added to this charted depth to find the true depth of water under a vessel at any particular time. In areas of uniform diurnal or semidiurnal tide patterns, the zero depth reference, or **tidal datum,** is usually equal to mean low water. The use of mean low water assures the sailor that the actual depth of the water is, in general, greater than that on the chart. In regions with mixed tides, mean lower low water is used as the tidal datum for the same reason. When the low-tide level falls below the mean value used as the tidal datum, a **minus tide** results. A minus tide can be a hazard to boaters, but it is cherished by clam diggers and students of marine biology, because it exposes shoreline usually covered by the sea.

As the water level along the shore increases in height, the tide is said to be rising or flooding; a rising tide is a **flood tide.** When the water level drops, the tide is falling or ebbing; a falling tide is an **ebb tide.**

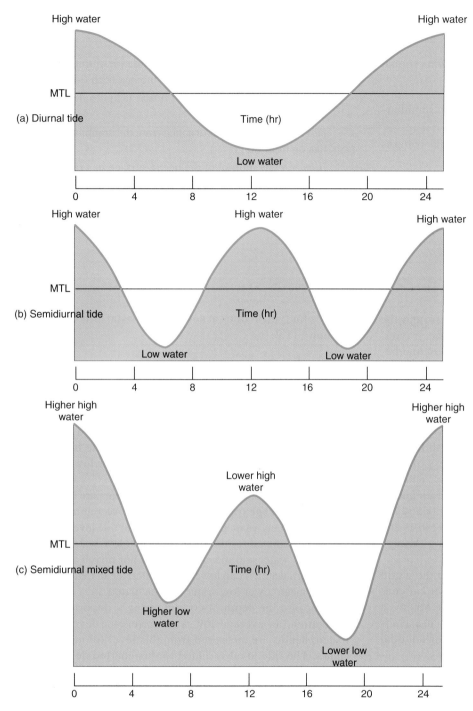

Figure 11.1 The three basic types of tides: (a) a once-daily, diurnal, tide; (b) a twice-daily, semidiurnal, tide; and (c) a semidiurnal mixed tide with diurnal inequality. MTL equals mean tide level.

11.3 Tidal Currents

Currents are associated with the rising and falling of the tide in coastal waters. These **tidal currents** may be extremely swift and dangerous as they move water into a region on the flood tide and out of the region on the ebb tide. When the tide turns, or changes from an ebb to a flood or vice versa, a period of **slack water** occurs during which the tidal currents slow and then reverse. Slack water may be the only time that a vessel can safely

navigate a narrow channel with swiftly moving tidal currents, sometimes in excess of 5 m/s (10 knots). The relationships of tidal currents to standing wave tides, to progressive tides, and to tidal current prediction are discussed in this chapter.

11.4 Equilibrium Tidal Theory

Oceanographers analyze tides in two ways. The tides are studied as mathematically ideal wave forms behaving uniformly in response to the laws of physics. This method is called **equilibrium tidal theory.** It is based on an Earth covered with a uniform layer of water, in order to simplify the relationships between the oceans and the tide-rising bodies, the Moon and the Sun. The tides are also studied as they occur naturally; this method is called **dynamic tidal analysis.** It studies the oceans' tides as they occur, modified by the landmasses, the geometry of the ocean basins, and Earth's rotation.

The effects of the Sun's and the Moon's gravity and the rotation of Earth on tides are most easily explained by studying equilibrium tides. In this discussion, Earth and the Moon act as a single unit, the Earth-Moon system that orbits the Sun (fig. 11.3). The Moon orbits Earth, held by Earth's gravitational force acting on the Moon (B in fig. 11.3a). There is also a force acting to pull the Moon away from Earth and send it out into space (B′ in fig. 11.3b). In the first part of this discussion, B′ is considered a **centrifugal force.** A centrifugal force is an apparent force present when one judges motion against a rotating frame of reference. Forces B and B′ must equal and opposite to keep the Moon in its orbit. Likewise, the Moon's gravitational force acting on Earth (C in fig. 11.3a) must be balanced by the apparent, or centrifugal, force (C′ in fig. 11.3b). B′ and C′ are present because the Earth-Moon system is rotating about a common axis at the system's center of mass, which is located 4640 km (2880 mi) from Earth's center along a line between Earth and the Moon. The Earth-Moon system is held in orbit about the Sun by the Sun's gravitational attraction (A in fig. 11.3c), while a centrifugal force again acts to pull the Earth-Moon system away from the Sun (A′ in fig. 11.3c). For the Earth-Moon system to remain in solar orbit, the gravitational forces must equal the centrifugal forces (fig. 11.3c).

Sir Isaac Newton's universal law of gravitation tells us that the force of attraction between any two bodies is proportional

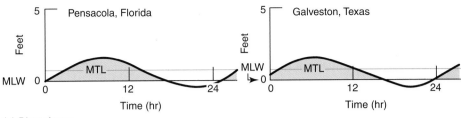

(a) Diurnal type

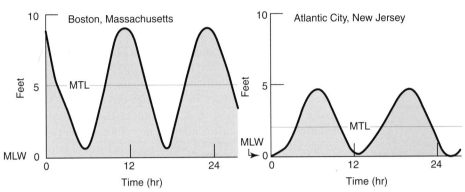

(b) Semidiurnal type

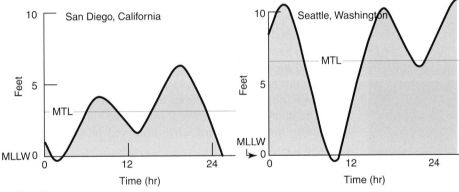

(c) Semidiurnal mixed type

Figure 11.2 Tide types and tidal ranges vary from one coastal area to another. The zero tide level equals mean low water *(MLW)* or mean lower low water *(MLLW)*, as appropriate. *MTL* equals mean tide level. All tide curves are for the same date. Tide types at any one location can change with time. These tide curves are intended to illustrate the characteristics of different tide types and should not be interpreted as indicating that these locations experience only one tide type.

If the gravitational forces are calculated for each unit mass of material at Earth's surface and at its center, the gravitational forces at Earth's surface change with location because of the changing distance between a unit mass on Earth's surface and the center of the Sun or Moon. The centrifugal force acting on each unit mass on Earth is constant and is equal and opposite to the gravitational force acting on a unit mass at Earth's center.

The difference in force per unit mass between the gravitational force of the Moon or the Sun and the centrifugal force at an Earth surface point is proportional to $G(M/R^3)$, where G is the gravitational constant, M is the mass of the Moon or the Sun, and R is the distance between the centers of Earth and the Moon or between the centers of Earth and the Sun.

The gravitational force exerted by the Moon on a unit mass is larger at the point on Earth's surface that is closest to the Moon than on a unit mass at Earth's center, because the distance between Earth's surface and the Moon is $R - r$. The quantity r is the radius of Earth. The gravitational force difference or excess acting toward the Moon pulls a unit mass away from Earth's center and produces a tide-raising force. At a point on the side of Earth opposite the Moon, the gravitational force is less than the constant centrifugal force because the distance is $R + r$. The centrifugal force difference or excess that pulls a unit mass away from Earth's center is also a tide-raising force. The calculation of these tide-raising forces is shown in appendix C.

A second way to understand tide-raising forces is to think of the Sun or the Moon as exerting a gravitational force that continuously attracts Earth toward the Sun or toward the Moon. In this case, the gravitational force, or **centripetal force,** of the Sun or the Moon holds Earth in orbit. The centripetal force is constant and equal to the average gravitational force of the Sun or the Moon acting at Earth's center. In this case, the difference between the average gravitational attraction and the gravitational attraction at individual points on Earth produces the tide-raising force. The force difference per unit mass between a surface point and Earth's center is proportional to $G(M/R^3)$ in this second case as well. The two approaches yield the same results. The derivation of $G(M/R^3)$ is found in appendix C.

In the Earth-Moon-Sun system, the mass of the Sun is very great, but the Sun is very far away. By contrast, the Moon is small, but it is close to Earth. Calculating the distribution of

to the product of the two masses divided by the square of the distance between the centers of the masses:

$$F = G\left(\frac{m_1 m_2}{R_2}\right)$$

where

G = universal gravitational constant, 6.67×10^{-8} cm³/g/s²
m_1 = mass of body 1 in grams
m_2 = mass of body 2 in grams
R = distance between centers of masses in centimeters

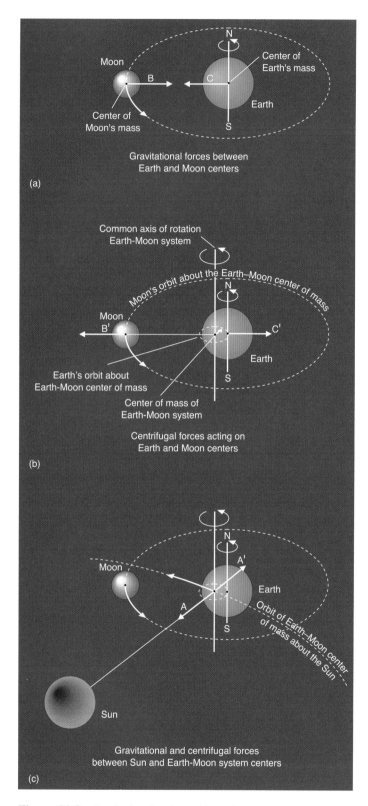

Figure 11.3 Gravitational and centrifugal forces act to keep the Earth-Moon system in balance.

these forces for each water particle at Earth's surface shows that the Moon has a greater attractive effect on the water particles than the Sun. In the following discussion, the effects of each tide-raising body are considered separately.

The Moon Tide

The water particles on the side of Earth facing the Moon are closest to the Moon and are acted on by the excess Moon gravitational force. Because the water is liquid and deformable, this force moves the water particles on Earth's surface toward a point directly under the Moon. This movement produces a bulge in the water covering. At the same time, the centrifugal force of the Earth-Moon system acting on the water particles at Earth's surface opposite the Moon creates an opposing bulge. This centrifugal force is equal to the magnitude of the excess gravity tide-raising force and is proportional to $-G(M/R^3)$. The minus sign indicates that this excess centrifugal force is acting opposite to the Moon's tide-producing gravity force. If we place the Moon opposite Earth's equator and then stop the Earth model and its Moon in space and time, we see the bulges in the water covering, as shown in figure 11.4. Remember that our model Earth initially had a water covering of uniform depth; therefore, as the two bulges are created, an area of low-water level is formed between the bulges. We now have a water covering with two bulges and two intervening depressions (or two crests and two troughs, or two high-tide levels and two low-tide levels) distributed around the equator.

Earth makes one rotation in about twenty-four hours, and the bulges (or crests) in the water covering tend to stay under the tide-producing body as Earth turns. As a result, a point on Earth that is initially at a crest (or high tide) passes to a trough (or low tide), to another high tide, to another low tide, and back to the original high tide as Earth completes one revolution. You can follow this process in figure 11.5. The effect created by the motion of Earth as it turns to the east when observed from space can also be interpreted as a wave form, for the bulges in the water covering move westward when observed relative to a fixed position on Earth.

The Tidal Day

While Earth turns eastward upon its axis, the Moon is moving in the same direction along its orbit about Earth. After twenty-four hours, the Earth point that began directly under the Moon is no longer directly under the Moon. Earth must turn for an additional fifty minutes, about 12°, to bring the starting point on Earth back in line with the Moon. Therefore, a **tidal day** is not twenty-four hours long but twenty-four hours and fifty minutes long. This difference also explains why corresponding tides arrive at any location about one hour later each day. This relationship is shown in figure 11.6.

The Tide Wave

The tides produced in this example are semidiurnal, with two highs and two lows each day. The tidal distortion of the model's water covering produces a wave form known as the **tide wave.** The crest of the tide wave is the high-water level, and the trough of the tide wave is the low-water level. The wavelength in this example is half the circumference of Earth,

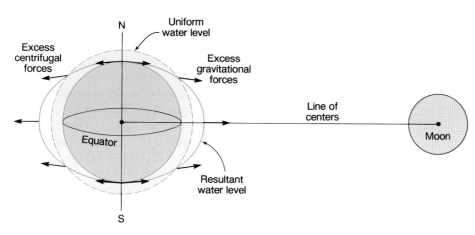

Figure 11.4　Distribution of tide-raising forces on Earth. Excess lunar gravitational and centrifugal forces distort the Earth model's water envelope to produce bulges and depressions. View is roughly in Earth's equatorial plane.

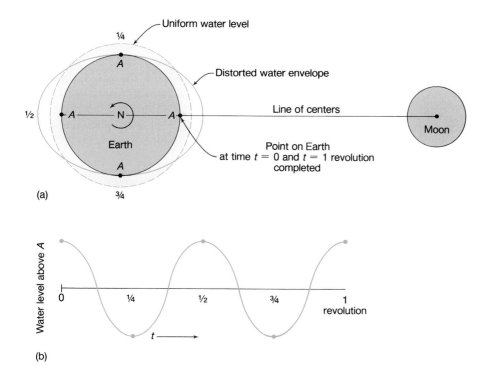

Figure 11.5　The change in water level at point *A* during one Earth rotation through the distorted water envelope [see fig. 11.4]. Fractions indicate portions of a revolution. View is down on Earth's North Pole, perpendicular to Earth's equatorial plane.

and the tide wave's period is about twelve hours and twenty-five minutes.

The Sun Tide

Although the Moon plays the greater role in producing the tides, the Sun produces its own tide wave. Despite the Sun's large mass, it is so far away from Earth that its tide-raising force is only 46% that of the Moon. The time required for Earth to revolve on its axis with respect to the Sun is on average twenty-four hours, not twenty-four hours and fifty minutes, as in the

case of the Moon tide. For this reason, the tide wave produced by the Moon is not only of greater magnitude than that produced by the Sun, but it also continually moves eastward relative to the tide wave produced by the Sun. Because the tidal forces of the Moon are greater than those of the Sun, the tidal period of the Moon is more important, and the tidal day is considered to be twenty-four hours and fifty minutes.

Spring Tides and Neap Tides

The Moon's orbit requires 29½ days relative to a point on Earth. During this period the Sun, Earth, and the Moon move in and out of phase with each other. At the new Moon, the Moon and Sun are on the same side of Earth, so the high tides, or bulges, produced independently by the Moon and the Sun coincide (fig. 11.7). Because the water level is the result of adding the two wave forms together, tides of maximum height and depression, or tides with the greatest **range** between high water and low water, are produced. These tides are known as **spring tides.** The vertical displacement, or amplitude, of the tide is one-half the range—the distance above or below mean tide level.

In a week's time, the Moon is in its first quarter; it has moved eastward along its orbit (about 12° per day) and is located approximately at right angles (or 90°) to the line between the centers of Earth and the Sun. The crest, or bulge, of the Moon tide is at right angles to the tide wave created by the Sun; the crests of the Moon tide will coincide with the troughs of the Sun tide, and the same will be true of the Sun's tide crests and the Moon's tide troughs (fig. 11.7). The crests and troughs tend to cancel each other out, and the range between high water and low water is small, producing low-amplitude **neap tides.**

At the end of another week, the Moon is full, and the Sun, Moon, and Earth are again lined up, producing crests that coincide and tides with the greatest range between high and low waters, or spring tides. These spring tides are followed by another period of neap tides, produced by the Moon in its last quarter when it again stands at right angles to the Sun-Earth line (fig. 11.7). The tides follow a four-week cycle of changing amplitude, with spring tides occurring every two weeks and a period of neap tides occurring in between. This progression can be seen in the portions of the tide records reproduced in figure 11.8. The effect occurs each lunar month and is the result of the Moon's tide wave moving around Earth relative to the Sun's tide wave.

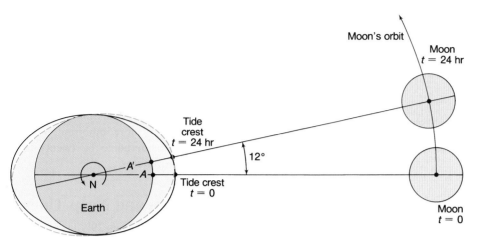

Figure 11.6 Point *A* requires twenty-four hours to complete one Earth rotation. During this time, the Moon moves 12° east along its orbit, carrying with it the tide crest. To move from *A* to *A′* requires an additional fifty minutes to complete a tidal day. View is down on Earth's North Pole, perpendicular to Earth's equatorial plane.

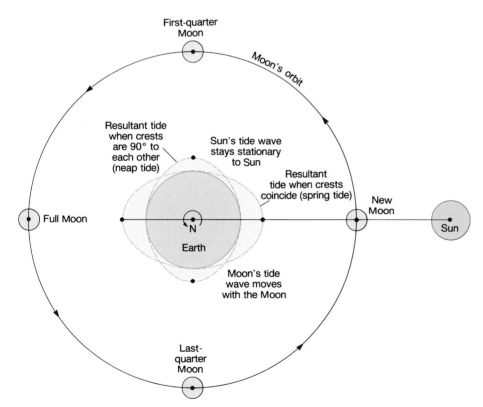

Figure 11.7 Spring tides result from the alignment of Earth, Sun, and Moon during the full Moon and the new Moon. During the Moon's first and last quarters neap tides are produced. The tidal range or vertical difference between high and low water is reduced during neap tides. View is down on Earth's North Pole, perpendicular to Earth's equatorial plane.

Declinational Tides

If the Moon or the Sun stands north or south of Earth's equator, one bulge, or high water, is in the Northern Hemisphere and the other is in the Southern Hemisphere (fig. 11.9). Under these conditions, a point at the middle latitudes on Earth's surface

passes through only one crest, or high tide, and one trough, or low tide, each tidal day. A diurnal tide, often called a **declinational tide,** is formed, because the Moon or the Sun is said to have declination when it stands above or below the equator.

Declinational (or diurnal) tides are influenced by both the Moon and the Sun. The Sun stands above 23½°N at the summer solstice and above 23½°S at the winter solstice. This variation causes the bulge created by the Sun to oscillate north and south of the equator in a regular fashion each year and tends to create more diurnal Sun tides during the winter and summer than during the spring and fall. The Moon's declination varies between 28½°N and 28½°S with reference to Earth. The Moon's orbit is inclined 5° to the Earth-Sun orbit, and it takes 18.6 years for the Moon to complete its cycle of maximum declination. When the Sun's and the Moon's declinations coincide, both tide waves become more diurnal. Each lunar month, the Moon travels from a declination of 5° above the Earth-Sun orbit plane to a declination of 5° below the plane and back.

Elliptical Orbits

The Moon does not move about Earth in a perfectly circular orbit, nor does Earth orbit the Sun at a constant distance. These orbits are elliptical, and therefore, Earth is closer to a certain tide-raising body at some times during an orbit than at other times. During the Northern Hemisphere's winter, Earth is closest to the Sun; therefore, the Sun plays a greater role as a tide producer in winter than in summer.

11.5 Dynamic Tidal Analysis

Equilibrium tidal theory helps us understand the distribution of wave-level changes and tide-raising forces, but it does not explain the tides as observed on the real Earth. Return to figure 11.2 and notice the variety of tidal ranges and tidal periods that appear at different locations on the same date. Refer also to figure 11.8, which shows different tides at different places during the same time period. The Sun-Moon-Earth system for all locations in the two figures is the same, but the equilibrium tidal theory does not explain natural tides at any particular location. Investigating the actual tides requires the dynamic approach, a mathematical study of tide waves as they occur.

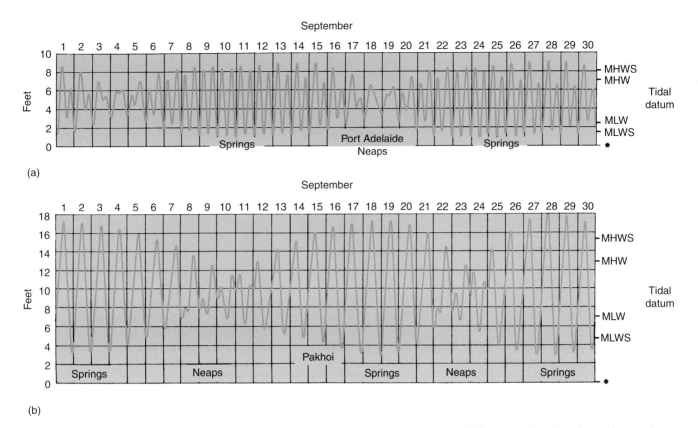

Figure 11.8 Spring and neap tides alternate during the tides' monthly cycle. *MHWS* is the mean high water spring tides; *MLWS* is mean low water spring tides. (a) A semidiurnal tide from Port Adelaide, Australia. (b) A diurnal tide from Pakhoi, China.

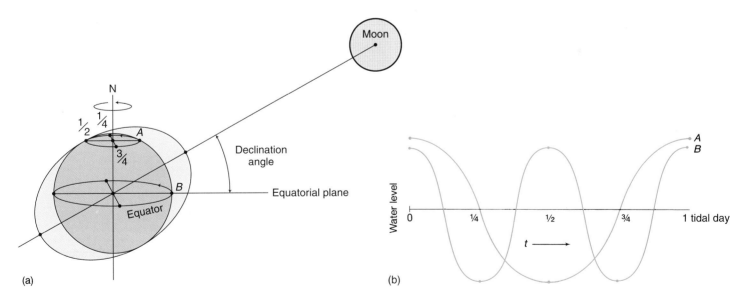

Figure 11.9 (a) The declination of the Moon produces a diurnal tide at latitude *A* and a semidiurnal tide at latitude *B*. (b) Fractions indicate portions of the tidal day. View is roughly in Earth's equatorial plane.

The Tide Wave

The behavior of the natural tide wave varies considerably from the tide wave of the water-covered model. Because the continents separate the oceans, the tide wave is discontinuous; the wave starts at the shore, moves across the ocean, and stops at the next shore. Only in the Southern Ocean around Antarctica

do the tide waves move continuously around Earth. A tide wave has a long wavelength compared to the depth of the oceans; therefore, it behaves like a shallow-water wave, with its speed controlled by the depth of the water. Because the wave is contained within the ocean basins, it can oscillate in the basin as a standing wave, and it is also reflected from the edge of the continents, refracted by changes in water depth, and diffracted as it passes through gaps between continents. In addition, the persistence of the tidal motion and the scale on which it occurs are so great that the Coriolis effect plays a role in the water's movement. All these factors together produce Earth's real tides. Because these interactions are complex, it is not possible to understand them together until each is first considered separately.

The tide wave's speed as a **free wave** moving across the water's surface is determined by the depth of the water. The tide wave moves as a free wave at about 200 m/s (400 mi/h) in a water depth of 4000 m (13,200 ft), but at the equator, Earth moves eastward under the tide wave at 463 m/s (1044 mi/h). This is more than twice the speed at which the tide can travel freely as a shallow-water wave. Under these conditions, the tide moves as a **forced wave** that is the result of the Moon's attractive force and Earth's rotation. Because Earth turns eastward faster than the tide wave moves freely westward, friction displaces the tide crest to the east of its expected position under the Moon. This eastward displacement continues until the friction force is balanced by a portion of the Moon's attractive force. These two forces, when balanced, hold the tide crest in a position to the east of the Moon rather than directly under it. This process is illustrated in figure 11.10.

Above 60°N or 60°S, the distance around a latitude circle is less than one-half the distance around the equator. Here, the free propagation speed of the tide wave equals the speed at which the sea floor moves under the wave form. Under these conditions, the crest of the tide stays aligned with the Moon, and less friction is generated between the rotating Earth and the moving wave form. Friction between the moving tide wave and the turning Earth also acts to slow the rotation rate of Earth, adding about 1½ milliseconds per 100 years to the length of a day.

Tidal motion extends to all depths in the oceans and can produce internal waves along water density boundaries between the upper and deeper layers in the oceans. Internal waves may become unstable and break, forming smaller waves and generating turbulence, which mixes water and dissipates energy. This tidal turbulence at mid-ocean depths becomes an important mechanism for resupplying nutrients from depth to the upper layers of the oceans.

Progressive Wave Tides

In a large ocean basin, the tide wave moving across the sea surface like a shallow-water wave is a **progressive tide.** Examples of progressive tides are found in the western North Pacific, the eastern South Pacific, and the South Atlantic Oceans. **Cotidal lines** are drawn on charts to mark the location of the tide crest at set time

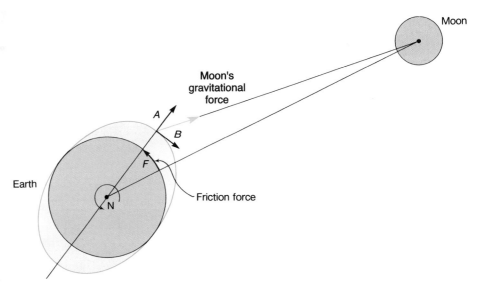

Figure 11.10   The crest of the tide wave is displaced eastward until the *B* component of the gravitational force balances the friction (*F*) between Earth and the tide wave. Component *A* of the gravitational force is the tide-raising force. Component *B* causes the tide wave to move as a forced wave. View is down on Earth's North Pole, perpendicular to Earth's equatorial plane.

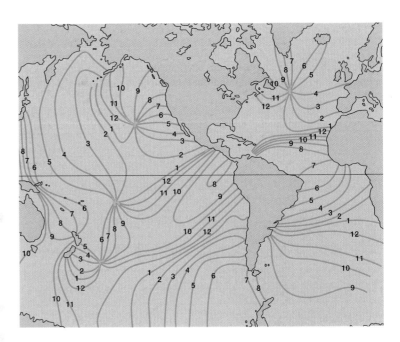

Figure 11.11   Cotidal lines for the world's oceans. The high-tide crest occurs at the same time along each cotidal line. Positions of the high tide are indicated for each hour over a twelve-hour period for semidiurnal tides.

intervals, generally one hour apart. The cotidal lines for the world's ocean tides are shown in figure 11.11 (see also box figure 1 in box titled "Measuring Tides from Space" in this chapter).

Because the tide wave is a shallow-water wave, the water particles move in elliptical orbits, and their motion extends to the sea floor. The horizontal component of the motion greatly exceeds the vertical motion. Because the time in which the water particles move in one direction is so long (one-half the tide period), the Coriolis effect becomes important. In the

Northern Hemisphere, the water particles are deflected to the right, and in the Southern Hemisphere, they are deflected to the left. This deflection causes a clockwise rotation of the water in the Northern Hemisphere and a counterclockwise rotation in the Southern Hemisphere. This circular (or rotary) movement is the oceanic tidal current described at the end of the "Standing Wave Tides" section.

Standing Wave Tides

In some ocean basins or parts of ocean basins, the tide wave is reflected from the edge of the continents, and a **standing wave tide** is produced (see chapter 10). Remember that if a container of water is tipped so that the water level is high at one end and low at the other, the water flows to the low end, raising the water level at that end as the water level at the high end drops. This movement produces a wave having a wavelength that is twice the length of the container, with antinodes at the ends of the basin and a node at the basin's center. This same process occurs in ocean basins but with some important modifications.

The high-water to low-water change in an ocean basin requires a long time, and the Coriolis effect must be included. The moving water, deflected to the right in the Northern Hemisphere, does not reach the low-tide end but instead is deflected to a position to the right of the initial high-tide position. This movement causes the tide crest to rotate counterclockwise around the basin in which it oscillates, but the tidal current rotates clockwise, because the current is deflected to the right in the Northern Hemisphere (see fig. 11.13). In the Southern Hemisphere, these directions are reversed.

In the **rotary standing tide wave,** the node becomes reduced to a central point, while the tide crest (shown as cotidal lines) progresses around the edges of the basin. (See figure 11.11 for a demonstration of this pattern in the northeastern and southwestern Pacific and in the North Atlantic.) The central point, or node, for a rotary tide is called the **amphidromic point.** The distance between low- and high-water levels, or the tidal range, for a rotary tide is shown on a chart by a series of lines decreasing in value as they approach the amphidromic point. The lines of equal tidal range are called **corange lines** (fig. 11.12). Near the amphidromic point the tidal range is small; the farther from the amphidromic point, the greater the range. Because the amphidromic point is located near the center of an ocean basin, many mid-ocean areas have small tidal ranges, while the shores of the landmasses forming the sides of the basins have larger tidal ranges. The value and position of corange lines in mid-ocean are not as well known as they are near shore.

The flow of water from the high-water side to the low-water side of a standing tide wave produces a rotating tidal current, as shown in figure 11.13. A rotating tidal current is also produced by the orbital motion of water particles in a progressive tide wave. If the tide wave is diurnal, the water particles travel in one complete circle in a tidal day. A

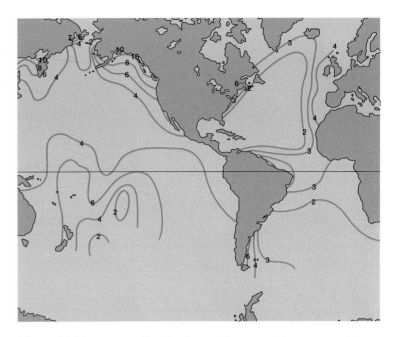

Figure 11.12   Corange lines for the world's oceans. These corange lines connect positions with the same spring tidal range. Open-ocean tidal ranges are less well known than nearshore ranges, where tide-level recorders are commonly used. Satellites now measure water elevation changes in mid-ocean; see the box titled "Measuring Tides from Space" in this chapter.

semidiurnal tide causes two circles, and a semidiurnal mixed tide produces two circles of unequal size. An example of a rotating semidiurnal mixed tidal current recorded at the Columbia River lightship in the North Pacific is presented in figure 11.14.

Rotary standing tides occur in basins in which the natural period of the basin approximates the tidal period. If the tidal period and the oscillation period of the basin coincide, the tides increase in amplitude. Table 11.1 relates basin depths and lengths or widths that produce natural oscillations equal to tidal periods. Remember from chapter 10 that the natural period of oscillation of a standing wave in a basin is

$$T = \left(\frac{1}{n}\right)\left(\frac{L}{\sqrt{gD}}\right)$$

where L, the wavelength, is twice the basin length, l, in closed basins and four times l in open-ended basins. A comparison of the values given in table 11.1 shows that deep-ocean basins must have great length to accommodate standing waves with tidal periods, whereas shallow basins may be much shorter. Most tides are semidiurnal, but dimensions of some basins cause the basin to resonate with a diurnal tidal period rather than a semidiurnal period. See the tide curves in figure 11.2 for Pensacola, Florida, and Galveston, Texas. In an open-ended tidal basin with a mixed tide, the semidiurnal portion of the tide may cause the basin to resonate with a node at the entrance to the basin and another node within the basin. The diurnal portion of the tide may have only a single node at the basin entrance. The result is a diurnal water-level pattern at the second node of the semidiurnal tide and a semidiurnal mixed pattern in all other parts of the basin. The harbor of Victoria, British Columbia, Canada, is located

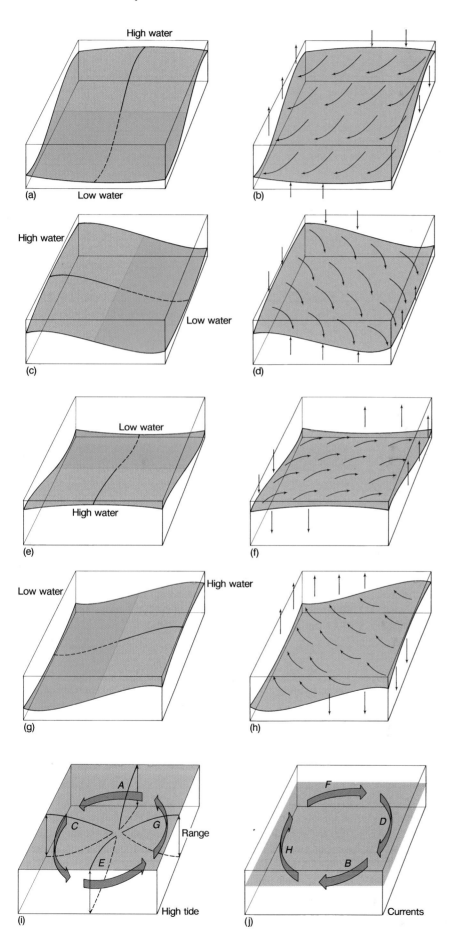

Figure 11.13 Rotary standing tide waves. At the instant high tide occurs at one side of the basin (a), the water begins to flow toward the low-tide side, creating a tidal current (b) that is deflected to its right in the Northern Hemisphere. The current displaces the high tide counterclockwise from (a) by way of (b) to (c). The process continues from (c) by way of (d) to (e), from (e) by way of (f) to (g), and from (g) by way of (h) to (a). This process results in a tide wave that rotates counterclockwise about the amphidromic point (i) and a tidal current that flows clockwise (j).

near a semidiurnal node so that it registers a diurnal tidal pattern in an inlet system of mixed tides.

Ocean tides are the result of combining progressive and standing wave tides with diurnal and semidiurnal characteristics. Different kinds of tides interact with each other along their boundaries, and the results are exceedingly complex.

Tide Waves in Narrow Basins

Unlike ocean basins, coastal bays and channels are often long and narrow with a length that is considerably greater than their width. These narrow basins have an open end toward the sea, so that the reflection of the tide wave occurs only at the head of the basin. To resonate with a tidal period, the length dimension need be only one-half the length cited for the closed basins in table 11.1. In this type of open basin, the node is at the entrance, and only one-fourth of the tide wave's form is present; an antinode is at the head of the bay. This situation is illustrated in chapter 10, figure 10.28. If the open basin is very narrow, oscillation occurs only along the length of the basin; there is no rotary motion because the basin is too narrow. For example, the Bay of Fundy in northeastern Canada has a tidal range near the entrance node of about 2 m (6.6 ft), whereas the range at the head of the bay is 11.7 m (35 ft). This particular bay has a natural oscillation period that is so well matched to the tidal period that every tidal impulse at its entrance creates a large oscillation at the head of the bay (fig. 11.15). In another bay along another coast, the shape of the basin may be such that it decreases rather than increases the tide's range. Every naturally occurring basin is unique in this regard.

11.6 Tidal Bores

In some areas of the world, large-amplitude tides cause large and rapid changes in water volume along shallow bays or river mouths. Under these conditions, the rising tide is forced to move toward

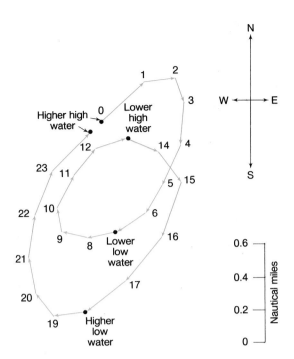

(a)

Figure 11.14 Ocean tidal currents are rotary currents. The *arrows* trace the path followed by water particles in a tide wave during a mixed tidal cycle. Two unequal tidal cycles are shown. *Numbers* indicate consecutive hours in each tidal stage. The Coriolis effect deflects the horizontal component of the water particles' orbital motion, causing them to move in a circular path.

Table 11.1	Dimensions of Closed Ocean Basins with Natural Periods Equaling Tidal Periods	
Tidal Period	**Depth (m)**	**Length or Width (km)**
Semidiurnal	4000	4428
12.42 h	3000	3835
44,712 s	2000	3131
	1000	2214
	500	1566
	100	700
	50	495
Diurnal	4000	8853
24.83 h	3000	7667
89,388 s	2000	6260
	1000	4427
	500	3130
	100	1399
	50	989

(b)

Figure 11.15 Low tide (a) and high tide (b) at the Rocks Provincial Park, Hopewell Cape, New Brunswick, Canada. The tidal range at the head of the bay exceeds 10 m (28 ft).

the land at a speed greater than that of the shallow-water wave, whose speed is determined by the depth of the water or the speed of the opposing river flow. When the forced tide wave breaks, it forms a spilling wave front that moves into the shallow water or up into the river. This wave front appears as a wall of turbulent water called a **tidal bore** and produces an abrupt change in water levels as it passes. A single bore may be formed, or a series of

bores may be produced. The bores are usually less than a meter in height but can be as much as 8 m (26 ft) high, as in the case of spring tides on the Qiantang River of China.

The Amazon, Trent, and Seine Rivers have bores. Fast-rising tides also send bores across the sand flats surrounding Mont-Saint-Michel in France and into Turnagain Arm of Cook's Inlet in Alaska. The bore in the Bay of Fundy in

Figure 11.16   The tidal bore moving up the Shubenacadie River in Nova Scotia, Canada.

Table 11.2	Predicted Times and Heights of High and Low Waters for July 6–9, 2006, San Diego, California, United States (32.7133° N, 117.1733° W)	
Day	**Time (h:min)[1]**	**Height (ft)**
6	01:26	0.80 low
	07:42	3.19 high
	11:47	2.46 low
	18:35	5.84 high
7	02:09	0.15 low
	08:39	3.43 high
	12:42	2.55 low
	19:17	6.26 high
8	02:49	−0.94 low
	09:23	3.66 high
	13:31	2.54 low
	19:59	6.69 high
9	03:27	−0.94 low
	10:02	3.87 high
	14:18	2.44 low
	20:42	7.05 high

1. hours:minutes. Time meridian 120°W. 00:00 is midnight; 12:00 is noon— Pacific Daylight Time. Heights are referred to mean lower low water, which is the chart datum of soundings.

Canada (fig. 11.16) has been reduced by the construction of a causeway. Towns in areas having tidal bores often post warnings; their turbulence can be a severe hazard because they suddenly flood areas that were open stretches of beach only minutes before.

11.7 Predicting Tides and Tidal Currents

Because of all the natural combinations of progressive and standing tides and the factors that affect them, it is not possible to predict Earth's tides from knowledge of the tide-raising bodies alone; equilibrium tidal theory is not adequate for the task. Accurate, dependable daily tidal predictions are made by combining actual local measurements with astronomical data.

The rise and fall of the tides are measured over a period of years at selected locations. Primary tide stations make these water-level measurements for at least nineteen years, to allow for the 18.6-year declinational period of the Moon. From these data, mean tide levels are calculated. Oceanographers use a technique called **harmonic analysis** to separate the tide record into components with magnitudes and periods that match the tide-raising forces of the Sun and the Moon. They are then able to isolate the effect of the local geography, known as the **local effect.** Tides for any location are predicted by combining the local effect with the predicted astronomical data. Complex and cumbersome mechanical computers or tide machines were once used to predict the tides, but today computers quickly and easily recombine the data and predict the time, date, and elevation of each high-water and low-water level.

Tide Tables

Tide and tidal current tables for North America were published by the National Ocean Survey (NOS) of the National Oceanic and Atmospheric Administration (NOAA) until

1996. These tables are now available on the Internet and published by private companies using NOAA data. Tide tables give the dates, times, and water levels for high and low water at primary tide stations (table 11.2). There are 196 primary tide stations in the United States, but many more locations require accurate tide predictions. The data for these auxiliary stations are determined by correcting nearby primary station data for time and tide height.

Tidal Current Tables

Tidal currents in the open ocean have been explained as rotary currents formed by the passing tide wave form and the deflection of water particles due to the Coriolis effect. Tidal currents in the deep sea are of scientific interest to oceanographers concerned with removing this circular motion from their data to obtain the net flows of the major ocean currents. Tidal currents in harbors and coastal waters are of major interest to commercial vessels and pleasure boaters because these currents can be very strong and must be taken into account by anyone who wants to navigate in such waters.

Like the tides, the tidal currents are first measured at selected primary locations in important inland waterways and channels. These current data are studied to determine how the speed and direction of the tidal current are related to the predicted tide-level changes. As before, the local effect is determined and is used to predict tidal currents on the basis of the tide tables.

Once tidal currents have been predicted for a location, they can be graphed with time and compared to the tidal height

Measuring Tides from Space

ntil recently, monitoring and recording of open-ocean tides had been done only by inference from measurements at coastal sites, some mid-ocean islands, and a few seafloor-mounted pressure gauges. Today's satellites, first *SEASAT,* then *GEOSAT,* and now the *ERS* series and *TOPEX/Poseidon,* can measure the absolute elevation of the sea surface by radar altimetry. The satellite uses a radar beam to measure the distance from the satellite to the sea surface, and this distance is compared to the distance between the satellite and Earth's center to obtain an elevation of the sea surface above Earth's center. *TOPEX/Poseidon* provides more accurate sea-level measurements than previous satellites, collecting ten measurements per second and requiring ten days to repeat its ground track between approximately 65°N and 65°S.

The changes in sea-surface elevation recorded by the satellite are caused by climate variations, water-density shifts due to temperature change, wind and atmospheric pressure fluctuations, and ocean current meanders, as well as by the passing tide wave.

Tidal elevations can be separated from other height changes, because of their definite and recognizable periods, allowing oceanic tidal maps based on satellite data to be drawn with high accuracy.

Oceanic tidal maps for specific tidal components have been derived from the *TOPEX/Poseidon* data. The astronomical tidal component with the largest effect on the tides is the M_2 or principal lunar semidiurnal component. Box figure 1 is a cotidal map of global M_2 tides derived from one year of *TOPEX/Poseidon* sea-surface elevation data that have been fitted to a computer model. As longer-term records are collected, it will be possible to clearly separate more of the tidal components, improve computer models, and predict the total ocean tide.

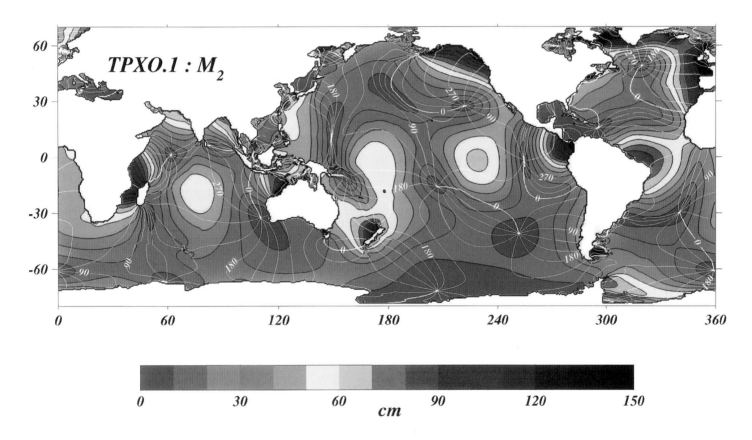

Box Figure 1 *TOPEX/Poseidon* data produced this cotidal chart of the principal lunar semidiurnal tidal component (M_2). The rotation of the tides about amphidromic points and areas of progressive tides are shown. Cotidal lines mark the position of the tidal wave crest, hour by hour; each starting point is a *thick white line* representing 0000 hour GMT. The *numbers* represent the degrees of rotation of the tidal crest about the amphidromic points. Cotidal amplitude (one-half the cotidal range) increases with distance from the amphidromic points. The color scale is in centimeters.

Table 11.3	Predicted Tidal Currents for August 1–4, 2006, Seymour Narrows, British Columbia, Canada (50.1333°N, 125.3500°W)		
Day	Slack Water Time (h:min)[1]	Maximum Current Time (h:min)[1]	Velocity (knots)[2]
1		01:38	−8.15
	04:59 flood tide begins		
		08:01	6.74
	11:25 ebb tide begins		
		14:14	−4.85
	17:07 flood tide begins		
		19:57	4.81
	22:42 ebb tide begins		
2		02:21	−7.46
	05:46 flood tide begins		
		08:59	6.83
	12:35 ebb tide begins		
		15:23	−4.21
	18:19 flood tide begins		
		20:55	3.56
	23:23 ebb tide begins		
3		03:14	−6.99
	06:39 flood tide begins		
		10:02	7.33
	13:49 ebb tide begins		
		16:43	−4.29
	19:46 flood tide begins		
		22:03	2.82
4	00:16 ebb tide begins		
		04:17	−6.91
	07:36 flood tide begins		
		11:04	8.30
	14:57 ebb tide begins		
		17:57	−5.20
	21:06 flood tide begins		
		23:13	2.79

1. hours:minutes. Time meridian 120°W. 00:00 is midnight; 12:00 is noon—Pacific Daylight Time.

2. Velocity is positive during flood tide and negative during ebb tide.

curves for that area. If maximum current times coincide with the times for either low or high water, the tide has a progressive wave form. If maximum current times coincide with mid-tide stages, the tide is a standing wave–type tide.

Tidal current data (table 11.3) are published in a format similar to that of the tide tables. The times of slack water, maximum flood currents, and maximum ebb currents, as well as the speed of the currents in knots and the direction of flow for ebb and flood currents are given for primary channel stations. Auxiliary tidal current stations are keyed to the primary current stations with correction factors to determine current speed, time, and direction at the secondary stations. This information allows the master of a vessel to decide at what time to arrive at a particular channel in order to find the current flowing in the right direction or how long to wait for slack

water before choosing to proceed through a particularly swift and turbulent passage.

11.8 Practical Considerations: Energy from Tides

Tidal energy was used to turn mill wheels in the coastal towns of northern Europe during the nineteenth century. The possibility of obtaining large amounts of energy from the tides still exists where there are large tidal ranges or narrow channels with swift tidal currents. Two systems can be used to extract energy from the rise and fall of the tide. Both require building a dam across a bay or an estuary so that seawater can be held in the bay at high tide. When the tide ebbs, a difference in water-level height is produced between the water behind the dam and the ebbing tide. When the elevation difference becomes sufficient, the seawater behind the dam is released through turbines to produce electrical power. The reservoir behind the dam is refilled on the next rising tide by opening gates in the dam. This single-action system produces power only during a portion of each ebb tide (fig. 11.17a). A tidal range of about 7 m (23 ft) is required for this system to produce power.

This same arrangement can be used as a double-action system. In this system, power is produced on both the ebbing and rising tide (fig. 11.17b). This system requires manipulating water heights across the dam and two-way flow through turbines.

These systems appear to be simple and cost-effective methods for producing electrical power, but there are few places in the world where the tidal range is sufficient and where natural bays or estuaries can be dammed at their entrances at reasonable cost and effort. Moreover, the appropriate tides and bays are not necessarily located near population centers that need the power. Installation and power-distribution costs, in addition to periodic low-power production because of the changing tidal amplitude over the tide's monthly cycle, make this type of power more expensive than that from other sources.

Since 1966, a 240-megawatt power plant alongside a dam on the Rance River Estuary in France has produced 5.4×10^{10} watt-hours of electricity per year. Present global energy demands could be satisfied by 250,000 plants of this capacity, but only about 255 sites have been identified around the world that have the potential for tidal energy development.

Tidal power was first considered for the Bay of Fundy in 1930. Canadian and American interest was casual because of the expense of the project until the rising cost of fossil fuels focused interest on alternative energy sources. The province of Nova Scotia commissioned a power station in the tidal estuary of the Annapolis River in 1984 (fig. 11.18), the world's largest straight-flow-rim–type single-effect scheme. Tidal ranges at the Annapolis site vary from 8.7 m (29 ft) during spring tides to 4.4 m (14 ft) during neap tides. The unit generates up to 20 megawatts of power from the head of water developed between the upstream basin and sea level downstream at low tide. Initially, this project was intended as a pilot to demonstrate the feasibility of a large-scale, straight-flow turbine in a tidal setting. The station has now been added to the province's principal electrical utility's hydrogenerating system. Annual production is

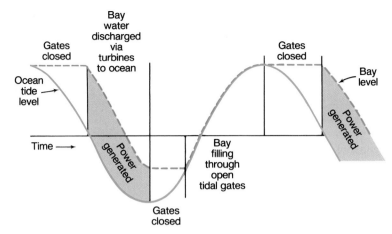

(a) Single-action power cycle; ebb only

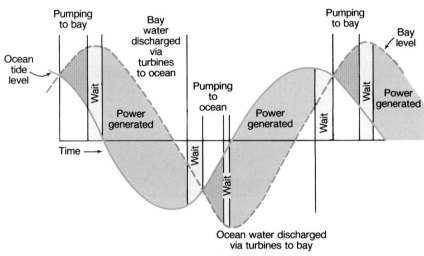

(b) Double-action power cycle; ebb and flood

Figure 11.17 Periods of power generation related to ocean tidal heights and water storage levels. (a) A single-action tidal power system. Power is generated on the ebb tide. (b) A double-action tidal power system. Power is generated on the ebb and flood tides.

Figure 11.18 The Annapolis River tidal power project in Nova Scotia (Bay of Fundy), Canada, is the first tidal power plant in North America.

$3–4 \times 10^{10}$ watt-hours. Power availability has been in excess of 95%. Similar plants exist in China and Russia.

Although tidal power does not release pollutants, it is not without environmental consequences. Dams isolate bays from the rivers and estuaries with which they had been connected. At present, the natural period of oscillation of the Bay of Fundy is about thirty minutes longer than the tidal period; these periods are sufficiently alike to cause the tides to resonate. Decreasing a bay's length with a dam shortens its period of oscillation. In the case of the Bay of Fundy, it is estimated that this could increase tidal ranges by 0.5 m (20 in) and tidal currents by 5% along the coast of Maine.

The damming of a bay or an estuary interferes with ship travel and port facilities. The dams are barriers to migratory species and alter the circulation patterns of the isolated basin. As the environmental impacts of dams have become better known, they have become far less popular as possible energy sources.

Swift tidal currents in inshore channels represent another possible energy source. Flowing water has been used for several centuries to turn the equivalent of windmills or water wheels for limited power. Because the tidal currents reverse with the tide, these "water mills" must operate with the current flowing in either direction.

Power generated by windmills depends on the density of the air, the blade diameter, and the cube of the wind speed. Water mills with a similar design depend on the density of the water, the blade diameter, and the cube of the current speed (table 11.4). A windmill in a 20-knot wind produces about the same power as a water mill with blades of the

Table 11.4	Estimated Power Generation (in kW) for Water Mills at Various Current Speeds		
	Current Speed		
Blade Diameter (m)	**5 Knots (2.5 m/s)**	**4.5 Knots (2.25 m/s)**	**3.5 Knots (1.75 m/s)**
2	8.5	5.5	3.7
5	53	35	23
10	210	140	90
15	480	310	205
20	850	550	370
30	1910	1250	820

same diameter in a 2-knot current because the density of water is about 1000 times the density of air:

$$0.001 \text{ g/cm}^3 (20 \text{ knots})^3 = 1 \text{ g/cm}^3 \times (2 \text{ knots})^3$$

$$\text{air density} \times (\text{speed})^3 = \text{water density} \times (\text{speed})^3$$

Tidal currents, although reversing, are regular and predictable. The currents are a natural renewable power source that is much easier to harvest than the energy from water waves. British and Norwegian engineering teams have begun looking at these strong currents for producing electrical energy. Propeller-driven turbine generators mounted on tripods have been designed to deliver 300 kilowatts of power per generating unit (fig. 11.19). In September 2002, a Norwegian energy research group (SINTEF, ABB and Hammerfest Strom) installed a prototype unit at the bottom of Norway's Kvalsundet Strait. The prototype is a tower with blades 20 m (60 ft) in diameter. The pitch of the blades can be changed to optimize the power as the tidal current varies. The turbines rotate 180° as the tidal current reverses. Power is sent to shore by cable. Another version of a turbine tidal generator is being developed by a British company, Marine Current Turbines, Ltd. This system employs two turbines mounted on either side of a steel piling set in the sea floor.

Figure 11.19   Computer image of a 15–16 m (50–53 ft) blade-driven turbine. This unit sits on the sea floor in an area of fast tidal flow. It is the marine counterpart of a modern windmill.

Summary

Diurnal tides have one high tide and one low tide each tidal day; semidiurnal tides have two high tides and two low tides. A semidiurnal mixed tide has two high tides and two low tides, but the high tides reach different heights and the low tides drop to different levels. For diurnal and semidiurnal tides, the greatest height reached by the water is high water, and the lowest point is low water. Mixed tides have higher high water, lower high water, higher low water, and lower low water. The zero depth on charts is referenced to mean low water or mean lower low water; low tides falling below these levels are minus tides. Rising tides are flood tides; falling tides are ebb tides.

Equilibrium tidal theory is used to explain the tides as a balance between gravitational and centrifugal forces. An equatorial tide is a semidiurnal tide, or a tide wave with two crests and two troughs. Because the Moon moves along its orbit as Earth rotates on its axis, a tidal day is twenty-four hours and fifty minutes long. The period of the semidiurnal tide is therefore twelve hours and twenty-five minutes.

The Sun's effect is less than half that of the Moon's, and the tidal day with respect to the Sun is twenty-four hours. Because the tidal force of the Moon is greater than that of the Sun, the tidal day is still considered to be twenty-four hours and fifty minutes.

Spring tides have the greatest range between high and low water; they occur at the new and full Moons, when Earth, the Sun, and the Moon are in line. Neap tides have the least tidal range; they occur at the Moon's first and last quarters, when the Moon is at right angles to the line of centers of Earth and the Sun.

When the Moon or Sun stands above or below the equator, equatorial tides become more diurnal. Diurnal tides are often called declinational tides. The elliptic orbits of Earth and the Moon also influence the tide.

The dynamic approach to tides investigates the actual tides as they occur in the ocean basins. The tide wave is discontinuous, except in the Southern Ocean. It is a shallow-water wave that oscillates in some ocean basins as a standing wave, and its motions persist long enough to be acted on by the Coriolis effect. The tide wave is reflected, refracted, and diffracted along its route. Because a point on Earth moves eastward faster at the equator than the tide wave progresses westward as a free wave, the tide wave moves as a forced wave, and its crest is displaced to the east of the tide-raising body. Above 60°N and 60°S latitudes, the crest is more nearly in line with the Moon, because the speeds of Earth and the tide wave match more closely.

In large ocean basins, the tide wave can move as a progressive wave. The Coriolis effect causes a rotary tidal current. Standing wave tides can also form in ocean basins; they rotate around an amphidromic point as they oscillate. Cotidal lines mark the progression of the tide crest, and regions of equal tidal range are identified by corange lines. The tidal range increases with the distance from an amphidromic point. Standing wave tides in narrow open-ended basins oscillate about a node at the entrance to the bay or basin; the antinode is at the head of the basin, as in the Bay of Fundy. There is no rotary motion when a bay is very narrow.

A rapidly moving tidal bore is caused when a large-amplitude tide wave moves into a shallow bay or river.

Tidal heights and currents are predicted from astronomical data and actual local measurements. NOAA provides data for annual tide and tidal current tables.

Single- and double-action dam and turbine systems extract energy from the tides. Tidal power plants are in use in France, Russia, and Canada. Few places have large enough tidal ranges and suitable locations for tidal dams. Tidal power has environmental drawbacks as well as high developmental costs. Tidal currents are another energy-producing possibility, but installation and service costs are considered high.

Key Terms

All key terms from this chapter can be viewed by term or definition when studied as flashcards on this book's website at www.mhhe.com/sverdrup10e.

diurnal tide, 271
semidiurnal tide, 271
semidiurnal mixed tide, 271
high water, 271
low water, 271
higher high water, 271

lower high water, 271
higher low water, 271
lower low water, 271
average tide/mean tide, 271
tidal datum, 271
minus tide, 271
flood tide, 271
ebb tide, 271
tidal current, 272
slack water, 272
equilibrium tidal theory, 272

dynamic tidal analysis, 272
centrifugal force, 272
centripetal force, 273
tidal day, 274
tide wave, 274
range, 275
spring tide, 275
neap tide, 275
declinational tide, 276
free wave, 278
forced wave, 278

progressive tide, 278
cotidal line, 278
standing wave tide, 279
rotary standing tide wave, 279
amphidromic point, 279
corange line, 279
tidal bore, 281
harmonic analysis, 282
local effect, 282

Study Questions

1. Distinguish between the terms in each pair:
   a. diurnal tide—semidiurnal tide
   b. tidal day—tidal period
   c. spring tide—neap tide
   d. flood tide—ebb tide
   e. cotidal lines—corange lines
2. What is the path of a water particle in the tidal current shown in figure 11.14 if a 1-knot current flowing south is also present?
3. Why is it more efficient to generate power by means of a tidal dam than to erect water mills in tidal currents?
4. Why are standing wave tides produced in small coastal basins as well as in large ocean basins? Use table 11.1.
5. Explain why it is necessary to have both a large tidal range and a relatively large volume of water behind a tidal dam to generate electrical power from the rise and fall of the tides.
6. Sketch each of the three different tidal patterns during a spring tide and a neap tide. Label the tide levels of the spring tide sequences.

7. Explain why a tide is a wave.
8. Explain the relationship between the tides and tidal currents.
9. Why does the tide act as a shallow-water forced wave in the ocean basins?
10. How do progressive tides and standing wave tides differ?
11. How is a tidal bore produced?
12. What would happen if there were no force counteracting the Sun's gravitational force on the center of mass of the Earth-Moon system?
13. Why are high tides higher during the winter at mid-latitudes in the Northern Hemisphere than they are in the summer?
14. Why must the tides be measured for approximately nineteen years at a location before the data can be used in tide forecasting?
15. Why are accurate satellite sea-surface elevation measurements over the entire Earth so important to oceanography?

Study Problems

1. How many days will pass before a high tide reoccurs at the same clock time?
2. Using the information in table 11.2, plot the San Diego, California, tide curves for July 8 and 9, 2006. What type of tide is the San Diego tide? Label the water levels on the curve.
3. Locate and label Mean Lower Low Water on the plot you constructed for problem 2.

4. Using the information in table 11.2, what are the maximum and minimum tidal ranges of the San Diego, California, tide during the time July 6 to 9, 2006?
5. At what time on August 4, 2006, should you arrive at Seymour Narrows to navigate the narrows at slack water between lunch and dinner? See table 11.3.

Rocky shoreline along the Pacific Ocean near Big Sur, California.

Chapter 12

Coasts, Beaches, and Estuaries

Chapter Outline

Learning Outcomes

After studying the information in this chapter students should be able to:

1. *recognize* different types of coasts when viewing images of them,
2. *describe* the processes that generate secondary coasts,
3. *label* features of the beach on a diagram of a typical beach profile,
4. *review* the seasonal movement of sand onto and off the beach,
5. *diagram* and *explain* the movement of sand in a coastal circulation cell,
6. *list* the different types of beaches,
7. *recognize* different types of coastal structures and *know* their intended purpose,
8. *locate* the National Marine Sanctuaries,
9. *characterize* the different types of estuaries, and
10. *sketch* the trend of mean sea level over the past fifteen years and label the axes.

Shores and beaches are the most familiar areas of the ocean for most people, but even the casual visitor senses changes in these areas. On any visit the tide may be high or low, the logs and drift may have changed position since the last visit, and the dunes and sandbars may have shifted since the previous summer. A visit to a beach farther along the coast or along a different ocean presents a different picture. The sand is a different color, or there is no sand at all; the waves break higher or lower across the beach; the slope of the beach is steeper or flatter; and so on. No beach is static, and no two beaches are exactly the same.

On our visits to the coast, we may stop to admire an inlet, photograph a quiet harbor, or visit a river estuary. These places are where the salt water from the oceans and the fresh water from the land meet. In some ways, all these areas behave like small oceans, but the frequent addition of fresh water gives them characteristics of their own. These areas share certain features, but, just as beaches differ, one estuary differs from another.

In this chapter, we study the types of coasts and beaches and the natural processes that create and maintain them. We also investigate the estuaries and partially enclosed seas of coastal areas and the circulation patterns that are unique to them. These are complex and sensitive parts of the ocean system; to preserve them we must understand them.

12.1 Major Zones

The **coasts** of the continents are the areas where the land meets the sea. The terms *coast, coastal area,* and *coastal zone* are used to describe these land edges that border the sea, including areas of cliffs, dunes, beaches, bays, coves, and river mouths. Some examples of the various kinds of coasts are presented in figures 12.1 and 12.2. The width of the coast, or the distance to which the coast extends inland, varies and is determined by local geography, climate, and vegetation, as well as by the perception of the limits of marine influence in social customs and culture. However, the coast is most generally described as the land area that is or has been affected by marine processes such as tides, winds, and waves, even though the direct effect of these processes may be felt only under extreme storm conditions. The seaward limit of the coast usually coincides with the beginning of the beach or shore but sometimes includes nearby offshore islands.

The term *coastal zone* includes the open coast as well as the semi-isolated and sheltered bays and estuaries that interrupt it. The coastal zone incorporates land and water areas and has become the standard term used in legal and legislative documents affecting U.S. coastal areas. In this context, the coastal zone's landward boundary is defined by a distance (often 200 ft or about 60 m) from some chosen reference (usually high water); the seaward boundary is defined by state and federal laws.

Coastal areas are regions of change in which the sea acts to alter the shape and configuration of the land. Sometimes these changes are extreme and occur rapidly (for instance, the damage caused by a hurricane). Sometimes the changes are subtle and so slow that they are not perceived by people during their lifetimes but are very impressive when considered over long periods (for example, the formation of the Mississippi River delta or the gradual erosion of Cape Hatteras, North Carolina). In general, coasts composed of soft, unconsolidated materials, such as sand, change more rapidly than coasts composed of rock. Over geologic time, coasts have changed dramatically as they have emerged and submerged with tectonic processes, climatic variations, and changing sea level; some remnants of these changes can still be seen in present coastlines.

The **shore** is a part of the coast; it is that region from the outer limit of wave action on the bottom (seaward of the lowest tide level) to the limit of the waves' direct influence on the land.

This land limit may be marked by a cliff or an elevation of the land above which the sea waves cannot break. Such features act as barriers to the wave-tossed drift of logs, seaweeds, and other debris. The **beach** is an accumulation of sediment (sand or gravel) that occupies a portion of the shore. The beach is not static but moving and dynamic, because the beach sediments are constantly being moved seaward, landward, and along the shore by nearshore wave and current action. Between the high-tide mark and the upper limit of the shore may be dunes or grass flats dotted with drift logs left behind by an exceptionally high tide or a severe storm (fig. 12.3).

(a)

(b)

(a)

(b)

(c)

(c)

Figure 12.1 (a) The Oregon coast is famed for its bold rocky headlands and intervening pocket beaches. (b) Bayou La Loutre at Yscloskey, Louisiana. The bayou is an old stream course of the Mississippi River. The salt content of the water changes as the wind drives fresh water from the river or salt water from the coast through the delta. (c) Glacial bays and fjords produce a rugged topography along the coast of Glacier Bay in southeastern Alaska.

Figure 12.2 (a) Cape of Good Hope, South Africa, where the Indian and Atlantic Oceans meet. (b) Sea stacks are common features along the southern coast of Australia, occurring in a wide variety of shapes and sizes. These sea stacks are part of a group called the Twelve Apostles. (c) These chalky cliffs along the Dorset coast of England were formed from the remains of microscopic, single-celled foraminifera, principally *Globigerina*.

Figure 12.3 Driftwood accumulates at the high-tide line in areas where timber is plentiful.

12.2 Types of Coasts

Any land form is the product of the processes that have changed it through time, and the study of land forms and the processes that have fashioned them is known as **geomorphology.** Coastal geomorphology considers tectonic processes; wave, wind, and current exposure; tidal range and tidal currents; sediment supply and coastal transport; and climate and climate changes.

Climate change can result in a worldwide, or **eustatic, change** in sea level that may submerge previous coasts or expose previous sea floor. During the last period of glaciation, sea level dropped by an estimated 100 m (328 ft) or more due to the storage of water on land as glacial ice and the cooling and subsequent decrease in volume of the remaining water in the ocean basins. With the global warming that signaled the end of the last ice age, sea level rose to its present height as land ice melted and the increase in average temperature of seawater caused a corresponding increase in volume. Sea level is continuing to rise even now as a result of global warming (see the discussion of greenhouse gases in chapter 7).

There are different ways to classify coasts for the purpose of describing and studying them. One is to classify them as erosional or depositional, depending on whether they predominantly lose or gain sediments, or shoreline. This is often influenced by whether the coast is located at the edge of the plate as part of an active, or leading, continental margin or away from the edge of the plate as part of a passive, or trailing, continental margin (the formation and characteristics of active and passive continental margins are discussed in chapter 3, section 3.4). In the United States, eastern coasts are part of a passive continental margin and are generally considered depositional. Western coasts are located along active continental margins and are generally considered erosional. These categories should not be confused with coastal erosion problems that are prevalent on both coasts.

Coasts of both types may be modified by eustatic sea-level changes. Some coasts (both erosional and depositional) have been eroded and their sediments swept away to be deposited elsewhere, while other coasts have received sediments from rivers and shore currents. In some parts of the world, coasts are modified by seasonal storms; other areas enjoy more benign weather. Coasts of polar regions are modified by their interaction with sea ice; tropical coasts are altered by reef-building corals. Each coast has its own identity, but different types of coasts can be recognized as the results of certain processes.

A system devised by the late Francis P. Shepard of the Scripps Institution of Oceanography organizes coasts into two process categories: (1) coasts that owe their character and appearance to processes that occur at the land-air boundary and (2) coasts that owe their character and appearance to processes that are primarily of marine origin. Further classification depends on whether their large-scale features are the products of tectonic, depositional, erosional, volcanic, or biological processes.

In the first category are coasts that have been formed by (1) erosion of the land by running surface water, wind, or land ice, followed by a sinking of the land or a rise in sea level; (2) deposits of sediments carried by rivers, glaciers, or the wind; (3) volcanic activity, including lava flows; and (4) uplift and subsidence of the land by earthquakes and associated crustal movements. Coasts formed by these processes are **primary coasts,** since there has not yet been time for the sea to substantially alter or modify their original appearance. The second category includes coasts formed by (1) erosion due to waves, currents, or the dissolving action of the seawater; (2) deposition of sediments by waves, tides, and currents; and (3) alteration by marine plants and animals. These coasts are **secondary coasts;** their character, even though it may have been originally land-derived, is now distinctly a result of the sea and its processes.

A relatively young coast may be rapidly modified by the sea, while a coast that is old may retain its land-derived characteristics. Keep in mind that absolute age is not really important because the classification is based only on whether the characteristics are derived from the land or from the sea, and it is possible to find both types of features along the same coast.

Primary Coasts

Coasts formed by subaerial erosion followed by sinking of the land or a rise in sea level include those that were covered by glaciers during the ice ages. During these glacial periods, sea level was lower than at present because much of the water was held as ice on the continents. Glaciers moved slowly across the land, scouring out valleys as they inched along to the sea. The weight of the ice caused the land to subside, and in some cases, when the ice began to melt, sea level rose faster than the land could rebound upward to its preglacial elevation. In other cases, glacial troughs were scoured below sea level and filled with seawater as the ice receded. The **fjords** of Norway, Greenland, New Zealand, Chile, and southeastern Alaska are the results of these processes (fig. 12.4). Fjords are long, deep, narrow channels with a U-shaped cross section. Where the glacier met the sea, there was often a collection of debris that formed a lip, creating a shallow entrance, or **sill.**

When a glacier or ice sheet ceases its forward motion and retreats, it leaves a mound of rubble, called a **moraine,** along the border of its greatest extension. If the glacier has reached the edge

Figure 12.4 The narrow channel of a fjord, Milford Sound, New Zealand.

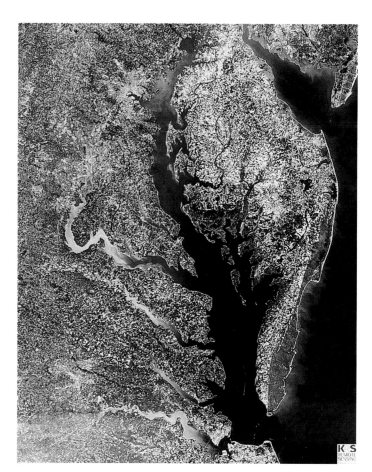

Figure 12.5 Delaware Bay *(upper right)* and Chesapeake Bay *(center)* are examples of drowned river valleys.

of a continent, this material becomes a part of the coastal area. Long Island, off the New York and Connecticut coasts, and Cape Cod, Massachusetts, are moraines. Moraines act as protective barriers to the continental coast. In some areas, land that was deeply covered by ice during the last ice age is still slowly rising; in Scandinavia, the rate of rise is about 1–5 cm (0.4–2 in) per year. Along other coasts, tectonic forces rather than loss of ice have caused uplift. Along the western coasts of North and South America, old wave-cut terraces can be identified now well above sea level.

When sea level was lowered during the ice ages, rivers flowed over the exposed shore to the sea. The rivers cut V-shaped channels into these areas, and in many cases, the channels had numerous side branches formed by feeder streams. As the sea level rose, these channels were filled with seawater, producing areas such as Chesapeake Bay and Delaware Bay on the eastern coast of the United States. A coast of this type is called a **drowned river valley,** or **ria coast,** shown in figure 12.5.

Rivers carrying extremely heavy sediment loads build deltas on top of the continental shelves. The **delta,** or deposit of river-borne sediment left at a river mouth, produces a flat,

fertile coastal area. Examples of these deposits are found at the mouths of the Mississippi, Ganges, Nile, and Amazon Rivers. A similar type of coast is produced when the eroded materials carried down from the hills by surface runoff and many small rivers join together to form an **alluvial plain.** The eastern seacoast of the United States south of Cape Hatteras was produced in this way.

It is estimated that every second, the rivers of the world carry 530 tons of sediment to the sea. This rate of removal is equal to the erosion of a layer 6 cm (2.4 in) thick from all land above sea level every 1000 years. This sediment helps to form and maintain the world's beaches, and much of it finally finds its way to the deep-ocean floor. All of it passes through the coastal zone and takes part in coastal processes. Refer to chapter 4 for sources of terrigenous material.

A **dune coast** is a wind-modified depositional coast. In Africa, the Western Sahara is gradually growing westward toward the Atlantic Ocean as the prevailing winds move sand from the inland desert to the coast. Along other dune coasts, for example, along central and northern Oregon (fig. 12.6), the winds move the sands inland to form dunes; elsewhere dunes driven by the wind migrate along the shore.

The Hawaiian Islands are the tops of large seamounts and have excellent examples of coasts formed by volcanic activity. Lava flows extending to the sea form black sand beaches or **lava**

Figure 12.6 Coastal sand dunes near Florence, Oregon.

Figure 12.7 A volcanic crater coast, Hanauma Bay, Hawaii, is a crater that has lost the seaward portion of its rim. Today, this crater coast is a park and marine preserve.

coasts. Craters formed by volcanic explosions became concave bays when they lost their rims on the seaward side, forming **cratered coasts** (fig. 12.7).

When tectonic activity results in faulting and displacement of Earth's crust, the coast is changed in characteristic ways. The California San Andreas Fault system lies along a boundary where crustal plates are moving parallel to each other along a transform fault. (Plate movements and transform faults are described in chapter 3.) Over long periods of time, faults formed in this area filled with seawater. The Gulf of California, also known as the Sea of Cortez, between Baja California and the mainland of Mexico, is found at the southern end of the fault system. At the northern end lie San Francisco and Tomales Bays, where the fault runs out into the Pacific Ocean. Tomales Bay is a particularly good example of a **fault bay.** See the satellite image of this fault system in figure 12.8. In other parts of the world, the Red Sea and the Scottish coast provide examples of **fault coasts.**

Secondary Coasts

Secondary coasts owe their present-day appearance to marine processes. As waves batter the coast, they constantly erode and grind away the shore; rocks and cliffs are undercut by the wave action and fall into the sea, where they are ground into sand. A coastal area formed of uniform material may have been irregular in shape with headlands and bays at one time, but the concentrated energy of the waves wears down the headlands more rapidly than the straighter shore and cove areas. With time, a more regular coastline results; examples are found in southern California, southern Australia, New Zealand (fig. 12.9), and England's cliffs of Dover. If the original coastline is made up of materials that vary greatly in composition and resistance to wave erosion, the result will be an increasingly irregular coastline of hard rock headlands separated by sandy coves. In some places, such as northern California, Oregon, Washington, Australia, and New Zealand, the headlands erode irregularly, and small rock islands and tall, slender pinnacles of resistant rock are formed; these are known as **sea stacks** (figs. 12.2b and 12.10). Headlands still connected to the mainland are undercut to produce sea caves or are cut through to produce arches or small windows.

In some cases, eroded materials are carried seaward by the waves and currents to areas just off the coast. If sufficient sand material is deposited in offshore shallows paralleling the beach, **bars** are produced. If still greater amounts of material are added, the bars may grow until they break the surface and form **barrier islands** (fig. 12.11). The barrier islands of the southeastern coast of the United States were formed during a period of rising sea level; seawater flooded low coastal areas, isolating the high dunes at the sea's edge and converting a primary coast to a secondary coast. Once a barrier island has formed, plants begin to grow, and the vegetation helps to stabilize the sand and increase

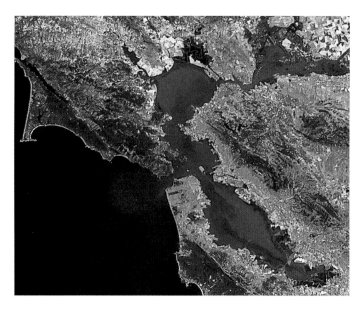

Figure 12.8   The San Andreas Fault runs along the California coast to the west of San Francisco, separating Point Reyes from the mainland and producing Bolinas and Tomales Bays. Tomales Bay is seen as a *dark line* at the *upper left.*

Figure 12.9   A cliffed coast made regular by erosion in New Zealand.

Figure 12.10   Sea stacks are common features along the coasts of northern California, Oregon, and Washington. They occur in a wide variety of shapes and sizes.

the island's elevation by trapping sediments and accumulating organic matter.

A line of barrier islands along a coast protects the shoreline behind the islands from storm waves and erosion, but the islands can sustain heavy damage. One of the greatest natural disasters to strike the United States was the September 8, 1900, Galveston hurricane. Galveston sits on Galveston Island, a barrier island along the Texas coast in the Gulf of Mexico. In 1900, the city had a population of 37,000, and the highest elevation on the island was 2.7 m (8.7 ft). Just before the full force of the hurricane struck, people were still playing on the beach, enjoying the novelty of unusually large waves. No one thought it would be a dangerous storm. By the time the citizens of Galveston realized how strong the hurricane was, it was too late to escape to the mainland. Winds estimated at 140 mph swept over the island, leaving devastation in their wake. The island was pounded by large waves, and a storm surge (chapter 7, section 7.9) 4.8 m high (15.7 ft) swept across it, destroying more than 3600 buildings in the city (fig. 12.12). An estimated 6000 to 8000 of the town's residents died. After the storm, Galveston constructed a seawall and raised the elevation of the island to protect it from future hurricanes. During 1989's Hurricane Hugo, Folly Island, along the coast of South Carolina, suffered extensive storm damage. Eighty-six of 290 oceanfront structures were more than 50% destroyed, and fifty were damaged beyond repair. The U.S. National Flood Insurance Program issued nearly $3 million in damage payments to Folly Island property owners. In 1997, Tropical Storm Josephine moved eastward across the Gulf of Mexico toward Florida. This storm did not develop hurricane-force winds, but it did cause extensive damage to coastal areas of Texas in the vicinity of Galveston. Josephine remained 482 km (300 mi) off the coast, producing steady, onshore winds of approximately 12 m/s (27 mi/h) for over a hundred hours. The winds raised the sea level at Galveston 80 cm (32 in) above normal high-tide levels and caused average wave heights of 3.5 m (11 ft) along the coast. The sustained period of high water and wave action brought by this storm produced severe beach erosion, loss of homes, and loss of the dunes that stood between the remaining homes and the water. Attempts are made to halt storm-caused erosion by building seawalls and beach-holding devices, but these do not always prove successful. Although well intentioned, they often result in aggravating the loss of material from barrier islands rather than halting it.

Sand spits and **hooks** are bars connected to the shore at one end (figs. 12.13 and 12.31). Spits and hooks may grow, shift position, wash away in a storm, or rebuild under more moderate conditions. The area between the mainland and these spits and hooks is protected from open ocean waves and is often the site of beach flats formed of sand or mud. If a spit grows sufficiently to close off the mouth of an inlet, a shallow lagoon is formed. Water percolates through the sand, and inside the lagoon, the water rises and falls with the tides.

In the tropical oceans, **reef coasts** result from the activities of sea organisms. Corals grow in the shallow, warm waters surrounding a landmass, and the small animals gradually build a fringing reef, which is attached directly to the landmass. In other places, a lagoon of quiet water may lie between the barrier reef and the land, and, in a few cases, the coral encircles and overlies a submerged seamount to form an atoll. The formation of these reef types is discussed in chapter 4, and the life forms that make up the reef are discussed in chapter 18.

Figure 12.11   Sea Island, Georgia, is a barrier island that has been extensively developed. The shallow water between the island and the mainland is visible on the *left*.

Figure 12.12   Residents of the city inspect the damage after the September 8, 1900, Galveston Island, Texas, hurricane. *Courtesy of the Rosenberg Library, Galveston, Texas.*

The Great Barrier Reef, stretching along the northeastern coast of Australia toward New Guinea, is the largest and most famous of the world's coral reefs. Coral atolls in the Pacific include Tarawa, Kwajalein, Enewetak, and Bikini. The reef-encircled islands of Iwo Jima and Okinawa are familiar as the sites of major battles in the Pacific during World War II.

Other marine animals form reeflike structures as their shells are deposited layer on layer, gradually building up a mass of hard material. There are large reef deposits of oyster shells in the Gulf of Mexico off the coasts of Louisiana and Texas. These reefs are so large that the shells are harvested commercially for lime production. Along the eastern coast of Florida, large populations of shell-bearing animals have contributed their shells directly to the shore; small shell fragments form the sand on the beaches.

Plants as well as animals may modify a coastal area. Along low-lying coasts in warm climates, mangrove trees grow in shallow water. Their great roots form a nearly impenetrable tangle, providing shelter for a unique community of other plants and animals (fig. 12.14). Coastal mangrove swamps are found along the Florida coast, northern Australia, the Bay of Bengal, and in the West Indies. In more temperate climates, low-lying protected coasts with areas of sand and mud are often thickly covered with grasses, forming another type of plant-maintained environment (fig. 12.15). These **salt marshes** may extend inland a considerable distance if the land is flat enough to permit periodic tidal flooding. Such marshes also form around protected bays and coves that have large changes in tide level. Salt marshes are extremely productive in terms of organic material and extend the shoreline seaward by trapping sediments. The effect of human activities on mangrove swamps and salt marshes is discussed in chapter 13.

Figure 12.13 A spit has formed partway across the entrance to Sequim Bay, Washington.

12.3 Anatomy of a Beach

The beach includes the sand and sediment lying along the shore and the sediments carried along the shore by the nearshore waves and currents. Let us consider first the beach where we walk, sunbathe, and relax. A specific terminology for beaches and beach features has developed from the many studies of beaches around the world. These terms allow each area to be described and prevent confusion when one beach is compared to another. A beach profile cuts through a beach perpendicular to the coast; the profile in figure 12.16 shows the major features that might appear on any beach. Use this figure while reading this section, but remember that a given beach may not have all these features, for each beach is the product of its specific environment.

The **backshore** is the dry region of a beach that is submerged only during the highest tides and severest storms. The **foreshore** extends out past the low-tide level, and the **offshore** includes the shallow-water areas seaward of the low-tide terrace to the limit of wave action on the sea floor.

Low terraces called **berms** appear on beaches in the backshore area. Berms are formed by wave-deposited material and have flat tops like the top of a terrace. Berms are recognizable by their slope, or rise in elevation. Some berms have a slight ridge crest that runs parallel to the beach; such a ridge is called the **berm crest.** When two berms are found on a beach (figs. 12.16 and 12.17*a*), the berm higher up the beach, or closer to the coastline, is the **winter,** or **storm, berm.** It is formed during severe winter storms when the waves reach high up the beach and pile up material along the backshore. The seaward berm, the berm closer to the water, is the **summer berm** formed by the gentle waves of spring and summer that do not reach as far up the beach. Once the winter berm is formed, it is not

Figure 12.14 Mangrove trees growing along the shore of Florida Bay.

disturbed by the less-intense storms during the year. The waves that produce a winter berm erase the summer berm, but after the storm season is over, a new summer berm is formed by the reduced wave action.

Figure 12.15   A coastal saltwater marsh in Gaspé, Québec, Canada.

Between the berm and the water may be a wave-cut **scarp** at the high-water level. The scarp is an abrupt change in the beach slope that is caused by the cutting action of waves at normal high tide (fig. 12.17*b*). Berms and scarps do not form in the foreshore because of the wave action and the continual rise and fall of the water. The foreshore area seaward of the low-tide level is often flat, forming a **low-tide terrace;** the upper portion has a steeper slope and is known as the **beach face.** The slope of the beach face is related to the particle size of the loose beach material and to the wave energy.

Seaward of the low-tide terrace, in the offshore region, may be **troughs** and bars that run parallel to the beach. These structures change seasonally as beach sediments move seaward, enlarging the bars during winter storms, and then shoreward in summer, diminishing the bars. When a bar accumulates enough sediment to break the surface and is stabilized by vegetation, it may become a barrier island.

Beaches do not occur along all shores. Along rocky cliffs, there may be no beach area between low and high tides. However, there may be small pocket beaches, each separated from the next by rocky headlands or cliffs (fig. 12.18).

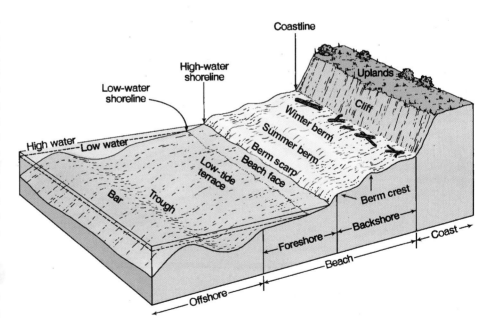

Figure 12.16   A typical beach profile with associated features.

12.4 Beach Dynamics

A sand beach exists because there is a balance between the supply and the removal of the sand. A sand beach that does not appear to change with time is not necessarily static (no new material supplied, no old material removed) but is more likely to represent a **dynamic equilibrium,** the supply of new sand

(a)   (b)

Figure 12.17   (a) Berms on a gravel beach. The winter, or storm, berm with drift logs is located high on the beach. The summer berm is between the winter berm and the beach face in the *foreground*. (b) A wave-cut scarp on a gravel beach.

Figure 12.18   A series of pocket beaches cut into an elevated marine terrace along the northern California coast. The elevated terrace was at one time formed just below sea level. Tectonic processes raised it to its present position.

equaling the loss of sand from the beach. The beach is continually changing, but it appears unchanged and remains in balance.

Natural Processes

The gentler waves of summer move sand shoreward and deposit it, where it remains until the winter. The large storm waves of winter remove the sand from a beach and transport it offshore, to a sandbar. These processes occur alternately, winter and summer, leaving the beach rocky and bare each winter and covered with sand during the summer (fig. 12.19). These seasonal changes are fluctuations about the equilibrium state of the beach each year. Other changes also appear in a cyclic fashion, such as the arrival of more sand during periods of late winter and spring river flooding. If a beach annually receives as much sand as it loses, the beach is in equilibrium and does not appear to change from year to year.

Single violent events may alter a beach. A great storm may arrive from such a direction that the usual beach current is reversed. A landslide may pile rubble across the beach and some distance into the water, interfering with the flow of sand along the beach. In either case, the beach changes because of changes in sediment supply, transport, and removal.

Waves moving toward a beach produce a current in the surf zone that moves water onshore and along the beach. This landward motion of water is called the **onshore current,** and this current transports sediment toward the shore in what is called **onshore transport.** Waves usually do not approach with their crests completely parallel to the beach but strike the shore at a slight angle. This pattern sets up a surf-zone **longshore current** that moves along the beach. Both onshore and longshore flow in the surf zone are illustrated in figure 12.20. The turbulence generated as the waves break in the surf zone tumbles the beach material in the water. The wave-produced longshore current moves this sediment parallel to the shore in the surf zone, producing **longshore transport.** Along both coasts of the North American continent, longshore transport generally moves the sediments in a southerly direction.

The up-rush, or **swash,** of water from each breaking wave moves the sand particles diagonally up and along the beach in the direction of the longshore current. The backwash from the receding wave moves downslope toward the surf zone, but the backwash is weaker than the swash because much of the receding water percolates down into the sand. The combined action of the swash and backwash moves particles in a zigzag, or sawtoothed, path along the swash zone as part of the longshore transport.

Evenly spaced crescent-shaped depressions called **cusps** are sometimes found along the sand and cobble beaches of quiet coves and longer, straighter coastlines (figs. 12.17b and 12.21). How cusps are formed is still somewhat of a mystery. They tend to form during neap tide periods, when the tidal range is at a minimum, and are often destroyed during periods of spring tide, when the tidal range is at a maximum. The size of cusps appears to be directly related to the amount of wave energy on the beach. Large cusps are associated with high wave energy, and smaller cusps seem to form where the wave energy is low. Cusp formation may be related to some form of wave interference as waves approach the beach.

Studies of sediment movements along coasts, into navigational channels, and around coastal structures provide important information to property owners, to engineers involved in the design and construction of shoreline installations—for example, piers, breakwaters, and seawalls—and to the military when planning troop and equipment movements in coastal areas.

The path traveled by beach sediments as the longshore current transports them from their source to their area of deposition is called a **drift sector.** Beaches within a drift sector usually remain in dynamic equilibrium and change little in appearance. Although the flow of material along the central portion of a drift

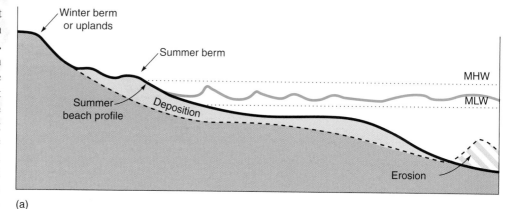

(a)

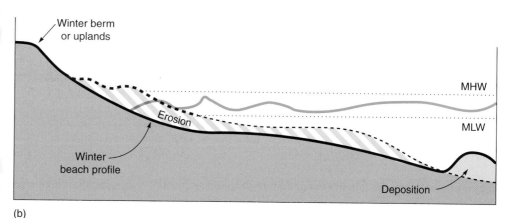

(b)

Figure 12.19 Seasonal beach changes. (a) The small waves and average tides of summer move sand from offshore to the beach and form a summer berm. (b) The high waves and storms of winter erode sand from the beach and store it in offshore bars. Winter conditions remove the summer berm, leaving only the winter berm on the beach.

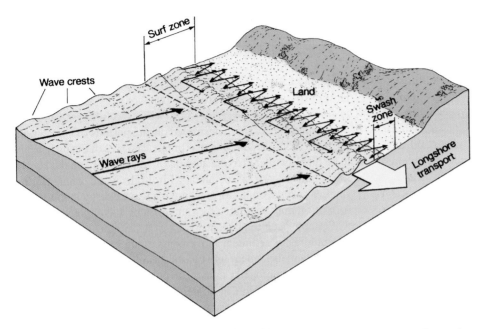

Figure 12.20 Waves in the surf zone produce a longshore current that transports sediments along the beach. *Arrows* indicate onshore and longshore water movement.

Figure 12.21 At Point Reyes National Seashore in California, wave action forms a series of cusps along a shore that has been made regular by wave erosion. How cusps form is not clearly understood.

sector maybe large, the amounts of sediment supplied and lost from the area are usually equal. At either end of the drift sector the beaches change with time. Areas at the sediment source are erosional; they are supplying or losing sediment. Areas at the depositional end of the drift sector are growing, or **accreting.** These processes are shown in figure 12.22. Although the longshore current is the usual transport mechanism within a drift sector, in some places tidal currents and coastal currents associated with large-scale oceanic circulation affect sediment transport.

Coastal Circulation

Onshore transport accumulates water as well as sand along a beach. This water must flow back out to sea and will do so one way or another. A headland may deflect the longshore current seaward, or the water flowing along the beach can return to the area beyond the surf zone in quieter water, such as areas over troughs or depressions in the sea floor. Regions of seaward return flow are frequently narrow and fast-moving. These are the rip currents that carry sediment through the surf zone (fig. 12.23); rip currents are also discussed in chapter 10 (see fig. 10.22).

Just seaward of the surf, the rip current dissipates into eddies. Some of the sediment is deposited here in quieter, deeper water and is lost from the down-beach flow. However, some is returned to shore with the onshore transport on either side of the rip current. The return of sediments to the beach, transport along the beach, transport back to sea in a rip current, and return again to the beach are processes that occur in a drift sector.

A series of these drift sectors, when linked together along a stretch of coast, forms a **coastal circulation cell.** In a major coastal circulation cell, the seaward transport of sediment results in deposits far enough offshore to prevent the sand from returning to the beach. At the end of a coastal circulation cell, the longshore transport is deflected away from the beach and the sand moves across the continental shelf and down a submarine canyon to the ocean basin floor. This sand has been removed from the coastal system, and sand beaches disappear beyond this point until a new source contributes sand to the next coastal circulation cell.

South of Point Conception in southern California, oceanographers recognize four distinct coastal circulation cells (fig. 12.24). Each cell begins and ends in a region of rocky headlands where beaches are sparse. Submarine canyons are found offshore in each cell. Beaches become wider as rivers contribute their sand to the longshore current, and in each cell, the current deposits the sediment in a submarine canyon, where it cascades to the ocean basin floor. Beaches just south of the canyon are sparse, and a new cell begins. Refer to chapter 4 for a discussion of submarine canyons.

Estimates of sediment transport rates along a beach are made by observing the rate at which sand is deposited, or accreted, on the upstream side of an obstruction or by observing the rate at which a sand spit migrates. Transport of sand along a section of coast varies between zero and several million cubic meters per year. Average values fall between 150,000 and 1,500,000 m^3 per year; 150,000 m^3 is more than 30,000 dump-truck loads. Once the enormous volume of the naturally moving sand is recognized, it becomes apparent why poorly designed harbors fill very quickly, beaches disappear, and spits migrate a considerable distance during a year. It also becomes obvious that the effort required to keep pace with the supply by dredging or pumping operations is enormously expensive and in many cases impossible.

The energy for the movement of beach materials comes from the waves and the wave-produced currents. Surface winds supply about 10^{14} watts of power to the ocean surface to produce the waves and currents. There are about 440,000 km (264,000 mi) of coastal zone in the world, and approximately half of this is exposed directly to the ocean waves. Ocean waves average 1 m (3.3 ft) in height and produce power that is equivalent to 104 watts for each meter of exposed coastline. Under storm conditions, the wave height averages about 3 m (10 ft), and a meter of coastline receives 10^5 watts. It is this supply of energy that causes beach erosion and the migration of sand along the narrow coastal zone.

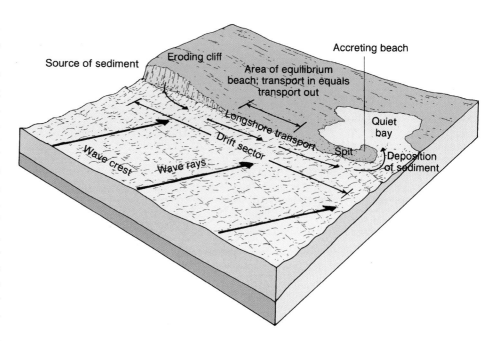

Figure 12.22 A drift sector extends along the coast from the sediment source to the area of deposit.

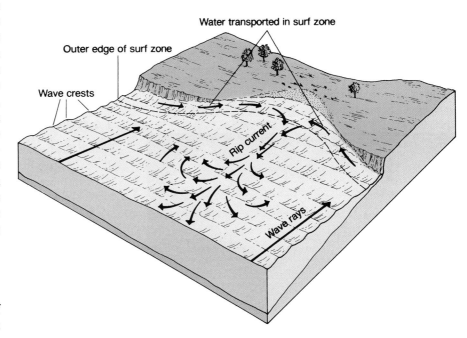

Figure 12.23 Rip currents form along a beach in areas of low surf, reduced onshore flow, and converging longshore currents.

12.5 Beach Types

Beaches are described in terms of (1) shape and structure, (2) composition of beach materials, (3) size of beach materials, and (4) color. In the first category, beaches are described as wide or narrow, steep or flat, and long or discontinuous (pocket beaches). A beach area that extends outward from the main beach and turns and parallels the shore is called a spit (see fig. 12.13). Spits frequently change their shape in response to waves, currents, and storms. In a wave-protected environment, a spit may extend from the shore to an offshore island. If this spit builds until it connects the rock or island to the shore, it forms what is called a **tombolo.** A spit extending offshore in a wide, sweeping arc bending in the direction of the prevailing current is called a hook. The sediment moves around the end of a hook and is deposited in the quiet water behind the point, often producing a broad, rounded, hook-shaped point. See figure 12.25 for an example of a tombolo and figure 12.31 for an example of a hook.

Materials that form beaches include shell, coral, minerals, rock particles, and lava. Shingles are flat, circular, smooth stones formed from layered rock when the beach slope, wave action, and stone size combine to slide the stones back and forth with the water movement. When stones roll rather than slide, rounded stones or cobbles are formed. The terms *sand, mud, pebble, cobble,* and *boulder* describe the size of beach particles (see chapter 4).

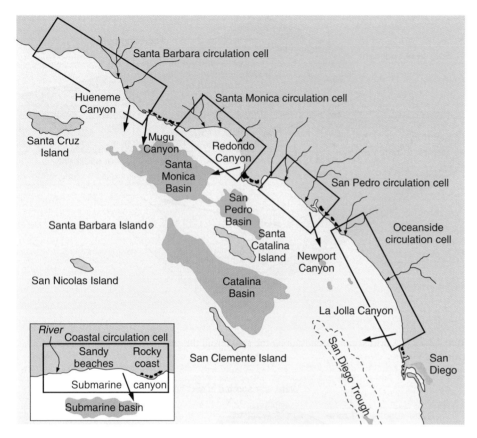

Figure 12.24 Major coastal sediment circulation cells along the California coast. Each cell starts with a sediment source and ends where beach material is transported into a submarine canyon.

Figure 12.25 A spit connected to an offshore island forms a tombolo. Tombolos are often formed by converging currents from nearshore eddies; the currents transport sediment toward the tombolo from both sides.

The composition and size of the beach material are related to the source of the material and the forces acting on the beach. Land materials are brought to the coast by rivers or are derived from the local cliffs by wave erosion. The sands of many beaches are the ground-up, eroded products of land materials rich in minerals such as quartz and feldspar. Small particles such as sand, mud, and clay are easily transported and redistributed by waves and currents; larger rocks are usually found close to their source, because they are too large to be moved any distance. The finer particles are carried away by the currents, and the larger rocks are left scattered on the beach. The Moeraki boulders (fig. 12.26), found along the eastern coast of New Zealand's South Island, are the world's largest examples of calcite-type mud balls. Some are 2 m (6.5 ft) in diameter and are eroded from the beach cliffs. In general, a beach littered with large blocks is an eroded beach, and the pebbles and boulders are called a **lag deposit.** If the lag deposit accumulates in sufficient quantity to protect a beach from further wave and water erosion, the beach is known as an **armored beach** (fig. 12.27).

Some beach materials come from offshore areas. Coral and shell particles broken by the pounding action of the waves are carried to the beaches by the moving water. A beach covered with uniform small particles of sand or mud is a depositional beach.

Some of the world's beaches have distinctive colors. In Hawaii, there are white sand beaches derived from coral and black sand beaches derived from basaltic lava. Green sands can be found in areas where a specific mineral (such as olivine or glauconite) is available in large enough quantities, and pink sands occur in regions with sufficient shell material.

12.6 Modifying Beaches

People like beaches and have great desires to own property along these sensitive fringes of land. Approximately 3.8 billion people (60% of the world's population) live within 100 km (60 mi) of a coastline. Population experts expect more than 6.3 billion

Figure 12.26 Mud balls are washed out of the cliffs of many beaches. These "Moeraki boulders," hardened by the deposition of calcite, are among the world's largest, approximately 2 m (6.5 ft) in diameter. The shape of the boulders is not due to wave action. They are found on the east coast of New Zealand's South Island.

people to live in coastal areas worldwide within the next fifty years. In the United States today, over one-half of the population lives within 80 km (50 mi) of the coasts (including the Great Lakes). By the year 2010, coastal population in the United States is estimated to grow to more than 127 million. The shore is affected directly and indirectly as increasing numbers of people make their homes in these vulnerable areas.

Coastal Structures

People change beaches by damming rivers to control floods and generate power. When a dam is built, a lake is formed behind the dam, and silt, sand, and gravel that once moved down the river to the coast are deposited in the lake behind the dam; an important source of sediment to the beaches has been removed. Although the sediment supply has been reduced, the longshore current continues to flow and carry away beach material. The net result is a loss of sediment and sand from the beaches, and the beaches erode.

Coastal zone engineering projects such as **breakwaters** and **jetties** are built to protect harbors and coastal areas from the force of the waves. Breakwaters are usually built parallel to the shore; jetties extend seaward to protect or partially enclose a water area or to stabilize a tidal inlet. The protected areas behind jetties and breakwaters are quiet, and the sediments suspended by the wave action and carried by the longshore current settle out in these quiet basins. The result is less material available for beaches farther down the coast, and, as the longshore current continues to flow, the next beach begins to disappear. Small-scale examples of this process are seen where **groins,** rock or timber structures, are placed perpendicular to the beach to trap sand carried in the longshore transport (fig. 12.28); sand is deposited on the up-current side of the groins, and sand is lost from the beach on the down-current side.

Individual owners, trying to stop the erosion of their property, build bulkheads and seawalls along their beaches. These expensive structures are made of timbers, concrete, or large boulders. The land is armored in hopes of preventing erosion by storm waves, high tides, and boat wakes. Many times these structures make the erosion problem worse because the waves expend their energy over a very narrow portion of the beach. If only a portion of the coastline is protected, wave energy may concentrate at the ends of the seawall and move behind it, or severe erosion may occur at the base of the seawall, and the wall may collapse. Seawalls can reflect waves to combine with incoming waves and cause higher, more damaging waves at the seawall or at some other place along the shore.

The natural shore allows waves to expend their energy over a wide area, eroding the land but maintaining the beaches, all a part of nature's processes of give and take. Time and time again, coastal facilities have been built in response to the demands of people, only to trigger a chain of events that results in newly created problems.

Figure 12.27 An armored beach. Lag deposits of large boulders and cobbles are left on an eroded beach. The rocks on this beach are eroded from glacial deposit cliffs that contain a large assortment of particle sizes, from clays to boulders.

Figure 12.28 These groins have been placed along the beach at Hastings in southern England to trap sand moving along the coast. Beach height difference at this groin is 4 m (12 ft).

Despite our efforts to protect them, beaches are disappearing because of rising sea level and coastal erosion (see the box "Rising Sea Level" in this chapter). Rising sea level has caused worldwide coastal land loss. More than 70% of the world's sandy beaches are eroding, and the percentage increases to nearly 90% for those beaches that have been carefully monitored. It is estimated that about 86% of the beaches along the East Coast from New York to South Carolina are being eroded. Beaches erode back from the waterline roughly 100 times faster than sea level rises vertically. Every millimeter rise in sea level results in 10 cm of shoreline retreat.

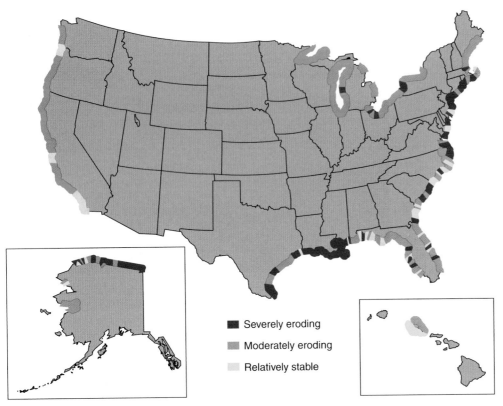

Figure 12.29 Shoreline erosion of the thirty-two coastal and Great Lakes states.

Legend:
- Severely eroding
- Moderately eroding
- Relatively stable

observed, and design modifications can be made before construction begins. Models of this type are expensive, but they are much less expensive than the costs of continually correcting the results of poorly designed structures that disturb the natural environment with undesired consequences.

The Santa Barbara Story

The Santa Barbara, California, harbor project is a classic example of interference with coastal zone processes. At Santa Barbara, a jetty and a breakwater were constructed to form a boat harbor. The jetty at the west side of the harbor juts out into the sea and turns southeastward, forming a breakwater that runs parallel to the coast (fig. 12.30). The longshore current and sediment transport move from west-northwest to east-southeast along this part of the Southern California coast (see fig. 12.24). This jetty-breakwater system creates a wave-sheltered area north and east of the structure and blocks the longshore current. Sand was deposited on the west side, where the longshore current was blocked, and the beach began to grow. The beaches to the southeast began to disappear because they were starved of sand.

When the beach to the west had grown until it reached the seaward limit of the jetty, the longshore current could again move sediments southeastward along the ocean side of the breakwater. When the longshore current and sediment reached the east end of the breakwater and the entrance to the harbor, the current formed an eddy that spiraled into the quiet water of the harbor. The sediment settled out and began to fill the harbor, forming a spit connected to the end of the breakwater. The sediment settling in the harbor deprived the beaches farther southeast of their supply but did not alter the forces acting to remove the sediment from these same beaches.

Today, a dredge pumps the sediments from the harbor through a pipe and back into the longshore current. In this way, the harbor remains open and the beaches to the southeast receive their needed supply of sand. Interference with a natural process requires the expenditure of much time, effort, and money to do the work nature did for no cost.

The History of Ediz Hook

Twenty-four hundred kilometers (1500 mi) north of Santa Barbara, another harbor has been altered through interference with the supply of sand. The harbor of Port Angeles, on the Strait of Juan de Fuca in Washington, is protected from storm waves by a naturally occurring, 5.6 km (3.5 mi) long

Sea-level rise results in beach erosion for several reasons. Deeper water decreases wave refraction (see chapter 10, section 10.7), which increases the capacity for longshore transport. Higher water level allows waves to break closer to shore and expend more energy on the beach. In addition, when sea level rises, the effects of waves and currents reach farther up the beach profile, causing a readjustment of the landward portion of the profile.

According to a U.S. Army Corps of Engineers study, more than 40% of the U.S. continental shoreline is losing more sediment than it receives (fig. 12.29). Of the 24,000 km (15,000 mi) of shoreline, 4300 km (2700 mi) are considered critical areas meriting public protection. Most of the areas designated critical are along the Atlantic and Gulf coasts, and many are on barrier islands. Ocean City, Maryland, has lost about 25% of its beachfront in recent years. Louisiana's Isles Dernieres barrier islands are losing up to 20 m (66 ft) of shore a year; as a result, the mainland suffers increasing coastal flooding and storm damage.

Designing coastal structures that correct or solve one problem without creating more problems is as much an art as a science. Measurements and calculations are made, but more is required, for there must be a thorough understanding of the natural processes in an area before the consequences of the new structure can be foreseen. A scale working model is a helpful aid in understanding the present state of a shore area and the results of future changes. Natural processes such as tides, waves, and currents are reproduced on a scale model, and then the proposed structure is introduced into the system. Its effect is

Figure 12.30 Santa Barbara harbor as seen from the south. In the *foreground,* a dredge removes the sediment that has built up in the protected, quiet water behind the jetty.

curving spit known as Ediz Hook (fig. 12.31). The hook is composed mainly of sand and gravel and protects an area large enough and deep enough to accommodate the largest and most modern commercial vessels. In recent years, the hook has undergone considerable erosion and is in danger of waves breaking through near where the hook is attached to the mainland.

To understand how the hook was formed and how its present condition has come about, one must go back about 124,000 years, to the time when the last glaciers retreated from this area. At that time, sea level was lower than it is now, and the Elwha River west of Port Angeles carried glacial sediments to the sea, forming a delta that spread to the east under the influence of local currents and waves. As sea level rose, the river, currents, and waves continued to carry sediment eastward, and the waves began to erode the cliffs, adding still more sediment to the longshore current. The result was a hook-type spit that built out into the water from where the shoreline makes an abrupt angle with the strait. Ediz Hook was gradually produced and was kept supplied with sand and gravel from cliff erosion and the river.

In 1911, the Elwha Dam was constructed to provide power and a freshwater reservoir. A second dam, the Glines Canyon Dam, was built farther upstream on the river in 1925. The sediments that had previously moved toward the hook from the river (estimated at 30,000 m^3 per year) were now being deposited behind the dams. In 1930, a pipeline to deliver fresh water from the Elwha Reservoir to Port Angeles was

constructed around the cliffs at beach level. Bulkheads built along the face of the cliffs to protect the pipeline cut off the hook's supply of cliff-eroded sediments, estimated at 380,000 m^3 per year.

Ediz Hook does more than protect the harbor; it is also the site of a large paper mill, a U.S. Coast Guard station, and several small harbor facilities. All are connected by a road running the length of the hook. In the 1950s, the city acted to protect the seaward side of the hook's base by armoring with large boulders and steel bulkheads. Cliff material was blasted into the water in hopes of supplying the required sediments. A constant battle between the people and the sea was fought. Nearly as fast as the armor was applied to the base of the spit, the waves tore away the protective barriers and removed the sediments. Only one-seventh of the total sediments once available to feed the spit remained in the longshore transport drift sector, and studies by the Army Corps of Engineers showed a possible loss of 270,000 m^3 of sand each year from the outside of the hook. The hook was sediment-starved.

In the winter of 1973–74, severe storms again damaged the hook, and a Corps of Engineers' program to strengthen the hook began. Over a fifty-year period, annual maintenance costs have been projected at about $30 million compared with revenue from the harbor of $425 million. Since the benefit-to-cost ratio is about 14:1, the project seems to be economically viable. During the summer of 2002, the Corps of Engineers maintenance work on Ediz Hook involved adding approximately 50,000 tons of rock along the hook.

Figure 12.31 Ediz Hook is a long spit forming a natural breakwater at Port Angeles, Washington. The protected harbor behind the spit is deep enough for mooring the largest supertankers (infrared false-color photo).

Since the two dams were built on the Elwha River early in the century, annual salmon runs declined from 380,000 to less than 3000 in the 1990s. Congress enacted a law in 1992 that authorized the removal of the dams to help salmon restoration. Once the dams are removed, the natural transport of sediment by the river to the coast will be restored, and Ediz Hook should no longer be as severely starved of sediment.

12.7 Estuaries

River mouths, fjords, fault bays, and other semienclosed bodies of salt water are all part of the coastal zone. One such body with a free connection with the ocean and fresh water diluting its average salinity to less than that of the adjacent sea is an **estuary.** Estuaries can be described by the geologic processes that formed them, and they can also be described by the dynamics of their freshwater and saltwater circulation.

Geomorphology may distinguish between estuary types in the same way that it is used to classify coasts. However, such a system does not consider the complex and dynamic combination of factors that come together when fresh water and salt water meet. In this chapter, estuaries are described mainly in terms of water exchange and salinity variation. Tides, river flow, and the geometry of an inlet are additional factors to be considered. Four basic estuary types are distinguishable.

Types of Estuaries

The simplest type of estuary is the **salt wedge estuary.** Salt wedge estuaries occur within the mouth of a river flowing directly into salt water. The fresh water flows rapidly out to sea at the surface, while the denser seawater flows upstream along the river bottom. The seawater is held back by the flow of the river, and an abrupt boundary forms, separating the intruding wedge of seawater and the fresh water moving downstream. A salt wedge estuary is shown in figure 12.32. The net seaward surface flow is almost entirely fast-moving river water because the large discharge of fresh water is forced into a surface layer above the salt wedge. The salt wedge moves upstream on the rising tide or when the river is at a low flow stage; the salt wedge moves downstream on the falling tide or when the river flow is high. The boundary between the salt water and the overriding river water is kept sharp by the rapidly moving river water. The moving fresh water erodes seawater from the face of the wedge. The waters of contrasting salinity mix and move upward into river water, raising the salinity of the seaward-moving fresh water. This one-way mixing process is called **entrainment;** very little river water is mixed downward into the salt wedge. Seawater from the ocean is continually added to the salt wedge to replace the salt water entrained into the seaward-flowing river water. In a salt wedge estuary, the circulation and mixing are controlled by the rate of river discharge; the influence of tidal currents is generally small compared to the influence of the river flow. Examples of salt wedge estuaries are found in the mouths of the Columbia, Hudson, and Mississippi Rivers. In the Columbia River, the salt wedge moves as far as 25 km (15 mi) upstream at times of high tide and low river flow, maintaining its identity as a sharp boundary between the two water types. Salt wedges also occur where rivers enter other types of estuaries: for example, at the mouth of the Sacramento River in San Francisco Bay.

Estuaries not of the salt wedge type are divided into three additional categories on the basis of their net circulation and the vertical distribution of salinity. These categories are the well-mixed estuary, the partially mixed estuary, and the fjord-type estuary.

Well-mixed estuaries (fig. 12.33) have strong tidal mixing and low river flow, creating a slow net seaward flow of water at all depths. Note that mixing is so complete that the salinity of the water is uniform over depth and decreases from the ocean to the river. There is little or no transport of seawater inward at depth; instead, salt is transferred inward by mixing and diffusion. The nearly vertical lines of constant salinity move seaward on the falling tide or when river flow increases, and they move landward on the rising tide or when river flow decreases. Many shallow estuaries, including the Chesapeake and Delaware Bays, are well-mixed estuaries.

Partially mixed estuaries (fig. 12.34) have a strong net seaward surface flow of fresh water and a strong inflow of seawater at depth. The seawater is mixed upward and combined with the river water by tidal current turbulence and entrainment to produce a seaward surface flow that is larger than that

of the river water alone. This two-layered circulation acts to rapidly exchange water between the estuary and the ocean. Salt moves into a partially mixed estuary by diffusion but also, and more important, by **advection,** the inflow of seawater at depth. Examples of partially mixed estuaries include deeper estuaries, such as Puget Sound, San Francisco Bay, and British Columbia's Strait of Georgia.

Fjord-type estuaries (fig. 12.35) are deep, small-surface-area estuaries with moderately high river input and little tidal mixing. This pattern occurs in the deep and narrow fjords of British Columbia, Alaska, the Scandinavian countries, Greenland, New Zealand, Chile, and other glaciated coasts. In these estuaries, the river water tends to remain at the surface and move seaward without much mixing with the underlying salt water. Most of the net flow is in the surface layer, and there is little influx of seawater at depth. Salt is supplied to the low-salinity surface layer at a slow rate by advective transport. The deeper water may stagnate because the entrance sill isolates it (see fig. 6.6). Little inflow occurs below the surface layer.

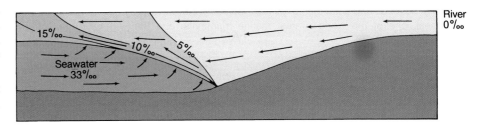

Figure 12.32   The salt wedge estuary. The high flow rate of the river holds back the salt water, which is entrained upward into the fast-moving river flow. Salinity is given in parts per thousand (‰).

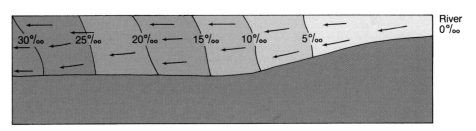

Figure 12.33   The well-mixed estuary. Strong tidal currents and wind-driven circulation distribute and mix the seawater throughout the shallow estuary. The net flow is weak and seaward at all depths. Salinity is given in parts per thousand (‰).

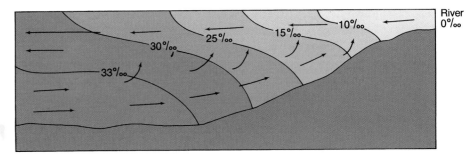

Figure 12.34   The partially mixed estuary. Seawater enters below the mixed water that is flowing seaward at the surface. Seaward surface net flow is larger than river flow alone. Salinity is given in parts per thousand (‰).

National Marine Sanctuaries

One hundred years after Yellowstone was established as our first national park, the U.S. Congress passed the Marine Protection, Research and Sanctuaries Act of 1972. The first sanctuary was Monitor National Marine Sanctuary, created in 1975 around the wreck of the Civil War ironclad *Monitor,* which lies in 70 m (230 ft) of water off Cape Hatteras, North Carolina. Today, there are thirteen national marine sanctuaries: five off the U.S. west coast, four off the U.S. east coast, including the Florida Keys, two in Hawaii, and one each in the Gulf of Mexico and American Samoa (box fig. 1). Together, these thirteen sites cover 46,000 km² (18,000 mi²). One more site is proposed: in Puget Sound, Washington, and still others are under consideration.

The newest marine sanctuary, created in June of 2006, is Papahānaumokuākea Marine National Monument. The Papahānaumokuākea Marine National Monument is the single largest conservation area in the United States, and the largest marine conservation area in the world. It encompasses 357,800 km² (137,000 mi²), an area larger than all the country's national parks combined. The region is home to more than 7000 marine species, about a quarter of which are found nowhere else on Earth. The islands are also a breeding site for millions of seabirds and endangered animals, including the Hawaiian monk seal, one of the world's rarest marine mammals, and the Hawaiian green sea turtle. Papahānaumokuākea is also of great cultural importance to Native Hawaiians with significant cultural sites found on the islands of Nihoa and Mokumanamana.

Monterey Bay National Marine Sanctuary stretches along the Pacific coast for 580 km (350 mi) and reaches seaward as much as 90 km (53 mi). This area supports a great diversity of marine life in various habitats, including a submarine canyon more than 3300 m (2 mi) deep. To the south, the Channel Islands National Marine Sanctuary is home to myriad seabirds, more than twenty kinds of sharks, thousands of sea lions, and northern fur seals, elephant seals, and the rare Guadalupe fur seal, which was hunted nearly to extinction in the nineteenth century. The edges of the islands are ringed with forests of giant kelp, seaweeds that provide food and shelter for a rich diversity of marine species.

North of Monterey Bay, the Gulf of the Farallones National Marine Sanctuary lies 50 km (30 mi) west of the Golden Gate. These islands provide a dwelling place for twenty-six species of marine mammals and the largest concentrations of breeding seabirds in the continental United States. More than a quarter of a million petrels, puffins, terns, auklets, murres, and other species visit this area every year. Still farther north, the beaches of the Olympic Coast National Marine Sanctuary remain almost as they

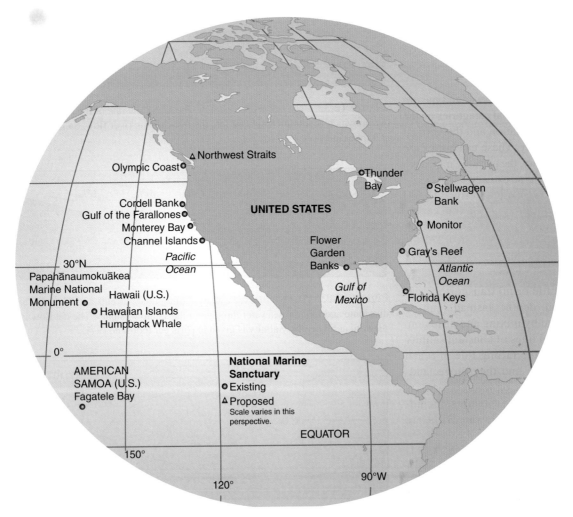

Box Figure 1   U.S. national marine sanctuaries.

Box Figure 2   Breaching humpback whales.

have always been, rich in marine life among rocky shores and tide pools.

Hawaiian Islands Humpback Whale National Marine Sanctuary is the breeding, calving, and nursing home for two-thirds of the North Pacific humpback whale population (box fig. 2). The very rare Hawaiian monk seal is also found here. On the U.S. Atlantic coast, Stellwagen Bank National Marine Sanctuary, a part of Massachusetts Bay east of Boston and north of Cape Cod, lies over an area of gravel and coarse sand left by receding glaciers. This is a very productive area and a feeding ground for humpback, fin, and northern right whales.

Sixty kilometers (100 mi) off the Texas-Louisiana coast in the Gulf of Mexico, a pair of salt domes support the northernmost coral reefs in North America. This is Flower Garden Banks National Marine Sanctuary, home to manta rays, hammerhead sharks, and loggerhead turtles. Off the tip of Florida, a 370 km (220 mi) arc of reefs, seagrass beds, and mangroves makes up the Florida Keys National Marine Sanctuary (box fig. 3).

These sanctuaries are national marine sanctuary multiple-use areas managed by the National Oceanic and Atmospheric Administration (NOAA), a branch of the Commerce Department. Commercial fishing, sport fishing, and spear fishing are permitted in the sanctuaries; so are boating, snorkeling, diving, and other kinds of recreation. Shipping traffic is not prohibited, but it is restricted to certain lanes and areas. In general, drilling, mining, dredging, dumping waste, and removing artifacts are forbidden.

Sanctuary designations have not been welcomed by all local populations. The bay surrounding Stellwagen Bank is heavily fished, and those who fish are opposed to further regulation even though catches continue to decline. Also, ships bound for Boston cross sanctuary waters, producing concern that they may interfere with whale migrations. The Florida Keys have a permanent population of more than 80,000 people. Many are economically dependent on the surrounding waters, and 2.5 million tourists come to snorkel, dive, and sport fish each year (bringing in more than $1 billion), so there

Box Figure 3   A diver watches a school of grunts in a coral garden in the Florida Keys.

has been significant resistance to the presence of NOAA. In the Channel Islands, a squid fishery is developing, and there is concern over the quantity of the catch, its effect on the food chains, and the numbers of other species being caught in the squid nets. How do we reconcile commercial fishing with its tradition of taking maximum harvests with the establishment of havens that seek to protect an entire sanctuary system?

To Learn More About National Marine Sanctuaries

Chadwick, D. 1998. Blue Refuges. *National Geographic*
    193 (3): 2–13.
Monterey Bay Aquarium
National Marine Sanctuaries—NOAA

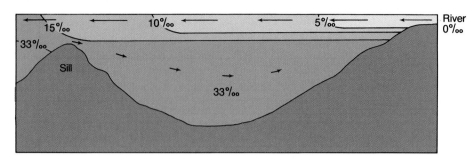

Figure 12.35 The fjord-type estuary. River water flows seaward over the surface of the deeper seawater and gains salt slowly. The deeper layers may become stagnant owing to the slow rate of inflow. Salinity is given in parts per thousand (‰).

The salt wedge estuary is a strongly stratified estuary. These estuaries are never very well mixed vertically, except above the boundary between the seaward-moving river water and the salt wedge. The other estuary types have various degrees of vertical stratification between the highly stratified and poorly mixed, and weakly stratified or well mixed. The processes that govern the degree of vertical mixing and stratification are the strength of the oscillatory tidal currents, the rate of freshwater addition, the roughness of the topography of the estuary over which these currents flow, and the average depth of the estuary. Tidal flows that change direction with the rise and fall of the tide produce energy for mixing; it does not usually result in a net directional flow and should not be confused with the net seaward or landward flow in each estuary.

Not all estuaries fit neatly into one of these categories. There are estuaries that fit between these types, and some estuaries change from one type to another seasonally with changes in river flow or weekly as the tides change from springs to neaps. Studying how an individual estuary compares to examples of the different types helps the oceanographer understand the processes in an estuary and the water exchange with the ocean.

Circulation Patterns

This discussion of estuary circulation is based on a partially mixed estuary, with its inflow from the ocean at depth and its surface outflow of mixed river and seawater. Tidal currents are produced as water flows into the estuary on the rising tide and flows out on the falling tide. Water in these currents moves back and forth in the estuary, but it does not immediately leave the system and enter the open sea. However, the surface water moves farther seaward on the falling tide than it moves inward on the rising tide, and a parcel of surface water is moved progressively farther seaward on each tidal cycle. In the same way, the deeper water from the sea moves into the estuary on the rising tide farther than it flows seaward on the falling tide. Averaged over many tidal cycles, this pattern produces a net movement seaward at the surface and a net movement landward at depth.

The **net circulation** of an estuary out (or seaward) at the top and in (or landward) at depth is of great importance because the seaward surface circulation carries wastes and accumulated debris seaward and disperses them in the larger oceanic system.

It is also through this circulation that organic materials and juvenile organisms produced in the estuaries and their marsh borderlands are moved seaward, while nutrient-rich water is brought inward at depth and replenishes the estuary's salt water.

Understanding the net circulation of an estuary and evaluating the flows of surface water seaward and the underlying ocean water landward can be a long process. To determine this pattern directly, the oceanographer installs recording current meters at various depths and at several cross-channel locations. Currents are measured over many tidal cycles for both spring and neap tides and at times of low, intermediate, and high freshwater discharge. The current records from each depth and location are averaged to discover the net current distribution for a given cross section in the estuary and the net flow of the estuary. Data are averaged over one-week periods to determine the importance of spring and neap tides, over monthly cycles to find changes in patterns due to seasonal river and climate fluctuations, and over several years to produce an annual pattern. The cost of installing and maintaining equipment, as well as the costs of processing the data, make this direct approach very expensive.

A less expensive, indirect approach is to assume a **water budget** with inflow equal to outflow and a constant estuary volume averaged over time; all processes that add or remove water must be considered. A **salt budget** is also assumed; salt added is equal to salt removed. At the entrance to the estuary, measurements of salinity are made over space and time, establishing the average salinity of the outflowing surface water, $\overline{S}_o$, and the average salinity of the deeper inflowing seawater, $\overline{S}_i$. The ratio

$$\frac{\overline{S}_i - \overline{S}_o}{\overline{S}_i}$$

is used to determine the fraction of river water in the seaward-moving surface layer at the estuary entrance.

If the average rate of river water inflow (R) is known for this same period, the volume rate of the surface layer's seaward flow (T_o) is found by using the following formulas:

$$\frac{\overline{S}_i - \overline{S}_o}{\overline{S}_i}(T_o) = R, \quad \text{or} \quad T_o = \frac{\overline{S}_i}{\overline{S}_i - \overline{S}_o}(R)$$

In a partially mixed estuary, the net seaward flow (T_o) is always larger than the river flow (R) and the river flow and the net saltwater inflow (T_i) combine to produce the seaward flow (T_o) and maintain the water budget. Inflow equals outflow.

$$T_o = T_i + R$$

Both evaporation, E, and precipitation, P, also remove and add water at the surface, as shown in figure 12.36. If both E and P are large, then R in the previous equations becomes ($R - E + P$).

This method assumes that, over the time period for which the calculations are used, the average water volume and the total salt content of the estuary remain constant. For more information on these equations, see appendix C.

Sea level is on the rise around the coasts of the world. Since August 1992, satellite altimeters have been measuring sea level on a global basis with unprecedented accuracy. Data from the *TOPEX/Poseidon* satellite (see chapter 1, section 1.10) provided observations of sea-level change from 1992 until 2005. *Jason-1*, launched in late 2001 as the successor to *TOPEX/Poseidon*, continues this record by providing an estimate of global mean sea level every ten days with an uncertainty of 3–4 mm. Over the past century, sea level has risen about 25 cm (10 in), but recently the rate of rise has increased. Sea level is presently rising at a rate of approximately 3.4 mm (0.13 in) per year (box fig. 1). It is difficult to predict rates of sea-level rise for the future, but current thought is that sea levels will rise at rates two to three times faster in the next 100 years than they did in the twentieth century. By the year 2100, sea level is expected to be 31–110 cm (1–3.5 ft) higher than at present; a good estimate is for an increase of 66 cm (2 ft).

Rising sea level is a global phenomenon, but the problems it causes are not distributed evenly around the world. Areas in mid-ocean such as small tropical atolls with elevations of only 1–2 m (3–7 ft) may be severely affected. Low coastal marshlands will undergo greater changes than coasts with steep, rocky coastlines. Each year, 100–130 km² (40–50 mi²) of Louisiana's Mississippi River wetland marsh disappear under the Gulf of Mexico (box fig. 2). Twelve thousand years ago, Louisiana's shoreline extended 200 km (120 mi) farther into the Gulf of Mexico. In the mid-1880s, Bailize was a busy river town at the tip of the delta; today, it is under 4.5 m (15 ft) of water.

Significant sea-level changes over hundreds to thousands of years are governed by variations in the amounts of water stored in Earth's reservoirs. Water is transferred from ocean to ice during glacial periods and back to the ocean during interglacial periods; this transference can account for a change in sea level of 100–200 m (300–650 ft).

Sediments accumulated in ocean basins also displace seawater. If the oceans of the world warmed by just 1°C, sea level would rise by about 60 cm (24 in) owing to the thermal expansion of seawater. Coasts may also rise or sink relative to sea level in response to isostatic forces.

The global average change in sea level may not be reflected in the gradual changes observed at any particular location. In the tropics, sea temperatures are nearly constant, so changes attributable to thermal expansion are small, but the seasonal oscillation of the

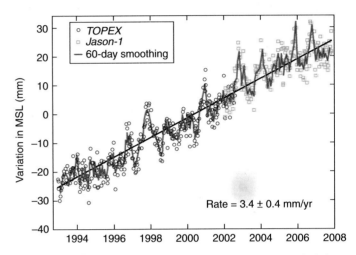

Box Figure 1   Temporal variations in global mean sea level (MSL) computed from *TOPEX/Poseidon* and *Jason-1* satellite altimeters. Each dot is a single ten-day estimate of the difference between MSL during those ten days and the mean MSL during the entire period from December 1992 through 2007.

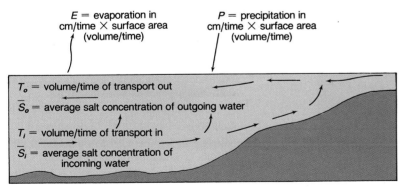

Figure 12.36   Water and salt enter and exit a partially mixed estuary.

Temperate-Zone Estuaries

In the middle latitudes of North America, most estuaries gain their fresh water from rivers, and evaporation from the estuary surface is minor or nearly balanced by direct precipitation. The salt concentration of the entering seawater, $\bar{S}_i$ is about 33‰, and $\bar{S}_o$ is about 30‰. When we use the preceding equation and the example in figure 12.37, T_o is shown to be about eleven times the rate

of river inflow, or R. T_i in this case is ten times R. The inflow rate of seawater at depth decreases as it moves into the estuary, and the surface flow increases seaward because of mixing and the incorporation of seawater from below. While this mixing is occurring, the average salt concentration of the seaward-moving surface flow is also increasing, and T_o is formed when 10 unit volumes of seawater have combined with the 1 unit volume of river water. These relatively large values for T_o and T_i compared to R make these net flows the key to understanding the exchange of estuary water with ocean water.

In some fjordlike estuaries, the surface layer is only as deep as the shallow sill, and the average salt content of the surface layer is very low, owing to weak mixing. In this case, T_o is approximately equal to R; T_i is negligible, and the deep water of the fjord stagnates.

12.8 High Evaporation Rates

Many bays located near 30°N and 30°S latitudes have low precipitation and high evaporation rates. Although rivers may flow into these bays, these bays are not estuaries if they do not

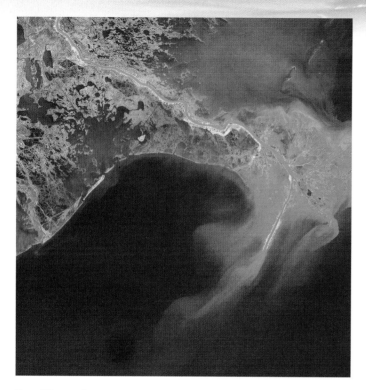

Box Figure 2 The Mississippi River delta. Twelve thousand years ago, Louisiana's shoreline extended 200 km (120 mi) farther into the Gulf of Mexico. Today, the river is kept in its channel by levees, and the sediments continue to the river's mouth. Sediments no longer replenish the delta and its marshes; instead, they are carried out into deep water.

meteorological equator (the intertropical convergence zone) between approximately 0° and 10°N causes sea level to vary in response to atmospheric pressure changes. Changes in sea-surface temperature associated with El Niño (see chapter 7) affect sea level, and in the North Atlantic and North Pacific, the annual seasonal shift from high- to low-pressure cells over the oceans causes a related change in sea level (rising with low atmospheric pressure and falling with high pressure). Onshore winds elevate and offshore winds depress the coastal water level. Ocean currents under the influence of the Coriolis effect create a sea-surface topography that varies by as much as a meter. Any climate-generated changes in speed or location of winds or currents alter sea level. Climate changes associated with atmospheric warming can be expected to alter the winds, the currents, and atmospheric pressure distributions as well as the thermal expansion of seawater. The atmospheric changes affect the sea surface, resulting in sea-level changes of as much as 1–2 m (3–7 ft).

To Learn More About Rising Sea Level

ATOC Consortium. 1998. Ocean Climate Change: Comparison of Acoustic Tomography, Satellite Altimetry and Modeling. *Science* 281 (5381): 1327–1332.

Nerem, R. S. 1995. Global Mean Sea Level Variations from TOPEX/Poseidon Altimeter Data. *Science* 268: 708–710.

Nerem, R. S., D. P. Chambers, E. W. Leuliette, G. T. Mitchum, and B. S. Giese, 1999. Variations in Global Mean Sea Level Associated with the 1997–1998 ENSO Event: Implications for Measuring Long-Term Sea-Level Change. *Geophysical Research Letters* 26 (19): 3005–3008.

Schneider, D. 1998. The Rising Seas. *Oceans, Scientific American Quarterly* 9 (3): 28–35.

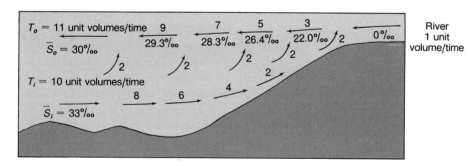

Figure 12.37 The seaward-moving surface water *(odd-numbered arrows)* increases in flow volume as the inflowing seawater *(even-numbered arrows)* moves upward. Upward mixing is indicated by the *vertical arrows*. For every 2 units of seawater mixed upward, the inflow decreases and the outflow increases by the same 2 units. Salinity is given in parts per thousand (‰).

experience net dilution. The contribution from rivers is usually minor when compared to the loss of water due to evaporation. The Red Sea and the Mediterranean Sea are examples. In these seas, evaporation increases the surface water salinity and density directly, causing surface water to sink and accumulate at depth, where it flows toward the open ocean. Ocean water flows into the sea at the surface, because it is less dense than the high-salinity outflow of deep water. The T_o and T_i flows have reversed, with T_o at depth and T_i at the surface (fig. 12.38). Because of this reversed circulation pattern, these bays are sometimes called **inverse estuaries.**

Compare figures 12.37 and 12.38. The evaporative bay in figure 12.38 shows a typical $\overline{S}_i$ value (36‰) for ocean-surface conditions at latitudes of 30°N and 30°S. The evaporative loss from the surface is 1 unit volume per time, resulting in a T_i flow of 20 units and a T_o flow of 19 units. The resulting $\overline{S}_o$ value is 37.9‰. The estuary in figure 12.37 has a river input of 1 unit volume per time and T_o and T_i flows of 11 and 10 units, respectively. In general, if the evaporation removal rate is similar to the river input of a temperate estuary, an inverse estuary or evaporative bay has a better rate of exchange with the ocean. Not only is T_o larger, but the exchange of water is more complete because the evaporative bay has a continuous overturn as surface water sinks and moves seaward.

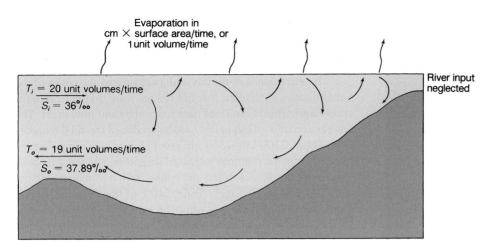

Figure 12.38 The evaporative sea or inverse estuary. Evaporation at the surface removes water (1 unit), and river inflow is negligible. Seawater flows inward at the surface ($T_i = 20$ units), and the seaward flow ($T_o = 19$ units) is at depth. Salinity is given in parts per thousand (‰).

12.9 Flushing Time

The mean volume of an estuary divided by T_o gives an estimate of the length of time required for the estuary to exchange its water. This is known as **flushing time.** If the net circulation is rapid and the total volume of an estuary is small, flushing is rapid, and the flushing time is short. A rapidly flushing estuary has a high carrying capacity for wastes because the wastes are moved rapidly out to sea and dispersed. A slowly flushing estuary risks accumulating wastes and building up high concentrations of land-derived pollutants. Population pressures on estuaries are heavy, for these areas are used as seaports, recreational areas, and industrial terminals as well as for their fishing resources. Understanding the circulation of estuary systems is essential to maintaining them as healthy, productive, and useful bodies of water.

If the waste products are associated only with the fresh water entering the estuary, it is possible to determine the ability of the fresh water to carry these products through the estuary without considering the action of the entire estuary. To do so the amount of fresh water in the estuary at any one time must be known. The freshwater volume is estimated from the average salinity of the estuary. For example, an estuary with an average salt concentration of 20‰ is one-third fresh water if the adjacent ocean salinity is 30‰. Dividing the freshwater volume by the rate of river water addition yields the flushing time for fresh water and its pollutants.

Some bays and harbors do not have a freshwater source; these areas are not estuaries and are flushed only by tidal action. On each change of the tide, a volume of water equal to the area of the bay multiplied by the difference between the high and low tide levels exits and enters the bay. When this volume, known as the **intertidal volume,** leaves the bay and enters the ocean, it may be displaced along the coast by a prevailing coastal current. On the next rising tide, an equivalent volume of different ocean water enters the bay. Under these conditions, the flushing time can be estimated by the number of tidal cycles required to remove and replace the bay volume. This number is found by dividing the mean bay volume by the intertidal volume exchanged per tidal cycle. However, the flushing is often not complete because currents along the coast do not always move the exiting intertidal volume a sufficient distance to prevent some portion of the same water from recycling into the bay. Therefore, the number of tidal cycles required for flushing may be greater than the calculated estimate.

The same conditions occur in estuaries that have a two-layered flow if mixing incorporates some of the exiting surface water into the deeper water flowing in from the sea. This mixing results in partial recycling of the T_o flow and modification of the T_i flow, and increases aeration of the deeper parts of the estuary.

Occasionally, the net circulation in an estuary is altered by ocean conditions. For example, if coastal upwelling occurs, the inflow of cooler, denser water to the estuary increases, and this increased inflow accelerates the estuary's circulation and causes the outflow to increase. When upwelling ceases and downwelling occurs, less-dense seawater is present at the estuary entrance, and the inflow is reduced. In these circumstances, the estuary's circulation may temporarily reverse as the denser, deeper water flows back to sea and the less-dense seawater moves inward at the surface.

Although each estuary has its own distinctive characteristics, the knowledge gained from studying one estuary can be used to understand other similar estuary systems. Understanding the circulation patterns and the processes that control these patterns allows us to judge the degree to which an estuary can be modified and used while still preserving its environmental and economic value. Because estuaries lie at the contact zone between the land and the sea, and because humans are inhabitants of the land, alterations have been made in these areas and probably cannot be entirely avoided in the future. But we owe it to ourselves and to those who come after us to make intelligent and knowledgeable uses of these areas, for they represent a great renewable natural resource if treated with care.

12.10 Practical Considerations: Case Histories

The Development of San Francisco Bay

The San Francisco Bay estuary (fig. 12.39) has a surface area of 1240 km² (480 mi²). Its river systems drain 40% of the area of California. When visited by the early Spanish missionaries in 1769, the bay was surrounded by wetlands, where the 10,000–15,000 native people gathered much of their food. Development came slowly until 1848, when gold was discovered, and the population of San Francisco increased from approximately 400 to 25,000 in 1850. Within fifty years of the initial gold boom, the marshes were nearly gone, the bay was shallowed, fresh water had been siphoned off for irrigation, and

Figure 12.39 The San Francisco Bay system. The city of San Francisco lies at the *bottom center.* The large, light-colored northern portion of the San Francisco Bay system is known as San Pablo Bay. The Sacramento River is joined by the San Joaquin River at the Y in the *upper center* of the photograph.

nonnative species of animals had been introduced, displacing native species.

Early in the bay's development, the fisheries resources (salmon, sturgeon, sardines, flatfish, crabs, and shrimp) were heavily exploited to feed the rapidly expanding population. By 1900, many of the fish, shellfish, and waterfowl stocks had been overharvested and depleted. Crabs were fished out in the 1880s, and the fishery was moved offshore, until its collapse in the 1960s. When the transcontinental railroad was completed in 1869, carload after carload of oysters was shipped from the East to mature on the bay's mudflats. The oysters did not become a self-reproducing stock, but many other small marine animals introduced with the oysters did adapt to the new environment, including the eastern soft-shell clam, the Japanese littleneck clam, and marine pests, the oyster drill and the shipworm. Even the striped bass, the bay's best-known sport fish, was introduced from the East Coast in 1879. The only commercially harvested fish left in the bay are herring and anchovy. For the story of more recent destructive invaders, see chapter 13.

Until 1884, hydraulic mining for gold in the surrounding watershed washed huge amounts of silt and mud directly into the streams and rivers. The silt and mud destroyed the salmon spawning areas. Much of the sand and mud reached the bay,

where it shallowed some areas, expanded the marshes, and altered the tidal flow. The rivers were flooded with the runoff from the barren land, and the resulting winter and spring floods produced changes in the estuaries, where the rivers enter the bay.

The marshlands of the bay and, particularly, the deltas of the Sacramento and the San Joaquin Rivers were diked, first to increase agricultural land and then for homes and industries. The conversion of wetland to dry land has reduced the tidal marshes from about 2200 km^2 (860 mi^2) to 125 km^2 (49 mi^2). The increased cropland required increased irrigation, and the state built a series of dams, reservoirs, and canals with a water-storage capacity of 20 km^3 (0.5 mi^3). Today, 40% of the Sacramento and San Joaquin Rivers' flow is removed for irrigation, and another 24% is sent by aqueduct to central and southern California.

The diversion of water so reduces the amount of river water during periods of low flow in the summer that the irrigation pumps cause the water to flow upstream from the bay, carrying with it hundreds of juvenile salmon and striped bass, which are drawn into the pumps and die on the fields or in the ditches. This is thought to have been a major cause of a drop in the population of striped bass since the 1960s. The loss of freshwater flow also changes the distribution and abundance of small, floating species on which the juvenile fish feed.

Intensive agricultural practices using fertilizers and pesticides, plus the runoff of water high in salts leached from the irrigated fields, have changed the quality of the fresh water entering the bay. An effort to lessen this effect involved directing runoff water from agriculture into holding reservoirs, some of which were also wildlife refuges. In 1982, deformed wildlife and depressed reproduction in bird species were discovered at the Kesterson Reservoir. Levels of the element selenium, 130 times normal, were found in plant and animal tissues. The selenium was being leached from the irrigated soils and concentrated in the reservoir. Kesterson Reservoir was closed in 1986. Recently, high concentrations of selenium have been found in south bay ducks.

Domestic and industrial wastes from urban areas also enter the bay. Residence time of water in north San Francisco Bay fluctuates between one day during peak river flow in the winter to two months during low summer flow. Winter flows carry fresh water into the south bay and increase its exchange with the central bay, decreasing south bay residence time from months to weeks at this period of the year. In the south bay, the city of San Jose has increased its discharge of treated wastewater as its population has grown. Since the 1980s, this freshwater increase has converted hundreds of acres of salt marsh to freshwater and brackish marsh, producing loss of habitat for native birds and small animals. San Jose has been ordered to create new habitat to offset this loss and to reclaim water from its treated sewage effluent in order to decrease the freshwater flow. If these improvements fail, a limit will be imposed on the volume of discharge, requiring the city to place limits on its population growth.

San Francisco Bay is considered the most modified major estuary in the United States. The city of San Francisco's population is roughly 775,000 in a metropolitan area of about 7 million. Despite all the factors involved in this increasingly crowded area, the bay appears to have suffered less total water-quality degradation than some other major estuaries. Water quality has been maintained, in part, because the greatest urbanization has occurred near the mouth of the bay, allowing wastes to exit quickly through the Golden Gate. Enhanced sewage treatment since the 1960s has improved conditions. There are patches of contamination from PCBs, oil, and chemical spills, but at this time, the bay does not seem to be suffering from an overproduction of marine plants or a depletion of oxygen. Shellfish collection has recently been permitted for the first time in decades. The pressure for increased growth, the recent years of drought, and the loss of fresh water to irrigation continue to affect bay plant and animal populations; brackish-water marshes disappear or convert to saltwater systems, displacing the native organisms. The delta smelt has declined by 90% from past levels, while adult striped bass have decreased from over 3 million in the past to 500,000 in 1991. In 1969, 118,000 winter chinook salmon made their way up the Sacramento River; in 1992, only 191 were counted. Since 1993, rainfall and river flow have improved in northern California, and in the summer of 1995, a large run of chinook salmon appeared in San Francisco Bay.

Local concern and five years of work resulted in the 1965 creation of the Bay Conservation and Development Commission (BCDC). Under the BCDC, public access to the bay's 276 miles of shoreline increased from 4 miles to 110 miles; the surface area of the bay increased by 800 acres, and filling was reduced from 2300 acres per year to nearly zero. The BCDC has permanently protected 99,000 acres of bay wetlands and has purchased an additional 10,000 acres in the north bay for wildlife habitat.

The Situation in Chesapeake Bay

The Chesapeake Bay estuary (fig. 12.40) is a shallow, drowned river valley with a surface area of 11,500 km² (7000 mi²) and an average depth of 6.5 m (21 ft). Chesapeake Bay flushes slowly; the residence time of the water is 1.16 years. The Chesapeake estuary has a long history as a major provider of oysters, blue crab, waterfowl, rockfish, and shad. It has also been under continuously increasing population pressure since colonial times and has been absorbing more and more discharges from the expanding urban centers of Washington, D.C., Baltimore, Norfolk, and dozens of towns in Maryland, Virginia, and Delaware. After heavy harvesting in the 1950s, the oyster catch dropped by two-thirds between 1960 and 1983. The oyster population at one time was large enough to filter a volume of water equal to the bay's volume of floating microscopic algae in about two weeks; the present low population takes about one year to filter an equivalent volume. Also between 1960 and 1983, the annual rockfish catch decreased from 6 million to 600,000 pounds, commercial fishing for shad has nearly disappeared in the upper Chesapeake, and the waterfowl population drop has been equally dramatic.

Although present harvests are low compared to those of the early years, the blue crab harvest for 1992 increased, and some areas report increased oyster catches. The bay's present total harvest of seafood is approximately 45 million kg per year

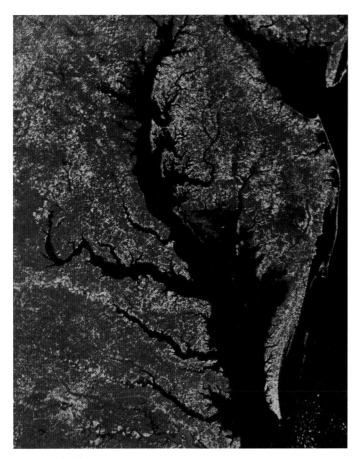

Figure 12.40 Chesapeake Bay is a shallow estuary located on the heavily populated eastern coast of the United States. The entrance to the estuary is at the *lower right.*

(99 × 10⁶ lb) with a dockside value of nearly $1 billion. The total value of the bay to Maryland and Virginia is estimated at $678 billion.

Partially treated and untreated sewage was recognized as the cause for outbreaks of typhoid fever among people who drank water and ate shellfish taken from the system in the early 1900s. Then, decaying untreated sewage robbed the water of oxygen; now, plant growth, stimulated by nutrients from the effluent of sewage treatment plants and the runoff from fertilized fields, decays and consumes the dissolved oxygen. Increased plant growth also reduces the clarity of the water and shades the bottom. The low oxygen levels and the reduction in bottom plant life due to decreased light alter the bottom habitat. In addition, wastes from some 5000 factories, from military bases and treated effluent from sewage plants in Pennsylvania, Maryland, Virginia, and southern New York find their way into the Chesapeake, adding conventional pollutants and toxic compounds that are trapped in the bottom sediments.

Efforts to improve water quality in Chesapeake Bay have produced nearly 4000 studies since the 1970s. In 1972, the Federal Clean Water Act provided enforcement standards and a system of permits governing the amount of pollutants that individual dischargers could dump into any body of water. Gradually, industrial and municipal dischargers were regulated, and efforts have reduced the input of phosphorus from sewage by 75% since 1981. At the same time, populations in all the areas of the Chesapeake have increased, and industries have expanded, causing the rate of effluent discharge to increase as well.

At the present time, it is estimated that industries and sewage plants along the shores of Maryland and Virginia discharge about 4 trillion gallons of treated wastewater annually, or about 20% of the water in the bay at any one time. An Environmental Protection Agency study costing $27 million over seven years launched the "Save the Bay" campaign in 1983. Scientists working for the campaign reported that industries in Maryland dump more than 2700 tons of heavy metals into the bay each year and that Virginia industries dump more than 400 tons during the same period. Much of the money has been spent to manage soil and fertilizer runoff from surrounding fields. About 41 million acres of the Chesapeake watershed are farmed by less than 3% of the watershed's population. Farming uses almost 700 million pounds of commercial fertilizer each year. Use of fertilizers has led to an increase in the ratio of nitrogen to phosphorus in bay waters, and many observers feel that this is the major cause of outbreaks of the toxic organism *Pfiesteria piscicida,* also known as "the cell from hell" (see chapter 16).

Critics charge that the Save the Bay program ignores industrial and municipal discharges in excess of the levels allowed under the current permit system. The federal permit system is a self-policing effort, with dischargers setting their own limits depending on available technology and cost. The permit system has not responded to new technologies and enforcement and, until recently, has not been independently monitored. Many industries trying to avoid the discharge permit process direct their discharges to municipal sewage systems that are principally equipped to treat domestic and organic wastes. Chesapeake Bay is shallow and has a large surface area; it has a very large ratio of watershed drainage area to volume of water. River systems that carry pollutants to the bay drain parts of New York, Pennsylvania, Delaware, Maryland, Washington, D.C., Virginia, and West Virginia. An unusual degree of cooperation must be achieved to solve the bay's problems.

While trying to understand the effects of human degradation of the Chesapeake, researchers have discovered that the natural changes in rainfall and temperature have large impacts as well. At irregular intervals, tropical storms and hurricanes produce flood conditions that have major effects on the water quality and the organisms. Cyclic variation in survival of eastern oysters and striped bass have been related to such weather changes. In 1972, 1975, and 1979, the annual freshwater inflow was double the average. In 1972 in particular, Tropical Storm Agnes dropped 25.5 cm (10 in) of rainfall in a short period of time. This sudden surge of fresh water reduced the bay's salinity to about one-fourth its normal minimum level and rinsed out the small floating organisms on which the larger animals feed. Severe winters, occurring at six- to eight-year intervals, promote heavy ice formation, which depresses the oxygen levels beneath the ice. These irregular but significant events superimposed on the increasingly degrading human influences make it difficult to determine exactly which factors are responsible for which changes in the Chesapeake estuary system.

Summary

The coast is the land area affected by the ocean. The shore extends from the low-tide level to the top of the wave zone. The beach is the accumulation of sediment along the shore.

Primary coasts are formed by nonmarine processes (for example, land erosion; river, glacier, or wind deposition; volcanic activity; and faulting). Primary coasts include fjords, drowned river valleys, deltas, alluvial plains, dune coasts, lava and cratered coasts, and fault coasts. Secondary coasts are modified by ocean processes (for example, ocean erosion; deposition by waves, tides, or currents; and modification by marine plants and animals). Erosion produces regular and irregular coastlines. Deposition of eroded materials creates sandbars, sand spits, and barrier islands. Barrier islands protect the continental coastline from storms, but the islands receive each storm's energy and destructive force.

Beaches exist in a dynamic equilibrium; supply balances removal of beach material. Gentle summer waves move the sand toward shore during onshore transport. In winter, high-energy storm waves scoop the sand off the beach and deposit it in a sandbar during offshore transport. In the surf zone, the breaking waves produce a longshore current. The longshore current moves the sediment along the shore as longshore transport. A beach accumulating as much material as it loses is in equilibrium. Rip currents move water and sediment seaward through the surf zone.

A series of drift sectors forms a coastal circulation cell that defines the paths followed by beach sediments from source to deposition. The energy for the movement comes from waves and the wave-produced currents.

A typical beach has features that include an offshore trough and bar and a beach area comprising a low-tide terrace, beach face, beach scarp, and berms. Winter, or storm, and summer berms are produced by seasonal changes in wave action. Beaches are described by their shape; the size, color, and composition of beach material; and their status as eroded or depositional beaches.

Dams, breakwaters, groins, and jetties interfere with the dynamic processes that create and maintain the oceans' beaches. When beaches are destabilized, they erode, and today, 40% of the U.S. continental shoreline loses more sediment than it receives. Santa Barbara harbor and Ediz Hook are examples of human interference with natural processes.

An estuary is a semi-isolated portion of the ocean that is diluted by fresh water. In a salt wedge estuary, seawater forms a sharply defined wedge moving under the fresh water with the tide; circulation and mixing are controlled by the rate of river discharge. In a shallow, well-mixed estuary, there is a net seaward flow at all depths due to strong mixing and low river flow. Salinity is uniform over depth but varies along the estuary. A partially mixed estuary has good mixing, seaward surface flow of mixed fresh water and seawater, and an inflow of seawater at depth. Salt is transported in this system by advection and mixing. Fjord-type estuaries are deep estuaries in which the fresh water moves out at the surface and there is little mixing or inflow at depth.

In a partially mixed estuary, progressive movement of surface water seaward and deep water inward occurs during each tidal cycle. The circulation of the estuary can be measured directly with current meters, a long and expensive process, or it can be measured indirectly by determining the water and salt budgets of the estuary.

In temperate latitudes, the volume transport of water between the estuary and the sea is much greater than the addition of fresh water. In fjords, the inward flow is small and the deep water tends to stagnate. In semienclosed seas with high evaporation rates, a net removal of fresh water occurs; the seaward flow is at depth and the ocean water enters at the surface.

Estuaries that flush rapidly have a higher capacity for dissipating wastes than those that flush slowly. In bays and harbors that are flushed only by tidal action, resident water and wastes can be recycled. Partial recycling can also happen in estuaries with two-layered flow as exiting surface water is mixed with incoming water.

Case histories of two estuaries, San Francisco Bay and Chesapeake Bay, are presented.

Key Terms

All key terms from this chapter can be viewed by term or definition when studied as flashcards on this book's website at www.mhhe.com/sverdrup10e.

coast, 289
shore, 289
beach, 289
geomorphology, 291
eustatic change, 291
primary coast, 291
secondary coast, 291
fjord, 291
sill, 291
moraine, 291
drowned river valley, 292
ria coast, 292
delta, 292
alluvial plain, 292

dune coast, 292
lava coast, 292
cratered coast, 293
fault bay, 293
fault coast, 293
sea stack, 293
bar, 293
barrier island, 293
sand spit, 294
hook, 294
reef coast, 294
salt marsh, 295
backshore, 296
foreshore, 296
offshore, 296
berm, 296
berm crest, 296
winter berm, 296

storm berm, 296
summer berm, 296
scarp, 297
low-tide terrace, 297
beach face, 297
trough, 297
dynamic equilibrium, 297
onshore current, 299
onshore transport, 299
longshore current, 299
longshore transport, 299
swash, 299
cusp, 299
drift sector, 299
accretion, 300
coastal circulation cell, 300
tombolo, 301
lag deposit, 302

armored beach, 302
breakwater, 303
jetty, 303
groin, 303
estuary, 306
salt wedge estuary, 306
entrainment, 307
well-mixed estuary, 307
partially mixed estuary, 307
advection, 307
fjord-type estuary, 307
net circulation, 310
water budget, 310
salt budget, 310
inverse estuary, 312
flushing time, 313
intertidal volume, 313

Study Questions

1. The processes that form a stretch of flat, uniform coast depend on whether the land is rising or sinking. Discuss the processes that form a flat coastal area under these two conditions.
2. Why are multiple berms more likely to be seen on a beach between March and August than between September and February?
3. Fjord coasts and drowned river valleys, or ria coasts, are primary coasts. Explain why their appearances are distinctly different. Give examples of each.
4. Describe the processes required to create a tombolo. Consider and discuss (a) the distribution of wave energy and (b) the longshore transport.
5. What conditions are required to maintain a beach with a constant profile and composition? Consider both a static and a dynamic environment.
6. Why does an eroding beach supplied with material from the cliffs of an old glacier deposit become an armored beach, while an eroding beach supplied by river sediments does not become armored?

7. How does the profile of a sand beach change during alternating seasonal periods of storm waves and more gentle small waves?
8. Sketch a beach section that is stable with respect to the supply and removal of beach sediments. Indicate the source of the sediments and the final deposition of sediments in the system. If the longshore transport or wave energy distribution is altered by a barrier perpendicular to the beach, what changes will occur?
9. What is causing the apparent rise of sea level around the world? Is the rise similar in all places?
10. What processes move beach sediments in a coastal circulation cell?
11. Compare the circulation of a semienclosed basin at 30°N with that of an estuary located at 60°N. Which is less likely to accumulate waste products at depth? Why?
12. Estuaries are classified by their net circulation and salt distribution in this chapter. What other features could be used to describe and classify them?

13. Why are estuaries with short flushing times less apt to degrade when used as the receiving water for urban runoff than estuaries with long flushing times?

14. Compare the circulations and histories of San Francisco Bay and Chesapeake Bay. How do these estuaries differ; how are they similar? What do you see in the future for each?

15. To prevent erosion of his beach, your neighbor wants to build a groin. What effect might the groin have on your neighbor's beach? On your beach if you live on the upstream side of the longshore current? On your beach if you live on the downstream side of the longshore current?

Study Problems

1. Determine the flushing time of an estuary in which $T_o = 9 \times 10^7$ m³/day and the volume of water is 30×10^8 m³.

2. An estuary has a volume of 50×10^9 m³. This water is 5% fresh water, and the fresh water is added at the rate of 6×10^7 m³/day. Why does the rate of addition of fresh water from the estuary to the ocean equal the input of fresh water from the land to the estuary? Consider the average salinity of the estuary as constant. What is the residence time of the fresh water?

3. Water entering an estuary at depth has a salinity of 34.5‰. The water leaving the estuary has a salinity of 29‰. The river inflow is 20×10^5 m³/day. Calculate T_o, the seaward transport.

4. A bay with no freshwater input can flush only by tidal exchange. If on each tidal cycle (ebb and flood) 10% of the bay's water volume is exchanged with ocean water, how much of the original water will still be in the bay after four tidal cycles? Solve for (1) no mixing between bay water and ocean water; (2) complete mixing between bay water and ocean water.

5. What is the flushing time of the bay in problem 4 for each case? Which case best duplicates natural conditions?

Oil platform in the North Sea.

Chapter 13

Environmental Issues and Concerns

Chapter Outline

Learning Outcomes

After studying the information in this chapter students should be able to:

1. *discuss* the problems posed by dumping solid wastes, sewage, and other toxicants in the oceans,

2. *describe* the origin of the Gulf of Mexico Dead Zone and the factors that control its size from year to year,

3. *evaluate* the hazards of plastic trash in the oceans,

4. *review* the trend in average number of marine oil spills per decade during the past four decades,

5. *explain* the importance of marine wetlands,

6. *review* the problem of marine invaders, and

7. *discuss* the problem of overfishing.

The oceans have always had a profound influence on the people who settled along their shores. They have provided food, clothing, and waste disposal. They have fascinated and inspired generations of artists, writers, poets, and musicians. They have provoked the imaginations of adventurers and explorers and stimulated the curiosity of scientists. For centuries, the oceans were safe from interference by humans because human populations and technology continued at low levels. But that is no longer so. In this chapter, we review ways in which our species affects the ocean: water pollution, uncontrolled harvesting, and the introduction of nonnative species.

Science documents changes in the marine environment and identifies causes, but science alone is not able to make the required corrections. Problems, once identified, may become issues that require policy changes, and changing policy may require making choices that are expensive and unpopular. Reaching a consensus in this arena is difficult and often complicated by political pressures. However, whether you live by the ocean or not, the ocean is a part of your world, and you need to consider the situations that are discussed in this chapter.

13.1 Water and Sediment Quality

Our growing populations, with their necessary industries, energy-generating facilities, and waste treatment plants, place a great burden on coastal zones. In the past, Earth's natural waters were assumed to be infinite in their ability to absorb and remove the by-products of human populations. However, too much use and too many wastes discharged into too small an area at too rapid a rate have produced problems that cannot be ignored. These discharges can and do exceed the ability of the natural systems to flush themselves and disperse their wastes into the open ocean.

Solid Waste Dumping

Using the sea as a dump for trash and garbage was and is a common practice around the world. Probably more than 25% of the mass of all material dumped at sea is dredged material from ports and waterways, and one of the major methods of industrial waste disposal is dumping at sea. After each modern war, obsolete military hardware and munitions have been disposed of at sea. The North Atlantic was the dumping ground for toxic gases confiscated from Germany at the end of World War II, and the bays and lagoons of many South Pacific islands were used for the disposal of Jeeps, tanks, bombs, and other items. After the United States left Vietnam, there was a similar unloading of vehicles and explosives in the waters around Southeast Asia. Today, Russia is living with the health and environmental problems caused by the disposal of nuclear wastes and power plants from obsolete nuclear submarines along the Arctic coast of northwest Russia.

Marine disposal of waste material and subsequent pollution of the marine environment must be regulated internationally. Once pollutants enter the marine environment, they can be transported anywhere in the oceans. This fact led to the development, beginning in the early 1970s, of a series of regional treaties and conventions governing marine pollution problems and more comprehensive international conventions establishing uniform standards to control global marine pollution. In 1972, the U.S. Congress passed the Marine Protection, Research, and Sanctuaries Act (MPRSA, also known as the Ocean Dumping Act) to regulate the intentional disposal of waste at sea and authorize research related to marine pollution. The MPRSA prohibits all ocean dumping, except as allowed by permits awarded by the Environmental Protection Agency (EPA), in any ocean waters under U.S. jurisdiction by any U.S. vessel or any vessel sailing from a U.S. port. It also bans any dumping of radiological, chemical, and biological warfare agents and any high-level radioactive waste, as well as medical wastes. The MPRSA also authorized the establishment of marine sanctuaries (see the box titled "National Marine Sanctuaries" in chapter 12).

One of the largest marine disposal sites in the United States, the New York Bight, is located off the mouth of the Hudson River, between New York and New Jersey. Street sweepings, garbage, dredge spoils, cellar dirt, and waste chemicals were dumped into this area beginning in the 1890s. Increasing quantities of floating debris found their way back to the beaches until 1934, when laws were passed to prohibit the dumping of "floatables." Refuse continued to be dumped, including building and subway construction debris, toxic wastes from industry, acids, and sewage sludge.

The dumping of toxic and oxygen-demanding wastes degraded the quality of the water and the sediments. Occasionally, this degraded water upwelled along the coast and killed the marine organisms in shallow water.

Table 13.1	U.S. Municipal Solid Waste (MSW) Generated and the Amount Recovered for Recycling, 2006 (weights in millions of tons)			
Type	**Weight Generated**	**Percentage of Total MSW**	**Weight Recovered**	**Percentage of Weight Generated That Was Recovered**
Paper	85.3	33.9	44.0	51.6
Yard trimmings	32.4	12.9	20.1	62.0
Food scraps	31.3	12.5	0.7	2.2
Plastics	29.5	11.7	2.0	6.9
Metals	19.1	7.6	7.0	36.3
Glass	13.2	5.3	2.9	21.8
Wood	13.9	5.5	1.3	9.4
Textiles	11.8	4.7	1.8	15.3
Rubber and leather	6.5	2.6	0.9	13.3
Other	8.3	3.3	1.1	13.6
Total	251.3	100.0	81.8	32.5

Data from United States Environmental Protection Agency.

Estimates of the amount of municipal solid waste (MSW) generated in the United States in 2006 are given in table 13.1. The United States generated approximately 251.3 million tons of MSW in 2006, or about 4.5 pounds per person per day. Including the composting of some solid waste, 81.8 million tons of MSW were recovered for recycling (32.5% of the total), leaving roughly 169.5 million tons of MSW that required disposal.

In 1986, amendments to the MPRSA directed that ocean disposal of all wastes cease at the New York Bight and be moved to a new site known as the 106-Mile Dumpsite, 171 km (106 mi) offshore in water approximately 2500 m (8200 ft) deep. In the most detailed investigation ever conducted on the impacts of ocean dumping, scientists have studied the effects of 42 million tons of sewage sludge dumped at the 106-Mile Dumpsite between 1986 and 1992. A survey of the marine organisms at the site found an incredible diversity of animals, including 798 different species, 171 families, and 14 phyla, most of which were previously unknown. The sludge material dumped at the site significantly affected the metabolism, diet, and variety of the organisms living there. Over the six-year period, urchins, starfish, and sea cucumbers increased tenfold at the dumpsite, and sea urchins were observed feeding on sludge-derived organic matter. This suggests that the long-term disposal of sludge into the food web at a particular site could change the species diversity of the organisms there, favoring species able to use the organic material available in sewage sludge as a food source. In addition, contaminants penetrated to a depth of 5 cm in the sediment as organisms living in the sediments burrowed through them. Sludge disposal at the 106-Mile Dumpsite was stopped in July 1992.

Because silver is used in photography, its concentration in sewage sludge is typically 150 times higher than in deep-sea sediments. Consequently, the presence of sewage sludge in marine sediments can be tracked by measuring the concentration of silver in the sediment. After the dumping stopped in 1992, silver levels in sediment samples from the site began to decline. In 1993, it was found that the number of organisms at the dumpsite was decreasing and that the ingestion of sewage-derived organic matter was also subsiding at the dumpsite. Occasional strong bottom currents capable of resuspending the contaminated sediment at the site and transporting it to the south were observed. The movement of contaminated sediments to the south was confirmed when sediment samples 50 nautical miles south of the dumpsite were found to have elevated levels of silver and increased densities of sediment-dwelling organisms were observed.

The problems associated with the production and disposal of waste materials will only increase in the future as global population grows from about 6.7 billion as of the middle of 2008 to a projected 10 billion before 2200.

Sewage Effluent

Most urban sewer systems in the United States were built in the late 1800s and early 1900s; these systems carried raw sewage to the closest body of water: river, lake, or ocean. Since then, the systems have grown by adding to or combining old systems, building new systems, and adding treatment facilities. Until 1972, the law allowed the discharge of anything into a body of water until the water was polluted; "polluted" was defined by the individual states. In 1972, the Federal Clean Water Act, in an effort to make U.S. water "fishable and swimmable," mandated the upgrade of sewage treatment to the secondary treatment level by 1977.

For years, Boston Harbor has received sewage effluent in various stages of treatment from an enormous urban area, and the harbor is seriously polluted. Under a provision of the Clean Water Act, Boston was able to apply for a series of treatment waivers because it discharged into marine waters rather than fresh water, and Boston continued to apply for waivers until 1985, when the EPA denied its last waiver application and Massachusetts ordered compliance with the Clean Water Act.

Construction of new sewage collecton systems and a sewage treatment plant began in 1989 at a cost of $3.8 billion. Important components of the new Deer Island Sewage Treatment Plant have been placed in operation in a series of steps since January 1995. The plant was constructed to remove human, household, business, and industrial pollutants from wastewater that originates in homes and businesses in forty-three different communities with a total population of roughly 2.5 million in the greater Boston area. It is the second-largest sewage treatment plant in the United States. The plant's peak capacity is 1270 million gallons of wastewater per day (mgd). Average daily flow is about 390 mgd. After the sewage is treated in the plant, it is discharged through an outfall tunnel 7 m (24 ft) in diameter beneath the sea floor that extends roughly 15.3 km (9.5 mi) offshore. The actual discharge of treated effluent is through a diffuser, which consists of more than fifty pipes that rise to the seabed over the last 2000 m (6600 ft) of the tunnel's length. Each pipe connects to a diffuser cap, which splits the flow into several streams. The purpose of the diffuser is to ensure the maximum practicable dispersion and dilution of the wastewater flow. Extensive monitoring plans are in place to evaluate what impact the discharge from the plant may have near the outfall site and further south along the coast where wind-driven currents may carry it.

The southern California coastal region from Los Angeles to San Diego is known as the Southern California Bight (SCB). The SCB was home to almost 20 million people in 2006. Ocean-related tourism in this area generates an estimated $9 billion annually. As important as this coastal region is to tourism, it is also used for a variety of other, seemingly incompatible purposes, many of which result in the discharge of pollutants to coastal waters. Sources of pollution include discharges from municipal wastewater facilities and power-generating stations, oil platforms, industrial effluents, and dredging operations. Nineteen municipal wastewater treatment facilities discharge treated water directly to the SCB; four of these are very large facilities discharging more than 100 million gallons of treated water per day. Historically, these four facilities have been the leading source of point source contaminants to the SCB. Over the past thirty years, the cumulative flow from these four facilities has increased an overall 16%, but the discharge of suspended solids, oxygen-consuming constituents (often referred to as "biological oxygen demand" constituents, or BODs), and oil and grease have decreased significantly (fig. 13.1). These reductions are due largely to increased and improved treatment, greater control of these materials at their source, and land disposal of some solids. Probably the most significant improvement in pollution control is due to the December 1998 upgrading of the Hyperion Treatment Plant

operated by the city of Los Angeles to provide full secondary treatment of wastewater. This upgrade cost over $1 billion and took more than ten years to complete. The emission of toxicants (substances that degrade the environment and are harmful to organisms, including heavy metals, chemical compounds, and excessive concentrations of nutrients) into the SCB has also declined dramatically and is discussed in the section "Toxicants."

Toxicants

Surface runoff from coastal urban areas feeds directly into the marine environment via storm sewers. Storm sewers carry a wide mix of materials, including silt, hydrocarbons from oil, residues from industry, pesticides and fertilizers from residential areas, and coliform bacteria from animal wastes. Even the chlorine added to drinking water and used to treat sewage effluent as a bactericide may form a complex with organic compounds in the water to produce chlorinated hydrocarbons that are toxic in the marine environment.

From rural and agricultural lands, runoff finds its way through lakes, streams, and rivers to the coast. This runoff supplies pesticides and nutrients, which can poison or overfertilize the waters. Excess nutrients can be destructive because the nutrients stimulate plant growth that eventually dies and decays, removing large quantities of oxygen. Lack of oxygen then kills other organisms, which in turn decay and continue to remove oxygen from the water. Animal wastes and failed septic tank systems also contribute contamination to rural runoff.

Many toxicants reaching the coast do not remain in the water but become adsorbed onto the small particles of matter suspended in the water column. These particles clump together and settle out because of their increased size, and toxicants become concentrated in the sediments. Analyses of core samples show changes in the concentration of toxicants in bottom sediments over time.

The concentration of heavy metals in marine sediments in the SCB generally increased from 1845 through 1970, until concentrations were two to four times their natural levels. Improved wastewater treatment led to a dramatic decrease in the discharge of most heavy metals to the SCB from 1970 through 1990, when emission levels reached a steady-state, low level (fig. 13.2*a–d*).

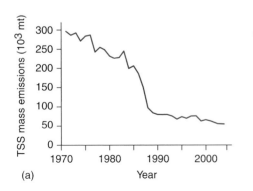

(a)

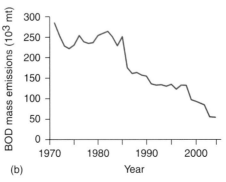

(b)

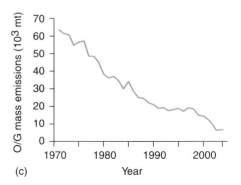

(c)

Figure 13.1 Estimated annual emissions from large municipal wastewater treatment facilities into the Southern California Bight between 1971 and 2004 of (a) total suspended solids (TSS), (b) biological oxygen demand substances (BOD), and (c) oil and grease (O/G) in thousands of metric tons.

Two additional toxicants that can have a major impact on the environment are the pesticide DDT (dichloro-diphenyl-trichloro-ethane) and PCBs (polychlorinated biphenyls). DDT came into wide agricultural and commercial usage in this country in the late 1940s. Over the next twenty-five years, approximately 675,000 tons were applied domestically. The peak year for use in the United States was 1959, when nearly 40,000 tons were applied. From that high point, usage declined steadily to about 6500 tons in 1971, after which DDT use in the United States was banned by the Environmental Protection Agency. PCBs are a group of over 200 synthetic compounds, some mutagenic (known to produce mutations in developing organisms) and carcinogenic (cancer-promoting). PCBs are among the most stable organic compounds known. Congress passed the Toxic Substances Control Act in 1976, which led to a ban in 1979 on the manufacture and use of PCBs at concentrations above 50 ppm. By 1984, additional restrictions were in place to extend the ban to most cases where the concentration was less than 50 ppm. Figure 13.3 shows concentrations of DDT and PCBs in sediments in Puget Sound from about 1890 to 1990. DDT and PCB contamination in the sediments peaked in 1960 and declined significantly thereafter through 1990, with the exception of a small peak in DDT measured in the 1986 sample. More

recent trends in the concentration of total PCBs in Puget Sound can be seen in data from the National Oceanic and Atmospheric Administration's Mussel Watch Project—a national program begun in 1968 to monitor pollutant concentrations in mussels at over 250 sites around the country. Mussel Watch data show that PCB inputs to the central part of Puget Sound leveled off during early 1990s, experienced a second small peak in 1998, and then declined again in 2000.

In 1998, a global pact (Protocol on Persistent Organic Pollutants) was negotiated by the United States and twenty-eight other countries; the pact requires most countries in western and central Europe, North America, and the former Soviet Union to ban production of compounds such as DDT and PCBs. Russia was granted a special exemption to continue production of PCBs until 2005 and to postpone destruction of stocks until 2020 while it converts its electrical power grid transformers to other insulating fluids.

Some of the particulate matter that adsorbs toxicants has a high organic content and forms a food source for marine creatures. In this way, heavy metals and organic toxicants associated with the particles find their way into the body tissues of organisms, where they may accumulate and be passed on to predators. Due to their mass, these particulars often deposit on the sea floor.

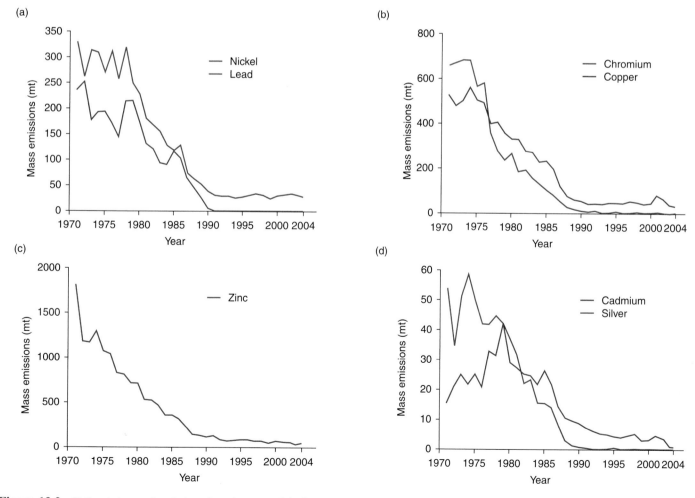

Figure 13.2 Estimated annual emissions from large municipal wastewater treatment facilities into the Southern California Bight between 1971 and 2004 of trace metals in metric tons: (a) nickel and lead, (b) chromium and copper, (c) zinc, and (d) cadmium and silver.

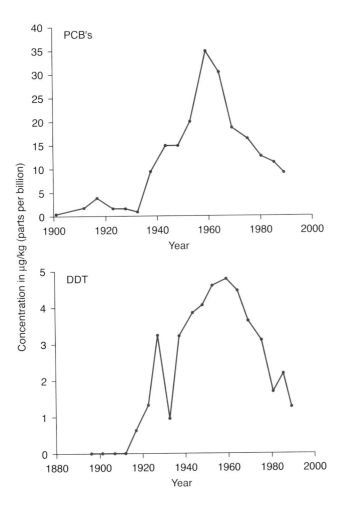

Figure 13.3 Concentration of PCBs and DDT in sediment cores from a deep-water Puget Sound station located due west of Seattle. Concentrations peaked in 1960 and then declined following the enactment of legislation controlling their use.

Table 13.2	Food Chain Concentration of DDT
	DDT Residues in Parts per Million (ppm)
Water	0.00005
Plankton	0.04
Silverside minnow	0.23
Sheepshead minnow	0.94
Pickerel (predator)	1.33
Needlefish (predator)	2.07
Heron (small animal predator)	3.57
Tern (small animal predator)	3.91
Herring gull (scavenger)	6.00
Osprey egg	13.8
Merganser (fish-eater)	22.8
Cormorant (fish-eater)	26.4

Reprinted with permission from George M. Woodwell, et al., *DDT Residues in an East Coast Estuary, Science* 156 (1967): 821–24. Copyright 1967 American Association for the Advancement of Science.

Thus, throughout the United States, toxic residues are found in estuarine bottom fish. Shellfish have been found to have concentrations of heavy metals at levels that are many thousands of times over the levels found in the surrounding waters. Scallops may have levels of cadmium about 2 million times over the water concentration, and oysters may have DDT levels 90,000 times over the water concentration. A classic 1967 study conducted after DDT had been sprayed on Long Island marshes to control mosquitoes documented the effect of this long-lived toxin as it was concentrated by the food chain (table 13.2).

A tragic example of what can happen when humans ingest organisms that had accumulated a toxin occurred between 1952 and 1968 in Minamata, Japan. An estimated 200–600 tons of waste mercury from an industrial source were released into the coastal bay from which villagers gathered their shellfish harvest. The mercury formed an organic complex that was readily taken up by the marine life; the accumulation of mercury in the shellfish led to severe mercury poisoning and death among those eating the shellfish. The physical and mental degenerative effects were especially severe on children whose mothers had eaten large amounts of shellfish during pregnancy. The condition produced is named Minamata disease.

This ability of adult mussels and oysters to absorb and concentrate contaminants has been put to use in the NOAA Mussel Watch Project. Since 1986, divers have collected samples from these shellfish to monitor the concentrations of some chemical contaminants in U.S. estuaries and coastal waters. Ten years after the first samples were taken, samples showed decreasing national trends for contaminants such as DDT, PCB, tin, and cadmium. More than 250 coast and estuary sites are sampled each year; almost half are located in waters near urban areas, within 20 km (12 mi) of population centers with more than 100,000 people. Three sites are located in Hawaii and two in Alaska. National mussel watch programs have been established in the United Kingdom and France, and there are multinational efforts in northern Europe. Mussel watches are being developed in the Asian Pacific Rim area.

Considering all the pathways, sources, and types of materials that can be classed as toxicants or pollutants, managing and eventually excluding them from the marine environment are extremely difficult and complicated. As seen in this discussion, efforts to reduce the discharge of toxic materials are showing signs of success, but it will take continued, long-term effort on the part of scientists, citizens, and governments to achieve a cleaner environment for the ocean world and the land world.

13.2 Gulf of Mexico Dead Zone

Each year off the mouth of the Mississippi River, a low-oxygen, or **hypoxic,** area forms in the northern Gulf of Mexico. It begins to appear in February, peaks in the summer, and dissipates in the fall, when storms stir up the gulf waters and increase mixing. It has been called the *dead zone* because the oxygen level in the water is too low to support most marine life. Fully oxygenated waters contain as much as 12 parts per million (ppm) of oxygen. Once oxygen falls to 5 ppm, fish and other aquatic animals can have trouble breathing. As the dead zone grows in size and oxygen levels drop, highly mobile organisms such as fish and shrimp flee the area. More slowly moving bottom-dwellers such as crabs, snails, clams, and worms can be overtaken by the hypoxic waters and suffocate. Sediment dwellers that can't leave a hypoxic zone begin dying at around 2 ppm. In some dead zones, oxygen levels remain near 0.5 ppm or lower for months. The low-oxygen water does not just hug the bottom; it can affect 80% of the water column in shallow areas, extending upward to within a few meters of the surface.

The dead zone is caused by the introduction of large amounts of nitrogen as nitrate, much of it from fertilizers, into the gulf by the Mississippi and Atchafalaya Rivers. The Mississippi River alone delivers an average 580 km^3 (140 mi^3) of fresh water to the Gulf of Mexico each year. Thirty-one states (41% of the area of the continental United States) and half of the nation's farmlands have rivers that drain into the Mississippi. Records show that from the 1950s onward, the use of fertilizers in these areas increased dramatically. The amount of nitrate from fertilizer that flowed into the gulf tripled between 1960 and the late 1990s, and the amount of another fertilizer, phosphate, doubled. It is estimated that about 30% of the nitrogen entering the gulf comes from fertilizers used in agriculture, another 30% comes from natural soil decomposition, and the remainder comes from a variety of sources, including animal manure, sewage treatment plants, airborne nitrous oxides from the burning of fossil fuels, and industrial emissions.

When nitrates are added to the water, they trigger the rapid growth of tiny marine algae. This phenomenon is commonly referred to as a bloom (see chapter 16). Massive blooms can produce a biological chain reaction in which there is a corresponding increase in small animals that feed on the algae, producing more plant and animal material than can be consumed by fish and other predators. Billions of these small organisms die and sink to the sea floor, where they decompose and are consumed by bacteria. These bacteria use up oxygen in the water in the process.

The seasonal appearance of the dead zone was first detected in 1972 but it was not until over twenty years later that scientists began to systematically map its location. The dead zone occupies different regions of the northern gulf from one year to the next, making it difficult to predict exactly what areas will be affected (fig. 13.4). The size of the dead zone seems to correlate with both the amount of agricultural fertilizer used in the Mississippi River drainage basin and with annual changes in rainfall, which cause variable water runoff into the

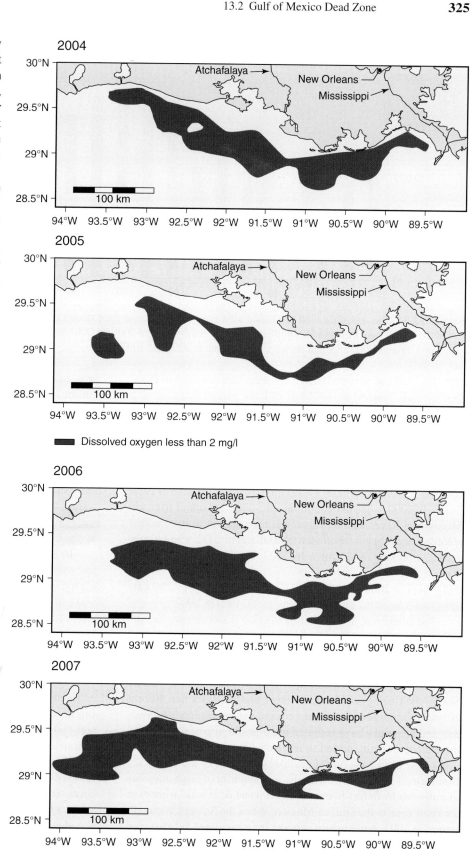

Figure 13.4 The location and size of the dead zone in the northern Gulf of Mexico vary from year to year. Examples of the size and shape of the dead zone (*brown*) as measured in July in the years 2004 through 2007. Research continues in an effort to better understand the factors controlling the pattern of hypoxic water. *Data Source: N. Rabalais, LUMCON.*

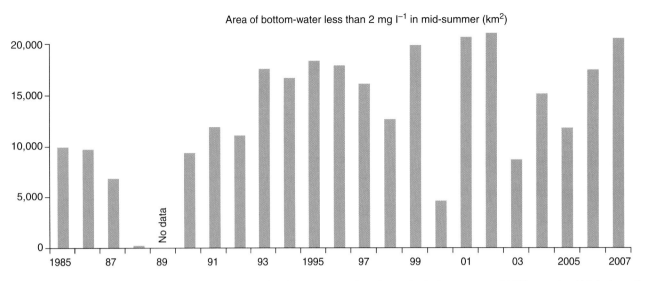

Area of bottom-water less than 2 mg l^{-1} in mid-summer (km^2)

Figure 13.5   The size of the Gulf of Mexico dead zone. The dead zone grew during the particularly wet year of 1993, remained high through the rest of the decade, and then shrank significantly during the dry year of 2000. In 2001 and 2002, it grew once again before shrinking to smaller sizes from 2003 through 2005 and then growing in size again in 2006 and 2007.

gulf. The smallest recorded size of the dead zone occurred in 1988. The hypoxia formed in 1988 but did not persist because of a drought and very little freshwater runoff into the Gulf. The dead zone covered 15,000–20,000 km^2 (5800–7700 mi^2) throughout much of the 1990s (fig. 13.5). In 1993, the Mississippi River flooded and the dead zone grew to an area of 17,500 km^2 (6750 mi^2), an area twice the size of Chesapeake Bay. In 1999, it covered nearly 20,000 km^2 (7700 mi^2), roughly the size of New Jersey. In 2000, when dry conditions prevailed and runoff from the Mississippi River was very low, the zone shrank to its smallest recorded size. Then it rebounded in 2001 and in 2002, when it covered 22,000 km^2 (8484 mi^2). In 2003, it shrank again to about 7000 km^2 (2700 mi^2) before growing again to roughly 15,500 km^2 (6000 mi^2) in 2004, 11,000 km^2 (4242 mi^2) in 2005, 17,280 km^2 (6662 mi^2) in 2006, and 20,500 km^2 (7903 mi^2) in 2007.

The dead zone in the Gulf of Mexico is not an isolated situation, but it is the second-largest human-caused coastal dead zone in the world. Similar problems occur in Chesapeake Bay, Japan, Australia, New Zealand, the Kattegat Strait between Denmark and Sweden, the northern Adriatic Sea, the East China Sea off the Chaingjang (Yantze River), and the Black Sea. Recent efforts to limit the use of fertilizers and restore wetlands in Sweden and Denmark have reduced the introduction of nutrients to the Kattegat Strait and led to increased levels of oxygen in the water. The dead zone on the northwestern shelf of the Black Sea first formed in the 1960s with the introduction of large amounts of nutrients from fertilizer runoff. At its peak, it was larger than the dead zone in the Gulf of Mexico. When the Soviet Union collapsed in 1991, government support for agriculture did also, and fertilizer usage dropped dramatically. By 1996, the dead zone in the Black Sea had disappeared for the first time in thirty years.

13.3 Plastic Trash

Walk any beach in any estuary or along any open coast and see the tide of plastics being washed ashore. It is estimated that every

year, more than 135,000 tons of plastic trash have been routinely dumped by naval, merchant, and fishing vessels. The National Academy of Sciences estimates that the commercial fishing industry yearly loses or discards about 149,000 tons of fishing gear (nets, ropes, traps, and buoys) made mainly of plastic and dumps another 26,000 tons of plastic packaging materials. Recreational and commercial vessels and oil and gas drilling platforms add their share. A total of 7 million tons of cargo and crew wastes were added during each year of the 1980s. Plastics are a worldwide problem; they come from many sources and are distributed by the currents to even the remotest ocean areas (fig. 13.6a).

In 1960, the U.S. production of plastics was 3 million tons; forty years later, the production was more than 50 million tons. Plastic is inexpensive, strong, and durable; these characteristics make it the most widely used manufacturing material in the world today and a major environmental problem. Nobody knows how long plastic stays in the marine environment, but an ordinary plastic six-pack ring could last 450 years.

Thousands of marine animals are crippled and killed each year by these materials. As many as 30,000 fur seals a year are estimated to become entangled in lost or discarded plastic fishing nets or to choke to death in plastic cargo straps. Lost lobster and crab traps made entirely or partially of plastic continue to trap animals; one year 25% of the 96,000 traps that had been set off of Florida's western coast were lost. Seabirds die entangled in six-pack rings and plastic fishing line; both seabirds and marine mammals swallow plastics. Porpoises and whales have been suffocated by plastic bags and sheeting, and the sheeting clings to coral and to rocky beaches, smothering plant and animal life. Fish are trapped in discarded netting; sea turtles eat plastic bags and die. Plastics are as great a source of mortality to marine organisms as oil spills, toxic wastes, and heavy metals (fig. 13.6b–d).

The Marine Plastic Pollution and Control Act of 1987 is the U.S. law that implements an international convention for the prevention of pollution from ships. This law prohibits the dumping

Figure 13.6 The impact of plastics on the marine environment. (a) Persistent litter includes plastic nets, floats, and containers. (b) A common murre entangled in a six-pack yoke. (c) A young gray seal caught in a trawl net. (d) A sea turtle with a partially ingested plastic bag. Photos (a), (b), and (c) are from Sable Island in the North Atlantic, approximately 240 km (150 mi) east of Nova Scotia, Canada.

(a)

(b)

(c)

(d)

of plastic debris everywhere in the oceans; other types of trash may be dumped at specific distances from shore, and ports are required to provide waste facilities for debris. The U.S. Coast Guard is the agency that must enforce this law in U.S. waters; no international enforcement exists.

Some manufacturers make biodegradable plastic by adding light-absorbing molecules that break down the plastic after a few months' exposure to sunlight. There is no evidence that this will solve the problem at sea; even at the surface, the water keeps the plastic cool, and it becomes coated with a thin film of organisms that shade it from the light. It may be that only people educated to act responsibly will be able to reduce the tide of plastics pollution rising around the world.

13.4 Ocean Waste Management Proposals

Expanding populations, industrialization, and economic growth have increased the amounts of our garbage and changed its characteristics. Landfill areas are being overwhelmed, and much of their contents is increasingly recognized as a threat to Earth's soil and freshwater reservoirs. Aluminum and plastics displace traditional materials such as paper, steel, and glass, and household cleaners, solvents, and pesticides are often environmentally harmful. Estimates in 1991 of world wastes produced each year are shown in table 13.3 and figure 13.7.

At the Woods Hole Oceanographic Institution in 1991, a group of scientists held a workshop to assess the potential of the deep ocean as an option for future waste management. They argued that controlled disposal in some ocean environments presents less risk to life than current projected land disposal, with its associated groundwater contamination, or incineration accompanied by atmospheric pollution. They pointed out that the abyssal plain of selected mid-ocean areas supports only sparse populations of marine organisms, and the slowly moving water

Table 13.3	Estimate of World Wastes Produced Each Year	
Type	**Millions of Metric Tons**	**Percentage of Total Waste**
Municipal solid waste	1500	36.2
Dry sewage sludge	65	1.6
Dredged material	1075	26.0
Industrial wastes	1500	36.2
Total	4140	100.0

above it stays at the bottom for thousands of years before it returns to the surface. They suggested that the deep-ocean floor far from shore may be a more appropriate site than coastal landfills that pollute inshore water and degrade beaches, and they proposed a large-scale experiment to assess the environmental effects of depositing 1 million tons of sewage sludge each year for ten years on the floor of the Hatteras Abyssal Plain in the Atlantic Ocean. Implementing such a project would require changes in the Ocean Dumping Act; this is considered unlikely at the present time. Some people feel that any experiments with ocean waste disposal will have to proceed by affiliation with other countries and should fall under UN regulations.

Another suggestion is to dispose of highly radioactive and long-lived nuclear wastes away from the continents in the muds of the deep-sea floor or in subduction zones, where the waste products will gradually recycle back into the mantle. Examining the wreckage of a Soviet nuclear submarine that sank in 1986 in the deep Atlantic with two nuclear reactors and thirty-four nuclear warheads has been suggested as a way to determine whether deep-sea muds can trap radioactive substances. Such proposals seek to avoid the possible contamination of land areas and freshwater supplies while eliminating the problem of long-term hazardous waste storage on land. No experiments involving any of these wastes are planned at this time.

Should we consider dumping to be an ocean resource that has a human and economic value just like any other resource? If so, do these suggestions deserve discussion? As the world's population and its refuse products continue to increase and available storage sites decrease, consideration of these questions may become inevitable.

13.5 Oil Spills

Humans' activities in the twenty-first century depend heavily on oil, and this dependence requires the bulk transport of crude oil by sea to the land-based refineries and centers of use. This transport creates the potential for accidents that release large volumes of oil and expose the world's coasts and estuaries to spills associated with vessel casualties and transfer procedures. Drilling offshore wells exposes marine areas to the risks of blowouts, spills, and leaks. Because industry, agriculture, and private and commercial transportation require petroleum and petroleum products, oil is constantly being released into the environment, to find its way directly or indirectly to the sea.

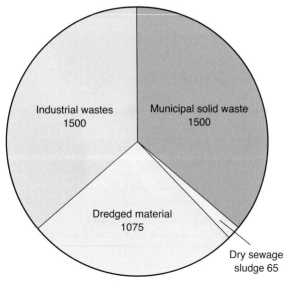

* Numbers in millions of metric tons

Figure 13.7 Estimate of world wastes produced each year

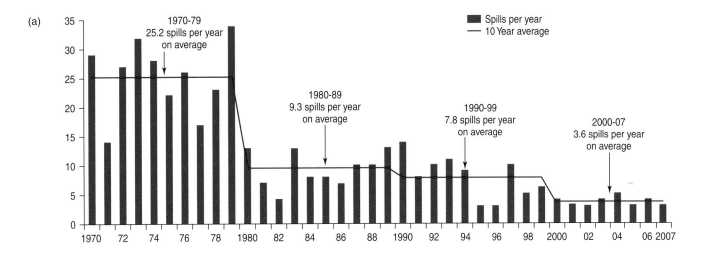

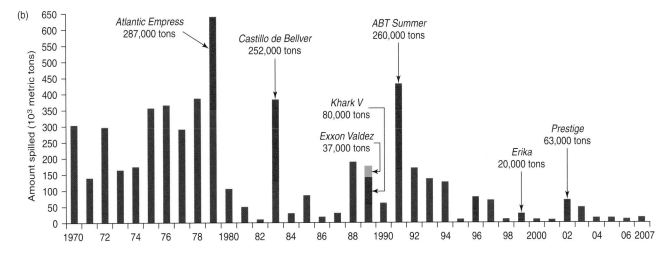

Figure 13.8   Annual (a) number of oils spills over 700 metric tons (205,800 gal) and (b) amount of oil spilled in thousands of metric tons in tanker accidents from 1970 to 2007.

Annual oil spills over 700 metric tons (205,800 gal) from tanker accidents from 1970 to 2001 are shown in figure 13.8a. The average number of large tanker spills per decade decreased significantly from over twenty-four per year in the 1970s to just over seven per year in the 1990s. The quantity of oil spilled annually during this time period is shown in figure 13.8b. Individual major spills are typically responsible for a high percentage of the total oil spilled. For example, from 1990 to 1999, 346 spills over 7 metric tons each occurred, totaling nearly 1.1 million metric tons (322.2 × 10^6 gal), but 830,000 metric tons (244 × 10^6 gal) (75% of the total) were spilled in just ten incidents (just over 1% of the accidents). Consequently, the volume of oil spilled in any given year can be strongly influenced by a single large spill, as seen in figure 13.8b. Table 13.4 lists selected marine oil spills from tanker accidents.

The damage caused by spills far out at sea is difficult to assess because no direct visual or economic impact occurs on coastal areas and damage to marine life cannot be accurately evaluated. Spills due to the grounding of vessels or accidents during transfer or storage happen in coastal areas, where the environmental degradation and loss of marine life can be observed. Spills that occur in estuaries and along coasts affect regions that are oceanographically complex, biologically sensitive, and

economically important. Three oil spills between 1978 and 1991 have become ecological landmarks; these are the sinking of the *Amoco Cadiz,* the grounding of the *Exxon Valdez,* and the discharge of crude oil into the Persian Gulf during the Gulf War.

In March 1978, the *Amoco Cadiz* lost its steering in the English Channel and broke up on the rocks of the Brittany coast of France (fig. 13.9a and b). Gale-force winds and high tides spread the oil over more than 300 km (180 mi) of the French coast; more than 3000 birds died; oyster farms and fishing suffered severely. Of the approximately 210,000 metric tons (61.7 × 10^6 gal) of oil spilled, roughly 30% evaporated and was carried over the French countryside; 20% was cleaned up by the army and volunteers; another 20% penetrated into the sand of the beaches to stay until winter storms washed it away; 10% (the lighter, more toxic fraction) dissolved in the seawater; and 20% sank to the sea floor in deeper water, where it continued to contaminate the area for an unknown period of time.

The United States experienced a large-volume oil spill in March 1989, when, 40 km (25 mi) out of Valdez, Alaska, the *Exxon Valdez,* loaded with 170,000 metric tons (50 × 10^6 gal) of crude oil, ran onto a reef, tearing huge gashes in its hull and spilling 35,000 metric tons (almost 10.3 × 10^6 gal) of oil into Prince William Sound. Local contingency plans were not adequate for

Table 13.4 **Selected Major Oil Spills from Tankers**

Ship	Year	Location	Oil Lost (metric tons)	Oil Lost (10⁶ gal)
Atlantic Princess	1979	Off Tabago, West Indies	287,000	84.4
ABT Summer	1991	700 nautical miles off Angola	260,000	76.4
Castillo de Beliver	1983	Off Saldanha Bay, South Africa	252,000	74.1
Amoco Cadiz	1978	Off Brittany, France	223,000	65.6
Haven	1991	Genoa, Italy	144,000	42.3
Odyssey	1988	700 nautical miles off Nova Scotia, Canada	132,000	38.8
Torrey Canyon	1967	Scilly Isles, United Kingdom	119,000	35.0
Sea Star	1972	Gulf of Oman	115,000	33.8
Urquiola	1976	La Coruna, Spain	100,000	29.4
Irenes Serenade	1980	Navarino Bay, Greece	100,000	29.4
Hawaiian Patriot	1977	300 nautical miles off Honolulu, Hawaii	95,000	27.9
Independenta	1979	Bosphorus, Turkey	95,000	27.9
Jakob Maersk	1975	Oporto, Portugal	88,000	25.9
Braer	1993	Shetland Islands, United Kingdom	85,000	25.0
Khark 5	1989	120 nautical miles off Atlantic coast of Morocco	80,000	23.5
Prestige	2002	150 nautical miles off the Atlantic coast of Spain	77,000	22.6
Aegean Sea	1992	La Coruna, Spain	74,000	21.8
Sea Empress	1996	Milford Haven, United Kingdom	72,000	21.2
Katina P.	1992	Off Maputo, Mozambique	72,000	21.2
Nova	1985	Off Kharg Island, Gulf of Iran	70,000	20.6
Assimi	1983	55 nautical miles off Muscat, Oman	53,000	15.6
Metula	1974	Magellan Straits, Chile	53,000	15.6
Wafra	1971	Off Cape Agulhas, South Africa	40,000	11.8
Exxon Valdez	1989	Prince William Sound, Alaska	37,000	10.9

cleaning up oil spills of this magnitude; delays, caused in part by out-of-service equipment, as well as a lack of equipment and personnel, the rugged coastline, the weather, and the tidal currents of the enclosed area in which the spill occurred, combined to intensify the problem. The oil spread quickly and was distributed unevenly over more than 2300 km² (900 mi²) of water, taking a severe toll on seabirds, marine mammals, fish, and other marine organisms. The oil moving out of Prince William Sound was not washed out to the open sea but was captured by the nearshore currents and moved westward parallel with the coast, repeatedly oiling the rocky wilderness beaches in the weeks that followed (fig. 13.9 *c* and *d*). In the subarctic conditions of Prince William Sound, the photochemical and microbial degradation of the oil proceeded more slowly than it would have in warmer temperate regions. Plant and animal populations also recover more slowly in the cold water, where organisms tend to live longer and reproduce more slowly. Researchers inspecting the area in 1992, three years after the disaster, reported that the area was recovering and repopulating with organisms as the oil aged and degraded. They also noted that the areas most intensively cleaned to remove oil are recovering more slowly and with less-balanced populations than the areas in which the oil has been allowed to degrade naturally. In the 1992 inspection, estimates were made of the dispersal of the spilled oil: 13% in the sediments, 2% in the beaches, 50% degraded naturally in the water, 20% degraded in the atmosphere, and 14% recovered and removed.

Much of the world's crude oil comes from wells of the Persian Gulf, and it must be transported the length of the gulf on its way to refineries around the world. Each year, a quarter of a million barrels of oil are routinely spilled into the gulf's shallow waters, making it one of the world's most polluted bodies of water. However, it is also an area of vigorous plant growth and supports fisheries of shrimp, mackerel, mullet, snapper, and grouper.

During the eight-year Iran-Iraq war, the bombing of oil facilities produced major spills, including one from an oil rig that poured out 172 metric tons (54,350 gal) per day for nearly three months, but the greatest catastrophe came in the 1991 Gulf War, when an estimated 800,000 metric tons (252.8 × 10⁶ gal) of crude oil gushed into gulf waters, the world's greatest oil spill to date. Some of this oil was deliberately released, some came from a refinery at a gulf battle site, and bombing contributed additional quantities. This spill was much larger than the estimated output of 475,000 metric tons of crude oil from Ixtoc 1, an offshore Mexican well that blew out in 1979 and was the previous largest spill on record.

The average depth of the Persian Gulf is 40 m (131 ft), and the circulation of its high-salinity water is sluggish. Changes in the wind in the weeks after the spill stopped the oil from drifting the entire length of the gulf; still, some 570 km (350 mi) of Saudi Arabia's shoreline were oiled. Oil spill experts from the United States, the United Kingdom, the Netherlands, Germany, Australia, and Japan rushed to help. They were able to protect the water-intake pipes of desalination plants and refineries, but other cleanup efforts were less successful (fig. 13.9*e*). About half of the oil evaporated, and about 300,000 metric tons (94.8 × 10⁶ gal) were recovered. Much sank to the bottom of the gulf, where oil may still be seeping from sunken tankers.

Figure 13.9 (a) The tanker *Amoco Cadiz* aground and broken in two off the coast of France in March 1978. (b) A bay along the French coast was fouled with oil. (c) Hot water under high pressure was used to clean Prince William Sound beaches after the 1989 *Exxon Valdez* oil spill. (d) Booms collect oil washed from a beach and hold it for pickup. (e) Oil remained along the Persian Gulf coast long after the 1991 Gulf War. (f) Cleaning up a 1993 oil spill along the beaches of Tampa Bay one more time in 1994.

The tragedies of these spills continue to demonstrate that no adequate technology exists to cope with large oil spills, particularly under difficult weather and sea conditions, along irregular coastlines, or far from land-based supplies. On the average, only between 8% and 15% of oil spilled is recovered, and these may be inflated estimates because the recovered oil has a high water content. The technology for oil cleanup at sea includes oil booms and oil skimmers (fig. 13.10). These devices are useful in confining and recovering small spills in protected waters, but they are not very effective in open and exposed ocean areas or in rough or stormy seas.

The onshore cleanup is often as destructive as the spill, especially in wilderness situations, where the numbers of people, their equipment, and their wastes further burden the environment. The toxicity of weathered crude oil is low, and many oil spill experts consider that cleanup efforts should be concerned not with removing oil from the beaches but with moving oil seaward to prevent it from reaching the beaches. Follow-up studies indicate that many of the cleanup methods used along shorelines cause more immediate and long-term ecological damage than leaving the oil to degrade naturally. Beach habitats treated with high-pressure, hot water take longer to recover than those left untreated; backhoeing and high-pressure washing (100 lb/in^2) destabilize gravel and sand beaches, killing animals and working the oil into the sediments (see fig. 13.9c). To prevent oil from reaching the beaches, some support newly developed, low-toxicity dispersants; the dispersed oil moves into deeper water, where it is diluted and its toxicity is lessened.

More data are needed to assess the effectiveness of attempts to speed up natural degradation by adding nutrients to the beaches to increase populations of biodegrading microorganisms. Releasing other strains of oil-degrading bacteria has been tested with limited success; the new organisms may not compete effectively with natural populations.

Once oil is removed from a beach, the first cleaning operation may not be the last. On the east coast of Florida, the beaches of Tampa Bay are lined with resorts and condominiums, and the economy depends heavily on these beaches for tourism. A freighter and two oil barges collided in August 1993. One barge carried diesel fuel and gasoline that caught fire and burned. The second barge carried heavy fuel oil; this barge sank and released the fuel oil to mix with the sediments on the bottom of the bay. Each time strong winds stir the bay, contaminated sediments are moved onshore to create a new oil contamination problem. Figure 13.9f shows the end of a costly January 1994 cleanup episode along St. Petersburg Beach.

On November 19, 2002, the tanker *Prestige* broke apart 150 nautical miles off the northwest coast of Spain. Like many tankers still at sea, the twenty-six-year-old *Prestige* was not built with the added safety of a double hull. When the ship suffered cracks in its single hull during a storm, it was not able to survive and eventually sank in roughly 3600 m (11,800 ft) of water. A month after the sinking, oil continued to leak from the wreck at a rate of about 125 metric tons (36,750 gal) per day. The French submersible *Nautile* was used to plug several cracks and holes in the vessel. Two months following the wreck, the leakage had been reduced to about 80 metric tons (23,500 gal) per day. Leakage also declined due to the cold temperature of the deep ocean. An estimated 63,000 metric tons (18.5 × 10^6 gal) of its cargo of 77,000 metric tons (22.6 × 10^6 gal) of fuel oil is believed to have been released.

After the *Exxon Valdez* spill, Congress enacted the Oil Pollution Act of 1990. Because of the law and subsequent international maritime regulations, nearly all vessels used to transport oil will have double hulls by the year 2020 to help protect against spills.

The immediate damage from a large spill is obvious and dramatic; by contrast, the effect of the small but continuous additions of oil that occur in every port and harbor are much more difficult to assess because they produce a chronic condition from which the environment has no chance to recover. Refined products such as gasoline and diesel fuel are more toxic to marine life than crude oil, but they evaporate rapidly and disperse quickly. Crude oil is slowly broken down by the action of water, sunlight, and bacteria, but the portion that settles on the sea floor moves down into the sediments, which it continues to contaminate for years.

Large volumes of crude oil also enter the marine environment from natural seeps. The current global rate of natural seepage of oil into the oceans is estimated to be 600,000 metric tons (176.4 × 10^6 gal), with a range between 200,000 and 2 million metric tons (58.8–588 × 10^6 gal) per year. Natural seeps are believed to be the single greatest source of oil in the marine environment.

13.6 Marine Wetlands

The value of shore and estuary areas as centers of productivity and nursery areas for the coastal marine environment is well known to oceanographers and biologists. Saltwater and brackish marshes and swamps, known as marine **wetlands,** border estuaries and provide nutrients, food, shelter, and spawning areas for marine species,

Figure 13.10   A catamaran oil skimmer. The vessel cruises at about 3 knots, guiding surface oil between the twin hulls. The oil adheres to a moving belt and is lifted into onboard storage tanks.

including such commercially important organisms as crabs, shrimp, oysters, clams, and many species of fish. Nevertheless coastal countries have long histories of filling wetlands and modifying coasts to provide croplands, port facilities, and industrial space for their growing populations (fig. 13.11). For examples of these trends, see the history of San Francisco Bay in chapter 12 and consider

Figure 13.11 The industrialized estuary of the Duwamish River at Seattle, Washington. The commercially developed flat areas adjacent to the river were once tidelands and salt marshes.

Holland, where the Dutch have been reclaiming wetlands for thousands of years, until at present about one-third of their land has been reclaimed from the North Sea. Although the Dutch will continue to develop parts of their coast, they have decided to return 15% of the reclaimed area to rivers and estuaries over the next twenty-five years because of the high costs of dike repair and pumping water, as well as their concern over wetland habitat loss and its impact on the Dutch national symbol, the stork.

Another type of wetland is being destroyed along the muddy shores of tropical and subtropical lagoons and estuaries where several species of mangrove trees grow. These salt-tolerant trees protect the shore from wave erosion and storm damage and provide specialized habitats for fish, crustaceans, and shellfish on and among their tangled roots. In recent years, mangroves have been logged for timber, wood chips, and fuel; mangrove swamps have been cleared and filled to provide land for crops, shrimp ponds, and resorts. Eighty percent of the mangroves in the Philippines, 73% of Bangladesh's trees, and more than 50% of coastal stands in Africa have been removed. Over 50% of the world's mangroves are gone.

In the United States, more than 139 million people, about 53% of the national total, live along the coast. This population is expected to increase by an average of 3600 people per day, reaching 165 million by the year 2015. This rate of growth is faster than that for the nation as a whole.

Areas of recreation and retirement replace wetlands with waterfront homes, each with its own individual pleasure-craft moorage (fig. 13.12). Between the

Figure 13.12 The wetlands of Barnegat Bay, New Jersey, were replaced by a housing and recreational complex.

Spartina: Valuable and Productive or Invasive and Destructive?

Spartina alterniflora, known as smooth cordgrass, many-spiked cordgrass, and saltmarsh cordgrass, is a deciduous, perennial flowering plant native to the Atlantic and Gulf Coasts of the United States. It is the dominant native species of the lower salt marshes along the Atlantic seaboard from Newfoundland to Florida and on the Gulf Coast from Florida to east Texas. It grows in the intertidal zone from mean higher high water to 1.8 m (6 ft) below mean higher high water.

These natural salt marshes are among the most productive habitats in the marine environment. Nutrient-rich water is brought to the wetlands during each high tide, making a high rate of food production possible. As the seaweed and marsh grass leaves die, bacteria break down the plant material, and insects, small shrimp-like organisms, fiddler crabs, and marsh snails eat the decaying plant tissue, digest it, and excrete wastes high in nutrients. Numerous insects occupy the marsh, feeding on living or dead plant tissue, and red-winged blackbirds, sparrows, rodents, rabbits, and deer feed directly on the cordgrass. Each tidal cycle carries plant material into the offshore water to be used by the subtidal organisms.

Spartina is an exceedingly competitive plant. It spreads primarily by underground stems; colonies form when pieces of the root system or whole plants float into an area and take root or when seeds float into a suitable area and germinate. *Spartina* establishes itself on substrates ranging from sand and silt to gravel and cobble, and is tolerant of salinities ranging from near fresh to salt water (35‰). *Spartina* is able to tolerate high salinities because salt glands on the surface of the leaves remove the salt from the plant sap, leaving visible white salt crystals. Because of the lack of oxygen, marsh sediments are high in sulfides that are toxic to most plants. *Spartina* has the ability to take up sulfides and convert them to sulfate, a form of sulfur that the plant can use; this ability makes it easier for the grass to colonize marsh environments. Another adaptive advantage is its biochemical photosyn-

Box Figure 1 A naturally occurring *Spartina* marsh.

thetic pathway, which uses carbon dioxide more efficiently than most other plants.

These characteristics make *Spartina* a valuable component of the estuaries where it occurs naturally. The plant functions as a stabilizer and sediment trap and as a nursery area for estuarine fishes and shellfishes. Once established, a stand of *Spartina* begins to trap sediment, changing the substrate elevation, and eventually the stand evolves into a high marsh system where *Spartina* is gradually displaced by higher-elevation, brackish-water species. As elevation increases, narrow, deep tidal channels form throughout the marsh (box fig. 1). Along the East Coast, *Spartina* is considered valuable for its ability to prevent erosion and marshland deterioration; it is also used for coastal restoration projects and the creation of new wetland sites.

---

mid-1950s and the mid-1980s, approximately 20,000 acres of coastal wetlands were lost each year in the contiguous United States. Estuarine wetland losses were greatest in six states: Louisiana, Florida, Texas, New Jersey, New York, and California. Much of the Louisiana loss is due to accelerated erosion and subsidence of Louisiana's coastal marshes; see the box titled "Rising Sea Level" in chapter 12.

In 1972, Congress enacted the first nationwide wetlands regulatory program in Section 404 of the Clean Water Act and focused attention on coastal wetlands by enacting the Coastal Zone Management Act. In the 1970s and 1980s, public concern led to the passage of tidal wetland protection laws in many coastal states and to stricter enforcement of federal laws. Each marine wetland is as individual as the estuary it borders, and the singular and distinctive nature is a part of the problem because the difficulty of protecting a wetland is

related to its legal definition. Under different legal interpretations and conditions, some wetlands escape protection while other, drier areas are included; in this case, one size or definition does not fit all.

13.7 Biological Invaders

A combination of physical and biological barriers sets the geographical limits of a species. These limits have changed with time as climate patterns have altered and as plate tectonics has shifted the configurations of the oceans and continents. For example, the opening of the Bering Strait between Asia and North America allowed interchange between the marine organisms of the North Pacific and the Arctic basins, and the separation of South America from Antarctica allowed the currents associated with the Southern Ocean's West

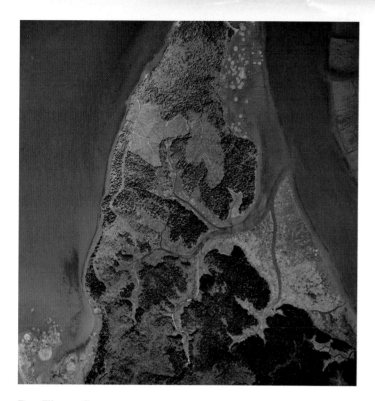

Box Figure 2 Circular patches of *Spartina* spread along the mudflats of Willapa Bay, Washington. The *large circle* at *lower left* is thought to be the original colony. This is a false-color image on infrared film.

steadily along Washington's tidal estuaries, crowding out the native plants and drastically altering the landscape by trapping sediment. *Spartina* modifies tidal mud flats, turning them into high marshes inhospitable to the many fish and waterfowl that depend on the mud flats. In 1945, *Spartina* covered 4.5 acres of Washington's Willapa Bay; it had spread to 432 acres in 1982, 2400 acres in 1992, and more than 3400 acres in 1995. State officials predict that by 2010, it will cover 30,000 of the bay's 80,000 acres if left unchecked (box fig. 2). It is already hampering the oyster harvest and the Dungeness crab fishery, and it interferes with the recreational use of beaches and waterfronts.

Spartina has been transplanted to England and New Zealand for land reclamation and shoreline stabilization. In New Zealand, the plant has spread rapidly, changing mud flats with marshy fringes to extensive salt meadows and reducing the number and kinds of birds and animals that use the marsh. Another species of *Spartina* (*S. maritima*) is native to marshes along the coasts of Europe and Africa. *S. alterniflora* was introduced into Great Britain from eastern North America in about 1800 and spread to form large colonies. It coexisted with the native species, but by 1870, a sterile hybrid that reproduced by underground stems appeared. In about 1890, a vigorous seed-producing form was derived naturally from this hybrid and spread rapidly along the coasts of Great Britain and northwestern France.

Efforts to control *Spartina* outside its natural environment have included burning, flooding, shading plants with black canvas or plastic, smothering the plants with dredged materials or clay, applying herbicide, and mowing repeatedly. Little success has been reported in New Zealand and England; Washington's management program has tried many of these methods and is presently using the herbicide glyphosate to control its spread. Work has begun to determine the feasibility of using insects as biological controls, but effective biological controls are considered ten years away. Even with a massive effort, it is doubtful that complete eradication of *Spartina* from nonnative habitats is possible, for it has become an integral part of these shorelines and estuaries during the last 100 to 200 years.

Spartina has been introduced to and naturalized in Washington, Oregon, California, England, France, New Zealand, and China. The plant was carried to Washington in packing material for oysters transplanted from the East Coast in 1894. Leaving its insect predators behind, the cordgrass has been spreading slowly and

---

Wind Drift to carry organisms from one cold-water ocean to another.

When humans began to cross oceans, they carried many organisms, plant and animal, intentionally and unintentionally. More than three and a half centuries ago, settlers and traders coming to the North American continent brought with them large communities of widely varied organisms that had attached to and bored into the wooden hulls of their ships in the harbors or bays where their journeys began. Many of the organisms transported across the oceans in this way were certainly swept from the ships' sides by the waves and currents, but a few invaders were present when the ships anchored at their journeys' ends. Hundreds of years later, there is no way of knowing when the first European barnacles or periwinkle snails began to

colonize the eastern coasts of the United States and Canada; these two organisms are now considered typical of this region.

The steel hulls of today's ships do not carry such communities for they are protected by antifouling paints and the ships move through the water at speeds sufficient to sweep away many of the organisms carried by wooden hulls. However, the ships of the present do carry ballast water that is loaded and unloaded to preserve the stability of the ship as it unloads and loads cargo. Tens of thousands of vessels with ballast tanks ranging in capacity from hundreds to thousands of gallons move across our oceans. These vessels rapidly transport this ballast water and its populations of small floating organisms across natural oceanic barriers; ballast water and living organisms may be released days or weeks later, thousands of kilometers from their point of origin.

Field Notes

Ecological Nowcasting of Sea Nettles in Chesapeake Bay

by Dr. Christopher Brown

Dr. Christopher W. Brown is an oceanographer at the Office of Research and Applications, National Environmental Satellite, Data and Information Service, National Oceanic and Atmospheric Administration. His research interests include biological patterns and their biogeochemical consequences in the ocean, and the remote detection, characterization, and prediction of marine organisms and populations.

One goal of biological oceanography is to understand and predict how the abundance and distribution of aquatic organisms will respond to changing environmental conditions. With recent advances in real-time observation and modeling, that goal now appears to be within reach.

Certain specific combinations of physical, chemical, and biotic conditions may lead to the development and persistence of biological events. For example, an algal bloom may form if a "seed" population exists when environmental conditions, such as light level and nutrient concentrations, are favorable for that species. By knowing the conditions necessary for an organism to thrive, we can predict the potential occurrences of the organism.

Advanced technology and telecommunications allow many environmental factors to be quickly measured, analyzed, and even forecast. For example, data can be derived from satellites and simulated by numerical models. The capability to retrieve, evaluate, and simulate environmental conditions rapidly offers the potential to predict or "nowcast" the presence of organisms as they appear, provided that the species' habitats can be defined.

My colleagues and I applied this ecological approach to nowcast the likelihood of encountering sea nettles (Chrysaora quinquecirrha), a stinging jellyfish, in Chesapeake Bay. Sea nettles are abundant every summer in the bay. They go through several stages of development during their life cycle (box fig. 1). Fertilized eggs form larvae that attach to hard surfaces such as oyster shells and grow into tiny polyps with tentacles. During the winter, polyps are in a dormant state attached to the bottom. From April through early summer, when temperature and salinity are favorable, the polyps change into immature medusae, called ephyrae, which measure approximately 0.1 cm in diameter. The ephyrae grow rapidly into medusae, the adult form of the jellyfish.

Sea nettles are biologically important in Chesapeake Bay and they interfere with recreational activities; thus knowledge of their distribution in the bay is valuable to the public and scientists. Nettles deter swimming and other water activities because contact with their stinging tentacles is painful, as their name implies. Sea nettles are also voracious predators, devouring copepods, fish eggs and larvae, and comb jellies. This predation affects food web dynamics and energy flow at several tropic levels and may impact finfish production in the bay. Consequently, the distribution of C. quinquecirrha could be used in ecosystem models to study how the nettles affect trophic-level energy flow and larval fish survival, and to alert swimmers of areas to avoid.

The procedure to nowcast the distribution pattern of C. quinquecirrha in Chesapeake Bay exploits their temperature

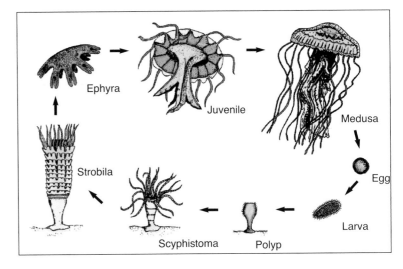

Box Figure 1 The life cycle of the sea nettle, *Chrysaora quinquecirrha.*

and salinity preferences and our ability to rapidly acquire these two environmental variables. Specifically, we generate a map that indicates the likelihood of encountering sea nettles in any part of the bay. We use a habitat model that relates the likelihood of sea nettle presence to estimated sea-surface temperature and salinity. Our weekly nowcast as well as information on the sea nettle and the procedures used in creating the maps are available on the Web at http://coastwatch.noaa.gov/seanettles.

The estimates of surface temperature and salinity are obtained with a hydrodynamic computer model used to study water quality in the bay. Our model uses near-real time data, such as river discharge volume and air temperature, as input, allowing us to simulate the salinity and temperature currently in the bay. We plan to acquire more accurate sea-surface temperatures measured by satellite in the future.

The habitat model was developed by analyzing coincident C. quinquecirrha population, salinity, and sea-surface temperature data collected during the spring, summer, and fall of 1987–2000 in surface waters (0–10 m deep) from the bay and selected tributaries. In situ nettle population density was measured by trawling with a relatively fine mesh net. As can be seen in box figure 2, C. quinquecirrha were found within a relatively narrow, well-defined range of temperature and salinity. Box figure 3 is a graphical representation of the habitat model.

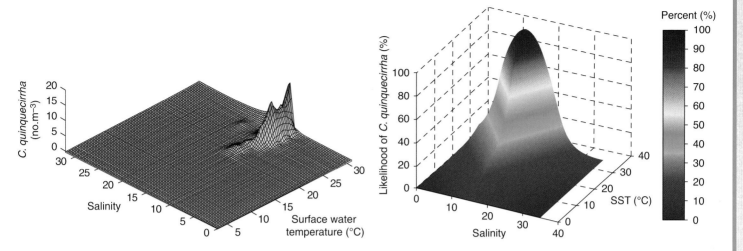

Box Figure 2 Salinity and sea-surface temperature values encompassing 95% of the sea nettle (*Chrysaora quinquecirrha*) population in surface waters of Chesapeake Bay. Maximum *C. quinquecirrha* density = 15 m^{-3}.

Box Figure 3 The likelihood of encountering sea nettles, *Chrysaora quinquecirrha*, in surface waters of the Chesapeake Bay as a function of salinity and surface temperature (SST).

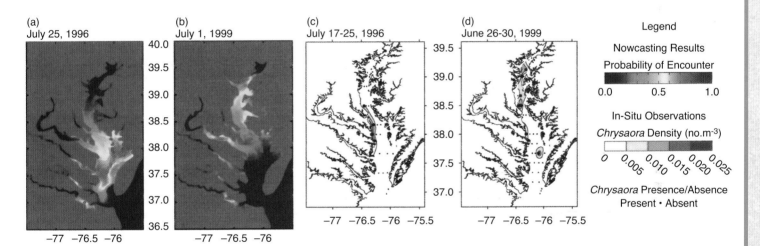

Box Figure 4 The probability of encountering sea nettles, *Chrysaora quinquecirrha*, in surface waters of the Chesapeake Bay for (a) July 25, 1996, and (b) July 1, 1999, and (c, d) observations of sea nettle population density from these two periods, collected by the Trophic Interactions in Estuarine Systems research program sponsored by the National Science Foundation.

Comparisons between observations and retrospective sea nettle predictions (box fig. 4a–d) indicate that the model performed well on a bay-wide scale. The relative downstream-upstream position of nettles in the bay was shown correctly for the summers of 1996 and 1999. In 1996, an extremely wet year, the model predicted that nettles should have been located principally in the lower reaches of the bay. In 1999, a more typical year for rainfall, the model predicted that nettles were positioned farther upstream in the bay and several tributaries. On a finer scale, however, the nowcasts fail to predict lateral variations in the bay, suggesting that additional factors influence sea nettle distribution. We will focus on adding other environmental variables known to influence sea nettle concentrations, such as bathymetry and wind speed.

Similarly, we will be able to predict the response of sea nettle distributions to climate variations if we know how changes in climate would impact salinity and surface temperature.

Our methodology to nowcast the distribution pattern of sea nettles can probably be applied to any organism. Many challenges remain, however. For example, the habitat of a target species may not be as easily defined as that of sea nettles, and we cannot routinely obtain key environmental indicators that are needed to define the habitats of many species. Nevertheless, we believe the approach described here, in conjunction with new technological capabilities, holds promise for achieving one of biological oceanography's age-old goals, that of predicting the distribution pattern of marine organisms.

J. T. Carlton and J. B. Geller sampled ballast water from 159 cargo ships from various Japanese ports that docked in Coos Bay, Oregon.* The ballast water contained members of all the major groups of floating organisms known as animal plankton: microscopic shrimplike copepods, marine worms, barnacles, flatworms, jellyfish, and shellfish. The copepod density was estimated to be greater than 1500 copepods per cubic meter; the density of juvenile forms or larvae of marine worms, barnacles, and shellfish was greater than 200 organisms per cubic meter. Whether organisms arrive in ballast water or in the packing of commercially harvested fish and shellfish, come attached to the floats of seaplanes, or are released from home aquariums, they all leave behind the natural controls of predators and disease found in their native environments. If these invaders are introduced into a hospitable new environment, they may flourish and severely disrupt its biological relationships, forcing out some species and destroying others.

Some biological invaders remain unnoticed for many years; others begin almost immediately to seriously disrupt the ecology of their adopted areas. In 1985, an Asian clam *(Potamocorbula amurensis)* (fig. 13.13) was discovered in a northern arm of San Francisco Bay. This clam was previously unknown in the area and appears to have arrived in its juvenile or larval form in the ballast water of a cargo ship from China. During the next six years, the clam spread southward into the bay and formed dense colonies, as many as 10,000 clams per square meter. The Asian clam feeds on plantlike, single-celled diatoms and the larvae of crustaceans. The food requirements of these huge populations reduce amounts available for native species, stressing the system's species balance and food chains. In 1990, a second intruder, the European green crab, or green shore crab *(Carcinus maenas)* (fig. 13.14), was recognized in the southern part of San Francisco Bay. This crab moved from Europe to the East Coast of the United States in the 1820s and to Australia in the 1950s. The green shore crab is less than 6.5 cm (3 in) broad and is voracious and belligerent. It feeds on clams and mussels and may spawn several times from a single mating. It feeds enthusiastically on the Asian clam, but it can also feed on the native species and can outcompete them for food. *Carcinus maenas* is now moving northward along the West Coast of the United States. In 1997, it had reached Coos Bay, Oregon; in 1998 was discovered in Willapa Bay and Grays Harbor, Washington; and in 1998 was found on the west coast of Vancouver Island, British Columbia.

In 1982, the ballast water of a ship from the coast of America carried a jellyfish-like organism known as a comb jelly *(Mnemiopsis leidyi)* into the Black Sea. From the Black Sea, it spread to the Azov Sea and has recently moved into the Mediterranean. It has no predator in these areas and devours huge quantities of plankton, small crabs, shrimp, fish eggs, and fish larvae.

Ballast water carrying the resting cells of Japanese organisms that cause red tides was responsible for shutting down the natural and aquaculture shellfish harvests of Tasmania and southern Australia in the 1980s; red tides are discussed in chapter 16. Fish can also be transported this way; for example,

*J. T. Carlton & J. B. Geller, 1993. Ecological Roulette: The Global Transport of Nonindigenous Marine Organisms. *Science:* 261 (5117):78–82.

Figure 13.13   Asian clam *(Potamocorbula amurensis).*

Figure 13.14   European green crab *(Carcinus maenas)*. The width of the crab's back (carapace) is about 4 cm (1.5 in). *Photo courtesy of Thomas M. Niesen, San Francisco State University*

Japanese sea bass were introduced into Australia's Sydney Harbor region in 1982–83. A small crab *(Hemigrapsus sanguineus)* common in Japanese waters was identified in New Jersey in 1988 and has since been found as far south as Chesapeake Bay and as far north as Cape Cod. The North American razor clam *(Ensis directus)* was detected in Germany in 1979, and it has since spread to France, Denmark, the Netherlands, and Belgium.

Not all invaders are animals. In the 1980s, an attractive, fast-growing, bright green tropical seaweed *(Caulerpa taxifolia)* was introduced into European aquariums. In 1985, some plants escaped into the Mediterranean Sea during a routine tank-cleaning at Monaco's aquarium. The seaweed rapidly spread and grew to cover more than 44.5 km^2 (11,000 acres) of the northern Mediterranean coastline by 1997. It has recently been reported off the coast of North Africa. In June 2000, *Caulerpa taxifolia* was discovered in a coastal lagoon just north of San Diego, California, and later in Huntington Harbor in Orange County, California, just 125 km (75 mi) further north. In areas where it becomes well established, it is capable of causing ecological and economic devastation by overgrowing and eliminating native species of seaweeds, seagrasses, reefs, and other organisms. In the Mediterranean, *Caulerpa taxifolia* has had a negative impact on tourism, recreational boating, and diving, and has harmed commercial fishing by causing changes in the distribution of fish and impeding net fishing. No effective method of removing it has yet been found.

As more and more alien species are recognized, biological oceanographers and marine biologists realize that an ecological revolution is taking place in estuaries and bays and along the rocky shores of all the continents. Many biologists refer to the introduction of these alien species as "biological pollution"; others are increasingly alarmed by the breakdown of natural barriers and the worldwide "biological homogenization" that may result.

In the case of ballast water, the introduction of zebra mussels into the Great Lakes and their subsequent invasion of the rivers and streams of the central United States sounded the alarm for the freshwater environment. In 1990, Congress enacted the Nonindigenous Aquatic Nuisance Prevention and Control Act (NANPCA). Under this law, the United States adopted voluntary regulations for exchanging ballast water on the high seas for vessels bound for the Great Lakes; this provision became law in 1993.

It has been calculated that, at any one time, in excess of 3000 species must be in motion in the ballast water of ocean-going ships around the world.* As yet no law comparable to NANPCA regulates ballast water in the marine environment, although both the United Nations and the International Council for the Exploration of the Sea have called for ballast water management. Exchange of ballast water at sea can be attempted, if there is no danger to ship or crew, but not all the organisms may be flushed out and the sediment in the bottom of the ballast tanks may not be removed. Proposals have included treating the ballast water by adding poisonous chemicals, heating the water, filtering the water, and exposing the water to ultraviolet radiation. None of the world's cargo vessels is designed for ballast management, however, and all these proposals require some redesign or refit of vessels.

Controlling ballast water will not close all doors to invading marine species and will be costly. However, it will lead to fewer foreign invasions. Keep in mind that "No introduced

*J. T. Carlton & J. B. Geller, 1993. Ecological Roulette: The Global Transport of Nonindigenous Marine Organisms. *Science:* 261 (5117):78–82.

marine organism, once established, has ever been successfully removed or contained, or the spread successfully slowed" (James T. Carlton, Maritime Studies Program, Williams College).

13.8 Overfishing and Incidental Catch

Around the world, too many fishing boats are taking too many fish, too fast. Diminishing fish populations are victims of relentless overfishing and management that fails to acknowledge declining stocks. Changes in world fish catches between peak fishing years and 1992 are given in table 13.5. Fish are being taken faster than the populations can reproduce. A 1995 report from the UN Food and Agriculture Organization (FAO) estimates that 70% of fish stocks worldwide are now overfished and depleted, concluding that world fisheries cannot be sustained at their present levels. According to the U.S. Office of Fisheries Conservation and Management, 41% of the species in U.S. waters are overfished. The 200-year-old Newfoundland cod fishery was closed by Canada in 1992, and in 1993, the National Marine Fisheries Service closed large parts of the U.S. cod fishery. The failure of the cod fishery is discussed further in chapter 17. North Atlantic swordfish landings in the United States declined 70% from 1980–90, and the average weight of the swordfish fell from 52 to 27 kg (115 to 60 lb). In 1991, several countries worked to reduce the Atlantic swordfish catch; Spain and the United States complied by reducing catches 15%. However, Japan increased its catch 70%; Portugal's catch rose 120%, and Canada's 300%. The Atlantic bluefin tuna population dropped 80% between 1970 and 1993; the bluefin population that spawns in the Gulf of Mexico has dropped 90% since 1975; and the Mediterranean population has declined by 50%. Many observers believe that this species is doomed, for it is the world's most valuable fish, selling for as much as $200/kg (nearly $91/lb) in Japan's specialty fish shops.

Fish populations are not the only casualties; other marine animals and marine birds are being affected as they compete for their share of the catch. In the Shetland Islands, nesting seabirds failed to breed in the mid- to late 1980s, apparently in response to lack of food when the sand eels in the area were overfished. In Kenya, overharvesting of the trigger fish on coral reefs

Table 13.5	Changes in Fishing Catch Between Peak Year and 1992						
Atlantic Ocean			**Pacific Ocean**			**Indian Ocean**[1]	
Area[2]	**Peak Year**	**Percent Change**	**Area**[2]	**Peak Year**	**Percent Change**	**Area**	**Percent Change**
NW	1973	−42	NW	1988	−10	West	+6
NE	1976	−16	NE	1987	−9	East	+5
WC	1984	−36	WC	1991	−2		
EC	1990	−20	EC	1981	−31		
SW	1987	−11	SW	1991	−2		
SE	1973	−53	SE	1989	−9		

1. The increase in Indian Ocean catches was due to implementation of more-sophisticated fishing techniques. Average annual growth is between 1988 and 1992.

2. NW, northwest; NE, northeast; WC, west-central; EC, east-central; SW, southwest; SE, southeast.

allowed abnormal growth of sea urchins, thereby damaging the reef ecosystem. Alaska's Steller sea lion populations have plummeted. An estimated 170,000 Steller sea lions existed in 1970. By 1990, the count was 60,000, and by 2000, only about 45,000 remained. Studies in the Gulf of Alaska and the Bering Sea show that more than 50% of the Steller sea lion's diet is pollock, but the heavily fished pollock stock in the Bering Sea was down from an estimated 12.2 million tons in 1988 to 6.5 million tons in 1995. The sea lions have to spend more time and energy obtaining the same amount of fish. The entire population has been listed as a threatened species since 1990, and the western U.S. stock was listed as endangered in 1997. Steller sea lions are protected under the Marine Mammal Protection Act, which prohibits the killing, harming, or harassing of any marine mammal, as well as the Endangered Species Act. With this federal protection, hope exists for the recovery of the Steller sea lion population.

Two hundred million people make their living by catching, processing, or selling fish, and millions more depend on fish as their source of protein. In Southeast Asia, more than 5 million people fish full time and contribute some $6.6 billion to the region's income. In northern Chile, fishing accounts for 18,000 jobs and 40% of the national income; Iceland depends on fishing for 17% of its income and 12%–13% of its employment.

In 1977, the United States extended control over its fisheries to an Exclusive Economic Zone (EEZ) that extends 200 miles out to sea and within which only U.S. boats are permitted to fish. Today, more than 122 other nations have claimed EEZs, and in most cases, the declaration of an EEZ has been followed by an increase in the country's fishing capacity. The world's fishing industry today has about twice the capacity it needs to make its annual catch. The world's fishing fleet doubled between 1970 and 1999, from 585,000 to 1.2 million large boats. China's fishing fleet is the largest in the world, with an estimated 450,000 vessels, roughly a third of the global fleet. The European Union nations are estimated to have fleets that are more than 40% larger than required. Overexpansion is encouraged by economic and political pressures that push artificial government supports to modernize fleets, supply fuel as well as sophisticated fishing gear, and prevent industry collapse, but these programs often undermine management programs and encourage overfishing.

The U.S. fishing fleet (fig. 13.15) has grown dramatically through a federal loan program encouraging the building of new boats. The new boats are extremely efficient, equipped with sonar and depth recorders and computers that remember the sites of previous catches and home in on these sites at a later date. Planes, helicopters, and even satellite data are used to find the fish. In addition, the fishing boat and gear, the crew's wages, and the fuel needed are all increasing the cost of commercial fishing, requiring the vessels to put even greater pressures on the shrinking schools of fish as the fishers strive to maintain their incomes.

Some countries are responding to the situation by consolidating their fishing fleets. Taiwan no longer issues licenses to boats under 1000 tons and has started a buy-back program for boats more than fifteen years old. Malaysia is cutting its

Figure 13.15 Seiners gather in Petersburg, Alaska, waiting for the salmon season to open.

number of fishers by 40% and favors more modern, higher-capacity boats. Iceland plans a 40% cut in its fishing capacity. These programs may decrease the number of fishing vessels, but they will not necessarily reduce the catch if the remaining vessels are larger and more efficient.

The oceanic drift-net fishery became a serious problem in the 1980s, laying out nets up to 65 km (40 mi) long each night. These gill nets of almost invisible nylon hang like walls in the water, trapping and killing nearly everything that swims into them. In 1990, the U.S. Marine Mammal Commission estimated the aggregate length of these nets at 40,000 km (25,000 mi), enough to ring Earth. Three Asian nations—Japan, Taiwan, and South Korea—have used the nets to catch squid and fish in the Pacific and Indian Oceans; in the Atlantic, several European nations have drift-netted, mainly for albacore tuna. These nets do not rot; sections torn free in storms float in the ocean for months, even years, as ghost nets catching everything they encounter; no estimate exists on the animal life they destroy. In the Pacific, Japan, Taiwan, and South Korea agreed to abide by a UN resolution to halt drift-net fisheries by the end of 1992. After four years of discussion, the European Union fisheries ministers voted to ban drift-net fishing in most of the northeast Atlantic beginning in 2002. The only countries to vote against the ban were France, which has the second-biggest drift-net fleet in the European Union, and Ireland. Italy, which has the biggest, abstained while the other twelve member states voted for the ban. The ban applies everywhere except in the Baltic waters where there are no dolphins. This concession was made in order to get the support of Denmark, Finland, and Sweden. Fishers in the Baltic use drift nets to catch salmon and sea trout. The decision ends a long campaign by environmentalists, including Greenpeace, which has been trying to convince the European Union to act for fifteen years.

Large numbers of marine animals die each year only because they are caught incidentally by people fishing for other species. **Incidental catch,** or **by-catch,** or what are often called "trash fish," represents a tremendous waste of marine resources. The FAO estimates that each year, 30 million tons of fish, about 25% of all reported commercial marine landings, are caught as

Table 13.6	**Global Incidental Catch**	
Fish Type	**Percent Discard of Landed Weight**	**Landed Weight (millions of metric tons)**
Shrimp and prawns	520	1.83
Crabs	249	1.12
Flounder, halibut, sole	75	1.26
Redfish, bass, conger eel	63	5.74
Lobster, spiny-rock lobster	55	0.21

by-catch and discarded. The world's shrimp fishery is estimated to have an annual catch of 1.8 million tons; the associated discarded by-catch is 9.5 million tons. Shrimp trawlers are estimated to catch more than 45,000 sea turtles each year; more than 12,000 of them do not survive. Alaskan trawlers for pollock and cod throw back to the sea some 25 million pounds of halibut, worth about $30 million, as well as salmon and king crab because they are prohibited from keeping or selling these fish. Another 550 million pounds of bottom fish are discarded in Alaskan waters to save space for larger or more valuable fish. See table 13.6.for global incidental catch rates for the most severely affected fisheries.

The discard rate on by-catch varies from place to place; if incidental catch does not bring a high enough price and if processors or markets are not available, these "trash fish" will be returned to the sea, usually dead. Whereas in the Gulf of Mexico 1 pound of shrimp results in 10.3 pounds of by-catch that is nearly all discarded, in Southeast Asia and other areas with local fisheries and fresh fish markets much of the by-catch is used.

13.9 Afterthoughts

Earth and its environment are works in progress. Since its beginning, Earth has constantly reworked and modified itself; living things have interacted with their environments, making changes and achieving new balances. Humans, however, have acquired the power to accelerate natural change and make fundamental environmental alterations for their own purposes. Few organisms compete with us successfully, and few environments are able to resist our presence. Humans will never be a zero impact factor, but we can challenge ourselves to understand and consider the implications of our choices. Science can help us understand the consequences resulting from our choices, but each of us, individually, must carefully define and protect the process of making the "best" choices.

We must also remember that *there is no away.* What we dispose of on this planet remains on this planet—out of sight may be out of mind but not out of our environment. In June 1995, after three years of deliberation, the Shell Oil Company planned to dump a decommissioned, floating 14,500-ton oil-storage tank 2000 m (6600 ft) down on the Atlantic's North Feni Ridge. Shell claimed that sea disposal of the tank, which contained an estimated 100 tons of heavy-metal sludge, was less hazardous than land disposal. The plan was stopped by adverse publicity and a huge public outcry; the tank was interred at a land site. Was the best choice made? Should the public be as concerned about the land burial as it was about ocean disposal?

The more we understand our Earth system, the better the choices that we can make. Each of us—by continuing our education (in school or out), by participating in the political process, by working with others—can help to make the intelligent, informed decisions that are required to maintain a healthy and productive planet.

Summary

Water quality is affected by dumping solid waste and liquid pollutants into coastal and offshore waters. Land runoff carries a mixture of toxicants, oil and gasoline residues, industrial wastes, pesticides, and fertilizers to estuaries and seacoasts. An expanding low-oxygen area in the Gulf of Mexico has been linked to increasing amounts of nitrates and phosphates flowing into the Gulf primarily from the Mississippi River. Toxicants are adsorbed onto silt particles and become concentrated in coastal sediments. Organisms further concentrate toxicants and pass them on to other members of marine food webs. Reducing the discharge of toxicants is showing success, with declines in lead, DDT, and PCBs.

Plastics are an increasing problem to marine life, killing thousands of fishes, birds, mammals, and turtles each year. Laws prohibit dumping of plastics at sea, but there is no international enforcement.

Because of the world's increasing quantities of solid wastes, the deep-ocean plains have been proposed for waste disposal by some scientists. It has also been suggested that long-lived nuclear wastes could be placed in deep-sea trenches for eventual recycling back into Earth's mantle.

Oil spills are a special problem for inshore waters. Three significant oil spills are the sinking of the *Amoco Cadiz,* the grounding of the *Exxon Valdez,* and the 1991 spills into the Persian Gulf. All have had devastating ecological effects that continue to demonstrate that no adequate cleanup technology exists at this time.

Wetlands border estuaries and coasts; they are important as areas of nutrients, food, and shelter for many marine species. Many wetlands have been filled, dredged, developed, and lost.

Organisms move across the oceans in the ballast water of thousands of cargo vessels, to be discharged into new environments far from their points of origin. Significant ecological disruptions have been found in San Francisco Bay, the Black Sea, the Azov Sea, coastal Australia, and coastal Europe. Many scientists refer to the introduction of alien species as biological pollution.

Overfishing in all the oceans is devastating the world's fisheries. People who make their living from fishing are losing their jobs, and the marine species that depend on the fish are declining as the fish populations decrease. The establishment of Exclusive Economic Zones has promoted overexpansion of fishing fleets. Incidental catch is the cause of a tremendous loss to fisheries and wildlife.

Key Terms

All key terms from this chapter can be viewed by term or definition when studied as flashcards on this book's website at www.mhhe.com/sverdrup10e.

hypoxic, 324
wetland, 332
incidental catch/by-catch, 340

Study Questions

1. If contaminated sediments are dredged from the floor of a harbor to use in a landfill, what hazards to the environment should be considered during the dredging, the transport, and the storage of the sediments?
2. Nutrient concentrations in coastal waters are increasing. Why? What is the result?
3. Why is toxicant concentration in sediments higher than toxicant concentrations in the overlying water? How are organisms in a polluted area affected?
4. Why are plastics such a problem in marine waters?
5. Why do some scientists think that ocean disposal of hazardous waste materials is safer than land disposal?
6. Compare the oil spills from the 1991 Gulf War and the *Exxon Valdez.* Consider the geography and climate of each area, the dispersal of the oil, the effect of the oil on the beaches and the organisms, the effect on people who gain their living from the sea, and the effectiveness of the cleanup.
7. What happens to oil when it has been spilled at sea? Into coastal waters?
8. Why are marine wetlands ecologically valuable, and why are they decreasing in all ocean shore areas?
9. Explain the consequences of ballast water transport of marine organisms.
10. How did the establishing of Exclusive Economic Zones affect coastal fisheries?
11. What is incidental catch, or by-catch, and what is its effect on ocean fisheries?
12. Why can a biological invader such as *Caulerpa taxifolia* spread so rapidly in a new environment?
13. How could the incidental catch, or by-catch, of the world's fisheries be used effectively?
14. What would have to happen socially or economically in order to effectively use the world's incidental catch?

Divers swim above a coral reef in the Cayman Islands.

Chapter 14

The Living Ocean

Chapter Outline

Learning Outcomes

After studying the information in this chapter students should be able to:

1. *Describe* how viscous and inertial forces influence how organisms of different sizes move through the water,

2. *Explain* differences and similarities between photosynthesis and chemosynthesis,

3. *Describe* how the biological pump increases the flux of atmospheric CO_2 into the ocean, and

4. *Explain* how different groups of organisms regulate their body temperatures and how this influences their ecological distributions.

The previous chapters combined basic principles of physics and chemistry to explain ocean processes. In this chapter and those that follow, we use these same principles to describe how and where organisms live in the oceans, and whether they live within water or ice, on the sea floor, or along the shores. Marine organisms have much in common with organisms on land, but the ocean environment also presents distinct challenges and survival problems. Scientists that study ocean life are referred to as either biological oceanographers or marine biologists. The distinctions between these two titles are subtle with no obvious boundary between them. Biological oceanographers tend to study how smaller organisms in the ocean affect, and are affected by, oceanographic processes. Marine biologists tend to study the biology of, and interactions between, larger organisms in the sea. Marine ecosystems are composed of an astonishing diversity of organisms, both large and small. Consequently, scientists that study ocean biology ultimately have to understand the interactions between different sizes and different groups of organisms. Regardless of whether scientists are referred to as marine biologists or biological oceanographers, they study the abundance and distribution of marine organisms and the relationships between the organisms and their environment.

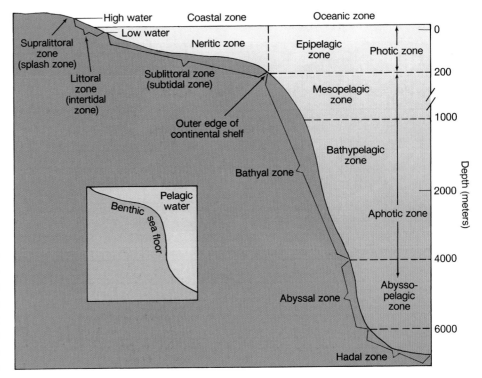

Figure 14.1 Zones of the marine environment.

14.1 Environmental Zones

The marine environment that organisms live within is vast and varied. It can be divided into two general zones: the **pelagic zone,** or water environment, and the **benthic zone,** or seafloor environment. The pelagic zone is further divided into the coastal or **neritic zone** (above the continental shelf) and the **oceanic zone** (open ocean away from the direct influence of land). Both the benthic and pelagic zones can be divided into additional zones based on different increments of depth, as illustrated in figure 14.1. The different benthic zones and the types of organisms that live within them will be discussed in more detail in section 14.5.

14.2 Considerations of Size

We begin our considerations of biology in the world's oceans with a discussion of size. The most abundant organisms in the sea are microbes (the terms "microbes" and "microorganisms" can be used interchangeably). These single-celled organisms carry out a variety of processes that have global ramifications, some of which will be discussed in more detail in chapters 15 and 16. The vast majority of microbes are generally only a few microns (a micron, μm, is 1/1000 of a millimeter) in size, much smaller than the width of a human hair. The organisms of the sea that can be seen easily, such as invertebrates, fish, or whales, vary dramatically in size. For example, a small marine worm may be less than an inch in length, and a blue whale may be almost 100 feet (33 m) in length. These organisms are all multicellular, which means they are composed of multiple cells that interact with one another to form structures such as tissues, organs, and limbs. Typically, the individual cells that make up a multicellular organism are more or less the same size (nerve cells are an exception to this rule), and they are comparable to the size of most microbes. About ten orders of magnitude separate the length of the smallest microbe from the length of the largest blue whale (~0.5 μm to ~33 m). In contrast, only about three orders of magnitude separate the size of the small 0.5 μm microbes from the size of large 100 μm cells. Why should this be so? Why aren't enormous organisms composed entirely of enormous cells? What determines the shape that organisms can assume, and how does shape relate to lifestyle? Why do some organisms have elaborate shapes and others have relatively simple shapes? As described below, the answers to these questions are derived from fundamental principles of physics and chemistry such as conservation of mass and energy covered in earlier chapters: organisms must accomplish tasks involving transport and retention of mass and energy, and factors such as size and shape affect how effectively they are able to do so. The field of research that looks for answers to how chemistry and physics affect basic biological characteristics such as size and shape is known as **biomechanics** and will serve as a starting point for thinking about life in the ocean.

Heat content, chemical constituents, and mass of living organisms differ from that of their environment. At the level of a cell, the main underlying mechanism used to maintain these differences is molecular transport. Molecular transport that acts upon mass is referred to as **diffusive,** transport that acts upon heat is referred to as **conductive,** and transport that acts upon momentum is referred to as **viscous.** These transport mechanisms all act across surfaces. Also, they act strongly across shorter distances and are weaker across longer distances. Think of how increasing the thickness of your clothes by putting on a sweater reduces the rate of heat conduction away from you when you are sitting in a cold room. Conduction alone can be used to dissipate heat from microscopic-sized organisms but cannot be used as the only means of dissipating heat from larger organisms composed of large numbers of closely packed cells. Similarly, diffusion alone can supply nutrients to and remove wastes from organisms the size of typical cells but not organisms the size of whales.

A fundamental difference between small organisms and large organisms is the ratio of their surface area to volume. Let's imagine a shrinking sphere. As the sphere shrinks from having a radius of 1000 units to a radius of only 1 unit the surface area decreases by the square of the radius (area = $4 \pi r^2$), but the volume decreases by the cube of the radius (volume = $4/3 \pi r^3$). Thus the surface area to volume ratio [$(4 \pi r^2 / 4/3 \pi r^3) = (3/r)$] will increase as the radius decreases. A smaller sphere has a larger surface area to volume ratio than a larger sphere (table 14.1). The same principle holds for organisms: a microscopic phytoplankton has a much larger surface area to volume ratio than a macroscopic whale. Different forces act on the surface of a cell than act on the volume of a cell. Organisms less than about a millimeter in size tend to be more impacted by viscosity, and larger organisms tend to be more impacted by gravity and inertia. The result of the different relative impacts of these forces is that small organisms live in a marine environment dominated by viscous forces and momentum is essentially non-existent. Microorganisms drag a layer of fluid with them as they move

Table 14.1	Relation Between Size and the Ratio of Surface Area to Volume		
Radius (r)	Surface Area ($4 \pi r^2$)	Volume ($4/3 \pi r^3$)	Surface Area: Volume (3/r)
1	12.56	4.188	3
10	1256	4188	0.3
100	125600	4188000	0.03
1000	12560000	4188000000	0.003

through water, and they do not glide once they stop swimming. Larger organisms live in a marine environment dominated by inertia; they leave behind the water they are swimming in, and they tend to continue to move for a period of time after they stop swimming. Gliding due to momentum and streamlining of body types is critical to how large organisms move through water.

The physical constraints associated with molecular transport and the relative impact of viscous versus inertial forces profoundly influences how organisms behave, how they move through their environment, how they sense their environment, how they find food to eat, and how they avoid being eaten. Cells are small because of the constraints set by molecular transport; diffusion alone simply cannot act fast enough to supply enough nutrients to an enormous cell. The shapes of small marine organisms can be very different from the shapes of large marine organisms because of the constraints set by the different forces that act upon them. As will be seen, there are apparently few biomechanical "solutions" to some of the "problems" faced by marine organisms. For example, bivalves, sponges, and microscopic ciliates are unrelated, and yet they all use cilia to filter the water for food (fig. 14.2); the bivalves and sponges use multiple ciliated cells, and the microscopic ciliates are a single-celled organism. The physical and chemical interactions between organisms and their environment and between each other ultimately determine ecosystem structure.

14.3 Groups of Organisms

Marine ecosystems are composed of diverse communities of interacting organisms. We just saw how cell size and the physical and chemical constraints of the environment influence interactions. Grouping organisms together based on either their relatedness or their function provides further insight into the complex interactions that occur within ecosystems. Taxonomic systems were developed about 250 years ago by the Swedish botanist Carolus Linneas and are still used today to name and classify related groups of organisms. **Taxonomy** is based primarily on shared morphology. For example, the genus and species name *Homo sapiens* is accepted in all countries as the name for humans, and all humans share similar, readily recognizable morphological characteristics. Taxonomic categories for some common marine organisms discussed in this chapter are listed in table 14.2. In recent years, the use of DNA sequence information has helped uncover evolutionary relationships among organisms that were not readily apparent based solely on morphology. The evolutionary connections between ancestor organisms and their descendants are referred to as **phylogenies.** Importantly, relatedness categories often do not

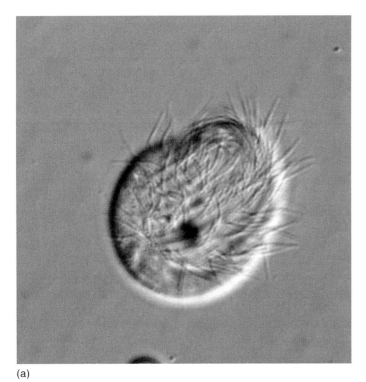

(a)

(b)

Figure 14.2 (a) A single-celled ciliate covered with cilia. The cilia are the small projections emanating from the cell surface. (b) Flame scallops feed on plankton that they filter from the seawater with cilia.

overlap with functional categories. Organisms that are not closely related to one another and/or organisms that are of very different sizes can carry out similar functions within an ecosystem. For

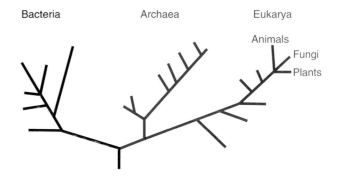

Figure 14.3 The tree of life is divided into three domains: Bacteria, Archaea, and Eukarya. The Bacteria and Archaea are single cells without nuclei. The Eukarya domain includes many single-celled organisms with nuclei as well as the animals, plants, and fungi.

example, the unrelated ciliates and bivalves described above both filter the water of microscopic organisms and thus impact the abundance of these organisms.

The broadest category of relatedness used to group different organisms is based on a combination of cell morphology/ structure and DNA sequences. This method of classification recognizes three basic domains of life: Bacteria, Archaea, and Eukarya (fig. 14.3). (Note that these names are capitalized when used to refer to the domain names; the names are not capitalized when used to refer to the common names bacteria, archaea, and eukaryotes.) Bacteria and Archaea belong to the most ancient groups of organisms, the **prokaryotes.** Fossil remnants of prokaryotes date back to about 3.5 billion years ago. Prokaryotes are unicellular, possess a simple cell architecture, and lack membrane-bound internal structures including a cell nucleus. Bacteria and Archaea share a similar morphology but can be readily distinguished from one another based on DNA sequences. Recent estimates suggest that as much as 40% of the prokaryotes in marine waters deeper than ~500 m may be archaea. **Eukaryotes** appear later in the fossil record, dating back to about 2.1 billion years ago. Eukaryotes are characterized by a complex cell architecture. All eukaryotes possess a **nucleus** that houses the majority of a cell's DNA. Most eukaryotes also contain **mitochondria** that are used to derive energy from food, and some eukaryotes contain **chloroplasts** that are used to harvest sunlight by photosynthesis (see below).

The broadest functional division between organisms is based on how they derive energy and carbon for growth (fig. 14.4). This method of classification recognizes two groups of organisms, **autotrophs** and **heterotrophs.** Autotrophs

Table 14.2	**Taxonomic Categories of Some Marine Organisms**				
	Sperm Whale	**Emperor Penguin**	**Chinook Salmon**	**Bay Scallop**	**Giant Kelp**
Phylum	Chordata	Chordata	Chordata	Mollusca	Phaeophyta
Class	Mammalian	Aves	Osteichthyes	Bivalvia	Phaeophycea
Order	Cetacea	Sphenisciformes	Salmoniformes	Ostreoida	Laminariales
Family	Kogiidae	Spheniscidae	Salmonidae	Pectinidae	Lessoniaceae
Genus	*Kogia*	*Aptenodytes*	*Oncorhynchus*	*Argopecten*	*Macrocystis*
Species	*sima*	*forsteri*	*tshawytscha*	*irradians*	*pyrifera*

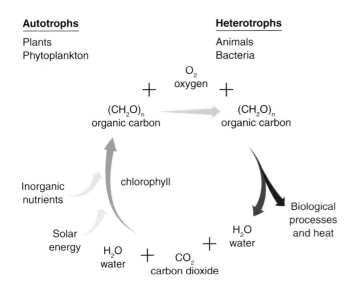

Autotrophs   **Heterotrophs**

Plants   Animals
Phytoplankton   Bacteria

Figure 14.4 Simplified schematic of connections between autotrophic and heterotrophic organisms. Photosynthetic organisms use solar energy to generate organic carbon from carbon dioxide and as a by-product, they generate oxygen. Heterotrophic organisms consume organic matter and breathe oxygen to drive biological processes and as a by-product, they generate carbon dioxide.

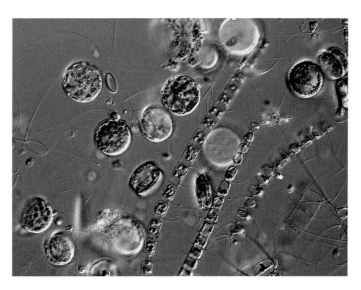

Figure 14.5 Phytoplankton sample collected by pulling a net with a mesh size of 20 µm through surface waters. Both chains and individual cells of different species of diatoms are present.

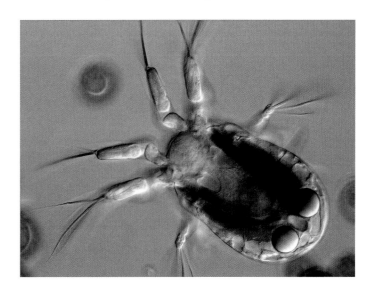

Figure 14.6 Juvenile stage of one of the crustacean zooplankton collected from coastal waters.

generate organic matter, such as the sugars and proteins needed for growth, from inorganic compounds such as nitrate or ammonium, phosphate, and carbon dioxide. The energy needed to drive formation of organic matter is derived either from sunlight or from chemical energy, and the carbon is derived from carbon dioxide. The organisms that harness light energy to generate organic matter are termed **photoautotrophs,** and the process they carry out is **photosynthesis.** An important by-product of photosynthesis is the generation of oxygen. The organisms that use chemical energy to drive production of organics are termed **chemoautotrophs,** and the process they carry out is **chemosynthesis.** Until recently, the number of different organisms able to chemosynthesize was believed to be relatively restricted. It now appears that a large percentage of the archaea that live in deep waters may carry out chemosynthesis. Autotrophs are said to occupy the base of the food web because they generate organic compounds from inorganic, non-living matter and because they fuel the growth of organisms that require organic matter as food. Heterotrophs consume the organic matter generated by autotrophs; they use the organic matter as both a carbon source and an energy source. Most heterotrophs use oxygen to oxidize the organic matter to derive energy. Heterotrophs generate carbon dioxide as a by-product of the oxidation of organic matter.

The functional categories of autotrophs and heterotrophs span across the relatedness categories of prokaryotes and eukaryotes. Prokaryotes can be photoautotrophic (cyanobacteria), chemoautotrophic (for example, the bacteria that live in worms from the hydrothermal vents and derive energy from the oxidation of hydrogen sulfide), or heterotrophic (for example, the bacteria such as *Escherichia coli* that live in our guts and feed on what we leave behind after digestion). Eukaryotes can be photoautotrophic (plants, algae) or heterotrophic (for example, humans, whales, fish). There are no known examples of eukaryotes that are chemoautotrophic. However, there are examples of eukaryotes,

such as the worm *Riftia pachyptila*, that rely solely on the organic material generated by the chemosynthetic bacteria that live in close association with them.

For life in the ocean, we can additionally group organisms based on habitat. **Plankton** are generally small (less than a few millimeters in size) and are unable to move faster than currents, so they drift with the water's overall movement. Some large organisms such as *Sargassum* seaweed and some jellyfish are also considered to be planktonic because they also drift with the currents. **Phytoplankton** photosynthesize and thus are photoautotrophic. They can be either prokaryotic or eukaryotic; they are almost always unicellular, although some species form chains of connected cells (fig. 14.5). **Zooplankton** consume organic matter and thus are heterotrophic. All zooplankton are eukaryotic, but they can be either unicellular or multicellular (fig. 14.6). Heterotrophic prokaryotes are members of the **bacterioplankton.** Some phytoplankton can

Figure 14.7 All fish are members of the nekton. Here, a diver observes Sargent Major fish at Santo, Vanuatu.

Figure 14.8 The benthic sun star, *Solaster,* typically has ten arms and a diameter of 25 cm (10 in).

both photosynthesize and consume organic matter depending on environmental conditions, and they are known as **mixotrophic plankton.** Organisms that are larger than the plankton and can swim faster than the currents are **nekton** (fig. 14.7). Both plankton and nekton live within the pelagic zone, or the water environment. They are found in both neritic and oceanic zones (fig. 14.1). Organisms that live on, in, or attached to the sea floor are the **benthos** (fig. 14.8), and they live in the benthic zone, which extends from the **supralittoral zone,** or splash zone, to the deepest depths of the ocean, or the **hadal zone** (fig. 14.1). The habitats of many organisms change as an individual matures during its life cycle. For example, the juvenile stage of crabs is part of the zooplankton, whereas the adult stage of crabs is part of the benthos.

14.4 Facts of Ocean Life

Various aspects of the marine environment pose challenges for the organisms that live there. As will be seen in subsequent chapters, there is an enormous diversity of organisms that live

in the sea, but as discussed in section 14.2, they all have to fit within the relatively limited number of potential biomechanical solutions to these challenges. What follows is a discussion of broad-scale chemical and physical environmental conditions that influence the distribution and abundance of marine life. In thinking about these different environmental conditions, remember that mass and energy are conserved. What this means is that molecules such as carbon or nitrogen can be found in a variety of different compounds, but in the end, there will always be a mass balance of the molecules. Also, organisms derive energy from their use of different compounds, but organisms do not create energy. These concepts are critical as we consider the various transformations that living organisms carry out in the world's oceans.

Light

The most abundant source of energy to the ocean is sunlight, which both heats the waters and supplies the light energy necessary for photosynthesis. The **euphotic zone** is the area of the ocean where there is sufficient sunlight for growth of photosynthetic organisms. The **aphotic zone** is the area of the ocean where no light penetrates (fig. 14.1). The depth to which light penetrates varies depending on a number of factors including (1) the angle at which the Sun's rays hit Earth's surface, (2) absorption of different wavelengths of light by water itself, and (3) the amount of particulate matter in water (review the discussion of light transmission in the oceans in chapter 5). The angle at which sunlight hits the ocean is determined by latitude, season, and time of day. Absorption of light by water molecules means that the total amount of light decreases exponentially with depth (fig. 14.9a); red wavelengths of light are absorbed at the shallowest depths, and blue wavelengths of light are absorbed at deeper depths (fig. 14.9b). The euphotic zone of coastal waters is shallower than the euphotic zone of the open ocean because more particles (both living and non-living) are found in water closer to land, and these particles absorb light. In the open ocean, the euphotic zone may extend to nearly 200 m (600 ft), whereas, in coastal regions, the euphotic zone may extend to only a few tens of meters. The penetration of sunlight is much greater in clear than in turbid seawater (fig. 14.9a).

Plants are the dominant photosynthetic organisms on land. Land plants remain fixed in one location during their life; only seeds disperse to new locations. Although plants may grow towards more sunlight, they cannot move to the sunnier side of the hill. Phytoplankton are the most abundant photosynthetic organisms in the oceans. Even those phytoplankton that are able to swim cannot move faster than the currents. Depending on the depth of ocean mixing, a community of phytoplankton may be mixed deep into colder, darker waters one day only to be returned to the warmer surface ocean the next day where they are confronted with bright sunlight. Phytoplankton must be able to adapt to wide changes in their physical environment over relatively short time periods.

The region shallower than the aphotic zone but deeper than the euphotic zone is sometimes referred to as the **twilight zone.** There is enough light in the twilight zone during the day for

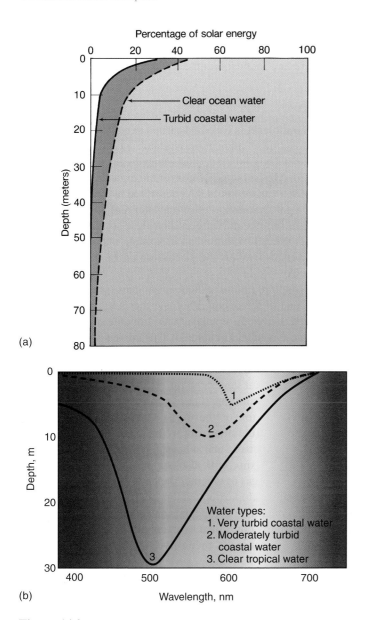

Figure 14.9   (a) The percentage of solar energy available at depth in clear and turbid water. (b) Depth penetration of different wavelengths of light in three different water types. Note that shorter wavelengths of light penetrate to deeper depths in clearer waters.

organisms to see objects to varying degrees; at the shallower depths of the twilight zone, distinct objects can be seen, but deeper in this zone only shadows are visible. With the advent of sonar during WWII, naval sonar researchers noticed that deep scattering layers—layers where sonar signals were heavily reflected—appeared to migrate up and down in the water column. During the day, the deep scattering layer is at mid-water depths, but at night, this "deep" layer moves to near the surface. These deep scattering layers are now known to result from high concentrations of zooplankton and some fishes or squid, whose bodies reflect the sonar signals. These organisms hide in deeper waters during the day to avoid visual predators and migrate to surface waters at night to feed.

Another source of light in the ocean is **bioluminescence.** This light is generated by the organisms themselves and in the sea is used primarily to avoid being eaten. On a dark night, the wake of a boat or moving fish or even an oar pulled through surface waters will cause the flashing of certain organisms, primarily a group of phytoplankton known as dinoflagellates, some jellyfish, shrimp, and squid. Bioluminescence is produced by cells when the enzyme luciferase acts on its substrate luciferin to produce light. Bioluminescence in the sea is commonly stimulated when an organism is physically agitated, which is why we see bioluminescence when something moves through the water. (This phenomenon is often incorrectly referred to as phosphorescence, but it has nothing to do with phosphorescence, which refers to the emission of light from individual excited molecules.) Bioluminescence also occurs on land (fireflies use flashes of light to identify mates), but bioluminescence is much more common in the ocean.

The most common function of bioluminescence in the ocean is thought to be predator avoidance. An attacking predator can be startled or distracted by the light produced by its prey, which may allow the prey to escape. For example, the arms of certain brittle stars (relatives of starfish) will begin to flash in repetitive patterns when the animal is attacked. If the attack continues, the brittle star loses the arm closest to the attacking animal. The discarded arm will continue to flash, holding the attention of the attacker, while the darkened animal crawls away, although with one less arm. The "burglar alarm" hypothesis has been developed to explain why some marine organisms produce light. If movement of water associated with the proximity of a predator causes the potential prey to flash, then repeated flashes act as a sort of burglar alarm, since they call attention to the predator. In this instance, rather than the individual that triggered the burglar alarm being arrested by police, this individual may instead be eaten by an even larger predator that can now see its newly illuminated prey!

Bioluminescence serves other functions besides predator avoidance. Bacteria that colonize marine snow (large particles of organic material that fall through the water column) often bioluminesce. Presumably fish are enticed to eat the glowing marine snow, and the bacteria find themselves once again in the food-rich environment of a fish gut. The use of submersibles and mid-water trawls suggests that a majority of deep-water fish and invertebrates can also bioluminesce. Angler fish use a lure filled with bioluminescent bacteria that dangles just above their mouth to lure their food. Flashlight fish have pouches beneath both eyes that are filled with bioluminescent bacteria that they use to see their prey. Many mid-water fish living in the "twilight zone" produce light on their underside, to avoid producing a shadow. In order to effectively mask themselves using this technique, the fish must be able to detect and match the incoming ambient light.

Color (or absence of color), instead of bioluminescence, is used by other organisms to avoid being eaten. Many zooplankton, including jellyfish, are nearly transparent and blend in well with their watery background. Many fish use color rather than transparency to blend in. In the relatively clear waters of the tropics, fish use bright colors to become almost invisible against their backdrops of colorful coral reefs. Other fish conceal themselves with bright color bands and blotches that disrupt the outline of the fish and may draw a predator's eye away from a vital spot. Bright colors are also commonly used for sexual recognition during breeding times. In coastal waters of temperate latitudes, less light penetrates to depth, and organisms use browns and grays to conceal themselves against rocks and in kelp beds. Bottom fish are commonly similar in color to the sea floor or are speckled with neutral colors. Flatfish have skin cells that expand and contract to produce color changes, and they can change their color to match almost any bottom type on which they live (fig. 14.10). Fish that swim near the ocean surface– such as herring, tuna, and mackerel—commonly have dark backs and light undersides. This pattern of coloring is known as countershading, and it allows the fish to blend in with the bottom when seen from above and with the surface when seen from below (fig. 14.11). Deep-water fish are usually small (rarely larger than 10 cm [4 in]), and they typically appear black at depth.

Figure 14.10   The peacock flounder can change its color and the pattern on its skin to match the sea floor.

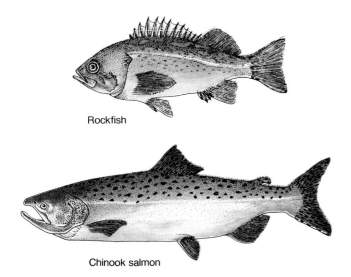

Rockfish

Chinook salmon

Figure 14.11   Viewed from above, the dark dorsal surface of the fish blends with the sea floor; viewed from below, the light ventral surface blends with the sea surface. This type of coloration is known as countershading.

Some species of deep-water shrimp are red when seen at the surface. At depths below the penetration of red light, these shrimp appear dark and inconspicuous.

Carbon

All life on our planet is based on carbon. Autotrophic organisms (both photosynthetic and chemosynthetic) utilize inorganic carbon dioxide (CO_2) as their carbon source to generate the organic carbon needed for growth. Heterotrophic organisms use organic carbon as their carbon source. As described in chapter 6, carbon dioxide is a gas that dissolves in seawater and is stored in large quantities in the ocean. Carbon dioxide also acts as a buffer to maintain the pH of the ocean in a range that is suitable for living organisms (review the discussion of the pH of seawater in chapter 6).

The growth of photosynthetic organisms is restricted to the euphotic zone of the ocean. Photosynthesis by phytoplankton and seaweeds drives the uptake of CO_2 from the surrounding waters as organic matter is generated. This creates a concentration gradient, and more CO_2 from the atmosphere diffuses into the surface waters to replace the CO_2 removed via photosynthesis. Most of the organic matter generated through photosynthesis is consumed by the heterotrophs that live within the euphotic zone. Only a small portion of the organic matter rains out of the surface waters to fuel other life in the ocean. This means that those areas of the ocean, such as coastal waters, that support more autotrophic organisms will also support more heterotrophic organisms, regardless of whether the heterotrophs are members of the bacterioplankton, zooplankton, nekton, or benthos. By the time carbon-containing compounds reach the deep sea, they will have been partially consumed by many different organisms, and the carbon compounds themselves will have undergone numerous transformations.

As heterotrophic organisms in deeper waters consume organic matter, they generate CO_2 and remove oxygen. The net effect of photosynthesis in the surface waters and the raining down of organic matter to deeper waters is that CO_2 is drawn down from the atmosphere and stored in deeper waters until ocean circulation brings the deep water back to the surface for equilibration with the atmosphere. This process of drawing CO_2 from the atmosphere into the ocean through the activity of biological processes is known as the **biological pump.** The biological pump plays a critical role in modulating atmospheric levels of CO_2. Estimates suggest that, if the biological pump was suddenly turned off, atmospheric levels of CO_2 would rise about 200 ppm, almost doubling preindustrial levels of atmospheric CO_2. We will look in more detail in the following chapters at how different organisms influence the efficiency of the biological pump.

Inorganic Nutrients

Autotrophic organisms require the same inorganic nutrients for growth that the plants in your garden require as fertilizer. All phytoplankton and seaweeds require the inorganic macronutrients phosphate (PO_4^{-3}) and nitrogen, most commonly in the form of nitrate (NO_3^-) or ammonium (NH_4^+). A small number of bacteria, including some prokaryotic phytoplankton known as cyanobacteria, can use N_2 gas as a source of nitrogen when NO_3^- or NH_4^+ concentrations are less than the cells' minimum requirements. One group of eukaryotic phytoplankton, diatoms (which will be covered in more detail in chapter 16), also require silicic acid ($SiOH_4$) for growth. Autotrophs additionally require micro-nutrients such as vitamins, iron, zinc, and manganese. The terms "macro-" and "micro-nutrients" refer to the relative concentrations of the nutrients required by autotrophs. Macro-nutrients are required in higher concentrations than micro-nutrients.

Inorganic nutrients are quickly consumed within the euphotic zone by the phytoplankton to generate nitrogen- and phosphate-containing organic matter. As this organic matter sinks through the water column, it is consumed by organisms and broken down into its inorganic constituents. Biologically required inorganic nutrients are found in the lowest concentrations in surface waters because phytoplankton use up the nutrients as they grow (fig. 14.12). Inorganic nutrients therefore display the opposite depth profile of sunlight, which has important implications for the distribution of phytoplankton species within the water column; the highest concentrations of nutrients are commonly found deeper in the water column where light levels are low. Nutrients enter surface waters through two main mechanisms: (1) nutrient run-off from land and (2) upwelling of nutrient-rich deep waters. Nitrogen gas (N_2), like CO_2, diffuses into seawater from the atmosphere, but as described earlier, relatively few species of bacteria can use N_2 as a nitrogen source, although use of this compound has important implications for the global nitrogen cycle. Geographically, the abundance of phytoplankton is greatest where the supply of nutrients is the highest, which occurs in coastal waters or in upwelling regions such as off the coast of Peru. It is important to note that the supply of a nutrient to surface waters can be high due to a process such as upwelling, but the amount of nutrient measured in those same waters can be low because of the rapid uptake of the nutrient by phytoplankton. If the concentration of a required nutrient falls below the minimum required concentration, phytoplankton growth slows, and some species will not be able to grow at all even if other nutrients are present at sufficient concentrations to support growth. A number of pioneering studies carried out over the last decade have shown that vast regions of the ocean (the Southern Ocean, Sub-Arctic Pacific, and Equatorial Pacific) have iron levels that are so low that the growth of phytoplankton in these regions is limited, even though the macro-nutrients nitrate and phosphate are in sufficient concentrations to support growth. When iron is added to surface waters in these areas, a bloom of phytoplankton occurs. These iron-limited areas are known as High Nitrate Low Chlorophyll (HNLC) regions. They will be discussed in more detail in chapter 15.

Oxygen

Oxygen enters the ocean in two ways: as a by-product of photosynthesis by marine photoautotrophs and through equilibration of surface waters with the atmosphere. Oxygen concentrations are highest in surface waters where they can reach saturating concentrations of 12 mg/l. Oxygen is removed from seawater predominantly through respiration, which is carried out by all living organisms. **Respiration** refers to the oxidation of organic matter to CO_2 to derive energy; oxygen is the most commonly used oxidant (fig. 14.4). The biologically mediated profiles of O_2 and CO_2 display opposite depth distributions: photosynthesis generates O_2 and removes CO_2; respiration removes O_2 and generates CO_2. Microorganisms and larvae with bodies less than a few millimeters in size depend on diffusion for oxygen uptake and CO_2 removal. Larger organisms such as worms, shellfish, and fish cannot rely solely on diffusion and instead use gills to take up oxygen from the seawater and circulating blood to exchange O_2 and CO_2 between the gills and other tissues. Marine mammals use lungs to take up oxygen from the air above the sea surface. Active species like fish and mammals require more oxygen than sessile species like sponges and barnacles.

Oxygen concentrations are highest in surface waters and decrease with depth. Oxygen is replenished in deeper waters through vertical circulation. The rate of removal of oxygen slows to a minimum at the oxygen minimum zone (about 800 m depth) because of the decreasing amounts of organic carbon available from the photic zone; below the oxygen minimum zone, oxygen concentration increases because of the vertical circulation of oxygenated waters. If the circulation of deep water is sluggish relative to the rate of removal of oxygen by respiration, oxygen levels in deeper waters may decrease to concentrations that are harmful to organisms. Respiration rates in deep waters may be enhanced when high concentrations of nutrients are introduced into stratified bodies of waters, as is currently being seen in many estuaries of the U.S. east coast and Gulf coast (review section 13.2). When nutrients are added to surface waters, more organic matter is

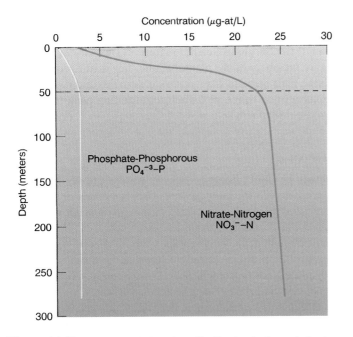

Figure 14.12 Nitrate and phosphate distribution in the main basin of Puget Sound in the late summer. The low surface values are the result of nutrient utilization by phytoplankton. The depth of the euphotic zone is indicated by the *dashed line*.

generated via photosynthesis. As the additional organic matter rains down to the sluggishly moving deeper waters, respiration can remove oxygen from the waters faster than it can be replenished via circulation. Fish require at least 5 mg/l of oxygen and below about 3 mg/L, many marine organisms cannot survive. **Hypoxia** occurs when oxygen concentrations fall below about 2 mg/L. Fish kills can result when bottom waters become hypoxic. **Anoxia** occurs when oxygen concentrations reach 0 mg/mL. Relatively few ocean environments are anoxic. Only **anaerobes** can survive anoxic conditions because they do not rely on oxygen to oxidize organic matter for energy. They instead use other oxidizing agents such as NO_3^- or SO_4^- that are present in the water.

Salinity

The salinity of surface waters from different regions of the ocean differs depending on climate, proximity to freshwater sources, season, temperature, and circulation patterns. The most dramatic differences occur near estuaries. The salinity of deep waters is relatively constant. Organisms must maintain balanced internal salt levels. To maintain a salt content that differs from surrounding waters, an organism actively removes or adds salt. The membranes that surround all cells are permeable only to certain molecules. **Osmosis** is a special type of diffusion in which water moves across a cell membrane from areas of low salinity (high water concentration relative to salt concentration) to areas of high salinity (low water concentration relative to salt concentration).

Different organisms maintain different internal salt concentrations. The body fluids of many bottom-dwelling creatures, such as sea cucumbers and sponges, contain the same salt content as seawater (fig. 14.13). There is no concentration gradient, so the water moves equally in both directions across cell membranes such that the salt content remains the same on both sides. In contrast, the internal salt concentration of most fish is lower than that of seawater, and marine fish tend to lose water from their tissues. Fish expend considerable energy to prevent dehydration, which would increase their salt content. Fish maintain a fluid balance by almost continuously drinking seawater and excreting salt. The outer skin of fishes is not entirely permeable to seawater, so the salt is excreted at the gills (fig. 14.13). Sharks and rays maintain high concentrations of urea in their tissues. The urea acts as a salt and eliminates the osmotic gradient and thus prevents water from leaving cells.

Species distributions are limited by salinity conditions. Organisms adapted to life in fresh waters are rarely able to live in ocean waters, and vice versa, organisms adapted to life in the ocean are rarely adapted to life in fresh water. One slight exception to this rule is that some fish and crustaceans use the low-salinity coastal bays and estuaries as breeding-grounds and nursery areas for their young. The adults then migrate further offshore into higher-salinity waters. Salmon and eels stand in stark contrast to most other animals because they can live in both fresh water and saltwater, depending on their life cycle stage. Salmon are **anadromous.** They spawn in fresh water, and after one to two years (depending on the species of

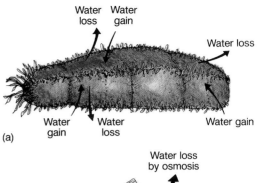

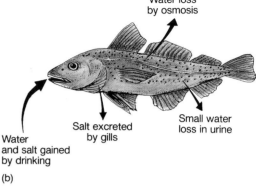

Figure 14.13 (a) The salt concentration of the seawater is the same as the salt concentration of the sea cucumber's body fluids (35‰). The water diffusing out of the sea cucumber is balanced by the water diffusing into it. (b) The salt concentration in the tissues of the fish is much lower (18‰) than that of the seawater (35‰). To balance the water lost by osmosis, the fish drinks salt water, from which the salt is removed and excreted.

salmon), the juvenile fish migrate down rivers to mature as adults in the open ocean. After several years, the adults return to their home streams to spawn, and the cycle begins again. The American and European eels are **catadromous** and display the opposite migratory pattern. They spawn in the open ocean of the Sargasso Sea but mature in fresh water. Juvenile eels are members of the plankton, and they drift north and east with the Gulf Stream. They return to estuaries and rivers where they live for up to ten years before migrating back to the Sargasso Sea to spawn.

Buoyancy

Whether an object floats or sinks in water depends on the difference between the buoyant force, which pushes an object up, and the gravitational force, which pushes an object down. A submerged object always displaces water upwards. If an object weighs more than the weight of the water displaced, then the object will sink. If an object weighs less that the weight of the water displaced, the object will float. Consequently, an object sinks if its density is greater than the density of water.

Organisms regulate their buoyancy through a variety of methods. Most plankton, despite their small size, are denser than seawater. Many species of plankton store oil droplets to decrease their density. They also possess various appendages such as spines and "feathers" that increase their surface area to volume ratio, further enhancing viscous forces and slowing sinking.

Larger organisms can enhance their buoyancy using trapped gases. For example, the Portuguese-man-of-war fills a bell with air, and many seaweeds use trapped gas in their fronds. Many fish maintain neutral buoyancy by using an internal swim bladder filled with gas. Some species fill their bladders by gulping air at the surface; others release gas from their blood through a gas gland to the swim bladder. When a fish changes depth, it adjusts the gas pressure in its swim bladder to compensate for pressure changes. You may have seen the effects of bringing a deep swimming fish to the surface too quickly: the bulging eyes and distended body result from the uncontrolled expansion of the swim bladder. Whales and seals decrease their density and increase flotation by storing large quantities of blubber. Sharks and other fish store oil in their livers and muscles; the giant squid excludes denser ions from its cells and replaces them with less dense ions. Seabirds float by storing fat, possessing light bones, and using air sacs developed for flight. Their feathers are waterproofed by an oily secretion called preen, which acts to seal air between the feathers and the skin. This helps to keep the birds warm and afloat.

Temperature

The hottest temperatures in the ocean occur at the black smokers of deep-sea hydrothermal vents (review the discussion of hydrothermal vents in chapter 3). The fluid exiting the vents is ~350°C (660°F). No organisms survive in 350°C water, but because the water that surrounds the vents is only about 2°C, this extraordinarily hot water is quickly cooled. Cool waters percolate through cracks in the sea floor near the vents, and the subsea floor is cooled to temperatures that, although still very hot, are also more amenable to life. Members of the domain Eukarya have the lowest temperature tolerance; the highest temperature that a eukaryote has been shown to survive is 60–62°C (140°F). Members of the domains Archaea and Bacteria, known as **hyperthermophiles,** can withstand temperatures greater than 90°C (194°F). The highest temperature for life has been reported for an archaea. This prokaryote, known as *Geobacter* strain 121, was isolated from near a hydrothermal vent and has been shown to survive in waters that are 121°C (250°F!) under pressures of 200 atm, which is the pressure of the deep-sea environment from which the prokaryote originated; only by maintaining the water under pressure is boiling prevented. Hypothermophiles are able to prevent their membranes and DNA from melting and their proteins from being inactivated at these very high temperatures. Think what happens when you boil an egg—the proteins unfold and the egg white and yolk solidify. The proteins of hypothermophiles are not inactivated at these high temperatures, and many biotechnologists are interested in understanding more about the molecular basis for this resistance to inactivation.

The temperature of 90% of the ocean is remarkably constant and varies from about −1°C to 4°C (30°F to 40°F). Surface waters are more variable and can range from about −1°C in Arctic and Antarctic waters to over 30°C (86°F) in the surface waters of the tropics and may reach even higher temperatures in shallow tidal pools. The temperature of surface waters is greatly

influenced by season and latitude. Annual changes in open-ocean surface temperatures are small at high and low latitudes, whereas the changes are much larger in the mid-latitudes. Temperature changes in surface waters are more extreme near the coast because this water is shallower and is influenced by temperature changes on land. Northerly winds that blow along the coast in the Northern Hemisphere and southerly winds in the Southern Hemisphere will cause deep, cold water to be upwelled and can make for cold surface waters near the coasts even on a warm, sunny day (review chapter 7).

About 20% of Earth's surface is frozen, including the sea ice in the Arctic and Southern Oceans. The lowest temperatures found anywhere in the oceans occur in the brine channels of winter sea ice. Brine channels are created when salts are expelled into the water as the ice crystals grow during sea ice formation. The temperature of these brine channels can drop to an amazing −35°C (−31°F) because the salinities in these channels can reach levels about six times higher than seawater. Cold-loving organisms that live in very cold environments are known as **psychrophiles,** and they have developed specialized attributes to allow them to exist in these environments. Many psychrophilic microorganisms possess **cryoprotectants,** such as dimethylsulphoniopropionate (DMSP), that lower the freezing point of their internal fluids. Many also possess ice-binding proteins that likely enhance cell survival during the cycling between winter freezes and summer thaws. These organisms also use different fatty acids to maintain flexible membranes even under the very cold and salty conditions of this environment.

Invertebrates and most fishes are **poikilotherms.** They possess no means of metabolically regulating their body temperature, and instead they rely solely upon heat conductance. Their internal temperature responds to the temperature of surrounding waters, and hence, their physiology is regulated by water temperature. Metabolic processes tend to proceed more rapidly when poikilotherms are in warmer water rather than in colder water. Cold-water species commonly grow more slowly, live longer, and attain a larger size. Fish that live in polar regions must prevent their blood from freezing. Without special adaptations, fish blood would freeze at about −0.8°C (31°F), which means that their blood would freeze in the approximately −2°C (28°F) waters of the poles. Antarctic fish possess anti-freeze proteins that lower the freezing point of their blood and allow them to survive. For some poikilotherms, a change in water temperature triggers spawning or dormancy. The geographical range of poikilotherms is largely restricted by water temperatures. This becomes apparent during major climate shifts as occurs during El Niño years (chapter 7). During the El Niño of 1983, surface water temperatures were significantly higher than usual in the eastern Pacific, and warm waters were found further north than usual. Species of fish normally considered warm-water fishes were found in Alaskan waters.

Seabirds and mammals are **homeotherms.** They can maintain a nearly constant body temperature, often well above the temperature of the surrounding seawater. For example, Weddell seals live in the sea ice around Antarctica, yet their body temperature is a balmy 36°C (remember, our body temperature is 37°C, or 98.6°F) despite the fact that they live in an environment where the air and water temperature is often far below 0°C. They use a thick layer of blubber to reduce heat loss. Emperor penguins use a kind of pouch to incubate their eggs; the temperature difference between the inside and outside of the pouch can be as much as 80°C (176°F!). Since the physiological capabilities of homeotherms are less regulated by water temperature, they often have wider geographical ranges. For example, whales annually migrate between polar and tropical waters. Some fish, notably tunas and lamnid sharks (such as the great white shark), are **endotherms.** They can maintain a higher temperature than the surrounding seawater but do not have the same level of temperature control as homeotherms. Endothermic fish use specialized arrays of blood vessels to prevent heat generated in muscles from being lost when blood circulates through the gills.

Pressure

The pressure at the sea surface is one atmosphere, which is equivalent to 14.7 pounds of air pressing on each square inch. Pressure increases rapidly with depth in the ocean because water is much denser than air, increasing by one atmosphere every 10 m. Therefore the pressure at a depth of 6000 m (19,685 ft, or ~3.7 mi) is about 600 atm, or the equivalent of almost 4.5 tons of weight pushing against each square inch! However, organisms are adapted to the pressure of their environment and they do not feel it; you probably did not even realize that you had 14.7 lbs of pressure pushing against every square inch of your body. Remember as well that water is not particularly compressible; the volume of water decreases by only about 4% per 1000 atm pressure.

Pressure, like temperature, affects a variety of cellular processes. Any biological process that depends on a change in volume is impacted by pressure, and organisms that live under high pressure possess attributes that uniquely adapt them to this life. Pressure-loving bacteria and archaea are known as **piezophiles.** Any biological process that requires volume expansion, such as the addition of gas to a fish's swim bladder to enhance buoyancy, will be more difficult under pressure. In contrast, processes that result in volume compression are easier under pressure. Both pressure and low temperatures tend to make a cell's membranes more rigid. Membrane functions include processes such as ion regulation, nutrient transport, and nerve impulses in multicellular organisms. Many deep-sea organisms maintain optimal membrane fluidity and function by changing the composition of their membranes.

Deep-living worms, crustaceans, bivalves, and sea cucumbers do not have gas-filled cavities or lungs that would collapse under pressure. Deep-living species of fish use swim bladders filled with oil rather than air to regulate buoyancy. Air-breathing marine mammals and birds can dive to great depths on a single breath of air—sperm whales can dive to depths greater than 2000 m (6561 ft, or 1.2 mi), and emperor penguins can dive to depths of more than 500 m (1640 ft). These deep-diving mammals and birds use high concentrations of a specialized form of the oxygen-binding protein hemoglobin known as myoglobin that releases the oxygen slowly from their blood as necessary.

Circulation

The ocean is in near constant motion, moved and mixed by currents (chapter 9), waves (chapter 10), and tides (chapter 11). Ocean circulation underlies the chemical and physical drivers of ocean biology discussed thus far. Circulation moves food and oxygen, replenishes nutrients, disperses plankton, and scatters the reproductive stages of both swimmers and attached organisms.

Plankton are moved both horizontally and vertically through the ocean by currents. In the open ocean, the distribution of different plankton is determined by large-scale circulation patterns such as the gyres. Vertical motions in the ocean are less than horizontal motions. The vertical movement of surface waters can become density stabilized due either to the interaction of freshwater and ocean waters or to the heating of surface waters. Many plankton can control their vertical distribution within the water column by swimming or by regulating their buoyancy. Some species of zooplankton migrate tens of meters each day, feed at the surface during the night, and hide from visual predators at depth during the day. Some species of phytoplankton, such as dinoflagellates, can also move vertically through the water column. They swim to deeper depths during the night where the higher concentration of nutrients are found and swim to the surface during the day to photosynthesize. Many planktonic juveniles that live in estuaries migrate through the water column according to the tidal cycle. In this way, they can avoid being flushed out of the estuary on an outgoing tide.

Water movement can also be important for concentrating organisms and particles at fronts or internal waves. A particle that is positively buoyant, or an organism that prefers to swim upwards, will be concentrated by downward flowing water at the boundary of two water masses. Neutrally buoyant water particles are not concentrated. Organisms that feed on plankton may take advantage of the ability of fronts to concentrate their food.

14.5 Bottom Environments

Our discussion thus far has focused mainly on the fluid environment of organisms. We now turn our focus to the solid surfaces of the ocean bottom. As described in section 14.1, the sea floor can be divided into different zones (fig. 14.1). The supralittoral zone, or splash zone, is covered by wave spray only during the highest spring tides. The organisms that exist within this zone must cope with extreme changes in their environment: hot versus cold, wet versus dry, pounding surf versus exposure to air, and coverage by salty ocean water versus fresh rain water. The **littoral zone,** or intertidal zone, is covered and uncovered with seawater once or twice a day as the water level changes between high and low tides. Conditions within the littoral zone vary greatly from site to site: changing from rock to sand, crashing waves to gentle surf, from high- to low-amplitude tides. Climatic conditions within the littoral zone vary depending on latitude and season. The **sublittoral zone,** or subtidal zone, is below the low-tide level and extends out over the continental shelf. The **bathyal** and **abyssal zones** are areas of complete darkness, without seasonal changes. The hadal zone lies beneath 6000 m (19,685 ft) and is associated with ocean trenches. Benthic organisms of the deep sea rely on the organic material that rains down from the euphotic zone. The organisms that live in the supralittoral and littoral zones are the most familiar to the average beach-goer.

In addition to the depth of the bottom environment, the base, or **substrate,** on or in which the organisms live is also critical. Rock, mud, sand, and gravel each provide a different type of food, shelter, or place for attachment that attracts different groups of organisms. The substrate of the sea floor is most variable along shallow coastal areas: sandbars, mud flats, rocky points, and stretches of gravel and cobble are frequently found along the same strip of coastline. At increased depths, the substrate is more uniform with a decrease in the particle size of the sediment.

Benthic organisms either live on or in the sea floor. **Epifauna** live on the surface, and **infauna** live within the sea floor. Epifauna are benthic animals and can be divided into two general categories: sessile and mobile. Suspension feeders such as barnacles, mussels, and some worms live attached to hard surfaces and feed by filtering the water for food particles. Because only a limited amount of food can be gained by filtering the water, all suspension feeders need to filter a maximum amount of water, while expending a minimum amount of energy. As noted earlier, all suspension feeders use cilia to bring water and food to them. The epifauna also include organisms that move across hard surfaces, including crabs, snails, and sea stars. These animals commonly "hunt" other animals as food, graze on algae or microbes growing on the sea floor, or scavenge detritus and dead animals that have sunk to the sea floor.

Epiflora are benthic algae. Seaweeds attach to rocks and grow up towards the light to photosynthesize. Seaweeds cannot grow attached to a sandy bottom. Many seaweeds use gas trapped in their fronds to keep the fronds afloat. There are limits to the depth of water that seaweeds can grow in, since they must be attached to the sea floor and must grow to near the surface.

Infauna burrow and live buried within the sea floor. Sediment type determines where infauna can live. For example, a burrowing worm from a mud flat or a shrimp from a sandy beach cannot live on a rocky reef. Many members of the infauna are deposit feeders. The sediment in which they live is a mixture of particulate organic and inorganic matter, dissolved inorganic and organic substances, and microorganisms. Different species of deposit feeders feed in different ways: some ingest the sediments, some have tentacles to gather particles, and some sort out the inorganic particles before ingesting what remains.

Organisms living attached to the sea floor often greatly modify their habitats. These organisms can be thought of as ecosystem engineers because they provide shelter and additional surfaces for attachment for other organisms, and they often affect the movement of sediments. Think of kelp forests and eel grass beds and the different environments they provide for other organisms. Some crabs moving across the sea floor carry sponges and barnacles attached to their shells. The most outstanding example of biological modification of a substrate is the coral reef. Here, the organisms create a specialized environment over the calcareous skeletons of other organisms.

14.6 Close Associations

Symbioses

Organisms interact with each other in a variety of ways. A **symbiosis** is a close ecological relationship between organisms of two different species. There are three different categories of symbioses. In **commensalism,** one partner benefits from the relationship and the other partner is unaffected. For example, barnacles that live on whales benefit from movement through the water, whereas the whale is thought to be unaffected. In **mutualism,** both partners benefit from the relationship. Various crabs carry sea anemones on their shells: the crabs provide the anemone with a place of attachment and scraps of its prey, and the anemone protects the crab with its stinging tentacles. The clownfish also provides protection to anemones, and the anemone's stinging tentacles provide protection for the clownfish (fig. 14.14). In **parasitism,** one partner lives at the expense of the other. Parasites obtain food and shelter by damaging, but not killing, their hosts. Parasitic worms are found in most fish.

Symbioses are very common and an integral part of the normal functioning of ocean ecosystems. Throughout this chapter, we've seen examples of different types of symbioses: the hydrothermal vent worm *Riftia* and its symbiotic chemosynthetic bacteria that provide organic matter, the angler fish and its symbiotic bacteria that provide light. Perhaps the best examples of symbiosis are tropical coral reefs. Coral reefs are formed by the symbiosis between coral polyps, an animal most closely related to sea anemones and jellyfish, and a unicellular alga. The algae photosynthesize and provide the corals with organic matter necessary for growth. The coral provides the algae with protection, access to sunlight, and some nutrients. The giant structures created by corals are themselves home for many organisms that are

Figure 14.14 Clownfish have a mutualistic relationship with the sea anemone.

Biodiversity is defined as the number of species, or the variety of life forms, found in a defined area. The oceans make up more than 90% by volume of the Earth's known biosphere (that part of the environment in which organisms are found). Therfore, understanding the importance of Earth's biodiversity, as well as maintaining this diversity, requires knowledge of those species present in the ocean. About 1.8 million species are currently known to science, with about 215,000 described species of marine animals. Estimates for the total number of species on Earth (both known and unknown) vary from a low of 3.6 million to an astronomical high of 100 million species. The most frequently accepted order-of-magnitude estimate is around 10 million species inhabit Earth. The smaller the organism, the less is known about it. The vast majority of organisms on Earth remain unknown to science.

Census of Marine Life (CoML) is a ten-year international project designed to determine the abundance, diversity, and distribution of marine life. Six years into the project, the Census consists of seventeen projects involving more than 2000 researchers from eighty nations. The Census will create the first comprehensive list of all forms of sea life, the first version of which will be released in 2010 as an online encyclopedia with a webpage for every discovered organism. The Census will also develop maps of the range of different species and their abundance. These organisms form interwoven ecosystems upon which we all depend.

The more heterogeneous and complex the physical environment, the more varied the environments that are available for organisms to colonize and the higher the species diversity. Consider the oceans: coral reefs, coastal areas, estuaries, open-ocean waters, deep sea floor, and hydrothermal vents are distinctly different ecosystems. Nearly 90% of the ocean's water lies below 100 m (330 ft), where it is cold, dark, and one of the most homogeneous environments on Earth. However, we now realize that this deep-water environment is actually extremely patchy in space and time. Nutrient patches caused by sinking phytoplankton blooms, fish and marine mammal carcasses, and other kinds of organic material provide environments for several hundred species of fish, small shrimp-like organisms, and squids. Small differences in the otherwise homogeneous deep-sea environment allow for greater biological diversification than might be expected. Competition for food and space, predation by other species, and natural disturbances all help to control the biodiversity of an area.

Geographical barriers divide the oceans into a series of environments. The arrangement of the continents and oceans combined with latitude, topography, and related climatic zones organizes the oceans into a series of areas with different patterns of circulation and different water properties. Boundaries in this water world exist in both the horizontal and the vertical and include water mass borders, changes in temperature, salinity, light, and density as well as isolating currents. Coral reefs provide a large number of microhabitats and are found only in tropical waters. As latitude decreases from the poles to the tropics, ocean species diversity tends to increase, just as it does on land. The Pacific has a greater number of species than the Atlantic because of its greater abundance of coral reefs. In the Pacific, coral reef diversity increases from all directions toward the Philippines and Indonesia.

Over the past few decades, concern has increased over the loss of species and the importance of conserving Earth's biodiversity. The scientist E. O. Wilson has coined the acronym HIPPO to describe the ways that biodiversity is currently being eroded. The letter H stands for habitat destruction, the letter I for invasive species, the two letter Ps stand for pollution and population growth of human communities, and the letter O stands for overharvesting. Estimates suggest that we are on a path to reducing half of Earth's plant and animal species to extinction or critical endangerment by the end of the century.

Yet to a very large degree, the human species depends on preservation of biological diversity for its own survival. The richer the diversity of life, the greater the opportunity for discovery of valuable drugs and other marine products. It has been said that conserving the variety of life is our insurance policy. In addition, the appreciation of biological diversity keeps us aware of the continuity of life, and through our concern for species, we gain an understanding of our biological heritage. However, our inadequate knowledge of biodiversity and its accurate measurement remain difficult problems. How much diversity is there? How fast is biodiversity being lost or degraded? What actions can slow or prevent an increasing rate of loss?

For centuries, it was thought that humans could not drive any ocean-living species to extinction. The sea was too big and too deep, its inhabitants too numerous, prolific, and widespread. While in the last 200 years only one marine mammal (Stellar sea cow) and four marine shellfish have become extinct, estimates of extinctions among small organisms in coastal areas and on coral reefs range from 100 to over 1000. Think about the possibility of marine extinctions; consider the diversity of the ocean's food webs, and remember that if you lose one component, you may upset an entire ecosystem and the other components suffer also. Think about losing strands in the web of life as you read chapters 15 through 18.

---

also part of symbioses. Reef health is maintained by fish whose nearly constant grazing prevents other algae from colonizing and overgrowing the reef.

14.7 Practical Considerations: Human Impacts on Marine Environments

Humans have brought great changes to the ocean environment, which has serious implications for ocean biology (chapter 13). First, a significant portion of the CO_2 generated from the burning of fossil fuels since the industrial revolution has made its way into the world's oceans, decreasing the pH of surface waters. If this enhanced influx of CO_2 into the oceans continues at the current rate, the ability of organisms, such as corals, to precipitate the calcium carbonate needed to build their reefs will be impacted. Second, an increasing number of people worldwide live within close proximity to the coast. With increased numbers of humans come increased levels of inorganic nutrients that are introduced into coastal waters. Enhanced nutrient input can expand regions of hypoxic waters, which in turn creates changes in ecosystem structure. Third, overfishing of and eventual extinction of numerous large predator species in marine waters are hypothesized to have dramatically changed ocean ecosystems. As top predators are removed from an ecosystem, the next most abundant fish is commonly harvested by humans in a process referred to as "fishing down the food web." Fourth, the average temperature of surface waters is estimated to have increased by ~0.6°C since the industrial revolution. Corals are particularly sensitive to sea-surface temperatures. Fifth, invasive species are being introduced into our coastal waters at increasing rates with subsequent disruptions to ecosystem structure. And finally, the combination of sea-level rise due to the increased water temperature and human activities is causing the loss of coastal wetlands. Wetlands not only host a diversity of organisms, but they also serve as barriers, or "sinks," for storm surges. The impact of the loss of wetlands was dramatically and tragically illustrated by the loss of lives and property associated with Hurricane Katrina in the summer of 2005.

Changes in the marine chemistry and physics of the world's oceans have profound impacts on ocean biology. In the coming chapters, we will look in more detail at the organisms that compose the ocean food web to understand how these organisms interact to form stable ecosystems and how these organisms respond to environmental changes.

Summary

Marine biologists and biological oceanographers study interactions between marine organisms and their environment. The marine environment is subdivided into different zones. The major environments are the pelagic and benthic zones, both of which have numerous subdivisions based on ocean depth.

Life in the oceans is incredibly diverse. The most abundant organisms in the ocean are microscopic in size, although the largest organisms on Earth are also found in the sea. The dominant physical forces that act on very small organisms differ from those that act on larger organisms. This in turn influences the shapes that organisms can assume.

Marine organisms can be categorized based on their carbon and energy sources and/or their relationships to one another. Unrelated organisms can carry out the same function within an environment. Marine organisms are commonly divided into plankton, nekton, and benthos.

Important physical and chemical factors that influence the distribution of marine organisms include light, temperature, pressure, salinity, and inorganic nutrients. The availability of light and inorganic nutrients limits the growth of phytoplankton populations. Depth profiles of light and biologically required inorganic nutrients display opposite patterns. Light levels are highest in surface waters, whereas inorganic nutrients are highest in deep waters. Phytoplankton produce oxygen and generate the organic matter that serves as food for other organisms. Regions of the ocean where the supply of nutrients to surface waters is high will support more phytoplankton, which in turn will support larger numbers of organisms that consume organic matter.

The vast majority of the ocean is relatively cold and dark, with nearly constant salinity and high pressures. The sediments of deeper benthic areas are generally composed of small-grained particles. The coldest temperature in the ocean is found in sea ice. The hottest temperature in the ocean is found at hydrothermal vents. The organisms that live under extreme temperatures or high pressures have developed specialized attributes that allow them to exist in these environments. Many organisms that live in the sea have also developed means of generating their own light in a process known as bioluminescence. Bioluminescence in the ocean is primarily used either to attract food or to avoid being eaten.

Near-shore environments are highly variable. The salinity and temperature of the water will change depending on depth and proximity to freshwater sources. The substrate that makes up near-shore benthic regions can also vary dramatically over relatively small distances. Substrate material determines the types of benthic organisms that can live in any given environment.

The physical and chemical conditions of the ocean are being impacted by humans, which in turn is impacting the organisms that live in these environments. Temperature changes impact the distribution of marine species. Changes in nutrient delivery to coastal waters will influence the abundance of phytoplankton. This, in combination with circulation patterns, can lead to the occurrence of low oxygen conditions. In addition, overfishing or the introduction of invasive species can have ripple effects through an ecosystem.

Key Terms

All key terms from this chapter can be viewed by term or definition when studied as flashcards on this book's website at **www.mhhe.com/sverdrup10e**.

pelagic zone, 344
benthic zone, 344

neritic zone, 344
oceanic zone, 344
biomechanics, 345
diffusive, 345
conductive, 345
viscous, 345
taxonomy, 345

phylogenies, 345
prokaryote, 346
eukaryote, 346
nucleus, 346
mitochondria, 346
chloroplasts, 346
autotrophs, 346

heterotrophs, 346
photoautotrophs, 347
photosynthesis, 347
chemoautotrophs, 347
chemosynthesis, 347
plankton, 347
phytoplankton, 347

Study Questions

1. Describe and compare two different benthic environments.
2. Describe how unrelated organisms can serve the same function in an ecosystem.
3. Compare the depth profiles of oxygen and carbon dioxide. Why do they differ?
4. Both light and inorganic nutrients are required for phytoplankton growth. Explain why they display different depth profiles.
5. Explain the difference between nutrient supply and nutrient concentration and the influence of these two parameters on phytoplankton abundance.
6. What does it mean that autotrophs are at the base of the food web?
7. Describe why small organisms are strongly influenced by viscosity and large organisms are more strongly influenced by inertia.
8. Describe what respiration means. Which organisms respire?
9. Why is a neritic habitat more variable than a deep-sea habitat?
10. Describe how a fish and a sea cucumber control their internal salt concentration.
11. How are commensalisms and mutualism the same? How are they different?
12. Describe how increased human populations in coastal areas can result in hypoxic bottom waters. Under which conditions can this occur?
13. Why does biodiversity in the oceans matter?
14. Describe the biological pump and how it impacts carbon dioxide concentrations in the atmosphere.
15. In what ways does upwelling contribute to increasing the numbers of organisms in surface waters?
16. Describe at least three reasons why a marine organism might produce light.

Rockfish swim in a forest of giant kelp *(Macrocystis pyrifera)* that reaches upward toward the sunlight.

Chapter 15

Production and Life

Chapter Outline

Learning Outcomes

After studying the information in this chapter students should be able to:

1. *Describe* the relationship between autotrophs and heterotrophs,

2. *Explain* the differences between bottom-up and top-down controls on population sizes,

3. *Explain* how phytoplankton biomass varies seasonally in temperate, tropical, and polar waters,

4. *Explain* the relation among abiotic factors (light, nutrient, temperature), phytoplankton productivity, and fisheries, and

5. *Compare* and *contrast* food web dynamics in open ocean versus coastal regions.

The focus of this chapter is photosynthetic autotrophs. These organisms use sunlight and inorganic compounds to generate the organic matter that serves as food for life in the sea. The vast majority of autotrophs in the ocean are phytoplankton, too small to be seen with the naked eye. These microorganisms occupy the base of the food web. Without phytoplankton, the ocean would be a very different place, with few (if any) of the more familiar large marine organisms. Understanding the variation in phytoplankton abundance and the rate at which organic matter is generated by phytoplankton is critical to understanding life in the ocean.

15.1 Primary Production

Gross and Net

Phytoplankton are single-celled microscopic organisms that carry out photosynthesis. The oxygen and organic carbon that phytoplankton generate through photosynthesis are critical for most life in the ocean. The absolute requirements of phytoplankton are relatively simple: sunlight, inorganic nutrients, and carbon dioxide. The required nutrients and carbon dioxide are dissolved in seawater. Growth of phytoplankton can be limited by temperature or by a lack of adequate amounts of either inorganic nutrients or sunlight; carbon dioxide is rarely limiting in the world's oceans.

Phytoplankton use **pigments** to absorb energy from sunlight. Photosynthetic organisms that produce oxygen possess the pigment **chlorophyll *a*,** regardless of whether they live on land or in the sea. Chlorophyll *a* absorbs light primarily in the blue and red regions of the spectrum. In the remarkable series of chemical reactions that occur during photosynthesis, absorbed light energy is converted into chemical energy, which in turn is used to drive chemical reactions such as the splitting of water to produce oxygen and the generation of organic carbon from carbon dioxide. The process of generating organic carbon from carbon dioxide is commonly referred to as **carbon fixation.**

Photosynthesis can be represented by the overall equation

$$6\,CO_2 \;+\; 12\,H_2O \;\xrightarrow[\text{chlorophyll } a]{\text{Solar energy}}\; C_6H_{12}O_6 \;+\; 6\,O_2 \;+\; 6\,H_2O$$

| 6 molecules carbon dioxide | + | 12 molecules water | | 1 molecule sugar | + | 6 molecules oxygen | + | 6 molecules water |

As seen in chapter 14 (see fig. 14.9), different wavelengths of light penetrate to different depths in the water column. Phytoplankton have evolved different types of pigments to allow them to utilize a broader spectrum of light than can be captured by chlorophyll *a* alone. For example, certain phytoplankton, such as diatoms, contain the pigment fucoxanthin (pronounced fu ko zan thin) in addition to chlorophyll *a;* other phytoplankton, such as cyanobacteria, contain the pigment phycoerythrin (pronounced fi ko e rith rin) in addition to chlorophyll *a* (fig. 15.1). Possession of additional pigments expands the spectrum of light that a phytoplankton species is able to absorb and thus extends the depths at which different phytoplankton are able to grow. It should be noted that not all the solar energy absorbed by the pigments in a phytoplankton cell is converted into chemical energy. In other words, the process is not 100% efficient. Some absorbed solar energy is lost as heat, and some is lost as **fluorescence.** Chlorophyll *a* fluorescence from individual phytoplankton cells can be viewed using a fluorescence-detecting microscope. An example of a fluorescent cell

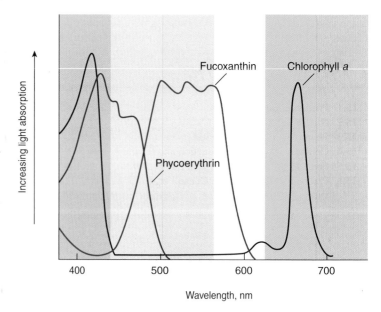

Figure 15.1 Light absorption by three photosynthetic pigments: chlorphyll *a*, fucoxanthin, and phycoerythrin. Note that possession of either phycoerythrin or fucoxanthin in addition to chlorophyll *a* expands the spectrum of light that can be absorbed.

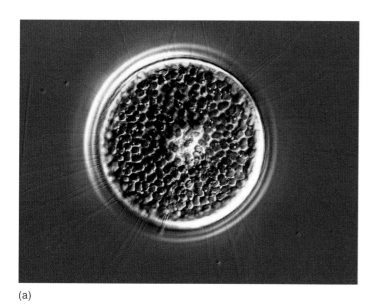

(a)

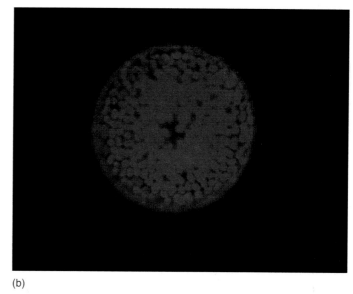

(b)

Figure 15.2 Micrographs of a diatom, a type of phytoplankton. The left image (a) was obtained by illuminating the cell with white light. The right image (b) was obtained by illuminating the cell with high-intensity blue light and the resulting red fluorescence from the chlorophyll *a* was photographed. In these types of cells, chlorophyll *a* is located within small structures known as chloroplasts.

is shown in figure 15.2. In the first image, white light illuminates a phytoplankton cell under high magnification. In the second image (taken at the same magnification), the cell has been illuminated with high-intensity blue light in excess of what can be used for photosynthesis. The resulting chlorophyll *a* fluorescence is detected as red light emitted by the cell.

Production of organic material from inorganic nutrients using light energy is termed **primary production.** The total amount of organic material produced through photosynthesis is **gross primary production.** However, not all the carbon fixed by phytoplankton is available for consumption by other organisms. The phytoplankton consume some of their organic matter through respiration because all organisms respire, regardless of whether they are able to breathe or not. As shown in the equation below, respiration produces chemical energy and generates carbon dioxide from organic carbon. Autotrophs respire the organic carbon obtained from fixation of carbon dioxide. Heterotrophs respire the organic carbon obtained from consumption of organic matter, commonly in the form of whole organisms (review section 14.2). Respiration is represented by the net overall equation

$$1\,C_6H_{12}O_6 + 6\,O_2 \longrightarrow 6\,CO_2 + 6\,H_2O + \text{chemical energy}$$

1	+	6		6	+	6	+	
molecule sugar		molecules oxygen		molecules carbon dioxide		molecules water		chemical energy

Net primary production is the gain in organic matter from photosynthesis by phytoplankton minus the reduction in organic matter due to respiration by phytoplankton. Net primary production reflects the gain in phytoplankton biomass that is available for consumption by heterotrophs. In marine ecosystems, the amount of net primary production determines the amount of food available for large organisms such as fish.

Primary productivity is typically expressed in units of carbon because this is the element that organic substances are based upon. It is also expressed in units of time because productivity is the rate at which phytoplankton biomass is produced due to photosynthesis. It can be expressed relative to area (per m^2) or relative to volume (per m^3). The amount of net primary productivity in coastal waters can be about 10 g C/m^2/day; in surface waters of the open ocean of the North Pacific, for example, net primary productivity is commonly less than 1 g C/m^2/day. Note that, although the terms "primary productivity" and "primary production" seem similar, they are not the same thing. "Primary productivity" refers to a rate and "primary production" refers to an amount.

Measuring Primary Productivity

Review the overall reaction of photosynthesis given in the previous section, and notice that the amount of oxygen produced is directly related to the amount of carbon dioxide taken up. Primary productivity can therefore be determined by measuring either the rate of uptake of carbon dioxide or the rate of production of oxygen. The most common method for estimating the rate of uptake of carbon dioxide relies upon measuring the uptake of a radioactive isotope, carbon-14. The carbon-14 acts as a tracer for carbon dioxide uptake. To measure primary productivity at a given site, water samples are commonly taken from different depths. Each water sample is then divided into multiple sub-samples. One set of triplicate samples (three replicate samples are used to determine the reproducibility of the analyses) are placed in clear bottles, or "light bottles," and a second set of triplicates are placed in light-proof, or "dark bottles." Carbon-14 is added to all bottles. The light and dark bottles are typically incubated with the carbon-14 for twenty-four hours by attaching them to a line suspended from a float or by incubating them in a flowing water incubator onboard a ship. When suspended from a float, the bottles are hung at the same depth they were sampled from; when incubated aboard ship, the light bottles are partially covered with screening to reduce the incident sunlight to levels similar to those that the samples were taken from. The incubation is ended by filtering the phytoplankton from the light and dark bottles. The amount of radioactivity present in the phytoplankton cells trapped on the filter is then measured. The dark bottles serve as a control for any background uptake of the radioactive label. Total net organic carbon production over time is calculated from the difference between the amount of carbon-14 taken up by the light and dark bottles from each depth after twenty-four hours.

These data can be used to compare primary productivity in different ecosystems and at different depths in the water column. An example of sample incubations and the resulting depth profile of primary productivity is illustrated in figure 15.3. Note that, in this example, net primary productivity is lower in surface waters than in midwaters. This suggests that some factor, such as low nutrient concentrations, limits primary productivity in the surface waters. The sub-surface maximum in primary productivity reflects a balance between sufficient sunlight and sufficient nutrients for the phytoplankton populations present in the water column. Note, also, that the carbon-14 method measures net primary productivity. Because phytoplankton respire, gross primary productivity is always higher than what is measured as net primary productivity (fig. 15.3). Deep in the water column, where light levels are low, net primary productivity is zero. The depth at which gross primary productivity and respiration rates are equal is the compensation depth (see also section 6.3). At the compensation depth, net primary productivity is zero.

A similar approach can be used to measure net primary productivity based on the rate of oxygen production, but without the requirement for use of radioactive compounds. Nonradioactive methods are commonly not as sensitive as radioactive methods but are more accessible to student researchers. Production of oxygen in sample bottles can be determined using a technique that relies on the addition of compounds that change color depending on oxygen concentration. These reagents are added to the bottles at the end of the twenty-four-hour incubation. Oxygen production is calculated as the difference between the oxygen concentration in light and dark bottles before and after the twenty-four-hour incubation. Careful technique is required to prevent oxygen from being accidentally introduced into a bottle during sample preparation.

The amount of carbon dioxide taken up and the amount of oxygen produced during photosynthesis reflect a chemical balance between starting material (carbon dioxide)

and product (oxygen). A similar relationship exists between the amounts of nitrogen and phosphorus removed from the water and the amount of organic matter produced. When all nutrients required by phytoplankton are abundant and the phytoplankton are growing optimally, the fixed ratios of the different elements by weight are

$$O_2 : C : N : P = 109 : 41 : 7.2 : 1$$

These ratios can be used to estimate primary productivity by measuring either the rate of nutrient uptake by phytoplankton or the rate of oxygen production. For example, if 10 mg of dissolved phosphorus (P) is removed by phytoplankton from a volume of water in a given amount of time, then 410 mg of carbon (C) as phytoplankton biomass will be produced. If the rate of delivery of nitrogen (N) or phosphorus into a region by upwelling and currents and the rate at which these compounds are removed by other currents are known, then the rate at which the nutrient is incorporated into phytoplankton biomass can be calculated and primary productivity can be determined. This approach allows oceanographers to estimate primary productivity

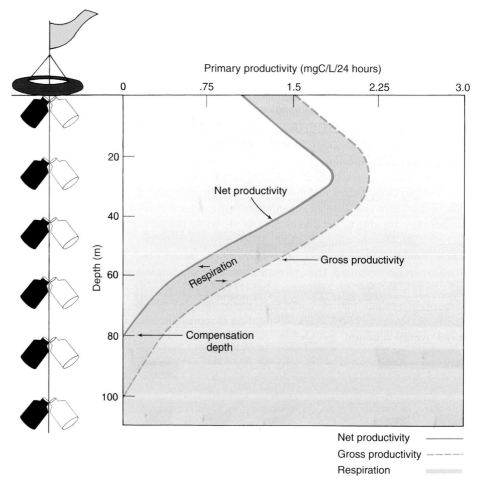

Figure 15.3 Results from a carbon-14 incubation experiment designed to determine net primary productivity with depth. In this example, productivity is measured in milligrams of carbon per liter of seawater per twenty-four-hour incubation. The compensation depth is where net primary productivity is zero. In this example, light, nutrients, and stability of the water column combine to provide conditions for the highest primary productivity at 30 m (100 ft).

over large areas of the ocean and to associate it with large-scale water movement and chemical cycles.

15.2 Nutrient Cycles

Inorganic nutrients are removed from seawater by phytoplankton in a ratio that reflects their biological demands. Under optimal growth conditions, the molar ratio (in contrast to the ratio based on molecular weight described above) of C:N:P in organisms is about 106:16:1. When organisms die and decompose, the nutrients are released back to the seawater in a similar ratio. This ratio is known as the **Redfield ratio** after the researcher who first reported the observation. The process of breaking down organic matter into inorganic constituents is referred to as **nutrient regeneration,** or **remineralization.** No significant amount of new matter comes to Earth from space. Therefore, inorganic nutrients must always be recycled to generate organic matter. The recycling of nutrients occurs to different extents in different ecosystems. The cycling of nutrients between inorganic and organic states is accomplished by complex interactions between different types of organisms, including microorganisms. These cycles of production and consumption form the basis for food web interactions discussed in section 15.5.

To understand how nutrients are recycled, we focus here on nitrogen and phosphorus. Inorganic nitrogen is dissolved in seawater primarily in three different forms: nitrate (NO_3), ammonium (NH_4^+), and nitrogen gas (N_2). Microorganisms carry out the conversions between these different forms of nitrogen. The most oxidized form of nitrogen is nitrate, and this is the form of inorganic nitrogen that is most abundant in deep waters. Inorganic phosphorus is dissolved primarily as PO_4^{3-} and it is also more abundant in deep than shallow waters. All living organisms require nitrogen and phosphorus to create essential compounds such as amino acids, proteins, and nucleic acids.

Nitrogen and phosphorus are continuously cycled between inorganic and organic states. Here, we enter the cycling loop with the phytoplankton because they remove inorganic nutrients from seawater during growth. Nitrate and ammonium are the two inorganic forms of nitrogen used by most phytoplankton. A relatively small number of cyanobacteria (prokaryotic phytoplankton) can use nitrogen gas. These cyanobacteria use nitrogen gas to generate nitrogen-containing organic matter in a process commonly referred to as **nitrogen fixation.** Although the nitrogen gas dissolved in seawater is plentiful, it is energetically costly for this subset of cyanobacteria to use it as a nitrogen source. Nitrogen gas is only used when other forms of inorganic nitrogen are below required concentrations. Through photosynthesis, inorganic nutrients are converted into organic matter that comprises the phytoplankton biomass. This biomass can be consumed by zooplankton, which in turn are consumed by other organisms. After each round of consumption, some of the nitrogen, phosphorus, and carbon is lost as waste. As different organisms consume the organic matter, some release regenerated inorganic nitrogen as nitrate and some release it as ammonium. Ultimately, as the organic matter generated by the phytoplankton either sinks or is taken deeper through the water column, it is converted into the

inorganic constituents through the activity of microorganisms. Deep in the water column, little digestible organic matter is left, and the waters contain high concentrations of nitrate and phosphate. When circulation brings this deep, nutrient-rich water back to the surface, the cycle begins again as the phytoplankton convert the inorganic nutrients into organic matter. Large fisheries are associated with upwelling regions (chapter 8) because nutrient-rich waters are upwelled to the surface and fuel the growth of phytoplankton, which in turn support the organisms that serve as food for other organisms.

15.3 Phytoplankton Biomass

The total biomass of the phytoplankton community at any instant in time is referred to as the phytoplankton **standing stock,** or **standing crop.** The standing stock is a product of two main processes. Growth and reproduction increase phytoplankton numbers and are dependent upon net primary productivity. The higher the rates of net primary production, the greater the amount of biomass generated through photosynthesis during a given period of time. Death and grazing of phytoplankton reduce phytoplankton numbers. The balance between production and consumption of phytoplankton biomass determines the standing stock.

There are a number of techniques available to determine phytoplankton biomass in a given water sample. One option is to count all the phytoplankton cells in a sample and multiply the number counted by the average amount of carbon per individual cell. Automated methods make this option feasible for some of the smaller phytoplankton. Another option relies on the fact that all phytoplankton possess chlorophyll *a*. Therefore, the concentration of chlorophyll *a* in a water sample is an estimate of the amount of photosynthetic biomass in the water sample. Note that chlorophyll *a* concentrations do not estimate the number of phytoplankton present in a water sample because different species of phytoplankton possess different amounts of chlorophyll *a* per cell. For example, species characterized by larger cells almost always have more chlorophyll *a* per cell than species characterized by smaller cells. Chlorophyll *a* concentrations at different depths in the water column are commonly measured using two different approaches. The most direct, but also the most labor-intensive, way is to filter the phytoplankton cells from a volume of water, extract the chlorophyll *a* from all the cells trapped on the filter, and determine the concentration of chlorophyll *a* in the extract. An indirect, but more rapid method relies on the fact that, when relatively strong intensities of blue light are absorbed by chlorophyll *a*, some absorbed light energy is released as fluorescence (see fig. 15.2). Higher concentrations of chlorophyll *a* in a body of water result in higher levels of chlorophyll *a* fluorescence. In addition to detecting chlorophyll *a* fluorescence with a microscope, fluorescence can also be measured electronically using an instrument known as a fluorometer. Fluorometers can be lowered over the side of a ship, or they can be mounted on robotic devices such as a Seasoar (see chapter 8). Both methods for measuring chlorophyll *a* provide estimates of phytoplankton biomass with depth through the water column. Depth profiles of chlorophyll *a* concentrations can be conducted at a relatively

Figure 15.4   Maps of seasonal distributions of chlorophyll *a* concentrations on land and in the oceans based on SeaWiFS satellite measurements of global color. In the ocean, *dark blue* indicates low concentration of chlorophyll *a*. Other colors (*light blue* to *green* to *yellow* to *orange*) indicate increasing concentrations. On land, *yellow* indicates low concentrations of chlorophyll *a* and *green* indicates high concentrations of chlorophyll *a*. Conditions are shown for April (a), July (b), October (c) in 1998, and January (d) in 1999.

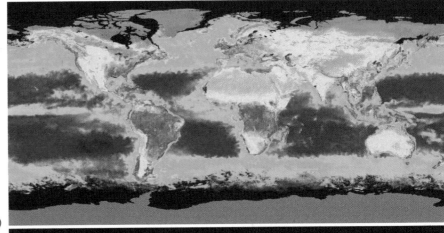

(a)

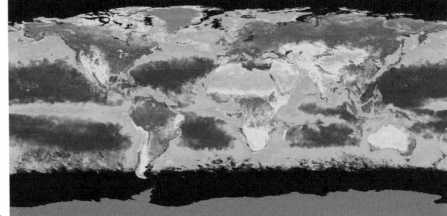

(b)

limited number of sites dependent on the duration of an oceanographic research cruise or on the spatial range of robotic sensors.

Satellites are used to determine the global distribution of chlorophyll *a* concentrations in surface waters. Satellites use a different method than fluorometers to estimate chlorophyll *a* concentrations. Satellites detect what is known as "water leaving radiance." To understand how satellites work, think about why leaves on a tree look green. Leaves possess chlorophyll *a*, and as stated before, chlorophyll *a* absorbs light primarily in the blue and red regions of the light spectrum. This means that we see leaves as green because these wavelengths of light are not absorbed by leaves and are instead reflected back to our eyes. Satellites act as our eyes and detect the wavelengths of light that leave surface waters. In other words, satellites detect **ocean color.** They measure ocean color for near-surface waters. SeaWiFS (Sea-viewing Wide Field-of-view Sensor) and other satellites can view every square kilometer of cloud-free ocean surface every forty-eight hours to provide incredibly detailed views of ocean color (fig. 15.4). The darker blue regions of the oceans (the cool colors) in these SeaWiFS images denote regions of low chlorophyll *a* concentrations and are related to features such as mid-ocean gyres (chapter 9). These are vast regions of the ocean where little upwelling occurs and thus nutrient concentrations, and resulting phytoplankton concentrations, are low. Higher concentrations of chlorophyll *a* in the oceans (indicated by the warmer colors, with the highest concentrations indicated by the reddest colors) are associated with the higher nutrient concentrations found close to land or at upwelling regions. Note that the colors used in these images are referred to as "false colors." These are not the true colors detected by the satellite but are simply a

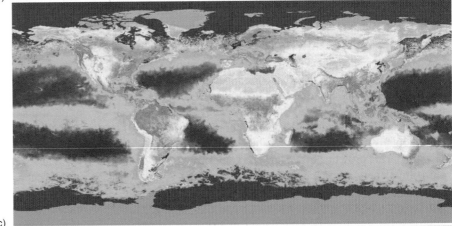

(c)

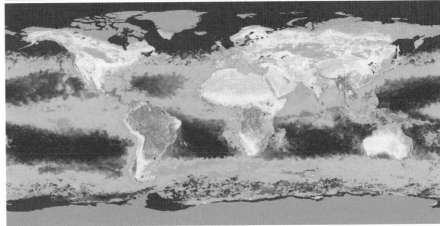

(d)

convention commonly used to denote high (warm colors for the ocean) or low (cool colors for the oceans) concentrations of chlorophyll *a*.

Blooms of phytoplankton occur when phytoplankton biomass increases more rapidly than it is consumed by grazers, primarily the zooplankton. Blooms commonly occur when growth conditions for the phytoplankton are favorable (sufficient nutrients and sunlight, for example), but zooplankton numbers are still low. Blooms can be detected as a rapid increase in chlorophyll *a* concentrations at a given site or a given depth using the methods described above. To understand blooms, imagine two hypothetical sites. At one site, the standing stock increases rapidly because grazing is at a minimum (fig. 15.5*a*). At the other site, the phytoplankton standing stock remains relatively low because grazers are able to consume the phytoplankton as quickly as new phytoplankton biomass is produced. At this site, chlorophyll *a* concentrations remain relatively low and constant over time (fig. 15.5*b*). At both sites, there is net primary production—new biomass is produced—but in one case, the standing stock or chlorophyll *a* concentration is low (fig. 15.5*a*), and in the other case the standing stock is high (fig. 15.5*b*).

The relation between primary production rates and phytoplankton biomass can be examined by dividing the primary productivity values by the biomass values. Biomass normalized primary production rates are given in units of mg C/mg phytoplankton biomass/day. Consider again the two sampling sites just described. Both have similar rates of primary production, but at one site the standing stock is high because the impacts of grazing are limited (fig. 15.5*a*). At the other site, the standing stock is low because grazing levels are high (fig. 15.5*b*). By normalizing primary productivity to phytoplankton biomass, the differences between these two sites become clearer (table 15.1). The phytoplankton at the site where biomass is low are actually growing more rapidly and fixing more carbon per unit time than the phytoplankton at the site where biomass is high. Another way to think about these two environments is in terms of the flow of carbon through the phytoplankton in a given amount of time. At the site where the bloom is occurring and phytoplankton biomass is high, most of the newly fixed carbon is present in the phytoplankton biomass, and little of the newly fixed carbon is present in the grazers. At the site where the standing stock is low despite high levels of primary productivity, the newly fixed carbon moves rapidly through the phytoplankton, and more is present in the grazers and other organisms that consume the newly fixed organic matter. The flow of carbon through the phytoplankton is much greater at the low biomass site than at the high biomass site.

(a)

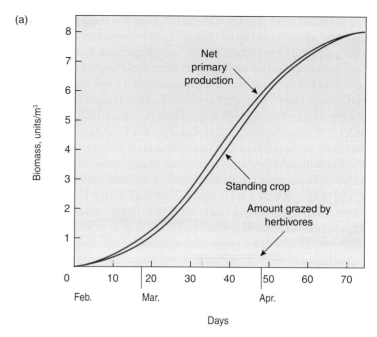

(b)

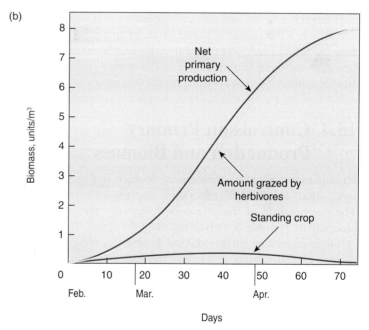

Figure 15.5 Biomass of phytoplankton and zooplankton during the spring. Net primary productivity is superimposed on the biomass plots. In both panels, net primary productivity increases over the seasons. In the top panel (a), grazing is low and standing stock increases. In the bottom panel (b), grazing is high and standing stock remains low.

Table 15.1	Relation Between Primary Productivity and Phytoplankton Biomass at Two Hypothetical Sites		
Site	Primary Productivity (mg C/m³/day)	Phytoplankton Biomass (mg phytoplankton biomass/m³)	Biomass Normalized Primary Productivity (mg C/mg phytoplankton biomass/day)
1	10	1	10
2	10	10	1

The examples described above illustrate that the flow of carbon through biomass can be very different between two sites even though primary productivity levels are comparable. These examples help demonstrate an important large-scale feature that characterizes life in the oceans and distinguishes it from life on land: the flow of carbon through the phytoplankton is more rapid than the flow of carbon through land plants. To understand how this works, let's compare large-scale standing stocks and primary productivity on land and in the oceans. Based on satellite images of chlorophyll *a* distributions across our planet, about 1000 times more carbon is stored in photosynthetic land-plant biomass at any given instant in time than is stored in phytoplankton biomass. In other words, the standing stock of plants is much greater than the standing stock of phytoplankton, yet the annual rate of net primary production in the ocean is comparable to the annual rate of net primary production on land. At the large scale, the situation in the ocean (on average) is analogous to the hypothetical site described previously where the standing stock is low. The situation on land (on average) is analogous to the hypothetical site described previously where the standing stock is high. In general, newly fixed carbon in the ocean moves rapidly from the phytoplankton to the heterotrophic consumers. In contrast, on land, newly fixed carbon is stored longer in land plants, and the flow of carbon from land plants to heterotrophic consumers is slower. A defining feature of life in the oceans is that carbon dioxide quickly becomes organic matter that serves as food.

15.4 Controls on Primary Production and Biomass

Phytoplankton primary productivity is controlled by temperature, light levels, and inorganic nutrient concentrations. The combination of these factors determines the maximum productivity possible in a given region of the ocean. The impact of these factors on phytoplankton biomass is referred to as bottom-up control. These factors determine how quickly phytoplankton grow and reproduce. Grazing by heterotrophic consumers, in addition to temperature, nutrients, and light, determines the maximum amount of biomass possible in a given region of the ocean. The influence of grazing on phytoplankton biomass is known as top-down control. The interactions between bottom-up and top-down controls determine phytoplankton standing stock. Review figure 15.5 with these concepts in mind. To illustrate how nutrients, light, and grazing impact the biomass and productivity of phytoplankton, three regions of the world's oceans will be considered: polar, temperate, and tropical regions.

Summer days are long in polar regions. Relative to lower latitudes, however, the intensity of light in polar regions is low, and the depth of light penetration is shallow. There is no sunlight for a significant portion of the year at the highest latitudes. Most of the Arctic is frozen throughout the winter

and only begins to thaw in the spring. Scientists must be very dedicated to work in the harsh environment of the poles. By summer, the days lengthen and light intensities reach a maximum. Much of the ice and snow melts, and nutrients are released into the waters. During this period of adequate sunlight and nutrients, the phytoplankton bloom and create high biomass. Growth of zooplankton lags behind growth of phytoplankton (fig. 15.6a). In late summer/early fall, temperatures and light intensities decrease rapidly, zooplankton abundance is at a maximum, and sea ice begins to fill the open water. Phytoplankton abundance during this period declines dramatically. A plot of biomass over time indicates that accumulation of phytoplankton biomass is restricted to the summer months (fig. 15.6a). Because the growing season is so short, required nutrients rarely fall below the level necessary for growth. At these latitudes, the availability of light controls phytoplankton growth. Even though the burst of primary productivity associated with the phytoplankton bloom is short-lived, the high amount of organic matter generated during this period supports numerous larger organisms, including seals, whales, and polar bears.

The situation in the subtropics and tropics (fig. 15.6b) is very different from that at the poles. Abundant high-intensity sunlight is available year-round in the tropics, but nutrient concentrations are rarely high enough to support high phytoplankton biomass. The water column in the tropics is density stratified due to surface warming. The phytoplankton remain within the stratified surface waters because they are not easily displaced into the underlying denser water. The same density stratification that keeps the phytoplankton in the surface waters means that the upwelling and mixing processes for renewing nutrient supply to surface waters are weak and localized. This poor nutrient supply to the euphotic zone limits phytoplankton production in the tropics and subtropics despite high levels of sunlight. Any increases in phytoplankton biomass are quickly consumed by the zooplankton. Compare the changes in biomass illustrated for the polar regions in figure 15.6a with the changes illustrated for subtropical regions in figure 15.6b.

The situation in temperate regions is more complicated than in polar regions or the tropics. In temperate regions, the intensity and duration of sunlight, as well as nutrient concentrations, vary with season. During the winter, solar irradiance is low, winter storms are relatively frequent, and the water column is generally well mixed. Temperate surface waters in winter are

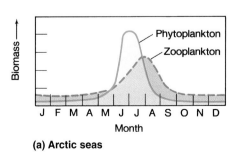

(a) Arctic seas

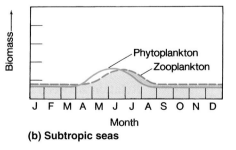

(b) Subtropic seas

Figure 15.6 Biomasses of the zooplankton and phytoplankton are closely coupled in (a) polar latitudes and (b) subtropical latitudes.

characterized by high concentrations of inorganic nutrients and low levels of solar energy. Because phytoplankton require both light and nutrients for growth, phytoplankton biomass is low during the winter months of temperate regions (fig. 15.7). During spring, solar radiation increases and warms the surface waters, which increases the density stability of the water column. Just as described above for tropical waters, density stratification in temperate waters also helps to maintain the phytoplankton in the surface waters. During spring in these waters, there is sufficient nutrients and sufficient sunlight for photosynthesis in the surface waters. Under these conditions, the phytoplankton bloom, and biomass increases dramatically. Phytoplankton bloom during the spring in temperate waters because they are able to grow faster than the zooplankton grazers in the water column can consume them. The downturn in the biomass curve during the later months of spring happens for two reasons. First, the amount of nutrients in the stratified surface waters declines as the phytoplankton use the nutrients for their growth. The density stratification also minimizes renewal of nutrients to the surface waters. Nutrient limitation slows the growth of the phytoplankton. As phytoplankton growth slows, the zooplankton grazers begin to "catch up," and their feeding causes a decline in phytoplankton biomass. Towards the end of summer, solar radiation begins to decline, surface waters cool, density stratification begins to break down, and nutrients are renewed in surface waters. A second, smaller bloom often occurs in the fall. Ultimately the fall bloom declines as winter sets in and light levels decrease. In addition, enhanced winter mixing means that the phytoplankton are mixed deep in the water column where they are unable to obtain sufficient light for photosynthesis.

Given this understanding of the relationship among seasons, nutrient supply, and solar irradiance, compare the conditions in temperate waters (fig. 15.7) with those in tropical waters. In the tropics, relatively low levels of surface mixing and overturn mean that nutrient levels remain relatively low throughout the seasons. Because solar irradiance remains high throughout the year, any increase in nutrients results in small blooms of phytoplankton that are quickly consumed by the zooplankton (fig. 15.8).

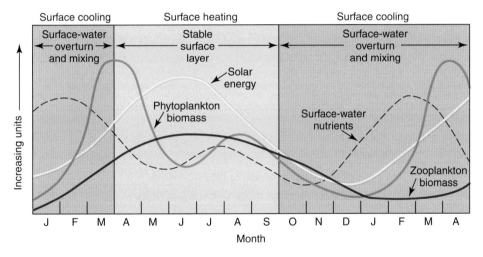

Figure 15.7 Phytoplankton and zooplankton biomass, nutrient supply, solar energy, and water column stability over the year in temperate waters in the Northern Hemisphere.

Global Primary Productivity

Global distributions of annual rates of primary production can be estimated based on primary productivity measurements conducted at discrete sites by scientists over the years and an understanding of the constraints on primary production in different ocean zones (fig. 15.9). When there is sufficient sunlight, regions that receive higher nutrient input display higher levels of net primary productivity. Coastal areas are generally much more productive than the open ocean. Rivers and land runoff supply nutrients to coastal waters and estuaries. Fresh water from rivers can also stabilize the water column by creating a low-density (less saline) surface layer that helps to maintain the phytoplankton in well-illuminated surface waters.

Narrow regions of particularly high productivity are found in the major upwelling zones along the west coasts of North and South America, the west coast of Africa, and the west side of the Indian Ocean (review chapter 9 on upwelling). The upwelled deep water brings nutrient-rich water to the surface, and when sufficient sunlight is available, phytoplankton blooms occur. The abundant populations of phytoplankton in

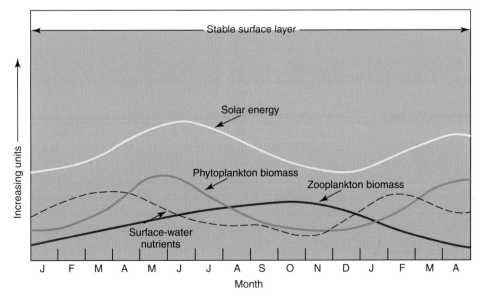

Figure 15.8 Lack of surface mixing and overturn in tropical waters results in relatively low phytoplankton and zooplankton biomass throughout the year.

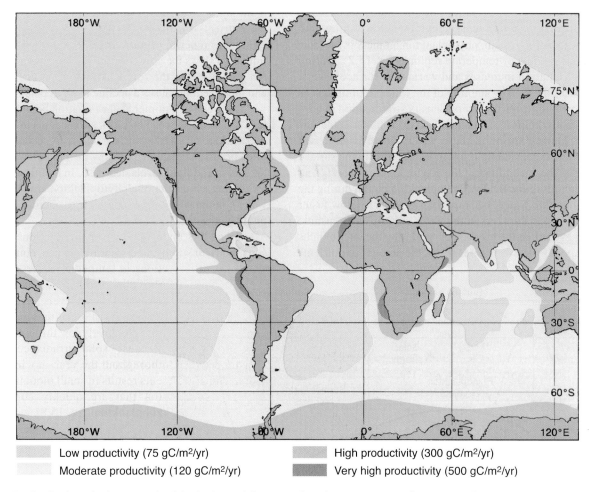

	Low productivity (75 gC/m²/yr)		High productivity (300 gC/m²/yr)
	Moderate productivity (120 gC/m²/yr)		Very high productivity (500 gC/m²/yr)

Figure 15.9　The distribution of primary productivity in the world's oceans, based on many years of ocean sampling.

Table 15.2	**World Ocean Primary Productivity**				
Area	**Mean Primary Productivity (gC/m²/yr)**	**World Ocean Area (km²)**	**(%)**		**Total Primary Productivity (metric tons carbon/yr)**
Upwelling	500	3.62×10^6	1.0		1.8×10^9
Coasts	300	36.2×10^6	10.0		10.8×10^9
Open oceans	130	322×10^6	89.0		40.3×10^9
All ocean areas	150	362×10^6	100.0		54.1×10^9

these regions form a first step in the food webs that result in large numbers of commercially valuable fish.

The Equatorial Pacific demonstrates the influence of open-ocean upwelling on primary productivity. The equatorial divergence is where upwelling occurs, bringing nutrient-rich waters to the surface. Phytoplankton abundance is much higher at the equatorial divergence than at other open-ocean areas that do not experience sustained upwelling. The same holds true for the surface divergence that surrounds Antarctica. In contrast, surface convergence results in downwelling and can create vast areas of low nutrient availability such as the open-ocean gyres. Review figure 15.4 and see that the open-ocean gyres correspond to regions of low chlorophyll *a* biomass, and the Equatorial Pacific corresponds to relatively high levels of chlorophyll *a* biomass.

Now compare the distribution of primary productivity illustrated in figure 15.9 with chlorophyll *a* distributions of figure 15.4 to see that regions with higher phytoplankton biomass tend to be regions with higher primary productivity.

On average, upwelling regions are almost twice as productive as coastal regions and about six times as productive as the open ocean (table 15.2). The total area covered by the open ocean is about ninety times greater than the area covered by upwelling regions and about nine times greater than the area covered by coastal regions (table 15.2). If we combine the two parameters (productivity and size), it becomes clear that most of the world's oceans are characterized by relatively low levels of productivity (table 15.2). Despite the low levels of productivity per unit area, the size of the ocean is so large—the ocean covers about 70% of

Table 15.3	**Gross Primary Productivity of Land and Ocean**				
Ocean Area	**Range (gC/m^2/yr)**	**Average (gC/m^2/yr)**		**Land Area**	**Amount (gC/m^2/yr)**
Open ocean	50–160	130 ± 35		Deserts, grasslands	50
Coastal ocean	100–500	300 ± 40		Forests, common crops, pastures	25–150
Estuaries	200–500	300 ± 100		Rain forests, moist crops, intensive agriculture	150–500
Upwelling zones	300–800	640 ± 150		Sugarcane and sorghum	500–1500
Salt marshes	1000–4000	2471			

Earth's surface—that about half of annual primary productivity on the planet is carried out by the microscopic phytoplankton.

High productivity regions and low productivity regions can be found both on land and in the oceans (table 15.3). For example, primary productivity per square meter in the open ocean is about the same as that of deserts on land. Upwelling regions are comparable to pastureland and lush forestland, and certain estuaries are comparable to the productivity of the most heavily cultivated land.

In the ocean, nitrogen is the nutrient that commonly limits net primary productivity, although as described in section 15.6, iron also limits productivity in some regions. In many lakes on land, phosphorus limits primary productivity. Therefore, nitrogen input into coastal waters and phosphorus input into lakes must be carefully controlled. Addition of a limiting nutrient can cause a bloom of phytoplankton. Review sections 13.2 and 14.3 to see that nutrient enrichment can result in production of high amounts of organic matter. Respiration of the enhanced amount of organic matter can result in dangerously low concentrations of dissolved oxygen in regions where water circulation is minimal.

15.5 Food Webs and the Biological Pump

Food webs describe the flow of nutrients and food between different groups of organisms. The majority of the organisms in any food web are less than a few centimeters in size (table 15.4). Photosynthetic phytoplankton form the base of the food web because they generate organic matter available for consumption by other organisms. A pyramid is the simplest way to visualize the flow of organic matter between organisms (fig. 15.10). Organic matter is generated at the base of the pyramid by the primary producers. A variety of zooplankton directly consume the phytoplankton, and they are referred to as herbivorous zooplankton by analogy to plant-eating animals on land. The zooplankton that consume the phytoplankton are known as primary consumers. The herbivorous zooplankton are in turn grazed upon by carnivorous zooplankton. The zooplankton that feed upon the herbivores are known as secondary consumers. The flow of energy and food between different groups of organisms in the pyramid construction is fairly direct: smaller organisms are consumed by bigger organisms, which in turn are consumed by even bigger organisms. The primary producers, primary consumers, secondary consumers, and so on represent different steps in the transfer of carbon and nutrients. Each step represents a different **trophic level.** Each

Table 15.4	**Relative Abundance and Size of Marine Organisms**	
Organic Form	**Size Range**	**Relative Abundance**
Fish	10–100 cm	0.01
Zooplankton	1 mm–10 cm	1.0
Phytoplankton	0.001–0.02 mm	10.0

transfer between trophic levels results in a loss of organic carbon and energy. Therefore, less total biomass is supported at each trophic level.

The yields of food resources for humans are highest when the marine species harvested are at the lowest trophic level possible. The relationship between phytoplankton production and theoretical fish production is illustrated for three regions of the ocean (table 15.5). The third column in this table represents estimated efficiency of energy transfer between trophic levels and the fourth column illustrates the trophic level at which humans commonly harvest food. Upwelling regions and coastal areas are nutrient-rich and support short, efficient food webs (fig. 15.11a and b). The high rates of primary production in these areas mean that food for herbivores is obtained easily when energy is efficiently transferred to the next trophic level. In contrast, open-ocean food webs are long and characterized by inefficient transfer of energy to higher trophic levels that humans harvest (fig. 15.11c).

A great deal of research over the years indicates that the flow of nutrients and food energy between different groups of organisms is actually more complicated than is depicted by a simple pyramid. All food webs reflect complex interactions between different groups of organisms. But some food webs are more complex than others. For example, the food web that supports herring is complicated because prey-predation levels change as the fish matures (fig. 15.12). As will be described in the following paragraphs, food webs in the open ocean are more like loops, since much of the material is "recycled" between the different groups, with little new input of nutrients into the ecosystem. These types of food webs are dominated by microbial processes and do not support many higher trophic levels. To illustrate different influences on food webs, the world's oceans will be divided into two general categories: high-nutrient input into surface waters as occurs in coastal waters and low-nutrient input into surface waters as occurs in the gyres of the open oceans.

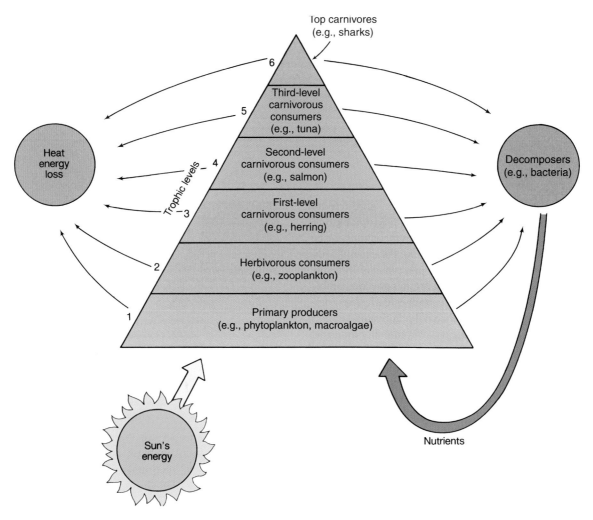

Figure 15.10 A trophic pyramid. Trophic levels are numbered from *base* to *top*. The first trophic level requires nutrients and energy. Nutrients are recycled at each level; energy is lost as heat at each level. Heat is lost from the ocean by radiation to a cooler atmosphere.

Table 15.5	Oceanic Food Production			
Area	Phytoplankton Production (metric tons of carbon/yr)	Efficiency of Mass and Energy Transfer per Trophic Level	Trophic Level Harvested	Estimated Fish Production (metric tons of carbon/yr)
Open ocean	40.3×10^9	10%	5	4.0×10^6
Coastal regions	10.8×10^9	15%	3	243×10^6
Upwelling areas	1.8×10^9	20%	2	360×10^6

Nitrate is the dominant form of nitrogen input into coastal environments and upwelling regions. These regions display some of the highest levels of primary productivity in the sea. The source of nitrate in these environments is either upwelled waters or run-off from land. High-nutrient regions are home to the largest size classes of phytoplankton. In general, coastal and upwelling food webs reflect a traditional trophic pyramid where larger phytoplankton are consumed by larger zooplankton, which are consumed by larger organisms such as juvenile fish. However, even in these high-nutrient regions, there are numerous examples of ways that the steady progression between different trophic levels is short-circuited. For example,

many zooplankton can consume phytoplankton that are many times larger than themselves, clearly breaking the traditional chain of big things eating smaller things. In addition, phytoplankton can excrete organic matter directly into seawater. This dissolved organic matter (DOM) is consumed and respired by the bacteria, again short-circuiting the traditional trophic pyramid. Despite these exceptions to the rule, coastal and upwelling food webs that begin with large phytoplankton are generally referred to as energy-efficient or short food webs because there are few transfers of organic matter from the autotrophic organisms to the largest heterotrophic organisms (table 15.5). Another way to say this is that short,

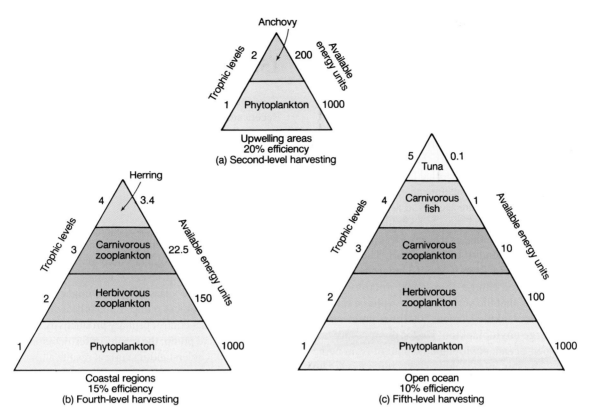

Figure 15.11 Trophic level efficiency varies among (a) upwelling areas, (b) coastal regions, and (c) the open ocean. The number of trophic levels and the level at which humans harvest differ with location.

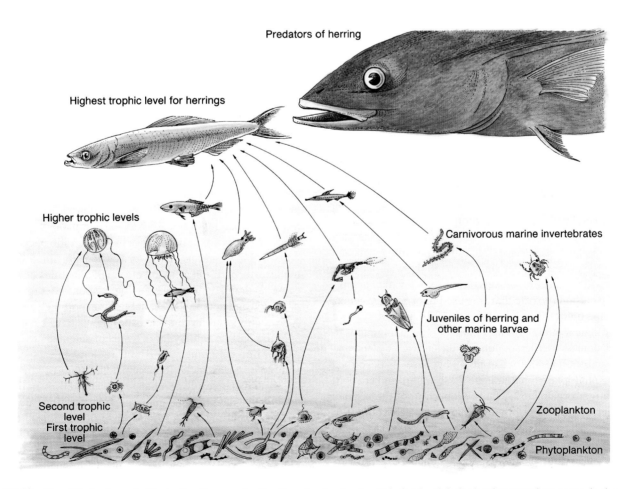

Figure 15.12 Simplified schematic illustrating the complexity of marine food webs. This food web is for herring at various stages in the herring's life.

energy-efficient food webs have few trophic transfers. The world's largest fisheries are found in nutrient-rich regions with energy-efficient food webs.

The vast majority of the world's oceans, such as the open-ocean gyres, are low-nutrient regions with low rates of primary productivity. The dominant form of nitrogen in these areas is ammonium because little nitrate enters the water through up-welling. Most of the ammonium is recycled between organisms. When zooplankton, for example, consume phytoplankton, some of the break-down products are released as ammonium. Ammonium excreted by zooplankton can be re-used by phyto-plankton as a source of inorganic nitrogen. Phytoplankton in these low-nutrient regions tend to be very small. The size of a typical open-ocean phytoplankton is about 1 micron. The size of a coastal phytoplankton can be almost 500 microns. Refer back to table 14.1 to see that the surface area to volume ratio of this typical small phytoplankton is about 500 times greater than that of the typical large phytoplankton. Under the low-nutrient conditions of the open-ocean, phytoplankton benefit from being small because their greater surface area allows them to take up more nutrients per cell volume. Smaller phytoplankton tend to be consumed by smaller zooplankton, and these smaller zooplankton in turn tend to be consumed by only slightly larger zooplankton. Numerous trophic transfers, with a loss of carbon and nutrients at each step, are required in open oceans to gen-erate large heterotrophic organisms (table 15.5). Much of the or-ganic and inorganic matter is instead recycled between the different small organisms. The open ocean environment is char-acterized by a tight loop of nutrient regeneration and nutrient up-take dominated by microbes. This type of food web is referred to as a **microbial loop.** Low productivity, multiple transfer food webs do not support large-scale fisheries.

The "biological pump" refers to the composite of food web processes that result in a draw-down, or transfer, of carbon dioxide from the atmosphere to the ocean due to phytoplankton photosynthesis and the generation of organic carbon that can be exported from surface waters to fuel life in the ocean. The way this works is that, as phytoplankton take up dissolved carbon dioxide for photosynthesis, more carbon dioxide from the atmo-sphere dissolves into the seawater. The organic matter generated from the phytoplankton is transferred through the food web, and a portion of the organic matter sinks through the water column to serve as food for other organisms. As the organic matter sinks to depth, the carbon dioxide that was used to originally gener-ate the organic matter is drawn-down to depth as well. A more efficient biological pump provides more organic compounds to deep waters than an inefficient pump. Nutrient-rich waters tend to support an efficient biological pump because, as described above, large phytoplankton dominate these waters and are con-sumed by larger organisms. A large proportion of the organic matter generated by large phytoplankton can be exported to deeper waters. The nutrient-poor waters of the open ocean tend to support an inefficient biological pump that is dominated by smaller phytoplankton; most of the organic matter and nutrients are recycled in the surface waters. Less of the organic matter generated by phytoplankton in the open ocean is exported to deeper waters.

15.6 Practical Considerations: Ocean Fertilization

Recent studies of ice cores indicate that atmospheric levels of car-bon dioxide varied between about 180 and 280 ppm for ~650,000 years prior to the industrial revolution. Current atmospheric levels of carbon dioxide are about 380 ppm and are predicted to continue to rise, perhaps reaching a doubling of pre-industrial levels by mid-century and as high as 800 ppm by the end of the century. The dra-matic rise in levels of carbon dioxide in the atmosphere has been linked to a potential for subsequent climate change. This has led some to consider fertilizing phytoplankton in the oceans as a means of enhancing the biological pump and slowing the rise in atmo-spheric carbon dioxide levels. A number of experiments have been conducted over the years to test the impact of fertilization on the open-ocean biological pump. By synthesizing our understanding of the factors that control primary productivity and the functioning of the biological pump, the results of fertilization experiments can be evaluated. Understanding current results allows informed deci-sions to be made about the feasibility of using ocean fertilization as a means of minimizing rising levels of atmospheric carbon dioxide.

Three main regions of the world's oceans are characterized by relatively high concentrations of nitrate and phosphate, but rela-tively low abundances of phytoplankton: the Southern Ocean, the Equatorial Pacific, and the North Pacific. Phytoplankton growth in these **high nitrate low chlorophyll** *a* **regions,** or **HNLC regions,** is not limited by light or macro-nutrients but is instead limited by low concentrations of the micro-nutrient iron. Iron is the fourth most abundant element on Earth, but the chemistry of iron in ocean waters means that most iron present in seawater is not available for use by phytoplankton. New inputs of iron are deliv-ered to surface waters in HNLC regions primarily in wind-blown dust or by weak upwelling. HNLC regions have been targeted as areas where phytoplankton growth could be enhanced by iron fer-tilization. Artificially fertilizing with iron would be significantly cheaper than fertilizing with nitrate because iron is required by phytoplankton at much lower concentrations. Review the Redfield ratio described in section 15.2 to see that the molar ratio of car-bon:nitrogen is 6.6:1. In contrast, the molar ratio of carbon:iron is 100,000:1! This means that, in areas where iron limits phytoplank-ton growth, addition of relatively small quantities of iron should cause a dramatic increase in the amount of organic carbon produced and potentially available for export to deep waters.

A number of large-scale experiments have been conducted over the past decade by large teams of scientists to test the im-pact of iron enrichment on HNLC regions. In each experiment, soluble iron was added to large patches (tens of km^2 in size) of ocean and the resulting changes monitored over time (up to twenty-five days). Three main research questions are addressed in these experiments: (1) does iron enrichment result in greater phytoplankton biomass in the fertilized HNLC regions, (2) how are different members of the food web impacted by iron addi-tions, and (3) does enhanced phytoplankton abundance in HNLCs lead to a greater export of organic carbon from surface waters and an enhanced draw-down of carbon dioxide from the atmosphere?

CalCOFI—Fifty Years of Coastal Ocean Data

*I*n the 1920s and 1930s, the coast of Southern California between Monterey and San Diego, was the site of a rich sardine fishery. During this period, more sardines were caught off the California coast than any other fish in North America. This was the fishery that supplied the twenty-four canneries along Monterey's Ocean View Avenue, made famous in John Steinbeck's novel *Cannery Row*. But in the 1940s, the fishery dwindled and collapsed. The catch of more than 550,000 metric tons in 1945 dropped to just 100,000 metric tons in 1947. In response to the fishery's failure, California Cooperative Fisheries Investigations (CalCOFI) was formed and began a series of ocean research cruises in 1949. During the thirty-five years between 1949 and 1984, more than 23,000 stations were sampled between the southern reaches of Baja California, the California-Oregon border, and several hundred kilometers offshore. Since 1984, another 7000 stations have been routinely sampled between San Diego and Point Conception.

Although the original purpose of the program was to look at the distribution and abundance of the sardine, it became clear that understanding the fluctuations in the sardine population required studying the links among the ocean, its currents, and the atmosphere, especially with respect to El Niños, La Niñas, and the even longer-term cycles that affect the Northeast Pacific.

Between 1958 and 1960, the warming of the California Current was related to an equatorial El Niño event; zooplankton and larval fish populations declined, and the harvest of coastal fish dropped from 114,000 metric tons in 1956 to 79,000 metric tons in 1960. Nearshore seaweed forests near San Diego also declined during this period. In the late 1970s, the California Current again became warmer and fresher, and the long-term decline in the commercial fish catch accelerated. At the same time, seabird counts decreased, seaweed forests declined, and changes were noted in the intertidal benthic and bird communities as southern species came to dominate communities. It is now recognized that an unexplained shift in Northeast Pacific conditions occurred in 1977, on which later El Niños and La Niñas were imposed.

During the 1983–84 El Niño, zooplankton, kelp forests, and fish again declined. The range of many fish populations and invertebrates shifted northward in 1983; seabirds along the Oregon coast had a bad breeding season; and the numbers of pups born to California sea lions and northern fur seals also dropped.

In 1985, sardine spawning reached 30,000 metric tons; by 1995, the sardine biomass reached 300,000 metric tons; and in 1999, the sardines surpassed 1 million metric tons for the first time since the mid-1940s. The stock has rebuilt and has extended its range northward, reaching Vancouver Island in 1998.

In 1998 and 1999, the area began a transition from warming to cooling, entering a La Niña cycle that peaked in 1999–2000. The phytoplankton population increased, with April 1999 chlorophyll measurements showing the highest values since 1984. It is thought that this may signal the end to the warming trend started in 1977. CalCOFI data banks now hold more than fifty years of coastal data

Box Figure 1 CalCOFI researchers use a plankton net known as a bongo net to collect duplicate samples of plankton.

concerning ocean currents, atmospheric conditions, and population levels of plankton, fishes, marine mammals, seabirds, and other marine organisms (box fig. 1).

In the California Current system, warming episodes that last one to two years affect the strength of the current and lower salinities—all linked to equatorial El Niños. Researchers investigating fish scales in sediment cores from the Santa Barbara basin have accumulated a 2000-year record that shows a pattern of alternating cycles of abundance between sardine and anchovy populations lasting not two but approximately thirty years. This cycle is thought to be related to the long-term Northeast Pacific trend that started in 1977. Sardines prefer warmer water and anchovies prefer cooler water, indicating some kind of long-term climatic oscillations that affect water temperature, productivity, and populations in this coastal area. Present indications are that the sardine is increasing in the California Current and the anchovy population is declining.

Another important use of CalCOFI data is associated with the interpretation of satellite data concerning phytoplankton. The *SeaWiFS* satellite measures phytoplankton absorption and reradiation of light, the changes in light leaving the water in response to concentrations of dissolved and suspended materials, and the fluorescence of photosynthetic pigments such as chlorophyll. These data are known as bio-optical data and are used to provide quantitative data on phytoplankton populations. To calibrate the satellite data, however, one must have direct measurements of populations at sea. CalCOFI has been collecting bio-optical data at more than 300 stations since 1993, and these data provide a key to interpreting the data from the *SeaWiFS* satellites.

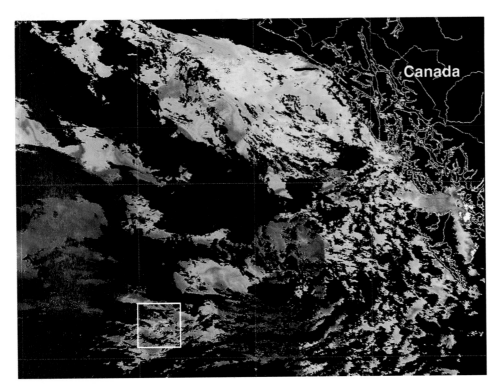

Figure 15.13 A SeaWiFS image of chlorophyll *a* distributions on day twenty of an iron fertilization experiment in the North Pacific. The warm colors (*reds* and *yellows*) indicate regions of high chlorophyll *a* concentrations, the cool colors (*blues*) indicate regions of low chlorophyll *a* concentrations. *Black areas* over the ocean indicate cloud cover, which prevents the satellite from detecting ocean color. The area highlighted with the *white box* is the site of the high chlorophyll *a* concentrations resulting from iron fertilization.

The answer to the first research question is clear. Within days of iron fertilization, chlorophyll *a* concentrations increase dramatically, in some experiments by as much as twenty times. The size of the fertilized patch can grow to as large as 1000 km². Frequently the fertilized patches are so vast that the chlorophyll *a* of the phytoplankton bloom can be seen in satellite images (fig. 15.13). The answer to the second question is also clear. The increase in chlorophyll *a* is accompanied by a shift in the food web toward large phytoplankton, such as diatoms, rather than the smaller phytoplankton that were present prior to the start of the fertilization. Remember, nutrient-limited open-ocean environments tend to be dominated by small phytoplankton about 1 micron in size. A dominant phytoplankton that tends to bloom after iron fertilization are needle-shaped diatoms about 100 microns in length.

The answer to the third research question of whether ocean fertilization enhances the biological pump is still not settled despite almost a decade of research by dedicated scientists. In all fertilization experiments, phytoplankton biomass and primary productivity increase. However, lower than predicted amounts of organic carbon are formed in response to these enrichments, and much less sinks out of the surface waters to fuel growth of organisms at depth. In other words, despite fertilization, the biological pump in HNLC regions appears to remain relatively inefficient. It is still not entirely clear why the efficiency of the biological pump is so much lower than expected. These experiments point out the difficulty in predicting how marine food webs will respond to changes in their chemical and physical environments. These experiments also illustrate that iron fertilization of the ocean would have to be carried out on a much larger scale than originally hypothesized to draw down significant quantities of carbon dioxide from the atmosphere, which would likely lead to large-scale changes in ocean food webs.

Summary

Phytoplankton are the dominant photosynthetic autotrophs in the sea. These microscopic organisms use sunlight and inorganic compounds to generate the organic matter that serves as food for life in the sea. The process of generating organic carbon from carbon dioxide is commonly referred to as carbon fixation. All photosynthetic organisms use the pigment chlorophyll *a* to absorb sunlight. Oxygen is formed as a by-product of photosynthesis. Organic matter is broken down through respiration to yield chemical energy, water, and carbon dioxide.

Gross primary productivity is the total amount of organic matter produced by photosynthesis per volume of seawater per unit of time. Net primary production is the gain in organic matter from photosynthesis by phytoplankton minus the reduction in organic matter due to respiration by phytoplankton. Primary productivity can be determined by measuring the rate of uptake of carbon dioxide or the rate of production of oxygen.

Phytoplankton remove required inorganic nutrients from seawater in a ratio that reflects their biological demands. When organisms die and decompose, the nutrients are released back into seawater in a similar ratio. Nutrients cycle between the land and the sea and through organic and inorganic compounds. Inorganic nutrient concentrations, sunlight, and temperature influence the rate of primary production. Phytoplankton blooms occur when phytoplankton reproduce more rapidly than they are consumed by zooplankton and other heterotrophs. Standing stock is the total phytoplankton biomass present at a given site at a given instant in time and is related to chlorophyll *a* concentrations.

Primary productivity is dependent on temperature, sunlight, and inorganic nutrients. At polar latitudes, the availability of light controls phytoplankton productivity; nutrients are not limiting. In the tropics, sunlight is available year-round, but the stability of the water column minimizes nutrient supply to surface waters. Phytoplankton in the tropics are commonly nutrient-limited. At temperate latitudes, light, nutrients, and water column stability vary over the seasons. Phytoplankton blooms in temperate regions commonly occur in the spring and fall.

Because high concentrations of required inorganic nutrients are delivered to the surface waters in coastal and upwelling regions, these areas are more productive than the open ocean, assuming sufficient sunlight is available. Upwelling regions are about twice as productive as coastal waters and about six times as productive as the open ocean. The size of the open ocean is about nine times as large as coastal environments, and coastal regions are about ten times larger than upwelling regions. Most of the ocean is characterized as nutrient-poor with low rates of primary productivity. However, the sizes of the world's oceans are so vast that about as much organic carbon is generated on a yearly basis by the phytoplankton of the sea as by the plants on land.

Food webs describe the interconnections between different types of organisms within an ecosystem. The phytoplankton are the primary producers. In the open ocean where nutrient concentrations are low, the dominant phytoplankton are small.

In nutrient-rich waters, the dominant phytoplankton are large. The organisms that consume the phytoplankton are the primary consumers. The organisms that consume the primary consumers are secondary consumers and so on. Trophic pyramids can be used to represent the relationships between producers and consumers in terms of organic matter and nutrients. Food web interactions influence the amount of organic matter that is available as food for organisms at deeper depths in the water column. The "biological pump" refers to the processes that result in the draw-down of carbon dioxide from the atmosphere to generate organic carbon that is available to rain down to depth. The efficiency of the biological pump influences the amount of carbon that is exported from surface waters. Open-ocean waters are characterized by a relatively inefficient biological pump. Nutrient-rich waters are characterized by a relatively efficient biological pump and more organic carbon is available for export to depth.

A decade's worth of large-scale experiments have shown that primary productivity in vast areas of the world's oceans is limited by iron. Fertilization of these areas with iron leads to rapid increases in phytoplankton abundance and shifts in the food web toward larger phytoplankton. The resulting efficiency of the biological pump is not enhanced as much as predicted, however. These experiments indicate that food webs are complicated, and the impacts of changes in the ocean environment on food webs are not easily predicted.

Key Terms

All key terms from this chapter can be viewed by term or definition when studied as flashcards on this book's website at www.mhhe.com/sverdrup10e.

pigments, 360
chlorophyll *a*, 360
carbon fixation, 360
fluorescence, 360
primary production, 361
gross primary production, 361

net primary production, 361
Redfield ratio, 363
nutrient regeneration/remineralization, 363
nitrogen fixation, 363
standing stock/standing crop, 363

ocean color, 364
bloom, 365
food webs, 369
trophic level, 369
microbial loop, 372
high nitrate low chlorophyll *a* (HNLC) regions, 372

Study Questions

1. Explain the general relationship between the abundance and size of organisms shown in table 15.4.
2. Distinguish between the terms in each pair:
   a. standing crop—biomass
   b. photosynthesis—respiration
   c. autotroph—heterotroph
   d. producer—consumer
   e. net productivity—gross productivity
   f. primary production—primary productivity
3. Compare primary productivity of polar, temperate, and tropical regions. What factors generally limit productivity in each region?

4. Trace a chain of organisms through the food web illustrated in fig. 15.12. Which organisms are involved and what trophic levels do they represent?
5. Which areas of the ocean are most productive and why?
6. Explain what is different between the flow of carbon through phytoplankton biomass in the ocean and the flow of carbon through plant biomass on land.
7. Explain why compensation depth is deeper in tropical waters with low productivity than it is in temperate waters with high productivity.
8. Explain the factors that result in the global differences in chlorophyll *a* concentrations illustrated in the satellite images of figure 15.4.

9. Explain the difference between primary productivity and phytoplankton biomass normalized primary productivity.

10. Explain three differences between open-ocean food webs and coastal food webs.

11. Explain why the biological pump is less efficient in the open ocean than in coastal or upwelling regions.

12. Why does the respiration of phytoplankton appear nearly constant with depth in figure 15.3?

13. Explain three ways to determine rates of primary production.

14. Explain why it was initially surprising that scientists found that large areas of the ocean are characterized by high concentrations of nitrate and phosphate but low concentrations of phytoplankton.

15. Explain why iron fertilization has been suggested as a means of reducing atmospheric levels of carbon dioxide.

16. Calculate how many square kilometers of (a) upwelling, (b) coastal, and (c) open ocean are required to produce 1 ton of fish.

The box jellyfish, *Chironex fleckeri*.

Chapter 16

The Plankton: Drifters of the Sea

Chapter Outline

Learning Outcomes

After studying the information in this chapter students should be able to:

1. *explain* under which conditions you would expect to find different groups of phytoplankton (e.g., cyanobacteria, diatoms, dinoflagellates),

2. *understand* the difference between meroplankton and holoplankton,

3. *explain* the concept of the microbial loop, and

4. *describe* what HABs are and why they are of increasing concern.

The word "plankton" comes from the Greek *planktos,* meaning to wander. All plankton are limited in their mobility, and they are moved along by currents. The phytoplankton, zooplankton, and bacterioplankton are the wanderers and drifters of the sea. The diversity of this group of organisms is astonishing, but they are largely unknown by most people because the majority of plankton are too small to be seen with the naked eye. Plankton are some of the more abundant organisms in the sea and are central to the functioning of all marine ecosystems. There are two general types of controls on plankton abundance. "Bottom-up control" refers to control of phytoplankton biomass through abiotic factors such as nutrient and light availability, and temperature. "Top-down control" refers to control of biomass through consumption of organisms. Keep these concepts in mind as you read about the different plankton and where they are found. Representatives of key members of the phytoplankton, zooplankton, and bacterioplankton will be discussed here.

16.1 Kinds of Plankton

In general, plankton are small organisms, commonly less than a few millimeters in size. Their small size means that they tend to be moved by currents from place to place while suspended in seawater. **Phytoplankton** use solar energy to generate oxygen and the organic food that fuels most of the rest of life in the seas. Phytoplankton form the base of most food webs. Cyanobacteria are the only bacterial members of the phytoplankton. All other phytoplankton are eukaryotic (review section 14.3 for definitions). **Zooplankton** are composed of unicellular as well as multicellular organisms. In general, zooplankton consume other organisms. Many juvenile stages of non-planktonic adults are also members of the zooplankton. Bacteria are not members of the zooplankton. Zooplankton contain the largest members of the plankton. For example, some jellyfish are huge, trailing 15 m (50 ft) tentacles. The **bacterioplankton** are composed of members of the domains Bacteria and Archaea. Bacterioplankton commonly rely upon inorganic and organic compounds that are dissolved in the seawater. Bacterioplankton play incredibly diverse roles in marine ecosystems.

Most plankton are small enough that they can be seen only with the aid of a microscope. Plankton are commonly categorized according to their size. **Picoplankton** are less than 2–3 microns in size. To get a sense of how small a picoplankton is, remember that the average human hair is about 100 microns in diameter. A 2 micron picoplankton is significantly smaller than the width of an average human hair! Picoplankton are composed of phytoplankton, unicellular zooplankton, and bacterioplankton. The extent of diversity of this group of organisms is only just beginning to be uncovered. The most abundant phytoplankton in the world's oceans is a group of cyanobacteria known as *Prochlorococcus.* These organisms are less than 1 micron in size and were only discovered by biological oceanographers about twenty years ago in 1988 (see boxed feature to learn more about the discovery of picoplankton). Within the last few years, there has been an explosion of new information about the diversity of the picoplankton because of the new ability to identify organisms based on their DNA sequences. Currently, only a few picoplankton can be cultured and maintained in the laboratory. **Nanoplankton** are between 2 and 20 microns in size. The picoplankton and nanoplankton dominate the plankton of open-ocean environments. **Microplankton,** or **net plankton,** range between 20 and 200 microns in size. The microplankton have been well studied over the years because it is relatively straightforward to capture and concentrate these organisms using a plankton net made of fine mesh that can be dragged through the water (see section 16.6). **Macroplankton** are 200 microns to 2 mm in size and can also be captured using a plankton net. The larger microplankton and macroplankton are visible members of coastal and upwelling environments. In some locations, their abundance is so great that their vertical movement through the water column can be detected with a depth recorder. This layering of organisms is referred to as the deep scattering layer (see chapters 5, 14).

16.2 Phytoplankton

Microscopic phytoplankton are often referred to as the "grasses of the sea." Just as a land without grass could not support the insects, small rodents, and birds that serve as food for larger organisms, a sea without phytoplankton could not support zooplankton and other larger organisms. As the British biological oceanographer Sir Alister Hardy said, "All flesh is grass."

By definition, phytoplankton are photosynthetic. They are members of a diverse group of organisms collectively referred to as algae. The vast majority of phytoplankton are unicellular. Some species of phytoplankton can form chains or aggregates of cells (fig. 16.1), but the cells still function independently of each other. *Sargassum* is the only large, multicellular planktonic alga. *Sargassum* can form large mats that provide shelter and food for a wide variety of specialized organisms, including certain fish and crabs, that occur nowhere else. The Sargasso Sea is a region of the large gyre in the North Atlantic and is well known for great floating masses of *Sargassum* seaweed. These floating masses frightened early mariners and became part of many sea stories. Lesser quantities of *Sargassum* are found elsewhere in warm-water environments.

Common members of marine phytoplankton are **diatoms, dinoflagellates, coccolithophorids, cyanobacteria,** and **green algae.** These different groups can be readily distinguished from one another based on morphology (such as aspects of their cell walls) and the suite of pigments they possess. Diatoms, dinoflagellates, and coccolithophorids are important members of the larger size classes—the nanoplankton and microplankton. Commonly, the presence of large phytoplankton in an ecosystem results in relatively efficient trophic transfers (see fig. 15.11). Green algae and cyanobacteria are important members of the picoplankton. They commonly dominate the open-ocean phytoplankton. There are many other groups of marine phytoplankton such as cryptomonads, silicoflagellates, and chrysomonads, but only the dominant players will be discussed here.

Diatoms are estimated to consist of tens of thousands of different species. Diatoms are found in both marine and freshwater environments in both pelagic and benthic realms. They come in a wide variety of sizes and shapes. Some species of diatom fall within the smallest range of the nanoplankton and some within the mid-range of the macroplankton. They are important members of coastal ecosystems, although they are also found in the open ocean. In some regions, diatoms generate as much as 90% of the organic matter produced during phytoplankton blooms. Diatoms tend to do best in high-nutrient, well-mixed waters. These are the regions where our greatest fisheries are found (see table 15.5). Overall, diatoms are estimated to be responsible for as much as 40% of marine primary productivity.

All diatoms possess the pigments fucoxanthin and chlorophyll *c* in addition to chlorophyll *a* (review fig. 15.1). The cell wall of diatoms is called the **frustule.** A defining feature of diatoms is that their frustule is composed primarily of silica that is embedded with a small amount of organic matter. The frustule displays elaborate nano-sized structures that are commonly used to define species. In fact, the structure of the diatom cell wall is so intricate that diatoms have long been used to test the optical resolution of microscopes (fig. 16.2). The frustule is surrounded by an organic matrix that prevents the silica from dissolving in seawater.

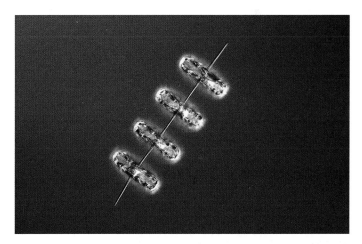

Figure 16.1 Most phytoplankton are unicellular. Some phytoplankton can form chains of connected cells. However, the individual cells within the chains act independently of each other. Two representative chain-forming diatoms are members of the genera *Melosira* (*left*) and *Thalassiosira* (*right*). The *Melosira* chain is composed of five linked cells; the *Thalassiosira* chain is composed of four linked cells.

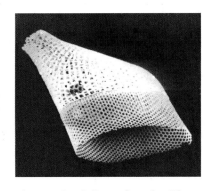
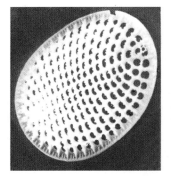
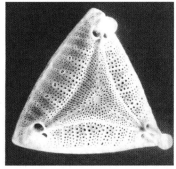

Figure 16.2 Scanning electron micrographs of diatom frustules. The organic casing that normally surrounds living diatoms has been removed, so the elaborate patterns of pores in the frustule can be seen.

Although there are other organisms in the sea that can use silicate, such as sponges and silicoflagellates (fig. 16.3), diatoms essentially control the cycling of biogenic silicate in the world's oceans.

Diatoms display two general shapes with either **radial** or **bilateral symmetry.** Radially symmetrical species are shaped more or less like Petri dishes or hat boxes and are members of the **centric** diatoms. Bilaterally symmetrical species are elongate and are shaped more or less like a cigar box and are members of the **pennate** diatoms. Micrographs of representative members of the centric and pennate diatoms are shown in figure 16.4, and drawings of representative species are shown in

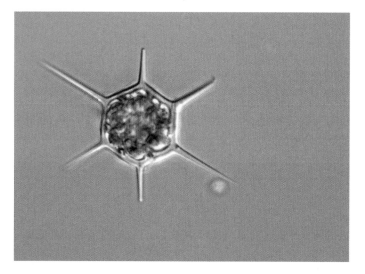

Figure 16.3 A representative coastal silicoflagellate. These members of the phytoplankton form cell walls of silica.

figure 16.5. The frustule of both pennate and centric species is composed of two halves (known as valves) held together by a series of bands also made of silica (think of wrapping tape around a Petri dish or hat box to hold the two halves together) (fig. 16.6). When a cell grows, the two halves slide apart due to the addition of new bands; the diameter of a valve cannot increase. All phytoplankton reproduce predominantly via asexual divisions: one cell divides to form two cells. When a diatom cell has grown large enough to undergo division, two new inner halves of the frustule are created, and the cell divides into two daughter cells. By this process, one daughter cell is the same size as the parent cell, and one is slightly smaller (fig. 16.6). Over successive generations, the mean cell size of the population will decrease. The most common means of escaping this cycle of diminishing cell size is through sexual reproduction. When a diatom decreases in size to about 40% of its original size, gametes can be produced (often in response to an environmental trigger), and upon successful mating, a zygote is formed. The zygote, also known as an **auxospore,** breaks free of its old cell wall and forms an entirely new and much larger frustule. The cycling over time between large and small cells within a given species of diatom will influence food web dynamics, since large cells may be grazed by different zooplankton than small cells.

The silica cell wall of diatoms is heavier than seawater, and unless a diatom actively controls its buoyancy, the cell will sink. Diatoms produce spines and other protrusions (fig. 16.4) that increase the surface area to volume ratio, slow sinking, and help maintain cells within surface waters (review section 14.2 on this topic). Eventually, however, either diatoms are eaten or they sink out of the euphotic zone to deeper waters. The zooplankton that consume diatoms produce fecal pellets that contain the heavy frustules,

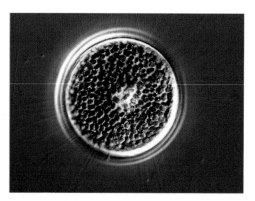

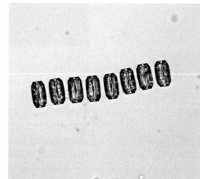

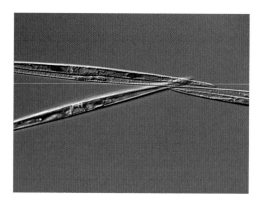

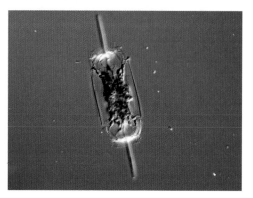

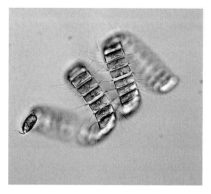

Figure 16.4 Representative members of coastal diatom communities of temperate waters. Four species of centric diatoms (*clockwise beginning upper left: Coscinodiscus, Thalassiosira, Chaetoceros,* and *Ditylum*) are shown in the *left panels* and one species of pennate diatom (*Pseudo-nitzschia*) is shown in the *right panel.*

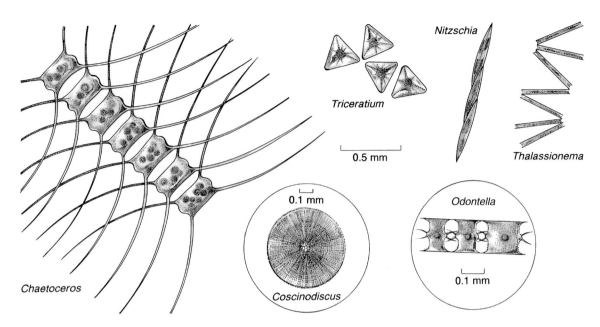

Figure 16.5 Drawings of centric and pennate diatoms.

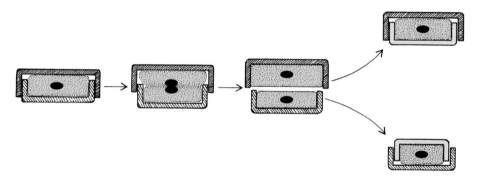

Figure 16.6 The division of a parent centric diatom into two daughter diatoms. The two halves of the pillboxlike cell separate, the cell contents divide, and a new inner half is formed for each pillbox. One daughter cell remains the same size as the parent; the other is smaller.

helping these organic matter–rich pellets to sink. As described in chapter 15, any organic matter generated by diatoms that sinks out of the surface water serves as food for organisms at depth. A small portion of the diatom-produced organic matter sinks all the way to the sea floor where it becomes embedded within the sediments. Over long periods of time, this organic matter contributes to petroleum deposits. The silica frustules can also accumulate on the sea floor where they form deposits known as diatomaceous earth. Sometimes, geologic processes lift these silica-rich sediments above sea level. These sites are mined for diatomaceous earth, which is used industrially to make filtration systems for swimming pools and as an abrasive for toothpaste and silver polish (review chapter 4, "Source and Chemistry" of sediments).

Dinoflagellates are another important member of the nano- and microplankton. Dinoflagellates tend to do well in relatively calm and well-stratified waters. A number of features distinguish the dinoflagellates from other phytoplankton. Their pigments include chlorophyll *a*, chlorophyll *c*, and peridinin. The cell wall of dinoflagellates is composed of cellulose and some species form plate-like structures as their cell wall. Dinoflagellates

assume a wide variety of shapes. One of the more intriguing aspects of dinoflagellates is that some species are autotrophic and possess pigments necessary for photosynthesis (fig. 16.7*a* and *b*); some species are heterotrophic (fig. 16.7*c* and *d*) and are incapable of photosynthesis; and some species are both autotrophic and heterotrophic, depending on food availability. Evolutionary biologists hypothesize that the ancestor of dinoflagellates was photosynthetic; over evolutionary time, however, some groups lost the ability to photosynthesize. The capacity to feed heterotrophically, either exclusively or as a supplement to photosynthesis, means that dinoflagellates are often found in low-nutrient environments, including those of the open ocean.

Dinoflagellates possess two flagella. One encircles the cell like a belt, and the other is located at a right angle to the first. The beating of the flagella makes the cells motile, and they spin like a top as they move through the water. Because the flagella are so fine, it is easiest to see the location of the flagella in drawings of dinoflagellates (fig. 16.8). Many species of dinoflagellates can swim toward the surface during the day to photosynthesize, and they can swim toward the more nutrient-rich deeper waters at night. Despite their ability to swim vertically, dinoflagellates cannot swim against the currents and so are members of the plankton.

For two reasons, dinoflagellates are probably the group of phytoplankton best known to the general public. First, some species of dinoflagellates bioluminesce. If you go swimming at night during the summer in temperate waters, for example, the glow you see as you move through the water is most likely due to dinoflagellate bioluminescence. They use the enzyme luciferase and the substrate luciferin to generate the light (see section 14.4 for a more detailed discussion of bioluminescence).

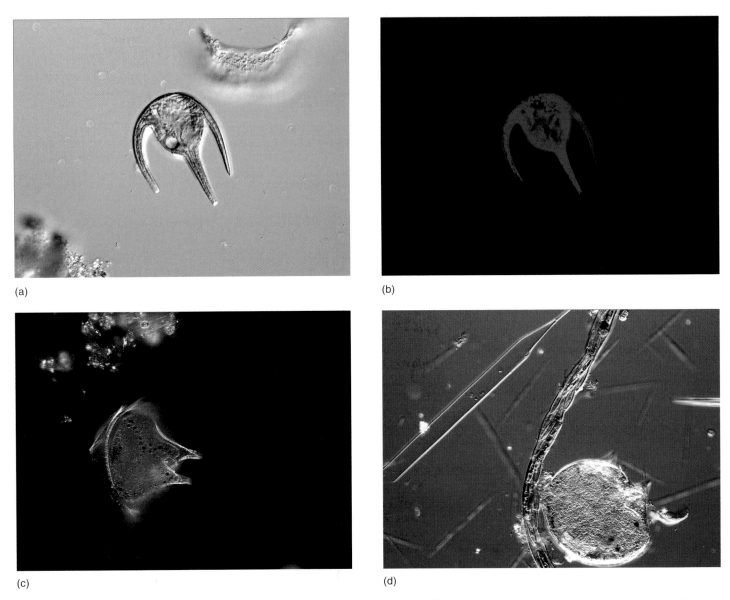

(a)

(b)

(c)

(d)

Figure 16.7 Two representative dinoflagellates from coastal environments. Dinoflagellates such as *Ceratium* (a) can be photosynthetic, and when illuminated with blue light, chlorophyll *a* fluorescence is detected as red light (b). Dinoflagellates such as *Protoperidinium* (c, d) are heterotrophic, and they use elaborate methods to capture their phytoplankton prey. Shown here is a *Protoperidinium* cell preparing to feed on chains of diatoms.

The second reason that dinoflagellates are well known is because some species produce toxins that are biomagnified through the food web. Consumption of toxin-contaminated food can impact the health of marine mammals and humans. Shellfish destined for human consumption are carefully screened for the presence of toxin to protect human health. If toxin is detected, shellfish beds are closed to harvesting. Toxic species of dinoflagellates will be further discussed in section 16.5.

Coccolithophorids are a third dominant member of the larger phytoplankton. They possess a cell wall embedded with calcium carbonate plates known as coccoliths (fig. 16.9*a*). Coccoliths are shed as cells grow and divide, and the coccoliths sink slowly through the water column. Coccoliths reflect light in a way that can be detected by satellites, which has greatly enhanced our knowledge of the bloom dynamics of these organisms. For example, during the late summer–early fall of 1997,

the water over most of the continental shelf in the eastern Bering Sea was colored aquamarine by a huge bloom of coccolithophorids (fig. 16.9*b*). The bloom was so large that it was visible from space and was recorded by the SeaWiFS sensor. A bloom this large had never been recorded before in the region and was attributed to unusual climate conditions.

Cyanobacteria are the most abundant phytoplankton in low-nutrient open-ocean environments. They also represent the most ancient lineage of oxygen-producing photosynthetic organisms. It is through their photosynthetic activity that early Earth became oxygenated (review chapter 2, "Age and Time"). The two most common groups of open-ocean cyanobacteria today are *Prochlorococcus* and *Synechococcus*. These organisms are about 1 micron in size and thus are members of the picoplankton. *Prochlorococcus* possesses special forms of the pigments chlorophyll *a* and *b; Synechococcus* possesses chlorophyll *a* and

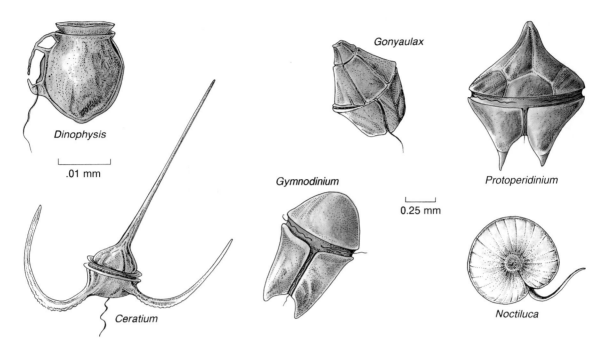

Dinophysis

.01 mm

Ceratium

Gonyaulax

Gymnodinium

0.25 mm

Protoperidinium

Noctiluca

Figure 16.8   Representative species of coastal dinoflagellates.

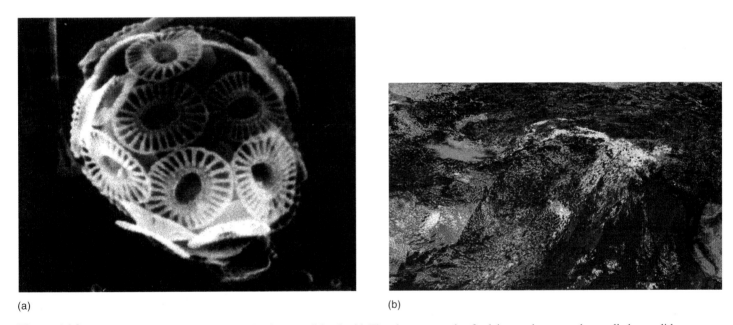

(a)

(b)

Figure 16.9   (a) A scanning electron micrograph of a coccolithophorid. The plates are made of calcium carbonate and are called coccoliths. (b) A true-color image showing the extent of a coccolithophorid bloom over the continental shelf of the eastern Bering Sea in September 1997. Image provided by the SeaWiFS Project, NASA/Goddard Space Flight Center. High concentrations of coccolith plates can accumulate on the sea floor as sediment and can eventually become a rock we call chalk.

phycoerythrin (see fig. 15.1). *Prochlorococcus* is globally distributed within tropical and sub-tropical waters at latitudes between 40°N and 40°S. *Prochlorococcus* dominates the euphotic zone of these open-ocean environments, particularly when surface waters are highly stratified and inorganic nutrients such as nitrate are depleted. In fact, in contrast to other phytoplankton, *Prochlorococcus* appears incapable of using nitrate as a nitrogen source and instead uses ammonium as a primary source of inorganic nitrogen. *Prochlorococcus* is estimated to be responsible

for between 10 and 50% of the net primary production of the open ocean depending on time of year. This single group of organisms plays an incredibly important role in the global carbon cycle. *Synechococcus* has a broader global distribution than *Prochlorococcus* and can use nitrate as a nitrogen source. *Synechococcus* blooms when the water column becomes mixed due to winter storms, and nitrate concentrations are higher.

Cyanobacteria also play an important role in nitrogen fixation—the use of nitrogen gas to generate nitrogen-containing

Field Notes

Discovery of the Role of Picoplankton
by Dr. E. Virginia Armbrust

Dr. E. Virginia Armbrust is a Professor of Oceanography at the University of Washington. Her research interests center on the biodiversity of microbial communities, particularly plankton communities.

About half of global primary productivity is due to photosynthesis by marine phytoplankton, despite the fact that phytoplankton biomass represents only about 0.2% of that of land plants. For most of the twentieth century, relatively large phytoplankton (~10 microns and larger) that could be easily viewed with conventional light microscopy were considered the major players in oceanic primary productivity. Scientists now know that the most abundant member of the phytoplankton community is so small that 100 cells could easily fit end-to-end across a human hair. Moreover, picoplankton, or phytoplankton less than 2 microns in size, represent about 50% of the total phytoplankton biomass in much of the open ocean. How is it possible then, that each year, these microscopic organisms produce as much organic carbon as trees, crops, and grasses combined? Phytoplankton, no matter how small, are part of a tightly coupled food web in which individual cells are consumed almost as quickly as they are produced, passing on their newly fixed organic carbon to different members of the heterotrophic community. Only during blooms does phytoplankton biomass reach a relatively high level, which is ultimately cropped by grazers.

Primary productivity by phytoplankton not only fuels marine food webs; it also plays a critical role in mitigating atmospheric carbon dioxide levels on our planet. Without phytoplankton and the transfer of their organic carbon to different trophic levels throughout the water column, atmospheric concentrations of CO_2 would likely more than double. As concern grows over global CO_2 concentrations, greater attention has been focused on understanding how different members of the phytoplankton community will respond to changing environmental conditions. The interactions among global climate change, primary productivity, and atmospheric CO_2 levels are complicated. A key step in beginning to define the various feedback loops is to identify which phytoplankton are present and what roles they play in different marine ecosystems. What follows is a minihistory of how the development of new scientific techniques has led to the discovery of entirely new groups of phytoplankton, which in turn provide insights into the complicated set of interactions that make up a marine ecosystem.

The first breakthrough in identifying picoplankton, which are the smallest members of the phytoplankton, came in the late 1970s, when two scientists—John Waterbury from Woods Hole

Box Figure 1 Two flow cytometers surround Rob Olson from Woods Hole Oceanographic Institution, at sea aboard the R/V *Oceanus* in March 2002.

Oceanographic Institution and John Sieburth from University of Rhode Island—identified bacteria-sized photosynthetic organisms now known as marine unicellular cyanobacteria of the genus *Synechococcus*. At the time of their discovery, the most common means of enumerating bacteria in the ocean was to stain cells with acridine orange, a fluorescent molecule that binds DNA. Fluorescent microscopy was then used to detect and count the stained bacterial cells. Waterbury noticed the presence of cells about 1 micron in size that fluoresced under the microscope even in the absence of stain. He realized that this fluorescence was due to the photosynthetic pigments known as phycobilins. These pigments characterize most cyanobacteria and allow them to absorb blue light, the wavelength of light that dominates deeper waters of the open ocean. Over the next ten years, the prokaryote *Synechococcus* was heralded as the world's most abundant phytoplankton, since it was distributed globally and appeared to dominate the low-nutrient waters of vast regions of the open ocean.

All this changed in the late 1980s, when Rob Olson of Woods Hole Oceanographic Institution and Penny Chisholm of Massachusetts Institute of Technology discovered an even smaller and more abundant cyanobacterium now known as *Prochlorococcus.* What made this new discovery possible was the use of flow cytometers at sea. Flow cytometry had been developed for biomedical research to rapidly detect cells stained with fluorescent compounds such as the DNA stain acridine orange. What distinguished flow cytometry from fluorescent microscopy was that thousands of cells in liquid could be analyzed for both cell size and fluorescent properties in a matter of seconds. Phytoplankton ecologists began taking flow cytometers to sea (see box fig. 1) in the early 1980s and using them to record the distribution of *Synechococcus.* The small size, high cell concentrations, and unique combination of fluorescent pigments made *Synechococcus* perfectly suited for detection by flow cytometry. During this period, scientists were working hard to optimize flow cytometers for use at sea to enable detection of even the dimmest fluorescent cells, particularly those found in surface waters. On a research cruise in the North Atlantic, Olson made the final modifications to his flow cytometer that allowed him to detect, for the first time, dimly fluorescing cells that were even smaller than *Synechococcus* and that did not display the characteristic fluorescence associated with phycobilins. This newly discovered genus of cyanobacteria, *Prochlorococcus,* contains chlorophyll *b,* which like the phycobilins, allows the cells to efficiently harvest blue light for photosynthesis. *Prochlorococcus* cells can be found even deeper in the water column than *Synechococcus* and have helped to change our notion of what light levels constitute the base of the euphotic zone. *Prochlorococcus* is routinely found at depths where the light levels are only 0.1% of surface light, a light intensity that approaches the limit of human perception. *Prochlorococcus* is now known to dominate temperate and tropical oceans. In the subtropical Pacific, for example, *Prochlorococcus* can represent 50% of phytoplankton biomass despite the fact that they are only about 0.6–0.8 μm in size. In the North Atlantic, cell concentrations of *Prochlorococcus* and *Synechococcus* oscillate with opposite patterns: *Synechococcus* blooms when *Prochlorococcus* is at its lowest cell concentrations, and *Prochlorococcus* blooms when *Synechococcus* is at its lowest cell numbers. Together, *Synechococcus* and *Prochlorococcus* are responsible for 30–80% of total primary productivity in the low-nutrient waters of the open oceans.

The discoveries of *Synechococcus* and *Prochlorococcus* both relied on the fact that these organisms contain a unique combination of photosynthetic pigments that can be detected with either fluorescent microscopy or flow cytometry. But what of the picoplankton that were still awaiting discovery that do not possess a defining pigment signature? The problem was simple: how do you know what you are missing if you cannot see it? The next breakthrough came in the mid-1990s, when oceanographers began using DNA-based techniques to identify previously unknown plankton in the ocean. This approach does not depend on culturing cells or even seeing them. Instead, DNA is isolated from an entire community of plankton of a particular size class. Polymerase chain reaction (PCR) is used to amplify specific genes from the mixed population. The DNA sequence of these targeted genes can be used to identify different groups of organisms based on similarity to DNA sequences present in a large DNA database known as Gen-Bank (http://www.ncbi.nlm.nih.gov). Using this approach, it now appears that the prokaryotic picoplankton are composed of different groups of *Synechococcus* and *Prochlorococcus* but nothing else. In contrast, new eukaryotic picoplankton are being discovered almost every time scientists sample different environments. One of the more interesting recently discovered eukaryotic photosynthetic picoplankton is *Ostreococcus.* This organism was first identified in a French lagoon and was determined to be the smallest known eukaryote at only 0.8 μm (micrometers) in size. *Ostreococcus* has now been observed in a wide variety of environments, including coastal waters and open oceans.

Now in the twenty-first century, studies of phytoplankton communities have once again been revolutionized by a new technique—this time, DNA sequencing of whole genomes. Fueled largely by support from the U.S. Department of Energy Microbial Genome Program (http://www.jgi.doe.gov), the complete DNA sequence has been obtained and is beginning to be deciphered for strains of *Prochlorococcus* and *Synechococcus,* strains of cyanobacteria that can fix atmospheric nitrogen, and multiple diatoms. The DNA sequence of these organisms will help define what genes they possess and thus the potential functions these organisms can carry out in the environment. For example, the presence or absence of particular genes in the *Prochlorococcus* and *Synechococcus* genomes provides insight into the oscillation in cell numbers observed in the North Atlantic. Sequenced strains of *Prochlorococcus* possess genes for ammonia utilization but are missing the genes necessary for nitrate utilization. In contrast, *Synechococcus* possesses both of these genes. During spring in the North Atlantic, deep mixing brings high levels of nitrate into the euphotic zone that only *Synechococcus* can use, allowing these cells to bloom. In the absence of these high nitrate concentrations, *Prochlorococcus* blooms. These kinds of insights into phytoplankton ecology will continue to emerge as more is learned about the types of organisms that inhabit the oceans and the various capabilities these organisms have stored in their genes.

Each discovery discussed here has resulted in an enhanced understanding of how oceanic ecosystems function. The goal now is to predict how resilient these communities will be to future environmental change because, without phytoplankton, life on Earth would be very different.

organic matter. Productivity in vast areas of the open ocean is limited by the availability of inorganic nitrogen. This suggests that any phytoplankton able to use nitrogen gas rather than other forms of inorganic nitrogen would have an advantage. However, as described in the previous chapters, relatively few prokaryotes, and no eukaryotes, can use nitrogen gas. Nitrogen gas is used by nitrogen-fixing organisms when the concentrations of other inorganic nitrogen sources are low. A great deal of research over the past decade has gone into determining which phytoplankton use nitrogen gas. Neither *Prochlorococcus* nor *Synechococcus* described above can use nitrogen gas. Instead, three general groups of nitrogen-fixing cyanobacteria have been identified. First, one subset of nitrogen-fixing cyanobacteria are symbionts of other organisms. For example, a stable symbiosis exists between the cyanobacterium *Richelia* and the open-ocean diatom *Rhizoselenia*. Second, the filamentous cyanobacterium *Trichodesmium* has long been considered the dominant nitrogen-fixing cyanobacterium in the open ocean. Surface blooms of *Trichodesmium* can be seen from on board the deck of a ship. Water samples containing these cells can be monitored for rates of nitrogen fixation. Recently, a third type of nitrogen-fixing cyanobacterium known as *Crocosphaera* was discovered by scientists using a combination of DNA technology and cell isolation. This nanoplankton-sized cyanobacterium likely plays an important role in nitrogen fixation in tropical open oceans.

Prasinophytes are a type of green algae that are the other important members of the picophytoplankton. The smallest size category of phytoplankton appears to be dominated almost exclusively by cyanobacteria and prasinophytes. Prasinophytes are eukaryotic phytoplankton that are closely related to land plants. Like land plants, they store starch and possess chlorophyll *a* and chlorophyll *b*. The smallest known eukaryotic cell—*Ostreococcus*—is a member of the prasinophytes. Another important member of this group is *Micromonas*. These organisms are found in both open-ocean and near-shore environments. Those ecosystems where the numerically dominant phytoplankton are either cyanobacteria or prasinophytes will be characterized by numerous trophic transfers to the highest trophic level (see fig. 15.11).

16.3 Zooplankton

Zooplankton are heterotrophs, and they consume other organisms. Zooplankton fit within two general categories of plankton: the holoplankton and the meroplankton. **Holoplankton** spend their entire lives as plankton. **Meroplankton** spend only a portion of their lives as plankton. The meroplankton include the eggs and the larval and juvenile stages of many organisms that spend most of their lives as either free swimmers (fish) or bottom dwellers (such as crabs and starfish). Some feed almost exclusively on phytoplankton and are herbivorous zooplankton. Others feed on other members of the zooplankton and are carnivorous. Those that feed on both phytoplankton and zooplankton are omnivorous. Many zooplankton can swim, and some can even dart rapidly over short distances in pursuit of prey or to escape predators. Even those zooplankton that can swim vertically are still transported by currents and therefore are members of the plankton. As you read about zooplankton, consider

the roles that different groups may play in different food webs. We begin this section with representatives of the holoplankton.

The life histories of different types of zooplankton are varied and show many strategies for survival in a world where reproduction rates are high and life spans are short. The unicellular zooplankton or protozoa grow rapidly, sometimes as fast as the phytoplankton because they, too, reproduce by cell division. In contrast, the multicellular zooplankton reproduce sexually and produce juvenile offspring that must mature. In warm waters where food supplies are abundant, some species may produce three to five generations of offspring a year. At high latitudes, where the season for phytoplankton growth is brief, zooplankton may produce only a single generation in a year. The voracious appetites, rapid growth rates, and short life spans of zooplankton lead to rapid liberation of nutrients that can be reused by phytoplankton.

Zooplankton exist in patches of high population density between areas that are much less heavily populated. The high population patches attract predators, and the sparser populations between the denser patches preserve the stock, as fewer predators feed there. Turbulence and eddies disperse individuals from the densely populated patches to the intervening sparser areas. Convergence zones and boundaries between water types concentrate zooplankton populations, which attract predators.

Plankton accumulate at the density boundaries caused by the layering of the surface waters; variation of light with depth and day-night cycles play additional roles. Some zooplankton migrate toward the sea surface each night to feed on the food-rich surface layer and descend during daylight to possibly reduce grazing by their predators or to follow their food resources. This daily migration may be as great as 500 m (1650 ft) or less than 10 m (33 ft).

Accumulations of organisms in a thin band extending horizontally along a pycnocline or at a preferred light intensity or food resource level are capable of partially reflecting sound waves from depth sounders. The zooplankton layer is seen on a bathymetric recording as a false bottom or a deep scattering layer, the DSL (see chapters 5 and 14). Echo-sound studies are used to record the vertical migration of this layer of plankton and to estimate abundance by measuring the vertical and horizontal extent of the layer.

Among the most well-studied and widespread zooplankton types worldwide are small **crustaceans** (shrimplike animals), the **copepods,** and **euphausiids** (fig. 16.10). These animals are basically herbivorous and consume more than half their body weight each day. They are found throughout the world. Copepods are a link between the phytoplankton, or producers, and first-level carnivorous consumers. Euphausiids are larger, move more slowly, and live longer than copepods. Euphausiids, because of their size, also eat some of the smaller zooplankton along with the phytoplankton that make up the bulk of their diet. Copepods and euphausiids both reproduce more slowly than diatoms, doubling their populations only three to four times a year. In the Arctic and Antarctic, the euphausiids are the **krill,** occurring in such quantities that they provide the main food for the **baleen** (or whalebone) whales. These whales have no teeth; instead, they have netlike strainers of baleen suspended from the roof of their mouth. After the whales gulp the water and plankton, they expel the water through the baleen, leaving the tiny krill behind. Whales of this type include the blue, right, humpback, sei, minke, and finback whales; whales are discussed

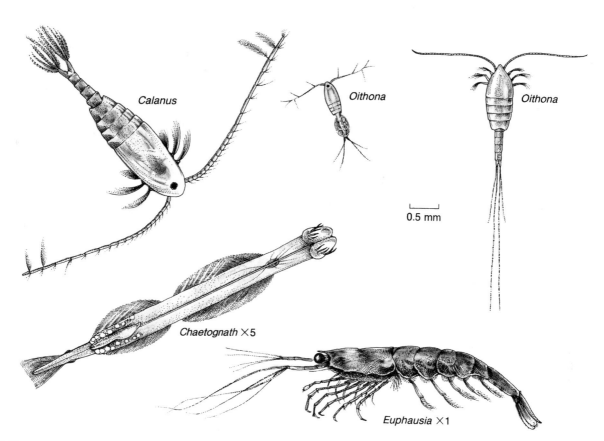

Figure 16.10   Crustacean members of the zooplankton and an arrowworm, or *chaetognath. Calanus* and *Oithona* are copepods. The shrimplike *Euphausia* is known as krill.

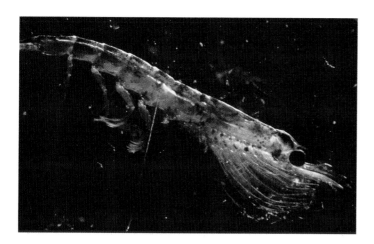

Figure 16.11   The Antarctic krill, *Euphausia superba,* dominates the zooplankton of the Antarctic Ocean.

in chapter 17. It is remarkable to consider that the largest organisms in the sea feed on some of the smallest organisms.

The Antarctic krill, *Euphausia superba* (fig. 16.11), is the most abundant of the eighty-five or so krill species. Antarctic krill are distributed patchily over an enormous area, perhaps as large as 36×10^6 km^2, which is about four times the size of the United States. The krill are circumpolar in distribution, but their concentration is not uniform; their greatest concentration occurs in the summer close to Antarctica where the highest abundance

of phytoplankton are found. In summer, krill live near the surface in huge swarms made up of a billion or more individuals; in the cold, dark winter, they are believed to live on the underside of the ice or perhaps dive down to the sea floor. Individuals are now known to live at least five years under natural conditions and up to nine years in the laboratory. In summer, females lay up to 10,000 eggs at a time, several times a season.

Early estimates of the total biomass of Antarctic krill ran as high as 6 billion metric tons; present estimates lie between 135 million and 1.35 billion tons. Because of the initial large biomass estimates, krill in the Southern Ocean was considered a potentially valuable international fishery. The first harvests were made by the former Soviet Union in the 1960s, and in later years, Japan, Korea, Poland, and Chile participated. The largest harvest, 529,000 metric tons, was made by Japan, Korea, Poland, and Chile during the Antarctic summer of 1980–81. During the next fifteen years, fewer vessels fished krill and the catch dropped: 81,000 metric tons in 1998 but increasing to 103,000 metric tons in 1999 and 104,000 metric tons in 2000. The most recently available numbers indicate that the world catch of krill continues to creep upward and reached 127,000 metric tons in 2005.

The high iodide content of krill meat prevents marketing for human consumption, and the economic future of the krill fishery is questionable. Krill is sold whole, as peeled tail meat, minced, or as a paste; it is used as livestock and poultry feed in Eastern Europe and as fish feed by the Japanese. Because krill deteriorate rapidly, krill used for consumption must be processed within three hours, limiting the daily harvest. The distance to the fishing grounds is long and the costs of vessels and fuel are high.

The Convention for the Conservation of Antarctic Marine Living Resources was established by treaty in 1981. The convention's goal is to keep any harvested population, including krill, from dropping below levels that ensure the replenishment of the adult population.

Arrowworms, or **chaetognaths** (see fig. 16.10), are abundant in ocean waters from the surface to great depths. These macroscopic (2–3 cm, or 1 in), nearly transparent, voracious carnivores feed on other members of the zooplankton. These are members of the trophic level corresponding to primary consumers. Several species of arrowworms are found in the sea, and in some cases, a particular species is found only in a certain water mass. The association between organism and water mass is so complete that the species can be used to identify the origin of the water sample in which it is found. In the North Atlantic, the arrowworm *Sagitta setosa* inhabits only the North Sea water mass and *Sagitta elegans* is found only in oceanic waters.

Foraminifera and **radiolarians** are microscopic, single-celled, amoeba-like protozoans; they are shown in figure 16.12. They feed primarily on phytoplankton but some can also eat small zooplankton. Foraminifera, such as the common *Globigerina,* are encased in a compartmented calcareous covering, or shell. Radiolarians are surrounded by a silica test, or shell. The radiolarian tests are ornately sculptured and covered with delicate spines. Openings in the test allow a continuity between internal protoplasm and an external layer of protoplasm. Pseudopodia (false feet), many with skeletal elements, radiate out from the cell. Radiolarians feed on diatoms and small protozoans caught in these pseudopodia. Both foraminifera and radiolarians are found in warmer regions of the oceans. After death, their shells and tests accumulate on the ocean floor, contributing to the sediments. Calcareous foraminifera tests are found in shallow-water sediments; the siliceous radiolarian tests, which are resistant to the dissolving action of the seawater, predominate at greater depths, commonly below 4000 m (13,200 ft) (see chapter 4). **Tintinnids** (fig. 16.12) are tiny protozoans with moving, hairlike structures, or **cilia.** These organisms are often called bell animals and are found in coastal waters and the open ocean.

Pteropods (fig. 16.13) are mollusks; they are related to snails and slugs. They have a foot that is modified into a transparent and gracefully undulating "wing." This grouping includes animals with shells and animals without shells. The shelled group are primarily filter feeders that capture plankton on a mucous net that they then consume. The group without shells specializes in eating

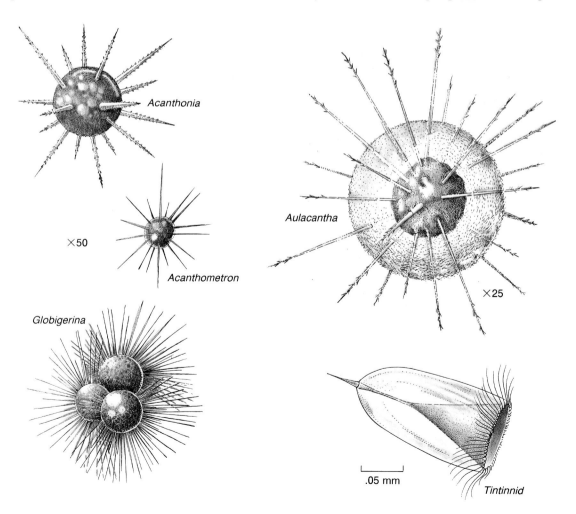

Figure 16.12   Selected radiolarians (*Acanthonia, Acanthometron,* and *Aulacantha*). A foraminiferan *(Globigerina)* is at *bottom left;* a *tintinnid* is at *bottom right.*

soft-bodied animals. Their hard, calcareous remains contribute to the bottom sediments in shallow tropical regions. Some pteropods are herbivores and some are carnivores.

Transparent, gelatinous, and bioluminescent, the **ctenophores,** or comb jellies (fig. 16.14), float in the surface waters. Some have trailing tentacles; all are propelled slowly by eight rows of beating cilia. The small, round forms are familiarly called sea gooseberries or sea walnuts; by contrast, the beautiful, tropical, narrow,

flattened Venus' girdle may grow to 30 cm (12 in) or more in length. A group of Venus' girdles drifting at the surface and catching the sunlight with their beating cilia is a spectacular sight from the deck of a ship. All ctenophores are carnivores, feeding on other zooplankton.

The tunicate, another transparent member of the zooplankton, is related to the more advanced vertebrate animals (animals with backbones) through its tadpolelike larval form. **Salps** (fig. 16.14) are pelagic tunicates that are cylindric and transparent; they are commonly found in dense patches scattered over many square kilometers of sea surface. Salps feed on phytoplankton and particulate matter.

Both ctenophores and pelagic tunicates, although jellylike and transparent, are not to be confused with jellyfish (fig. 16.15). True jellyfish, or sea jellies, come from another and unrelated group of animals, the **Coelenterata,** also called **Cnidaria.** Some jellyfish, such as the common *Aurelia* and the colorful *Cyanea,* with its trailing, stinging tentacles that capture zooplankton, spend their entire lives as drifters. Others, such as *Gonionemus,* a small jellyfish of the Atlantic and Pacific Oceans, and *Aequorea,* found in many temperate waters, are members of the plankton for only a portion of their lives; they eventually settle and change to a bottom-dwelling, attached form similar to a sea anemone. Another group of unusual jellyfish are **colonial organisms,** including the Portuguese man-of-war, *Physalia,* and the small by-the-wind-sailor, *Velella.* Both are

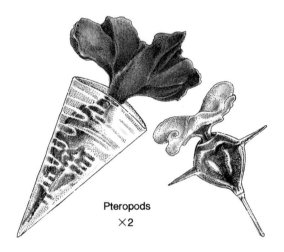

Figure 16.13   The pteropods are planktonic mollusks.

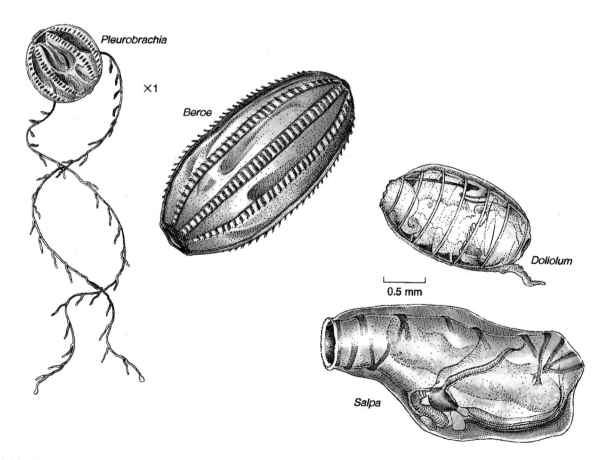

Figure 16.14   The comb jellies (ctenophores) *Pleurobrachia* and *Beroe. Pleurobrachia* is often called a sea gooseberry. *Salpa* and *Doliolum* are tunicates.

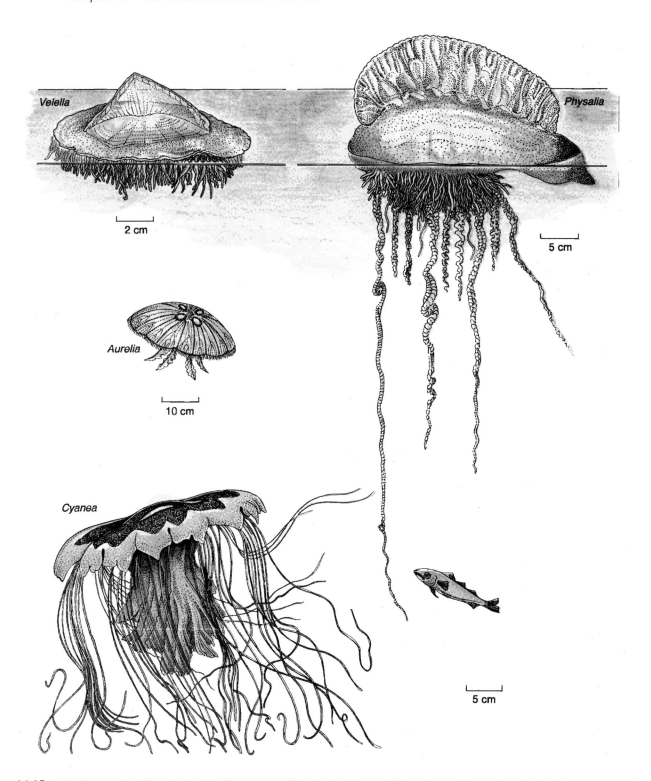

Figure 16.15 Jellyfish belong to Coelenterata, or Cnidaria. *Velella,* the by-the-wind-sailor *(top left),* and *Physalia,* the Portuguese man-of-war *(top right),* are colonial forms.

collections of individual but specialized animals. Some gather food, reproduce, or protect the colony with stinging cells, and others form a float.

A small fishery for sea jellies or jellyfish exists off the Atlantic coast of the United States; total catch reported for 1999 was 578 metric tons. Caught off the coast of Georgia, *Stomolophus*

melagris, known as cannonball, jelly ball, or cabbage head jelly-fish, are dehydrated in brine, then sliced, diced, and packaged for Asian markets. *S. melagris* has a solid structure, no stinging cells, and measures up to 20 cm (8 in) in diameter.

In contrast to the holoplankton that spend their entire lives as members of the plankton (fig. 16.16), the meroplankton

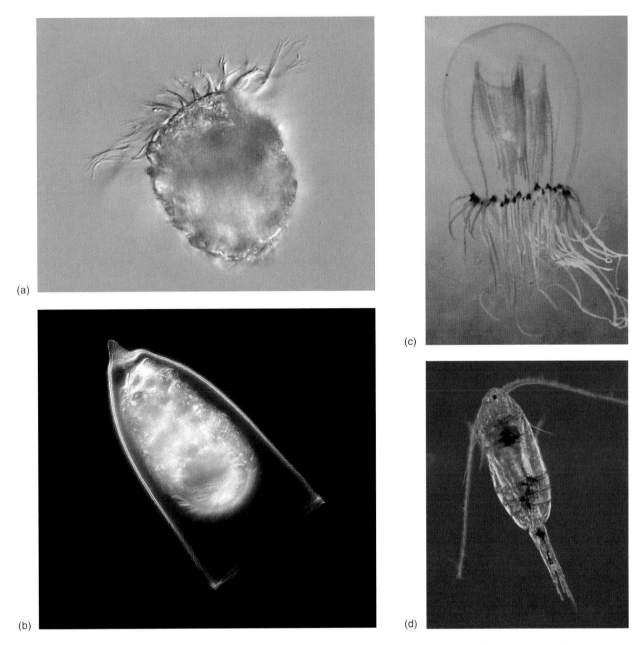

Figure 16.16 Organisms that spend their whole lives in the zooplankton are known as holoplankton and include single-celled protozoa such as (a) a ciliate and (b) a tintinnid, and multicellular oganisms such as (c) the jellyfish *(Polyorchis penicellatus)* and (d) a copepod.

spend only part of their lives as members of the plankton. For a few weeks, the **larvae** (or young forms) of oysters, clams, barnacles, crabs, worms, snails, starfish, and many other organisms are a part of the zooplankton. The currents carry these larvae to new locations, where they find food sources and places to settle. In this way, areas in which species may have died out are repopulated and overcrowding in the home area is reduced. Sea animals produce larvae in enormous numbers, so these meroplankton are an important food source for other members of the zooplankton and other animals. The parent animals may produce millions of spawn, but only small numbers of males and females must survive to adulthood to guarantee survival of the stock.

Larvae often look very unlike the adult forms into which they develop (fig. 16.17). Early scientists who found and described these larvae gave each a name, thinking they had discovered a new type of animal. We keep some of these names today, referring, for example, to the trochophore larvae of worms, the veliger larvae of sea snails, the zoea larvae of crabs, and the nauplius larvae of barnacles.

Other members of the meroplankton include fish eggs, fish larvae, and juvenile fish. The young fish feed on other larvae until they grow large enough to hunt for other foods. Some large seaweeds release **spores,** or reproductive cells, that drift in the plankton until they are consumed or settle out to grow attached to the sea bottom.

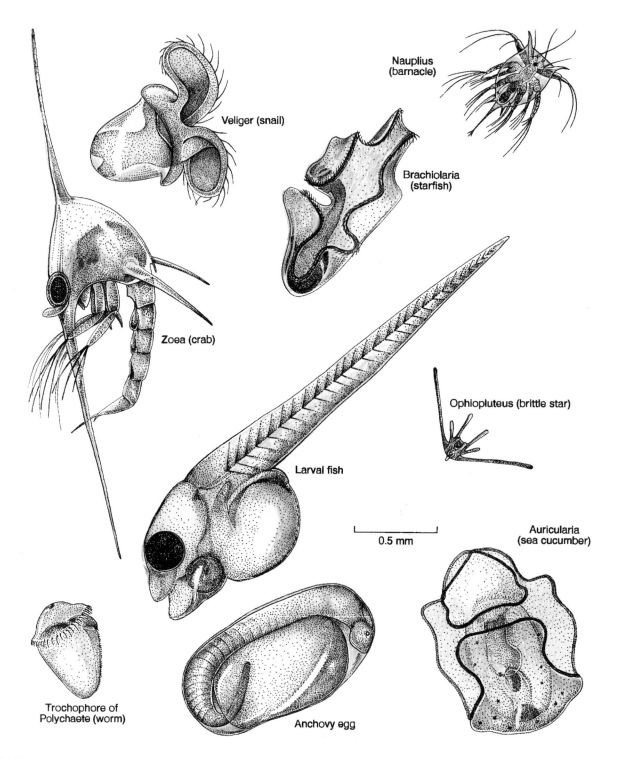

Veliger (snail)

Nauplius
(barnacle)

Brachiolaria
(starfish)

Zoea (crab)

Ophiopluteus (brittle star)

Larval fish

0.5 mm

Auricularia
(sea cucumber)

Trochophore of
Polychaete (worm)

Anchovy egg

Figure 16.17 Members of the meroplankton. All are larval forms of nonplanktonic adults.

16.4 Bacterioplankton

Bacterioplankton are composed of members of the two most ancient domains of life—Bacteria and Archaea. Bacterioplankton share a similar morphology: they are unicellular, they are commonly around 1–2 microns in size, and they possess no internal membrane-bound cell structures. Members of the Bacteria and Archaea can be distinguished from one another based on DNA sequences. It is estimated that about 1×10^{29} bacterioplankton exist within marine environments.

Bacterioplankton have an enormous impact on biogeochemical cycles on our planet. They are directly involved in recycling nutrients, some can use nitrogen gas and they therefore influence nitrogen cycles, and they play important roles in the global carbon cycle. Photosynthetic and nitrogen-fixing cyanobacteria were covered in section 16.2. Here we will discuss heterotrophic and chemosynthetic members of the bacterioplankton.

Most marine bacterioplankton exist either as single cells that float freely in the water column or as colonies of cells attached to sinking particles. Marine bacteria can be a significant food source for planktonic larvae and single-celled zooplankton and mixotrophic algae. A film of bacteria is commonly found on particles of floating organic matter. The attached bacteria help break down the organic matter, regenerating inorganic nutrients. The small size of some particles, with their attached microbial populations, makes them ideal food for small zooplankton. Bacterioplankton also utilize dissolved organic matter (DOM) that is released by phytoplankton and that is also a product of zooplankton feeding and excretion. Overall, about half the organic matter generated by phytoplankton is consumed as DOM by the bacterioplankton. "The microbial loop" refers to the processes that convert DOM into biomass that can be consumed by other organisms (fig. 16.18).

Our view of the composition of bacterioplankton is changing rapidly. Traditionally, marine bacteria were mainly studied by first isolating them from seawater using nutrient-enriched seawater media. The capabilities of these isolated microbes were then determined in laboratory studies. It is now clear that most bacterioplankton in the world's oceans have been largely resistant to these attempts at isolation and subsequent cultivation in the laboratory. With the advent of DNA-based identification technologies, knowledge of bacterioplankton diversity has increased exponentially. However, most members of the Bacteria and Archaea in the sea are still known only by a small segment of DNA sequence (the 16S rRNA gene). The inability to study these DNA-identified microbes in the laboratory means that it is still difficult to know exactly what processes they carry out within the world's oceans. This creates a problem: scientists know that a particular type of bacteria is present within the water column or sediment sample based on its DNA sequence, but they don't know exactly what processes the organism carries out. To address this quandary, marine microbiologists have developed new techniques to isolate the bacterioplankton adapted to the low-nutrient environments characteristic of most of the world's oceans. Bacteria that live in these environments are known as oligotrophs. They do not grow at the high-nutrient concentrations characteristic of near-shore environments, which were used in the early isolation attempts. Use of these new low-nutrient culturing techniques led to the recent isolation of one of the most abundant bacteria in the sea—*Pelagibacter ubique*. This organism grows slowly on open-ocean seawater with no added nutrients and does not increase its growth rate in response to the addition of organic carbon. The entire DNA se-

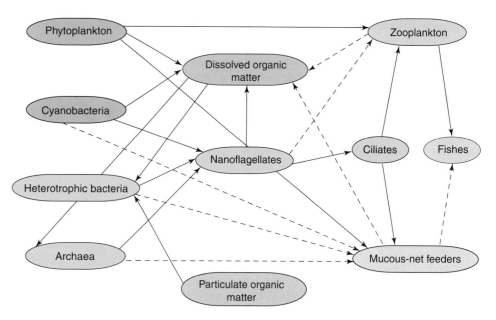

Figure 16.18   Simplified schematic of a marine food web highlighting the importance of the microbial loop composed of photosynthetic *(green boxes)* and heterotrophic *(yellow boxes)* microorganisms that convert dissolved organic matter and decaying particulate organic matter *(brown boxes)* into biomass that can be consumed by larger organisms *(blue boxes)*. Dominant pathways are indicated with *solid arrows*. More minor pathways are indicated with *dashed arrows*.

Figure 16.19   The perforated probe was driven into the seafloor crust by the ROV *Jason II*. The probe allows the sampling of fluids circulating in the upper oceanic crust. *Photo courtesy of Dr. H. Paul Johnson, School of Oceanography, University of Washington.*

quence of *Pelagibacter* has recently been obtained, which should lead to further insights into how these cells function.

Another breakthrough has come from an increased understanding of Archaea. Members of the Archaea were originally considered to be restricted to extreme environments such as the hydrothermal vents (fig. 16.19) where vent water can be acidic and extremely hot. The use of DNA-based technologies now indicates that Archaea are more widespread than originally realized. One type of Archaea, known as marine group I, comprises about 40% of the bacterioplankton within mesopelagic waters, making them one of the more abundant groups of

Extremophiles

Extremophiles are microorganisms that thrive under conditions that would be fatal to other life forms: extreme temperatures (hot and cold), high levels of acid or salt, no oxygen, no sunlight. Extremophiles not only flourish under these severe conditions but may actually require them to reproduce. Although some extremophiles have been known for forty years or more, scientists have been discovering more and more of these organisms in environments that were once thought lifeless.

These single-celled microorganisms resemble bacteria, for they have no membrane-bounded nucleus. However, when their genes were compared to the genes of bacteria, it was discovered that they are distinctly different. In fact, they appear to share a common ancestor with the eukaryotes, organisms such as ourselves with a membrane-bounded nucleus and cellular functional units, or organelles. This discovery opened for review the basic categories of all living organisms, and the result is a reorganization of life categories into three major domains: Bacteria, Archaea (the extremophiles), and Eukarya (all nuclei-containing organisms) (see fig. 14.3). This discovery was made by Professor Carl Woese at the University of Illinois; it has completely revised the way we think about microbial biology and the organization of life on Earth.

In the oceans, heat-loving members of the Archaea require temperatures in excess of 80°C (176°F) for maximum growth. *Pyrolobus fumarii* is an extreme example. It was found at a depth of 3650 m (12,000 ft) in a hot vent in the mid-Atlantic Ridge southwest of the Azores. Its name means "fire lobe of the chimney" from its shape and the black-smoker vent where it was found. *P. fumarii* stops growing below 90°C (194°F) and reproduces at temperatures up to 113°C (235°F). It uses hydrogen and sulfur compounds as sources of energy and can also use nitrogen gas. It is able to live with or without oxygen. Other Archaea are commonly found in the plumes of hot water that occur after an undersea eruption; it is unclear how deep into Earth's crust these microorganisms are able to exist (box fig. 1).

Another surprise was finding that close relatives of the hot-vent Archaea are common and abundant components of the marine plankton, living in the cold, oxygenated waters off both coasts of North America. They have since been found at all latitudes in water below 100 m (330 ft), in the guts of deep-sea cucumbers, as well as in marine sediments. A microorganism found in Antarctic sea ice grows best at 4°C (39°F) and does not reproduce at temperatures above 12°C (54°F). And still other extremophiles have been found living in the salt ponds constructed for the evaporation of seawater.

Scientists in the United States, Japan, Germany, and other countries have a particular interest in the enzymes of extremophiles. Enzymes are required in all living cells to speed up chemical reactions without being altered themselves. Standard enzymes stop working when they are exposed to heat or other extremes, so those used in research and industrial processes must

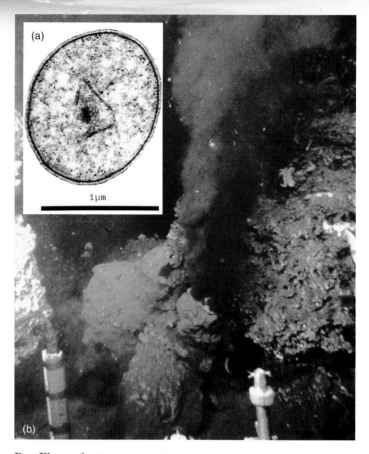

Box Figure 1 The extremophile *Pyrococcus endeavorii* (a) was isolated from an East Pacific Rise black smoker (b). This microorganism grows at temperatures that exceed 100°C (212°F).

be protected during reactions and while in storage. Heat-loving extremophile enzymes are already being used in biological and genetic research, forensic DNA testing, and medical diagnosis and screening procedures for genetic susceptibility to some diseases. In industry, these enzymes have increased the efficiency of compounds that stabilize food flavorings and reduce unpleasant odors in medicines. Enzymes that work at low temperatures may be useful in food processing where products must be kept cold to prevent spoilage. An important result of research with enzymes from extremophiles is learning how to redesign conventional enzymes to work under harsher conditions.

The discovery of microorganisms where none was assumed to exist and the recognition of a new branch of life are astonishing. It is likely that more discoveries and new technologies based on such discoveries will continue for some time to come.

microorganisms on the planet. A member of the marine group I was recently isolated and studied in the laboratory. This organism does not rely on solar energy to drive the formation of organic carbon. Instead it relies on chemosynthesis to derive energy from chemicals dissolved in seawater. This organism fixes carbon dioxide into organic matter and oxidizes ammonia to form nitrate. These organisms are hypothesized to play important roles in both carbon and nitrogen cycles.

Many of the bacterioplankton at the hydrothermal vents also carry out chemosynthesis to generate organic matter. They oxidize the hydrogen sulfide present in high concentrations in the hydrothermal vent water. There are worms that live at the hydrothermal vents that do not possess guts or mouths but instead harbor chemosynthetic microorganisms that generate organic carbon for them. These worms do not require sunlight or food generated in surface waters to survive.

16.5 Viruses

A **virus** is a noncellular particle composed of genetic material surrounded by a protein coat. Viruses are obligate parasites and are only able to carry out their metabolic activities and replicate themselves inside a host. Viruses likely infect all living organisms in the sea, from bacteria to whales. The marine environment is also a potential reservoir for disease-causing viruses. For example, some viruses are thought to cycle between marine and terrestrial mammals. In general, viral abundance is strongly correlated with planktonic biomass in the oceans. Viral abundance is highest where bacteria and phytoplankton abundances are highest; viral abundance decreases with depth and distance from shore. It is estimated that there are about 3×10^6 viruses in a milliliter of deep seawater and about 1×10^8 viruses in a milliliter of productive coastal waters. If we assume that most of the ocean is similar to the deep sea, then on average there are about 3×10^9 viruses per liter of seawater. If we estimate the volume of the oceans at about 1.3×10^{21} liters, then there are about 4×10^{30} viruses in the world's oceans. In fact, viruses are so abundant that scientists estimate that if they were all laid out end to end they would extend even further into outer space than the nearest sixty galaxies! Viruses are the most abundant biological entity in seawater.

Viruses are major pathogens of marine microbes. Some estimate that viruses have the same impact on microbial communities as zooplankton grazing. Others estimate that viruses kill as much as 40% of marine bacteria on a daily basis. It is likely that the extent of viral-induced death of microbes varies depending on the environment under study, but it is clear that viruses have a huge impact on microbial communities. When viruses kill a microbe, the cell is lysed, and the cell's organic matter is released as DOM. This DOM is available for utilization by heterotrophic bacteria. Viruses enhance recycling of nutrients and likely impact each interaction of the microbial loop depicted in figure 16.18.

During replication within a host cell, a virus may acquire some of the host's genes. When the host cell lyses and virus particles are released, some of these newly acquired genes may be transported to another cell upon the next round of infection. This can lead to an exchange of genetic material between organisms. The high abundance of viruses in the ocean and the apparently routine infection of microbes suggest that this transfer of genetic material between organisms may have been occurring naturally for a very long time.

16.6 Sampling the Plankton

The biological oceanographer needs to know which species make up the plankton in a given geographical area, their abundance, and where in the water column they are located.

Figure 16.20   After a plankton tow, the net is rinsed to wash the plankton down into the sample cup at the narrow end of the net.

Traditionally, plankton are sampled by towing fine-mesh, cone-shaped nets (fig. 16.20) through the sea behind a vessel or by dropping a net straight down over the side of a non-moving vessel and pulling it up like a bucket. After the net is returned to the deck, it is rinsed carefully, and the catch is collected. The total water volume sampled is measured by placing a flow meter in the mouth of the net. Multiplying water flow by the cross-sectional area of the net yields the sampled volume. It is also possible to raise and lower the net as it is being towed horizontally. This motion allows sampling to be averaged over both horizontal distance and depth. If the sample is to be taken at a specific depth, the net may be lowered closed and opened only when the desired depth is reached. After the towing operation, the net is again closed before it is brought up through the shallower water.

Multiple-net systems may be mounted on a single frame; the nets are opened and closed on command from the ship. The frame also carries electronic sensors that relay data on salinity, temperature, water flow, light level, net depth, and cable angle to the ship's computer.

Plankton tows must be rapid enough to catch the organisms but slow enough to let the water pass through the net. If the tow is too fast, the water is pushed away from the mouth of the net, and less water than expected is filtered. The size of the mesh from which the net is made also influences the catch; a fine-mesh net clogs rapidly and a large-mesh net loses organisms. Plankton may also be filtered from samples taken by a water bottle or a submersible pump.

Whether a plankton sample is taken by net, bottle, or pump, the next step is to determine the number and kinds of plankton in the sample. Many different approaches can be used. A portion of the sample can be examined under the microscope. It is labor-intensive to manually count all the organisms within a sample, and usually only a few types are targeted for counting. A portion of the sample can also be examined using flow cytometry, an instrument described in the box on picoplankton. Flow cytometers detect fluorescence from individual cells. The fluorescence detected can be due to phytoplankton pigments or to stains added to the cells. Viral, bacterioplankton, picoplankton, and nanoplankton abundances can be estimated with this approach. The plankton can also be collected on a filter, and DNA can be extracted from all the cells. Special techniques are then used to identify which organisms were trapped on the filter based on their DNA sequences. The most spectacular example of this approach was an experiment conducted recently in the Sargasso Sea. Two hundred liters of seawater were processed through filters designed to capture as many bacterioplankton as possible. The DNA from the organisms captured was extracted and over a million new genes were sequenced. Based on this work, scientists estimate that as many as 1800 new species were identified in this one experiment.

Sonar may be used to determine the quantity of zooplankton directly; the echo returned to the ship is related to the density of the zooplankton but does not determine the species present. The abundance of phytoplankton can also be determined by dissolving the chlorophyll *a* pigment out of the sample and measuring the pigment concentration. New optical instrumentation has been designed to measure the natural fluorescent signal coming from chlorophyll *a* in phytoplankton cells. These measurements are made at sea, directly and instantaneously at depth, and then are related to phytoplankton abundance.

None of these methods tells us how the various plankton species behave. A video plankton recorder designed at the Woods Hole Oceanographic Institution uses a strobe light with four video cameras at four different magnifications to photograph the plankton. The strobe flashes sixty times a second, capturing images of the organisms that researchers hope will enable them to learn about the swimming, feeding, and reproduction of the plankton. Eventually, they hope to program the system to recognize different kinds of plankton, enabling them to acquire species and population information while the system is running at sea.

16.7 Practical Considerations: Marine Toxins

Harmful Algal Blooms

The term **harmful algal blooms,** or **HABs,** is used to describe both toxic and nuisance phytoplankton blooms. HAB, rather than **red tide,** has become the preferred scientific term because

Figure 16.21 This nontoxic red tide of the dinoflagellate *Noctiluca* extended for 10 km (6 mi) along the shore of Puget Sound, Washington, in 1996. The bloom persisted for a week.

the blooms have nothing to do with the tides, and they may or may not color the water red. During a toxic bloom, particular species of phytoplankton produce a toxin that can be biomagnified through the food web. The production of toxin is commonly only detected once it has reached measurable levels in an organism of direct interest to humans such as shellfish or marine mammals. A nuisance bloom is not associated with toxin production and may include the following types of events: production by certain phytoplankton of such high quantities of mucous-like substances that fishermen's nets are clogged; blooms of a species of the diatom *Chaetoceros* that has long, sharp spines that irritate the gills of pen-reared salmon—the salmon cannot escape, and they suffocate because of the mucus produced in response to the irritation; phytoplankton blooms that result from nutrient enrichment—these blooms may make recreational water activities unpleasant. The picoplankton *Aureococcus anophagefferens* and *Aureoumbra lagunensis* also do not produce a toxin, but they can reach such high cell concentrations that some shellfish stop feeding. Some phytoplankton may bloom and color the water red but are not harmful to humans and other organisms (fig. 16.21). The Red Sea received its name because of dense blooms of nontoxic cyanobacteria that have large amounts of red pigment. The Gulf of California has been called the Vermillion Sea for the same reason. The focus for the rest of this chapter will be toxic phytoplankton blooms.

Only a few species of phytoplankton produce toxins. The toxins associated with HABs (table 16.1) are often powerful nerve poisons that in the most extreme cases can cause paralysis, memory loss, or death. For example, one set of algal toxins known as saxitoxins is fifty times more lethal than strychnine. It is still not clear exactly which factors cause a limited number of species of phytoplankton to produce toxins. The syndromes in humans associated with consumption of toxin-contaminated food are paralytic shellfish poisoning (PSP), neurotoxic shellfish poisoning (NSP), diarrhetic shellfish poisoning (DSP), ciguatera fish poisoning, and

Table 16.1	Phytoplankton Toxins		
Toxin	**Condition Produced**	**Phytoplankton Responsible**	**Characteristics**
Brevetoxin	Neurotoxic shellfish poisoning (NSP)	*Karenia brevis;* dinoflagellate	Affects nervous system; respiratory failure in fish and marine mammals; food poisoning symptoms in humans
Ciguatoxin	Ciguatera fish poisoning	*Gambierdiscus toxicus;* dinoflagellate	Affects nervous system; human symptoms variable
Domoic acid	Amnesiac shellfish poisoning (ASP)	*Pseudo-nitzschia;* diatom	Acts on vertebrate nervous system
Okadaic acid	Diarrhetic shellfish poisoning (DSP)	*Dinophysis* and *Prorocentrum;* dinoflagellates	Affects metabolism, membrane transport, cell division
Exotoxins		*Pfiesteria piscicida;* dinoflagellate	Mode of action unknown; produces mortality in fish, neurotoxic symptoms in humans
Saxitoxin	Paralytic shellfish poisoning (PSP)	*Alexandrium, Gonyaulax,* and *Gymnodinium;* dinoflagellates	Causes paralysis and respiratory failure in humans

amnesiac shellfish poisoning (ASP). As described below, the first four syndromes are caused by toxin-producing dinoflagellates; ASP is caused by toxin-producing diatoms. These syndromes all have the term "shellfish" or "fish" in their name because consumption of toxin-containing shellfish or fish is the most common route of toxicity in humans. The shellfish, for example, feed on toxin-producing phytoplankton and concentrate the toxin in their tissues. Humans commonly consume multiple shellfish at a time, further concentrating the toxin. The toxins are not affected by heat, so cooking the shellfish or fish does not neutralize the toxin. In addition to their impact on human health, the toxins can kill fish, shellfish, marine mammals, birds, and other animals that may consume toxin-contaminated seafood. In recent years, a number of research centers have been developed to understand the impacts of ocean processes such as HABs on human health. These research centers bring together public health researchers and oceanographers to address these questions.

The frequency and geographical distribution of HAB incidents have increased over the last few decades. Part of the observed increase is due to enhanced monitoring systems; more toxic events are reported because better detection methods are now in place. However, there are also documented cases of expansion of toxic blooms to previously unaffected areas. Researchers are currently trying to determine the role that human activities may play in the expansion of toxic blooms. Some see the expansion of harmful and toxic algal blooms as an indication of large-scale marine ecologic disturbances. Possible causes for the expansion of HAB events are as follows. Toxic species can be transported between impacted and unimpacted areas via ship ballast water transport or shellfish transplantation. Some species of HABs may be stimulated by the continuous addition of nutrients to coastal waters from sewage, agriculture, and aquaculture ponds. Large-scale climate fluctuations, such as those caused by El Niño, impact circulation patterns. This can lead to the transport of HABs to previously unaffected regions. For example, the frequency and persistence of NSP blooms on the west coast of Florida do not appear to have changed over the last 120 years despite increases in

nutrient enrichment of coastal waters. Instead, major range expansions of these organisms have coincided with unusual climatic events.

Paralytic Shellfish Poisoning (PSP)

Three genera of dinoflagellates—*Alexandrium, Gymnodinium,* and *Pyrodinium*—produce the toxins known as saxitoxins that cause PSP. PSP occurs seasonally along the northeastern and western coasts of the United States. Prior to the 1970s, PSP was documented only in North America, Europe, and Japan. Currently, PSP events have also been reported in South America, Australia, Southeast Asia, and India. In the United States' waters, PSP is associated primarily with *Alexandrium.* Part of the *Alexandrium* life cycle includes formation of dormant cysts that facilitate over-wintering of the cells. The cysts return to active growth once environmental conditions are favorable. Cyst formation means that repeated toxic events are almost inevitable once a bay experiences a toxin-producing *Alexandrium* event. Humpback whales have been poisoned by feeding on mackerel that had fed on toxic *Alexandrium.* The earliest documented cases of PSP in North America occurred in 1793 among Captain George Vancouver's crew. Five crew members became ill and one died after eating mussels gathered along the British Columbia coast.

A recent study examined the impact of saxitoxins on shellfish. Saxitoxin causes toxicity by binding to a molecule (the sodium channel) that is required for nerve and muscle activity; saxitoxin binding results in neuromuscular paralysis. Scientists found that when softshell clams (*Mya arenaria*) were fed on saxitoxin-producing *Alexandrium* for just twenty-four hours, some clams displayed signs of distress (reduced burrowing and impaired siphon retraction), and others appeared to be unaffected. Saxitoxin was concentrated to much higher levels in the unaffected, apparently resistant, clams. The clams that were sensitive to the toxin had been collected from areas with no previous reports of PSP outbreaks. The clams that were resistant to the toxin were collected from areas with repeated PSP outbreaks. The researchers found that a mutation in the clam's sodium channel conveyed resistance to the toxin. In nature, sensitive clams that do not possess the mutation are likely killed by exposure to saxitoxin. Only those clams within a population that possess the mutation are resistant to exposure, and these resistant clams are the ones that will survive and be able to reproduce in an area. The implications of this study are that repeated exposures to toxic outbreaks result in selection for clams that possess the mutation necessary to survive exposure to the toxin. This study also implies that repeated exposure to saxitoxin results in clams that concentrate higher levels of toxin and are more dangerous for human consumption.

Neurotoxic Shellfish Poisoning (NSP)

A single species of dinoflagellate—*Karenia brevis*—produces the toxin known as brevitoxin that causes NSP. *Karenia brevis* is present in the Gulf of Mexico, and historically, NSP was restricted to the west coast of Florida. The Florida Current and the Gulf Stream can transport cells out of the Gulf of Mexico and into Atlantic waters. A toxic event was reported in 1987 along the coast of North Carolina, likely due to a shoreward intrusion of Gulf Stream water carrying *K. brevis* cells. *Karenia brevis* is easily lysed in turbulent water and releases the toxin directly into seawater. The toxin can be aerosolized by wave action and can be inhaled, which results in respiratory problems and eye irritation. Fish may be rapidly killed when *K. brevis* cells are lysed as they pass through the gills, directly transferring the toxin to the gill tissue. At least 149 manatees died in 1996 due to exposure to *K. brevis*.

Diarrhetic Shellfish Poisoning (DSP)

Dinoflagellates of the genus *Dinophysis* and the species *Prorocentrum lima* produce the toxin okadaic acid. The gastrointestinal symptoms associated with okadaic acid are the mildest of toxins described here. DSP is widespread with reported cases from Europe, Japan, North and South America, South Africa, New Zealand, Australia, and Thailand.

Ciguatera Fish Poisoning

Ciguatera poisoning was first documented in the early sixteenth century, but it has probably been present in the tropics for much longer. It is estimated that each year, between 10,000 and 50,000 people eating fish in tropical regions are affected by ciguatera poisoning. More than 400 species of fish have been found to be affected, and there is no way to prepare affected fish to make it safe to eat. In the United States, over 2000 cases of ciguatera poisoning may occur each year, clustered in Florida, Hawaii, the Virgin Islands, and Puerto Rico. Symptoms of ciguatera poisoning are extremely variable; they may include headaches, nausea, irregular pulse beat, reduced blood pressure, and, in severe cases, convulsions, muscular paralysis, hallucinations, and death.

Several dinoflagellates are associated with ciguatera poisoning, but the dinoflagellate *Gambierdiscus toxicus* is most often associated with the problem. This dinoflagellate was first isolated in 1976 by scientists in Japan and French Polynesia. A toxin isolated at the University of Hawaii was named ciguatoxin; more toxic compounds have since been identified. Ciguatoxic dinoflagellates live attached to many types of seaweeds and appear to need nutrients exuded by the seaweeds. The seaweeds are eaten by herbivorous fish, and the ciguatoxins move through the food web stored in the fishes' liver. Initially, only a few species are toxic, while at the peak of the outbreak, almost all reef fish become toxic, and in the final stages, only large eels and certain snappers and groupers remain toxic. This cycle appears to take at least eight years and may last for as long as thirty years. Unfortunately, no easy way has been found to routinely measure ciguatoxin levels in any seafood. Outbreaks of ciguatera seem to be preceded by a disruption of the reef system triggered by both natural phenomena such as monsoons and seismic events and human activities such as construction of piers and jetties. However, not all reef disruptions are followed by an increase in fish toxicity.

Amnesiac Shellfish Poisoning (ASP)

In 1987, on Prince Edward Island, Canada, three people died and more than a hundred others became sick from eating shellfish contaminated with domoic acid. Improved chemical detection allowed scientists to trace the toxin to a bloom of *Pseudo-nitzschia*, a diatom, not a dinoflagellate. Diatoms had not been known to produce toxins until this outbreak occurred. Domoic acid poisoning, or amnesiac shellfish poisoning (ASP), may cause short-term memory loss in humans. Other symptoms include nausea, muscle weakness, disorientation, and organ failure. A second recorded case of ASP occurred in September 1991 in California's Monterey Bay, where more than 100 brown pelicans and cormorants died or showed unusual neurological symptoms after eating anchovies that had been feeding on *Pseudo-nitzschia*. That same year and the next, razor clam and crab fisheries along the Oregon and Washington coasts were closed because of high levels of domoic acid. In 1998, more than 100 sea lions died off the California coast from feeding on anchovies and sardines that had fed on the diatoms. The reasons for the sudden appearances of toxic levels of domoic acid are unknown.

Pfiesteria

The dinoflagellate *Pfiesteria piscicida*, was first identified in North Carolina in 1993. Since then, *P. piscicida* and other *Pfiesteria*-like organisms have been reported as the cause of major fish kills in estuaries and coastal waters of the Mid-Atlantic and southeastern United States. Cells may form among amoeboid and flagellated swimming forms that stun and poison fish. *P. piscicida* is considered a voracious predator on other marine microorganisms and has been linked to human symptoms that include sores, severe memory loss, nausea, respiratory distress, and vertigo. Whether fish kills are the product of the toxin released into the environment or whether *P. piscicida* kills by direct means is not clearly understood.

Cholera

Vibrio cholerae is a bacterium, but its occurrence in marine waters is associated with phytoplankton blooms. These areas are a reservoir for new environmentally hardy variants of the cholera bacterium (*Vibrio cholerae*) that adhere to members of the phytoplankton and zooplankton. In 1991, an epidemic of a new cholera variant broke out along the Peruvian coast; over a fifteen-month period, more than half a million people became ill and nearly 5000 people died in nineteen Latin American countries. *V. cholerae* was isolated from phytoplankton in the upwelling zone along the Peruvian coast. The disease is thought to have crossed the Pacific in the bilge water of ships from Bangladesh. Since 1960, seasonal outbreaks of cholera in Bangladesh have been related to coastal phytoplankton blooms; the bacterium is found associated with cyanobacteria, diatoms, dinoflagellates, seaweeds, and most heavily with zooplankton egg sacs.

Summary

Plankton are the drifting organisms. The microscopic plankton are divided into groups by size. Phytoplankton are autotrophic single cells or filaments. Their abundance can be controlled by nutrient availability, light, and temperature. These abiotic factors represent bottom-up controls. Abundance is also controlled through consumption by other organisms in a series of processes known as top-down controls.

Sargassum is the only large planktonic seaweed. The diatoms are found in cold, upwelled water; they are yellow-brown, with a hard, transparent frustule, and they store oil, which increases their buoyancy. Centric forms are round; pennate forms are elongate. Diatoms reproduce rapidly by cell division and make up the first trophic level in marine ecosystems.

Dinoflagellates are single cells with both autotrophic and heterotrophic capabilities. Their cell walls are smooth or are heavily armored with cellulose plates. They, too, reproduce by cell division. These organisms are responsible for much of the bioluminescence in the oceans. Coccolithophorids and silicoflagellates are small autotrophic members of the phytoplankton. Cyanobacteria and green algae are important members of picophytoplankton communities. Some species of cyanobacteria can also use nitrogen gas as a source of inorganic nitrogen.

Some herbivorous zooplankton reproduce several times a year; others reproduce only once, depending on the water temperature and phytoplankton food supply. Carnivorous zooplankton are important in the recycling of nutrients to the phytoplankton. High concentrations of zooplankton are found at convergence zones and along density boundaries. Zooplankton migrate toward the sea surface at night and away from it during the day, forming the deep scattering layer.

Zooplankton that spend their entire lives in the plankton are called holoplankton. The copepods and euphausiids are the most common members of the holoplankton. Euphausiids are also known as krill; they form a basic food of the baleen whales. Krill is the zooplankton base for all the Antarctic food webs; recently, it has been harvested for human consumption with mixed success.

Other small members of the holoplankton are the carnivorous arrowworms; the calcareous-shelled foraminifera; the delicate, silica-shelled radiolarians; the ciliated tintinnids; and the swimming snails, or pteropods. Large zooplankton include the comb jellies, salps, and jellyfish; all are nearly transparent, but each belongs to a different zoologic group.

The meroplankton are the juvenile (or larval) stages of nonplanktonic adults. This group comprises fish eggs, very young fish, and the larvae of barnacles, snails, crabs, starfish, and many other nonplanktonic animals. The spores of seaweeds are also planktonic.

Great numbers of bacterioplankton live free in seawater and coat every available surface. They are important as food sources and agents of decay and breakdown of organic matter. Dissolved organic material (DOM) moves through the microbial loop and is converted into particulate organic matter that is consumed by zooplankton. Viruses are noncellular, parasitic particles that are abundant in seawater. Viruses can infect bacteria and other marine organisms.

Plankton sampling is usually done with a plankton net or with a water bottle. The kinds of organisms in a sample are determined microscopically; the numbers of organisms can be counted using flow cytometry. Organisms can also be identified based on their DNA sequences.

Blooms of dinoflagellates and some other kinds of phytoplankton produce harmful algal blooms (HABs). Some HABs are toxic; others are not. The toxin is concentrated in shellfish and produces paralytic and other types of shellfish poisoning in humans and sometimes in other animals. HABs appear to be triggered by particular combinations of environmental factors, including addition of increasing amounts of nutrients to coastal waters. *Pfiesteria*-like dinoflagellates have complex life cycles, cause fish kills, and affect humans. Domoic acid is produced by diatoms. Its presence causes shellfish-harvest closures and has affected birds and marine mammals, as well as humans. Ciguatoxin is another dinoflagellate product that affects humans and hampers fishery development around the world. Warm, nutrient-rich coastal waters have been associated with outbreaks of new cholera variants.

Key Terms

All key terms from this chapter can be viewed by term or definition when studied as flashcards on this book's website at www.mhhe.com/sverdrup10e.

phytoplankton, 378
zooplankton, 378
bacterioplankton, 378
picoplankton, 378
nanoplankton, 378
microplankton/net plankton, 378
macroplankton, 378

diatom, 379
dinoflagellate, 379
coccolithophorid, 379
cyanobacteria, 379
green algae, 379
frustule, 379
radial symmetry, 380
bilateral symmetry, 380
centric, 380
pennate, 380
auxospore, 380
holoplankton, 386

meroplankton, 386
crustacean, 386
copepod, 386
euphausiid, 386
krill, 386
baleen, 386
chaetognath, 388
foraminafera, 388
radiolarian, 388
tintinnid, 388
cilia, 388
pteropod, 388

ctenophore, 389
salp, 389
Coelenterata/Cnidaria, 389
colonial organism, 389
larva, 391
spore, 391
extremophile, 394
virus, 395
harmful algal bloom (HAB), 396
red tide, 396
ciguatera, 398

Study Questions

1. Why does a pycnocline located above the compensation depth promote a phytoplankton bloom?
2. Why are meroplankton produced in such large numbers?
3. Even after toxin-producing phytoplankton are no longer found at a particular site, the shellfish may still not be safe to eat. Explain.
4. Patches with abundant populations of zooplankton are frequently found separated by patches with sparse populations. How does this pattern help to ensure survival from predators?
5. If the krill of the Southern Ocean were heavily harvested for human consumption, explain the possible effects on the rest of the organisms in that area.
6. Describe four ways to subdivide the plankton.
7. Consider temperate coastal waters. Why do diatoms tend to bloom in the spring and dinoflagellates tend to bloom in the summer?
8. Discuss what happens to a diatom population if no auxospores form.
9. Why is a planktonic stage important to a nonplanktonic adult?

10. Discuss how plankton may maintain themselves in a given region of the ocean even though currents are flowing through that region.
11. When you are sampling plankton with a plankton net, how can you determine the quantity of the plankton in a volume of water? Assume that you know (a) the cross-sectional area of the net and (b) the length of time you towed the net and the speed at which you towed the net, or you know (a) the cross-sectional area of the net and (b) the distance you towed the net.
12. Distinguish between diatoms and dinoflagellates and between euphausiids and copepods. Which are heterotrophs and which are autotrophs?
13. What is the microbial loop, and why is it important?
14. Episodes of toxic phytoplankton blooms appear to be increasing along the world's coasts. (a) List possible reasons for this increase. (b) Distinguish among PSP, NSP, ASP, and ciguatera.
15. Toxic materials, such as oil, may form a thin layer at the sea surface where many plankton also accumulate. How might this affect populations of the nekton and the benthos?

School of blue stripe sea perch or snapper.

The Nekton: Free Swimmers
of the Sea

Chapter Outline

Learning Outcomes

After studying the information in this chapter students should be able to:

1. *explain* the concept of keystone predators and how they impact species diversity within an ecosystem,

2. *explain* the impacts of top-down and bottom-up controls on marine mammals,

3. *compare* and *contrast* how threats to marine mammals and to commercial fisheries have changed over the past century, and

4. *explain* the relation of climate patterns (El Niño and Pacific decadal oscillation) on fisheries.

The nekton are the free swimmers of the oceans. Thousands of species of nekton swim freely through the pelagic and neritic regions of the oceans. Nekton are fishes, marine mammals, ocean-living reptiles, and squid. Members of the nekton are able to move toward their food and away from their predators; many occupy the top trophic levels of marine food webs. Sizes range from the smallest fish of the tropical reefs to the largest animal ever to have existed on Earth, the blue whale. This chapter examines representative swimmers in coastal waters and in the open sea; it contains additional information on their harvests, including quantities taken, methods used, and current population status.

17.1 Mammals

Marine **mammals** are homeotherms, which means that they can maintain a near-constant body temperature, frequently well above water temperatures (review section on temperature, chapter 14). Marine mammals also breathe air, although they can make dives of spectacular depths and duration. Record diving depths reported for some marine mammals and humans are given in table 17.1. The best diving mammals have streamlined shapes that reduce the drag on their bodies as they move through the water (review section on considerations of size, chapter 14). This lessens their swimming efforts and lowers their oxygen consumption. The best divers also show significant differences in the concentrations and distribution of the protein myoglobin when compared to terrestrial mammals. Myoglobin helps regulate oxygen concentrations and is found primarily in muscle tissue. Diving animals (and birds) appear to rely on oxygen bound to myoglobin while diving and are able to regulate its flow as needed during the dive. The concentration of myoglobin in marine mammals is three to ten times higher than that found in their terrestrial relatives.

Some marine mammals spend their entire lives at sea; others return to land to mate and give birth. Their young are born live (not as eggs) and are nursed by their mothers. Included in the marine mammals are large and small whales (including porpoises and dolphins), seals, sea lions, sea otters, and sea cows.

Whales

Whales belong to the mammal group called **cetaceans;** see table 17.2 for information about representative whales. Some

Table 17.1	Record Diving Depths of Mammals	
Species	**Dive Depth (m)**	**(ft)**
Human (*Homo sapiens*)	105	347
California sea lion (*Zalophus californianus*)	250	825
Common porpoise (*Delphinus delphis*)	260	858
Killer whale (*Orcinus orca*)	260	858
Bottle-nose dolphin (*Tursiops truncatus*)	535	1766
Pilot whale (*Globicephala melanena*)	610	2013
Beluga whale (*Delphinopterus leucas*)	650	2145
Weddell seal (*Leptonychotes wedellii*)	>700	>2310
Elephant seal (*Mirounga angustiorostris*)	>1500	>4950
Sperm whale (*Physeter catodon*)	>2000	>6600

cetaceans are toothed, pursuing and catching their prey with their teeth and jaws (for example, the killer whale, the sperm whale, and the small whales known as dolphins and porpoises). Other whales instead have mouths fitted with strainers of **baleen,** or whalebone, through which they filter the seawater to capture krill and other plankton. The blue, finback, right, sei, gray, and humpback whales are baleen whales. The mouths of toothed and baleen whales are compared in figure 17.1. The blue, finback, and right whales possess baleen and swim open-mouthed to engulf water and plankton. The tongue pushes the water through the baleen, and the krill are trapped. The sei whale swims with its mouth partly open and uses its tongue to remove the organisms trapped in the baleen. The humpbacks circle an area rich in krill and expel air to form a circular screen, or a net of bubbles. The krill bunch together toward the center of this net, and the whales pass through the dense cloud of krill and scoop them up. The gray whale feeds mainly on small bottom crustaceans and worms.

Some whales migrate seasonally over thousands of miles; other whales stay in cold water and migrate over relatively short distances. The California gray whale and the humpback whale make long migratory journeys. Gray whales are found only in the North Pacific and adjacent seas. They are the most coastal of all great whales and are now composed of two populations—a large Eastern Pacific Stock and a small remnant Western Pacific Stock (table 17.3). In summer, the California gray whale is found in the shallow waters of the Bering Sea and the adjacent Arctic Ocean, where they feed all summer, building up layers of fat and blubber. In October, when the northern seas begin

Table 17.2	**Principal Characteristics of Some Whales**				
Common Name	**Scientific Name**	**Distribution**	**Weight (tons)**	**Length (m)**	**Food**
Toothed					
Sperm	*Physeter catadon*	Worldwide	35	18	Squid, octopus, deep-sea fish
Narwhal	*Monodon monoceros*	Artic Circle	2	5	Fish, squid, shrimp
Killer	*Orcinus orca*	Worldwide	11	10	Mammals, fish, squid
Baleen					
Blue	*Balaenoptera musculus*	Worldwide; north-south migrations	80–150	33	Krill
Fin	*Balaenoptera physalus*	20°–75°N, 20°–75°S; north-south migrations	50–70	24–27	Krill, fish
Humpback	*Megaptera novaeangliae*	Worldwide; north-south migrations along coasts	35	11–17	Krill, copepods, fish, squid
Southern right	*Eubalaena australis*	Southern Hemisphere, 20°–55°S	80	17	Copepods, krill
Sei	*Balaenoptera borealis*	Mid-latitudes; seasonal migrations	30	15–18	Copepods, other plankton
Gray	*Eschrichtius robustus*	North Pacific; commonly within few kilometers of shore	35	11–15	Benthic invertebrates: amphipods, polychaete worms
Bowhead	*Balaena mysticetus*	Arctic and subarctic	100	18–20	Krill, copepods
Antarctic minke	*Balaenoptera bonaerensis*	Southern Hemisphere, 10°S to ice pack	14	9–11	Krill

(b)

(a)

(c)

Figure 17.1 (a) The killer whale *(Orcinus orca)* is a toothed whale. (b) The humpback whale *(Megaptera novaeangliae)* is a baleen whale. These two humpbacks with their mouths open illustrate the location of the baleen. (c) Bowhead whales *(Balaena mysticetus)* have the longest baleens of any whales and they are shown clearly here. This dead bowhead whale was hauled onto the ice and is shown lying on its back.

to freeze, the whales begin to move south, and in December, the first gray whales arrive off the west coast of Baja California. Here, they spend the winter in the warm, calm waters of sheltered lagoons, where the gray whales calve and mate but find little food; by the time they leave for their northward migration in February and March, they have lost 20–30% of their body weight. The animals move north singly or in twos or threes, sometimes in groups of ten to twelve, up the western coast of the United States, Canada, and Alaska. Moving at about 5 knots day and night, the whales make their annual 18,000 km (11,000 mi) migratory journey to link areas that provide abundant food with

areas that ensure reproductive success (fig. 17.2a). Protection of the gray whales' calving and winter grounds in Baja California has allowed the population to build, and in 1994, the gray whale became the first marine mammal to be removed from the U.S. endangered species list.

The humpback whale also has well-defined migration patterns. Humpbacks are found in three geographically and reproductively isolated populations: in the North Pacific, North Atlantic, and Southern Ocean and a poorly studied population in the Arabian Sea. The North Pacific humpback spends the summer feeding in the Gulf of Alaska, along the northern islands

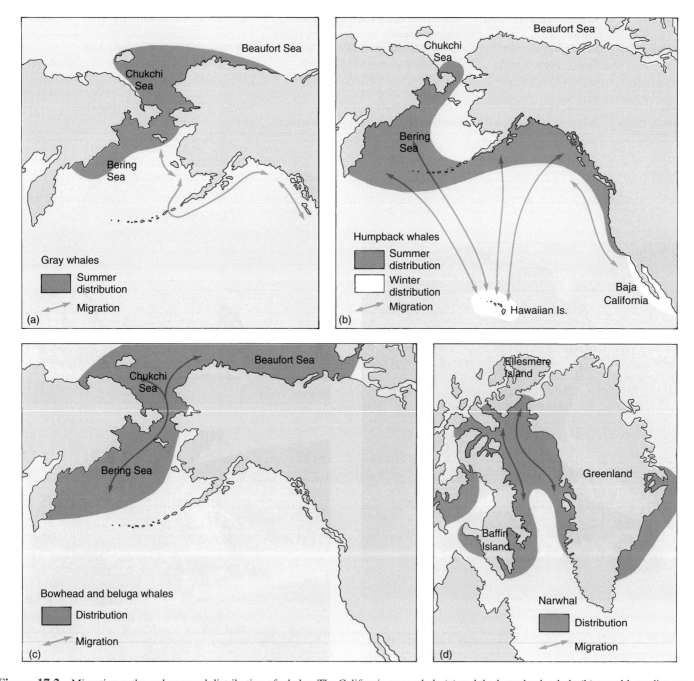

Figure 17.2 Migration paths and seasonal distribution of whales. The California gray whale (a) and the humpback whale (b) travel long distances between cold-water feeding and warm-water calving and mating areas. The bowhead and beluga whales (c) and the narwhal (d) remain in cold water and migrate over short distances.

Table 17.3	Recent World Population Estimates of Representative Whale Populations Based on IWC Assessments			
Population		**Year(s) to Which Estimate Applies**	**Estimated Population Size**	**Status**
Minke				Most abundant whale
	Southern Hemisphere	1982/83–1988/89	761,000	
	North Atlantic	1996/2001	174,000	
	West Greenland	2005	10,800	
Fin				Endangered
	North Atlantic	1996/2001	30,000	
	West Greenland	2005	3,200	
Gray				Endangered
	Western North Pacific	2007	121	
	Eastern North Pacific	1997/98	26,300	
Bowhead				Endangered
	Berin-Chukchi-Beaufort Seas	2001	10,500	
	West Greenland	2006	1,230	
Humpback				Endangered
	North Pacific	2007	At least 10,000	
	Southern Hemisphere	1997/98	42,000	
Right				Endangered
	Western North Atlantic	2001	300	
	Southern Hemisphere	1997	7,500	

of Japan, and in the Bering Sea; in winter, these North Pacific humpbacks migrate to the Mariana Islands in the west Pacific, the Hawaiian Islands in the central Pacific, and along the west coast of Baja California in the eastern Pacific. At these warmer latitudes, calves are born and mating takes place (fig. 17.2b). Globally, there may be as many as 40,000 humpback whales (table 17.3). Some populations are showing signs of recovery from exploitation.

The bowhead, the beluga, and the single-tusked narwhal remain in cold water but still migrate over short distances each year. The beluga and narwhal are toothed whales; the bowhead is a baleen whale (fig. 17.1c). The largest population of bowhead whales is found in the Bering, Chukchi, and Beaufort Seas; this whale spends nearly all its life near the edge of the Arctic ice pack. Bowheads, singly or in pairs, often accompanied by belugas, migrate north from the Bering Sea to feed in the Beaufort and Chukchi Seas as the ice recedes in the spring, returning south to the Bering Sea in groups of up to fifty as the ice begins to extend in the winter. Their mating and reproductive cycles are not well known, but they probably mate during the spring migration and calve sometime during April and May of the following year (fig. 17.2c).

The narwhal is the most northerly whale and is found only in Arctic waters, most commonly on both sides of Greenland. This toothed whale possesses only two teeth. In males, one tooth grows out through the front of the head and becomes a tusk up to 3 m long. These tusks are thought to result from sexual selection, similar to the plume of male peacocks. In summer, these whales move north along the coasts of Ellesmere and Baffin Islands, and in the autumn, they return south to the waters along the Greenland coast (fig. 17.2d).

The whales of stories and songs are the "great whales": the blue, sperm, humpback, finback, sei, and right whales (fig. 17.3 and table 17.2); they are also the whales of the whaling industry.

Whaling

The earliest-known European whaling was done by the Norse between A.D. 800 and 1000. The Basque people of France and Spain hunted whales first in the Bay of Biscay, and then in the 1500s, the Basque whalers crossed the Atlantic to Labrador. They set up whaling stations along the Labrador coast to process the blubber of bowhead and right whales into oil for transport back across the Atlantic. In Red Bay, Labrador, the operation reached its peak in the 1560s and 1570s, when 1000 people gathered seasonally to hunt whales and produce 500,000 gallons of whale oil each year. By 1600, whaling had become a major commercial activity among the Dutch and the British, and at about the same time, the Japanese independently began harvesting whales. In the 1700s and 1800s, whalers from the United States, Great Britain, and the Scandinavian and other northern European countries pursued the whales far from shore, hunting them for their oil and baleen. Kills were made using handheld harpoons, and the whales were cut up and processed on land or onboard the ships at sea. Long voyages, intense effort, and dangerous combat between whalers and whales characterized these whale hunts.

In 1868, Svend Føyn, a Norwegian, invented the harpoon gun with its explosive harpoon and changed the character of whaling. Ships were motorized, and in 1925, harvesting was increased further by the addition of great factory ships, to which the small, high-speed whale-hunting vessels brought the dead whales for processing. This system freed the fleets, centered in the Antarctic, from dependence on shore stations. These methods continued into the twentieth century, greatly increasing the efficiency of the hunt and rapidly depleting the whale stocks of the world.

In the 1930s, the annual blue whale harvest reduced the population to less than 4% of its original numbers, threatening the

Figure 17.3 Relative sizes of baleen and toothed whales.

species with extinction. In 1946, representatives from Australia, Argentina, Britain, Canada, Denmark, France, Iceland, Japan, Mexico, New Zealand, Norway, Panama, South Africa, the former Soviet Union, and the United States met in Washington, D.C., to establish the International Whaling Commission (IWC), which was set up under the International Convention for the Regulation of Whaling. The purpose of the convention was to ensure proper conservation of whale stocks and thus make possible the orderly development of the whaling industry. Regulations prohibited the killing of the blue, gray, bowhead, and right whales and of cows with calves. Opening and closing dates for whaling and minimum size data were set for each species harvested. Although each factory ship carried an observer, the individual could only report offenses and recommend disciplinary action; the government of the country registering the ship involved was responsible for any penalties. With IWC and its regulations in effect, 31,072 whales were killed in 1951, more than 50,000 in 1960, and over 66,000 in 1962. The most recent assessment of whale stocks provided by the IWC is given in table 17.3. Note that in 1989, the IWC decided to provide numbers only for populations that had been assessed in detail.

As the 1970s drew to a close, the era of commercial whaling appeared to be closing as well. In 1979, the IWC placed a moratorium on all whaling in the Indian Ocean and outlawed the use of factory ships as floating bases from which to send out hunter vessels, but whaling continued from land bases in Antarctica. In the spring of 1982, the IWC voted a moratorium on commercial harvesting of whales, except dolphin and porpoise, in order to study the whale populations and assess their ability to recover. The moratorium began in 1985–86.

In 1992, whaling nations reminded the IWC that the original intent of the organization was to "conserve" whales to protect the stocks as a harvestable natural resource. Citing the large population of North Atlantic minke whales, Norway chose to set its own national catch limits and resumed whaling in 1993; its 2000 harvest was 487 minkes. In 1994, the IWC voted twenty-three to one (Japan) and with several abstentions to establish a whale sanctuary in Antarctic waters south of 55°S. In 1999, Iceland left the IWC and returned to whaling. In recent years, much attention has focused on countries that issue permits for killing whales for scientific purposes. In 2006/2007, Japan and Iceland used special permits to catch 505 Antarctic minke and 253 common minke whales, 105 sei, fifty Bryde's, and six sperm whales. The issue of special-permit whaling remains controversial within the IWC.

The IWC permits whaling by native peoples (in Alaska, Greenland, and the former Soviet Union) in order to balance the conservation of whales with the cultural and subsistence needs of these peoples. In 1997, the Native American Makah tribe of Washington, citing an 1895 treaty with the United States, received permission to harvest five gray whales from the annual limit of 140 gray whales harvested in the North Pacific for traditional aboriginal subsistence needs. One whale was harvested by the Makahs in 1999 and another in the early summer of 2001.

While populations of California gray whales have recovered and some populations of other whale species are slowly increasing, the majority of whale species are still present in low numbers when compared to their estimated original populations. The eastern North Pacific right whale was nearly exterminated during the intensive whaling of the 1940s–60s, and it is considered the most endangered population of the large whales. Some researchers believe that the failure of whale populations to recover is due to the difficulty of finding mates in such small populations; the possibility also exists that the noise produced by increasing ship traffic interferes with whale communication. Other scientists are concerned that krill harvesting and the global depletion of fish species are affecting the whale populations; pollution may also play a role.

The intertwined history of whales and people has not yet ended. In the words of Herman Melville from the pages of *Moby Dick:*

> The moot point is whether the Leviathan can long endure so wide a chase and so remorseless a havoc; whether he must not at last be exterminated from the waters, and the last whale, like the last man, smoke his last pipe, and then himself evaporate in the final puff.

Dolphins and Porpoises

Dolphins and porpoises are small, toothed whales (fig. 17.4*e*). Bottle-nose dolphins *(Tursiops truncates)* are the most familiar of all cetaceans. These are the dolphins commonly seen in zoos and aquaria. They inhabit temperate and tropical coastal waters. The behavior and social system of these animals are diverse. In the open ocean, dolphins and porpoises are observed traveling at high speeds and in large schools. Porpoises have been timed swimming at speeds in excess of 30 knots, a feat that interests scientists and researchers. Strong bonds have been observed between mothers and calves. It is essential to the survival of young calves that they stay in close proximity to their mothers during travel. The calf swims close to the mother near her dorsal fin in a position that provides hydrodynamic benefits to the calf. The calf swims significantly faster with less effort when positioned next to the mother. The calf is able to glide about a third of the time, something that is not possible when swimming separately from the mother. Their ability to communicate and their intelligence, as demonstrated by their learning and recall abilities, are under study. They have been trained to help divers and to act as messengers between divers working underwater and surface vessels. They have been used by the U.S. Navy to locate underwater mines.

These smaller cetaceans are found in both tropical and temperate waters. Although marine, they will go up rivers and channels into shallow brackish waters. They have been seen moving across the very shallow lakes and canals of the Mississippi delta region in water barely deep enough to support their high-speed swimming.

The last two decades have placed great pressure on the world's dolphins and porpoises. A 1990 United Nations–sponsored symposium reported that more than a million of these mammals die each year in nets, usually the unwanted by-catch of those fishing for other species. Two small species, the Mexican porpoise and the black dolphin of the Chilean coast, may be endangered. Only a few hundred Mexican dolphins are left in the Gulf of California.

Seals and Sea Lions

Seals and sea lions belong to the **pinnipeds,** or "feather-footed" animals, so named for their four characteristic flippers. Representative pinnipeds are shown in figures 17.4 and 17.5. Both seals and sea lions spend most of their time at sea, hunting for fish and squid, but need to rest and breed on land.

(a)

(b)

(c)

(d)

(e)

Figure 17.4   Marine mammals. (a) The harbor seal *(Phoca vitulina)* is a friendly, curious animal that coexists well with humans. (b) This large male northern fur seal *(Callorhinus ursinus)* stands vigilant over his harem of females in the Pribilof Islands. (c) A sea otter *(Enhydra lutris)* and her pup dine on crab. (d) The walrus *(Odobenus rosmarus)* feeds from the bottom and relaxes on the Arctic ice floes. (e) Pacific white-sided dolphins *(Lagenorhynchus obliguidens)* keep pace with the bow wave of a research vessel.

To stay warm, they have hair and blubber, although not nearly as much blubber as whales. They are also large animals with a relatively small surface area to volume (review section "Considerations of Size," chapter 14). Seals are divided into true seals and eared seals. True seals have small ear holes and their rear flippers cannot be moved forward. Therefore, they do not get around well on land but move quickly through the water. Common true seals include the harbor seal *(Phoca vitulina)*, the harp seal *(Pagophilus groenlandicus)*, the Weddel *(Leptonychotes weddellii)* and leopard *(Hydrurga leptonyx)* seals of the Antarctic, and the northern elephant seal

Whale Falls

*T*he death of a whale suddenly sends a huge, localized source of food to the sea floor. In 1987, Craig Smith and colleagues from the University of Hawaii accidentally discovered the carcass of a blue whale on the floor of the Santa Catalina basin off California (box fig. 1a). Their studies show that a whale carcass, or whale fall, supports a large community of organisms. First, the scavengers such as hagfish, crabs, and sharks reduce the body to bones in as little as four months, and then the bacteria take over. Whalebones are rich in fats and oils that give the animal buoyancy in life; after death, they provide nutrition for anaerobic bacteria. These bacteria decompose the fatty substances and generate hydrogen sulfide and other compounds that diffuse out through the bone. Chemosynthetic bacteria then metabolize the sulfides and form bacterial mats over the skeleton. The bacterial mats are grazed by worms, mollusks, crustaceans, and other organisms. Other animals may be attracted, and they get their nutrition from the sulfides, the bacterial mats, the fatty substances in the bones, or other animals at the site.

The number of species found on a single skeleton is surprising: 5098 animals from 178 species were isolated from five vertebral bones recovered from one whale. Small mussels are shown on recovered whalebone in box figure 1b. The bone surface area totaled 0.83 m^2 (9 ft^2). Ten of these species, including worms and limpets, have been found only associated with whale skeletons. By comparison, the most fertile hydrothermal vent areas have yielded only 121 species and hydrocarbon seeps just thirty-six.

How abundant are these whale falls? At this time, that is difficult to say, for they can be anywhere on the ocean floor and are difficult to locate. In 1993, the U.S. Navy searched 20 km^2 (8 mi^2) of the Pacific Missile Range off California with side-scan sonar while seeking a lost missile. Eight whale falls were videotaped, and one of them was subsequently located and examined by Smith and his team using a submersible. Scientists in Japan, New Zealand, and Iceland are also looking for whale-fall sites. With permission from the National Marine Fisheries Service, Smith has taken two dead stranded whales out to sea and sunk them to observe the colonizing of the carcasses and learn more about the diversity of the organisms associated with whale falls.

As many as fifteen of the whale-fall species have also been found in other sulfide-rich habitats such as the deep-sea hydrothermal vents. Smith has suggested that the whale falls may serve as "stepping-stones" for the dispersal of organisms that depend on chemosynthesis, as from one hydrothermal vent to another. Objections to this theory include that the number of species that have been found to overlap whale falls and vents is small, that vents are found in more than 1500 m (5000 ft) of water, while most whales live and die in shallower water along the edge of the continental shelf, and that the increasing number of vents being found indicates that dispersal from vent to vent is not a problem. Whether or not the

(a)

(b)

Box Figure 1   (a) The skull, jawbones, and vertebrae of a 21 m (70 ft) blue whale on the floor of the Santa Catalina basin off southern California, depth 1240 m (4000 ft). The skull is about 1.5 m (5 ft) long, and each vertebra is about 40 cm (16 in) long. (b) Sulfide-loving mussels (each about 1 cm [0.4 in] long) clustered on a recovered whalebone. Up to 178 species of animals have been found living on a single carcass.

whale falls act as stepping-stones, it is of great interest that these whale-fall communities exist, that hydrothermal vents are not as isolated as was thought, and that another new kind of community has been discovered on the sea floor, a place once thought to be cold, dark, and unsuited to life.

(*Mirounga angustirostris*), whose males can reach 3 tons. Common eared seals include the California sea lion (*Zalophus californianus*) and the northern fur seal (*Callorhinus ursinus*), once almost hunted to extinction for their fur. Eared seals have external ears and they can move their rear flippers forward so they can move quickly on land and can rear up into a partially erect position.

Pinnipeds display similar mating habits. Some breed on ice and some on land. They frequently migrate long distances to arrive at the breeding grounds. Commonly, the largest males arrive on shore first to establish territories, which they defend with fierce fighting. The females and subordinate males arrive later. Dominant males can have harems of as many as fifty females that they mate with. Females give birth to their pups on

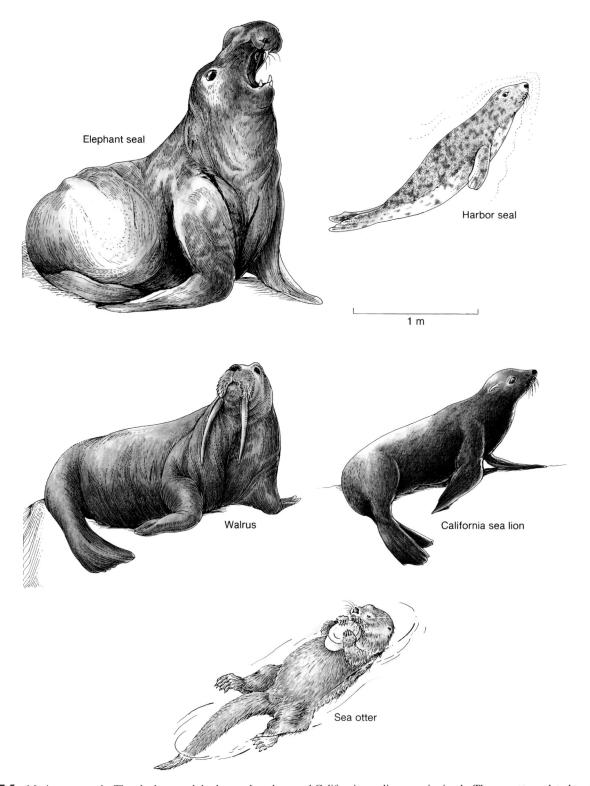

Figure 17.5 Marine mammals. The elephant seal, harbor seal, walrus, and California sea lion are pinnipeds. The sea otter, related to river otters, feeds on clams and sea urchins while floating on its back.

shore, as the pups cannot swim and must be nursed. The mothers leave them for times to hunt offshore. The breeding grounds are referred to as rookeries.

Some pinnipeds are currently experiencing a period of relative peace compared to the sealing days of the nineteenth and early twentieth centuries. Between 1870 and 1880, hunters reduced the population of northern elephant seals *(Mirounga angustirostris)* to as few as fifty animals. Because these animals spend most of their lives at sea, their entire populations were not wiped out. The most recent population estimates suggest population sizes of as many as 150,000 animals. Their rookeries in Mexico and U.S. waters are now fully protected. The Guadalupe fur seal (*Arctocephalus townsendi*) was also hunted to near extinction, until in 1892 only seven animals were thought to survive. Population size has recovered to about 10,000 animals, although almost all pups appear to be raised on a single island close to highly developed areas. Intensive hunting of seal populations can be considered as examples of the top-down control of organism abundance discussed in chapters 15 and 16. In these examples, humans are the predators.

Population sizes of the northern fur seal over the years have been controlled with both top-down and bottom-up processes. Heavy hunting had reduced population sizes of fur seals in the Pribilof Islands in the southern Bering Sea to about 200,000–300,000 animals by 1910. In the 1950s, population sizes appeared to reach about 2.5 million animals. But, today, population sizes are estimated at about 1.5 million and appear to be continuing to decline. One factor that has been suggested as part of the explanation for the continued decline is a reduced prey base. Walleye pollock is a major food source for these seals. Pollack is also one of the largest commercial fisheries in the world. Reduced food supply would represent a bottom-up control of population size of the northern fur seals.

Sea Otters

Sea otters (see figs. 17.4*c* and 17.5) are related to river otters but they live in salt water. Sea otters inhabit coastal areas, taking shellfish and other foods from the bottom in relatively shallow waters. They differ from seals and whales in having no insulating layer of blubber beneath their skin. Sea otters spend almost all their time in the water, where they breed and give birth. Sea otters live in and around kelp beds, where they often use rocks to help them break open huge quantities of sea urchins, abalone, mussels, and crabs that they consume for food. They often float on their backs and use their bodies as a table to work on while they eat. An average male sea otter that weighs about 80 lbs will consume almost 20 lbs of food each day, about 20–25% of his body weight. That amount of food would be comparable to a 120 lb human consuming 30 lbs of food each day without gaining any weight. Sea otters are the smallest marine mammals. They depend on their thick fur that traps air to keep them warm. This soft, thick fur is the reason sea otters were almost hunted to extinction. The sea otter was brought under protection in 1911, although many scientists at the time doubted the species could survive. Today's population consists of about 108,000 individuals.

Sea otters are considered keystone predators in kelp forests. A keystone predator is one that maintains species diversity within an ecosystem. An important food source for sea otters is the sea urchins that graze on kelp. If the numbers of sea otters decline, the numbers of sea urchins increase. Increased numbers of sea urchins impact the ability of kelp forests to grow. Kelp forests serve as nurseries for small fish and other

animals and a reduced density of kelp reduces the size of fish nurseries. This example of a keystone predator illustrates the complicated interactions that define food webs (review food webs in chapter 16).

Walrus

The walrus (see figs. 17.4*d* and 17.5) is placed in a separate taxonomic subgroup. It has no external ears and is able to rotate its hind flippers so that it can walk on a hard surface. Its heavy canine teeth (or tusks) are unique, and both males and females have them. These tusks help the walrus haul itself out of the water onto the ice and are probably used to glide over the bottom, like sled runners, while the animals forage for clams with their heavy, muscular whisker pads.

The Pacific walrus (*Odobenus rosmarus*) lives in the waters and on both coasts of the Bering and Chukchi Seas. It is thought that the population before the eighteenth-century exploitation was 200,000–250,000 animals. The 1950 population was 50,000–150,000. Between 1975 and 1990, aerial surveys of walruses were carried out jointly between the United States and Russia every five years. In 1990, surveys of the Pacific walrus population were suspended because of problems with the survey methods such that they were no longer considered reliable. The last survey in 1990 estimated a population size of 201,000 individuals. The current population size of walruses is unknown.

Sea Cows

Manatees and **dugongs** are also known as **sea cows** (fig. 17.6); they are members of the order **Sirenia** and are thought by some to be the source of mermaid stories. Manatees and dugongs are the world's only herbivorous marine mammals. Manatees are found in the brackish coastal bays and waterways of the warm southern Atlantic coasts and in the Caribbean, and dugongs are found in the seas of Southeast Asia, Africa, and Australia. Like whales, they never leave the water.

The dugongs have disappeared from much of the Indian Ocean and South China Sea due to degradation of habitat. This 3 m (10 ft) long animal weighing about 360 kg (800 lbs) depends on seagrass beds for food, and these beds are being cleared for development and smothered by the silt and mud from eroding, overgrazed, and deforested lands. The dugongs are also hunted for their tusks, which are considered to have aphrodisiac properties and are used to make amulets. Many of the animals are also injured or killed by boat propellers.

Manatees in the coastal water of the Caribbean and South Atlantic are frequently injured and killed by collisions with the propellers of large and small vessels. Known manatee mortality from all causes between 1979 and 1992 was more than 1700 animals, 26% from boat collisions. In 1996, 600 manatees died, and at least 158 of their deaths were due to the neurotoxin produced by the phytoplankton *Karenia brevis* (see chapter 16). There were 416 manatee deaths from all causes in 2006. The

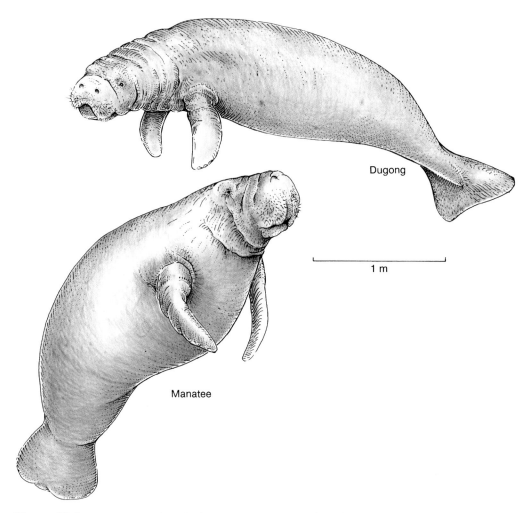

Figure 17.6   The manatee of the Caribbean and the dugong of Southeast Asia are herbivorous marine mammals.

cows were slow-moving, docile, totally unafraid of humans, and present in only limited numbers. These characteristics, coupled with a low reproduction rate, made them unable to withstand the human hunting pressure. The last Steller sea cow was killed for its meat in about 1768.

Polar Bears

The polar bear *(Ursus maritimus)* is the top predator in the Arctic (fig. 17.7). They are long-lived (up to twenty-five years) carnivores with dense fur and blubber for insulation. They live only in the Northern Hemisphere and in a region where winter temperatures can be as low as −50°F; they maintain a body temperature of 98.6°F. Polar bears actually have more problems with overheating than in getting too cold. Adult males weigh 250–800 kg (550–1700 lbs) and females weigh 100–300 kg (200–700 lbs). They feed primarily on ringed seals. They hunt seals by waiting for them to emerge from the openings in the ice that the seals make so they can breathe or climb out to rest. The polar bear may wait hours for the seal to emerge. Polar bears typically only eat the skin and blubber of the seal. The remaining seal meat serves as food for other Arctic animals such as the Arctic fox. Polar bears also feed on the carcasses of dead whales or walruses.

Polar bears travel long distances over the ice, 30 km (19 mi) or more each day. They are also strong swimmers, swimming continuously for 100 km (62 mi). Polar bears commonly breed in April. The males spend the winter on the pack ice; pregnant females dig large dens either on the mainland or on sea ice to spend the winter. They commonly give birth to two cubs in December or January. In spring, mother and cubs migrate to the coast near the open sea. Cubs remain with their mother for at least two years.

The world's total population of polar bears is estimated at 20,000 to 25,000. They are found in the United States, Canada, Greenland, Norway (Svalbard Islands), and Russia. The Alaskan population is about 4700 individuals. On January 9, 2007, the U.S. Fish and Wildlife Service proposed adding the polar bear to the federal list of threatened and endangered species, in May of 2008, the polar bear was officially listed as threatened. The primary threat to polar bears is the decrease in sea ice coverage that is occurring. These bears are completely dependent on sea ice for all aspects of their lives.

Figure 17.7   A polar bear on the ice in the central Arctic Ocean.

destruction of Florida's manatee habitat is also accelerating as salt marshes, seagrass beds, and mangrove areas are drained, reclaimed, or otherwise destroyed.

Long extinct, the Steller sea cow existed in the shallow waters off the Commander Islands in the Bering Sea. These sea

Russia and Norway prohibit all hunting of polar bear. The U.S. Marine Mammal Protection Act of 1972 prohibits the killing of all marine mammals, including polar bear, except by native people for subsistence purposes.

Marine Mammal Protection Act

In 1972, the U.S. Congress established the Marine Mammal Protection Act, Public Law 92–522. This act includes a ban on the taking or importing of any marine mammals or marine mammal product. "Taking" is defined as the act of harvesting, hunting, capturing, or killing any marine mammal or attempting to do so. The act covers all U.S. territorial waters and fishery zones. It is also unlawful "for any person subject to the jurisdiction of the United States or any vessel or any convoy once subject to the jurisdiction of the United States to take any marine mammals on the high seas," except as provided under preexisting international treaty.

The act effectively removed the animals and their products from commercial trade in the United States. Only under strict permit procedures and with the approval of the Marine Mammal Commission can a few individual marine mammals be caught for scientific research and public display.

In 1994, the Marine Mammal Protection Act was amended to provide for exemption for Native subsistence, for permits for scientific research, for a program to authorize and control the taking of marine mammals incidental to commercial fishing operations, for the preparation of stock assessments for all marine mammals in U.S. waters, and for studies of fishery/pinniped interactions.

An important overall goal of the act is to maintain marine mammals as a "significant functioning element in the ecosystem of which they are a part." The act has dramatically reduced the death and injury of marine mammals. A recent report by the National Academy of Sciences indicates that today, the biggest threats to marine mammals are habitat degradation and the cumulative effects of harassment. Many effects of human activities on marine mammals can occur over dramatically different time scales. Individual animals may be affected immediately or over the course of years; populations may be affected over the course of years to generations; and ecosystem effects may not become obvious until generations or even centuries have passed. As has been seen in previous sections, trends in population sizes are difficult to determine—many of these animals spend a majority of their lives underwater or away from easy sightings by humans.

Communication

Many marine mammals use sound to communicate with each other and sound instead of sight to picture their underwater environment. The best-known communication between marine mammals are the "songs" of the male humpback whales. Different humpback populations have different songs, and the songs are transmitted from one individual to another within the population. Songs last up to thirty minutes and are changed and modified during each breeding season. These songs are thought to be announcements of presence and territory, although some scientists believe the singing is a secondary sexual characteristic of males in the breeding season. Female gray whales stay in contact with their calves by a series of grunts, and Weddell seals are known to communicate by audible squeaks.

Using sound to picture the environment is known as **echolocation.** The ability to make the sharp sounds required to produce the echoes that allow marine mammals to orient themselves and locate objects is suspected in all toothed whales, some pinnipeds (Weddell seal, California sea lion), and possibly the walrus. A few baleen whales—the gray, blue, and minke—also have this capacity. Although these animals can produce a range of sounds, the most useful sound for echolocation appears to be clicks of short duration released in single pulses or trains of pulses. The bottle-nose dolphin produces clicks in frequencies audible to the human ear and higher, each lasting less than a millisecond and repeated up to 800 times per second. When each click hits its target, part of the sound is reflected back; the animal continually evaluates the time and direction of return to learn the speed, distance, and direction of the reflecting target (fig. 17.8). Low-frequency clicks are used to scan the general surroundings, and higher frequencies are used for distinguishing specific objects.

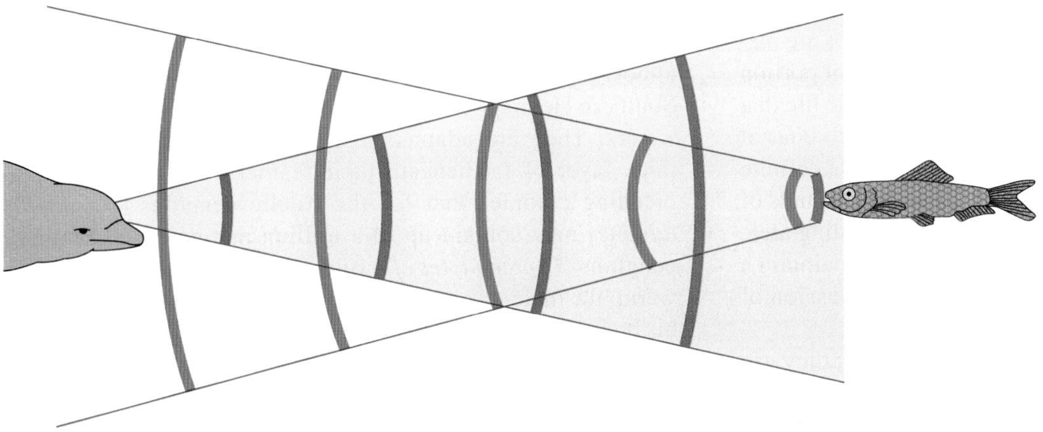

Porpoises and dolphins move air in their nasal passages to vibrate the structures that produce the clicks; the whistles and squeaks are made by forcing air out of nasal sacs. The bulbous, fatty, rounded forehead of the porpoise acts as a lens to concentrate the clicks into a beam and direct them forward. Sperm whales produce shorter, more powerful, long-range pulses at lower frequencies; these sounds may travel several kilometers. Each pulse from a sperm whale is compound, made up of as many as nine

Figure 17.8   The pattern of clicks produced by a porpoise for echolocation is shown in *orange*. The reflected echoes by which the porpoise determines the speed, distance, and direction of the target are shown in *green*.

separate clicks. The sperm whale's massive forehead is filled with oil that may be used to focus the sound pulses.

These animals must be able to pick up the faint incoming echoes of their own clicks and screen out the louder outgoing clicks and other sea noises. Sounds enter through the lower jaw and travel through the skull by bone conduction. Within the lower jaw, fat and oil bodies channel the sound directly to the middle ear. Areas on each side of the forehead are also very sensitive to incoming sound. The hearing centers in the brains of marine mammals are extremely well developed, presumably to analyze and interpret returning sound messages. Their vision centers are less developed, and they are believed to have no sense of smell.

In recent years, there has been much debate on effects of human-generated noise on whale behavior. The navy uses sonar, which can travel for miles underwater, to detect the presence of submarine and other potential underwater hazards. A number of years ago, fourteen beaked whales were found beached on islands in the Bahamas about thirty-six hours after navy vessels used sonar for training exercises, causing many to speculate that the navy sonar was harmful to whales. The navy has developed research programs with scientists to determine how best to avoid harming these animals. Reported strandings appear to be localized to tropical and subtropical regions with steep cliffs on the sea floor. Researchers hypothesize that the beaked whales may be particularly sensitive. Beaked whales dive deeper than any other air-breathing mammal. During normal dives their lungs collapse when they descend to depths greater than 70 m. This physiology prevents too much nitrogen from entering their bloodstream, a cause of decompression sickness. The researchers hypothesize that naval sonars may mimic sounds made by killer whales, their natural predator. This sound may stimulate the beaked whales to dive repeatedly to depths shallower than the depth at which their lungs collapse, eventually causing the whales to experience decompression sickness, which can be lethal. Proposed possible solutions to these effects of human-generated noise on whales are to avoid training exercises in regions where whales are known to migrate and when possible to limit the duration of the exercises.

17.2 Marine Birds

About 3% of the estimated 9000–10,000 species of birds are marine species. Seabirds are birds that spend a significant portion of their lives at sea. Some are so well adapted to oceanic life that they come ashore only to reproduce; others move into coastal waters to feed; but all return to shore to nest. Most have definite breeding sites and seasons, and they may migrate thousands of kilometers as they travel from feeding ground to breeding area. Birds are homeotherms, which means that they can maintain a constant body temperature (see chapter 14 for a discussion of temperature).

Seabirds swim at the sea surface and underwater; they use their webbed feet, their wings, or a combination of wings and feet. They float by using fat deposits in combination with light bones and air sacs developed for flight. Their feathers are waterproofed by an oily secretion called preen, and the air trapped under their feathers helps keep the birds afloat, insulates their bodies, and prevents heat loss. When diving, the birds reduce their buoyancy by exhaling the air from their lungs and air sacs and pulling in their feathers close to their bodies to squeeze out the trapped air. The underwater swimmers such as cormorants and penguins have thicker, heavier bones, and penguins have no air sacs.

The eyesight of seabirds is highly developed, for they depend on their vision to locate fish in the water. Their senses of hearing and smell are less developed, and their least developed sense is taste; they have few taste buds and swallow quickly. They drink seawater or obtain water from their food. Excess salt is concentrated by a gland over each eye, and a salty solution drips down or is blown out through their nasal passages. To conserve water, the birds reduce and concentrate their urine, forming uric acid, a nearly nontoxic white paste that is mixed with feces and excreted. Because of their great activity and high metabolic rate, birds are huge eaters. Seabirds feed on fish, squid, krill, egg masses, and bottom invertebrates, as well as on carrion and garbage. They are most plentiful where food is abundant, and their presence is a good indicator of high productivity in surface water.

There are four general related groups of seabirds: (1) albatrosses, petrels, and shearwaters; (2) penguins; (3) pelicans, cormorants, boobies, and frigate birds; and (4) gulls, terns, and auks. The wandering albatross *(Diomedea exulans)* of the southern oceans is the most truly oceanic of marine birds with the largest wing spread of all birds, 3.5 m (11 ft). These great white birds with black-tipped wings spend four to five years at sea before returning to their nesting sites. The smaller, North Pacific, black-footed albatross *(Phoebastria nigripes)* is a pelagic scavenger, searching the sea surface for edible refuse, including scraps from ships and fishing boats. Its primary foods are squid, crab, and surface fish. The smallest of the oceanic birds, Wilson's petrel *(Oceanites oceanicus)*, is a swallowlike bird that breeds in Antarctica and flies 16,000 km (10,000 mi) along the Gulf Stream to Labrador during the Southern Hemisphere winter, returning to the Southern Hemisphere for the southern summer, another 16,000 km.

Penguins (fig. 17.9a) cannot fly. Their wings have been modified into flippers that allow them to swim underwater, using their feet for steering. Their underwater swimming speed is almost 10 mph. All but one species of penguin live in the Southern Hemisphere, primarily in Antarctic and Subantarctic waters. They are adapted to life in the cold. They have a thick layer of fat beneath their feathers. Penguins establish breeding colonies that for the Adélie penguins *(Pygoscelis adeliae)* may contain up to a million mating pairs. Emperor penguins *(Aptenodytes forsteri)* huddle together to keep warm; the male incubates the single laid egg on his feet against his body for over two months while the female returns to sea to hunt. Chicks are born in the spring when the plankton blooms. The Galápagos penguin *(Spheniscus mendiculus)* lives along the equator, far from Antarctica. This penguin is restricted to regions where cold water upwells.

(a)

Pelicans and cormorants are large fishing birds with big beaks. They are strong fliers found mostly in coastal areas, but some venture far out to sea. A pelican (fig. 17.9b) has a particularly large beak from which hangs a pouch used in catching fish. White pelicans of North America *(Pelecanus erythrorhynchos)* fish in groups, herding small schools of fish into shallow water and then scooping them up in their large pouches. The Pacific's brown pelican *(Pelecanus occidentalis)* does a spectacular dive from up to 10 m (30 ft) above the water to capture its prey. Cormorants are black, long-bodied birds with snakelike necks and moderately long bills that are hooked at the tip. They settle on the water and dive from the surface, swimming primarily with their feet but also using their wings. Because they do not have the water-repellent feathers of other seabirds, cormorants must return to land periodically to dry out.

Terns (fig. 17.9c) and gulls are members of the fourth group of seabirds and are found all over the world except in the South

(b)

(c)

(d)   (e)

Figure 17.9   (a) These Magellanic penguins are part of a 500,000-bird colony at Punta Tombo, Argentina. (b) A brown pelican takes off from the water. (c) Crested terns face the wind on the beach of Heron Island, a part of the Great Barrier Reef, Australia. (d) A mixed flock of dunlins and sanderlings take flight along a sandy beach. (e) Kittiwakes (gull-like birds) and common murres nesting on an island in Kachemak Bay, Alaska.

Pacific between South America and Australia. Gulls are strong flyers and will eat anything; they forage over beach and open water. The terns are small, graceful birds with slender bills and forked tails. The Arctic tern *(Sterna paradiseae)* breeds in the Arctic and each winter migrates south of the Antarctic Circle, a round trip of 35,000 km (20,000 mi).

Puffins, murres, and auks are heavy-bodied, short-winged, short-legged diving birds. They feed on fish, crustaceans, squid, and some krill. All are limited to the North Atlantic, North Pacific, and Arctic areas, where most nest on isolated cliffs and islands. The great auk was a large, slow, flightless bird, 0.6 m (2 ft) high, that provided food for generations of sea travelers. It was killed for both its meat and its feathers (for stuffing mattresses), and as its numbers dwindled, museums and private collectors paid more and more for each bird. The last two were killed on a small island off Iceland on June 2, 1844.

Shorebirds differ from seabirds in that they do not swim much and thus have a greater dependency on land. They include sandpipers, plovers, stilts, avocets, snipes, oystercatchers, turnstones, and phalaropes (fig. 17.9*d*). Many migrate long distances between winter feeding grounds and spring nesting grounds. For example, the semipalmated sandpiper *(Calidris pusilla)* flies about 2000 miles without stopping from the North Atlantic coast to Suriname in South America. The red knot *(Calidris canutus)* flies from the Arctic Circle to Tierra del Fuego at the tip of South America. Shorebirds commonly feed on small crustaceans, clams, and snails using a variety of styles: running and stabbing consists of capturing prey by stabbing it; chiseling and hammering describe how bivalves are opened; and pecking and probing are used to find prey beneath the mud or sand surface—the length of the bill determines what can be gathered as food. The great blue herons *(Ardea herodias)* and snowy egrets *(Egretta thula)* are waders; their long necks enable them to strike at small fish in the water.

17.3 Marine Reptiles

Although many land reptiles visit the shore to feed, mainly on crabs and shellfish, few reptiles are found in today's seas. Examples of marine reptiles are shown in figure 17.10. All reptiles are poikilotherms, which means their metabolism rates vary with temperature (see chapter 14 on temperature). The only modern marine lizard is the big, gregarious marine iguana of the Galápagos Islands. It lives along the shore and dives into the water at low tide to feed on the algae. This iguana has evolved a flattened tail for swimming and has strong legs with large claws for climbing back up on the cliffs. It regulates its buoyancy by expelling air, allowing it to remain underwater. The large monitor lizards of the Indian Ocean islands, known as the Komodo dragons, are capable of swimming but do so only under duress.

Alligators may enter shallow shore water, and crocodiles are known to go to sea. The estuarine crocodile of Asia is found in the coastal waters of India, Sri Lanka, Malaysia, and Australia. The Indian gavial and false gavial are slender-nosed, fish-eating crocodilians found in nearshore waters.

Sea Snakes

About fifty kinds of sea snakes are found in the warm waters of the Pacific and Indian Oceans; there are no sea snakes in the Atlantic Ocean. Sea snakes are extremely poisonous; they have small mouths, flattened tails for swimming, and nostrils on the upper surface of the snout that can be closed when the snakes are submerged. The sea snakes' skin is nearly impervious to salt, but it is permeable to gases, passing nitrogen gas as well as carbon dioxide and oxygen. The snake is able to lose nitrogen gas to the water, allowing it to dive as deep as 100 m (300 ft), stay submerged as long as two hours, and surface rapidly with no decompression problems. Sea snakes eat fish, and most reproduce at sea by giving birth to live young.

Sea Turtles

Sea turtles can be distinguished from land turtles because sea turtles can not retract their heads into their shells and because their front limbs have been modified into flippers for swimming. Male and female turtles mate offshore and then the females return to shore to lay their eggs; females may return to shore several times during breeding season, each time laying as many as 100 eggs that are buried in the sand. The five species of sea turtles found in U.S. and Caribbean waters are the green *(Chelonia mydas)*, hawksbill *(Eretmochelys imbricate)*, loggerhead *(Caretta caretta)*, Kemp's ridley turtle *(Lepidochelys kempii)*, and leatherback turtle *(Dermochelys coriacea)*.

Green turtles (fig. 17.10) are generally found in tropical and subtropical waters between 30°N and 30°S. They appear to obtain sexual maturity when they reach anywhere between twenty and fifty years old. Once they reach sexual maturity, females begin returning to the beaches where they were born. Green turtles occupy three different types of habitats. Nesting occurs on high-energy beaches, juveniles congregate in open-ocean convergence zones (review chapter 9, convergence zones), and the adults occupy shallow coastal areas. They are primarily herbivorous, feeding mainly on seagrass and algae, although the juveniles can also feed on small animals. Major nesting colonies in the United States are on the east coast of Florida, in the U.S. Virgin Islands, and in Puerto Rico. Hawksbill turtles are found throughout the tropical oceans. Their name refers to their prominent hooked beak. They reach sexual maturity when they are over thirty-five years old. Major nesting colonies in the United States are in Puerto Rico, the Virgin Islands, Florida, and Hawaii. As with green turtles, hawksbill juveniles are found in oceanic convergence zones, where they feed on algae. As they mature, they feed primarily on sponges found around coral reefs or rocky outcroppings. Loggerhead sea turtles are distributed throughout subtropical and temperate waters, primarily on continental shelves and in estuaries. They become sexually mature at thirty to forty years of age. Loggerhead hatchlings move offshore to convergence zones where *Sargassum* is located. After a few years at sea, they move into continental shelf regions to feed on crabs and shellfish. Nesting is concentrated in the temperate zone and subtropics. In U.S. waters, nesting sites are found along the Atlantic coast of Florida to North Carolina. The Kemp's ridley turtle is found primarily in the Gulf of Mexico. They

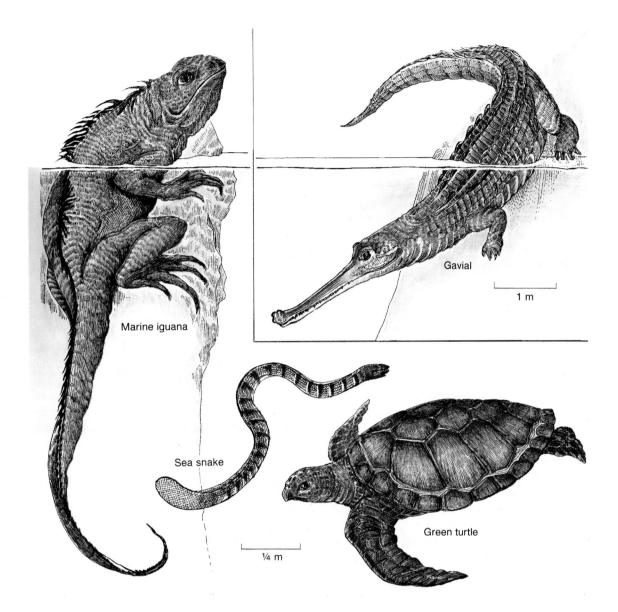

Figure 17.10 Marine reptiles. The marine iguana is the only modern marine lizard; it inhabits the Galápagos Islands. The gavial is a fish-eating crocodilian from India. Sea snakes are venomous; they are found only in the Pacific and Indian Oceans. The green turtle feeds on seagrasses and seaweeds in tropical coastal waters.

become sexually mature at seven to fifteen years old. They nest in large aggregations in Rancho Nuevo, on the northeastern coast of Mexico. The hatchlings also congregate on *Sargassum*. Adults appear to feed on crabs and mollusks.

The leatherback sea turtle is distinctive from other sea turtles. It is the largest of all sea turtles, with an average weight of about 500 kg (0.5 ton); the largest known leatherback was 3 m from beak tip to tail and weighed 1 ton. These turtles also do not have a hard shell as other sea turtles. Instead they have a shell of distinctive, thick, leathery skin. The leatherback is almost entirely pelagic and can be found from the topics to the poles. Adults can migrate across entire ocean basins. They appear to reach sexual maturity at eight to fifteen years old, much younger than the green, hawksbill, and loggerhead sea turtles. Nesting occurs only in the tropics, which is the main time that these animals enter coastal waters. They feed primarily on jellyfish,

which they suck into their mouths with a large amount of water; specialized spines in the throat allow the leatherback to expel the water but retain the jellyfish.

Sea turtle population numbers are being dramatically reduced due to pollution, habitat degradation, and the demand for turtle products. Turtle eggs and turtle meat are prized throughout the Pacific, and turtle nests are raided by poachers. Sea turtle eggs incubate for up to a few months, during which time they are vulnerable to harvesting by humans, dogs, rats, or other carnivores. In addition, as discussed previously, sea turtles reach sexual maturity at a relatively late age. All turtles listed above are on the U.S. endangered species list. Placing sea turtles on the endangered list makes it illegal to import sea turtle products into this country. The Kemp's ridley sea turtle is the most endangered of all, primarily because the females nest in large groups at the single known nesting site in Rancho Nuevo and are therefore particularly vulnerable to disruption. In 1947, 42,000 females were detected nesting there; in 1985, the number of identified nests was only 740. The Mexican government now protects many of the nests at Rancho Nuevo and in 2000, over 6000 nests were found, suggesting that this species may be recovering. Another major threat to sea turtles is incidental capture, injury, or mortality during fishing operations. A turtle excluder device, or TED, was developed to minimize the incidental capture of turtles when fishing for shrimp. A TED is a grid of bars with an opening either at the top or at the bottom of the trawl net. The grid is fitted into the neck of a shrimp trawl. Small animals such as shrimp pass through the bars and are caught in the bag end of the trawl, whereas larger animals, such as sea turtles or sharks, strike the grid bars and are ejected through the opening.

17.4 Squid

Squid (fig. 17.11) are abundant in the world's oceans. Until recently, little has been known about squid at all ocean depths. The use of video cameras in submersibles and robots is providing researchers with a record of rarely seen, deep-water squids, many never before seen alive. They are elusive, swim rapidly, and are now known to not only float with neutral buoyancy but also to rest on the bottom. Their wide range of bioluminescence and coloration allows them to change color and disappear rapidly.

Squid have eight short arms and two thin, long tentacles; the tentacles have a number of suckers at their ends, and in some species, bioluminescent lures to attract prey. Squid can remain motionless or move backwards or forwards. They range from a few centimeters in length to the giant squid (*Architeuthis*), a deep-water species that may be 20 m (65 ft) long and weigh nearly 1 ton. Scientists know little about these huge organisms, for no giant squid has yet been captured alive.

17.5 Fish

Fish dominate the nekton. Fish are found at all depths and in all the oceans. As described in chapter 15, their distribution patterns are determined directly or indirectly by their dependency on the ocean's primary producers. Fish are concentrated in upwelling areas, shallow coastal areas, and estuaries. The surface waters support much greater populations per unit of water volume than the deeper zones, where food resources are sparser.

There are three major groups of fish: jawless fishes, which include hag fishes and lampreys; cartilaginous fishes, which include sharks, rays, skates, and ratfish; and bony fishes, which include the great majority of fishes. Sharks are considered apex predators (see fig. 15.10) because they prey on many species of lower trophic levels and they have few natural predators. Fish come in a wide variety of shapes related to their environment and behavior. Some are streamlined, designed to move rapidly through the water (tuna and marlin); some are laterally compressed and swim more slowly around reefs and shorelines (snapper and tropical butterflyfish), others are bottom-dwelling fish that are flattened for life on the sea floor (sole and halibut); while still others are elongated for living in soft sediments and under rocks (some eels). Fins provide the push or thrust for locomotion and occur in a variety of shapes and sizes. Fins are used to change direction, turn, balance, and brake. Flying fish use their

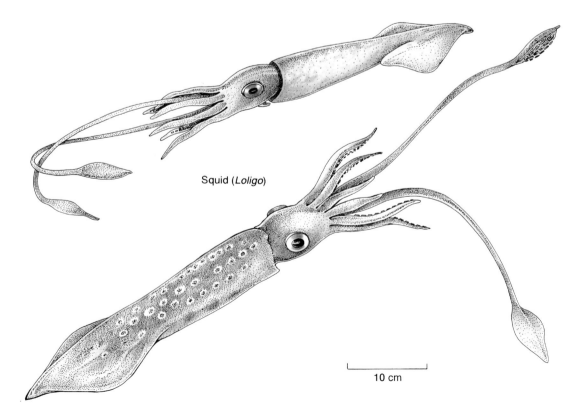

Squid (*Loligo*)

10 cm

Figure 17.11   The squid is a swimming mollusk and a member of the nekton.

fins to glide above the sea surface; mudskippers and sculpins walk on their fins. A sample of the variety of marine fishes is shown in figure 17.12.

Schooling is common among certain types of fish (herring, mackerel, menhaden). Schools may consist of a few fish in a small area, or they may cover several square kilometers; for example, herring in the North Sea have been seen in schools 15 km long and 5 km wide (9 × 3 mi). Usually, the fish in a school are of the same species and are similar in size. Fish schools have no definite leaders, and the fish change position continually. Schooling fish keep their relationship to one another constant as the school moves or changes direction. Most schooling fish have wide-angle eyes and the ability to sense changes in water displacement, which allow them to keep their place with respect to their neighbors. Schooling probably developed as a means of

protection; each fish has less chance of being eaten in the school than alone. The school may also keep reproductive members of the population together.

Ocean fish are divided into two groups: (1) fish with skeletons of cartilage and (2) fish with skeletons of bone. Cartilaginous fish—sharks and their relatives, the skates, rays, and ratfish—are considered more primitive than bony fish. The fish that are used for food are mainly those with skeletons of bone.

Sharks and Rays

The shark is an ancient fish; it predates the mammals, having first appeared in Earth's oceans 450 million years ago. Unlike most fishes, most sharks bear live young. They also differ from other fish by their skeletons of cartilage rather than bone and by

(a)

(b)

(c)

(d)

Figure 17.12 The bony fishes dominate the world's aquatic environments. Their diversity is enormous, and they are found in almost every conceivable aquatic environment. Two of the thousands of species of reef fish: an emperor angel (a) and a butterflyfish (b). The wolf eel (c) is a carnivore with strong jaws and long teeth. The scales of the seahorse (d) are modified to form protective armor. Seahorse populations are declining, and their slow reproduction rate is not likely to support increasing harvests for aquariums, traditional medicine, and gourmet dishes. The red snapper (e) inhabits rocky coastal areas; it is used as a food fish.

(e)

their toothlike scales. Shark scales have a covering of dentine similar to that on the teeth of vertebrates, and they are extremely abrasive; sharkskin has been used as a sandpaper and polishing material. The shark's teeth are modified scales; they are replaced rapidly if lost, and they occur in as many as seven overlapping rows. Sharks are actively aware of their environment through good eyesight; excellent senses of smell, hearing, and mechanical reception; and electrical sense.

Sharks see well under dim-light conditions. They have the ability to sense chemicals in their environment through smell, taste, a general chemical sense, and unique pit-organs distributed over their bodies. These pit-organs contain clusters of sensory cells resembling taste buds. The shark's sense of smell is acute; it has a pair of nasal sacs located in front of its mouth, and when water flows into the sacs as the shark swims, it passes over a series of thin folds with many receptor cells. Sharks are most sensitive to chemicals associated with their feeding, and all are able to detect such chemicals in amounts as dilute as one part per billion.

Receptors along the shark's sides are sensitive to touch, vibration, currents, sound, and pressure. The movement of water from currents or from an injured or distressed fish are sensed by the shark's lines of receptors, which communicate with the watery environment by a series of tubes; water displacement stimulates nerve impulses along these systems. The shark uses its hearing in the location of its prey. Pores in the shark's skin, especially around the head and mouth, are sensitive to small electrical fields. Fish and other small marine organisms produce electrical fields around themselves, and the shark uses its electroreception sense to locate prey and recognize food. As a shark swims through Earth's magnetic field, an electrical field is produced that varies with direction, giving the shark its own compass.

More than 350 species of shark are known, and scientists are still discovering new species. Sharks are widely spread through the oceans and are found in rivers more than a hundred miles from the sea, and one species, the Lake Nicaragua shark, is found landlocked in fresh water. Some of these sharks are shown in figure 17.13. The whale shark (*Rhicodon typus)* is the world's largest fish, reaching lengths of more than 15 m (50 ft). This shark filters plankton from the water and is harmless to other fish and mammals. The basking shark (*Cetorhinus maximus*), 5–12 m (15–40 ft) long, is another plankton feeder. It is found commonly off the California coast and in the North Atlantic, where it has been harvested for its oil-rich liver. A third species of plankton-feeding shark, 4 m (14 ft) long and weighing approximately 680 kg (1500 lbs), nicknamed megamouth *(Megachasma pelagios)*, was discovered in 1976. This shark is hypothesized to migrate vertically through the water column over the day/night cycle, spending the day in deep waters (~150 m) and at shallower depths (~15 m) at night. This pattern of migrating to shallow depths at night is also seen with the deep scattering layer of zooplankton and small fish (chapters 14 and 16).

Many sharks are swift and active predators, attacking quickly and efficiently, using their rows of serrated teeth to remove massive amounts of tissue or whole limbs and body portions. They also play an important role as scavengers and, like wolves and the large cats on land, eliminate the diseased and

aged animals. Sharks can and do attack humans, although the reasons for these attacks and the periodic frenzied feeding observed in groups of sharks are not understood. A human swimming inefficiently at the surface may look like a struggling, ailing animal and be attacked; a diver swimming completely submerged may appear as a more natural part of the environment and be ignored.

Skates and rays (fig. 17.14) are flattened, sharklike fish that live near the sea floor; there are some 450–550 species of skates and rays. They move by undulating their large side fins, which gives them the appearance of flying through the water. Their five pairs of gill slits are on the underside of their bodies, not along their sides. The large manta rays are plankton feeders, but most rays and skates are carnivorous, eating fish but preferring crustaceans, mollusks, and other benthic organisms. Their tails are usually thin and whiplike, and in the case of stingrays, carry a poisonous barb at the base. Some skates and a few rays have shock-producing electric organs that can deliver shocks of up to 200 volts; these are located along the side of the tail in skates and on the wings of the rays. Their purpose appears to be mainly defensive. Like the sharks, most rays bear their young live. Skates enclose the fertilized eggs in a leathery capsule called a sea purse or mermaid's purse that is ejected into the sea and from which the young emerge in a few months.

Commercial Species of Bony Fish

Bony fish can be categorized based on body shape. The fast swimmers such as tuna and marlin have streamlined bodies that allow them to move quickly through water (see "Considerations of Size" in chapter 14); fish such as angelfishes have laterally compressed bodies that allow them to maneuver around coral reefs or through kelp beds; fish such as flounders and halibuts are flat and adapted to living on the bottom—they lie on one side with both eyes on top. Fish that live on or near the bottom are known as **demersal fish.**

Most commercially valuable food fish are bony fish (fig. 17.14) and are found between the surface and depths of about 200 m (600 ft). Among the most important fisheries are those for the enormously abundant small, herring-type fish, such as sardine, anchovy, menhaden, and herring. These fish feed directly on plankton and are components of short, efficient food webs (see chapter 16). They are found in large schools and areas of high primary productivity such as upwelling zones.

Deep-Sea Species of Bony Fish

The fish of the deep sea are not well known because the depths below the epipelagic zone are difficult and expensive to sample. Representative fish are shown in figure 17.15.

In the dim, transitional mesopelagic layer, between 200 and 1000 m (660 and 3300 ft), the waters support vast schools of small, luminous fish. Fishes of the mesopelagic zone are small, with large mouths, hinged jaws, and needlelike teeth. *Cyclothone* is believed to be the most common fish in the sea. Each of its species lives at a relatively fixed depth; the

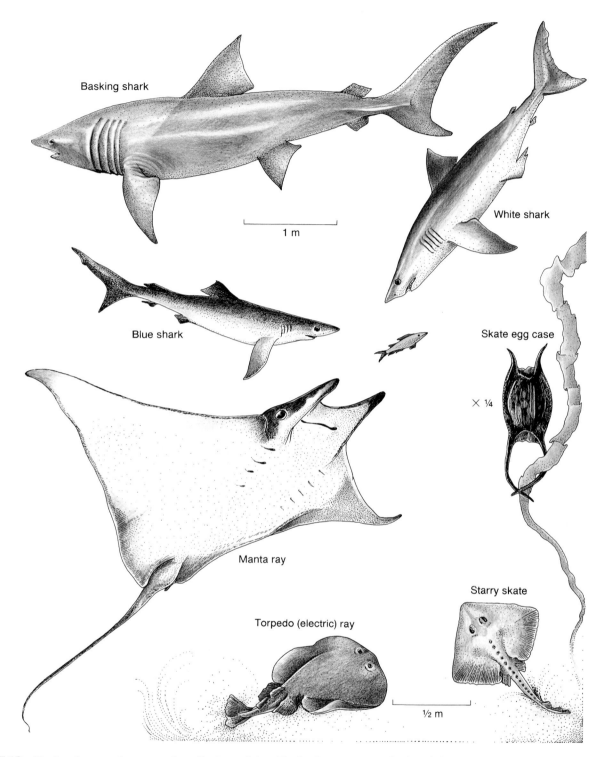

Figure 17.13   Sharks, skates, and rays are all cartilaginous fishes. The leathery egg case of a skate is known as a mermaid's purse.

deeper-living species are black, and the shallower-living species are silvery to blend with the dim light. At this depth, the lanternfish has a worldwide distribution; some 200 species are distinguished by the pattern of light organs along their sides. They are a major item in the diet of tuna, squid, and porpoises. *Stomias,* a fish with a huge mouth, long, pointed teeth, and light organs along its sides, and the large-eyed hatchetfish

prey on the great clouds of euphausiids and copepods found at the mesopelagic depth.

In the perpetually dark bathypelagic zone, there is little food; only about 5% of the food produced in the photic zone is available at these depths. Here, the fish are small, between 2 and 10 cm (1 and 5 in), and have no spines or scales but are fierce and monstrous in appearance. They are mostly black in color

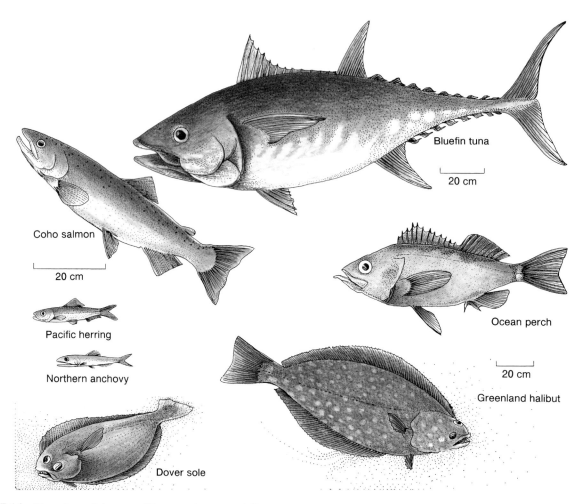

Figure 17.14 These bony fishes are all harvested commercially.

with small eyes, huge mouths, and expandable stomachs. They breathe slowly, and the tissues of their small bodies have a high water and low protein content. Floating at constant depth without using energy for swimming, these fish go for long periods between feedings by using their food for energy rather than for increased tissue production. Most have bioluminescent organs or photophores on their undersides that allow them to blend with any light from above. These photophores may show different patterns among species and between sexes; they may also identify predators and are used as lures to attract prey (review section 14.4 on light). Some species have large teeth that incline back toward the gullet so the prey cannot escape, and others have gaping mouths with jaws that unhinge to allow the catching and eating of fish larger than themselves.

Among the most famous of these predators are *Macropharynx longicaudatus* and the related *Saccopharynx,* which have funnel-like throats and tapering bodies ending in whiplike tails. When the stomach is empty, the fish appears slender, but it expands to accept anything the great mouth can swallow. The female anglerfish, *Ceratias halboelli,* has a dorsal fin modified into a fishing rod, which slides along a canal in the back of the fish; an illuminated lure dangles from the tip of the rod. Other fish are attracted to the lure, which the anglerfish moves forward along her back until the bait is just above her jaws.

17.6 Practical Considerations: Commercial Fisheries

Global fishery statistics are maintained by the Food and Agricultural Organization of the United Nations (FAO). The latest report was released in 2006 and covered data up to 2004. In 1950, the total world catch was approximately 21 million metric tons. Over the next fifty years, the fishing effort by all nations intensified and the technology and gear used to catch fish improved dramatically (see chapter 13). In 1960, the catch was 40 million metric tons and in 1970, the catch was 70 million metric tons. In 2004, the catch reached 86 million metric tons. Aquaculture (discussed in more detail in section that follows on fish farming) is the most rapidly growing food sector and in 2004, marine aquaculture contributed about 18 million metric tons to the total amount. The greatest amount of aquaculture is freshwater production of fish such as carp. Overall, it is estimated that capture fisheries and aquaculture from both marine and freshwater environments provided 2.6 billion people with at least 20% of their average animal protein intake. Preliminary estimates from 2005 indicate that the total fishery production available for human consumption increased by about 1 million metric tons to 107 metric tons. However, the global supply per person remained about the same because of population

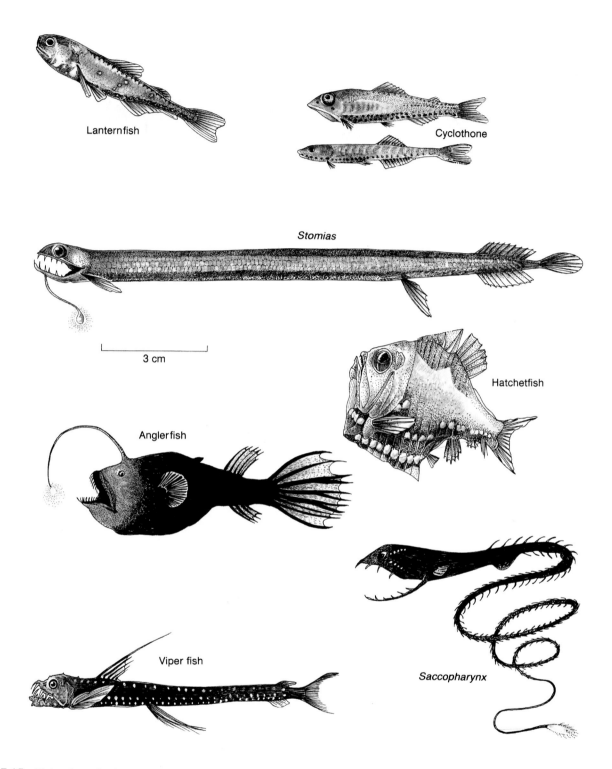

Figure 17.15 Fishes from the deep sea.

growth. The Alaskan pollock fishery is an example of a recent fishery that has responded to consumer demand. Pollock is a bottom fish; it is processed to remove the fats and oils that actually give the fish its flavor; and a highly refined fish protein called **surimi** is produced. Surimi is processed again and flavored to form artificial crab, shrimp, and scallops. The processing of pollock to form surimi is shown in figure 17.16.

World capture fishery production has remained relatively stable for the past decade except for the dramatic fluctuations caused by differences in the catches for Peruvian anchoveta. This species feeds directly on plankton and is particularly susceptible to the El Niño Southern Oscillation (see chapter 7). For example, catches for this species varied from 1.7 million metric tons in 1998 (an El Niño year) to 11.3 metric tons in 2000

(a)

(b)

(c)

Figure 17.16 (a) Freshly caught pollock are poured from a trawl net onto the deck of a seafood-processing vessel. (b) Fishes are cleaned and boned below decks in the processing plant. After the fishes have been washed to remove blood and oils and then dried, the fish meat becomes a flavorless, odorless fish product, surimi. (c) The surimi base is made into analogs such as crab shapes.

(a non–El Niño year). Fluctuations in other fisheries seem to compensate for one another so that the total remains relatively constant. The FAO report concludes that over the past ten to fifteen years, the proportion of overexploited and depleted stocks have remained relatively constant. It was estimated that in 2005, about one-quarter of the fish stocks monitored by the FAO are underexploited or moderately exploited and could potentially produce more; about half the fish stocks are fully exploited, with no room for further expansion; and about one-quarter of the fish stocks are overexploited, depleted, or recovering from depletion. This last category appears particularly serious for fisheries exploited solely or partially in the open ocean or for highly migratory sharks. The FAO suggests that the maximum wild capture potential has probably been reached for the world's oceans.

Five fisheries are discussed in the following section; each illustrates the difficulties faced by those fishing, the fishery managers, the fishing nations, and the fish consumers as the catches decline and the costs rise.

Anchovies

The anchovy fishery is concentrated in the upwelling zone off the coast of Peru and has produced the world's greatest fish catches for any single species. Anchovies are small, fast-growing fish that feed directly on the phytoplankton in the upwelling zone and travel in dense schools that are easy to net in large quantities. These fish are primarily processed into fish meal that is exported as feed for domestic animals.

The fishery began in 1950 with a harvest of 7000 metric tons. By 1962, the harvest had risen to 6.5 million metric tons, and as world demand for fish meal rose, the anchovy fishing intensified. In 1970, a peak fishing year, about 12.3 million metric tons were taken. As the fishing increased, the average size of the harvested fish decreased, and more and more fish (smaller and younger) were required to make up the same catch. During periods in which the Peruvian coast is visited by El Niños (see chapter 7 and fig. 17.17), the general wind pattern reverses, the

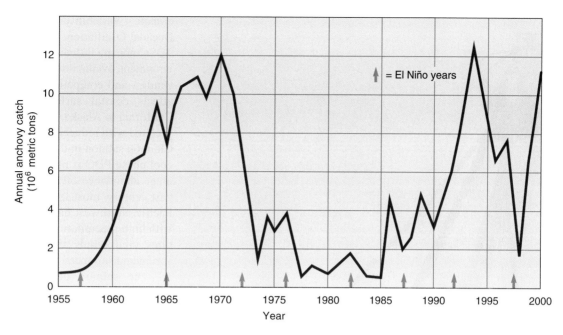

Figure 17.17 The anchovy catch of Peru and Ecuador by calendar year, given in millions of metric tons. Years of El Niño are 1957, 1965, 1972, 1976, 1982–83, 1992–93, and 1997–98.

winds become westerly, diminishing the upwelling and moving warmer, nutrient-poor surface water eastward into the areas of normally cold and productive upwelled water. The decrease in nutrients and the increase in temperature result in low phytoplankton abundance, the effects of which ripple through the whole ecosystem. El Niño devastates the fishing, and large numbers of seabirds that feed on the schools of anchovies starve.

Follow the changes in the anchovy catch as related to El Niños in figure 17.17. Heavy fishing in 1970–71 combined with the severe effect of a 1972 El Niño resulted in only 2 million metric tons harvested in 1973. With government quotas in place, the catch increased slowly in 1974 and 1975, but El Niño struck again in 1976 and in 1982–83. The catch stabilized in the years 1988–90 and continued to increase through the weak El Niño year of 1992–93. The 1994 catch of 12.5 million metric tons is the greatest catch yet taken from this area. Catches in 1995–97 held steady between 8.6 million and 7.6 million metric tons. The impact of the 1997–98 El Niño was dramatic; only 1.7 million metric tons were harvested in 1998. In this fishery, overfishing combined with environmental changes intensify the effect of overharvesting. By 1999, the catch had increased to 8.7 million metric tons and by 2000, 11.3 million metric tons.

Tuna

Top predators such as the Atlantic bluefin tuna *(Thunnus thynnus)* have been considered overexploited since the early 1980s. These fish reach lengths of 9 feet and can weigh as much as three-quarters of a ton. They are endothermic and can thus range from subtropical spawning areas to subarctic feeding grounds. The fact that they are endothermic also makes them strong swimmers. There are two recognized breeding grounds for these fish—the Gulf of Mexico and the Mediterranean Sea. The regulatory board for this fishery—the International Commission for the

Conservation of Atlantic Tunas (ICCAT)—recognizes two management units, fish from the west and from the east Atlantic Ocean, defined by the 45°W meridian. Over the past thirty years, the western Atlantic breeding populations have declined dramatically in numbers. In 1996, electronic tags that archive and transmit data to satellites began being used to track the behavior of these highly migratory fish. The use of these sophisticated tagging devices has made it clear that these fish can move back and forth across the Atlantic in a single year. This indicates that the metabolic cost of swimming across an ocean basin must be relatively low compared to the ecological benefits derived from this migration. Bluefin tuna appear to stay primarily within the upper 300 m of the water column, with occasional dives to as deep as 1000 m. The West Atlantic bluefin tuna were documented moving to the Gulf of Mexico and the eastern Mediterranean Sea during the breeding season, indicating that both regions need to be protected for the health of the western Atlantic stock. The use of these tags can help determine whether fish return to breeding grounds to spawn every year and potential impacts of oceanographic conditions on their migratory patterns.

Salmon

Another fishery in which the United States has an important role and investment is the salmon industry of the Pacific Northwest and Alaska (fig. 17.18). There are seven species of salmon in these regions: pink *(Oncorhynchus gorbuscha)*, sockeye *(Oncorhynchus nerka)*, chum *(Oncorhynchus keta)*, Chinook *(Oncorhynchus tshawytscha)*, coho *(Oncorhynchus kisutch)*, steelhead *(Oncorhynchus mykiss)*, and sea-run cutthroat *(Oncorhynchus clarki)*. Salmon spawn in fresh water and remain there as juveniles for a year or less, depending on the species; sockeye are distinctive because they require a lake for part of their life cycle. All salmon eventually journey to the sea, where

Figure 17.18   A salmon purse seiner hauling in its net.

climate variability termed the Pacific Decadal Oscillation, or PDO (review chapter 9). Salmon are influenced by their physical environment, availability of food, predators of juveniles, and competitors for food. The PDO affects coastal surface temperatures from California to Alaska. Warm phase PDO is associated with reduced abundance of coho and Chinook salmon in the Pacific Northwest and cool phase PDO is associated with above average abundance of these two fish. Over the past century, most salmon populations in the Pacific Northwest have fared best in periods with high precipitation, cool coastal temperatures, and winds that favor upwelling in spring and summer.

The salmon depend on high-quality, pollution-free fresh water and clean, gravel-bottomed, shady, cool streams for successful natural spawning. With increasing numbers of people, there is increasing pressure on salmon habitat. We dam our rivers for power usage and flood control, cutting off the salmon from their home streams; we harvest trees up to the stream edge, removing shade and allowing mud and silt eroded from the exposed land to cover the streambed. We use streams to carry away our wastes, degrading the water quality and damaging both the juvenile fish and the returning runs. In many areas, these activities are being corrected, but the naturally spawning fish are now fewer in number, and we must pay both the cleanup costs and the costs of hatchery programs to supplement the dwindling natural stocks. Different populations of Chinook, coho, steelhead, and sockeye are now listed as endangered.

Federal, state, and privately owned hatcheries harvest eggs from mature fish, raise the juveniles, and hold them until an optimal time for release. These juveniles go to sea, and their survivors return to their natal streams at maturity. This practice can be called ocean or sea ranching. Because the eggs and the juveniles are held in controlled environments, hatchery fish have a survival rate much higher than that of wild fish. At sea, the hatchery and wild fish mix together and return to the same streams. If commercial and sport fishing quotas are based on total numbers of returning fish, the wild stocks will become depleted as the production of hatchery fish increases.

Atlantic Cod

The Atlantic cod *(Gadus morhua)* is a demersal fish found on both sides of the Atlantic. In U.S. waters, cod are managed as two stocks: Gulf of Maine and Georges Bank southward. In

they live and grow for one to four years. Much less is known about the saltwater phase of their life cycle. Fully mature salmon return to spawn in the freshwater streams, rivers, or lakes that they were born in (review section 14.4 on salinity). This means that fish spawn in fresh water under one governmental jurisdiction and are caught as they move through coastal waters of other governmental jurisdictions. This complicates negotiations to determine the length of fishing periods and the number of fish to be caught.

Each year's salmon fishery depends on the number of fish returning to spawn in previous years. In 1997, scientists recognized a link between salmon abundance and large-scale

1992, Canada banned cod fishing in the waters off Newfoundland and in 1993 and 1994 extended the ban south into the Gulf of St. Lawrence, effectively shutting down its historic northwest Atlantic cod fishery. Iceland warned its fishing industry that the cod stock could collapse unless the annual catch was cut back by 40%. Also in 1994, the U.S. New England Fishery Management Council stopped the fishing for cod on Georges Bank.

Cod are bottom-dwellers, or ground fish; they feed on small fish, crab, squid, and clams. They live twenty to twenty-five years, mature in three to seven years, and produce several million eggs per spawning fish. How could the fish that had fed the Pilgrims and produced catches of 50,000 metric tons by hand fishing a hundred years ago have failed so completely?

For nearly 400 years, cod were fished with hand lines; these gave way to longlines with hundreds of hooks and net traps in the nineteenth century. By 1895, trawl nets 45–60 m (150–200 ft) long were in use, and in the 1950s, huge factory trawlers from Great Britain, the former Soviet Union, Germany, and other nations joined the fishing. In 1968, the combined fleet caught 810,000 metric tons of cod, nearly three times more than had been caught in a single year before 1954. By the 1970s, the cod catch had dropped to less than 200,000 metric tons, and both Canada and the United States extended their territorial waters out 320 km (200 mi) to exclude foreign vessels. Recommended cuts in harvest quotas were not made, and to maintain catch rates, fishing effort was intensified. Between 1968 and 1992, the catch declined 69%, and between 1993 and 1996, another drop of the same magnitude occurred. More than 30,000 people lost their jobs with the closure of the Atlantic cod fishery. The Grand Banks area off Newfoundland remains closed to cod fishing.

The New England fishery followed the same course—enlarging the fleet from 825 boats to 1423 between 1977 and 1982. Although the catches had begun their decline, the New England Fishery Council was unwilling to control the fishing and eliminated catch quotas in 1982. In 1991, the National Marine Fisheries Service (NMFS) was sued to reinstate the regulations, and by 1993, the NMFS announced that 55% of the New England cod population had been caught and that only closing the fishery could save the stock. The 1999 cod catch was 9700 metric tons, and the 2005 catch was 6900 metric tons, compared with 43,700 metric tons in 1990. The fishery remains tightly controlled, with fishing confined to certain areas and a limit of 10,000 metric tons.

Every year since 2001, European fisheries scientists have advised the European Union to ban cod fishing in the North Sea. Again in 2006, it was recommended that the fishery be closed for at least two years to allow North Sea stocks to rise to a minimum sustainable level. Instead, the 2008 allowable catch for North Sea cod has been increased by 11% due to a higher than expected detection of juvenile fish.

Sharks

As overfishing takes its toll on familiar commercial pelagic species such as tuna and swordfish, the fishing pressure is transferred to other species. Most sharks are vulnerable to overfishing because they are long-lived, take many years to reach sexual maturity, and only have a few young at a time. Recovery from overfishing can take years or decades for many shark species. Until recently, sharks were only fished for sport and local markets. When U.S. commercial shark fishing began in the late 1970s, 21 metric tons of thresher shark were brought to California docks; by 1982, the thresher catch was 1100 metric tons. In 1989, the catch fell to 300 metric tons, and the Pacific thresher fishery collapsed.

The Magnusun-Steven Fishery Conservation and Management Act requires overfished shark stocks to be rebuilt and for healthy shark populations to be maintained. In the Atlantic, Gulf of Mexico, and Caribbean Sea, the large coastal sharks such as tiger (*Galeocerdo cuvier*) and hammerhead (*Sphyrna mokarran*) are overfished; the status is unknown for highly migratory pelagic sharks such as blue (*Prionace glauca*) and shortfin mako (*Isurus oxyrinchus*). The current status of most sharks in the Pacific is unknown. The practice of removing the fin(s) from a shark and discarding the remainder of the shark at sea is prohibited for any person in U.S. waters or onboard a U.S. fishing vessel.

In the United States, the NMFS has set strict quotas on commercial shark fishing in the Atlantic, Gulf of Mexico, and Caribbean (fig. 17.19); recreational fishing was restricted and finning banned. In 2002, the United Nations Convention on the International Trade in Endangered Species placed basking and whale sharks, the world's largest fish, on its list of fish that can be hunted and traded only within strict limits. Concern continues over the loss of the ocean's most widespread top predator; no one knows how many sharks are in the world's oceans, and no one knows how their loss will affect ocean food webs.

Fish Farming

The world's demand for seafood is increasing at the same time that the ocean harvest of wild fish is decreasing. An alternative way to increase the fish harvest is by fish farming, known as

Figure 17.19 National Marine Fisheries Service scientists sample a shark population to evaluate the effects of commercial fisheries.

aquaculture, the growing or farming of animals and plants in a water environment. The term **mariculture** is used for the growing or farming of marine plants and animals. Currently, aquaculture provides 30% the world's fish market. The UN Food and Agricultural Organization projects that by 2010, aquaculture will provide 35% of the world's fish for human consumption. Global aquaculture harvests more than doubled in weight between 1986 and 1996. At the present time, more than 25% of all fish consumed by humans is raised in aquaculture ponds.

Farming the water began in China some 4000 years ago. The Chinese were culturing common carp in 1000 B.C., and in 500 B.C., a book was written giving directions on fish farming, including methods for building the pond, selecting the stock, and harvesting it. In China, Southeast Asia, and Japan, fish farming in fresh and salt water has continued to the present as a practical and productive method of raising large quantities of fish such as carp, milkfish, *Tilapia,* and catfish. Most of these farms are family-run, labor-intensive operations. Fish farmers may practice **monoculture,** raising fish as a single species, or **polyculture,** raising more than one species as surface feeders and bottom feeders to make use of the total volume of the pond.

In the United States, fish farming produces only about 2% of fishery products, but that amount includes 50% of the catfish and nearly all the trout. To be successful and profitable in the United States, fish farming requires a proven market and large-scale operation combined with the development of technology to reduce costs. The species chosen for farming must reproduce easily in captivity, have juvenile forms that survive well under controlled conditions, gain weight rapidly, eat cheap and available food, and fetch a high market price.

As commercial fishing has declined, mariculture has responded by more than tripling its worldwide output. In 1987, mariculture provided 4 million metric tons of fish and shellfish; in 2005, it supplied about 19 million metric tons. Marine fish are grown in floating cages or pens kept in calm, shallow, protected, coastal areas, which are becoming less and less available as shorelines are being increasingly built over and used for recreation. The lack of suitable sites, the impact of fish pens on water quality (increasing nutrients, decreasing oxygen), the release of antibiotics into the marine environment, and the possibility of the transmission of fish disease to wild populations are pushing the development of offshore, submersible systems that are more efficient and that distribute waste and excess food over larger areas. Additional research is going into high-energy feeds that promote faster growth and therefore decrease the time needed to bring the fish to market size.

Farming of Atlantic and Pacific salmon has surged past the catch of wild salmon. About half of store-bought salmon is now farmed. Since the late 1970s over 1 million metric tons of farmed salmon have been produced per year. A majority of the farmed salmon is raised in open net pens (fig. 17.20) in coastal areas near migration paths of wild salmon as they move to and from the ocean. A number of reports has suggested that farmed salmon can infect wild salmon with sea lice and in early 2008, a virus known as infectious salmon anemia was documented to be killing Chilean farmed salmon. Farmed salmon depend on a diet that is ~45% fish meal and 25% fish oil, commonly derived from menhaden and anchoveta. On average, every pound of farmed salmon requires 3 pounds of wild-caught fish as food, which acts to increase pressures on wild fisheries rather than decrease pressures.

The Canadian government has adopted a national policy of encouraging salmon farms. In British Columbia, the fish farmed are not the native Pacific salmon species but Atlantic salmon, and there is concern over escapees competing with the Pacific salmon. Populations of parasitic sea lice increase dramatically in salmon pens. Farming large numbers of fish requires introducing antibiotics and produces significant quantities of fish waste and uneaten food that deplete the water of oxygen. A typical farm may raise 400,000 to 1 million fish. The native British Columbia pink salmon run has declined from 3.6 million to 147,000 fish in two years, and many blame the salmon farms. In the United States, Maine, Washington, California, and Oregon allow salmon farming, but the permit procedure is long and complex. Mariculture research is focusing on several other fish species for marketing by 2010. In Hawaii, pen-reared mahi mahi now grow to 1.5 kg (3 lb) size in 150 days and are expected to be the first to enter the market. Norway has the technology to grow Atlantic cod, but the cost is still too high for the commercial market. The United States and Canada are experimenting with the Atlantic halibut.

Figure 17.20 An aerial view of salmon pens, in which fish are raised to market size. Typical net pens are 100 m across and 30 m deep. *By Sea Farm Washington, Inc.*

Summary

The nekton swim freely and independently. The marine mammals include whales, seals, sea lions, walruses, sea otters, and sea cows. Marine mammals breathe air and maintain near-constant body temperatures. The U.S. Marine Mammal Protection Act was passed in 1972 to protect all marine mammals and to prohibit commercial trade in marine mammal products. Today, some of the biggest threats to marine mammals are habitat degradation and the cumulative effects of harassment.

Whales belong to the cetaceans. Some whales such as sperm whales and killer whales, dolphins, and porpoises possess teeth. Others whales such as humpbacks and blue whales possess baleen that they use to filter the water for plankton. Some whales migrate seasonally over thousands of miles. Many whales are endangered and the International Whaling Commission regulates whaling on a voluntary basis; a moratorium on commercial whaling is in effect, but scientific whaling continues.

Seals, walruses, sea lions, and sea otters belong to the pinnipeds. They spend most of their time at sea, hunting for food, but still need to come ashore to rest and breed. Sea otters are considered keystone predators in kelp forests. Fur seals and sea otters were hunted heavily during the nineteenth and twentieth centuries, but population numbers have recovered somewhat. Manatees and dugongs feed on seagrass and are found in warm waters of the Indian and Atlantic Oceans where they are caught between increasing human contacts and loss of habitat; the Steller sea cow is extinct. The polar bear is the top predator of the Arctic's marine food web and is completely dependent on the presence of sea-ice. The primary threat to polar bears is diminishing sea-ice coverage.

Only 3% of total bird species are marine; these birds are specialized for life at sea. Many marine birds are migratory and travel long distances each year. Seabirds spend a significant portion of their lives at sea; shorebirds do not swim much and have a greater dependence on land.

Sea snakes, the marine iguana, seagoing crocodiles, and sea turtles are the reptile members of the nekton. Many sea turtles are endangered. Female sea turtles return to specific beaches to lay their eggs. The buried eggs incubate for up to a few months. During this time they are susceptible to hunting and poaching.

Fish are found at all depths in all oceans. Sharks, rays, and skates are primitive fish with cartilaginous skeletons. All other fish have bony skeletons, including the commercially fished species and the highly specialized types of the deeper ocean.

World fish catches have increased dramatically since 1950 due to increased fishing effort and improved technology. Although stock abundance has declined, new fisheries have developed, and the fishing effort has continued high to keep the catch stable. Currently, about a quarter of fish stocks are overexploited. The FAO suggests that the maximum wild capture of fish has been reached for the world's oceans.

The problems of the traditional commercial fisheries include overfishing, increased regulation, and increased costs. The Peruvian anchovy fishery had produced the world's greatest fish catches until it was hit by the effects of overfishing and El Niño. The Pacific Northwest and Alaskan salmon fisheries are experiencing difficulty due to the degradation of the salmon's freshwater environment and the loss of natural salmon runs. The Atlantic cod fishery collapsed due to overfishing. World shark fisheries are depleting shark populations in all the oceans.

Aquaculture and mariculture are increasing worldwide, and they have proved to be practical and productive in much of Asia. Norway, Chile, and Scotland produce most of the farmed salmon. In the United States, marine aquaculture is a small industry that requires large-scale operations, proven markets, and cost-cutting technology to succeed. Undesirable side effects include degraded water quality and possible disease transfer to the native stocks.

Key Terms

All key terms from this chapter can be viewed by term or definition when studied as flashcards on this book's website at www.mhhe.com/sverdrup10e.

mammal, 402	dugong, 411	surimi, 423
cetacean, 402	sea cow, 411	aquaculture, 428
baleen, 402	Sirenia, 411	mariculture, 428
pinniped, 407	echolocation, 413	monoculture, 428
manatee, 411	demersal fish, 420	polyculture, 428

Study Questions

1. The Marine Mammal Protection Act of 1972 has allowed the recovery of certain populations to such an extent that the marine mammals are now competing for food resources of economic value to humans. Under these circumstances, should any adjustments be made to this law? Explain why.
2. What sensory abilities help a shark to find its prey?
3. In the last fifty years, the world's marine fish catch has grown from 40 million metric tons to more than 80 million metric tons. During this period, the catch from many large fisheries has declined. How do you explain this contradictory situation?
4. How have the problems of migratory seabirds changed during the last hundred years?
5. What uses do marine mammals make of sound?
6. Why has the sea otter population diminished in the area of the Aleutian Islands over the last five years?
7. How can the ability to predict El Niño help manage the Peruvian anchovy fishery?
8. Why are more toothed whales found at low latitudes and why do more baleen whales reside in temperate and polar latitudes? Consider food requirements only.
9. Compare the problems of the Atlantic cod fishery and the Pacific salmon fishery. Are there any factors in common to both fisheries? How do these fisheries differ?
10. Sea snakes are thought to have their origin in the land snakes of India and Southeast Asia. Why are there no sea snakes in the Atlantic Ocean? What might be the effect of a sea-level canal through the isthmus of Panama?
11. Explain the status of sea turtle populations at the present time.
12. Discuss the impact of humans on populations of walrus, fur seals, and sea otters.
13. Compare fishing as practiced in the United States and in less-developed countries.
14. Compare fish farming in the United States with fish farming in other countries.
15. What characteristics do deep-sea fish have in common? How do these characteristics enhance fish survival in this environment?

Sea stars and other organisms on a rocky shore at low tide.

Chapter 18

The Benthos: Dwellers of the Sea Floor

Chapter Outline

Learning Outcomes

After studying the information in this chapter students should be able to:

1. *describe* which factors limit where seaweeds can grow,

2. *understand* the role of keystone predators in intertidal regions,

3. *describe* the factors that dictate where benthic animals grow, and

4. *understand* the role of symbioses in different ecosystems.

The animals and algae that live on the sea floor or in the sediments are members of the benthos. The algae and plants are found in the sunlit shallow coastal areas and in the intertidal zones, while the benthic animals are found at all ocean depths. The benthos include remarkably rich and diverse groups of organisms, among them the luxuriant, colorful tropical coral reefs, the newly discovered organisms surrounding deep-sea hydrothermal vents, the great cold-water kelp forests, and the hidden life below the surface of a mud flat or a sandy beach. Benthic organisms are important food resources and provide valuable commercial harvests: for example, oysters, clams, crabs, and lobsters. In this chapter, we first present an overview of the benthos by group and habitat, with special focus on the world's more intriguing benthic communities, and then we consider the harvesting of the benthos, its problems, and its potential. This chapter is intended to serve only as an introduction to this topic, for no single chapter can do justice to the diversity of organisms present in this group.

18.1 Algae and Plants

Seaweeds are members of a large group called **algae.** The unicellular planktonic algae were considered in chapter 16. In this chapter, we consider the large, benthic, multicellular algae, or seaweeds, that are conspicuous members of coastal intertidal and subtidal communities. Like plants, algae photosynthesize and contain pigments, including chlorophyll *a,* to absorb solar energy. Algae are distinguished from plants by body form, reproduction, accessory pigments, and storage products. Algae have simple tissues; they do not produce flowers or seeds, and they do not have roots.

General Characteristics of Benthic Algae

Seaweeds are benthic organisms; they grow attached to rocks, shells, or any solid object. Seaweeds are attached by a basal organ known as a **holdfast** that anchors the seaweed firmly to a solid base or substrate. The holdfast is not a root; it does not absorb water or nutrients. Above the holdfast is a stemlike portion known as the **stipe.** The stipe may be so short that it is barely identifiable, or it may be up to 35 m (115 ft) in length. It acts as a flexible connection between the holdfast and the **blades,** the alga's photosynthetic organs. Seaweed blades are thin and are

bathed on all sides by water. They serve the same purpose as leaves, but they do not have the specialized tissues and veins of leaves because algae do not require that water be conducted from the ground, up a stem, and through the veins to leaf cells. The blades may be flat, ruffled, feathery, or even encrusted with calcium carbonate. The general characteristics of a benthic alga are shown in figure 18.1.

Seaweeds are not found in areas of mud or sand where their holdfasts have nothing to which they can attach. Sometimes during a storm seaweeds are dislodged, taking with them the rocks to which the holdfasts cling. Seaweeds carrying their rock anchors, may drift for some time before they sink back to the bottom. If they sink too deep for sufficient light to reach them, they die, but if their blades remain in the photic zone, they will continue to grow. Because the benthic algae are dependent on sunlight, they are confined to the shallow depths of the ocean, where they are surrounded by water, dissolved carbon dioxide, and nutrients. They are efficient primary producers, exposing a large blade area to both the water and the Sun.

In the sea, the quality and quantity of light change with depth. The red end of the visible light spectrum is quickly absorbed at the surface, and the blue-green light penetrates to the greatest depths (review fig. 14.9). On land and at the sea surface, algae receive the full spectrum of visible light. The green algae have the same chlorophyll pigments as land plants, and they are able to absorb the same wavelengths of light and they are found in shallow water. Algae found at moderate depths have a brown pigment that is more efficient at trapping the shorter wavelengths. At maximum growing depths, the algae are red, for the red pigment can best absorb the remaining blue-green light. The characteristic pattern for seaweed growing on a rocky shore is the green algae in shallow water, then the brown algae, and the red algae at greater depths. Few large red algae are seen at low tide, for they are primarily sublittoral species.

Seaweeds provide food and shelter for many animals. They act in the sea much as forests and shrubs do on land. Some fish and other animals, such as sea urchins, limpets, and some snails, feed directly on the algae; other animals feed on shreds and pieces of algae as they settle to the bottom. Some organisms use large seaweeds as a place of attachment; some of the smaller algae grow on the larger forms. Algae that produce calcareous outer coverings are important in the building of tropical coral reefs; these are discussed in section 18.4 of this chapter.

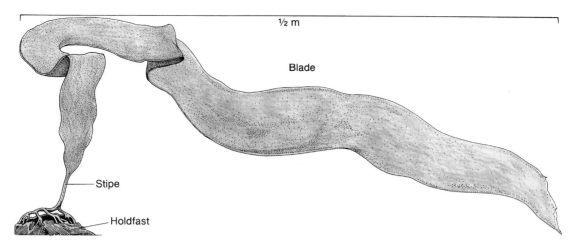

Figure 18.1   Benthic algae are attached to the sea floor by a holdfast. A stipe connects the holdfast to the blade. *Laminaria* is a genus of kelp and a member of the brown algae.

Kinds of Seaweeds

As just described, algae can be divided into different groups based on the combination of pigments they possess. The pigments that the green, brown, and red algae contain provide them with their characteristic color. Review figure 15.1 to see how different pigments influence the spectrum of light that algae can absorb. Remember that leaves on a tree appear green, for example, because their combination of pigments absorbs mainly blue and red light but not green light. The color a plant or an algae appears are the colors of light that are not well absorbed by their pigments. These colors are instead reflected back to our eyes. As described below, categorization of algae is based on additional characteristics besides their suite of pigments. Therefore, relying on visible color alone to categorize the algae can sometimes be misleading. Some red algae appear brown, green, or violet, and some brown algae appear black or greenish. Representative algae are shown in figure 18.2.

Green algae are moderate in size and may form fine branches or thin, flat sheets. They are mostly freshwater organisms, but a small number occur in the sea, including *Ulva,* the sea lettuce, and *Codium,* known as dead man's fingers. Green algae are related to the land plants; they have the same green chlorophyll pigments as land plants and they also store starch as a food reserve.

All brown algae are marine and range from simple microscopic chains of cells to the **kelps,** which are the largest of the algae. Kelps have more tissue structure than most algae but are much simpler than flowering land plants. The kelps have strong stipes and effective holdfasts that allow them to colonize rocky points in fast currents or heavy surf, a habitat favored by the sea palm, *Postelsia.* Other kelps grow with holdfasts well below the depth of wave action and float their blades at the surface supported by gas-filled floats; for example, the bull kelp, *Nereocystis,* is especially abundant along the Alaskan, British Columbian, and Washington coasts, and the great kelp, *Macrocystis,* is found along the California coast. Other species are found off Chile, New Zealand, northern Europe, and Japan. Their brown color comes from the pigment fucoxanthin, the same pigment found in diatoms (fig. 15.1). Storage products include the carbohydrates laminarin and mannitol but not starch.

The red algae are almost exclusively marine; they are the most abundant and widespread of the large algae. Their body forms are varied and often beautiful, flat, ruffled, lacy, or intricately branched. Their life histories are specialized and complex, and they are considered the most advanced of the algae. All contain the pigments phycoerythrin and phycocyanin (review fig. 15.1). Their storage product is floridean starch.

Although the diatoms were discussed as part of the plankton (see chapter 16), there are also benthic diatoms. These diatoms are usually of the pennate type and grow on rocks, muds, docks, and blades of kelp, where they produce a slippery brown coating.

Marine Plant Communities

A few flowering plants with true roots, stems, and leaves have made a home in the sea. Seagrasses grow, often completely submerged, in patches along muddy beaches, where they help to stabilize the sediments. Eelgrass, with its strap-shaped leaves, is found on mud and sand in the quiet waters of bays and estuaries along the Pacific and Atlantic coasts, and turtle grass is common along the Gulf Coast. Surf grass flourishes in more turbulent areas exposed to waves and tidal action. Seagrasses are important primary producers, and their decomposing leaves add large quantities of vegetative material and nutrients to shores and estuaries. They provide a place of attachment for sponges, small worms, and tunicates and a feeding ground for many benthic organisms.

Temperate-area salt marshes are dominated by marsh grasses able to tolerate the brackish water. Marsh grasses are partly consumed by marsh herbivores, but much of the vegetation breaks down in the marsh and is washed into the estuaries by tidal creeks. There the vegetative remains are broken down by bacteria that release nutrients to the water to be reused by the bacteria. See the box on *Spartina* salt marshes in chapter 13.

The mangroves (see fig. 12.14) grow in the intertidal zone along humid, tropical coasts. They are salt-tolerant, woody trees with special adaptations that allow them to thrive in oxygen-deficient muds. Their waxy leaves reduce water loss, and they

Figure 18.2   Representative benthic algae. *Ulva* and *Codium* are green algae. *Postelsia, Nereocystis,* and *Macrocystis* are kelps. The kelps and *Fucus* are brown algae. The red algae are *Corallina, Porphyra,* and *Polyneura. Corallina* has a hard, calcareous covering.

excrete excess salt through glands located on their leaves. Some mangroves grow prop roots from overhanging branches to extend the root system above the water; others send up vertical roots from roots below the water's surface. Birds, insects, and other animals live in the mangrove's leafy canopy, and its intertwining prop roots provide shelter for small marine organisms. The roots also trap sediments and organic material, which help to build and extend the shore seaward.

18.2 Animals

Benthic animals, unlike benthic algae, are found at all depths and are associated with all substrates. More than 150,000 benthic species exist, as opposed to about 3000 pelagic species. About 80% of the benthic animals belong to the **epifauna;** these animals live on or attached to rocky areas or firm sediments. Animals that live buried in the substrate belong to the **infauna** and are associated with soft sediments such as mud and sand. There are only about 30,000 known species of infauna, as compared with 120,000 species of epifauna.

Some animals of the sea floor are **sessile,** or attached to the sea floor, as adults (for example, barnacles, sea anemones, and oysters), while others are motile all their lives (for example, crabs, starfish, and snails). Although most benthic forms produce motile larval stages that spend a few weeks of their lives as members of the meroplankton (see section 16.3), the planktonic larval stage is of particular importance to the sessile species. The mobility of the juvenile stages allows these species to relieve overcrowding and to colonize new areas. Sessile adults must wait for their food to come to them, either under its own power or carried in by waves, tides, and currents; motile organisms are able to pursue their prey, scavenge over the bottom, or graze on seaweed-covered rocks.

The distribution of benthic animals is not governed by any single factor such as light or pressure. Animal distribution is controlled by a complex interaction of factors, creating living conditions that are extremely variable. The substrate may be solid rock, movable cobbles, shifting sand, or soft mud. Temperature is nearly constant in deep water but changes abruptly in shallow areas covered and uncovered by the daily tidal cycle; salinity, pH, exposure to air, oxygen content of the water, and water turbulence also change abruptly in the intertidal zone. Benthic animals exist at all depths and are as diverse as the conditions under which they live. Their lifestyles are related to their varied habitats, and the following sections provide an overview of habitats and lifestyles.

Animals of the Rocky Shore

The rocky coast is a region of rich and complex algal, plant, and animal communities living in an area of environmental extremes. It is a meeting place between the more variable land conditions and the more stable sea conditions. As the water moves in and out on its daily tidal cycle, rapidly changing combinations of temperature, salinity, moisture, pH, dissolved oxygen, and food supply are encountered. At the top of the littoral zone, organisms must cope with long periods of exposure, heat, cold, rain, snow, and predation by land animals and seabirds as well as turbulence of waves and seaward water flow. At the littoral zone's lowest reaches, organisms are rarely exposed but have their own problems of competition for space as well as predation. The exposure endured by marine life at different levels in the littoral zone is shown in figure 18.3.

The distribution of the algae and animals is governed by their ability to cope with the stresses that accompany exposure, turbulence, and loss of water, as well as food web dynamics. Along the rocky coasts of such areas as North America, Australia, and South Africa, biologists have noted that patterns form as the algae and animals sort themselves out over the intertidal zone. This grouping is called zonation; **vertical zonation,** or **intertidal zonation,** is shown in figure 18.4. The distribution of the seaweeds, with the green algae in shallow water, the brown algae in the intertidal zone, and the red algae in the subtidal area, is an example of such vertical zonation.

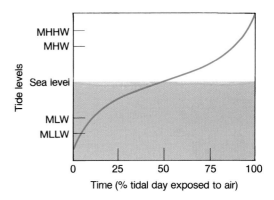

Figure 18.3 The time of exposure to air for intertidal benthic organisms is determined by their location above and below the sea level and by the tidal range. (MLLW = mean lower low water; MLW = mean low water; MHW = mean high water; MHHW = mean higher high water.)

In chapter 17, the concept of a keystone predator was introduced. The example discussed was the role of the sea otter as a keystone predator, whose presence maintains diversity within kelp forests. Another example of a keystone predator comes from the classic work of Robert Paine, who studied intertidal community structure. Paine and his students set out to understand why different species are distributed in different zones within the intertidal along the coast of Washington. They knew that the dominant mussels *(Mytilus californianus)* in this region are susceptible to desiccation and need to be covered by water for most of the day. Therefore, water height determined the upper limit of the mussels. The lower limit of the mussels was determined by their predator, the starfish *(Pisaster ochraceous)*. This starfish is larger than mussels and even more sensitive to desiccation so its upper limit is shallower than that of the mussel. Therefore a narrow band of mussels grows in this intertidal region, with the upper limit determined by water height and the lower limit determined by predators. Another interesting part of the experiment was that Paine and his students removed by hand all the starfish from the zone below the mussels. Fairly soon thereafter, the mussels invaded the lower zone and grew so rapidly that they eliminated other species within the area and the biodiversity of the region dropped dramatically. This drop in biodiversity with removal of a single predator is the foundation of the concept of keystone predator. Keep this concept in mind when considering distributions of organisms along the rocky shore.

The following is a general discussion of the more conspicuous life forms found in these rocky intertidal zones. It is not a guide to the marine life of rocky shores, as substantial differences occur from region to region. If you wish to explore the rocky shores in your area, obtain a manual written specifically for your location.

In the supralittoral (or splash) zone, which is above the high-water level and is covered with water only during storms and the highest tides, the animals and algae occupy an area that is as nearly land as it is ocean bed. The width of the zone varies with the slope of the rocky shore, variations in light and

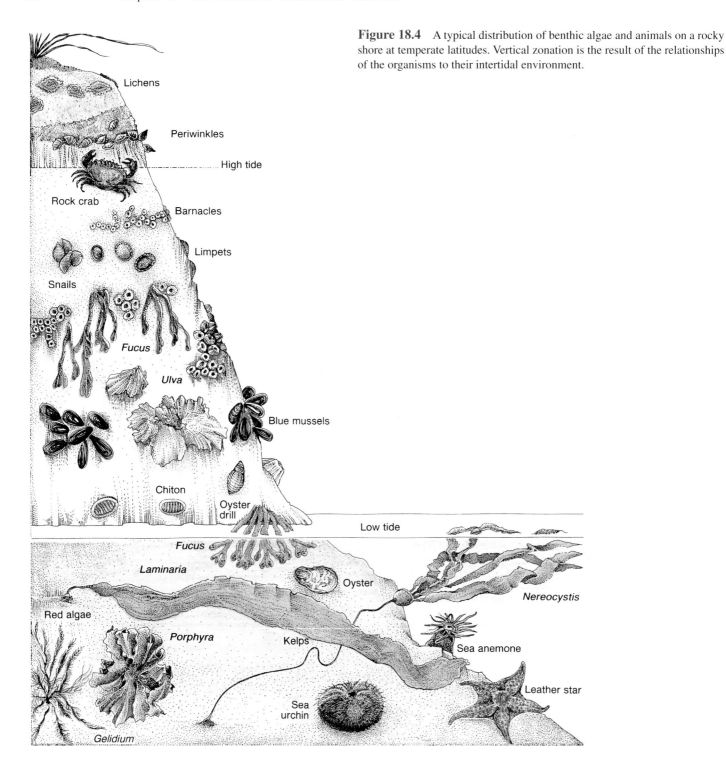

Figure 18.4 A typical distribution of benthic algae and animals on a rocky shore at temperate latitudes. Vertical zonation is the result of the relationships of the organisms to their intertidal environment.

shade, exposure to waves and spray, tidal range, and the frequency of cool days and fogs. At the top of this area, patches of dark lichens and algae appear as crusts on the rocks; these crusts are often nearly indistinguishable from the rock itself. Scattered tufts of algae provide grazing for the small herbivorous snails and limpets of the supralittoral zone. The periwinkle snail *Littorina* is well adapted to an environment that is more dry than wet; it is an air-breather, and some species will drown if caught underwater. At low tide, snails withdraw into their shells to prevent moisture loss and seal themselves off from the air with

a horny disk called an operculum. Limpets are able to use their single muscular foot to press themselves tightly to the rocks to prevent drying out.

Just below the zone of snails and limpets, the small acorn barnacles filter food from the seawater and are able to survive even though they are covered with water only briefly during the few days of the spring tides each month. Barnacles are crustaceans, related to crabs and lobsters but cemented firmly in place. They have been described as animals that lie on their backs and spend their lives kicking food into their mouths with

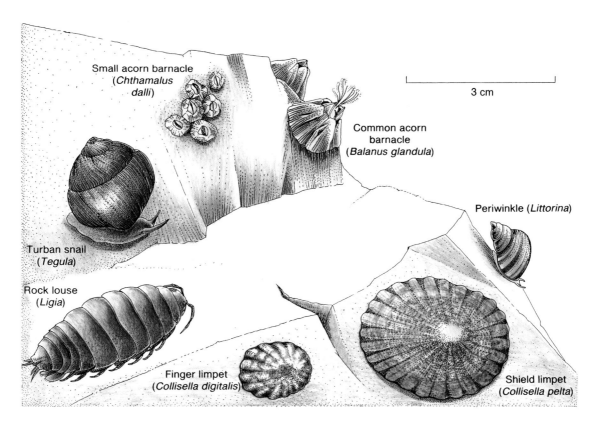

Figure 18.5 Organisms of the supralittoral zone. The limpet and the snail *Littorina* are herbivores. The barnacles feed on particulate matter in the water. *Ligia* is a scavenger.

their feet. In some areas, the rocky splash zone is the home of another crustacean, the large (3–4 cm, or 1–2 in) isopod *Ligia*. Organisms of the supralittoral zone are shown in figure 18.5.

Conspicuous members of the upper midlittoral zone are illustrated in figure 18.6. These organisms include several other species of barnacles, limpets, snails, and two other **mollusks:** the bivalved (or two-shelled) mussels and chitons, which may appear to resemble limpets but on closer inspection will be seen to have shells of eight separate plates. Chitons, like limpets, are grazers that scrape algae from hard surfaces. Mussels are filter feeders; food strained from the water is trapped in a heavy mucus and moved to the mouth by liplike palps. A muscular foot anchors chitons and limpets; strong cement secures barnacles; and special threads attach mussels to the rocks. Tightly closed shells protect many of the organisms from drying out during periods of low tide, and their rounded profiles present little resistance to the breaking waves. Species of brown algae in this zone, typically rockweed (*Fucus*), have strong holdfasts and flexible stipes.

Gooseneck barnacles are found attached to rocks where wave action is strong. They have evolved an interesting feeding style, facing shoreward and feeding by taking particulate matter from the runback of the surf rather than facing the sea, as might be expected. Mussel beds promote shelter for less conspicuous animals, such as the segmented sea worm *Nereis* and small crustaceans. Shore crabs of varied colors and patterns are found in the moist shelter of the rocks. The crabs are active predators as well as important scavengers. Small sea anemones, huddled together in large groups to conserve moisture, are also found in the higher regions of the midlittoral zone.

The area is crowded; the competition for space appears extreme. The free-swimming juveniles (or larval forms) of these animals settle and compete for space, each with its own special set of requirements. New space in an inhabited area becomes available as the whelks, which are carnivorous snails, prey on some species of thinner-shelled barnacles, and the sea stars move up with the tide to feed on the mussels. This predation restricts the thinner-shelled barnacles to the upper levels of the littoral zone, which are too dry for the predator snails. Other species of barnacles inhabit the lower midlittoral zone in association with the predator snails, which cannot pierce the heavier plates of the mature individuals. Seasonal die-offs of algae and the battering action of strong seas and floating logs also act to clear space for newcomers.

A selection of organisms from the lower littoral zone is found in figure 18.7. The larger anemones are common inhabitants of the midlittoral and lower littoral zones. These delicate-looking, flowerlike animals attach firmly to the rocks and spread their tentacles, loaded with poisonous darts called nematocysts. The darts are fired when small fish, shrimp, or worms brush the tentacles. The prey is paralyzed, the tentacles grasp, and the prey is pushed down into the anemone's central mouth. A finger inserted into the tentacles of most anemones will feel only a stickiness and a tingling sensation, but some large tropical species can produce a painful injury. Some snails and sea slugs are unaffected by the nematocysts and prey on the anemones. Certain sea slugs store the anemone's nematocysts in their tissues and use them for their own defense.

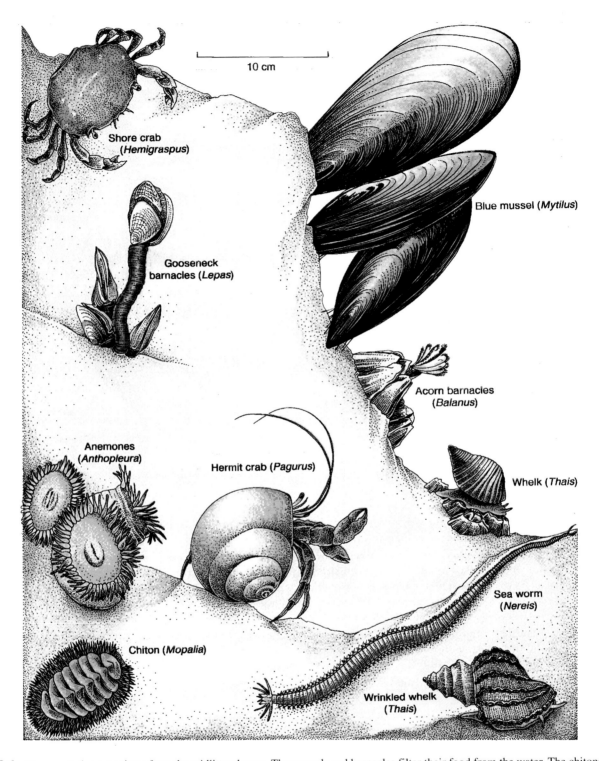

Figure 18.6 Representative organisms from the midlittoral zone. The mussels and barnacles filter their food from the water. The chitons graze on the algae covering the rocks. *Thais,* a snail, is a carnivore. Small shore crabs and hermit crabs are scavengers. *Balanus cariosus* is a larger and heavier barnacle than the barnacles of the supralittoral zone. *Nereis* is often found in the mussel beds. The anemones huddle together to keep moist when exposed.

Starfish of many colors and sizes make their home in the lower littoral zone; these slow-moving but voracious carnivores prey on shellfish, sea urchins, and limpets. Their mouths are on their undersides, at the center of their central disks, surrounded by strong arms that are equipped with hundreds of tiny suction cups, or tube feet. The tube feet are operated by a water-vascular system, a kind of hydraulic system that attaches the animal very firmly to a hard surface. In feeding, the tube feet attach to the shell of the intended prey and, by a combination of holding and pulling, open the shell sufficiently to insert the starfish's

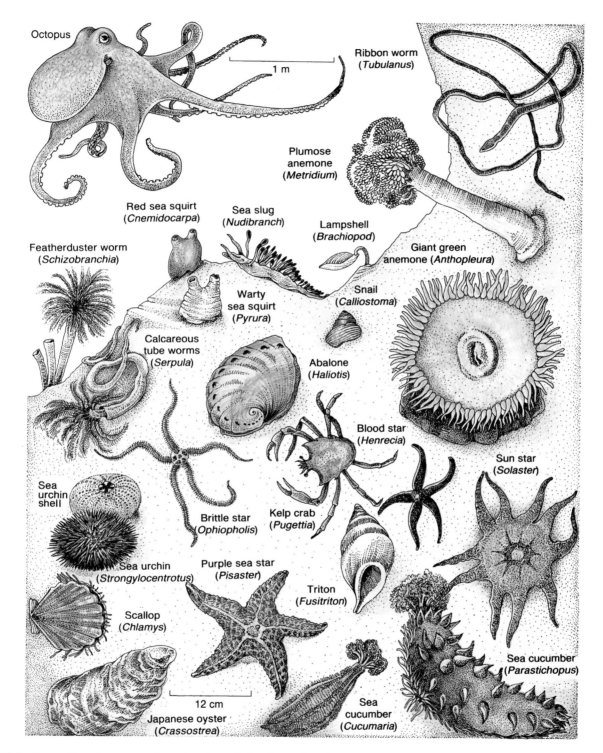

Figure 18.7   Lower littoral zone organisms. A variety of related organisms inhabit the area. The starfish feed on the oysters; the related sea urchins are herbivores; and the sea cucumbers feed on detritus suspended in the water. Among the mollusks are the oysters, scallops, snails, abalone, nudibranchs, and octopuses. The oysters and scallops are filter feeders. *Calliostoma* and the abalone are grazers; the nudibranch, octopus, and triton snail are predators. Note the size of the anemones in this zone.

stomach, which can be extruded through its mouth; enzymes are released, and digestion begins. Shellfish sense the approach of a starfish by substances it liberates into the water, and some execute violent escape maneuvers. Scallops swim jerkily; clams and cockles jump away; even the slow-moving sea urchins and

limpets move as rapidly as possible. Starfish are so effective as predators that oyster farmers must take care to exclude them from their oyster beds.

The filter-feeding sponges, some flat and some vase-shaped, encrust the rocks. On a minus tide, delicate, free-living flatworms and long **nemerteans,** or ribbon worms, armed with poison-injecting mouthparts, are found keeping moist under the mats of algae. Snails and crabs inhabit this zone; the scallop, another filter-feeding bivalve, and red algae are found as well, along the beds of kelp, eelgrass, and surf grass. Calcareous red algae encrust some rocks and are seen as tufts on others. Occasionally, **brachiopods,** or lampshells, are found in the lower littoral zone. They resemble clams but are completely unrelated to them. Their shells enclose a coiled ridge of tentacles used in feeding.

Beautiful, graceful, and colorful sea slugs, or **nudibranchs,** are active predators, feeding on sponges, anemones, and the spawn of other organisms. Although soft-bodied, they have few if any enemies because they produce poisonous acidic secretions. Herbivores are also present in the lower intertidal region; species of chitons and limpets as well as the sea urchin graze on the algae covering the rocks. Sea cucumbers are found wedged in cracks and crevices; some types are identifiable by their brightly colored tentacles, which act as mops to remove food particles from the water and thrust them into the animal's mouth. Tube worms secrete the leathery or calcareous tubes in which they live and extend only their graceful, feathered tentacles to strain their food from the water.

Octopuses are seen occasionally from shore on very low tides. These eight-armed carnivorous animals are soft-bodied mollusks. They feed on crabs and shellfish and live in caves or dens identifiable by the piles of waste shells outside. They are known for their ability to flash color changes and move gracefully and swiftly over the bottom and through the water. The world's largest octopus is found in the coastal waters of the eastern North Pacific. It commonly measures 2–3 m (10–16.5 ft) in diameter and weighs 20 kg (45 lb), but specimens in excess of 7 m (23 ft) and 45 kg (100 lb) have been observed. They are shy and nonaggressive, although they are curious and have been shown to have learning ability and memory.

Tide Pools

The zonation of benthic forms varies with local conditions; zones are generally narrow where the beach is steep and the tidal range is small, while zones are wide in areas where the beach is flat and the range of tides is large. The orientation of the shore to sunlight and shade, wave action, and beach topography often results in the apparent displacement of benthic zones. A tide pool formed from water left by the receding tide in a rock depression or basin provides a habitat for animals common to the lower littoral zone. Some small, isolated tide pools provide a very specialized habitat of increased salinity and temperature due to evaporation and solar heating. On a summer day, the water in a tide pool may feel quite warm to the touch. Other tide pools act as catch basins for rainwater, lowering the salinity of the water and its temperature in the fall and winter. Isolated tide pools often support blooms of microscopic algae that give the water the appearance of pea soup, and in various parts of the world, a tiny, bright-red copepod, *Tigripus,* is also found.

The deeper the tide pool and the greater the volume of water, the more stable the environment during isolation by the receding tide. The larger the tide pool, the more slowly it will change temperature, salinity, pH, and the carbon dioxide–oxygen balance. Subtidal animals such as starfish, sea urchins, and sea cucumbers can only survive in large, deep tide pools. These animals do not tolerate significant changes in their chemical and physical environment. A few fish species, such as the small sculpins, can be found in tide pools. These fish are patterned and colored to match the rocks and the algae within the pool. They spend much of their time resting on the bottom, swimming in short spurts from one resting place to another. Each tide pool is a specialized environment populated with organisms that are able to survive under the conditions established in that particular pool.

The bottom of the littoral (or intertidal) zone merges into the beginning of the sublittoral (or subtidal) zone extending across the continental shelf. If the shallow areas of the subtidal zone are rocky, many of the same lower littoral zone organisms will be found. When soft sediments begin to collect in protected areas or deeper water, the populations change, and animals of the rocky bottom are replaced by those of the mud and sand substrates.

Animals of the Soft Substrates

The distribution of life in soft sediments is shown in figure 18.8. A selection of animals from this region is found in figure 18.9. Along exposed gravel and sand shores, waves produce an unstable benthic environment. Few algae can attach to the shifting substrate, and few grazing animals are found. Sands and muds deposited in coves and bays with reduced water motion provide a more stable habitat. Here, the size and shape of the sediment particles and the organic content of the sediment determine the quality of the environment. The size of the spaces between the particles regulates the flow of water and the availability of dissolved oxygen. Beach sand is fairly coarse and porous, gaining and losing water quickly, while fine particles of mud hold more water and replace the water more slowly. Sand beaches exchange water, dissolved wastes, and organic particles more quickly than mud flats. The finer the mud particles, the tighter they pack together and the slower the exchange of water; oxygen is not resupplied quickly, and wastes are removed slowly. Digging into the mud will generally show a black layer 1 or 2 cm (0.5 or 1 in) below the surface. Above this layer, the water between the sediment particles contains dissolved oxygen; below the black layer, organisms (mainly bacteria) function without oxygen, producing hydrogen sulfide (the rotten egg smell). Lack of oxygen restricts the depth to which infauna species can be found, but some animals, such as clams, live below the oxygenated level. Clams use long extensions called siphons to obtain food and oxygen from the water above the sediments.

In locations protected from waves and currents, eelgrass and surf grass help stabilize the small-particle sediments and provide shelter, substrate, and food, creating a special community of plants and animals. (See "Marine Plant Communities" in section 18.1 of this chapter.)

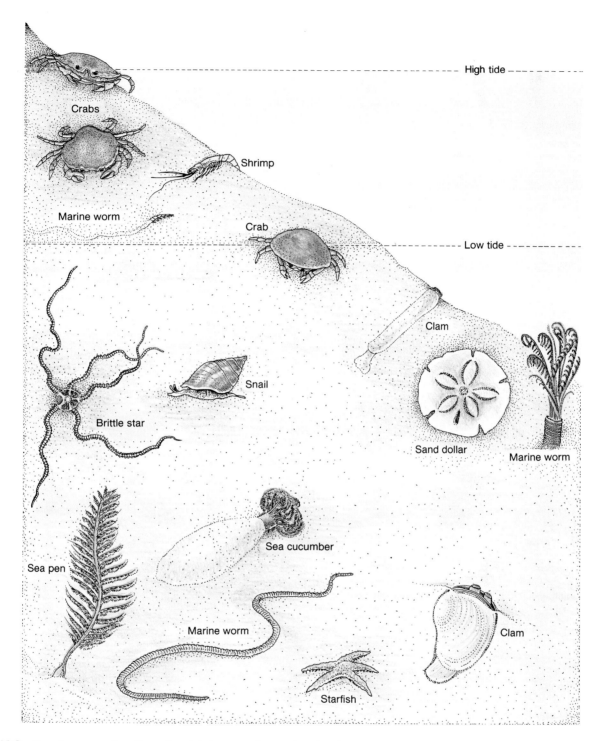

Figure 18.8 Zonation on a soft-sediment beach is less conspicuous than that found on a rocky beach. Animals living at the higher tide levels burrow to stay moist.

Most sand and mud animals are **detritus** feeders. Most detritus is formed from plant and algal material that is degraded by bacteria and fungi. The sand dollar feeds on detritus particles found between sand grains. Clams, cockles, and some worms are filter feeders, feeding on the detritus and microscopic organisms suspended in the water. Other animals are deposit feeders that engulf the sediment and process it in their gut to extract organic matter in a manner similar to that of earthworms. These deposit feeders are usually found in muds or muddy sands that have a high organic content: for example, burrowing sea cucumbers and the lugworm *Arenicola,* which produces the coiled castings seen outside its burrow. The process of sediment disruption by feeding or burrowing organisms is known as **bioturbation.** Small crustaceans, crabs, and some worm species are scavengers, preying on any available algal or animal material, while still other worms and snails are carnivores. The Moon snail is a clam eater that drills a hole in the shell of its prey and then sucks out the flesh.

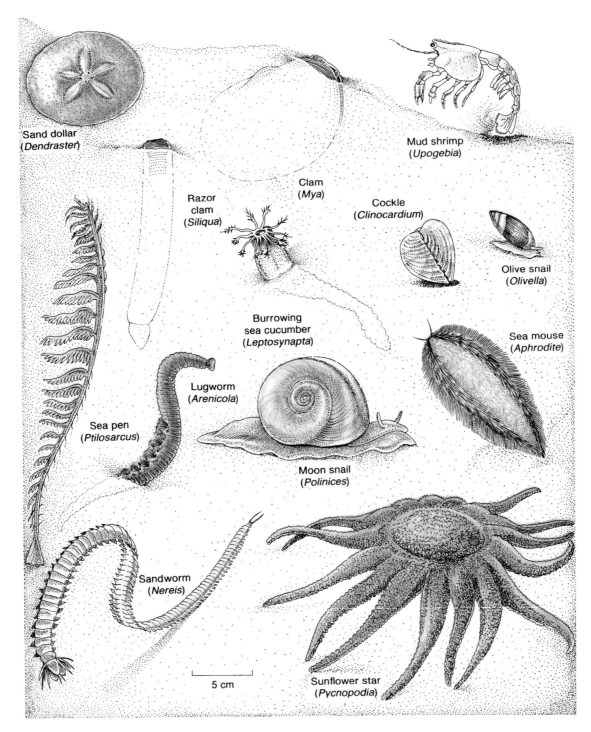

Figure 18.9 Organisms of the soft sediments. Infauna types include the shrimp *(Upogebia),* lugworm *(Arenicola),* clam, cockle, and burrowing sea cucumber. The sand dollar feeds on detritus; the Moon snail drills its way into shellfish; the sea pens feed from the water above the soft bottom. The sea mouse, like *Nereis,* is a polychaete worm.

Bacteria not only play the major role in the decomposition of organic material; they also serve as a major protein source. It is estimated that 25–50% of the material the bacteria decompose is converted into bacterial cell material, which is consumed by other microorganisms, which in turn serve as food for tiny worms, clams, and crustaceans. Areas of mud that are high in organic detritus produce large quantities of bacteria.

The intertidal area of a soft-sediment beach shows some zonation of benthic organisms, but it is not nearly as clear-cut as the zonation along a rocky cliff. In temperate latitudes, small crustaceans called sandhoppers are found at the high intertidal region; they are replaced by ghost crabs in the tropics. Lugworms, mole crabs, and ghost shrimp occupy the midbeach area, while clams, cockles, polychaete worms, and sand dollars are found in the lower intertidal region. The subtidal zone is home to sea cucumbers, sea pens, more crabs and clams, and some species of worms, snails, and sea slugs.

Animals of the Deep-Sea Floor

The deep-sea floor includes the flat abyssal plains, the trenches, and the rocky slopes of seamounts and mid-ocean ridges (see chapter 3). The sea-floor sediments are more uniform and their particle size is smaller than those of the shallower regions close to land sources. The environment of the bathyal, abyssal, and hadal zones is uniformly cold and dark.

The stable conditions of the deep-sea floor appear to have favored deposit-feeding infaunal animals of many species. Many members of the deep-sea infauna are small. They are known as the **benthic meiofauna** and measure 2 mm or less. This group of organisms include nematode worms, burrowing crustaceans, and segmented worms. At 7000 m (23,000 ft), tusk shells lie buried in the ooze with tentacles at the sediment surface to feed on the foraminifera. Acorn worms are found frequently in samples taken at 4000 m (13,000 ft). Hagfish burrow into the sediment at the

2000 m (6600 ft) depth. Detritus-eating worms and bivalved mollusks have been found on the sea floor in all the oceans.

Among the epifauna, protozoans are abundant and are widely distributed. Glass sponges attach to the scattered rocks on oceanic ridges and seamounts (fig. 18.10), as do sea squirts and sea anemones. Their stalks lift them above the soft sediments into the water, where they feed by straining out organic matter. Stalked barnacles attach to the stalks of glass sponges and sea squirts as well as to shells and boulders. Tube worms are common, ranging in size from a few millimeters to 20 cm (8 in), and sea spiders with four pairs of very long legs that span up to 60 cm (27 in) are found at depths to 7000 m (23,000 ft). Snails are found to the greatest of depths; those in the deepest trenches frequently have no eyes or eye-stalks.

The beard worms, or **pogonophora,** are found in more-productive areas at depths to 10,000 m (33,000 ft). They secrete a close-fitting tube and stand erect, with only their lower portion buried in the sediment. They have no mouth, no gut, and no anus and absorb their needed molecules through their skin. The pogonophora associated with the vent communities are discussed in section 18.6 of this chapter.

(b)

(a)

(c)

Figure 18.10   (a) A vase-shaped glass sponge 640 m (2100 ft) under the surface on the Brown Bear Seamount in the northeastern Pacific Ocean. (b) A deep-sea crab photographed at a depth of 2000 m (6550 ft) on the Juan de Fuca Ridge. (c) A group of deep-sea sponges and an anemone are seen at 684 m (2244 ft) on the Brown Bear Seamount.

(a)

(b)

Figure 18.11 Unprotected wood, such as driftwood, is subject to destruction by marine borers. (a) This beach log has been riddled by shipworms. (b) The small holes surrounding a shipworm hole are bored by a crustacean known as a gribble.

Horny corals, or sea fans, which resemble algae more than animals, grow at depths of 5000–6000 m (16,000–20,000 ft); so do solitary stone corals, which grow larger at these depths than do the coral organisms in the surface waters. Sea lilies, or crinoids, which are related to starfish, are also found at this depth, as are brittle stars and sea cucumbers. Sea cucumbers live in areas of sediments that are rich in organic substances; they are a dominant and widely spread organism of the deep-sea floor.

Fouling and Boring Organisms

Organisms that settle and grow on pilings, docks, and boat hulls are said to foul these surfaces. Fouling organisms include barnacles, anemones, tube worms, sea squirts, and algae. When these organisms grow on a vessel's hull, they slow its movement through the water and add to the costs in shipping time and haul-out for cleaning. Fouling organisms also make it difficult to operate equipment in the ocean environment (see Field Notes,

"Biofouling"), and much research goes into experiments with paints and metal alloys to discover methods of discouraging and controlling these fouling organisms.

Other organisms naturally drill or bore their way into the substrate. Sponges bore into scallop and clam shells; snails bore into oysters; some clams bore into rock. Organisms that bore into wood are costly, for they destroy harbor and port structures as well as wooden boat hulls. Two organisms are responsible for most of the wood boring: the shipworm (*Teredo*) and the gribble. *Teredo* is a wormlike mollusk with one end covered by a bivalved shell. These mollusks secrete enzymes that break down and at least partially digest wood fibers scraped away by the shell's rocking and turning movement as they bore deep and extremely destructive holes all through a piece of wood. The gribble is a crustacean; it gnaws more superficial and smaller burrows, but it, too, is very destructive (fig. 18.11).

The benthos is a remarkably diverse grouping of algae and animals (fig. 18.12). It is easily accessible in the shallow littoral zone, where it has been studied by scientists and students for hundreds of years, and today, submersibles, remotely operated vehicles (ROVs), and autonomous underwater vehicles (AUVs) survey previously unsampled marine habitats. As we continue to explore, we may expect to make discoveries as surprising and as unexpected as the vent communities in the rift areas. When we consider the uses we make of the oceans at present and the uses for which we may need the oceans in the future, we must consider their impact on the benthic environment and its inhabitants. We gain little if we use one resource at the expense of another. The balance within the marine environment is fragile, easy to degrade, and difficult (possibly impossible) to reconstruct. The health of the ocean's plants and animals is important to us as a sign of the health of our planet. We are all linked, and each of us affects the others.

18.3 High-Energy Environments

Recent work has shown that intertidal communities constantly battered by the waves are more productive than the world's lush, green rain forests. On average, waves deliver 0.335 watts of energy per square centimeter of coastline, about fifteen times more energy than comes from the Sun. Even during calm periods, wave energy is 100% greater than solar energy. The algae have little woody tissues; instead, the kelps of the rocky intertidal have 2.5 times more photosynthetic area per square meter of growing surface and are from two to ten times more productive than rain forest vegetation. The mussels, which are consumers, have been found to match or exceed the rain forest productivity when growing in areas of high wave action.

The method of harnessing the energy is indirect. Wave action reduces predators, such as starfish and sea urchins, allowing more mussels and more kelp to live in a unit area. The moving water brings a constant supply of nutrients to the algae and keeps their blades in motion, so they are never in the shade for long. Last, the waves can dislodge mussels, allowing more kelp to move in. The waves allow the primary producers to become highly concentrated, so the consumers can grow and expand their populations as well.

(a)

(b)

(c)

(d)

(e)

Figure 18.12 The benthos is a large, varied group of animals living on or in the sea floor, especially the organisms of the tide pools and the intertidal areas of the rocky coasts. (a) The anemone *Tealia crassicornis* is a sessile carnivore. The graceful and beautiful nudibranchs, or sea slugs, include (b) the white *Dirona albolineata,* (c) the orange-flecked *Triopha carpenteri,* and (d) *Hermissenda crassicornis* with its orange-and-white-striped tips. Starfish come in a remarkable diversity of shapes and sizes: (e) the bright orange bloodstar *(Henrica levinscula),* the slender-armed *Evasterias troschelli,* the many-armed *Solaster dawsoni, Mediaster aequalis* with its wide disk and broad arms, the leather star *(Dermasterias imbricata),* and the purple, rough-skinned *Pisaster ochraceus.*

(continued)

(f)

(g)

(h)

(i)

Figure 18.12 (continued)   (f) All starfish are carnivores and use their tube feet to hold and open the shellfish on which they feed. (g) The purple sea urchin *(Strongylocentrotus purpuratus)* and the green urchin *(S. droebachiensis)* are closely related to the sea stars but are herbivores, clipping off the algae with their specially constructed mouthparts (h). (i) The pink sea scallop *(Chlamys hastata hericia)* lies open as it filters organic particles from the seawater.

18.4 Coral Reefs

Tropical coral reefs are the most diverse and structurally complex of all marine communities (fig. 18.13). Coral reefs fringe one-sixth of the world's coastlines and provide habitat for tens of thousands of fish and other organisms. The largest coral reef in the world, the Great Barrier Reef, stretches more than 2000 km (1200 mi) from New Guinea southward along the eastern coast of Australia. Reef-building corals require warm, clear, shallow, clean water and a firm substrate to which they can attach. Because the water temperature must not go below 18°C and the optimal temperature is 23°–25°C, their growth is restricted to tropical waters between 30°N and 30°S and away from cold-water currents. Most Caribbean corals are found in the upper 50 m (150 ft) of lighted water; Indian and Pacific corals are found to depths of 150 m (500 ft) in the more transparent water of those oceans. Reefs usually are not found where sediments limit water transparency.

Tropical Corals

Corals are colonial animals, and individual coral animals are called **polyps** (fig. 18.14). A coral polyp is very similar to a tiny sea anemone with its tentacles and stinging cells, but, unlike the anemone, a coral polyp extracts calcium carbonate from the water and forms a calcareous skeletal cup. The corals of the curio shop and jeweler are the calcareous skeletons of the coral polyps. Large numbers of these polyps grow together in colonies of delicately branched forms or rounded masses.

Clear, shallow water is required by the reef-building coral, because within the tissues of the polyps are masses of single-celled dinoflagellate algae called **zooxanthellae** that require light for photosynthesis and therefore are limited to the photic zone (review section 16.2 on phytoplankton). Polyps and zooxanthellae have a symbiotic relationship (review dicussion of symbioses, chapter 14), in which the coral provides the algal cells with a protected environment, carbon dioxide, and nitrate

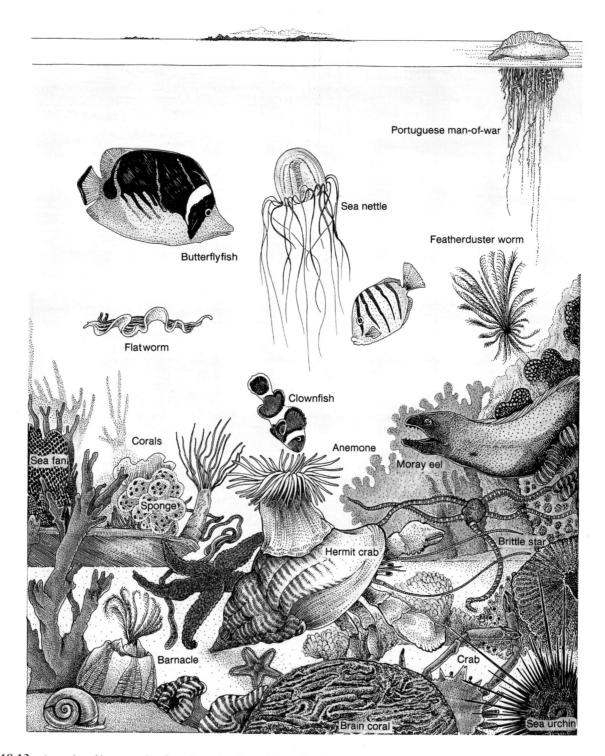

Figure 18.13 A coral reef is a complex, interdependent but self-contained community. Members include corals, clams, sponges, sea urchins, anemones, tube worms, algae, and fish.

and phosphate nutrients, and the algal cells photosynthesize, returning oxygen and removing waste. The zooxanthellae supply the corals with substantial amounts of their photosynthetic products; some coral species receive as much as 60% of their nutrition from the algae. Zooxanthellae also enhance the ability of the coral to extract the calcium carbonate from the seawater and increase the growth of their calcareous skeletons. The degree of interdependence between zooxanthellae and coral is thought to vary from species to species. The polyps feed actively at night, extending their tentacles to feed on zooplankton, but during the day, their tentacles are contracted, exposing the outer layer of cells containing zooxanthellae to the sunlight.

Field Notes

Biofouling

by Dr. Francisco Chavez

Dr. Francisco Chavez is a senior scientist at the Monterey Bay Aquarium Research Institute in Moss Landing, California. He is a marine biologist who conducts research on interactions among climate, ocean physics, marine chemistry, and ocean ecosystem variations on global to regional scales.

Competition between humans and marine organisms has been a part of ocean life for centuries. Marine organisms inevitably colonize submerged man-made structures and surfaces, for example ship hulls, buoys, instruments and nets. Microorganisms such as bacteria or micro-algae and larger organisms such as sea anemones, barnacles, and mollusks may dominate such populations. There is a succession of organisms (from smaller to larger) in a way equivalent to the colonization of an empty field, and there are also larger organisms such as gooseneck barnacles that do not depend on the smaller organisms for survival. Barnacles require a substrate for attachment and as attached adults feed primarily on planktonic organisms in the waters that surround them. This unwanted surface growth is known as biofouling. Biofouling colonies increase the drag on ships, cause increased corrosion, inhibit the functioning of instrument sensors, and foul nets. The removal of fouling organisms from marine equipment costs millions of dollars each year.

Once a surface of any kind is introduced into the ocean, it forms a substrate that is soon coated by inorganic and organic polymers (molecules made up of repeating structural units) from the seawater. This is a conditioning process that depends on environmental factors such as surface roughness, chemical composition of the material, the pH of seawater, its temperature, nutrient levels, and current velocity. After the initial conditioning viruses, bacteria, algae, and fungi colonize the surface. This process may occur within minutes or it may take weeks. The result is a slimelike matrix (box fig. 1). This matrix enhances further population growth and settlement. This slimy organic coating has a spongelike structure that contains relatively large amounts of water. The water channels and voids of the coating provide access to fresh nutrients and oxygen, also opportunities for microbes to exchange genes that fuel productivity and growth. The result is an alteration of the original surface by these

Box Figure 1 A slimelike matrix with an associated population of viruses, bacteria, and fungi that has formed on an instrument attached to a marine cable.

biochemical processes. Such stable, multilayered, biofilm environments are an ideal location for the growth of macro-organisms.

Micro-colonies flourish in this network and so do protozoan predators and larger planktonic plants and animals. Growing biodiversity eventually allows macro-foulers to gain a strong foothold. Barnacles are probably the most successful macro-organisms to colonize surfaces in oceanic conditions (box fig. 2).

Natural floating objects such as tree trunks often serve as substrates for these communities. Whales are also colonized and

Tropical Coral Reefs

The corals require a firm base to which they can cement their skeletons. The classic reef types of the tropical sea—fringing reefs and barrier reefs—are attached to existing islands or landmasses. Atolls are attached to submerged seamounts. (For a review of reef formation around a seamount, see chapter 4.) Corals are slow-growing organisms; some species grow less than 1 cm (0.4 in) in a year, and others add up to 5 cm (2 in)

a year. The same corals may be found in different shapes and sizes, depending on the depth and wave action. Environmental conditions vary over a reef, forming both horizontal and vertical zonation patterns that are the product of wave action and water depth, as shown in figure 18.15. On the sheltered (or lagoon) side of the reef, the shallow **reef flat** is covered with a large variety of branched corals and other organisms. Fine coral particles broken off from the reef top produce sand, which fills the sheltered lagoon floor. On the reef's windward

(a)

(b)

Box Figure 2 (a) The underside of a buoy before deployment with an acoustic doppler current meter attached to measure horizontal currents. (b) The same instrument after having been deployed for a period of months now encrusted with gooseneck barnacles.

have been observed scratching their backs on boats and moorings trying to rid themselves of their unwanted hitchhikers. Such hitchhikers require a near-surface environment to which they can attach and from which they can feed. For example, in the equatorial Pacific, thousands of miles from land, it takes only a few months for moored instrument-buoys to collect hundreds of attached barnacles that feed on plankton from the currents. As the barnacles grow they often cause expensive scientific instruments to malfunction and corrode at accelerated rates. Also the growing community of attached organisms attracts other feeders such as fish. The fish in turn attract fishers who sometimes accidentally damage the buoys.

A number of methods have been tested and developed over the years to reduce and discourage biofouling. Most methods use extremely toxic pesticides and biocides. These toxins are usually mixed into paints that are easily applied. The toxins initially reduced biofouling, but the toxicity of the paint decreases

over time and the paint must be frequently reapplied. Sometimes the combination of surface texture and chemicals encourages the microorganisms to flourish. Unfortunately the impact of these chemicals on the ocean is not well understood. Biofouling may also be discouraged through the use of materials that deter biofilm growth—for example, copper. Shutter mechanisms have been used to protect instrument surfaces from continuous exposure to seawater and marine growth. However, moving parts exposed to seawater are susceptible to mechanical failure. In recent years, antifouling research has focused on understanding the interaction between the biofilm and the microorganisms that attach, and these studies are looking for natural chemicals and bacteria that act antagonistically toward higher organisms. Currently there are no completely effective solutions for preventing biofouling, and the competition between marine life and human use of man-made surfaces in the ocean continues.

side, the reef's highest point, or **reef crest,** may be exposed at low tide and is pounded by the breaking waves of the surf zone. Here, the more massive rounded corals grow. Below the low-tide line to a depth of 10–20 m (35–65 ft) on the seaward side is a zone of steep, rugged buttresses, which alternate with grooves on the reef face. Masses of large corals grow here, and many large fish frequent the area. The buttresses dissipate the wave energy, and the grooves drain off fine sands and debris, which would smother the coral colonies. At depths of

20–30 m (65–100 ft), there is little wave energy, and the light intensity is only about 25% of its surface value; still, it is adequate to support reef algae and corals. The corals are less massive at this depth, and more delicately branched forms are found here. Between 30 and 40 m (100 and 130 ft), the slope is gentle and the level of light is very reduced; sediments accumulate at this depth, and the coral growth becomes patchy. Below 50 m (165 ft), the slope drops off sharply into the deep water.

Figure 18.14 A colony of star coral polyps on a reef in the Caribbean Sea, Bonaire, Lesser Antilles. Most reef-building corals are colonies of interconnected polyps that use their tentacles to capture food.

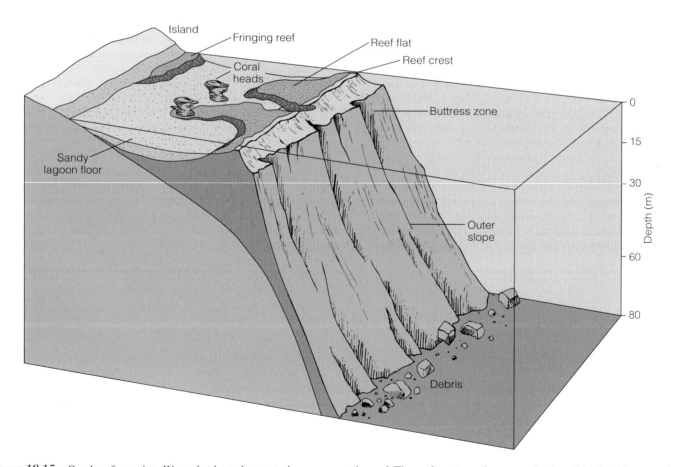

Figure 18.15 Coral reef zonation. Water depths and wave action vary over the reef. The reef crest may be exposed at low tide. Sand from coral debris fills the lagoon and drains off down the grooves of the outer slope, where buttresses dissipate wave energy from the open sea.

Figure 18.16 Coral reefs are rich communities of organisms. (a) Corals and a sea fan at Mana Island, Fiji. (b) A butterflyfish *(left)* and a moorish idol *(right)* at Soma Soma Straits, Taveuni Island, Fiji.

Coral reefs are complex assemblages of many different types of algae and animals (see figs. 18.13 and 18.16), and competition for space and food is intense. Algae, sponges, and corals are constantly overgrowing and competing with each other. Some species are active only at night, such as some fishes, snails, shrimp, the octopus, fireworm, and moray eel. During the day, other species depend on color and vision to make their way. It has been estimated that as many as 3000 animal species may live together on a single reef. The giant reef clam *Tridacna* can measure up to a meter in length and weigh over 150 kg (330 lb). It also possesses zooxanthellae in large numbers in the colorful tissues that line the edges of the shell. Crabs, moray eels, colorful reef fish, poisonous stonefish, long-spined sea urchins, seahorses, shrimp, lobsters, sponges, and many more organisms are all found living here together. On some reefs, the zooxanthellae have been shown to produce several times more organic material per unit of space than the phytoplankton, probably owing to the rapid recycling of nutrients between the corals and the zooxanthellae.

The reefs are not formed exclusively from the calcium carbonate skeletons of the coral. Encrusting algae that produce an outer calcareous covering also contribute; so do the minute shells of foraminifera, the shells of bivalves, the calcareous tubes of polychaete worms, and the spines and plates of sea urchins. All are compressed and cemented together to form new places for more organisms to live. At the same time, some sponges, worms, and clams bore into the reef; some fish graze on the coral and the algae; and the sea cucumbers feed on the broken fragments, reducing them to sandy sediments.

Coral reefs provide another example of the importance individual species can play in maintaining ecosystem diversity. The sea urchin *Diadema antillarum* plays a critical role in maintaining species diversity within coral reefs. This urchin is an herbivore that grazes algae, preventing it from growing over the corals. The sea urchin can also graze certain areas so intensively that algal free areas are created that allow coral larvae to settle and develop. In 1983, sea urchins in the Caribbean suddenly underwent mass mortaliy, potentially due to infection by a water-borne pathogen. In some areas, over 93% of the sea urchins died. The loss of sea urchins resulted in a dramatic increase in algal coverage of the affected reefs and a decrease in coral coverage. More than two decades later, recovery of the reefs and the sea urchin populations is still low.

Coral Bleaching

Oceanographers, marine biologists, and all those who have enjoyed the experience of a tropical coral reef have become increasingly concerned with the health of these biologically rich but delicately balanced regions of the oceans. Coral bleaching episodes, in which corals expel their zooxanthellae, exposing the coral's white calcium carbonate skeleton, have occurred from time to time, but unless the episodes were especially severe, the corals regained their algae and recovered. Between 1876 and 1979, only three bleaching events occurred, but over sixty bleach events were documented between 1979 and 1990. In the last two decades, this bleaching has become more frequent and more severe, and not all reefs have recovered. Large amounts of bleaching occurred at the time of the 1982–83 El Niño, during which shallow reefs of the Java Sea lost 80–90% of their living coral cover. In 1990–91, large-scale bleaching occurred in both the Caribbean and French Polynesia. During the 1997–98 El Niño event, which lasted for over a year, surface waters reached the highest temperatures on record throughout the Indian and western Pacific Oceans. 2005 saw the most extreme coral bleaching event to hit the Caribbean and Atlantic coral reefs. However, in 2005, an extensive monitoring system coordinated through the Global Coral Reef Monitoring Network was in place. The Coral Reef Watch of the National Oceanic and Atmospheric Administration (NOAA) utilizes remote sensing and in situ tools to monitor and report on the physical environmental conditions of coral reefs. Both groups provide nearly real-time information to coral reef managers to help them monitor their local reefs.

Predation and Disease

Periodically, the population of the sea star *Acanthaster* known as the crown of thorns increases dramatically; this sea star feeds on the coral polyps. Fossil evidence points to outbreaks having occurred along the Great Barrier Reef for at least the last 8000 years. Why the *Acanthaster* population increases so rapidly is unknown, but evidence points to a correlation between rainy weather (low salinity) and runoff (increased nutrients), allowing large numbers of *Acanthaster* larvae to survive. The clearing of land for agriculture and the development of coastal areas may also be related. Concern also exists that the harvesting of the large conchs that prey on the starfish may upset the population balance. One outbreak began in 1995 and by 1997 was affecting 40% of the reef. Outbreaks in individual reefs can last one to five years; outbreaks throughout reef systems can last up to a decade or longer. It takes ten to fifteen years for reef areas to recover from an *Acanthaster* outbreak, but the opening up of areas on the reef by the starfish may also allow slower-growing coral species to expand.

Reef-building algae as well as corals have recently been found under attack by several previously unknown diseases. One of these is coralline lethal orange disease, known as CLOD. CLOD is caused by a bright orange bacterial pathogen that is lethal to the encrusting red algae (corallines) that deposit calcium carbonate on the reefs. These algae cement together sand, dead algae, and other debris to form a hard, stable substrate. The disease was initially found in 1993 in the Cook Islands and Fiji; by 1994, it had spread to the Solomon Islands and New Guinea; and by 1995, it was found over a 6000 km (3600 mi) range of the South Pacific. It is unclear whether CLOD has been recently introduced from some obscure location or whether it has been present on the reefs but has now evolved into a more virulent form.

Human Activities

Humans and their activities are among the greatest threats to the reefs. Reefs are damaged by careless sport divers trampling delicate corals and by boats grounding or dragging their anchors. Deforestation and development increase land runoff, carrying sediments, nutrients, and human and animal waste into coastal waters. Additional nutrients favor the growth of algae; the algae out-compete the corals for space, smothering the existing corals and preventing new coral colonies from starting.

Coral reefs are despoiled by shell collectors and mined for building materials. In French Polynesia, Thailand, and Sri Lanka, tons of coral are used as construction material each year. Low-lying coastal areas that lose their coral reefs have no protection against storm surges that accompany extreme weather events. This is particularly important to island nations in the Pacific.

Where catching reef fish with a net is difficult, fishers may use sodium cyanide and explosives. Whether it is cyanide squirted into the water or explosives detonated in the water, the fish are temporarily stunned, making them easy to capture. These methods also kill many other reef organisms, including the reef-building organisms. Although banned in Indonesia and the Philippines, these methods are often the first choice of those in the very lucrative international aquarium and restaurant businesses.

At present, the most seriously threatened areas are considered to be the coasts of East Africa, Indonesia, Philippines, south China, Haiti, and the U.S. Virgin Islands. Ensuring the continued existence of these beautiful and productive areas requires an increased understanding of the complex nature of reef communities and the development of policies designed to protect them from human interference. The United Nations Environment Program, the Intergovernmental Oceanographic Commission, the World Meteorological Organization, and scientists from many different countries are involved in monitoring stations that cover the world's major reefs. The NOAA program, Coral Reef Watch, is developing a long-term coral reef monitoring system able to predict coral bleaching in all U.S. coral reef areas. The data from these stations are used with satellite sea-temperature data and biological monitoring to provide a coral reef early warning system (CREWS). Additional marine protected areas, increased scientific monitoring, and more global observer systems to regularly collect data are needed to ensure the survival of these complex and fragile areas.

Deep-Water Corals

Not all corals are found in shallow, warm tropical waters. Deep-water corals, or "cool corals," lack zooxanthellae; they form large reefs at temperatures down to 4°C and at depths of up to 2000 meters. Researchers believe these corals feed on periodic drifts of dead plankton or sift food from passing currents. Some reefs appear to be hundreds of years old. Reefs have been located off Norway, Scotland, Ireland, Nova Scotia, and Alaska's Aleutian Islands. There is concern that modern fishing with bottom trawling nets may crush these formations; in some places, newly discovered reefs had already been damaged by nets. Canada created a coral preserve in 2001, and Norway has barred trawling in several areas; other nations are considering similar plans.

18.5 Deep-Ocean Chemosynthetic Communities

Hot Vents

In March 1977, an expedition from Woods Hole Oceanographic Institution using the research submersible *Alvin* discovered densely populated communities of animals living around hydrothermal vents along the Galápagos Rift of the East Pacific Rise. Before then, all deep-sea benthic communities were assumed to consist of small numbers of deposit-feeding, slow-growing animals living on and in the soft sediments of the sea floor and depending for nutrition on the slow descent of decayed organic material from the surface layers. Vent communities have now been found scattered throughout the world at seafloor spreading centers, and nearly 300 new species have been identified. Vent communities are known between 9°N and 21°N latitude along the East Pacific Rise, in the Mariana and Okinawa troughs and North Fiji basin of the western Pacific, along the Mid-Atlantic Ridge, in the east Pacific from 49°–22°S, and on the Gorda and Juan de Fuca Ridges of the northeast Pacific.

The animals living in these deep-sea vent areas include filter-feeding clams and mussels, in addition to anemones, worms, barnacles, limpets, crabs, and fish. The clams are very large and show one of the fastest growth rates known for deep-sea animals, up to 4 cm (2 in) per year. The tube worms (fig. 18.17) are startling in size, up to 3 m (10 ft). They have been placed in the phylum Vestimentifera (a phylum is the most basic of taxonomic categories). These crowded communities have a biomass per unit area that is 500–1000 times greater than the biomass of the usual deep-sea floor.

The tube worms in the vent areas have no mouths and no digestive systems; the soft tissue mass of their internal body cavity is filled with bacteria. The tube worms have the ability to transport sulfide made nontoxic by a binding protein, carbon dioxide, and oxygen to the bacteria held in their body tissues; the synthesis of organic molecules is done within the bacteria. These kinds of bacteria are the primary producers in the communities

Figure 18.17 Tube worms crowd around deep-sea hydrothermal vents. The internal body cavity of each worm is filled with bacteria that synthesize organic molecules by chemosynthesis. *Photo © Richard A. Lutz.*

surrounding vents. They use chemosynthesis to fix carbon dioxide into organic molecules (review section 14.3). Despite their size and abundance, these tube worms are completely dependent on an internal symbiosis with the bacteria for their nutrition. Both clams and mussels have large numbers of bacteria in their gills. The mussels have only a rudimentary gut, and the clams and mussels have red flesh and red blood; the color is due to the oxygen-binding molecule hemoglobin. The oxygen is needed for oxidation of the hydrogen sulfide and to maintain body tissues and high growth rates.

The vent plankton and the bacteria in the water provide nutrition for a variety of filter feeders. Other bacteria form mats that surround the vents, and these bacterial mats are the food for snails and other grazers. The vent microorganisms are the base of a broad-based, self-contained trophic system in which nutrition passes from animal to animal by symbiosis, grazing, filtering, and predation.

During studies of the large shrimp populations surrounding some vent areas, researchers noticed a reflective spot just behind the head of the shrimp. The reflective spot consists of two lobes, each connected to the brain by a large nerve cord; light-sensitive pigment related to visual pigment is also present. This system allows the shrimp to detect the radiation associated with hot vents and may allow the shrimp to orient themselves with the vents and their food supply.

In 1991, a submersible monitoring vents on the East Pacific Rise, west of Acapulco and at a depth of 2500 m (1.5 mi), came upon a vent area immediately after a volcanic eruption that killed the larger organisms and left a white carpet of bacteria over the sea floor. Returning in 1992, researchers found the bacterial mats being grazed by large groups of fish, crabs, and other species. By 1993, giant tube worms were more than a meter (3.3 ft) high. Growing at nearly 1 m per year, these tube worms are another extremely fast-growing marine invertebrate. At this time, researchers found that metal-rich sulfide deposits had grown into 10 m (30 ft) chimneys. Previously, geologists had believed that such formations took decades to form. Between 1995 and 1997, the number of species more than doubled, from

Deep-Sea Ice Worms

Gas hydrates are icelike, crystalline deposits that form at low temperatures and high pressures; they are composed primarily of methane gas with some hydrogen sulfide gas. Most gas hydrate deposits are found on the sea floor buried under heavy loads of sediments (see chapter 4). In 1997, a research cruise in the Gulf of Mexico found a gas hydrate mound free of sediments at a depth of 540 m (1800 ft). The mound showed two distinct color bands, yellow and white; the yellow band contained oil and the white band did not. More surprisingly, the entire exposed surface of the hydrate was covered with worms, 2500 per square meter (box fig. 1). This was the first time nonmicrobial life had been seen on gas hydrates.

The worms live in individual oval depressions in the ice, and in more than 95% of the depressions, only one worm was visible. The worms are 2–4 cm (about 1–1.5 in) in length, and each worm is pink with a red blood vessel down the back and a digestive tract. Its eyes are reduced, and the sexes were found to be separate. The researchers believe that the worms feed on the bacteria living on the surface of the ice. The worms can move their lateral extensions (parapodia), setting up water currents that raise the oxygen level for the worms and bacteria. This may also be the way in which the depressions are formed. The worms are a new species (*Hesiocaeca methanicola*) of the polychaete family of annelid or segmented worms.

The researchers also found the same worms living under as much as 10 cm (4 in) of sediment. This raises the question of how much deeper in the sediments the worms can colonize.

H. methanicola does not appear to suffer from predators, although potential predators such as fish and large crustaceans were present on the hydrate surface. Revisiting the hydrate mound one month later, researchers found no empty depressions, and the density of worms had increased to 3000 per square meter. Perhaps the worms are unpalatable.

Box Figure 1 Methane ice worms are found in depressions on the exposed surface of a gas hydrate. The worms are believed to create the depressions.

---

twelve to twenty-nine. Mussels and small worms called serpulids moved in and so did clouds of tiny crustaceans, but no giant white clams had yet arrived.

How vents are first colonized by various animals is still not fully understood. Initially, the distances between vent areas appeared to be a barrier, but as more and more vents are discovered, distance seems less of an obstacle when combined with the fast maturation of the organisms and the large numbers of free-swimming larvae they produce. For another way in which organisms might move between vents, see the box titled "Whale Falls" in chapter 17.

In December 2000, researchers using the submersible *Alvin* and the imaging vehicle *Argo II* discovered a new type of hydrothermal vent field. This field, called the Lost City, was discovered on a terrace 700–800 m (2300–2600 ft) down in the North Atlantic, just off the Mid-Atlantic Ridge. Lost City has no black smokers but instead has steep-sided, white carbonate pinnacles 10–30 m (33–99 ft) tall. These carbonate chimneys support dense microbial populations of both Archaea and Bacteria. There are no large populations of larger organisms surrounding the vents; crabs and sea urchins are rare, sponges and corals more abundant. See chapter 3 for more information about the Lost City.

Cold Seeps

Off the coasts of Florida, Oregon, and Japan, communities based on chemosynthesis have been found associated with cold seepage areas. Bacteria are the primary producers of the communities, using methane and hydrogen sulfide. Along the continental slope of Louisiana and Texas in the Gulf of Mexico, faulting has fractured the sea bottom and produced environments where oil and gas seep up and onto the surface of the sea floor. Clams, mussels, and large tube worms were first collected from these sites (fig. 18.18) in 1985. A variety of other animals, including fishes, crustaceans, and mollusks, commonly found along the continental shelf are abundant around these seeps, attracted by the food supply.

In 1990, salt seeps were reported on the floor of the Gulf of Mexico. The escape of gas through surface sediments has formed depressions that filled with salt brine more than 3.5 times the usual salinity of seawater. These extremely dense

Figure 18.18 Photographs from Gulf of Mexico oil and gas seeps. (a) Mussels and tube worms. (b) Tube worms. (c) Seeping gas and oil, bacterial mats, and chemosynthetic communities in the Green Canyon oil lease area. Note sampling devices and gas bubbles.

brine lakes also contain methane and are surrounded by large communities of mussels.

18.6 Sampling the Benthos

At low tide, beaches all over the world are sampled by researchers armed with buckets and shovels. To understand the relationships between the organisms and their environment, one needs to know where the algae and animals are found on the beach with respect to the tide levels, or the vertical zonation pattern. This pattern is established by determining the slope of the beach and marking locations relative to mean sea level; this is done by surveying a **beach transect line** directly up the beach from low-tide level to

a point above high tide. Trenches paralleling the transect line are dug on sandy beaches to reveal the infauna populations below a measured surface area, and on a rocky shore, surface counts of individuals within an area of specific size are made along the transect line. Once the transect line is established, it is possible to return to the same place season after season or year after year to study seasonal changes in the populations or to determine the number of juveniles added yearly. Another type of study compares a natural area to an area stripped of all benthic organisms. The rate of repopulation and the sequence in which the algae and animals return give information on the relationships among species as well as the recovery rates in the event of a catastrophe due to human activity or natural occurrences.

Figure 18.19 A biological dredge used to collect epibenthic organisms from rocky substrate.

The classic method of obtaining information on subtidal species is by using a bottom dredge towed by a slowly moving ship. A dredge is a metal frame to which a heavy net bag is attached; it is dragged over the sea floor, scraping up the organisms in its path (fig. 18.19). This is a strictly qualitative sampling method, as the number and kinds of organisms cannot be accurately related to the area sampled. It is possible to attach a measuring wheel to the dredge frame, so that when the dredge is on the bottom, the wheel measures the distance over which it is dragged; then, knowing the distance and the width of the dredge, researchers can compute the approximate size of the area sampled, assuming the dredge did not bounce or skip over a rough bottom.

Soft bottoms can be sampled with a bottom grab or a box corer (see chapter 4). The sediment collected is washed through a series of mesh screens of varying sizes, and the organisms are collected and counted.

Divers wearing scuba gear can sample and photograph bottom populations directly in depths to approximately 35 m (105 ft). A diver can place a frame of a specific area on the bottom, identify the species within it, and count individuals of each type. If the bottom is not disturbed, it is possible to return to the sample plot and check it again. Other methods of observing the bottom without disturbing it include underwater photography and television cameras that are operated remotely from a ship or from a submersible.

In the past, sampling benthos of the bathyal, abyssal, and hadal zones required large research vessels and extended periods of time at sea to sample small, widely separated areas. Today, submersibles, ROVs, and AUVs provide new ways to reach previously unvisited parts of the deep ocean and view its inhabitants, undisturbed and in their natural world.

18.7 Practical Considerations: Harvesting the Benthos

The Animals

Benthic animals are a valuable part of the seafood harvest; they include crustaceans (crabs, shrimp, prawns, and lobsters) and mollusks (shellfish—clams, mussels, and oysters—as well as squid). The world's wild harvests for the year 2000 were catches of 7.2 million metric tons of shellfish (including squid) and another 6 million tons of crustaceans. Because of the demand for shellfish and crustaceans, the catches are more important in dollar value than their weight would suggest. In the United States, oysters are harvested in the Gulf of Mexico, southern New England, Chesapeake Bay, and Puget Sound; lobsters are fished in New England; crabs and clams are caught along all U.S. coasts; and shrimp are caught in the Gulf area and off other coasts as well. In many cases, these fisheries make important contributions to the local economy.

Many of the problems of the finfish fisheries have been repeated in the benthic fisheries. For example, between 1975 and 1980, more and more boats entered the king crab fishery of the Bering Sea (fig. 18.20). The catch of 1979 (70,000 metric tons) had dwindled to 5000 metric tons in 1994. The rapid decrease in catch was apparently due to overfishing combined with an insufficient knowledge of the king crab's natural history. King crabs migrate across the floor of the Bering Sea, but without better knowledge of crab populations and migration patterns, effective fishery conservation programs are difficult to implement. Bristol Bay is the most productive region of the Bering Sea, but in 1994 and 1995, the fishery was closed because of low numbers of female crabs in the Bristol Bay population. This Bristol Bay fishery reopened in 1996, and by 1998 the catch had reached 7500 metric tons, but the catch declined in succeeding years, and the 2001 quota was set at 3200 metric tons.

Attempts to increase the harvest of crustaceans and mollusks have focused on expanding world aquaculture or mariculture efforts. In 2000, the combined mariculture harvest of crustaceans and mollusks was more than 10 million metric tons. Oysters, mussels, scallops, and clams are raised on mariculture farms around the world. In Asia and Europe, raft culture of mollusks is popular. The larvae attach to ropes trailing below rafts, and this attachment keeps the shellfish in the water column, with abundant food and few predators. Mariculture in Japan, the Republic of Korea, France, and Spain is dominated

Figure 18.20   Alaska king crab being unloaded from a crab pot.

Figure 18.21   Commerical raft culture of bay mussels in a cove on Whidbey Island in Puget Sound. This mussel farm was the first mussel aquaculture venture in the United States.

by oysters and mussels. Raft culture in Japan is responsible for a harvest of about 250,000 metric tons of oysters each year. Spain produces over 248,000 metric tons of mussels by raft culture. The Japanese also culture scallops in hanging net cages as well as on the bottom, yielding a yearly harvest of 211,000 metric tons. Scallop culture in China increased to nearly 1 million metric tons in 2000. Mussel and oyster farmers in the United States (fig. 18.21) face many of the difficulties encountered by fish farmers. Costs, licensing policies, technology to replace hand labor, and research to improve diet and disease control require attention if the U.S. seafood harvest is to increase in the same way.

World shrimp and prawn production from mariculture ponds was over 1 million metric tons in 1999 and 2000. Crustaceans contribute 5% of the world's aquaculture production by weight, but their value is 25% of the total value. Shrimp feed is about 30% fish meal and 3% fish oil; intensive shrimp aquaculture adds to the pressure on wild fisheries (see also the discussion of salmon aquaculture in chapter 17). The United States is the world's largest market for shrimp, approaching 50% of all seafood imports, and U.S. corporations are taking advantage of the lower costs in Latin America with large-scale investments in shrimp farming.

The Algae

Seaweeds are gathered from the wild in northern Europe, Japan, China, and southeast Asia; they are also an important part of the Japanese aquaculture/mariculture industry. The 1999 world estimate for total seaweed harvests, mainly from mariculture, was 9 million metric tons. The 1999 and 2000 algal mariculture harvests in metric tons were 5.0 million for brown algae, 1.9 million for red algae, and less than 1.0 million for green algae.

Algae are good sources of vitamins and minerals but not of food calories, because most of the cellular material is indigestible. Certain species of green algae, known as sea lettuces, are used in seaweed soups, in salads, and as a flavoring in other dishes. A species of kelp, *Laminaria japonica,* is the **kombu** of Asia. Its blade is used in soups and stews, and it is used fresh, dried, pickled, and salted. It is also sweetened and shredded for use in candy and cakes. Another species of *Laminaria* is used in Europe in the same way. Historically along coastal areas, kelp was used as winter fodder for sheep and cattle and to mulch and fertilize the fields. Along the northwest coast of the United States and Canada, the stipes of bull kelp are made into pickles. The red alga *Porphyra* is the **nori** of Japan and the **laver** of the British Isles. It has been cultivated in Japan since 1700. Nori is used in soups and stews and is rolled around portions of rice and fish for flavor. Laver is fried or used in salads.

There are also important industrial uses for some algal products. **Algin** is extracted from brown algae, and **agar** and **carrageenan** are obtained from red algae. In California, kelp has been harvested for the production of algin. Algin

Genetic Manipulation of Fish and Shellfish

Fisheries scientists in North America, Japan, and northern Europe are using gene-transfer and chromosome manipulation techniques to keep coastal waters and fish farms stocked with species of rapid development and high growth rate. Chromosome manipulation is being used to produce sterile fish, fish of selected gender, and fish known as **triploid,** which carry an extra set of **chromosomes** (box fig. 1).

Triploid salmon and trout are produced by exposing the eggs to temperature, pressure, or chemical shock shortly after fertilization. The shock interferes with the division of the egg nucleus; the egg retains an extra, or third, set of chromosomes. The third set is retained throughout the fishes' development and inhibits their sexual maturation. Salmon with normal numbers of chromosomes grow, mature sexually, spawn, and die, while triploid fish do not mature sexually and reach a larger size, since they are not spending energy on egg or sperm production. The improved growth and survival of these fish has led to widespread production of farmed triploid trout in the United Kingdom.

This same technique allows hybridization of salmon and trout. Whereas normal chromosome-number hybrids between different species of salmon and trout do not survive, triploid hybrids do. In Idaho, a hybrid between rainbow trout and coho salmon has been shown to resist a common viral disease that causes serious economic loss to trout producers.

The gender of fish can be determined by chromosome manipulation and hormone treatment. Using a technique called gynogenesis, researchers expose fish sperm to ultraviolet light, denaturing its chromosomes. Eggs fertilized with the inactivated-chromosome sperm are briefly chilled, causing the eggs to develop with only the female's chromosomes. The result is all female offspring. Many female fish grow bigger and live longer than males. Female flounder grow to twice the size of males; female coho salmon have firmer and more flavorful flesh; and an all-female brood of sturgeon would bring higher profits to caviar producers. An alternative to gynogenesis is androgenesis, in which the egg chromosomes are inactivated and the sperm chromosomes are doubled.

Genes for a specific characteristic can be extracted from a species of fish, or an artificial gene that codes for a specific trait can be developed to enhance fish stocks. Either can be transferred into an egg or sperm of a different species of fish to produce a **transgenic** fish that may have faster growth, greater disease resistance, or more efficient food conversion than its parents. The Food and Drug Administration has been asked to approve the marketing of farmed Atlantic salmon that grows to market size in about half the time (it takes under wild conditions). The freezing temperatures in parts of Canada have been a major obstacle to the farming of Atlantic salmon. An antifreeze protein that inhibits ice crystal formation in fish has been successfully transferred to Atlantic salmon stock and their offspring.

Summer mortality and fluctuations in marketability have kept the oyster industry seasonal. Normal oysters enter a summer reproductive phase during which their meat is poor in quality and unmarketable. Triploid oysters, or all-season oysters, do not usually produce sperm or eggs but continue to grow steadily, becoming significantly larger than normal oysters and ready for the summer market. Triploid Pacific oysters have become an important part of U.S. West Coast aquaculture, where they account for one-third to one-half of the total production. Tetraploid oysters have also been produced. These oysters have four sets of chromosomes, and they are fertile. When they are mated with normal diploid oysters (two sets of chromosomes), only triploid oysters are produced.

The era of commercial transgenic fish and shellfish has arrived. There are no regulations in either the United States or Canada concerning transgenic fish, and researchers go to great pains to prevent their experimental stock from escaping to the wild environment, rigging pens with fences, screens, and alarm systems. However, occasional pen failures do allow transgenic fish to escape into the wild. If mixing occurs between experimental and normal fish, can it endanger the spawning population? Will the rapid inbreeding that is possible produce populations more susceptible to disease or environmental problems? How will the higher growth rate of sterile fish affect a mixed population? Although these problems and questions must be addressed, chromosome and gene manipulation is a useful tool in aquaculture and fish management.

Box Figure 1 (a) Chromosomes of the Pacific oyster *(Crassostrea gigas)* seen through a compound microscope. The diploid oyster has two sets of ten chromosomes. (b) The triploid oyster has three sets of ten chromosomes.

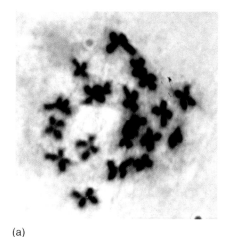

(a)

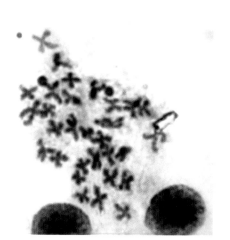

(b)

derivatives are used as stabilizers in dairy products, paints, inks, and cosmetics and to strengthen ceramics, improve the consistency of plaster, thicken jams, and put a longer-lasting head on beer. Agar-producing algae are harvested in Japan, Africa, Mexico, and South America. Agar is used as a medium for bacterial culture in laboratories and hospitals, in addition to serving as an ingredient in desserts and in pharmaceutical products. Algae rich in carrageenan are gathered in the wild in New England and northeastern Canada; it is a stabilizer and emulsifier used to prevent separation in ice creams, salad dressings, soups, puddings, cosmetics, and medicines. About 455 metric tons (1 million pounds) of agar and 4550 metric tons (10 million pounds) of carrageenan are used in the United States each year.

Biomedical Products

Many benthic organisms produce biologically active compounds that have therapeutic use. Extracts of some sponges yield anti-inflammatory and antibiotic substances, and an anticoagulant has been isolated from red algae. Some corals produce antimicrobial compounds, and the sea anemone *Anthopleura* provides a cardiac stimulant. Extracts of abalone and oyster act as antibacterial agents, and a muscle relaxant has been isolated from the snail *Murex*. The adhesive secreted by mussels is used to secure dental fillings and crowns, and good results have been obtained using this substance to repair the cornea and retina of the eye.

Thousands of active compounds from marine organisms have been screened since the mid-1980s for their anticancer, anti-inflammatory, antitumor, immune suppressant, and antiviral possibilities. The tools of molecular biology have made it possible to screen substances rapidly: for example, to test natural compounds for their effect on enzymes responsible for the growth of cancer cells. Discodermolide is an anticancer compound derived from a marine sponge. The sponge was first collected in 1987; the compound was isolated in 1990; and a license for development and manufacture was granted in 1998.

Most of the substances studied come from soft-bodied invertebrate organisms that rely on toxins for defense; many also live in dense populations and under very crowded conditions, such as on coral reefs and pier pilings. These animals have developed an array of compounds that prevent predation and give them an advantage in the competition for space. Toxic compounds poison predators or adjacent organisms, providing space for growth, and nontoxic substances make the animals unpalatable. Since the mid-1980s, more than 5000 compounds of this type have been reported.

Collecting, extracting, identifying, testing, and evaluating active natural compounds from the world's benthos are time-consuming and expensive endeavors. From discovery to pharmacy shelf takes ten to fifteen years and may cost hundreds of millions of dollars. If the process is successful, there is the additional problem that the demand for these organisms might exceed their supply until the substance is successfully synthesized.

Summary

Benthic algae are anchored to firm substrates. These algae have a holdfast, a stipe, and photosynthetic blades but no roots, stems, or leaves. Algal growth along a rocky beach ranges from green algae at the surface to brown algae at moderate depths to red algae, which are found primarily below the low-tide level. Each group's pigments trap the available sunlight at these depths. Algae are generally classified by their principal pigment. The brown algae include the large kelps. Seaweeds provide food, shelter, and substrate for other organisms in the area. There are also benthic diatoms and a few seed plants, including seagrasses and mangroves.

Benthic animals are subdivided into the epifauna, which live on or attached to the bottom, and the infauna, which live buried in the substrate. Animals that inhabit the rocky littoral region are sorted by the stresses of the area into a series of zones. Organisms that live in the supralittoral (or splash) zone spend long periods of time out of water. The animals of the midlittoral zone experience nearly equal periods of exposure and submergence. These animals have tight shells or live close together to prevent drying out. The area is crowded, and competition for space is great. The lower littoral zone is a less stressful environment. It is home to a wide variety of animals. The organisms of the littoral zone are herbivores and carnivores, and each has its specialized lifestyle and adaptations for survival.

The zonation of the organisms in the benthic region varies with local conditions. Tide pools provide homes for lower littoral zone organisms; they can also become extremely specialized habitats.

Mud, sand, and gravel areas are less stable than rocky areas. The size of the spaces between the substrate particles determines the porosity of the sediments. Some beaches have a higher organic content than others. Few algae can attach to soft sediments; thus, few grazers are found here. Eelgrass and surf grass provide food and shelter for specialized communities. Most organisms that live on soft sediments are detritus feeders or deposit feeders. Zonation patterns are not conspicuous along soft bottoms. Bacteria play an important role in the decomposition of plant material and its reduction to detritus. The bacteria themselves represent a large food resource.

The environment of the deep-sea floor is very uniform. The diversity of species increases with depth, but the population density decreases. The microscopic members of the deep-sea infauna are the benthic meiofauna. Larger burrowers such as sea cucumbers continually rework the sediments. The organisms of the epifauna are found at all depths.

Some organisms specialize in attaching to surfaces, and others bore into them. Wood-borers are very destructive.

High-energy coastal benthic environments are two to ten times more productive than rain forest vegetation. Tropical coral reefs are specialized, self-contained systems. The coral animals require warm, clear, clean, shallow water and a firm substrate. Photosynthetic dinoflagellates, called zooxanthellae, live in the cells of the corals and giant clams. The reef exists in a complex but delicate biological balance that can easily be upset. Reefs have a typical zonation and structure associated with depth and wave exposure. Coral reefs are presently under great stress from coral bleaching, predation, and disease. Human activities are among the greatest threats; damage is due to overfishing, pollution, and the use of dynamite and poison. An international network of monitoring stations is being implemented.

A number of deep, cold-water coral reefs have been located. These corals lack zooxanthellae, and much has yet to be learned about them.

Self-contained deep-ocean benthic communities of large, fast-growing animals depend on chemosynthetic bacteria as the first step in the food web. Communities are associated with hot-water vents and cold seeps. Dense bacterial communities have been found associated with carbonate chimneys in the North Atlantic.

Benthic organisms are sampled by hand in the intertidal zone; deeper samples are obtained with dredges, grabs, corers, ROVs, AUVs, and submersibles. Sampling is both qualitative and quantitative.

Shellfish and crustaceans are valuable world food resources. Large-scale aquaculture is being used to increase harvests of oysters, mussels, and shrimp.

Algae are gathered in many countries and cultivated in Japan. Some are used directly as food; others are used as stabilizers and emulsifiers in foods and other products. Biologically active substances with potentially practical uses have been isolated from benthic organisms. Thousands of active compounds are being screened for anticancer, antitumor, and other properties.

Key Terms

All key terms from this chapter can be viewed by term or definition when studied as flashcards on this book's website at www.mhhe.com/sverdrup10e.

algae, 432
holdfast, 432
stipe, 432
blade, 432

kelp, 433
epifauna, 434
infauna, 434
sessile, 435
vertical zonation, 435
intertidal zonation, 435
mollusk, 437
nemertean, 440
brachiopod, 440

nudibranch, 440
detritus, 441
bioturbation, 441
benthic meiofauna, 443
pogonophora, 443
polyp, 446
zooxanthellae, 446
reef flat, 448
reef crest, 449

beach transect line, 455
kombu, 457
nori/laver, 457
algin, 457
agar, 457
carrageenan, 457
triploid, 458
chromosome, 458
transgenic, 458

Study Questions

1. Explain the relationship among sulfides, bacteria, tube worms, and clams in a hydrothermal vent environment.
2. In what ways are the benthic algae (seaweeds) adapted for life in the littoral and sublittoral zones? Consider their structure, pigments, and life requirements.
3. In what ways are the benthic algae important in the ocean environment?
4. Discuss the food-gathering strategies of motile and sessile organisms in the littoral and sublittoral zones.
5. Discuss the factors that are responsible for the littoral zonation of marine organisms along a rocky shore.
6. Design an original organism to inhabit the supralittoral, the littoral, or the sublittoral zone. Consider its requirements for food, shelter, and protection from predators, its adaptations to its environment, and its life history.
7. Why are some subtidal organisms found in a tide pool high on a rocky beach, while other subtidal forms are not?
8. Why do few benthic organisms live on a beach made of non-cohesive sediments in a wave and surf area?
9. Discuss the importance of bacteria to benthic organisms.
10. Compare a square-meter area of deep-sea floor with a square-meter area of the rocky intertidal zone. What differences do you expect to find? Consider biomass, species abundance, and substrate.
11. How are coral reefs able to support a rich and varied population when the water surrounding the reef is clear and devoid of planktonic primary producers?
12. What is coral bleaching? Why does it happen? What is its result?
13. Compare the organisms found growing around deep-ocean hot-water vents and the organisms found around gas and oil seeps.
14. Discuss the genetic manipulation of fish and shellfish. Do the advantages of such techniques outweigh the possible disadvantages?
15. Compare photosynthesis and chemosynthesis; how are they similar and how are they different?

Appendix A

Scientific (or Exponential) Notation

Writing very large and very small numbers is simplified by using exponents, or powers of 10, to indicate the number of zeroes required to the left or to the right of the decimal point. The numbers that are equal to some of the powers of 10 are as follows:

$$1,000,000,000. = 10^9 = \text{one billion}$$
$$1,000,000. = 10^6 = \text{one million}$$
$$1000. = 10^3 = \text{one thousand}$$
$$100. = 10^2 = \text{one hundred}$$
$$10. = 10^1 = \text{ten}$$
$$1. = 10^0 = \text{one}$$
$$0.1 = 10^{-1} = \text{one tenth}$$
$$0.01 = 10^{-2} = \text{one hundredth}$$
$$0.001 = 10^{-3} = \text{one thousandth}$$
$$0.000001 = 10^{-6} = \text{one millionth}$$
$$0.000000001 = 10^{-9} = \text{one billionth}$$

149,000,000 is rewritten by moving the decimal point eight places to the left and multiplying by the exponential number 10^8, to form 1.49×10^8. In the same way, 605,000 becomes 6.05×10^5.

A very small number such as 0.000032 becomes 3.2×10^{-5} by moving the decimal point five places to the right and multiplying by the exponential number 10^{-5}. Similarly, 0.00000372 becomes 3.72×10^{-6}.

To add or subtract numbers written in exponential notation, convert the numbers to the same power of 10. For example,

$$
\begin{array}{ccccc}
1.49 \times 10^3 & & 14.90 \times 10^2 & & 1.490 \times 10^3 \\
+6.05 \times 10^2 & = & 6.05 \times 10^2 & = & 0.605 \times 10^3 \\
\hline
& & 20.95 \times 10^2 & & 2.095 \times 10^3
\end{array}
$$

$$
\begin{array}{ccccc}
2.36 \times 10^3 & & 23.60 \times 10^2 & & 2.360 \times 10^3 \\
-1.05 \times 10^2 & = & -1.05 \times 10^2 & = & -1.05 \times 10^3 \\
\hline
& & 22.55 \times 10^2 & & 2.255 \times 10^3
\end{array}
$$

To multiply, the exponents are added and the numbers are multiplied:

$$
\begin{array}{r}
4.6 \times 10^3 \\
\times 2.2 \times 10^2 \\
\hline
10.12 \times 10^5 = 1.012 \times 10^6
\end{array}
$$

To divide, subtract the exponents and divide the numbers:

$$\frac{6.0 \times 10^8}{2.5 \times 10^3} = 2.4 \times 10^5$$

The following prefixes correspond to the powers of 10 and are used in combination with metric units:

Exponential Notation

Exponential Value	Prefix	Symbol
10^{18}	exa	E
10^{15}	peta	P
10^{12}	tera	T
10^{9}	giga	G
10^{6}	mega	M
10^{3}	kilo	k
10^{2}	hecto	h
10^{1}	deka	da
10^{-1}	deci	d
10^{-2}	centi	c
10^{-3}	milli	m
10^{-6}	micro	μ
10^{-9}	nano	n
10^{-12}	pico	p
10^{-15}	femto	f
10^{-18}	atto	a

Appendix B

SI Units

The _Système international d' unités,_ or International System of Units, is a simplified system of metric units (known as SI units) adopted by international convention for scientific use.

Basic SI Units

Quantity	Unit	Symbol
length	meter	m
mass	kilogram	kg
time	second	s
temperature	Kelvin	K

Derived SI Units

Quantity	Unit	Symbol	Expression
area	meter squared	m^2	m^2
volume	meter cubed	m^3	m^3
density	kilogram per cubic meter	kg/m^3	kg/m^3
speed	meter per second	m/s	m/s
acceleration	meter per second per second	m/s^2	m/s^2
force	newton	N	$(kg)(m)/s^2$
pressure	pascal	Pa	N/m^2
energy	joule	J	(N)(m)
power	watt	W	J/s; (N)(m)/s

Length: The Basic SI Unit Is the Meter

Unit	Metric Equivalent	English Equivalent	Other
meter (m)	100 centimeters 1000 millimeters	39.37 inches 3.281 feet	0.546 fathom
kilometer (km)	1000 meters	0.621 land mile	0.540 nautical mile
centimeter (cm)	10 millimeters 0.01 meter	0.394 inch	
millimeter (mm)	0.1 centimeter 0.001 meter	0.0394 inch	
land mile (mi)	1609 meters	5280 feet	0.869 nautical mile
nautical mile (nm)	1852 meters	1.151 land miles 6076 feet	1 minute of latitude
fathom (fm)	1.8288 meters	6 feet	

Area: Derived from Length

Unit	Metric Equivalent	English Equivalent	Other
square meter (m^2)	10,000 square centimeters	10.76 square feet	
square kilometer (km^2)	1,000,000 square meters	0.386 square land mile	0.292 square nautical mile
square centimeter (cm^2)	100 square millimeters	0.151 square inch	

Volume: Derived from Length

Unit	Metric Equivalent	English Equivalent	Other
cubic meter (m^3)	1,000,000 cubic centimeters 1000 liters	35.32 cubic feet 264 U.S. gallons	
cubic kilometer (km^3)	1,000,000,000 cubic meters	0.2399 cubic land mile	0.157 cubic nautical mile
liter (L) or (l)	1000 cubic centimeters	1.06 quarts, 0.264 U.S. gallon	
milliliter (mL) or (ml)	1.0 cubic centimeter		

Mass: The Basic SI Unit Is the Kilogram

Unit	Meteric Equivalent	English Equivalent
kilogram (kg)	1000 grams	2.205 pounds
gram (g)		0.035 ounce
metric ton, or	1000 kilograms	2205 pounds
tonne (t)	1,000,000 grams	
U.S. ton	907 kilograms	2000 pounds

Time: The Basic SI Unit Is the Second

Unit	Metric and English Equivalent	
minute	60 seconds	
hour	60 minutes; 3600 seconds	
day	86,400 seconds 24 hours	} mean solar day
year	31,556,880 seconds 8756.8 hours 365.25 solar days	} mean solar year

Temperature: The Basic SI Unit Is the Kelvin

Reference Point	Kelvin (K)	Celsius (°C)	Fahrenheit (°F)
absolute zero	0	−273.2	−459.7
seawater freezes	271.2	−2.0	28.4
fresh water freezes	273.2	0.0	32.0
human body	310.2	37.0	98.6
fresh water boils	373.2	100.0	212.0
conversions	$K = °C + 273.2°$	$°C = \dfrac{(°F - 32)}{1.8}$	$°F = (1.8 \times °C) + 32$

Speed (Velocity): The Derived SI Unit Is the Meter per Second

Unit	Metric Equivalent	English Equivalent	Other
meter per second (m/s)	100 centimeters per second 3.60 kilometers per hour	3.281 feet per second 2.237 land miles per hour	1.944 knots
kilometer per hour (km/h)	0.277 meter per second	0.909 foot per second	0.55 knot
knot (kt)	0.51 meter per second	1.151 land miles per hour	1 nautical mile per hour

Acceleration: The Derived SI Unit Is the Meter per Second per Second

Unit	Metric Equivalent	English Equivalent
meter per second per second (m/s^2)	100 centimeters per second per second 12,960 kilometers per hour per hour	3.281 feet per second per second 8048 miles per hour per hour

Force: The Derived SI Unit Is the Newton

Unit	Metric Equivalent	English Equivalent
newton (N)	100,000 dynes	0.2248 pound force
dyne (dyn)	0.00001 newton	0.000002248 pound force

Pressure: The Derived SI Unit Is the Pascal

Unit	Metric Equivalent	English Equivalent	Other
pascal (Pa)	1 newton per square meter 10 dynes per square centimeter		
bar	100,000 pascals 1000 millibars	14.5 pounds per square inch	0.927 atmosphere 29.54 inches of mercury
standard atmosphere (atm)	1.013 bars 101,300 pascals	14.7 pounds per square inch	29.92 inches of mercury or 76 cm of mercury

Energy: The Derived SI Unit Is the Joule

Unit	Metric Equivalent	English Equivalent
joule (J)	1 newton-meter 0.2389 calorie	0.0009481 British thermal unit
calorie (cal)	4.186 joules	0.003968 British thermal unit

Power: The Derived SI Unit Is the Watt

Unit	Metric Equivalent	English Equivalent
watt (W)	1 joule per second 0.2389 calorie per second 0.001 kilowatt	0.0569 British thermal unit per minute 0.001341 horsepower

Density: The Derived SI Unit Is the Kilogram per Cubic Meter

Unit	Metric Equivalent	English Equivalent
kilogram per cubic meter	0.001 gram per cubic centimeter (g/cm^3)	0.0624 pound per cubic foot

To learn more about metrics, go to the U.S. Metric Association, Inc., http://lamar.colostate.edu/~hillger/.

Equations and Quantitative Relationships

Proportion Ratios

Modeling: $A / B = A'/B'$

Scaling and vertical distortion: $(D/L) \times V$. Distortion $= D'/L'$

Distance, Rate, Time Problems

Distance = speed × time

Area = area spreading rate × time

Volume = volume flow rate × time

Mass transferred = mass transfer rate × time

Rotation = rotation rate × time

Volume flow rate = current speed × area

Equilibrium, Steady-State Problems

Volumes, masses, and energy are held constant in time

All inflows equal ouflows. $(\text{rate} \times \text{time})_{in} = (\text{rate} \times \text{time})_{out}$

Residence time = mass or volume / supply or removal rate

Exponential Problems

$X = Y\, e^{\pm(rz)}$ Requires math tables or calculator

In time-dependent problems z = time, r = (A/time), a rate

X is the value of Y after a given time. If rz is +, then X > Y

If rz is –, then X < Y

When X = 0.5 (Y), (rz is negative) and (rz) = 0.693

Radioactive decay example. Half-life calculation:

$$0.5\,(Y) = 1\,(Y)\,e^{-(rt)}$$

Light attenuation example:

$$I_z = I_o\, e^{-kz}$$

where I_o = the light intensity at the sea surface, I_z = the light intensity at a depth z, and k is the attenuation coefficient (see discussion in chapter 5). The attenuation coefficient, k, is about $1.7/D$, where D is the Secchi disk depth. k can be determined for total available light or for individual wavelengths of light.

Population example:

$$P_{\text{time}=t} = P_{t=0}\, e^{+(rt)}$$

P is the number of individuals or mass. r in this case is a combination of reproduction rate minus the (death rate + grazing rate).

Three-Dimensional Problems

Volume = length × width × depth

Area = length × width

Slope Problems

Slope = rise/run

Horizontal distance × slope = elevation or height change

Small-Particle Settling Velocity: Stokes Law

$V = (2/9)\,[g\,(\rho_1 - \rho_2)/\mu]r^2$

V = settling speed, terminal velocity in cm/s

2/9 = shape factor constant for all small spheres

g = acceleration due to Earth's gravity, 981 cm/s^2

ρ_1 = particle density (of quartz), g/cm^3

ρ_2 = seawater density, g/cm^3

μ = viscosity of seawater

r = radius of particle in cm

$$V_{\text{cm/s}} = (2.62 \times 10^4)\,(r\text{cm})^2$$

Used to determine settling rate of small particles with diameters less than 0.125 mm. Can be used in reverse to determine size of small particles by measuring settling rates.

Heat Problems

Latent heat of fusion of water at 0°C = 80 cal/g

Latent heat of vaporization of water at 100°C = 540 cal/g

Heat capacity of water = 1 cal/g/°C

Calories = mass in g × heat capacity × temperature change, °C

Interpolation of Data Tables

Known values are axes values (a, c, d, and f) and corresponding table values (P, R, V, and X). Chosen axes values are b (between a and c) and e (between d and f). Find the table value T at chosen axes values b and e.

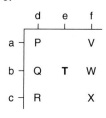

Step 1: Find the table value Q for axes values b and d.

$$Q = P + (b - a)(R - P)/(c - a)$$

Step 2: Find the table value W for axes values b and f.

$$W = V + (b - a)(X - V)/(c - a)$$

Step 3: Find the table value T for axes values b and e.

$$T = Q + (e - d)(W - Q)/(f - d)$$

Seawater, Constancy of Composition

$$Cl_1‰/Cl_2‰ = S_1‰/S_2‰ = ionA_1/ionA_2 = ionB_1/ionB_2$$

Subscripts 1 and 2 refer to two different concentrations.

Hydrostatic Pressure

$$P = \rho g z$$

P is pressure, ρ is fluid density, g is Earth's gravity, and z is the height of the fluid.

Progressive Surface Water Waves

Simple sinusoidal wave theory:

Wave speed, $C = L/T$, where L = wavelength and T, the wave period, is conserved.

$C^2 = (g/2\pi) L \tanh[(2\pi/L)D]$, where g = Earth's gravity, D = water depth, L = wavelength, and $\tanh[\emptyset]$ = the hyperbolic tangent of angle $[\emptyset]$. Here, $\emptyset$ equals $(2\pi/L)D$ in radians. See math tables or use calculator.

Deep-water approximation:

When $D > L/2$, $\tanh[(2\pi/L)D] \approx 1.0$, $C^2 = (g/2\pi) L$

Shallow-water approximation:

When $D < L/20$, $\tanh[(2\pi/L)D] \approx (2\pi/L)D$, $C = \sqrt{gD}$

Internal Waves

The celerity, C, or wave speed of internal waves along a boundary of a two-layer system is determined by a relationship between the density of the layers and their thicknesses. When the wavelength, L, is long compared with the water depths:

$$C^2 = (g/2\pi)L\left[\frac{\rho - \rho'}{\rho\cotanh(2\pi h/L) + \rho'\cotanh(2\pi h'/L)}\right]$$

h and ρ are the thickness and density of the denser lower layer, and h' and ρ' are the thickness and density of the less-dense upper layer. Cotanh is the hyperbolic cotangent. When L is small compared to the water depths, cotanh $(2\pi h/L)$ and cotanh $(2\pi h'/L)$ approach the value of 1.0 and

$$C^2 = (g/2\pi)L\left[\frac{\rho - \rho'}{\rho + \rho'}\right]$$

This equation relates to the deep-water wave equation at the sea surface, which is the boundary of two fluids—air and water. In this later case, ρ' of air is about 1/1000 the value of ρ of water and is neglected.

Standing Waves, Seiches

Based on $L/T = \sqrt{gD}$, shallow-water wave condition.
Closed basin:

The period of oscillation is $T = (1/n)(2l/\sqrt{gD})$ where n = the number of wave nodes, $L = 2 \times$ basin length, l; $L = 2l$, and D = basin depth.

Open-end basin:

The period of oscillation is

$T = (1/n)(4l/\sqrt{gD})$. $L = 4 \times$ basin length, $L = 4l$

Refraction of Light, Sound, and Waves

Snell's Law:
The change of speed of propagation causes refraction, bending, of wave rays and wave fronts.

$$C_1 \sin(a_2) = C_2 \sin(a_1)$$

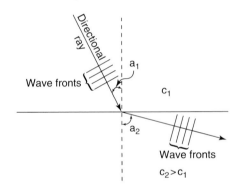

Tide-Raising Forces

The force between two masses due to their mutual gravitational attraction.

$$F = G (M_1 M_2)/R^2$$

where M_1 = mass 1, M_2 = mass 2, R = distance between centers of masses M_1 and M_2, and G = Newton's constant of gravitation = 6.67×10^{-8} cm³/g/s² = 6.67×10^{-8} dyne $\cdot$ cm²/g².

The gravitational force per unit mass on Earth, M_1, caused by the Sun or the Moon, mass M_2, is,

$$F/M_1 = G (M_2/R_2)$$

where R is the distance between the unit Earth mass M_1 and the Sun or the Moon center.

The tide-raising force, $\Delta F/M_1$, is the difference between the gravitational force acting on a unit mass at Earth's surface directly in line with the tide-raising body and a unit mass at Earth's center. (F/M_1)sur. $- (F/M_1)$cen. $= \Delta F/M_1$

$$\Delta F/M_1 = [G M_2/(R - r)^2] - [GM_2/(R)^2]$$

where R is the distance between the center of Earth and center of the Sun or Moon and r is Earth's radius.

$$\Delta F/M_1 = (-GM_2/R^2)\,[1 - (1/(1 - r/R)^2)]$$
$$= (-GM_2/R^2)\,[-2r/R + (r/R)^2]/(1 - r/R)^2$$

If the magnitude of r/R and $(r/R)^2$ is calculated, then $(r/R)^2 \ll 2r/R$ and $(r/R)^2$ is neglected in the numerator; $(r/R) \ll 1.0$ and (r/R) is neglected in the denominator.

$$\Delta F/M_1 = (-GM_2/R^2)\,[(-2r/R)/1] = 2GM_2\,r/R^3$$

Calculating Great Circle Distance

The great circle distance between two points on Earth can be calculated knowing the latitude and longitude of each point, and the average radius of Earth.

Let (ϕ_1, θ_1) and (ϕ_2, θ_2) be the geographical latitude and longitude expressed in radians of two locations. Let the difference in their longitude be $\Delta\theta$ expressed in radians.

Then the geocentric angular distance between the two points expressed in radians, $\Delta\sigma$, can be calculated from the following formula:

$$\Delta\sigma = \arctan\left(\frac{\sqrt{(\cos\phi_2\sin\Delta\theta)^2 + (\cos\phi_1\sin\phi_2 - \sin\phi_1\cos\phi_2\cos\Delta\theta)^2}}{\sin\phi_1\sin\phi_2 + \cos\phi_1\cos\phi_2\cos\Delta\theta}\right)$$

The great circle distance in kilometers between the two points is equal to the geocentric angular distance between the two points multiplied by Earth's average radius of curvature, which is about 6372.8 km.

For example, calculate the distance between Los Angeles International Airport (LAX, point 1) in Los Angeles, CA (N 33° 56.4′, W 118°24.0′) and O'Hare airport (ORD, point 2) in Chicago, IL (N 41°58.7′, W 87°54.3′).

First, convert these coordinates to decimal degrees (remember that North latitude is positive and West longitude is negative); then convert them to radians by multiplying the decimal degrees by $(\pi/180)$.

LAX (point 1): $\phi_1 = +33.94°$ or 0.5924 rad
                  $\theta_1 = -118.40°$ or -2.0665 rad

ORD (point 2): $\phi_2 = +41.98°$ or 0.7327 rad
                  $\theta_2 = -87.91°$ or -1.5343 rad

then: $\Delta\theta = (-2.0665) - (-1.5343) = -0.5322$

Input these coordinates in radians into the geocentric angular distance formula to calculate $\Delta\sigma \approx 0.4406$. Multiply this angular distance in radians by the average radius of Earth to obtain the distance between the two points.

$$(0.4406) \times (6371\text{ km}) = 2807\text{ km (1744 mi)}$$

Water and Salt Budgets for Partially Mixed Estuaries

Water in = water out = water budget, volume constant

Salt in = salt out = salt budget, salt content constant

T_o = volume transport of mixed water out, vol/time

T_i = volume transport of seawater in, vol/time

R = river flow into estuary, vol/time

S_o = average salt content of T_o water, mass/vol

$\bar{S}_i$ = average salt content of T_i seawater, mass/vol

Water budget: $T_o = T_i + R$

Salt budget: $T_o\,\bar{S}_o = T_i\,\bar{S}_i$

Combined budgets: $T_o = [\bar{S}_i/(\bar{S}_i - \bar{S}_o)]\,R$

Estuary flushing time or estuary water residence time equals estuary volume/To.

$(\bar{S}_i - \bar{S}_o)/\bar{S}_i$ = fraction of fresh water in T_o type flow

$(\bar{S}_i - \bar{S}_a)/\bar{S}_i$ = fraction of fresh water in estuary volume

where $\bar{S}_a$ = average salt concentration of estuary water

$[(\bar{S}_i - \bar{S}_a)/\bar{S}_i] \times$ estuary volume = freshwater volume stored in the estuary

(Freshwater volume of estuary/R) = residence time of fresh water in the estuary

Plant Production of Carbon by Phytoplankton

Production of carbon (C), release of oxygen (O_2), and use of the nutrients nitrogen (N) and phosphorus (P) by phytoplankton in photosynthesis occurs with fixed ratios on a mass basis. If the use of or production rate of one of the terms is known, the others are calculated from mass-to-mass ratios.

$$O_2{:}C{:}N{:}P = 109{:}41{:}7.2{:}1$$

Glossary

A

absorption taking in of a substance by chemical or molecular means; change of sound or light energy into some other form, usually heat, in passing through a medium or striking a surface.

abyssal pertaining to the great depths of the ocean below approximately 4000 m.

abyssal clay lithogenous sediment on the deep-sea floor composed of at least 70% clay-sized particles by weight.

abyssal hill low, rounded submarine hill less than 1000 m high.

abyssal plain flat ocean basin floor extending seaward from the base of the continental slope and continental rise.

abyssopelagic oceanic zone from 4000 m to the deepest depths.

accretion natural or artificial deposition of sediment along a beach, resulting in the buildup of new land.

acoustic profiling the use of seismic energy to measure sediment thickness and layering on the sea floor.

active margin *see* leading margin.

adsorption attraction of ions to a solid surface.

advection horizontal or vertical transport of seawater, as by a current.

agar substance produced by red algae; the gelatin-like product of these algae.

algae marine and freshwater organisms (including most seaweeds) that are single-celled, colonial, or multicelled, with chlorophyll but no true roots, stems, or leaves and with no flowers or seeds.

algin complex organic substance found in or obtained from brown algae.

alluvial plain flat deposit of terrestrial sediment eroded by water from higher elevations.

amphidromic point point from which cotidal lines radiate on a chart; the nodal, or low-amplitude, point for a rotary tide.

amplitude for a wave, the vertical distance from sea level to crest or from undisturbed sea level to trough, or one-half the wave height.

anadromous migratory pattern in which juvenile fish migrate down rivers to mature as adults in the open oceans.

anaerobic living or functioning in the absence of oxygen.

andesite a volcanic rock intermediate in composition between basalt and granite; associated with subduction zones.

anion negatively charged ion.

anoxic deficient in oxygen.

Antarctic Circle *see* Arctic and Antarctic Circles.

antinode portion of a standing wave with maximum vertical motion.

aphotic zone that part of the ocean in which light is insufficient to carry on photosynthesis.

aquaculture (mariculture) cultivation of aquatic organisms under controlled conditions.

Arctic and Antarctic Circles latitudes 66 1/2°N and 66 1/2°S, respectively, marking the boundaries of light and darkness during the summer and winter solstices.

armored beach a beach that is protected from wave and water erosion by coarse-size lag deposits.

aseismic ridge *see* transverse ridge.

asthenosphere upper, deformable portion of Earth's mantle, the layer below the lithosphere; probably partially molten; may be site of convection cells.

atmospheric pressure pressure, at any point on Earth, exerted by the atmosphere as a consequence of gravitational force exerted on the column of air lying directly above the point.

atoll ring-shaped coral reef that encloses a lagoon in which there is no exposed preexisting land and which is surrounded by the open sea.

attenuation decrease in the energy of a wave or beam of particles occurring as the distance from the source increases; caused by absorption, scattering, and divergence from a point source.

autotrophic pertaining to organisms able to manufacture their own food from inorganic substances. *See also* chemosynthesis and photosynthesis.

autumnal equinox *see* equinoxes.

auxospore naked cell of a diatom, which grows to full size and forms a new siliceous covering.

B

backshore beach zone lying between the foreshore and the coast, acted on by waves only during severe storms and exceptionally high water.

bacterioplankton composed of members of the domains Bacteria and Archaea.

baleen whalebone; horny material growing down from the upper jaw of plankton-feeding whales; forms a strainer, or filtering organ, consisting of numerous plates with fringed edges.

bar offshore ridge or mound of sand, gravel, or other loose material that is submerged, at least at high tide; located especially at the mouth of a river or estuary or lying a short distance from and parallel to the beach.

barrier island deposit of sand, parallel to shore and raised above sea level; may support vegetation and animal life.

barrier reef coral reef that parallels land but is some distance offshore, with water between reef and land.

basalt fine-grained, dark igneous rock, rich in iron and magnesium, characteristic of oceanic crust.

basin large depression of the sea floor having about equal dimensions of length and width.

bathyal pertaining to ocean depths between approximately 1000 and 4000 m.

bathymetry study and mapping of seafloor elevations and the variations of water depth; the topography of the sea floor.

bathypelagic oceanic zone from 1000–4000 m.

beach zone of unconsolidated material between the mean low-water line and the line of permanent vegetation, which is also the effective limit of storm waves; sometimes includes the material moving in offshore, onshore, and longshore transport.

beach face section of the foreshore normally exposed to the action of waves.

Beaufort scale scale of wind forces by range of velocity; scale of sea state created by winds of these velocities.

benthic of the sea floor, or pertaining to organisms living on or in the sea floor.

benthos organisms living on or in the ocean bottom.

berm nearly horizontal portion of a beach (backshore) with an abrupt face; formed from the deposition of material by wave action at high tide.

berm crest ridge marking the seaward limit of a berm.

bilateral symmetry having right and left halves that are approximate mirror images of each other.

biodiversity the number of species in an area compared to the number of individuals.

biogenous sediment sediment derived from organisms.

biological pump photosynthetic transfer of carbon as CO_2 from the atmosphere to the ocean in the form of organic molecules; carbon is transferred to intermediate and deep-ocean water when organic material sinks and decays.

bioluminescence production of light by living organisms as a result of a chemical reaction either within certain cells or organs or outside the cells in some form of excretion.

biomass the total mass of all or specific living organisms, usually expressed as dry weight in grams of carbon per unit area or unit volume.

biomechanics the field of research that looks for answers to how chemistry and physics affect basic biological characteristics such as size and shape.

bioturbation reworking of sediments by organisms that burrow into them and ingest them.

blade flat, photosynthetic, "leafy" portion of an alga or a seaweed.

bloom high concentration of phytoplankton in an area, caused by increased reproduction; often produces discoloration of the water. *See also* red tide.

breaker sea surface water wave that has become too steep to be stable and collapses.

breakwater structure protecting a shore area, harbor, anchorage, or basin from waves; a type of jetty.

buffer substance able to neutralize acids and bases, therefore able to maintain a stable pH.

bulkhead structure separating land and water areas; primarily designed to resist earth sliding and slumping or to reduce wave erosion at the base of a cliff.

buoy floating object anchored to the bottom or attached to another object; used as a navigational aid or surface marker.

buoyancy ability of an object to float due to the support of the fluid the body is in or on.

by-catch *see* incidental catch.

C

caballing mixing of two water types with identical densities but different temperatures and salinities; the resulting mixture is denser than its components.

calcareous containing or composed of calcium carbonate.

calcareous ooze fine-grained deep-ocean biogenous sediment containing at least 30% calcareous tests, or the remains of small marine organisms.

calorie amount of heat required to raise the temperature of 1 g of water 1°C.

calving breaking away of a mass of ice from its parent glacier, iceberg, or sea-ice formation.

capillary wave wave with wavelength less than 1.5 cm in which the primary restoring force is surface tension.

carbonate a sediment or rock formed from the accumulation of carbonate minerals ($CaCO_3$) precipitated organically or inorganically.

carbonate compensation depth (CCD), also known as the calcite compensation depth, is the depth at which the amount of calcium carbonate ($CaCO_3$) preserved falls below 20% of the total sediment. This is also commonly defined as the depth at which the amount of calcium carbonate produced by organisms as skeletal material in the overlying water column is equal to the rate at which it is dissolved in the water. No calcium carbonate will be deposited below this depth.

carbon fixation process of generating organic carbon from carbon dioxide.

carnivore flesh-eating organism.

carrageenan substance produced by certain algae that acts as a thickening agent.

catadromous migratory pattern in which juvenile fish spawn in the open ocean but mature in fresh water.

cation positively charged ion.

cat's-paw patch of ripples on the water's surface, related to a discrete gust of wind.

centrifugal force outward-directed force acting on a body moving along a curved path or rotating about an axis; an inertial force.

centripetal force inward-directed force necessary to keep an object moving in a curved path or rotating about an axis.

chaetognaths free-swimming, carnivorous, pelagic, wormlike, planktonic animals; arrowworms.

chemoautotrophs organisms that use chemical energy to drive production of organics.

chemosynthesis formation of organic compounds with energy derived from inorganic substances such as ammonia, methane, sulfur, and hydrogen.

chloride atom of chlorine in solution, forming an ion with a negative charge.

chlorinity (Cl‰) measure of the chloride content of seawater in grams per kilogram.

chlorophyll group of green pigments that are active in photosynthesis.

chloroplasts used to harvest sunlight by photosynthesis in cells.

chromosome one of the bodies in a cell that carries the genes in a linear order.

chronometer portable clock of great accuracy used in determining longitude at sea.

ciguatera toxin found in fish of tropical regions; produced by dinoflagellates.

cilia microscopic, hairlike projections of living cells that beat in coordinated fashion and produce movement.

cluster a group of galaxies. A cluster may contain thousands of galaxies.

coast strip of land of indefinite width that extends from the shore inland to the first major change in terrain that is unaffected by marine processes.

coastal circulation cell (drift sector, littoral cell) longshore transport cell pattern of sediment moving from a source to a place of deposition.

coccolithophorid microscopic, planktonic alga surrounded by a cell wall with embedded calcareous plates (coccoliths).

cohesion molecular force between particles within a substance that acts to hold the particles together.

colonial organism organism consisting of semi-independent parts that do not exist as separate units; groups of organisms with specialized functions that form a coordinated unit.

commensalism an intimate association between different organisms in which one is benefited and the other is neither harmed nor benefited.

compensation depth depth at which there is a balance between the oxygen produced by algae through photosynthesis and that consumed through respiration; net oxygen production is zero.

condensation process by which a vapor becomes a liquid or a solid.

conduction transfer of heat energy through matter by internal molecular motion.

conservative constituent component or property of seawater whose value changes only as a result of mixing, diffusion, and advection and not as a result of biological or chemical processes; for example, salinity.

consumer animal that feeds on plants (primary consumer) or on other animals (secondary consumer).

continental crust crust forming the continental land blocks; mainly granite and its derivatives.

continental drift the movement of continents; the name of Alfred Wegener's theory, preceding plate tectonics.

continental margin zone separating the continents from the deep-sea bottom, usually subdivided into shelf, slope, and rise.

continental rise gentle slope formed by the deposition of sediments at the base of a continental slope.

continental shelf zone bordering a continent, extending from the line of permanent immersion to the depth at which there is a marked or rather steep descent to the great depths.

continental shelf break zone along which there is a marked increase of slope at the outer margin of a continental shelf.

continental slope relatively steep downward slope from the continental shelf break to depth.

contour line on a chart or graph connecting points of equal elevation, temperature, salinity, or other property.

convection transmission of heat by the movement of a heated gas or liquid; vertical circulation resulting from changes in density of a fluid.

convection cell circulation in a fluid, or fluidlike material, caused by heating from below. Heating the base of a fluid lowers its density, causing it to rise. The rising fluid cools, becomes denser, and sinks, creating circulation.

convergence situation in which substances come together, usually resulting in the sinking, or downwelling, of surface water and the rising of air.

convergent plate boundary a boundary between two plates that are converging or colliding with one another.

copepod small, shrimplike member of the zooplankton; in the class Crustacea.

coral colonial animal that secretes a hard outer calcareous skeleton; the skeletons of coral animals form in part the framework for warm-water reefs.

corange lines in a rotary tide, lines of equal tidal range about the amphidromic point.

core vertical, cylindric sample of bottom sediments, from which the nature of the bottom can be determined; also the central zone of Earth, thought to be liquid or molten on the outside and solid on the inside.

corer device that plunges a hollow tube into bottom sediments to extract a vertical sample.

Coriolis effect apparent force acting on a body in motion, due to the rotation of Earth, causing deflection to the right in the Northern Hemisphere and to the left in the Southern Hemisphere; the force is proportional to the speed and varies with latitude of the moving body.

cosmogenous sediment sediment particles with an origin in outer space; for example, meteor fragments and cosmic dust.

cotidal lines lines on a chart marking the location of the tide crest at stated time intervals.

covalent bond chemical bond formed by the sharing of one or more pairs of electrons.

cratered coast a primary coast formed when the seaward side of a volcanic crater is eroded away or is blown away by a volcanic eruption; opening the interior of the crater to the sea and creating a concave bay.

cratons large pieces of Earth's crust that form the centers of continents.

crest *see* berm crest, reef crest, wave crest.

crust outer shell of the solid Earth; the lower limit is usually considered to be the Mohorovičić discontinuity.

crustacean member of a class of primarily aquatic organisms with paired jointed appendages and a hard outer skeleton; includes lobsters, crabs, shrimps, and copepods.

cryoprotectants materials that lower the freezing points of organisms' internal fluids.

ctenophore transparent, planktonic animal, spherical or cylindrical with rows of cilia; comb jelly.

Curie temperature temperature at which the magnetic signature is frozen into an igneous rock during cooling.

current horizontal movement of water.

current meter instrument for measuring the speed and direction of a current.

cusp one of a series of evenly spaced, crescent-shaped depressions along sand and gravel beaches.

cyanobacteria member of the phytoplankton that can dominate in open-ocean environments.

cyclone *see* typhoon, hurricane.

D

deadweight ton (DWT) capacity of a vessel in tons of cargo, fuel, stores, and so on; determined by the weight of the water displaced.

declinational tide *see* diurnal tide.

decomposer heterotrophic; microorganisms (usually bacteria and fungi) that break down nonliving organic matter and release nutrients, which are then available for reuse by autotrophs.

deep exceptionally deep area of the ocean floor, usually below 6000 m.

deep scattering layer (DSL) layer of organisms that move away from the surface during the day and toward the surface at

night; the layer scatters or returns vertically directed sound pulses.

deep-water wave wave in water, the depth of which is greater than one-half the wavelength.

degree an arbitrary measure of temperature. Temperature is measured in degrees using one of three different scales: Fahrenheit (°F), Celsius (°C), and Kelvin (K). See appendix B for conversions among these scales.

delta area of unconsolidated sediment deposits, usually triangular in outline, formed at the mouth of a river.

demersal fish fish living near and on the bottom.

density property of a substance defined as mass per unit volume and usually expressed in grams per cubic centimeter or kilograms per cubic meter.

depth recorder *see* echo sounder.

desalination process of obtaining fresh water from seawater.

detritus any loose material, especially decomposed, broken, and dead organic materials.

dew point highest temperature at which water vapor condenses from an air mass. At the dew point temperature, the relative humidity is 100%.

diatom microscopic unicellular alga with an external skeleton of silica.

diatomaceous ooze sediment made up of more than 30% skeletal remains of diatoms.

diffraction process that transmits energy laterally along a wave crest.

diffusion movement of a substance from a region of higher concentration to a region of lower concentration (movement along a concentration gradient may be due to molecular motion or turbulence).

dinoflagellate member of the plankton. Some species are photosynthetic and others are not.

dipole a magnetic field like Earth's, with two opposite poles.

dispersion (sorting) sorting of waves as they move out from a storm center; occurs because long-period waves travel faster in deep water than short-period waves.

diurnal inequality difference in height between the two high waters or two low waters of each tidal day; the difference in speed between the two flood currents or two ebb currents of each tidal day.

diurnal tide (declinational tide) tide with one high water and one low water each tidal day.

divergence horizontal flow of substances away from a common center, associated with upwelling in water and descending motions in air.

divergent plate boundary a boundary between two plates that are diverging or moving apart from one another.

Dobson unit ozone unit, defined as 0.01 mm thickness of ozone at Standard Temperature and Pressure (0°C and 1 atmosphere pressure). If all the ozone over a certain area is compressed to 1 atmosphere pressure at 0°C, it forms a slab of a thickness corresponding to a number of Dobson units.

doldrums nautical term for the belt of light, variable winds near the equator.

domain the highest level for grouping organisms. There are three domains of life—Bacteria, Archaea, and Eukarya.

downwelling zone sinking of water, usually the result of a surface convergence or an increase in density of water at the sea surface.

dredge cylindric or boxlike sampling device made of metal, net, or both that is dragged across the bottom to obtain biological or geologic samples.

drift bottle bottle released into the sea for use in studying currents; contains a card identifying date and place of release and requesting the finder to return it with date and place of recovery.

drift sector *see* coastal circulation cell.

dugong *see* sea cow.

dune wind-formed hill or ridge of sand.

dune coast a primary coast formed by the deposition of sand in dunes by the wind.

dynamic equilibrium state in which the sums of all changes are balanced and there is no net change.

E

Earth sphere depth uniform depth of Earth below the present mean sea level, if the solid Earth surface were smoothed off evenly (2440 m).

ebb current movement of a tidal current away from shore or seaward in a tidal stream as the tide level decreases.

ebb tide falling tide; the period of the tide between high water and the next low water.

echolocation use of sound waves by some marine animals to locate and identify underwater objects.

echo sounder (depth recorder) instrument used to measure the depth of water by measuring the time interval between the release of a sound pulse and the return of its echo from the bottom. *See also* precision depth recorder (PDR).

ecology the study of interactions between organisms and their environment and of interactions among organisms.

ecosystem the organisms in a community and the nonliving environment with which they interact.

eddy circular movement of water.

Ekman spiral in a theoretical ocean of infinite depth, unlimited extent, and uniform viscosity, with a steady wind blowing over the surface, the surface water moves 45° to the right of the wind in the Northern Hemisphere. At greater depths, the water moves farther to the right with decreased speed, until at some depth (approximately 100 m), the water moves opposite to the wind direction. Net water transport is 90° to the right of the wind in the Northern Hemisphere. Movement is to the left in the Southern Hemisphere.

electrodialysis separation process in which electrodes of opposite charge are placed on each side of a semipermeable membrane to accelerate the diffusion of ions across the membrane.

electromagnetic radiation waves of energy formed by simultaneous electrical and magnetic oscillations; the electromagnetic spectrum is the continuum of all electromagnetic radiation from low-energy radio waves to high-energy gamma rays, including visible light.

electromagnetic spectrum *see* electromagnetic radiation.

El Niño wind-driven reversal of the Pacific equatorial currents, resulting in the movement of warm water toward the coasts of the Americas, so called because it generally develops just after Christmas.

endotherms maintain a higher temperature than the surrounding seawater but do not have the same level of temperature control as homeotherms.

entrainment mixing of salt water into fresh water overlying salt water, as in an estuary.

epicenter point on Earth's surface directly above an earthquake location, specified by identifying the latitude and longitude of the earthquake. *See also* focus, hypocenter.

epifauna animals living attached to the sea bottom or moving freely over it.

epipelagic upper portion of the oceanic pelagic zone, extending from the surface to about 200 m.

episodic wave abnormally high wave unrelated to local storm conditions.

equator 0° latitude, determined by a plane that is perpendicular to Earth's axis and is everywhere equidistant from the North and South Poles.

equilibrium tide theoretical tide formed by the tide-producing forces of the Moon and Sun on a nonrotating, water-covered Earth.

equinoxes days of the year when the Sun stands directly above the equator, so that day and night are of equal length around the world. The vernal equinox occurs about March 21, and the autumnal equinox occurs September 22–23.

escarpment nearly continuous line of cliffs or steep slopes caused by erosion or faulting.

estuary semi-isolated portion of the ocean that is diluted by freshwater drainage from land.

eukaryote grouping of organisms based on cell morphology. Eukaryotes are either unicellular or multicellular. All eukaryotic cells contain internal membrane-bound cell structures, including the nucleus that contains cellular DNA.

euphausiid planktonic, shrimplike crustacean. *See also* krill.

euphotic zone depth of the water column where there is sufficient sunlight for growth of photosynthetic organisms.

eustatic change global change in sea level that affects all of the world's coastlines.

evaporation process by which liquid becomes vapor. Returns moisture to the atmosphere in the hydrologic cycle; water warmed by the Sun's heat evaporates, becomes vapor or gas, and rises into the atmosphere.

evaporite deposit of minerals left behind by evaporating water, especially salt.

evapotranspiration the combined effect of evaporation and transpiration.

extremophile microorganism that thrives under extreme conditions of temperature, lack of oxygen, or high acid or salt levels; these conditions kill most other organisms.

F

fast ice sea ice that is anchored to shore or the sea floor in shallow water.

fathom a unit of length equal to 1.8 m (6 ft); used to measure water depth.

fault break or fracture in Earth's crust in which one side has been displaced relative to the other.

fault bay a bay formed by faulting along a primary coast.

fault coast a primary coast formed by tectonic activity and faulting.

fetch continuous area of water over which the wind blows in essentially a constant direction.

filament chain of living cells.

fjord narrow, deep, steep-walled inlet formed by the submergence of a mountainous coast or by the entrance of the ocean into a deeply excavated glacial trough after the melting of the glacier. A deep, small-surface-area estuary with moderately high river input and little tidal mixing.

flagellum long, whiplike extension from a living cell's surface that by its motion moves the cell; plural, flagella.

floe discrete patch of sea ice moved by the currents or wind.

flood current movement of a tidal current toward the shore or up a tidal stream as the tide height increases.

flood tide rising tide; the period of the tide between low water and the next high water.

flushing time length of time required for an estuary to exchange its water with the open ocean.

focus the location of an earthquake within Earth. Focus is specified by identifying latitude, longitude, and depth of the earthquake. *See also* epicenter.

fog visible assemblage of tiny droplets of water formed by condensation of water vapor in the air; a cloud with its base at the surface of Earth.

food chain sequence of organisms in which each is food for the next higher members in the sequence. *See also* food web.

food web complex of interacting food chains; all the feeding relations of a community taken together; includes production, consumption, decomposition, and the flow of energy.

foraminifera minute, one-celled animals that usually secrete calcareous shells.

foraminifera ooze sediment made up of 30% or more of skeletal remains of foraminifera.

forced wave wave generated by a continuously acting force and caused to move at a speed faster than it freely travels.

foreshore portion of the shore that includes the low-tide terrace and the beach face.

fouling attachment or growth of marine organisms on underwater objects, usually objects that are made or introduced by humans.

fracture zone long, linear zone of irregular bathymetry of the sea floor, characterized by asymmetric ridges and troughs; commonly associated with fault zones.

free wave wave that continues to move at its natural speed after its generation by a force.

friction resistance of a surface to the motion of a body moving (for example, sliding or rolling) along that surface.

fringing reef reef attached directly to the shore of an island or a continent and not separated from it by a lagoon.

frustule siliceous external shell of a diatom.

G

gabbro a coarse-grained, dark igneous rock, rich in iron and magnesium, the slow-cooling equivalent of basalt.

galaxy a huge aggregate of stars held together by mutual gravitation.

generating force disturbing force that creates a wave, such as wind or a landslide entering water.

geomorphology study of Earth's land forms and the processes that have formed them.

geostrophic flow horizontal flow of water occurring when there is a balance between gravitational forces and the Coriolis effect.

glacier mass of land ice, formed by the recrystallization of compacted old snow, flowing slowly from an accumulation area to an area of ice loss by melting, sublimination, or calving.

Global Positioning System (GPS) a worldwide radio-navigation system consisting of twenty-four navigational satellites and five ground-based monitoring stations. GPS uses this system of satellites as reference points for calculating accurate positions on the surface of Earth with readily available GPS receivers.

Gondwanaland an ancient landmass that fragmented to produce Africa, South America, Antarctica, Australia, and India.

graben a portion of Earth's crust that has moved downward and is bounded by steep faults; a rift.

grab sampler instrument used to remove a piece of the ocean floor for study.

granite crystalline, coarse-grained, igneous rock composed mainly of quartz and feldspar.

gravitational force mutual force of attraction between particles of matter (bodies).

gravity Earth's gravity; acceleration due to Earth's mass is 981 cm/s^2; g is used in equations.

gravity wave water wave form in which gravity acts as the restoring force; a wave with wavelength greater than 2 cm.

great circle the intersection of a plane passing through the center of Earth with the surface of Earth. Great circles are formed by the equator and any two meridians of longitude 180° apart.

greenhouse effect the gradual increase in average global temperature caused by the absorption of infrared radiation from Earth's surface by "greenhouse gases" in the atmosphere such as water vapor and carbon dioxide.

Greenwich Mean Time (GMT) solar time along the prime meridian passing through Greenwich, England; also known as Universal Time or Zulu Time.

groin protective structure for the shore, usually built perpendicular to the shoreline; used to trap littoral drift or to retard erosion of the shore; a type of jetty.

group speed speed at which a group of waves travels (in deep water, group speed equals one-half the speed of an individual wave); the speed at which the wave energy is propagated.

guyot submerged, flat-topped seamount. Also known as a tablemount.

gyre circular movement of water, larger than an eddy; usually applied to a bigger system.

H

habitat place where a plant or animal species naturally lives and grows.

hadal pertaining to the greatest depths of the ocean.

half-life time required for half of an initial quantity of a radioactive isotope to decay.

halocline water layer with a large change in salinity with depth.

harmful algal bloom (HAB) term used to describe both toxic and nuisance phytoplankton blooms.

harmonic analysis process of separating astronomical tide-causing effects from the tide record, in order to predict the tides at any location.

heat a measure of total kinetic energy of atoms and molecules in a substance.

heat budget accounting for the total amount of the Sun's heat received on Earth during one year as being exactly equal to the total amount lost because of radiation and reflection.

heat capacity the quantity of heat required to produce a unit change of temperature in a unit mass of material. *See also* specific heat.

herbivore animal that feeds only on plants.

heterotrophic pertaining to organisms requiring organic compounds for food; unable to manufacture food from inorganic compounds.

higher high water higher of the two high waters of any tidal day in a region of mixed tides.

higher low water higher of the two low waters of any tidal day in a region of mixed tides.

high nitrate low chlorophyll *a* (HNLC) regions regions of the surface ocean where there are high nitrate concentrations, but iron limits phytoplankton productivity.

high water maximum height reached by a rising tide.

holdfast organ of a benthic alga that attaches the alga to the sea floor.

holoplankton organisms living their entire life cycle in the floating (planktonic) state.

homeotherm organism with a body temperature that varies only within narrow limits.

hook spit turned landward at its outer end.

horse latitudes regions of calms and variable winds that coincide with latitudes 30°–35°N and S.

hot spot surface expression of a persistent rising plume of hot mantle material.

hurricane severe, cyclonic, tropical storm at sea, with winds of 120 km (73 mi) per hour or more; generally applied to Atlantic Ocean storms. *See also* typhoon.

hydrogen bond in water, the weak attraction between the positively charged hydrogen end of one water molecule and the negatively charged oxygen end of another water molecule.

hydrogenous sediment sediment precipitated from substances dissolved in seawater.

hydrologic cycle movement of water among the land, oceans, and atmosphere due to vertical and horizontal transport, evaporation, and precipitation.

hydrothermal vent seafloor outlet for high-temperature groundwater and associated mineral deposits; a hot spring.

hypocenter *see* focus.

hypothermophile microorganism that can grow at temperatures greater than 90°C. Many microorganisms that live at the hydrothermal vents are hyperthermophiles.

hypothesis initial explanation of data that is based on well-established physical or chemical laws.

hypoxic having low oxygen levels in the water; organisms may find survival in a hypoxic environment difficult or impossible.

hypsographic curve graph of land elevation and ocean depth versus area.

I

iceberg mass of land ice that has broken away from a glacier and floats in the sea.

igneous rock rock formed by solidification of a molten magma.

incidental catch the portion of any catch or harvest taken in addition to the targeted species.

inertia the property of matter that causes it to resist any change in its motion.

infauna animals that live buried in the sediment.

inner core the innermost region of Earth. It is solid and consists primarily of iron with minor amounts of other elements that likely include nickel, sulfur, and oxygen.

internal wave wave created below the sea surface at the boundary between two density layers.

international date line an imaginary line through the Pacific Ocean roughly corresponding to 180° longitude, to the east of which, by international agreement, the calendar date is one day earlier than to the west.

intertidal *see* littoral.

intertidal volume in an embayment, the volume of water gained or lost owing to the rise and fall of the tide.

inverse estuary an embayment, often located in arid climates, with high evaporation and little freshwater input. Circulation is seaward at depth and inward at the surface, opposite that of a typical estuary. Salinity is generally higher than average seawater salinities.

ion positively or negatively charged atom or group of atoms.

ionic bond the electrostatic force that holds together oppositely charged ions.

island arc system chain of volcanic islands formed above the sinking plate at a subduction zone.

isobar line of constant pressure.

isobath contour of constant depth.

isohaline having a uniform salt content.

isopycnal having a uniform density.

isostasy mechanism by which areas of Earth's crust rise or subside until their masses are in balance, "floating" on the mantle.

isothermal having a uniform temperature.

isotope atoms of the same element having different numbers of neutrons.

J

jellyfish or sea jelly semitransparent, bell-shaped pelagic organism, often with long tentacles bearing stinging cells.

jet stream (polar) a stream of air, between 30° and 50°N and S and about 12 km above Earth, moving from west to east at an average speed of 100 km/h.

jetty structure located to influence currents or to protect the entrance to a harbor or river from waves (U.S. terminology). *See also* breakwater, groin.

K

kelp any of several large, brown algae, including the largest known algae.

kinetic energy energy produced by the motion of an object.

knot a unit of speed equal to 0.51 m/s or 1 nautical mile per hour.

krill small, shrimplike crustaceans found in huge masses in polar waters and eaten by baleen whales.

L

lag deposits large particles left on a beach after the smaller particles are washed away.

lagoon shallow body of water that usually has a shallow, restricted outlet to the sea.

Langmuir cells shallow wind-driven circulation; paired helixes of moving water form windrows of debris along convergence lines.

La Niña condition of colder-than-normal surface water in the eastern tropical Pacific.

larva immature juvenile form of certain animals.

latent heat of fusion amount of heat required to change the state of 1 g of water from ice to liquid.

latent heat of vaporization amount of heat required to change the state of 1 g of water from liquid to gas.

latitude distance north or south of the equator. Latitude is the angle between the equatorial plane and a line drawn outward from the center of Earth to a point on the surface of Earth. Latitude varies from 0° to +90° north of the equator and 0° to −90° south of the equator. Together with longitude, it specifies the location of a point on the surface of Earth.

Laurasia an ancient landmass that fragmented to produce North America and Eurasia.

lava magma, or molten rock, that has reached Earth's surface; the same material solidified after cooling.

lava coast a primary coast formed by active volcanism producing lava flows that extend to the sea.

leading margin (active margin) the edge of the overriding plate at a trench or subduction zone.

lee shelter; the part or side sheltered from wind or waves.

light-year the distance light travels in one year. A light-year is equal to 9.46×10^{12} km, or 5.87×10^{12} mi.

lithification the conversion of loose sediment to solid rock.

lithogenous sediment sediment composed of rock particles eroded mainly from the continents by water, wind, and waves.

lithosphere outer, rigid portion of Earth; includes the continental and oceanic crusts and the upper part of the mantle.

littoral area of the shore between mean high water and mean low water; the intertidal zone.

longitude distance east or west of the prime meridian. Longitude is the angle in the equatorial plane between the prime meridian and a second meridian that passes through a point on the surface of Earth whose location is being specified.

Longitude may be specified in one of two ways; either from 0° to 360° east of the prime meridian, or 0° to 180° east and 0° to 180° west. Together with latitude, it specifies the location of a point on the surface of Earth.

longshore current current produced in the surf zone by the waves breaking at an angle with the shore; the current runs roughly parallel to the shoreline.

longshore transport (littoral drift) movement of sediment by the longshore current.

loran navigational system in which position is determined by measuring the difference in the time of reception of synchronized radio signals; derived from the phrase "long-range navigation."

lower high water lower of the two high waters of any tidal day in a region of mixed tides.

lower low water lower of the two low waters of any tidal day in a region of mixed tides.

low-tide terrace flat section of the foreshore seaward of the sloping beach face.

low water lowest elevation reached by a falling tide.

lunar month time required for the Moon to pass from one new Moon to another new Moon (approximately twenty-nine days).

lysocline depth at which calcareous skeletal material first begins to dissolve.

M

magma molten rock material that forms igneous rocks upon cooling; magma that reaches Earth's surface is referred to as lava.

magnetic pole either of the two points on Earth's surface where the magnetic field force lines are vertical.

manatee *see* sea cow.

manganese nodules rounded, layered lumps found on the deep-ocean floor that contain, on average, about 18% manganese, 17% iron, with smaller amounts of nickel, cobalt, and copper and traces of two dozen other metals; a hydrogenous sediment.

mantle main bulk of Earth between the crust and the core; increasing pressure and temperature with depth divide the mantle into concentric layers.

mariculture *see* aquaculture.

maximum sustained yield maximum number or amount of a species that can be harvested each year without steady depletion of the stock; the remaining stock is able to replace the harvested members by natural reproduction.

meander turn or winding curve of a current.

mean Earth sphere depth the depth below sea level of the surface of the solid Earth if it

was perfectly smooth with no variation in elevation. This is 2403 m (7884 ft) below present sea level.

mean ocean sphere depth the depth of the ocean if the solid Earth was perfectly smooth with no variation in elevation. This is 2646 m (8682 ft).

mean sea level average height of the sea surface, based on observations of all stages of the tide over a nineteen-year period in the United States.

meiobenthos very small animals living buried in the sediments of the sea floor.

Mercator projection a map projection in which the surface of Earth is projected onto a cylinder. Distortion is great at high latitudes and the poles cannot be shown. Mercator projections are frequently used for navigation because a straight line drawn on them is a line of true direction or constant compass heading.

meridian circle of longitude passing through the poles and any given point on Earth's surface.

meroplankton floating developmental stages (eggs and larvae) of organisms that as adults belong to the nekton and benthos.

mesopelagic oceanic zone from 200–1000 m.

mesosphere either the layer of the atmosphere above the stratosphere extending from about 50–90 km, or the region of the mantle beneath the asthenosphere.

metamorphic rock a rock created by the alteration of a preexisting rock by high temperature and/or pressure.

microbial loop component of marine food webs in which dissolved organic material cycles through bacteria and nanoplankton then back to small members of the zooplankton.

microplankton net plankton, composed of individuals from 0.07–1 mm in size but large enough to be retained by a small mesh net.

minus tide low-tide level below the mean value of the low tides or the zero tidal depth reference.

mitigation coastal management concept requiring developers to replace developed areas with equivalent natural areas or to reengineer other areas to resemble areas prior to development.

mitochondria used to derive energy from food in cells.

mixed tide type of tide in which large inequalities between the two high waters and the two low waters occur in a tidal day.

mixotrophic plankton phytoplankton that can both photosynthesize and consume organic matter depending on environmental conditions.

Mohorovičić discontinuity (Moho) boundary between crust and mantle, marked by a rapid increase in seismic wave speed.

mollusks marine animals, usually with shells; includes mussels, oysters, clams, snails, and slugs.

monoculture cultivation of only one species of organism in an aquaculture system.

monsoon name for seasonal winds; first applied to the winds over the Arabian Sea, which blow for six months from the northeast and for six months from the southwest; now extended to similar winds in other parts of the world; in India, the term is popularly applied to the southwest monsoon and to the rains that it brings.

Moon tide portion of the tide generated solely by the Moon's tide-raising force, as distinguished from that of the Sun.

moraine glacial deposit of rock, gravel, and other sediment left at the margin of an ice sheet.

mutualism an intimate association between different organisms in which both organisms benefit.

N

nannoplankton plankton that passes through an ordinary plankton net but can be removed from the water by centrifuging water samples.

nautical mile unit of length equal to 1852 m, or 1.15 land miles or 1 minute of latitude.

neap tides tides occurring near the times of the first and last quarters of the Moon, when the range of the tide is least.

nebula a large dense cloud of gas and dust in space.

nekton pelagic animals that are active swimmers; for example, adult squid, fish, and marine mammals.

neritic zone shallow-water marine environment extending from low water to the edge of the continental shelf. *See also* pelagic.

net circulation the long-term transport of water out of an estuary at the surface and into the estuary at depth averaged over many tidal cycles.

net plankton *see* microplankton.

nitrogen fixation process of using nitrogen gas as a source of inorganic nitrogen.

node point of least or zero vertical motion in a standing wave.

nonconservative constituent component or property of seawater whose value changes as a result of biological or chemical processes as well as by mixing, advection, and diffusion; for example, nutrients and oxygen in seawater.

normalization standard approach dividing primary productivity values by biomass values.

nucleus part of a cell that houses the majority of the cell's DNA.

nudibranchs soft-bodied, gastropod mollusks; sea slugs.

nutrient in the ocean, any one of a number of inorganic or organic compounds or ions used primarily in the nutrition of primary producers; nitrogen and phosphorus compounds are examples.

O

oceanic pertaining to the ocean water seaward of the continental shelf; the "open ocean." *See also* pelagic.

oceanic crust crust below the deep-ocean sediments; mainly basalt.

oceanic zone open ocean away from the direct influence of land.

offshore direction seaward of the shore.

offshore current any current flowing away from the shore.

offshore transport movement of sediment or water away from the shore.

onshore direction toward the shore.

onshore current any current flowing toward the shore.

onshore transport movement of sediment or water toward the shore.

oolith small, rounded accretionary body of calcium carbonate. Created by precipitation of calcium carbonate in shallow, warm seawater due to a change in water temperature or acidity.

ooze fine-grained deep-ocean sediment composed of at least 30% calcareous or siliceous remains of small marine organisms, the remainder being clay-size particles.

orbit in water waves, the path followed by the water particles affected by the wave motion; also, the path of a body subjected to the gravitational force of another body, such as Earth's orbit around the Sun.

orographic effect precipitation patterns caused by the flow of air over mountains.

osmosis tendency of water to diffuse through a semipermeable membrane to make the concentration of water on one side of the membrane equal to that on the other side.

osmotic pressure pressure that builds up in a confined fluid because of osmosis.

outer core a region surrounding the inner core. It is liquid and consists primarily of iron with minor amounts of other elements that likely include nickel, sulfur, and oxygen.

overturn sinking of denser water and its replacement by less-dense water from below.

oxygen minimum zone in which respiration and decay reduce dissolved oxygen to a minimum, usually between 800 and 1000 m.

ozone a form of oxygen that absorbs ultraviolet radiation from the Sun; O_3.

P

paleoceanography the study of ocean characteristics and processes in the past.

paleomagnetism study of ancient magnetism recorded in rocks; includes study of changes in location of Earth's magnetic poles through time and reversals in Earth's magnetic field.

Pangaea an ancient landmass that consisted of all of the present-day continents; it fragmented into Laurasia and Gondwanaland.

parallel circle on the surface of Earth parallel to the plane of the equator and connecting all points of equal latitude; a line of latitude.

parasitism an intimate association between different organisms in which one is benefited and the other is harmed.

partially mixed estuary is one with a strong net seaward flow of fresh water at the surface and a strong inward flow of seawater at depth.

passive margin *see* trailing margin.

pelagic zone primary division of the sea, which includes the whole mass of water subdivided into neritic and oceanic zones; also pertaining to the open sea.

period *see* tidal period, wave period.

pH measure of the concentration of hydrogen ions in a water solution; the concentration of hydrogen ions determines the acidity of the solution; $pH = -\log_{10}(H^+)$, where H^+ is the concentration of hydrogen ions in gram atoms per liter.

phosphorite a sedimentary rock composed largely of calcium phosphate and largely in the form of concretions and nodules.

photic zone layer of a body of water that receives ample sunlight for photosynthesis; usually less than 100 m deep.

photoautotrophs organisms that harness light energy to generate organic matter.

photo cell a device that converts light energy to electrical energy; used to determine solar radiation below the sea surface.

photophore luminous organ found on fish.

photosynthesis manufacture by plants of organic substances and release of oxygen from carbon dioxide and water in the presence of sunlight and the green pigment chlorophyll.

phylogenies evolutionary connections between ancestor organisms and their descendents.

physiographic map portrayal of Earth's features by perspective drawing.

phytoplankton microscopic algal and photosynthetic forms of plankton.

piezophile microorganisms that grow at high pressures such as those that live in the deep sea.

pinniped member of the marine mammal group that is characterized by four swimming flippers; for example, seals and sea lions.

plankton passively drifting or weakly swimming organisms.

plate one of roughly a dozen segments of lithosphere that move independently on the asthenosphere. Plates consist of oceanic and/or continental crust fused to the rigid upper mantle overlying the asthenosphere.

plate tectonics theory and study of Earth's lithospheric plates, their formation, movement, interaction, and destruction; the attempts to explain Earth's crustal changes in terms of plate movements.

poikilotherm organism with a body temperature that varies according to the temperature of its surroundings.

polar easterlies winds blowing from the poles toward approximately 60°N and 60°S; winds are northeasterly in the Northern Hemisphere and southeasterly in the Southern Hemisphere.

polar molecule a molecule with an unevenly distributed electrical charge. One end of the molecule has a slight positive charge and the other end has a slight negative charge.

polar reversal the periodic reversal of Earth's magnetic field where the north magnetic pole becomes the south magnetic pole and vice versa.

polar wandering curve a plot of the apparent location of Earth's north magnetic pole as a function of geologic time.

Polaris also known as the North Star, is located less than 1° from the celestial pole, a line corresponding to the extension of Earth's axis of rotation into the sky from the north geographic pole. The angular elevation of Polaris above the horizon corresponds to the latitude of an observer in the Northern Hemisphere.

polychaetes marine segmented worms, some in tubes, some free-swimming.

polyculture cultivation of more than one species of organism in an aquaculture system.

polyp sessile stage in the life history of certain members of the phylum Coelenterata (Cnidaria); sea anemones and corals.

potential energy energy that an object has because of its position or condition.

precipitation falling products of condensation in the atmosphere, such as rain, snow, or hail; also the falling out of a substance from solution.

precision depth recorder (PDR) instrument used to obtain a continuous pictorial record of the ocean bottom by timing the returning echoes of sound pulses. *See also* echo sounder.

primary coast coastline shaped primarily by terrestrial processes rather than marine processes.

primary production amount of living matter, or biomass, that is produced by photosynthetic or chemosynthetic organisms, usually expressed in grams of carbon per volume of seawater.

primary productivity rate at which biomass is produced by photosynthesis and chemosynthesis, usually expressed as grams of carbon produced per volume of seawater per day.

prime meridian meridian of 0° longitude, used as the origin for measurements of longitude; internationally accepted as the meridian of the Royal Naval Observatory, Greenwich, England.

producer an organism that manufactures food from inorganic substances.

progressive tide tide wave that moves, or progresses, in a nearly constant direction.

progressive wind wave wave that moves, or progresses, in a certain direction.

projection a system of projecting lines of latitude and longitude onto a plane surface to create a map with specific physical properties; *see also* Mercator projection.

prokaryote grouping of organisms based on morphology. All prokaryotes are unicellular with no internal membrane structures. Prokaryotes are composed of members of the domains Archaea and Bacteria.

protozoa minute, mostly one-celled animals.

pseudopodium flowing, temporary extension of the protoplasm of a cell, used in locomotion or feeding.

psychrophile microorganisms that grow in cold temperatures such as those that live in sea ice.

pteropod pelagic snail whose foot is modified for swimming.

pteropod ooze sediment made up of more than 30% shells of pteropods.

P-wave (or primary wave) a type of seismic wave in which material is alternately compressed and stretched in the direction of propagation of the wave.

pycnocline water layer with a large change in density with depth.

R

radar system of determining and displaying the distance of an object by measuring the time interval between transmission of a radio signal and reception of the echo return; derived from the phrase "radio detecting and ranging."

radial symmetry having similar parts regularly arranged around a central axis.

radiation energy transmitted as electromagnetic rays or waves without the need of a substance to conduct the energy.

radiolarian ooze sediment made up of more than 30% skeletal remains of radiolarians.

radiolarians single-celled protozoans with siliceous tests.

radiometric dating determining ages of geologic samples by measuring the relative abundance of radioactive isotopes and comparing isotopic systems.

rafting transport of sediment, rock, silt, and other land matter out to sea by ice, logs, and the like, with the deposition of the rafted material when the carrying agent melts or disintegrates.

rain shadow an area of low precipitation on the leeward (or sheltered) side of an island or mountain. Precipitation occurs on the windward side as the air is forced to ascend the side of the island or mountain so the descending air on the leeward side is dry.

range *see* tidal range.

red clay red to brown fine-grained lithogenous deposit, of predominantly clay size, which is derived from land, transported by winds and currents, and deposited far from land and at great depth; also known as brown mud and brown clay.

red tide red coloration, usually of coastal waters, caused by large quantities of microscopic organisms (generally dinoflagellates); some red tides result in mass fish kills, others contaminate shellfish, and still others produce no toxic effects.

reef offshore hazard to navigation, made up of consolidated rock or coral, with a depth of 20 m or less.

reef coast a secondary coast formed by reef-building corals in tropical waters.

reef crest highest portion of a coral reef on the exposed seaward edge of the reef.

reef flat portion of a coral reef landward of the reef crest and seaward of the lagoon.

reflection rebounding of light, heat, sound, waves, and so on after striking a surface.

refraction change in direction, or bending, of a wave.

relict sediments sediments deposited by processes no longer active.

reservoir a source or place of temporary residence for water, such as the oceans or atmosphere.

residence time mean time that a substance remains in a given area before being replaced, calculated by dividing the amount of a substance by its rate of addition or subtraction.

respiration metabolic process by which food or food-storage molecules yield the energy on which all living cells depend.

restoring force force that returns a disturbed water surface to the equilibrium level, such as surface tension and gravity.

reverse osmosis the process of producing fresh water from seawater by forcing the water molecules through a semipermeable membrane, leaving behind salt ions and other impurities.

ria coast a primary coast formed when rising sea level, caused by the melting of glaciers and ice sheets following the last ice age, flooded coastal river valleys.

ridge long, narrow elevation of the sea floor, with steep sides and irregular topography.

rift valley trough formed by faulting along a zone in which plates move apart and new crust is created, such as along the crest of a ridge system.

rift zone a region where the lithosphere splits and separates, allowing new crustal material to intrude into the crack or rift.

rip current strong surface current flowing seaward from shore; the return movement of water piled up on the shore by incoming waves and wind.

rise long, broad elevation that rises gently and generally smoothly from the sea floor.

rotary current tidal current that continually changes its direction of flow through all points of the compass during a tidal period.

rotary standing tide tide that is the result of a standing wave moving around the central node of a basin.

S

salinity measure of the quantity of dissolved salts in seawater. It is formally defined as the total amount of dissolved solids in seawater in parts per thousand (‰) by weight when all the carbonate has been converted to oxide, all the bromide and iodide have been converted to chloride, and all organic matter is completely oxidized.

salinometer instrument for determining the salinity of water by measuring the electrical conductivity of a water sample of a known temperature.

salt budget balance between the rates of salt addition to and removal from a body of water.

salt marsh a relatively shallow coastal environment populated with salt-tolerant grasses. These are often found in temperate climates. They are extremely productive biologically and extend the shoreline seaward by trapping fine sediment.

salt wedge intrusion of salt water along the bottom; in an estuary, the wedge moves upstream on high tide and seaward on low tide.

sand spit *see* spit.

satellite body that revolves around a planet; a moon; a device launched from Earth into orbit around a planet or the Sun.

saturation concentration the maximum amount of any substance that can be held in solution without any of the substance coming out of solution. When a dissolved substance has reached its saturation concentration, the rate at which it enters into the solution will equal the rate at which it leaves the solution.

scarp elongated and comparatively steep slope separating flat or gently sloping areas on the sea floor or on a beach.

scattering random redirection of light or sound energy by reflection from an uneven sea bottom or sea surface, from water molecules, or from particles suspended in the water.

sea same as the ocean; subdivision of the ocean; surface waves generated or sustained by the wind within their fetch, as opposed to swell.

sea cow (dugong, manatee) large herbivorous marine mammal of tropical and subtropical waters; includes the manatee and the dugong.

seafloor spreading movement of crustal plates away from the mid-ocean ridges; process that creates new crustal material at the mid-ocean ridges.

sea level height of the sea surface above or below some reference level. *See also* mean sea level.

seamount isolated volcanic peak that rises at least 1000 m from the sea floor.

sea smoke type of fog caused by dry, cold air moving over warmer water.

sea stack isolated mass of rock rising from the sea near a headland from which it has been separated by erosion.

sea state numerical or written description of the roughness of the ocean surface relative to wave height.

Secchi disk white, or white and black, disk used to measure the transparency of the water by observing the depth at which the disk disappears from view.

secondary coast coastline shaped primarily by marine forces or marine organisms.

sediment particulate organic and inorganic matter that accumulates in loose, unconsolidated form.

sedimentary rock a rock formed by the cementation of mineral grains accumulated by wind, water, or ice transportation to the site of deposition or by chemical precipitation at the site.

seiche standing wave oscillation of an enclosed or semienclosed body of water that continues, pendulum fashion, after the generating force ceases.

seismic pertaining to or caused by earthquakes or Earth movements.

seismic sea wave *see* tsunami.

seismic tomography the use of seismic data to produce computerized, detailed, three-dimensional maps of the boundaries between Earth's layers.

seismic waves elastic disturbances, or vibrations, that are generated by earthquakes.

semidiurnal tide tide with two high waters and two low waters each tidal day.

semipermeable membrane membrane that allows some substances to pass through but restricts or prevents the passage of other substances.

sessile permanently fixed or sedentary; not free-moving.

set direction in which the current flows.

shallow-water wave wave in water whose depth is less than one-twentieth the average wavelength.

shingles flat, water-worn pebbles, or cobbles, found in beds along a beach.

shoal elevation of the sea bottom comprising any material except rock or coral (in which case it is a reef); may endanger surface navigation.

shore strip of ground bordering any body of water and alternately exposed and covered by tides and waves.

sidereal day time period determined by one rotation of Earth relative to a far-distant star, about four minutes shorter than the mean solar day.

sigma-t abbreviated value of the density of seawater; at a given temperature and salinity, neglecting pressure; $\sigma_t =$ (density $- 1) \times 1000$.

siliceous containing or composed of silica.

siliceous ooze fine-grained deep-ocean biogenous sediment containing at least 30% siliceous tests, or the remains of small marine organisms.

sill shallow area that separates two basins from one another or a coastal bay from the adjacent ocean.

slack water state of a tidal current when its velocity is near zero; occurs when the tidal current changes direction.

slick area of smooth surface water.

sofar channel natural sound channel in the oceans in which sound can be transmitted for very long distances; the depth of minimum sound velocity; derived from the phrase "sound fixing and ranging."

solar constant rate at which solar radiation is received on a unit surface that is perpendicular to the direction of incident radiation just outside Earth's atmosphere at Earth's mean distance from the Sun; equal to 2 cal/cm^2/min.

solar day time period determined by one rotation of Earth relative to the Sun; the mean solar day is twenty-four hours.

solstice times of the year when the Sun stands directly above 23 1/2°N or 23 1/2°S latitude. The winter solstice occurs about December 22, and the summer solstice occurs about June 22.

sonar method or equipment for determining, by underwater sound, the presence, location, or nature of objects in the sea; derived from the phrase "sound navigation and ranging."

sorting *see* dispersion.

sounding measurement of the depth of water beneath a vessel.

sound shadow zone area of the ocean into which sound does not penetrate because the density structure of the water refracts the sound waves.

Southern Oscillation a periodic reversal of the low- and high-pressure areas that typically dominate the eastern and western equatorial Pacific, respectively.

specific gravity ratio of the density of a substance to the density of 4°C water.

specific heat ratio of the heat capacity of a substance to the heat capacity of water.

spectrum the regular ordering of wave phenomena such as electromagnetic radiation by wavelength or frequency. A familiar spectrum is the regular ordering of visible light from the long-wavelength red to the short-wavelength violet when it passes through a prism.

sphere depth thickness of a material spread uniformly over a smooth sphere having the same area as Earth.

spicules small calcareous or siliceous skeletal structures found in sponges.

spit (sand spit) low tongue of land, or a relatively long, narrow shoal extending from the shore.

splash zone *see* supralittoral.

spoil dredged material.

spore minute, unicellular, asexual reproductive structure of an alga.

spreading center region along which new crustal material is produced.

spreading rate the rate at which two plates move apart. Spreading rates are generally between about 2 and 10 cm (0.8 and 4 in) per year.

spring tides tides occurring near the times of the new and full Moon, when the range of the tide is greatest.

standing crop biomass present at any given time.

standing wave type of wave in which the surface of the water oscillates vertically between fixed points called nodes, without progression; the points of maximum vertical rise and fall are called antinodes.

steepness of wave *see* wave steepness.

stipe portion of an alga between the holdfast and the blade.

storm berm *see* winter berm.

storm center area of origin for surface waves generated by the wind; an intense atmospheric low-pressure system.

storm tide (storm surge) along a coast, the exceptionally high water accompanying a storm, owing to wind stress and low atmospheric pressure, made even higher when associated with a high tide and shallow depths.

stratosphere the layer of the atmosphere above the troposphere where temperature is constant or increases with altitude.

subduction zone plane descending away from a trench and defined by its seismic activity, interpreted as the convergence zone between a sinking plate and an overriding plate.

sublimation transition of a substance from its solid state to its gaseous state without becoming a liquid.

sublittoral benthic zone from the low-tide line to the seaward edge of the continental shelf; the subtidal zone.

submarine canyon relatively narrow, V-shaped, deep depression with steep slopes, the bottom of which grades continuously downward across the continental slope.

submersible a research submarine, designed for manned or remote operation at great depths.

subsidence sinking of a broad area of the crust without appreciable deformation.

substrate material making up the base on which an organism lives or to which it is attached.

subtidal *see* sublittoral.

summer berm a seasonal berm that is built by low-energy waves during the summer and removed by high-energy waves in the winter.

summer solstice *see* solstice.

Sun tide portion of the tide generated solely by the Sun's tide-raising force, as distinguished from that of the Moon.

supersaturation when the concentration of a dissolved substance is higher than its normal saturation value.

supralittoral benthic zone above the high-tide level that is moistened by waves, spray, and extremely high tides; also called splash zone.

surf wave activity in the area between the shoreline and the outermost limit of the breakers.

surface tension tendency of a liquid surface to contract owing to bonding forces between molecules.

surimi refined fish protein that is used to form artificial crab, shrimp, and scallop meat.

swash the up-rush of water onto the beach from a breaking wave.

swash zone beach area where water from a breaking wave rushes.

S-wave (or secondary wave) a type of seismic wave in which material is sheared from side to side, perpendicular to the direction of propagation of the wave.

swell long and relatively uniform wind-generated ocean waves that have traveled out of their generating area.

symbiosis living together in intimate association of two dissimilar organisms.

T

taxonomy scientific classification of organisms.

tectonic pertaining to processes that cause large-scale deformation and movement of Earth's crust.

tektites particles with a characteristic round shape, formed from rock melting during meteorite impact.

temperature a measure of the rate of atomic or molecular motion in a substance.

terranes fragments of Earth's crust bounded by faults, each fragment with a history distinct from each other fragment.

terrigenous sediments of the land; sediments composed predominantly of material derived from the land.

test the shell of an organism.

theory a tested, reliable, and precise statement of the relationships among reproducible observations.

thermocline water layer with a large change in temperature with depth.

thermohaline circulation vertical circulation caused by changes in density; driven by variations in temperature and salinity.

thermosphere the layer of the atmosphere above the mesosphere; extends from 90 km to outer space.

tidal bore high-tide crest that advances rapidly up an estuary or river as a breaking wave.

tidal current alternating horizontal movement of water associated with the rise and fall of the tide.

tidal datum reference level from which ocean depths and tide heights are measured; the zero tide level.

tidal day time interval between two successive passes of the Moon over a meridian, approximately twenty-four hours and fifty minutes.

tidal period elapsed time between successive high waters or successive low waters.

tidal range difference in height between consecutive high and low waters.

tide periodic rising and falling of the sea surface that results from the gravitational attractions of the Moon and Sun acting on the rotating Earth.

tide wave long-period gravity wave that has its origin in the tide-producing force and is observed as the rise and fall of the tide.

tombolo deposit of unconsolidated material that connects an island to another island or to the mainland.

topography general elevation pattern of the land surface (or the ocean bottom). *See also* bathymetry.

toxicant substance dissolved in water that produces a harmful effect on organisms, either by an immediate large dose or by small doses over a period of time.

trace element an element dissolved in seawater at a concentration less than one part per million.

trade winds wind systems that occupy most of the tropics and blow from approximately 30°N and 30°S toward the equator; winds are northeasterly in the Northern Hemisphere and southeasterly in the Southern Hemisphere.

trailing margin (passive margin) the continental margin closest to the mid-ocean ridge.

transform fault fault with horizontal displacement connecting the ends of an offset in a mid-ocean ridge. Some plates slide past each other along a transform fault.

transform plate boundary a boundary between two plates that are sliding past one another. This boundary is marked by a transform fault.

transgenic describing an organism that contains hereditary material from another organism incorporated into its genetic material.

transpiration process by which plants return moisture to air. Plants take up water through roots and lose water through pores in their leaves. An actively growing plant daily transpires five to ten times as much water as it holds at one time.

transverse ridge ridge running at nearly right angles to the main or principal ridge.

trench long, deep, and narrow depression of the sea floor with relatively steep sides, associated with a subduction zone.

triploid condition in which cells have three sets of chromosomes.

trophic relating to nutrition; a trophic level is the position of an organism in a food chain or food (trophic) pyramid.

Tropics of Cancer and Capricorn latitudes 23 1/2°N and 23 1/2°S, respectively, marking the maximum angular distance of the Sun from the equator during the summer and winter solstices.

troposphere the lowest layer of the atmosphere, where the temperature decreases with altitude.

trough long depression of the sea floor, having relatively gentle sides; normally wider and shallower than a trench. *See also* wave trough.

T-S diagram graph of temperature versus salinity, on which seawater samples taken at various depths are used to describe a water mass.

tsunami (seismic sea wave) long-period sea wave produced by a submarine earthquake, volcanic eruption, sediment slide, or seafloor faulting. It may travel across the ocean for thousands of miles unnoticed from its point of origin and build up to great heights over shallow water at the shore.

tube worm any worm or wormlike organism that builds a tube or sheath attached to a submerged substrate.

turbidite sediment deposited by a turbidity current, showing a pattern of coarse particles at the bottom, grading gradually upward to fine silt or mud.

turbidity loss of water clarity or transparency owing to the presence of suspended material.

turbidity current dense, sediment-laden current flowing downward along an underwater slope.

twilight zone region of the water column where there is some light but not enough for photosynthesis.

typhoon severe, cyclonic, tropical storm originating in the western Pacific Ocean, particularly in the vicinity of the South China Sea. *See also* hurricane.

U

ultraplankton plankton that are smaller than nannoplankton; they are difficult to separate from water.

Universal Time solar time along the prime meridian passing through Greenwich, England; also known as Greenwich Mean Time (GMT) or Zulu Time.

upwelling rising of water rich in nutrients toward the surface, usually the result of diverging surface currents.

V

vernal equinox see equinoxes.

vertebrates animals with backbones or a spinal column.

virus a noncellular infectious agent that reproduces only in living cells.

viscosity property of a fluid to resist flow; internal friction of a fluid.

W

Wadati-Benioff zone dipping patterns of earthquake activity that descend into the mantle along convergent plate boundaries.

water bottle device used to obtain a water sample at depth.

water budget balance between the rates of water added and lost in an area.

water mass body of water identified by similar patterns of temperature and salinity from surface to depth.

water type body of water identified by a specific range of temperature and salinity from a common source.

wave periodic disturbance that moves through or over the surface of a medium with a speed determined by the properties of the medium.

wave crest highest part of a wave.

wave height vertical distance between a wave crest and the adjacent trough.

wavelength horizontal distance between two successive wave crests or two successive wave troughs.

wave period time required for two successive wave crests or troughs to pass a fixed point.

wave ray line indicating the direction waves travel; drawn at right angles to the wave crests.

wave steepness ratio of wave height to wavelength.

wave train series of similar waves from the same direction.

wave trough lowest part of a wave.

well-mixed estuary is one in which there is strong wind-driven and tidal mixing. The salinity of the water in the estuary is relatively constant with depth and decreases from the ocean to the river.

westerlies wind systems blowing from the west between latitudes of approximately 30°N and 60°N and 30°S and 60°S; southwesterly in the Northern Hemisphere and northwesterly in the Southern Hemisphere.

wind wave wave created by the action of the wind on the sea surface.

winter berm a relatively permanent berm that is formed by high-energy waves in the winter.

winter solstice see solstice.

Z

zenith point in the sky that is immediately overhead.

zonation parallel bands of distinctive plant and animal associations found within the littoral zones and distributed to take advantage of optimal conditions for survival.

zooplankton animal forms of plankton.

zooxanthellae symbiotic microscopic organisms (dinoflagellates) found in corals and other marine organisms.

Zulu Time solar time along the prime meridian passing through Greenwich, England; also known as Greenwich Mean Time (GMT) or Universal Time.

Credits

Photo

Chapter 1

Opener: © Brand X RF; **1.1a–c:** Courtesy of the Burke Museum of Natural History and Culture, Seattle, WA (cat 1-11433, 1-794, 1-158); **1.2:** © Anders Ryman/Corbis; **1.3:** Courtesy of Sea and Shore Museum, Port Gamble, WA; **1.5:** From Johannes van *Keulen's Sea Atlas,* 1682–84; **1.8:** © National Maritime Museum, London; **1.10:** Franklin Folger, 1769; **1.11:** NASA; **Page 10, Box Fig. 1:** © Mary Rose Trust; **Page 10, Box Fig. 2:** © Hans Hammerskiold for Vasa Museum, Stockholm, Sweden; **Page 11, Box Fig. 3:** © Henry Thomason, Texas Historical Commission; **Page 11, Box Fig. 4:** © Quest Group, Ltd.; **1.13a, c–i:** *Challenger Expedition*: Dec. 21, 1872–May 24, 1876. Engravings from *Challenger Reports*, vol. 1, 1885; **1.14:** © The Norwegian Maritime Museum; **Page 16, Box Fig. 1 and Page 17, Box Fig. 2:** Courtesy Marcia McNutt; **Page 18, Box Fig. 3:** © M. Leet for MBARI; **1.15:** © The Norwegian Maritime Museum; **1.16a:** © Scripps Institution of Oceanography, University of California, San Diego; **1.16b:** © D. Weisman/ Woods Hole Oceanographic Institution; **1.17a:** Courtesy of Joe Creager, University of Washington; **1.17b:** © Ocean Drilling Team, Texas A&M; **1.18:** CZCS Project/Goddard Space Flight Center; **1.19a:** Photo courtesy of the International Argo Steering Team; **1.20:** Japan Marine Science and Technology Center and the Integrated Ocean Drilling Team.

Chapter 2

Opener, 2.1: Image by Reto Stockli, NASA/Goddard Space Flight Center, Enhancements by Robert Simmon; **2.2:** © AP/Wide World Photos; **2.3:** NASA; **Page 30, Box Fig. 1:** © James A. Sugar/Corbis; **Page 31, Box Figs. 2 & 3:** © L.A. Frank/University of Iowa; **2.4:** NASA; **2.10:** © National Maritime Museum, London; **2.14:** Courtesy Stephen P. Miller, University of California, Santa Barbara; **2.17:** Department of Defense.

Chapter 3

Opener: NOAA; **3.3:** NASA; **3.7:** Paleogeographic Maps by Christopher R. Scotese © 2003 PALEOMAP Project (www.scotese.com); **3.8:** B. Heezen & M. Tharp, World Ocean Floor, © Marie Tharp, 1977 South Nyack, NY 10960. Reproduced by permission; **3.11:** Dr. Robert M. Kleckhefer, Chevron Overseas Petroleum, Inc., San Ramon, CA; **3.14:** Courtesy Ocean Drilling Program/Texas A&M University, College Station, TX; **3.26:** © OAR/National Underseas Research; **3.32:** © P.W. Lipman, U.S. Geological Survey; **Page 76:** The NEPTUNE Project (www.neptune.washington.edu); **Page 77, Box Figs. 1 & 2:** Courtesy John Delaney, School of Oceanography, University of Washington; **3.36, 3.37:** Paleogeographic Maps by Christopher R. Scotese © 2003 PALEOMAP Project (www.scotese.com); **3.39:** © News Office/Woods Hole Oceanographic Institution, Photo by Rod Catanech; **3.40a, c, 3.41:** Courtesy John Delaney, School of Oceanography, University of Washington; **Page 86, Box Fig. 1:** © Alyn & Alison Duxbury; **Page 87, Box Figs. 2 & 3:** Courtesy Deborah Kelley, University of Washington; **Page 87, Box Fig. 4:** © Alyn & Alison Duxbury; **Page 87, Box Fig. 5:** Courtesy Deborah Kelley, University of Washington.

Chapter 4

Opener: © AGE FotoSearch RF; **4.2:** Courtesy John Delaney, School of Oceanography, University of Washington; **Page 93, Box Fig. 2:** Courtesy Williamson & Associates, Seattle, WA; **4.3:** Courtesy Walter H. F. Smith/NOAA; **4.9:** © Hugo Ortner; **4.10:** Official U.S. Navy photo by R.F. Diel; **Page 104:** © Ed Seibel, 2000 MBARI; **4.19:** Courtesy Williamson & Associates, Seattle, WA; **4.21a–b:** Ken Adkins, School of Fisheries, University of Washington; **4.21c:** Courtesy of Steve Nathan and R. Mark Leckie; **4.22:** Courtesy Dr. Virgil E. Barnes; **4.23:** Courtesy of Joe Creager, University of Washington; **4.24:** NOAA; **4.26a:** Courtesy of Dean McManus, University of Washington; **4.26b:** Courtesy of Dawn Wright, Oregon State University; **4.27:** Courtesy of Kathy Newell, School of Oceanography, University of Washington; **4.28c:** Courtesy of Sara Barnes, School of Oceanography, University of Washington; **4.28d:** © Alyn & Alison Duxbury; **4.28e:** Courtesy Dr. Richard Sternberg, School of Oceanography, University of Washington; **4.29:** Courtesy Chevron Corporation.

Chapter 5

Opener: © Getty R.F.; **5.4:** © Sandra Hines, News and Information, University of Washington; **5.16:** © Alyn & Alison Duxbury; **5.17:** Courtesy of Dennis K. Clark, MOBY Project Team Leader, NOAA; **5.19:** © Alyn & Alison Duxbury; **5.22a:** Courtesy Kathy Newell, School of Oceanography, University of Washington; **5.22b:** Courtesy of Roger Anderson, Polar Science Center, Applied Physics Laboratory, University of Washington; **5.22c:** © Captain Lawson W. Brigham, retired, USCG; **5.23a–b:** J. Zwally, Goddard Space Flight Center; **5.23c–d:** D.J. Cavaneri & P. Gloersen, Goddard Space Center; **5.24a:** Edward Joshberger, U.S. Geological Survey, Department of the Interior; **5.24b:** Courtesy Kathy Newell, School of Oceanography, University of Washington; **5.25:** Earth Observatory Satellite Co., Lanham, MD, © 1991 Discover Magazine; © William Rosenthal/ Jeroboam; **5.26:** © William Rosenthal/Jeroboam.

Chapter 6

Opener: © Tom Stack & Therisa Stack/Tom Stack & Associates; **Page 159, Box Fig. 1a:** © Peter Grootes, Quarterly Research Center, University of Washington; **Page 159, Box Fig. 1b–c:** Courtesy of Christine Massey, University of Washington; **Page 159, Box Fig. 1d:** © Peter Grootes, Quarterly Research Center, University of Washington; **6.8:** Courtesy of John Conomos, U.S. Geological Survey.

Chapter 7

Opener: © AGE Fotostock RF; **7.7:** M. Chahine, Jet Propulsion Lab, Goddard Space Flight Center; **7.13:** NASA; **7.21:** Courtesy of W. Timothy Liu, Jet Propulsion Laboratory, California Institute of Technology; **7.28:** NOAA satellite photo; **7.31:** Scientific Visualization Studio, NASA, Goddard Space Flight Center; **Page 192:** LuAnne Thompson; **7.33:** NASA; **7.35:** U.S. Army Corps of Engineers photo by Alan Dooley; **7.36:** Photo taken on September 19, 2005, by Mark Wolfe/FEMA; **7.37:** Photo taken on October 4, 2005, by John Fleck/FEMA.

Chapter 8

Opener: © Anne Keiser/National Geographic Image Collection; **Page 207, Box Figs. 1 & 2:** Courtesy of Captain Lawson W. Brigham, retired, USCG; **Page 208, Box Fig. 3:** Courtesy of Sandra Hines, University of Washington; **Page 208, Box Figs. 4 & 5:** Courtesy James Johnson, Applied Physics Lab, University of Washington; **8.10:** James R. Postel, School of Oceanography, University of Washington; **8.11:** Courtesy Chelsea Instruments Ltd.; **8.12:** © News Office, Woods Hole Oceanographic Institution; **Page 213, Box Fig. 1 and Page 214, Box Fig. 2:** © Charles C. Eriksen; **8.13c:** Courtesy Lockheed Missiles and Space Co., Sunnyvale, CA.

18.14: © David Hall/Photo Researchers, Inc.; **18.16:** Courtesy Kathy Newell, School of Oceanography, University of Washington; **18.17:** Courtesy Richard A. Lutz; **Page 454, Box Fig. 1:** Courtesy Dr. Charles Fisher, Professor of Biology, Penn State; **18.18:** Courtesy Dr. James Brooks, Geochemical and Environmental Research Group, College Station, TX; **18.19:** Courtesy Kathy Newell, School of Oceanography, University of Washington; **18.20:** Courtesy Brad Matsen, Seattle, WA; **18.21:** Courtesy Ken Chew, School of Fisheries, University of Washington; **Page 458, Box Fig. 1:** Reprinted by permission from Washington Sea Grant Program, University of Washington.

Line Art

Chapter 2

Table 2.4: Data from H. W. Menard and S. M. Smith. 1966. Hypsometry of Ocean Basin Provinces. *Journal of Geophysical Research,* 71 (18): 4305–25.

Chapter 3

Figure 3.1: Courtesy of Berkeley Seismological Laboratory, University of California Berkeley. Reprinted with permission. **Figure 3.10:** Earthquake Epicenters, 1961–1967. From Barazangi and Dormar, *Bulletin of Seismological Society of America,* 1969. Two maps depicting earthquake epicenters. Seismological Society of America, El Cerrito, CA. **Figure 3.12:** Data from J. G. Sclater and J. Crowe, "On the Variability of Oceanic Heat Flow Average" in *Journal of Geographical Research,* 81:17 (June 1976), p. 3004. American Geophysical Union, Washington, D.C. **Figure 3.16:** From D. H. Tarling and J. C. Mitchell, "The Earth's Magnetic Polarity as a Function of Millions of Years Before Present" in *Geology,* 4.3 (1976). The Geological Society of America. Reprinted with permission. **Figure 3.17:** Modified with permission from R. L. Larson and W. C. Pitman, III, 1972, Geological Society of America Bulletin. **Figure 3.19:** From The Bedrock Geology of the World by R. Larson, et al. © 1985 by R. L. Larson and W. C. Pitman. Used with permission of W. H. Freeman and Company. **Figure 3.31:** From Robert J. Stern, "A Subduction Prime for Instructors of Introductory-Geology Courses and Authors of Introductory-Geology Textbooks," *Journal of Geoscience Education* 46:221-28, 1998. Reprinted with permission. **Figure 3.33:** Courtesy of NASA. **Figure 3.40b:** Courtesy of Veronique Robigou, University of Washington. Reprinted with permission.

Chapter 4

Figure 4.8: From pp. 321–322 from SUBMARINE GEOLOGY, 3rd ed. By Francis P. Shepard. Copyright © 1948, 1963. Reprinted by permission of Pearson Education, Inc. **Field Notes Figure 1:** Figure is modified from Figure 1 in Clauge, D. A., and J. G. Moore, 2002. "The Proximal Part of the Giant Submarine Wailau Landslide, Molokai, Hawaii." *Journal of Volcanology and Geothermal Research* 113: 259–87. **Field Notes Figure 2:** Figure is modified from Figure 1 in Moore, J. G., D. A. Clague, R. T. Holcomb, P. W. Lipman, W. R. Normark, and M. E. Torresan. 1989. Prodigious Submarine Landslides on the Hawaiian Ridge. *Journal of Geophysical Research* 94: 17465–84. **Figure 4.20:** From *Sedimentary Geology: An Introduction to Sedimentary Rocks and Stratigraphy,* 2/e by Donald R. Prothero and Fred Schwab, p. 198. © 1996, 2004 by W. H. Freeman and Company. Used with permission. **Table 4.3:** Reprinted from *Chemical Oceanography* 2nd Edition, Vol. 1, 1975, Riley & Skirrow, "Major constituents of seawater," with permission of Elsevier. **Table 4.4:** Reprinted from *Chemical Oceanography* 2nd Edition, Vol. 1, 1975, Riley & Skirrow, "Major constituents of seawater," with permission of Elsevier. **Figure 4.25 a–b:** From M. N. Hill, *The Sea,* Vol. 2. © 1963. Reprinted with permission of John Wiley & Sons.

Chapter 5

Table 5.5: Reprinted from *Chemical Oceanography,* Vol. 8, 1983, Riley & Chester, with permission of Elsevier. **Figure 5.11:** From Sverdrup/Johnson/Fleming, *The Oceans,* 1942, renewed 1970, p. 105. Adapted by permission of Prentice Hall, Inc., Upper Saddle River, N.J.

Chapter 6

Table 6.1: From J. P. Riley and G. Skirrow, *Chemical Oceanography,* Vol. 1, 2nd edition. Copyright 1975, Elsevier Science. Reprinted by permission. **Table 6.2:** From J. P. Riley and R. Chester, *Chemical Oceanography,* Vol. 8. Copyright 1983, Elsevier Science. Reprinted by permission. **Table 6.3:** Data from U.S. Geological Survey. **Figure 6.6:** Modified from Carbon Cycle, Wikipedia (http://en.wikipedia.org/wiki/carbon-cycle).

Chapter 7

Figure 7.3: Source: NOAA Meteorological Satellite Laboratory. **Figure 7.9:** Source: NOAA Marine Climate of Washington. **Figure 7.11a:** Scripps Institution of Oceanography/ UCSD **Figure 7.11b:** Source info: Scripps Institution of Oceanography, UC San Diego. Reprinted with permission. **Figure 7.12:** Source: Modified from National Oceanic and Atmospheric Administration: www.cpc.ncep.noaa.gov. **Figure 7.30:** From *The Open University, Open Circulation,* 2nd edition. Reprinted with permission of Elsevier. **Figure 7.32:** Data from NOAA-CIRES Climate Diagnostic Center, University of Colorado at Boulder. **Figure 7.34:** Image

courtesy of Robert Leben, Colorado Center for Astrodynamics Research, University of Colorado, Boulder. Reprinted with permission.

Chapter 8

Figure 8.14: Adapted from the National Renewable Energy Laboratory.

Chapter 9

Box 1, Figure 2: Source: Curtis E. Ebbesmeyer and W. James Ingraham, Jr., "Shoe Spill in the North Pacific," in American Geophysical Union EOS, *Transactions,* Vol. 73, No. 34, August 25, 1992, pp. 261, 365. **Box 1, Figure 3:** Source: Curtis E. Ebbesmeyer and W. James Ingraham, Jr., "Shoe Spill in the North Pacific," in American Geophysical Union EOS, *Transactions,* Vol. 73, No. 34, August 25, 1992, pp. 261, 365. **Figure 9.12:** From Sverdrup/Johnson/Fleming, *The Oceans,* 1942, renewed 1970, p. 105. Adapted by permission of Prentice Hall, Inc., Upper Saddle River, N.J.

Chapter 10

Figure 10.26: Source: www.ndbc.noaa.gov/dart.shtml. **Figure 10.28a:** Source: Adapted from "Oceanic internal waves" found at http://www.ifm.uni-hamburg.de/ers-sar/Sdata/oceanic/intwaves/intro/index.html. **Figure 10.28c:** From Alpers, W., P. Brandt, and A. Rubino, Internal Waves Generated in the Straits of Gibraltar and Messina: Observations from Space, in V. Barale and M. Gade (Editors), "Remote Sensing of the European Seas," Springer, p. 321, 2008. With kind permission of Springer Science+Business Media.

Chapter 11

Table 11.2: Source: www.tidepredictor.com, University of South Carolina. **Table 11.3:** Source: www.tidepredictor.com, University of South Carolina.

Chapter 12

Figure 12.24: D. L. Inamn and J. D. Frautschy, *Littoral Process and the Development of Shorelines,* 1965, pp. 511–536. Coastal Engineering, ASCE, New York. Reprinted with permission of ASCE. **Figure 12.29:** Image Courtesy of NASA. **Box 1, Figure 1:** From Leuliette, E. W., R. S. Nerem, and G. T. Mitchum, 2004: Calibration of TOPEX/Poseidon and Jason altimeter data to construct a continuous record of mean sea level change. *Marine Geodesy,* 27(1–2), 79–94. http://sealevel.colorado.edu. Reprinted with permission from the Colorado Center for Astrodynamics Research.

Chapter 13

Table 13.1: Data from U.S. Environmental Protection Agency. **Table 13.2:** From George M. Woodwell, et al., DDT Residues in an East Coast Estuary, *Science* 156 (1967): 821–24.

Chapter 15

Chapter 16

Index

Note: Page references followed by the letters *f* and *t* indicate figures and tables, respectively.